DRUG DISCOVERY RESEARCH

BICENTENNIAL 1807 WILEY 2007 BICENTENNIAL

The Wiley Bicentennial–Knowledge for Generations

Each generation has its unique needs and aspirations. When Charles Wiley first opened his small printing shop in lower Manhattan in 1807, it was a generation of boundless potential searching for an identity. And we were there, helping to define a new American literary tradition. Over half a century later, in the midst of the Second Industrial Revolution, it was a generation focused on building the future. Once again, we were there, supplying the critical scientific, technical, and engineering knowledge that helped frame the world. Throughout the 20th Century, and into the new millennium, nations began to reach out beyond their own borders and a new international community was born. Wiley was there, expanding its operations around the world to enable a global exchange of ideas, opinions, and know-how.

For 200 years, Wiley has been an integral part of each generation's journey, enabling the flow of information and understanding necessary to meet their needs and fulfill their aspirations. Today, bold new technologies are changing the way we live and learn. Wiley will be there, providing you the must-have knowledge you need to imagine new worlds, new possibilities, and new opportunities.

Generations come and go, but you can always count on Wiley to provide you the knowledge you need, when and where you need it!

William J. Pesce
President and Chief Executive Officer

Peter Booth Wiley
Chairman of the Board

DRUG DISCOVERY RESEARCH

New Frontiers in the Post-Genomic Era

Edited by

ZIWEI HUANG

San Diego, California

WILEY-INTERSCIENCE
A JOHN WILEY & SONS, INC., PUBLICATION

Published by John Wiley & Sons, Inc., Hoboken, New Jersey
Published simultaneously in Canada

Library of Congress Cataloging-in-Publication Data:

ISBN 978-0-471-67200-5

Printed in the United States of America
10 9 8 7 6 5 4 3 2 1

CONTENTS

PART II. *CHEMICAL AND SYNTHETIC APPROACHES IN DRUG DISCOVERY*

*denotes the author(s) whom should be contacted for all correspondences regarding the specific chapter.

FOREWORD

NIGHT THOUGHTS OF A CHEMIST BEMUSED BY BIOTECHNOLOGY

This book covers a wide variety of topics of interest to the pharmaceutical industry, and does so very well. The writing is usually clear, the illustrations apt, and the references generous. There is some sense, however, of desperation, which probably reflects the awareness of the impossible task now facing those in that industry. No matter what they do, or how well, it will never be enough, for either the patients or the company profits. There are good reasons for such uneasiness. The task, as now posed, may be impossible.

Biology, and therefore human biomedicine, is a product of evolution, and is thus a palimpsest of all that has gone before. Each variation, genetic or epigenetic, has consequences, throughout the organism, as it modifies catalytic and structural proteins and the various feedbacks loops in which they are imbedded. If, on balance, the various changes leave the organism more fit to survive and reproduce in the environment at that time, the change may become widespread, even if some of its effects are deleterious. The effects produce cascades of small differences in different populations, leading to a balanced diversity in their genomes and in metabolisms. This means that each drug acts against a different background. Those that act on the genome or on its expression or, nominally, one simple receptor, similarly change the physiology and its potential response to further intervention. As a result of this variability and plasticity of the matrix, a drug, in extreme cases, can kill one person and cure another, so that clinical trials are always veiled by some level of statistical mystification.

What can be done to ameliorate this problem, which is rapidly making a "red queen's race" of the industry? Bacterial infections were, and are, relatively straightforward problems, because bacteria are relatively far phylogenetically from eukaryotes and hence some key features of their metabolism are strongly uncoupled those of polycellular organisms. But metabolic and genetic aberrations in patients are very similar to those in normal tissues, and hence their selective targeting is far more subtle. Can any response but a deadly game of cut and try be used in such circumstances? Perhaps not easily.

Because biological evolution in general is exceedingly conservative and modular, the histories of biomedical cycles and genetic systems are represented, often in only slightly distorted forms, in contemporary organisms. That information, and the connections within the systems themselves, can lead to a deeper understanding of its responses to perturbations. Obtaining and using this information will not be a quick or easy task. Some aspects may be very generally applicable, however. The lifespan of a mouse is very different from that of an elephant, but would a mouse altered to live for many decades still be a mouse? Only a very deep analysis can tell us how to do new things and what the side-effects might be. The kind of work described in this book is laying the foundation for such understanding and for the skills required to make use of it.

Paul C. Lauterbur*
Urbana, Illinois
March 2007

*Note by the Editor: Shortly after writing this Foreword, Professor Paul C. Lauterbur, 2003 Nobel laureate in medicine, passed away on March 27, 2007. This book is dedicated to the memory of this great man who has had a life-long passion in applying scientific innovations to alleviate profound human sufferings.

INTRODUCTION

This book is a product of the collective effort and contribution of 70 scholars and researchers from 25 universities, institutes and companies around the world who are experts in a wide range of different disciplines related to biomedical and drug discovery research but share a common interest in translating research discoveries into new therapies for human diseases. In this book, they have shared their knowledge, recent research discoveries, and future perspectives of the various aspects of drug discovery research.

The recent completion of the human genome project has ushered in a new era of post-genomic scientific research and biomedical applications. As shown in the central illustration on the front cover of the book, the rapid advances in many scientific disciplines such as genomics, structure, chemistry, computer-aided drug design (CADD), biology, and pharmacology and their integration are changing the face of drug discovery research and facilitating the translation of the vast information from the human genome to the understanding of the biology of the human body and development of new medicine. As such, the book focuses on the interplay of the research in these fields for the discovery of new drugs and describes some of the most fascinating and cross-disciplinary frontiers in post-genomic biomedical and biopharmaceutical science, as written by experts working in such research frontiers. The intent of the book is to provide both graduate students and researchers who study, work, or have an interest in biotechnology and pharmaceutical science with a general introduction to and practical examples of some of the many methods and techniques that can be used in biopharmaceutical research in the post-genomic era.

The book has a total of 20 chapters, divided into three different sections: Part I, Computational and Structural Approaches in Drug Discovery (Chapters 1–5); Part II, Chemical and Synthetic Approaches in Drug Discovery (Chapters 6–14); and Part III, Biological Approaches and Translational Research in Drug Discovery (Chapters 15–20). Their arrangement and order reflect the overall scheme of modern drug discovery research shown on the cover illustration, which involves structure-based and computer-aided drug design, chemical synthesis and optimization of drug molecules, and biological and preclinical evaluation. The five chapters in Part I describe the recent advances in the development of computational methods and their application in virtual screening to identify new ligands or inhibitors of a protein target. In addition to receptor binding or so-called molecular docking simulations of ligands by virtual screening (Chapters 1–4), this section describes the development and use of computational techniques to predict other properties such as metabolism and toxicity of drug candidates (Chapter 3) or to design novel combinatorial compound libraries to be synthesized (Chapter 5). Part II has a collection of nine chapters devoted to the advances in chemical synthesis and chemical strategies for the development or delivery of both small and large molecule drugs or probes of protein functions. Chapters 6–11 cover a wide range of chemistries of small molecules of various natures and their application for the discovery of drugs and research probes. As small molecules have been a major source of drugs for the pharmaceutical industry, these chapters provide an important account of the state-of-the-art chemical methods in the research and discovery of these small molecule drugs or probes. Besides small molecules, synthetic molecules of larger sizes – such as chemically modified proteins or peptides – are becoming another important class of therapeutic and research agents, as described in Chapter 12 which uses synthetic chemokine analogs as an example. Chapter 13 reviews the chemical strategies for studying post-translational modification proteomics which has a significant implication on the development of protein-based therapeutics. The merging of nanotechnology and biomedicine is becoming an important trend. The role of chemistry in nano-medicine is highlighted in Chapter 14, which presents the development of nano-materials for the delivery and controlled release of drugs, using antibiotics as the example. Part III focuses on the biology of drug discovery, with six chapters reviewing biological and translational research with the goal of bringing novel laboratory research agents to the clinic. Chapters 15–18 provide an account of some of the most remarkable achievements and successful case stories in modern research and development of anticancer drugs and drugs for treating neurodegenerative disease. Chapter 19 reviews the role of animal models in drug discovery. Last but not least, research and applications of stem cells are becoming a highly promising new avenue for biomedical and drug discovery research, as described in Chapter 20.

It should be pointed out that the classification of the chapters into the three different sections and the overtly simplified summary as described above are limited and partial. As readers will find out when reading each of the 20 chapters in detail, many chapters are actually multidisciplinary and cross the boundaries of these sections. Indeed, one salient feature that this book wishes to present is the cross-disciplinary and collaborative nature of modern drug discovery research as demonstrated by these 20 chapters written by biologists, chemists, and clinicians whose works are highly interrelated to each other. Another important feature of the book is its focus on post-genomic drug discovery research in the United States and China. As China is becoming a global powerhouse in research, this book hopes to showcase some of the latest and most exciting research in drug discovery in China, together with the work of colleagues in the United States.

Readers of this book should note that a corresponding author is denoted for each chapter and his/her e-mail address provided in the Contributors pages. Readers who have questions or inquiries about a particular chapter are encouraged to communicate with the corresponding author directly through the e-mail address provided. A complete list of authors, corresponding authors, and affiliations can be found in the Table of Contents. Please also note that many chapters have color figures. All color figures are placed together in the center of the book. While black-and-white copies (so stated specifically in the figure captions) of these color figures are provided within each chapter, readers should refer to the color versions for better comprehension of these figures.

I would like to thank all those who have helped with and contributed to this book. Most importantly, I want to acknowledge all my co-authors of this book. I won't name them one by one here as there are 69 of them! Not only are they the pioneers, leaders, or experts in their fields of research, many also are or have become good friends and colleagues through our work on this book. It has been a great experience for me to work with these outstanding individuals and bring their fascinating research to the readers. The knowledge, insights, and perspectives shared by these wonderful co-authors are truly valuable to me and, I hope, to the readers. The final piece of the book was completed by my former colleague at the University of Illinois at Urbana-Champaign (UIUC), Professor Paul C. Lauterbur, who kindly wrote the Foreword. The idea of putting together this book was first conceived in 2003 when I was at UIUC where Paul and I were both faculty members of the Department of Chemistry and the Center for Biophysics and Computational Biology. The book was started there and continued in San Diego with the moving of my laboratory to San Diego in 2004. In this sense, it is personally satisfying to see that, in a somewhat serendipitous way, this book began and ended at Illinois. Recognized by the 2003 Nobel Prize in Physiology or Medicine, Paul Lauterbur's pioneering research has led to the development of modern magnetic resonance imaging (MRI), which represents one of the most significant medical diagnostic discoveries of the 20th century and has benefited patients throughout the world. This classic work demonstrates the power of cross-disciplinary research to discover innovative medical applications, which is the theme and goal of this book.

In addition to the science, many people deserve recognition for their efforts and help in other aspects of the book. Cindy Cook, Jonathan Rose and Danielle Lacourciere have tirelessly worked on the editing and production of the book. Sameer Kawatkar, Serena Chen, Pak-Nei Hon, Mi Youn Lim and Harsukh Gevariya have helped with the cover illustrations and design.

As the editor and on behalf of my 69 co-authors, I hope that this book can be a helpful tool and reference for your study and research of post-genomic drug discovery, an endeavor filled with difficult challenges for scientific intellect and great promises for human health.

Ziwei Huang
San Diego, California
March 2007

CONTRIBUTORS

Jing An, University of Illinois at Urbana-Champaign, Urban, IL 61801; The Burnham Institute for Medical Research, La Jolla, CA 92037; and Raylight Corporation, Chemokine Pharmaceutical Inc., La Jolla, CA 92037, USA

Tiba Aynechi, Department of Pharmeceutical Chemistry, University of California, San Francisco, San Francisco, CA 94143, USA

Jeremy L. Baryza, Department of Chemistry, Stanford University, Stanford, CA 94305, USA

Natasja Brooijmans, Graduate Program in Chemistry and Chemical Biology, University of California, San Francisco, San Francisco, CA 94143, USA

Dennis A. Carson, Moores Cancer Center, University of California, San Diego, La Jolla, CA 92093, USA; dcarson@ucsd.edu

Grace Shiahuy Chen, Department of Applied Chemistry, Providence University, Shalu, Taiwan, Republic of China

Huei-Sheng Vincent Chen, The Burnham Institute for Medical Research, University of California, San Diego, La Jolla, CA 92037, USA; hsv_chen@burnham.org

Ling Chen, Guangzhou Institute of Biomedicine and Health, Chinese Academy of Sciences, Guangzhou, People's Republic of China; chen_ling@gibh.ac.cn

Sai-Juan Chen, Shanghai Institute of Hematology (SIH) and State Key Lab of Medical Genomics, Rui Jin Hospital, School of Medicine, Shanghai Jiao Tong University, Shanghai, 200025, People's Republic of China

Zhu Chen, Shanghai Institute of Hematology (SIH) and State Key Lab of Medical Genomics, Rui Jin Hospital, School of Medicine, Shanghai Jiao Tong University, Shanghai, 200025, People's Republic of China; zchen@stn.sh.cn

Ji-Wang Chern, School of Pharmacy, National Taiwan University, Taipei 100, Taiwan, Republic of China; chern@jwc.mc.ntu.edu.tw

Won-Tak Choi, Department of Biochemistry, University of Illinois at Urbana—Champaign, Urbana, IL 61801, USA

Howard B. Cottam, Moores Cancer Center, University of California, San Diego, La Jolla, CA 92093, USA

Sutapa Ghosh, The Burnham Institute for Medical Research, La Jolla, CA 92037

Zhen Gong, Drug Discovery & Design Centre, Shanghai Institute of Materia Medica, Chinese Academy of Sciences, Shanghai 201203, People's Republic of China

Yanting Guo, Key Laboratory of Bioorganic Phosphorus Chemistry and Chemical Biology, Department of Chemistry, Tsinghua University, Beijing 100084, People's Republic of China

Andreas H. Guse, University Hospital Hamburg—Eppendorf, Center of Experimental Medicine, Institute of Biochemistry and Molecular Biology 1: Cellular Signal Transduction, 20246 Hamburg, Germany

Andrew D. Hamilton, Department of Chemistry, Yale University, New Haven, CT 06520, USA; andrew.hamilton@yale.edu

Paul J. Hergenrother, Department of Chemistry, Roger Adams Laboratory, University of Illinois, Urbana, IL 61801, USA; hergenro@uiuc.edu

Michael K. Hilinski, Department of Chemistry, Stanford University, Stanford, CA 94305, USA

Joshua C. Horan, Department of Chemistry, Stanford University, Stanford, CA 94305, USA

Jia Hu, Key Laboratory of Bioorganic Phosphorus Chemistry and Chemical Biology, Department of Chemistry, Tsinghua University, Beijing 100084, People's Republic of China

Ziwei Huang, University of Illinois at Urban-Champaign, Urbana, IL 61801; The Burnham Institute for Medical Research and University of California, San Diego, La Jolla, CA 92037, USA; ziweihuang@burnham.org

Marc Hyer, The Burnham Institute for Medical Research, La Jolla, CA 92037, USA

Jamie Imitola, Program in Stem Cells & Regeneration, Center for Neuroscience and Aging, Burnham Institute for Medical Research, La Jolla, CA 92037, USA

Hualiang Jiang, Drug Discovery & Design Centre, Shanghai Institute of Materia Medica, Chinese Academy of Sciences, Shanghai 201203; and School of Pharmacy, East China University of Science and Technology, Shanghai 200237, People's Republic of China; hljiang@mail.shcnc.ac.cn

Xuliang Jiang, Serono Research Institute, Rockland, MA 02370, USA; xuliang.jiang@serono.com

Cindy Kan, Department of Chemistry, Stanford University, Stanford, CA 94305, USA

Yohichi Kumaki, Raylight Corporation, Chemokine Pharmaceutical Inc., La Jolla, CA 92037, USA

I. M. Krishna Kumar, The Burnham Institute for Medical Research, La Jolla, CA 92037, USA

Irwin D. Kuntz, Department of Pharmaceutical Chemistry, University of California, San Francisco, San Francisco, CA 94143, USA; kuntz@cgl.ucsf.edu

P. Therese Lang, Graduate Program in Chemistry and Chemical Biology, University of California, San Francisco, San Francisco, CA 94143, USA

Gui-In Lee, Department of Chemistry, Yale University, New Haven, CT 06520, USA

Jean Pyo Lee, Program in Stem Cells & Regeneration, Center for Neuroscience and Aging, The Burnham Institute for Medical Research, La Jolla, CA 92037, USA

Jian Li, Drug Discovery & Design Centre, Shanghai Institute of Materia Medica, Chinese Academy of Sciences, Shanghai 201203, People's Republic of China

Song Li, Beijing Institute of Pharmacology and Toxicology, Haidian District, Beijing 100850, People's Republic of China; lis@nic.bmi.ac.cn

Yanmei Li, Key Laboratory of Bioorganic Phosphorus Chemistry and Chemical Biology, Department of Chemistry, Tsinghua University, Beijing 100084, People's Republic of China; liym@mail.tsinghua.edu.cn

Guoqiang Lin, Shanghai Institute of Organic Chemistry, Chinese Academy of Sciences, Shanghai 201203, People's Republic of China; lingq@mail.sioc.ac.cn

Stuart A. Lipton, The Burnham Institute for Medical Research, University of California, San Diego, La Jolla, CA 92037, USA; slipton@burnham.org

Hong Liu, Drug Discovery & Design Centre, Shanghai Institute of Materia Medica, Chinese Academy of Sciences, Shanghai 201203, People's Republic of China

Zhaowen Luo, Serono Research Institute, Rockland, MA 02370, USA

Malcolm J. McGregor, Serono Research Institute, Rockland, MA 02370, USA

Demetri Moustakas, Joint Graduate Program in Bioengineering, University of California, Berkeley, Berkeley, CA 94720, USA

Franz-Joseph Mueller, Program in Stem Cells & Regeneration, Center for Neuroscience and Aging, The Burnham Institute for Medical Research, La Jolla, CA 92037, USA

Aihua Nie, Beijing Institute of Pharmacology and Toxicology, Haidian District, Beijing, People's Republic of China, 100850; and The Burnham Institute for Medical Research, La Jolla, CA 92037, USA

Amanda C. Nottbohm, Department of Chemistry, Roger Adams Laboratory, University of Illinois, Urbana, IL 61801, USA

Connie M. Oshiro, Roche, Palo Alto, CA 94304, USA

Vaclav Ourednik, Program in Stem Cells & Regeneration, Center for Neuroscience and Aging, The Burnham Institute for Medical Research, La Jolla, CA 92037, USA

Kook In Park, Program in Stem Cells and Regeneration, Center for Neuroscience and Aging, The Burnham Institute for Medical Research, La Jolla, CA 92037, USA

Prithi Rajan, Program in Stem Cells & Regeneration, Center for Neuroscience and Aging, The Burnham Institute for Medical Research, La Jolla, CA 92037, USA

John C. Reed, The Burnham Institute for Medical Research, La Jolla, CA 92037, USA; jreed@burnham.org

Douglas D. Richman, Department of Molecular Pathology, University of California, San Diego, La Jolla, CA 92037, USA

Brian Shoichet, Department of Pharmaceutical Chemistry, University of California, San Francisco, San Francisco, CA 94143, USA

Evan Snyder, Program in Stem Cells & Regeneration, Center for Neuroscience and Aging, The Burnham Institute for Medical Research, La Jolla, CA 92037, USA; esnyder@burnham.org

Joseph G. Sodroski, Dana-Farber Cancer Institute, Harvard Medical School, Boston, MA 02115, USA

Yang D. Teng, Program in Stem Cells & Regeneration, Center for Neuroscience and Aging, The Burnham Institute for Medical Research, La Jolla, CA 92037, USA

Jian-Hua Tong, Shanghai Institute of Hematology (SIH) and State Key Lab of Medical Genomics, Rui Jin Hospital, School of Medicine, Shanghai Jiao Tong University, Shanghai, 200025, People's Republic of China

Vishal A. Verma, Department of Chemistry, Stanford University, Stanford, CA 94305, USA

Yen Wei, Department of Chemistry, Drexel University, Philadelphia, PA 19104, USA; weiyen@drexel.edu

Kate Welsh, The Burnham Institute for Medical Research, La Jolla, CA 92037, USA

Paul A. Wender, Department of Chemistry, Stanford University, Stanford, CA 94305, USA; wenderp@stanford.edu

Chi-Wai Wong, Guangzhou Institute of Biomedicine and Health, Chinese Academy of Sciences, Guangzhou, People's Republic of China

Junhai Xiao, Beijing Institute of Pharmacology and Toxicology, Haidian District, Beijing, 100850, People's Republic of China

Hang Yin, Department of Chemistry, Yale University, New Haven, CT 06520, USA

Houping Yin, Department of Chemistry, Drexel University, Philadelphia, PA 19104, USA

Dayong Zhai, The Burnham Institute for Medical Research, La Jolla, CA 92037, USA

Jiange Zhang, Department of Medicinal Chemistry, School of Pharmacy, Zhengzhou University, Zhengzhou 450052, People's Republic of China

Lihe Zhang, State Key Laboratory of Natural and Biomimetic Drugs, School of Pharmaceutical Sciences, Peking University, Beijing 100083, People's Republic of China; zdszlh@bjmu.edu.cn

Laura Zheng, Department of Chemistry, Drexel University, Philadelphia, PA 19104, USA

Weiliang Zhu, Drug Discovery & Design Centre, Shanghai Institute of Materia Medica, Chinese Academy of Sciences, Shanghai 201203, People's Republic of China

PART I

COMPUTATIONAL AND STRUCTURAL APPROACHES IN DRUG DISCOVERY

CHAPTER 1

MOLECULAR DOCKING AND STRUCTURE-BASED DESIGN

P. Therese Lang, Tiba Aynechi, Demetri Moustakas, Brian Shoichet, Irwin D. Kuntz[*], Natasja Brooijmans, and Connie M. Oshiro

1.1 INTRODUCTION

The discovery of new drugs is a complex process. It generally starts with the identification of compounds that bind to a target or show efficacy in a simple screen. Molecules that show good affinity are called "hits." The next step is to find compounds that have attractive pharmaceutical properties—for example, low toxicity and sufficient aqueous solubility to be orally active. Such compounds are often called "leads." Traditionally, "hits" have been found by screening, while "leads" are developed from "hits" through chemical synthesis. Screening normally involves large numbers of compounds from natural products, corporate databases, or organic chemistry companies that can be examined for biological activity in high-throughput assays. Commercial systems can process millions of tests per day for enzyme targets. The best compounds are moved forward in a process aimed at modifying their chemical structure to improve potency, specificity, and *in vivo* activity while lowering toxicity and side effects. Synthetic methods include combinatorial chemistry and library synthesis (Fig. 1.1).

[*]Corresponding author: kuntz@cgl.ucsf.edu

Drug Discovery Research: New Frontiers in the Post-Genomic Era, Edited by Ziwei Huang

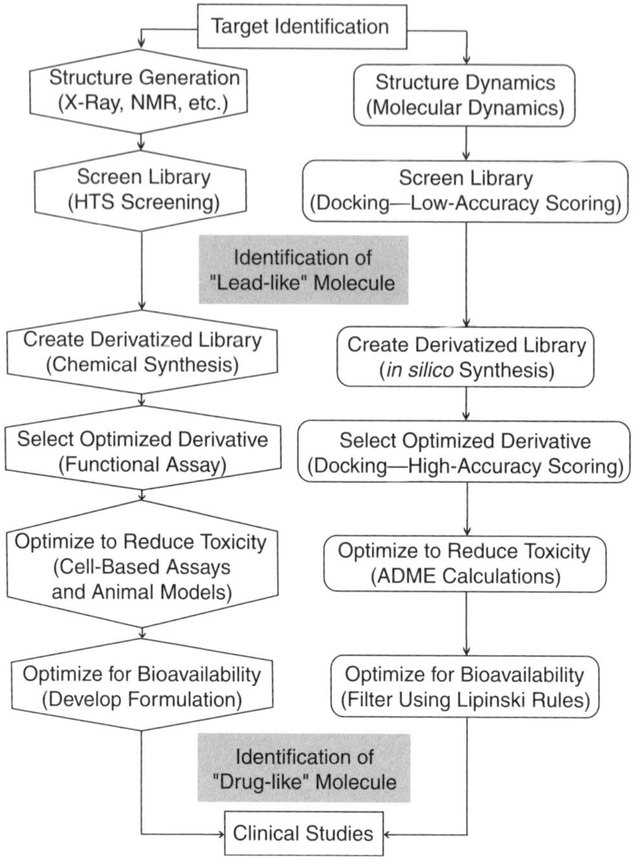

Figure 1.1 Example flow of drug design process from both experimental (left) and computational (right) perspectives.

Computational methods have proved useful in many aspects of the discovery process (Alvarez, 2004). A variety of strategies are available. If an active "lead" is known, it is straightforward to query a database for molecules with similar properties using pharmacophore searches or quantitative structure–activity relationships (QSAR). These methods typically use information available for the ligand, inhibitor, or substrate rather than the receptor. However, there is growing interest in mapping out receptor properties—either from known family relationships with other members of the receptor family or through pharmacophore strategies applied directly to the receptor structures (Arnold et al., 2004; Hajduk et al., 2005). This chapter focuses on a set of strategies in which direct knowledge of the receptor structure is used to identify or design ligands that possess good steric and chemical complementarity to specific sites on the target macromolecule. This process is referred to as "structure-based drug design" (Brooijmans and Kuntz, 2003).

The structure-based drug design paradigm is analogous to experimental screening. Structures for the receptor or target are obtained either from the literature or from in-house operations. These structures come from crystallography or nuclear magnetic resonance (NMR) experiments, but there is increasing interest in high-quality homology-modeled structures (Chance et al., 2004). Computer analogs of ligands are generated. These families are often called "virtual libraries" and may consist of compounds from corporate, academic, or commercial holdings (Laird and Blake, 2004; Webb, 2005). A virtual library might also include molecules that are not physically available but might be obtained through chemical synthesis, perhaps using combinatorial chemistry (Jorgensen, 2004; Kick et al., 1997). Screening of the virtual library against the target structure involves some form of positioning the putative ligand in three-dimensional space and evaluating the intermolecular interactions for that particular geometry. Typically, the process is an iterative one: The ligand is moved, and the new geometry is evaluated. This cycle is repeated until some "best scoring" geometry is identified for the particular ligand under test. Then, the next ligand in the list is chosen, and the whole procedure begins again. The goal of virtual screening is to identify the best binding candidates from the library for experimental testing. Because the virtual libraries can be huge—upwards of a billion compounds—this triage procedure is a critical step.

Once a binding candidate has been found, structure-based design can be used to optimize binding affinity. For this operation, one starts with a "hit" or "lead" with a known activity. Often, a structure for the ligand–target complex is available. There are many computational methods available for evaluating chemical variants of the "lead" that offer suggestions about the direction for the next round of synthesis. Ideally, a number of such variants are prepared, and their properties and structures are obtained so that a selection of molecules are available to take into further biological testing. Optimization methods are typically much more computationally intensive than other virtual screening approaches.

There has recently been a merging of these two ideas. Virtual libraries containing 1000–10,000 molecular fragments (sometimes called "anchors" or "scaffolds") are used in the initial screening. The most promising are then expanded using computer synthesis in a combinatorial fashion (Jorgensen, 2004; Kick et al., 1997; Miranker and Karplus, 1991).

In this chapter, we will focus on a particular subset of molecular design strategies called "docking," in which candidate molecules are matched to receptor structures and evaluated for chemical and geometric complementarity. We will not discuss the broad field of quantitative structure–activity relationships (QSAR) that focus on the chemical structures of ligands alone (Bender and Glen, 2005).

1.2 MOLECULAR DOCKING

1.2.1 Overview

The basis of molecular docking is the calculation or estimation of the free energy of binding of a ligand to a specific receptor site in a fixed environment. The free energy of binding yields, directly, an equilibrium binding constant and, indirectly, the preferred binding mode of the ligand–receptor complex. There are important scientific and mathematical issues involved. For example, it is currently much easier to calculate the energy/enthalpy of interaction than to obtain the free energy because we lack efficient ways to obtain the entropic contributions. Second, the interactions of the ligand and receptor with the solvent (water, salts, and other components) are not easy to quantitate. Third, searching through the large number of conformations of the receptor and the ligand and their relative positions are difficult computer science problems. The need to repeat these calculations for large numbers of putative ligands and many possible targets requires serious attention to the algorithms. Docking protocols have adopted a variety of heuristics to make useful calculations with the knowledge that a complete high-level calculation is not feasible for the systems of biological or therapeutic interest.

What can we expect from current approaches? The best-case calculations are accurate to within approximately 0.5 kcal/mol of experimental results, but these are generally free energy differences obtained using perturbation techniques on a related family of ligands (Jorgensen, 2004). Routine results are rarely within 1 kcal/mol of experimental results, and library searches of diverse chemical types have larger inaccuracies. Work continues on improving the force fields that model the enthalpic terms (Bernacki et al., 2005). Estimates of entropic contributions are empirical, and adequate sampling of configurations and conformations search is a complex combinatorial problem. Consequently, searching large databases for new leads requires protocols that deal with four specific tasks:

1. Receptor site identification
2. Receptor site characterization
3. Orientation of the ligand within the site
4. Evaluation of the ligand

These steps are described in turn in the succeeding sections, and examples are given.

1.2.2 Receptor Site Identification

With the sequencing of the human genome and recent advances in structural techniques, the number of publicly available biomolecular structures has exploded over the last few years with over 40,000 in the Protein Data Bank (Berman et al., 2000). The two main experimental sources for three-dimensional (3D) structures of biomolecules are X-ray crystallography and high-resolution NMR spectroscopy. X-ray crystallography provides structures of biomolecules in the crystalline states, and with the crystallization of the ribosome, the upper limit of the experiment has been extended to the 1000-kDa range. NMR spectroscopy is limited to approximately 50 kDa, but the method has the advantage of providing additional information about the dynamics of the structure. As an alternative, if the structure has not been solved experimentally, computational techniques such as homology modeling can be used to predict 3D structures. We next discuss some pros and cons of each of these sources for structure-based drug design.

For X-ray crystal structures to be sufficiently accurate for drug designing purposes, a resolution of approximately 2 Å, an R-factor below 20%, and an R_{free} factor below 30% are preferred. It is important to note that the majority of 3D crystal structures of biomolecules do not have hydrogens or highly flexible residues included in the file. The missing atoms must be considered before structure-based drug design can begin. In addition, crystal packing forces may locally influence protein conformation, particularly for nucleic acids and surface active sites.

The result of structure determination with NMR is an ensemble of structures that agree equally well with experimental data. Although an averaged structure can be derived, it has been shown that the entire ensemble provides a more complete description of the system from an experimental perspective (Staunton et al., 2003). For structure-based drug design purposes, though, there are several methods for choosing the appropriate structure, including selecting the member of the ensemble closest to the average as measured by some distance metric or cross-docking to all members of the ensemble. Unfortunately, there is no generally accepted standard of accuracy for NMR-generated structures. As a rough rule of thumb, a high-resolution NMR structure should preferably have approximately 20 (distance or dihedral) restraints per residue (Berman and Westbrook, 2004).

If no experimental structural information is available for the target biomolecule, homology modeling can provide structures to guide the search for novel lead compounds. It should be noted that, depending on the method, homology modeling yields average errors of 3-Å root-mean-square deviation of proteins with greater than 50% sequence similarity, with larger errors for increasing sequence dissimilarity (Nissen et al., 2000). Nevertheless, homology modeling has proven to be successful in several cases, including discoveries of a highly potent DNA methylation inhibitor and a compound that discriminates between two voltage-gated ion channels with 20-fold accuracy (Staunton et al., 2003).

A number of more general issues associated with the selection and preparation of a receptor structure should be noted. In many structures, ions are required for structural or functional purposes. However, modeling this type of chemistry is often difficult because the formal charge and associated desolvation energy of ions are extremely complicated to compute accurately. The protonation states of residues such as histidine, lysine, glutamic acid, or aspartic acid are highly dependent on the environment in the active site and may even change in response to the ligand (Hensen et al., 2004). In particular for crystal structures, critical water molecules may be present in the active site, and it is often difficult to predict whether the water can be replaced or should be included in the model of the receptor. All these issues can affect the quality of the model and should be carefully considered.

1.2.3 Receptor Site Characterization

Once an accurate structure has been determined for the target, ligands are typically restricted to lie within one geometric region of the macromolecule, generally known as the "binding site." This region is generally selected because, upon ligand binding, normal function is altered. A given receptor can have one or more binding sites, such as the active sites of enzymes, allosteric sites, the binding or recognition sites of receptors, or even a dimer interface. The exact location of the binding site may be well known through experiment. However, if the binding site is not known, automated methods exist to identify potentially interesting regions.

Experimental data that indicate the binding site is the best source, if available. Experimentally derived structures of the biomolecule complexed with the natural substrate or a known inhibitor directly indicate the binding site. The Protein Data Bank and the Nucleic Acid Data Bank contain a large number of these types of structures (Berman et al., 1992, 2000). The Cambridge Crystallographic Data Centre/Astex has compiled a subset of biologically relevant protein–ligand complexes identified as being reliable for structure-based drug design purposes (Nissink et al., 2002). If direct observations are not available, binding regions can be identified through mutational experiments (such as alanine screening) or other biochemical assays.

There are cases where binding site information is not available. In these instances, computational tools can be used to indicate probable binding areas. We will describe two methods: The first is based upon geometric features, and the second uses chemical functionality of the receptor surface. For illustration, we consider SPHGEN from the DOCK suite of programs and the Multiple Copy Simultaneous Search (MCSS) approach (Ewing et al., 2001; Kuntz et al., 1982; Meng et al., 1992; Miranker and Karplus, 1991). In addition, an interesting statistical characterization has been recently described (Hajduk et al., 2005).

SPHGEN automatically identifies a target site by computing a set of site points or sphere centers, which serves to create a negative image of the surface. The algorithm begins by mapping the geometric features of the receptor surface, as defined by Lee and Richards, using the dms program (Richards, 1977). Then, spheres of varying radii are analytically generated to touch the molecular surface at two points, with the sphere center lying along the surface normal and with no portion of the sphere intersecting with a receptor atom. These overlapped spheres indicate various surface features, including invaginations and clefts. A clustering protocol, using radial overlap as a metric, is then used to indicate potential areas for ligand binding. The largest cluster is generally used as the binding site, and, once generated, the cluster is used as a template for possible ligand atom positions.

MCSS identifies binding sites by mapping the chemical properties of the biomolecular surface. Thousands of copies of specific molecular fragments are distributed in the target region of the protein. Then, energy minimization is performed on the ensemble, creating distinct local minima for each fragment. This process is repeated on a variety of chemical functionalities until the surface is adequately described. Although this method does not capture every geometric detail of the binding site, it does provide a basic pharmacophore that can be used in later studies (Arnold et al., 2004; Miranker and Karplus, 1991).

1.2.4 Orientation of the Ligand in the Target Site

There are two basic strategies for exploring the orientational degrees of freedom for putative ligands. The first uses a search grid for both the translational space and the euler angle space. This brute-force method is feasible if one is studying a few ligands in a restricted site. It can be extended to larger libraries if parallel processing is available (Jorgensen, 2004). Alternatively, protocols have been developed to prescreen orientation space. For example, the DOCK program uses a geometric pairing algorithm, matching the sphere centers (described above) with ligand centers (usually ligand atoms). The matching criterion is based on a comparison of intersphere and interatom distances. Exhaustive or selective searches can be done over the match matrix. Careful placement of the spheres is an important step in getting good-quality results. The second type of method selectively samples all of orientation and conformation space using search engines. For example, the Metropolis algorithm and simulated annealing are used in QXP, and the genetic algorithm has been implemented an Autodock (McMartin and Bohacek, 1997; Morris et al., 1998). These algorithms have been studied extensively for many applications. Their strengths and limitations are well understood.

1.2.5 Evaluation of Ligand Orientations

The many configurations (orientations and conformations) of the ligand need to be evaluated with a scoring function to identify the energetically most favorable ligand binding pose. Ideally, the scoring function would calculate the ligand free energy of binding in aqueous solution (Beveridge and Dicapua, 1989; Kollman, 1993). However, the large computational expense of these calculations leads to the introduction of scoring functions that calculate a range of simplifications of the ligand binding free energy.

Scoring functions can be broadly classified into two categories: those based on first-principles-derived molecular mechanics force fields, and those based on functions fit to empirically derived binding data. For the purposes of this review, scoring functions that employ quantum mechanics are not considered as the extreme computational cost of these calculations make them prohibitive for use during small-molecule docking.

Of the first-principles-derived scoring functions, the most computationally efficient are those that approximate the binding free energy (ΔG_{bind}) as the molecular mechanics protein–ligand interaction energy. Molecular mechanics treats the molecule as a collection of atoms governed by a set of classical mechanical potential functions (Weiner and Kollman, 1981). Parameters for these potentials are derived from small-molecule experiments and refined to yield correct structural and thermodynamic quantities such as bond stretching frequencies or heats of formation. The primary DOCK energy scoring function approximates the ligand–receptor binding energy using the AMBER molecular mechanics intermolecular interaction energy, a sum of the Lennard-Jones 6–12 van der Waals (vdW) potential, and the Coulombic potential, given in equation (1.1):

$$E = \sum_{i}^{\text{Lig}} \sum_{j}^{\text{Rec}} \left(\frac{A_{ij}}{r_{ij}^{12}} - \frac{B_{ij}}{r_{ij}^{6}} + K \frac{q_i q_j}{D r_{ij}} \right) \tag{1.1}$$

where i indexes the ligand atoms and j indexes the receptor atoms; A and B are the vdW attractive and dispersive parameters, respectively; q is the partial charge on the atom; K is the scaling constant that converts electrostatic energy into kcal/mol; D is the dielectric constant of the medium; and r_{ij} is the distance between ligand atom i and receptor atom j (Pearlman et al., 1995). This scoring function is limited by its use of a distance-dependent dielectric screening function to mediate all charge–charge interactions. This dielectric treatment assumes that the dielectric value of the solvent is uniform between all charge pairs.

A class of scoring functions has been developed that utilizes implicit models of solvation to calculate the electrostatic component of the molecular mechanics intermolecular interaction energy in a more sophisticated fashion than the simple Coulombic approach previously described. Both the Generalized Born (GB) and Poisson–Boltzmann (PB) terms have been combined with an empirically derived surface area (SA) term to include the energy of desolvating nonpolar atoms, and the resulting GB/SA and PB/SA methods have been implemented into molecular dynamics and docking methods (Feig and Brooks, 2004; Feig et al., 2004; Honig and Nicholls, 1995). While these functions are more computationally intensive than the Coulombic electrostatic energy functions, their proper treatment of solvation effects yields more accurate energy scores, and they are therefore frequently used in a hierarchical fashion to rescore docked ligand poses. DOCK 5 implements a GB/SA scoring function that is recommended for use in a rescoring capacity (Zou et al., 1999).

The most computationally intensive class of first-principles-derived scoring functions combine molecular dynamics (MD) simulations with implicit or explicit solvation to average the interaction energies from a Boltzmann-weighted ensemble of complex structures, yielding accurate estimates of the binding free energy that takes into account protein flexibility. The implicit solvation methods, MM-PB/SA and MM-GB/SA, perform a short, explicit solvent MD simulation from which a set of snapshots of the protein–ligand complex structure are saved (Gohlke and Case, 2004; Wang et al., 2001; Zhou and Madura, 2004). These snapshots, representing a Boltzmann-weighted ensemble of complex structures, are rescored with either the PB/SA or GB/SA scoring functions, and the average interaction score of the snapshots is taken as the free energy of binding for the ligand. Several research groups in academia and industry use these methods in a hierarchical fashion to rescore ligands identified as potential binders (Kollman et al., 2000).

The second major class of scoring functions models the binding free energy as a weighted sum of several different types of interaction energies, with and without explicit vdW and electrostatic terms. Many of these functions are based upon a comparison of receptor–ligand complexes and experimental binding data (K_i). Programs such as AutoDock, FlexX, GOLD, and Glide implement a variety of empirically derived energy score functions, including the well-known ChemScore and PLP functions (Eldridge et al., 1997; Friesner et al., 2004; Gehlhaar et al., 1995; Halgren et al., 2004; Hoffmann et al., 1999; Morris et al., 1998; Verdonk et al., 2003).

1.3 LIGAND STRUCTURE GENERATION

To identify virtual screening "hits," databases consisting of three-dimensional structures of putative ligands are searched using the methodology described above. How are these structures obtained? In some cases, structures can be taken directly from a database of experimentally determined structures; for example, the Cambridge Crystallographic Database contains over 360,000 small-molecule crystallographically determined structures. In cases where only the two-dimensional information is known, one can create computer representations of the covalent structure using conformation generation programs such as Concord (Tripos), ChemX (Accelrys), Rubicon (Daylight), and Omega (OpenEye). These programs provide one or more conformers consistent with the chemical connectivity and general rules of physical organic chemistry. Often, these conformers will be ranked by some energy formula. Finally, some vendors make libraries of three-dimensional structures directly available, including the Advanced Chemical Development (www.acdlabs.com), MDL Drug Data Report (www.mdl.com), National Cancer Institute Open Database Compounds (cactus.nci.nih.gov), Tripos Discovery Research Screening Libraries (www.tripos.com), InfoChem GmbH database (www.infochem.de), Thomson Index Chemicus database (scientific.thomson.com), and ZINC (blaster.docking.org/zinc).

Two strategies have emerged to study flexible ligands. The first, generally referred to as incremental construction, breaks down the ligand into smaller pieces and then rebuilds it during the docking calculation. One example of this technique starts with a fragment of the compound (an "anchor") and then adds atoms in layers during a docking or an optimization cycle. This approach has been called "anchor and grow." In the alternate method, conformers for each compound can be pregenerated, stored in a database, and then rigidly docked. Molecular dynamics and Monte Carlo techniques offer a combination strategy where the starting point is a single conformation of the ligand that then explores alternatives during the dynamics phase. There have been some tests of the two strategies, but there is no strong consensus of which is better (Lorber and Shoichet, 2005; Moustakas et al., 2006).

Other important issues that influence ligand structure and docking are choices of partial charges, tautomer preferences, and pK_a values. A new database (ZINC) that deals with many of these concerns is now available through the Shoichet group at UCSF (Irwin and Shoichet, 2005).

1.4 DESCRIPTION OF DOCKING PROGRAMS

While we do not have the space to provide descriptions of the many different approaches to molecular docking, Table 1.1 summarizes features of several of the most frequently used programs. We also present a brief synopsis of DOCK, developed at UCSF.

The current version of the DOCK program, is written in C++ and provides an object-oriented implementation in which each major component of the DOCK algorithm is a class with a documented interface, allowing these DOCK functions to be modified or replaced easily. As a result, it has been possible to independently validate and optimize the rigid body sampling, the flexible sampling, the energy scoring functions, and our minimizers. DOCK features an energy scoring function based on a molecular mechanics force field, solvation corrections using implicit solvent models, integration with the complete AMBER force field score, rigid body docking, ligand conformational searching, binding pose cluster analysis, and local minimization methods and also includes support for parallel computing using the MPI standard.

TABLE 1.1 Examples of Commonly Used Structure-Based Drug Design Packages

Method	Ligand Sampling Method[a]	Receptor Sampling Method[a,b]	Scoring Function[b,c]	Solvation Scoring[b,d]	Reference(s)
AutoDock	GA	SE	MM + ED	DDD, DS	Morris et al., 1998; Osterberg et al., 2002
DOCK 3	CE	SE	MM	PBE, DS	Lorber & Shoichet, 1998; Wei et al., 2004
DOCK 4/5	IC	SE	MM	DDD, GB, PB	Ewing et al., 2001; Knegtel et al., 1997 Moustakas et al., 2006
EUDOCK	CE	CE	MM	DDD	Pang et al., 2001a
FlexX/FlexE	IC	SE	ED	NA	Claussen et al., 2001
Glide	CE + MC	TS	MM + ED	DS	Eldridge et al., 1997; Friesner et al., 2004; Halgren et al., 2004; Sherman et al., 1996
GOLD	GA	NA	MM + ED	NA	Jones et al., 1997; Verdonk et al., 2003
ICM-Dock	MC	MC	MM + ED	DDD, PBE, DS	Abagyan et al., 1994; Totrov & Abagyan, 1997
MM-PBSA	MD	MD	MM	GB, PB	Kollman et al., 2000
QXP	TS + MC	MD	MM + ED	DDD	McMartin & Bohacek, 1997

[a]Sampling methods are defined as Genetic Algorithm (GA), Conformational Expansion (CE), Monte Carlo (MC), Molecular Dynamics (MD), Incremental Construction (IC), Merged Target Structure Ensemble (SE), and Torsional Search (TS); Section 1.3 for more information.

[b]If the package does not accommodate this option, the symbol NA (not available) is used.

[c]Scoring functions are defined as either empirically derived (ED) or based on molecule mechanics (MM); see Section 1.2.5 for more information.

[d]Additional accuracy can be added to the scoring function using implicit solvent models. The most commonly used options are Distance-Dependent Dielectric (DDD), Poisson Boltzmann Dielectric (PBE), a parameterized desolvation term (DS), Generalized Born (GB), and linearized Poisson Boltzmann (PB); see Section 1.2.5 for more information.

1.5 TESTS OF DOCKING AND STRUCTURE-BASED DESIGN

Despite well-known methodological weaknesses, structure-based screening using molecular docking has had important successes. Pragmatically, docking has suggested new, nonobvious ligands for multiple targets; these have been subsequently tested and shown to bind experimentally. Hugo Kubinyi, in a recent review, describes over 50 macromolecular targets for which ligands have been discovered using docking-based approaches (see Table 1.2 for a partial list) (Kubinyi, 2006). Most of these projects used experimental X-ray structures to represent the protein. In several cases, homology-modeled structures were employed (Evers and Klebe, 2004; Schapira et al., 2003).

TABLE 1.2 Recent Examples of Novel Inhibitor Discovery Using Molecular Docking

Target	Representative Hit	Lead Inhibitor IC_{50}	Follow-Up Inhibitor IC_{50}	Complex Structure
p56 Lck SH2 domain (Huang et al., 2004)	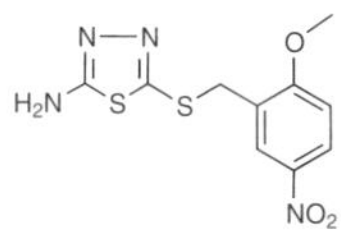	10 μM	NR	No
Neurokinin-1 receptor (Evers and Klebe, 2004)	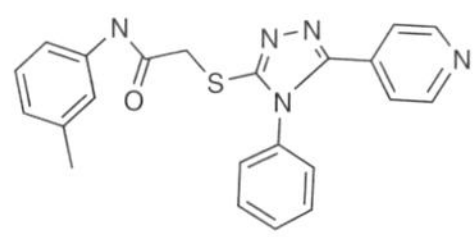	0.25 μM	NR	No
AICAR transformylase (Li et al., 2004)	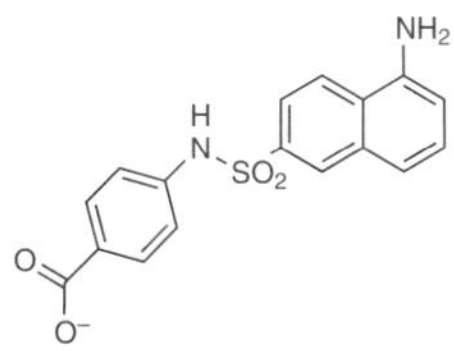	0.15 μM	NR	No
IMPDH (Pickett et al., 2003)	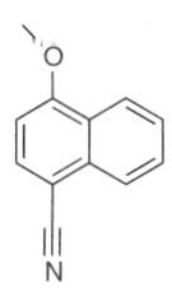	31 μM	NR	No
Checkpoint kinase 1 (Lyne et al., 2004)	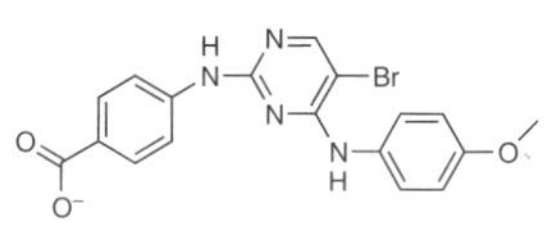	0.11 μM	NR	No
DNA gyrase (Boehm et al., 2001)	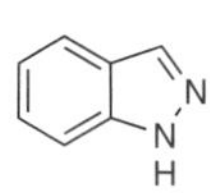	10,000 μM	0 μM	Yes
Aldose reductase (Iwata et al., 2001)	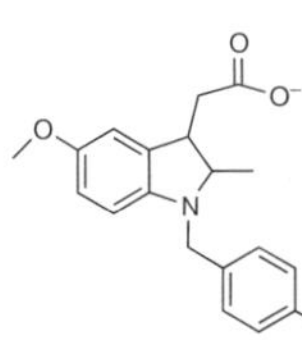	4.3 μM	0 μM	No
CDK4 (Honma et al., 2001)	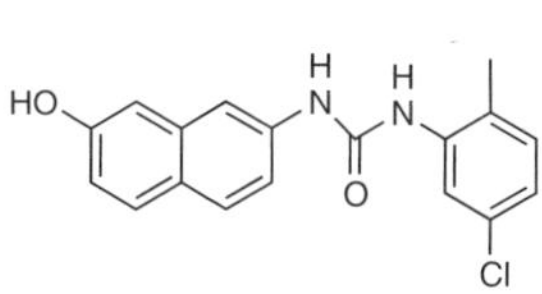	44 μM	0 μM	Yes

(Continued)

TABLE 1.2 *(Continued)*

Target	Representative Hit	Lead Inhibitor IC_{50}	Follow-Up Inhibitor IC_{50}	Complex Structure
Matriptase (Enyedy et al., 2001)	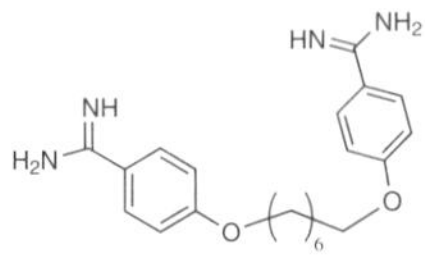	0.92 μM	0 μM	No
Bcl-2 (Enyedy et al., 2001)	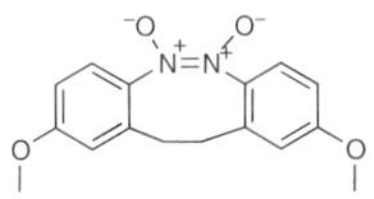	10.4 μM	NR	No
Adenovirus protease (Abarbanel 1984; Pang et al., 2001b)	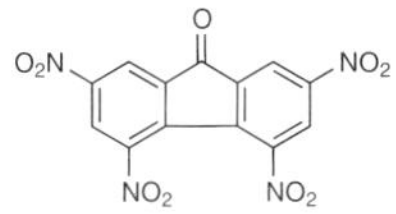	3.1 μM	NR	No
AmpC β-lactamase (Powers et al., 2002)	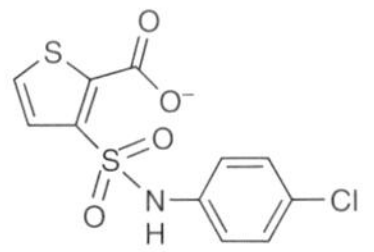	26 μM	1 μM	Yes
retinoic acid receptor (Schapira et al., 2001)	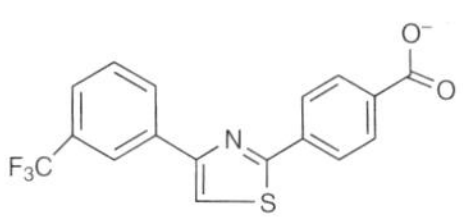	2 μM	NR	No
TGT (Gradler et al., 2001)	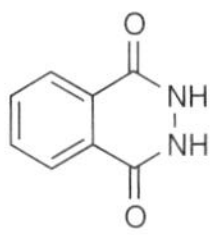	8.3 μM	0 μM	Yes
carbonic anhydrase II (Gruneberg et al., 2001)	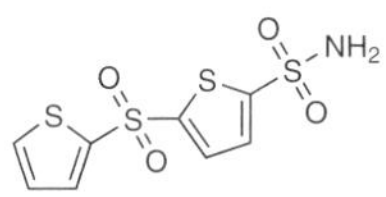	0.0008 μM	NR	Yes
HPRTase (Freymann et al., 2000)	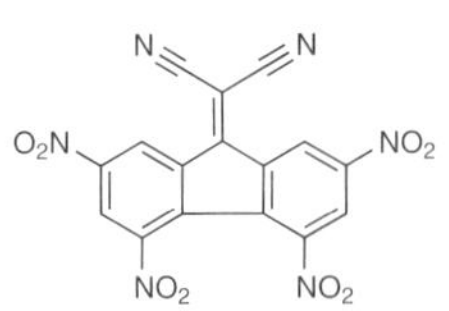	2.2 μM	NR	No
dihydro-dipicolinate (Paiva et al., 2001)	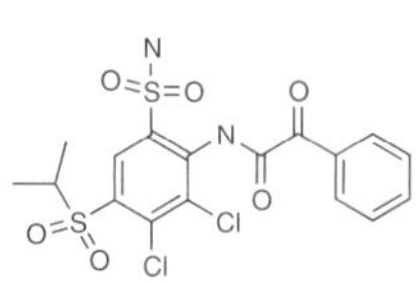	7.2 μM	NR	No

In recent work, the structures of known ligands in complex with their receptors have been predicted by docking, beginning with the structures of the independent molecules (Fig. 1.2) (Rizzo et al., 2000; Rosenfeld et al., 2003). In these studies, where the binding affinity is known but the structure of the complex is not, the docking predictions have been relatively accurate. The caveat to this is that there are many cases where docking mispredicts geometries in retrospective tests. Still, in published cases where the goal was genuine prediction, the docked geometry has often turned out to correspond closely to the subsequent experimental result.

A more difficult test is comparing the predicted geometries of novel ligands that emerge from the docking screens themselves. There are many examples of such predictions of ligand and geometry from docking screens against simple model cavity sites. These sites are small, completely enclosed by the protein, and dominated by one particular type of interaction, such as hydrophobicity, a single hydrogen bond acceptor, or a single electrostatic interaction. These features have allowed for multiple predictions of new ligands that are tested experimentally, often including structure

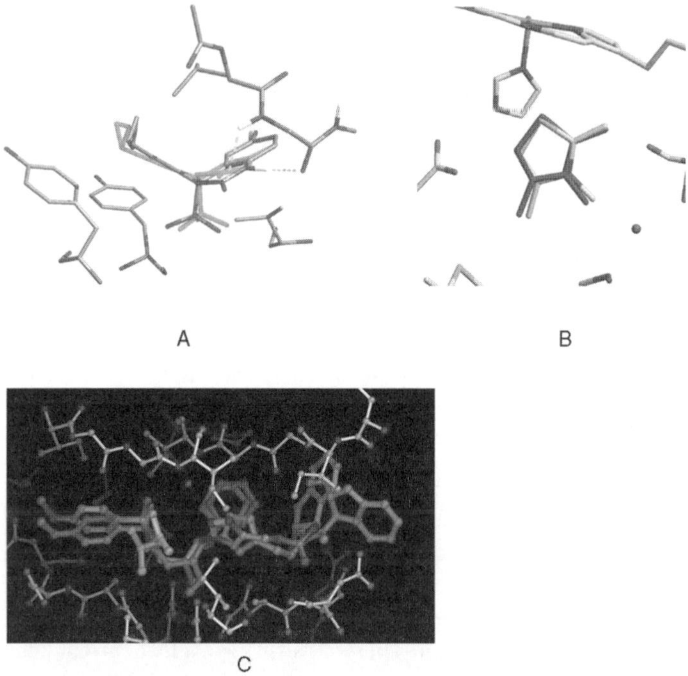

A B

C

Figure 1.2 Predicted complexes versus X-ray crystallographic structures that were subsequently determined. (**A**) Predicted (carbons in gray) and experimental (green) structures for sustiva in HIV reverse transcriptase (Rizzo et al., 2000; Rosenfeld et al., 2003). (**B**) Predicted (magenta) and experimental (white carbon atoms) structures of 2,3,4-trimethylthizole in the W191G cavity of cytochrome c peroxidase (Rizzo et al., 2000; Rosenfeld et al., 2003). (**C**) Predicted (green) and experimental structure (carbons in gray) of an HIV protease inhibitor (ligands with thick bonds, enzyme residues with thin bonds) (Brik et al., 2003, 2005; courtesy of Art Olson, TSRI). See color plates.

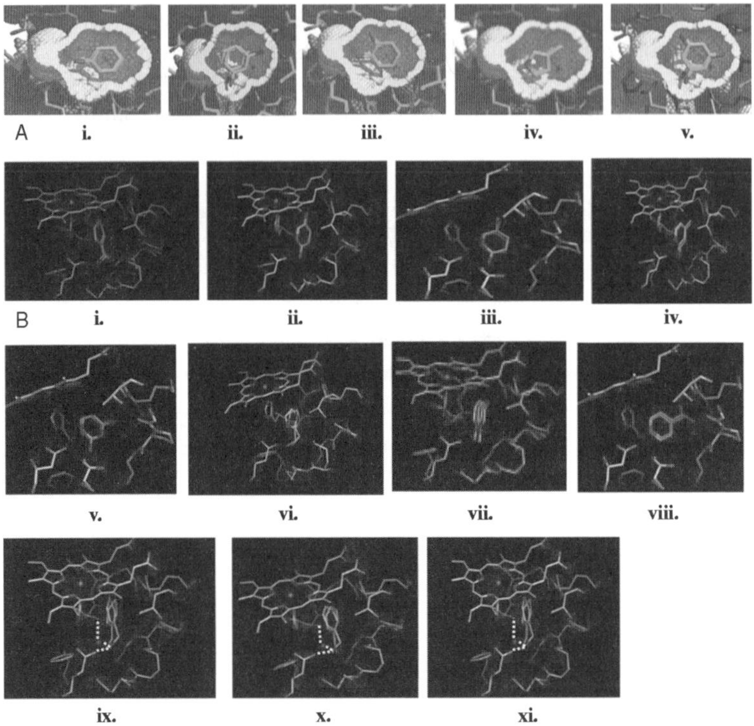

Figure 1.3 Docking predicted ligands from virtual screening against simple cavity sites. (**A**) The docked prediction (carbons in green) superposed on the crystallographic result (carbons in cyan). The surface of the L99A/M102Q cavity of T4 lysozyme (yellow) is cut away to reveal the complex. (i) Phenol, (ii) Chlorophenol, (iii) fluoroaniline, (iv) methylpyrrole, (v) difluorophenol. (**B**) The docked prediction (carbons in green) superposed on the crystallographic result (carbons in yellow) in the W191G cavity of cytochrome c peroxidase. (i) Thiophene-amidine, (ii) diaminopyridine, (iii) 2-amino-5-methylpyridine, (iv) 2-amino-4-methylpyridine, (v) diaminopyrimidine, (vi) hydroxymethyl-imidazole, (vii) 3-methyl-*n*-methylpyridine, (viii) 4-hydroxymethly-pyridine, (ix) aminomethyl-cyclopentane, (x) aminomethyl-benzene, (xi) aminomethyl-furan. See color plates.

determination (Fig. 1.3) (Brenk et al., 2006; Graves et al., 2005; Wei et al., 2002, 2004). Thus, X-ray crystal structures have been determined for about 25 ligands bound to three different cavities; in every case, the docking prediction corresponds closely to the X-ray crystallographic result. These results suggest that current docking algorithms are adequate to capture first-order determinants of binding fidelity (Fig. 1.4) (Gradler et al., 2001; Gruneberg et al., 2002; Powers et al., 2002; Wei et al., 2002).

How does performance in simple sites translate into larger, more drug-like sites? The consensus of many retrospective and prospective docking screens is that the ability to predict ligands and their geometries diminishes considerably in biology-relevant targets. In most cases, this failure reflects the increased complexity of the binding sites and the greater opportunities to find decoy ligand geometries. Nevertheless, there are examples of successful ligand prediction followed by structural

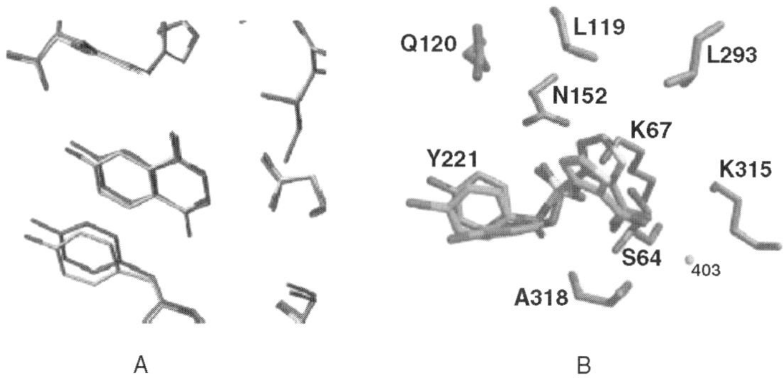

Figure 1.4 Predicted versus experimental structures from virtual screening. (**A**) The docked (carbons in orange) versus the crystallographic structure of the 4- aminophthalhydrazide bound to tRNA guanine transglycosylase (Gradler et al., 2001). (**B**) The docked (carbons in green) versus the crystallographic structure (carbons in orange) of 3-((4-chloroanilino)-sulfonyl)-thiophene-2-carboxylate bound to β-lactamase (enzyme carbons in gray) (Powers et al., 2002). See color plates.

determination, and in these cases the docking prediction is often close to the experimental result (Fig. 1.4) (Gradler et al., 2001; Gruneberg et al., 2002; Powers et al., 2002; Wei et al., 2002). These studies suggest that when the method does correctly predict a new ligand, even for a complicated, drug-like binding site, it does so for the right reasons.

An important question is whether structure-based screening is worth the effort, assuming that groups have access to high-throughput screening for ligand discovery. The two types of screens have been compared only a few times publicly, though rather more in unpublished industrial work. In the few published studies, the virtual screens had "hit rates" 10- to 1700-fold higher than the empirical screens (Table 1.3) (Doman et al., 2002; Kick et al., 1997; Oshiro et al., 2004; Paiva et al., 2001; Wyss et al., 2003). In the case with the best hit-rate enhancement, that of the diabetes-associated enzyme PTP-1B, the comparison was an imperfect one. Here, different libraries were targeted by the virtual and high-throughput screen, and a slightly different assay was used. In very recent work, Eric Brown and colleagues at McMaster University challenged the virtual screening community to predict the affinities of 50,000 molecules, none of which had been tested before but which were

TABLE 1.3 Hit Rates and Drug-Like Properties for Inhibitors Discovered with High-Throughput and Virtual Screening Against the Enzyme PTP-1B (Doman et al., 2002)

Technique	Number Compounds Tested	Hits with $IC_{50} < 100\ \mu M$	Hits with $IC_{50} < 10\ \mu M$	Rule of Five Compliant Hits[a]	Hit Rate[b]
HTS	400,000	85	6	23	0.021%
Docking	365	127	18	73	34.8%

[a]Number of 100 μM or better inhibitors that passed all four "rule of five" criteria (Paiva et al., 2001).
[b]The number of compounds experimentally tested divided by the number with IC_{50} values of 100 μM or less.

about to be tested in a high-throughput screen against dihydrofolate reductase (DHFR). In this experiment, the docking and HTS libraries were precisely the same, as were the experimental conditions. One of the startling results of this experiment was the very small number of hits to emerge from the screen. Indeed, whereas several groups were able to enrich putative inhibitors among their high-scoring molecules, the experimental group eventually concluded that they had no reliable hits at all (Elowe et al., 2005; Lang, et al., 2005). Intriguingly, several of the computational groups were able to indicate the lack of binders as part of their predictions (Brenk et al., 2005). Whereas the lack of experimental hits prevents definitive conclusions from this study, what does seem clear is that there is room for more of these comparative studies and "competitions."

1.6 CONCLUSIONS AND FUTURE DIRECTIONS

We have described the general steps involved in docking as well as typical algorithms for addressing each stage of the methodology. We have also listed examples in which prepackaged programs have been successfully used to discover novel inhibitors of a wide range of medically relevant applications. In the future, these methods will become progressively more integrated in the drug design process.

However, there are still several components of docking that need improvement. The two most-debated open questions in the field involve improving the scoring functions and developing algorithms for receptor flexibility. For scoring functions, research focuses on improving the treatment of solvent and the effects of entropy loss upon binding. Most docking approaches currently include drastic approximations of both of these properties, which have shown improvements over older methods. However, it is necessary to develop new schemes that treat these issues more accurately while preserving the speed of the calculation. Configurational entropy contributions are also difficult to calculate. Techniques that use molecular dynamics simulations to generate ensembles of ligand positions that are then rescored with high-accuracy scoring functions generate very accurate free energies of binding; however, they are computationally expensive (Kollman et al., 2000). It will be necessary to develop sampling techniques able to generate Boltzmann-weighted ensembles of ligand poses without requiring the use of expensive molecular dynamics calculations. Similarly, for receptor flexibility, many prepackaged programs have recently begun to develop new algorithms that allow for some measure of induced fit in the binding site. Several of these methods show great promise and will be further perfected with time, allowing for more and more structural rearrangement.

ACKNOWLEDGMENTS

We thank the sponsors of this work: the National Institute of General Medical Sciences of the National Institutes of Health through grants GM 56531 (P. O. Montellano, PI to IDK) and GM 59957 (BKS) as well as the AFPE fellowship to PTL.

REFERENCES

Abagyan, R., Totrov, M., and Kuznetsov, D. (1994). ICM—A new method for protein modeling and design: Applications to docking and structure prediction from the distorted native conformation. *J Comput Chem* **15**, 488–506.

Abarbanel, R. M. (1984). Protein structural knowledge engineering [dissertation]. University of California, San Francisco.

Alvarez, J. C. (2004). High-throughput docking as a source of novel drug leads. *Curr Opin Chem Biol* **8**, 365–370.

Anderson, A. C. (2003). The process of structure-based drug design. *Chem Biol* **10**, 787–797.

Arnold, J. R., Burdick, K. W., Pegg, S. C. H., Toba, S., Lamb, M. L., and Kuntz, I. D. (2004). Siteprint: Three-dimensional pharmacophore descriptors derived from protein binding sites for family based active site analysis, classification, and drug design. *J Chem Inf Comp Sci* **44**, 2190–2198.

Bender, A., and Glen, R. C. (2005). Discussion of measures of enrichment in virtual screening: Comparing the information content of descriptors with increasing levels of sophistication. *J Chem Inf Model* **45**, 1369–1375.

Berman, H. M., and Westbrook, J. D. (2004). The impact of structural genomics on the protein data bank. *Am J Pharmacogenomics* **4**, 247–252.

Berman, H. M., Olson, W. K., Beveridge, D. L., Westbrook, J., Gelbin, A., Demeny, T., Hsieh, S. H., Srinivasan, A. R., and Schneider, B. (1992). The nucleic acid database. A comprehensive relational database of three-dimensional structures of nucleic acids. *Biophys J* **63**, 751–759.

Berman, H. M., Westbrook, J., Feng, Z., Gilliland, G., Bhat, T. N., Weissig, H., Shindyalov, I. N., and Bourne, P. E. (2000). The protein data bank. *Nucleic Acids Res* **28**, 235–242.

Bernacki, K., Kalyanaraman, C., and Jacobson, M. P. (2005). Virtual ligand screening against escherichia coli dihydrofolate reductase: Improving docking enrichment using physics-based methods. *J Biomol Screen* **10**, 675–681.

Beveridge, D. L., and Dicapua, F. M. (1989). Free-energy via molecular simulation—Applications to chemical and biomolecular systems. *Annu Rev Biophys Biophys Chem* **18**, 431–492.

Boehm, H. J., Boehringer, M., Bur, D., Gmuender, H., Huber, W., Klaus, W., Kostrewa, D., Kuehne, H., Luebbers, T., Meunier-Keller, N., and Mueller, F. (2000). Novel inhibitors of DNA gyrase: 3d structure based biased needle screening, hit validation by biophysical methods, and 3d guided optimization. A promising alternative to random screening. *J Med Chem* **43**, 2664–2674.

Brenk, R., Irwin, J. J., and Shoichet, B. K. (2005). Here be dragons: Docking and screening in an uncharted region of chemical space. *J Biomol Screen* **10**, 667–674.

Brenk, R., Vetter, S., Boyce, S., Goodin, D. B., and Shoichet, B. K. (2006). Probing molecular docking in a charged model binding site. *J Mol Biol*, **5**, 1449–1470.

Brik, A., Muldoon, J., Lin, Y. C., Elder, J. H., Goodsell, D. S., Olson, A. J., Fokin, V. V., Sharpless, K. B., and Wong, C. H. (2003). Rapid diversity-oriented synthesis in microtiter plates for *in situ* screening of HIV protease inhibitors. *ChemBioChem* **4**, 1246–1248.

Brik, A., Alexandratos, J., Lin, Y. C., Elder, J. H., Olson, A. J., Wlodawer, A., Goodsell, D. S., and Wong, C. H. (2005). 1,2,3-Triazole as a peptide surrogate in the rapid synthesis of HIV-1 protease inhibitors. *ChemBioChem* **6**, 1167–1169.

Brooijmans, N., and Kuntz, I. D. (2003). Molecular recognition and docking algorithms. *Annu Rev Biophys Biomol Struct* **32**, 335–373.

Chance, M. R., Fiser, A., Sali, A., Pieper, U., Eswar, N., Xu, G. P., Fajardo, J. E., Radhakannan, T., and Marinkovic, N. (2004). High-throughput computational and experimental techniques in structural genomics. *Genome Res* **14**, 2145–2154.

Claussen, H., Buning, C., Rarey, M., and Lengauer, T. (2001). Flexe: Efficient molecular docking considering protein structure variations. *J Mol Biol* **308**, 377–395.

Doman, T. N., McGovern, S. L., Witherbee, B. J., Kasten, T. P., Kurumbail, R., Stallings, W. C., Connolly, D. T., and Shoichet, B. K. (2002). Molecular docking and high-throughput screening for novel inhibitors of protein tyrosine phosphatase-1b. *J Med Chem* **45**, 2213–2221.

Eldridge, M. D., Murray, C. W., Auton, T. R., Paolini, G. V., and Mee, R. P. (1997). Empirical scoring functions. 1. The development of a fast empirical scoring function to estimate the binding affinity of ligands in receptor complexes. *J Comput Aided Mol Des* **11**, 425–445.

Elowe, N. H., Blanchard, J. E., Cechetto, J. D., and Brown, E. D. (2005). Experimental screening of dihydrofolate reductase yields a "Test set" of 50000 small molecules for a computational data-mining and docking competition. *J Biomol Screen* **10**, 653–657.

Enyedy, I. J., Ling, Y., Nacro, K., Tomita, Y., Wu, X., Cao, Y., Guo, R., Li, B., Zhu, X., Huang, Y., Long, Y. Q., Roller, P. P., Yang, D., and Wang, S. (2001). Discovery of small-molecule inhibitors of bcl-2 through structure-based computer screening. *J Med Chem* **44**, 4313–4324.

Evers, A., and Klebe, G. (2004). Ligand-supported homology modeling of g-protein-coupled receptor sites: Models sufficient for successful virtual screening. *Angew Chem Int Edit Engl* **43**, 248–251.

Ewing, T. J., Makino, S., Skillman, A. G., and Kuntz, I. D. (2001). Dock 4.0: Search strategies for automated molecular docking of flexible molecule databases. *J Comput Aided Mol Des* **15**, 411–428.

Feig, M., and Brooks, C. L. (2004). Recent advances in the development and application of implicit solvent models in biomolecule simulations. *Curr Opin Chem Biol* **14**, 217–224.

Feig, M., Onufriev, A., Lee, M. S., Im, W., Case, D. A., and Brooks, C. L. (2004). Performance comparison of generalized born and poisson methods in the calculation of electrostatic solvation energies for protein structures. *J Comput Chem* **25**, 265–284.

Freymann, D. M., Wenck, M. A., Engel, J. C., Feng, J., Focia, P. J., Eakin, A. E., and Craig, S. P. (2000). Efficient identification of inhibitors targeting the closed active site conformation of the hprt from trypanosoma cruzi. *Chem Biol* **7**, 957–968.

Friesner, R. A., Banks, J. L., Murphy, R. B., Halgren, T. A., Klicic, J. J., Mainz, D. T., Repasky, M. P., Knoll, E. H., Shelley, M., Perry, J. K., Shaw, D. E., Francis, P., and Shenkin, P. S. (2004). Glide: A new approach for rapid, accurate docking and scoring. 1. Method and assessment of docking accuracy. *J Med Chem* **47**, 1739–1749.

Gehlhaar, D. K., Verkhivker, G. M., Rejto, P. A., Sherman, C. J., Fogel, D. B., Fogel, L. J., and Freer, S. T. (1995). Molecular recognition of the inhibitor AG-1343 by HIV-1 protease—Conformationally flexible docking by evolutionary programming. *Chem Biol* **2**, 317–324.

Gohlke, H., and Case, D. A. (2004). Converging free energy estimates: MM-PB(GB)SA studies on the protein–protein complex ras-raf. *J Comput Chem* **25**, 238–250.

Gradler, U., Gerber, H. D., Goodenough-Lashua, D. M., Garcia, G. A., Ficner, R., Reuter, K., Stubbs, M. T., and Klebe, G. (2001). A new target for shigellosis: Rational design and crystallographic studies of inhibitors of tRNA-guanine transglycosylase. *J Mol Biol* **306**, 455–467.

Graves, A. P., Brenk, R., and Shoichet, B. K. (2005). Decoys for docking. *J Med Chem* **48**, 3714–3728.

Gruneberg, S., Wendt, B., and Klebe, G. (2001). Subnanomolar inhibitors from computer screening: A model study using human carbonic anhydrase ii. *Angew Chem Int Ed Engl* **40**, 389–393.

Gruneberg, S., Stubbs, M. T., and Klebe, G. (2002). Successful virtual screening for novel inhibitors of human carbonic anhydrase: Strategy and experimental confirmation. *J Med Chem* **45**, 3588–3602.

Hajduk, P. J., Huth, J. R., and Fesik, S. W. (2005). Druggability indices for protein targets derived from nmr-based screening data. *J Med Chem* **48**, 2518–2525.

Halgren, T. A., Murphy, R. B., Friesner, R. A., Beard, H. S., Frye, L. L., Pollard, W. T., and Banks, J. L. (2004). Glide: A new approach for rapid, accurate docking and scoring. 2. Enrichment factors in database screening. *J Med Chem* **47**, 1750–1759.

Hensen, C., Hermann, J. C., Nam, K. H., Ma, S. H., Gao, J. L., and Holtje, H. D. (2004). A combined qm/mm approach to protein–ligand interactions: Polarization effects of the hiv-1 protease on selected high affinity inhibitors. *J Med Chem* **47**, 6673–6680.

Hoffmann, D., Kramer, B., Washio, T., Steinmetzer, T., Rarey, M., and Lengauer, T. (1999). Two-stage method for protein–ligand docking. *J Med Chem* **42**, 4422–4433.

Honig, B., and Nicholls, A. (1995). Classical electrostatics in biology and chemistry. *Science* **268**, 1144–1149.

Honma, T., Hayashi, K., Aoyama, T., Hashimoto, N., Machida, T., Fukasawa, K., Iwama, T., Ikeura, C., Ikuta, M., Suzuki-Takahashi, I., Iwasawa, Y., Hayama, T., Nishimura, S., and Morishima, H. (2001). Structure-based generation of a new class of potent cdk4 inhibitors: New *de novo* design strategy and library design. *J Med Chem* **44**, 4615–4627.

Huang, N., Nagarsekar, A., Xia, G., Hayashi, J., and MacKerell, A. D., Jr. (2004). Identification of non-phosphate-containing small molecular weight inhibitors of the tyrosine kinase p56 lck sh2 domain via *in silico* screening against the py + 3 binding site. *J Med Chem* **47**, 3502–3511.

Irwin, J. J., and Shoichet, B. K. (2005). Zinc—A free database of commercially available compounds for virtual screening. *J Chem Inf Model* **45**, 177–182.

Iwata, Y., Arisawa, M., Hamada, R., Kita, Y., Mizutani, M. Y., Tomioka, N., Itai, A., and Miyamoto, S. (2001). Discovery of novel aldose reductase inhibitors using a protein structure-based approach: 3d-database search followed by design and synthesis. *J Med Chem* **44**, 1718–1728.

Jones, G., Willett, P., Glen, R. C., Leach, A. R., and Taylor, R. (1997). Development and validation of a genetic algorithm for flexible docking. *J Mol Biol* **267**, 727–748.

Jorgensen, W. L. (2004). The many roles of computation in drug discovery. *Science* **303**, 1813–1818.

Kick, E. K., Roe, D. C., Skillman, A. G., Liu, G., Ewing, T. J., Sun, Y., Kuntz, I. D., and Ellman, J. A. (1997). Structure-based design and combinatorial chemistry yield low nanomolar inhibitors of cathepsin d. *Chem Biol* **4**, 297–307.

Knegtel, R. M. A., Kuntz, I. D., and Oshiro, C. M. (1997). Molecular docking to ensembles of protein structures. *J Mol Biol* **266**, 424–440.

Kollman, P. (1993). Free-energy calculations—Applications to chemical and biochemical phenomena. *Chem Rev* **93**, 2395–2417.

Kollman, P. A., Massova, I., Reyes, C., Kuhn, B., Huo, S., Chong, L., Lee, M., Lee, T., Duan, Y., Wang, W., Donini, O., Cieplak, P., Srinivasan, J., Case, D. A., and Cheatham, T. E., 3rd. (2000). Calculating structures and free energies of complex molecules: Combining molecular mechanics and continuum models. *Acc Chem Res* **33**, 889–897.

Kubinyi, H. (2006). Success stories of computer-aided design. In S. Eakins (Ed.), *Computer Applications in Pharmaceutical Research and Development*. New York: Wiley, pp. 000–000.

Kuntz, I. D., Blaney, J. M., Oatley, S. J., Langridge, R., and Ferrin, T. E. (1982). A geometric approach to macromolecule–ligand interactions. *J Mol Biol* **161**, 269–288.

Laird, E. R., and Blake, J. F. (2004). Structure-based generation of viable leads from small combinatorial libraries. *Curr Opin Drug Discovery Dev* **7**, 354–359.

Lang, P. T., Kuntz, I. D., Maggiora, G. M., and Bajorath, J. (2005). Evaluating the high-throughput screening computations. *J Biomol Screen* **10**, 649–652.

Li, C., Xu, L., Wolan, D. W., Wilson, I. A., and Olson, A. J. (2004). Virtual screening of human 5-aminoimidazole-4-carboxamide ribonucleotide transformylase against the nci diversity set by use of autodock to identify novel nonfolate inhibitors. *J Med Chem* **47**, 6681–6690.

Lingham, R. B., Silverman, K. C., Jayasuriya, H., Kim, B. M., Amo, S. E., Wilson, F. R., Rew, D. J., MD, M. D. S., Bergstrom, J. D., Koblan, K. S., Graham, S. L., Kohl, N. E., Gibbs, J. B., and Singh, S. B. (1998). Clavaric acid and steroidal analogues as ras- and fpp-directed inhibitors of human farnesyl-protein transferase. *J Med Chem* **41**, 4492–4501.

Lorber, D. M., and Shoichet, B. K. (1998). Flexible ligand docking using conformational ensembles. *Protein Sci* **7**, 938–950.

Lorber, D. M., and Shoichet, B. K. (2005). Hierarchical docking of databases of multiple ligand conformations. *Curr Top Med Chem* **5**, 739–749.

Lyne, P. D., Kenny, P. W., Cosgrove, D. A., Deng, C., Zabludoff, S., Wendoloski, J. J., and Ashwell, S. (2004). Identification of compounds with nanomolar binding affinity for checkpoint kinase-1 using knowledge-based virtual screening. *J Med Chem* **47**, 1962–1968.

McMartin, C., and Bohacek, R. S. (1997). Qxp: Powerful, rapid computer algorithms for structure-based drug design. *J Comput Aided Mol Des* **11**, 333–344.

Meng, E. C., Shoichet, B. K., and Kuntz, I. D. (1992). Automated docking with grid-based energy evaluation. *J Comput Chem* **13**, 505–524.

Miranker, A., and Karplus, M. (1991). Functionality maps of binding sites: A multiple copy simultaneous search method. *Proteins* **11**, 29–34.

Morris, G. M., Goodsell, D. S., Halliday, R. S., Huey, R., Hart, W. E., Belew, R. K., and Olson, A. J. (1998). Automated docking using a lamarckian genetic algorithm and an empirical binding free energy function. *J Comput Chem* **19**, 1639–1662.

Moustakas, D., Lang, P.T., Pegg, S., Pettersen, E., Kuntz, I., Brooijmans, N., and Rizzo, R. (2006). Development and validation of a modular, extensible docking program: Dock 5, *J Comput Aided Mol Des* **20**, 601–619.

Murcko, M. A. (1997). An introduction to *de novo* ligand design. In P. S. Charifson (Ed.), *Practical Application of Computer-Aided Drug Design*. New York: Marcel Dekker, p. 395.

Nissen, P., Hansen, J., Ban, N., Moore, P. B., and Steitz, T. A. (2000). The structural basis of ribosome activity in peptide bond synthesis. *Science* **289**, 920–930.

Nissink, J. W., Murray, C., Hartshorn, M., Verdonk, M. L., Cole, J. C., and Taylor, R. (2002). A new test set for validating predictions of protein–ligand interaction. *Proteins* **49**, 457–471.

Oshiro, C., Bradley, E. K., Eksterowicz, J., Evensen, E., Lamb, M. L., Lanctot, J. K., Putta, S., Stanton, R., and Grootenhuis, P. D. (2004). Performance of 3d-database molecular docking studies into homology models. *J Med Chem* **47**, 764–767.

Osterberg, F., Morris, G. M., Sanner, M. F., Olson, A. J., and Goodsell, D. S. (2002). Automated docking to multiple target structures: Incorporation of protein mobility and structural water heterogeneity in autodock. *Proteins* **46**, 34–40.

Paiva, A. M., Vanderwall, D. E., Blanchard, J. S., Kozarich, J. W., Williamson, J. M., and Kelly, T. M. (2001). Inhibitors of dihydrodipicolinate reductase, a key enzyme of the diaminopimelate pathway of *Mycobacterium tuberculosis*. *Biochim Biophys Acta* **1545**, 67–77.

Pang, Y. P., Perola, E., Xu, K., and Prendergast, F. G. (2001a). Eudoc: A computer program for identification of drug interaction sites in macromolecules and drug leads from chemical databases. *J Comput Chem* **22**, 1750–1771.

Pang, Y. P., Xu, K., Kollmeyer, T. M., Perola, E., McGrath, W. J., Green, D. T., and Mangel, W. F. (2001b). Discovery of a new inhibitor lead of adenovirus proteinase: Steps toward selective, irreversible inhibitors of cysteine proteinases. *FEBS Lett* **502**, 93–97.

Pearlman, D. A., Case, D. A., Caldwell, J. W., Ross, W. S., Cheatham, T. E., Debolt, S., Ferguson, D., Seibel, G., and Kollman, P. (1995). Amber, a package of computer programs for applying molecular mechanics, normal-mode analysis, molecular dynamics and free-energy calculations to simulate the structural and energetic properties of molecules. *Comput Phys Comm* **91**, 1–41.

Pickett, S. D., Sherborne, B. S., Wilkinson, T., Bennett, J., Borkakoti, N., Broadhurst, M., Hurst, D., Kilford, I., McKinnell, M., and Jones, P. S. (2003). Discovery of novel low molecular weight inhibitors of impdh via virtual needle screening. *Bioorg Med Chem Lett* **13**, 1691–1694.

Powers, R. A., Morandi, F., and Shoichet, B. K. (2002). Structure-based discovery of a novel, noncovalent inhibitor of ampc beta-lactamase. *Structure* **10**, 1013–1023.

Richards, F. M. (1977). Areas, volumes, packing, and protein-structure. *Annu Rev Biophys Bioeng* **6**, 151–176.

Rizzo, R., Wang, D., Tirado-Rives, J., and Jorgensen, W. (2000). Validation of a model for the complex of hiv-1 reverse transcriptase with sustiva through computation of resistance profiles. *J Am Chem Soc* **122**, 12898–12900.

Rosenfeld, R. J., Goodsell, D. S., Musah, R. A., Morris, G. M., Goodin, D. B., and Olson, A. J. (2003). Automated docking of ligands to an artificial active site: Augmenting crystallographic analysis with computer modeling. *J Comput Aided Mol Des* **17**, 525–536.

Schapira, M., Raaka, B. M., Samuels, H. H., and Abagyan, R. (2001). *In silico* discovery of novel retinoic acid receptor agonist structures. *BMC Struct Biol* **1**, 1.

Schapira, M., Raaka, B. M., Das, S., Fan, L., Totrov, M., Zhou, Z. U., Wilson, S. R., Abagyan, R., and Samuels, H. H. (2003). Discovery of diverse thyroid hormone receptor antagonists by high-throughput docking. *Proc Natl Acad Sci USA* **100**, 7354–7359.

Sherman, W., Day, T., Jacobson, M., Friesner, R., and Farid, R. (1996). Novel procedure for modeling ligand/receptor induced fit effects. *J Med Chem*

Staunton, D., Owen, J., and Campbell, I. D. (2003). Nmr and structural genomics. *Acc Chem Res* **36**, 207–214.

Totrov, M., and Abagyan, R. (1997). Flexible protein–ligand docking by global energy optimization in internal coordinates. *Proteins* 215–220.

Verdonk, M. L., Cole, J. C., Hartshorn, M. J., Murray, C. W., and Taylor, R. D. (2003). Improved protein–ligand docking using gold. *Proteins* **52,** 609–623.

Wang, J. M., Morin, P., Wang, W., and Kollman, P. A. (2001). Use of mm-pbsa in reproducing the binding free energies to hiv-1 rt of tibo derivatives and predicting the binding mode to hiv-1 rt of efavirenz by docking and mm-pbsa. *J Am Chem Soc* **123**, 5221–5230.

Webb, T. R. (2005). Current directions in the evolution of compound libraries. *Curr Opin Drug Discovery Dev* **8**, 303–308.

Wei, B. Q., Baase, W. A., Weaver, L. H., Matthews, B. W., and Shoichet, B. K. (2002). A model binding site for testing scoring functions in molecular docking. *J Mol Biol* **322**, 339–355.

Wei, B. Q., Weaver, L. H., Ferrari, A. M., Matthews, B. W., and Shoichet, B. K. (2004). Testing a flexible-receptor docking algorithm in a model binding site. *J Mol Biol* **337**, 1161–1182.

Weiner, P. K., and Kollman, P. A. (1981). Amber: Assisted model building with energy refinement. A general program for modeling molecules and their interactions. *J Comp Chem* **2**, 287.

Wyss, P. C., Gerber, P., Hartman, P. G., Hubschwerlen, C., Locher, H., Marty, H. P., and Stahl, M. (2003). Novel dihydrofolate reductase inhibitors. Structure-based versus diversity-based library design and high-throughput synthesis and screening. *J Med Chem* **46**, 2304–2312.

Zhou, Z. G., and Madura, J. D. (2004). Relative free energy of binding and binding mode calculations of hiv-1 rt inhibitors based on dock-mm-pb/gs. *Proteins: Struct Funtct Bioinf* **57**, 493–503.

Zou, X., Sun, Y., and Kuntz, I. D. (1999). Inclusion of solvation in ligand binding free energy calculations using the generalized-born model. *J Am Chem Soc* **121**, 8033–8043.

CHAPTER 2

RECENT ADVANCES IN DRUG DISCOVERY RESEARCH USING STRUCTURE-BASED VIRTUAL SCREENING TECHNIQUES: EXAMPLES OF SUCCESS FOR DIVERSE PROTEIN TARGETS

*Sutapa Ghosh, Aihua Nie, Jing An, and Ziwei Huang**

2.1 INTRODUCTION

The identification of a proper lead compound for a given molecular target is a critical step in the process of drug discovery. Traditionally, high-throughput screening (HTS) of large chemical libraries has been a primary source of identification of novel lead compounds. In recent years, the rapid progress in the human genome project has provided an ever-increasing number of potential drug targets to be screened (Westbrook et al., 2003). Moreover, advances in high-throughput (HTP) strategies for structure determination (Blundell and Abell, 2002) have yielded a large number of three-dimensional (3D) structures of therapeutically relevant targets. To this end, computational tools are becoming increasingly important to efficiently integrate

*Corresponding author: Ziweihuang@burnham.org

Drug Discovery Research: New Frontiers in the Post-Genomic Era, Edited by Ziwei Huang

this rapidly increasing wealth of structural data with the more traditional lead optimization techniques. Amongst them, virtual screening is bringing a more focused approach to drug discovery by using high-performance computing to analyze chemical databases and prioritize compounds for synthesis and analysis.

2.2 VIRTUAL SCREENING OF CHEMICAL LIBRARIES

Virtual screening is a computational filter to reduce the size of a chemical library to be screened experimentally and offers an opportunity to drastically reduce the time and effort associated with lead identification. The benefits are a focused subset with enhanced hit rates and a prioritized library for screening and synthesis. There are two fundamental approaches for virtual screening: a ligand-based approach (Bajorath, 2002) and a receptor-based approach (Gohlke et al., 2000). The ligand-based approach aims to identify molecules with physical and chemical similarities (pharmacophore-based, descriptor-based) to known ligands that are likely to interact with the target. This type of approach limits the diversity of the hits as they are biased by the properties of known ligands. Receptor-based virtual screening (protein–ligand docking, active site-directed pharmacophores) uses knowledge of the target protein's 3D structure to impose a structure-based filter on a chemical database to select candidate compounds that are likely to interact favorably with the protein's active site residues. This is a more open-ended approach that allows the identification of structurally novel ligands that may have similar interactions like known ligands or may have different interactions with other parts of the binding site. Receptor-based virtual screening (RBVS) and some of its recent applications will be discussed here.

2.2.1 Receptor-Based Virtual Screening

Molecular recognition (Fische, 1894) is the fundamental basis for drug action in which drug molecules exhibit pharmacological activity by binding to a target protein and forming a stable protein–ligand complex. The major factors contributing to the stability of such complexes include shape complementarity of the ligand to the ligand-binding site, formation of key hydrogen bonds, and good hydrophobic and electrostatic interactions with the active site residues as well as stable conformation of the bound ligand when the complex is formed. Receptor-based virtual screening (RBVS) aims to exploit the molecular recognition between a ligand and a target protein to select chemical entities that bind strongly to the active sites of biologically relevant targets for which the three-dimensional structures are known or inferred. This approach uses docking and scoring (Muegge and Enyedy, 2004) to sort the candidates in a virtual library. The docking algorithms (Halperin et al., 2002; Brooijmans and Kuntz, 2003) deal with the prediction of ligand conformation and orientation (or pose) within the targeted active site of the receptor. The scoring methods (Wang et al., 2003) evaluate the binding interactions between the target and the small molecule and aim to predict the biological activity of the compound based on the computed binding interactions.

In RBVS, one starts with a 3D structure of a target protein and a 3D database of ligands and uses virtual filtering to dock and score compounds as a means to identify potential lead candidates for further analysis and improvement. Some of the most widely used software programs in high-throughput RBVS are DOCK (Ewing et al., 2001), FlexX (Rarey et al., 1996), GOLD (Jones et al., 1997), Glide (Halgren et al., 2004), ICM (Totrov and Abagyan, 1997), FRED (McGann et al., 2003), and AutoDock (Goodsell et al., 1993). All such software programs have docking and scoring algorithms to generate subsets of a compound collection with higher affinity against a target by predicting their binding mode (by docking) and affinity (by scoring) and retrieving those with the highest scores. These programs can dock single or multiple conformers of small-molecule libraries. Although docking algorithms have matured significantly over the years, some challenges still remain, such as the treatment of protein-flexibility issues arising from induced fit or other conformational changes of the protein that occur upon ligand binding and the participation of water molecules in protein–ligand interactions. In some cases, limited resolution of available crystal structures of target proteins also affect the outcome of the virtual screening effort.

Receptors are usually represented in three forms in the different docking algorithms. They are treated as either atoms, surfaces, or potential energy grids (Halperin et al., 2002). Ligand flexibility is also treated in three different ways (Brooijmans and Kuntz, 2003) in these algorithms. The three ligand flexibility treatment methods are: (a) systemic search (Leach, 1996), using conformational search or incremental construction; (b) random or stochastic search (Kitchen et al., 2004), using Monte Carlo (Hart and Read, 1992) genetic algorithm (Oshiro et al., 1995), or tabu search (Baxter et al., 1998); and (c) simulation methods using molecular dynamics (Di Nola et al., 1994). The treatment of protein flexibility (Carlson and McCammon, 2000) in docking algorithms is less advanced than that of ligand flexibility and approaches such as rotamer libraries (Leach, 1994), molecular dynamics, and Monte Carlo simulations, have, in some cases, been applied to a limited region of the protein binding site. Docking programs such as DOCK, FlexX, and Glide use the systematic approach for treating small-molecule flexibility; AutoDock, GOLD, and MOE use the random/stochastic method for treating ligands; and DOCK, Glide, MOE (2003), and AutoDock incorporate the simulation option for ligand flexibility treatment.

The evaluation and ranking of the predicted ligand poses are critical steps in the success of the receptor-based virtual screening technique. The scoring methods (Ewing et al., 2001) implemented in various docking software programs are empirically or semiempirically derived, and these functions attempt to estimate the tightness of target–ligand interactions in bound complexes. Most scoring functions implemented in docking programs make various assumptions in the evaluation of the docked complex and oversimplify certain physical characteristics (such as hydrophobic, entropic, and solvation effects) that determine molecular recognition. Three major categories of scoring functions are currently used in the major docking programs; empirical, force-field based, or knowledge-based scoring functions.

The various force-field scoring functions are based on the different molecular mechanics of the force-field parameters used in the different programs. These scoring functions typically measure only the potential energy of the system by quantifying the sum of the energies between the receptor–ligand interaction energy and the internal ligand energy induced by binding. Most force-field scoring functions consider only one protein conformation, which greatly simplifies the actual scenario. The major drawback of this scheme is that it often relies on nonbonded interaction energy terms of standard force fields formulated to model gas-phase contributions to structure and energetics and do not include solvation and entropic terms (Kitchen et al., 2004). Some examples of force-field-based scoring functions include: G-score (Kramer et al., 1999a,b), DOCK energy score (Meng et al., 1992), and GOLD score (Verdonk et al., 2003).

The empirical scoring functions are fit to reproduce experimental data such as binding energy as the sum of individual uncorrelated terms. These coefficients were determined by either regression analysis or partial least-squares fitting on a large training set of experimentally available receptor–ligand complexes (Bohm, 1994, 1998). The success of these scoring functions are measured by comparing the predicted to experimental binding affinity. The strength of these types of functions is that they are trained on a large number of receptor–ligand complexes available in the PDB. The weakness is that because they are constructed empirically, they are sensitive to the complexes represented in their training sets and therefore are unsuitable for VS projects, where these functions are likely to encounter novel ligands and novel ligand orientations. Some examples of empirical scores include: LUDI (Bohm, 1992), Chemscore (Eldridge, 1997), SCORE (Tao and Lai, 2001), and *F*-Score (Rarey et al., 1996).

Knowledge-based scoring functions are designed to reproduce experimental structures rather than binding energies. In this case, the structural information from a large number of complexes in the PDB was used to construct atomic interaction potentials, where a number of atom-type interactions are defined depending on their molecular environment. Similar to the empirical scoring functions, the knowledge-based functions are essentially biased toward the limited information that is implicitly encoded in their derivation. However, their computational simplicity allows them to be effectively used in virtual screening and in scoring large chemical databases. Some examples of knowledge-based scores include PMF (Muegge and Martin, 1999; Muegge, 2000, 2001) and DrugScore (Gohlke et al., 2000).

Regardless of the choice of docking software used and even when the binding conformations are predicted, the major drawback in the RBVS procedure is the unavailability of appropriate scoring functions to differentiate between correct and incorrect poses of bound ligands and to identify false-negative and false-positive hits. Given the limitations of the current scoring functions, a recent trend in this field has been the use of consensus scoring schemes (Charifson et al., 1999) and visual inspection to select likely candidates. Consensus scoring combines the information from different scores to balance errors in individual scores, reduces the number of false positives identified by individual scoring functions, and improves the odds of identifying the true ligands.

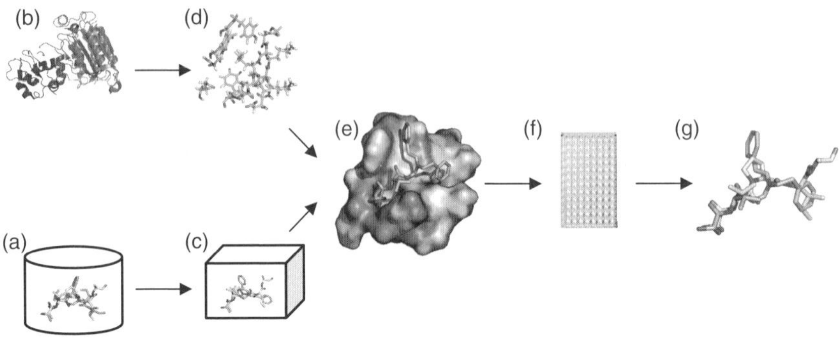

Figure 2.1 Typical receptor-based virtual screening scheme consisting of many steps in repeated cycles. **(A)** Small-molecule database preparation: Choose library, strip counterions, add H-atoms, check/fix valency problems, protonate at physiological pH, calculate 2D properties, convert to 3D, energy minimize small molecule. **(B)** Target structure selection: X-ray, NMR or homology model, quality of structure. **(C)** Filter library: Lipinski's Rule of Five, pharmacophore model. **(D)** Binding site definition: Ligand binding site, catalytic site, protein–protein interaction site. **(E)** Docking to predict ligand conformation and orientation at binding site. Scoring to evaluate the interaction energy between the target and the ligand. Visual inspection to select molecules showing good interactions with binding site residues. **(F)** Testing of virtual hits in biological assay for activity. **(G)** Lead optimization: Compounds showing biological activity further modified using medicinal chemistry, SAR, structural studies to get second-generation compounds. See color plates.

The output of a docking-based screen is a set of 3D models of the predicted binding mode of each compound against the receptor, together with a ranking that is a measure of the quality of fit. Thus, it represents the most detailed and relevant model for identifying a receptor-focused subset of ligands. A few important factors to consider when selecting docking software are the capability of the software for iterative refinement of docking parameter/protocol based on new results, adaptability to additional scoring functions, pre- and/or post-docking filters, design, components, and results of validation studies, user learning curve, customer support, cost, speed, user interface, input/output structural file formats, code availability, and upgrading possibility. Like any HTP technology, RBVS workflows are designed to balance accuracy and time commitment. A typical RBVS scheme is shown in Fig. 2.1. Over the last few years, a number of laboratories have published virtual screening studies using a variety of workflow and docking programs, and the number is steadily rising, demonstrating the level of interest in this approach and its broad applicability.

For the small-molecule compound database, it is desirable to have maximum structural diversity in the virtual library so as to maximize the chances of finding a hit for the target macromolecules. Some of the commonly used small-molecule databases in virtual screening are large public databases such as ZINC (3.3 million commercially available compounds; free), Available Chemicals Directory (ACD, 4 million entries; not free), National Cancer Institute compound database (NCI, 400,000 entries; free), and MDDR (MDL Drug Data Report, >147,000 entries).

Other possibilities for small-molecule collection include CMC (Comprehensive Medicinal Chemistry, >8600 entries), CSD (Cambridge Structural Database), Beilstein, and SciFinder. Large pharmaceutical companies have corporate databases of a few million compounds.

2.2.2 Recent Successes from RBVS

2.2.2.1 Bcl-2 Family Proteins

Proteins in the Bcl-2 family are central regulators of programmed cell death, and members that inhibit apoptosis, such as Bcl-x_L and Bcl-2, are overexpressed in many cancers, making them important targets for anticancer drug development (Shangary and Johnson, 2003). The development of inhibitors of these proteins as potential anticancer therapeutics has been first explored in our laboratory by virtual screening of a chemical database using RBVS techniques. At the time the study was carried out, no experimental structure of Bcl-2 was available. Because Bcl-2 and Bcl-x_L have 47.2% sequence identity, a homology model of Bcl-2 was constructed based on the NMR structure of Bcl-x_L complexed with the Bak peptide (Sattler et al., 1997). The Bak BH3 peptide binds to a largely hydrophobic binding pocket formed by the BH1, BH2, and BH3 domains of in Bcl-2 as well as in Bcl-x_L. This binding pocket in Bcl-2 appears to be essential for its anti-apoptotic function since mutations at this site abolish its function (Yin Oltvai et al., 1994; Cosulich et al., 1997). Using the Bak binding surface pocket in the homology model of Bcl-2 as the targeted active site, virtual screening of the Available Chemicals Directory (ACD) database of 193,833 compounds was carried out (Wang et al., 2000) using the program DOCK3.5. Fifty-three candidate molecules that displayed relatively lower binding energy, favorable shape complementarity, and/or potential of forming hydrogen bonds with the Bcl-2 protein residues were picked for the next phase of biological testing. Twenty-eight of these compounds with diverse scaffold and possible druglike properties were actually obtained for biological testing. Subsequent fluorescence polarization assays verified that the most potent compound, HA14-1, competed with Bak BH3 peptide for binding to Bcl-2 ($IC_{50} = \sim 9\ \mu M$). The structure of this compound is shown in Table 2.1, and a structural model for the complex of HA14-1 with the Bcl-2 surface pocket as predicted by computer docking calculations is shown in Fig. 2.2. Apoptosis induced by HA14-1 in HL-60 cells is characterized by activation of caspase protease and loss of $\Delta\psi_m$. This was the first small-molecule ligand identified for this family of proteins, and this identification opened new avenues for the development of anticancer agents targeting members of this family of proteins.

Another small organic molecular inhibitor of Bcl-2, named compound 6 (structure shown in Table 2.1), was discovered by virtual screening in another laboratory (Enyedy et al., 2001), starting with a homology model of Bcl-2. The binding site was identified as all residues within 8 Å of the Bak BH3 peptide binding site of Bcl-2. The National Cancer Institute database of 206,876 compounds was virtually screened against the modeled structure of Bcl-2 using the program DOCK3.5. The top 500 compounds with the best scores were further filtered based on those with nonpeptidic character. Thirty-five compounds were

TABLE 2.1 Some Examples of Recent Success in RBVS

Target Protein	Disease Treatment Area	Small Moleule Library	VS Software Used	Potency in Binding Assay	Structure of a Representative Hit Compound
Bcl-2 (Wang et al., 2000)	Cancer	ACD	DOCK3.5	$IC_{50} = 9\ \mu M$	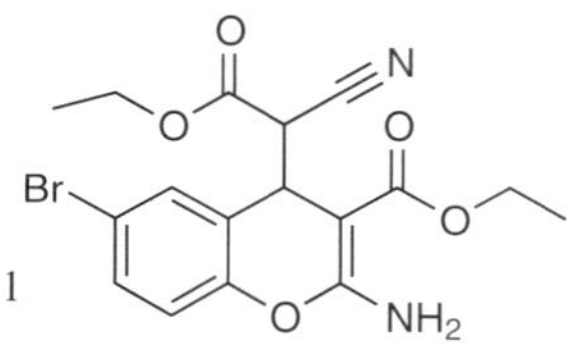 HA14-1
Bcl-2 (Enyedy et al., 2001)	Cancer	NCI DB	DOCK	$IC_{50} = 4\ \mu M$	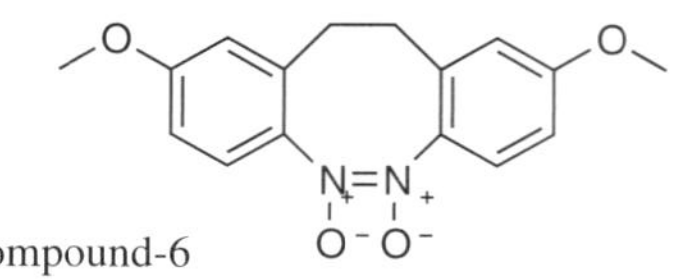 Compound-6
Bcl-2 (Lugovskoy et al., 2002)	Cancer	Chembridge Chemnavigator	TreeDOCK	$IC_{50} = 8.5 \pm 0.6\ \mu M$	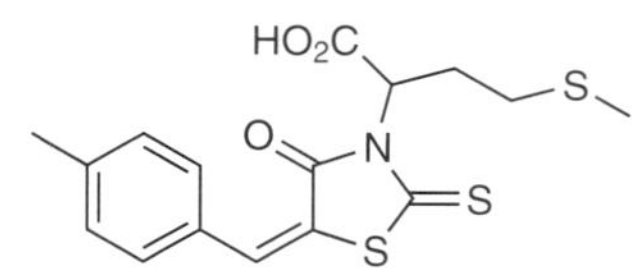 BH3I-ISCH$_3$
XIAP (Nikolovska-Coleska et al., 2004)	Cancer	TCM-3D	DOCK	$IC_{50} = 4.1 \pm 1.1\ \mu M$	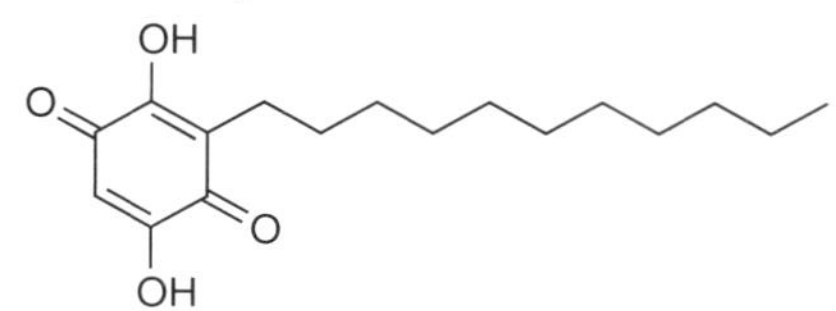 Embelin

Acetyl Cholinesterase (Mizutani and Itai, 2004)	Alzheimer's disease	ACD & Maybridge	ADAM & EVE	$IC_{50} = 590$ nM	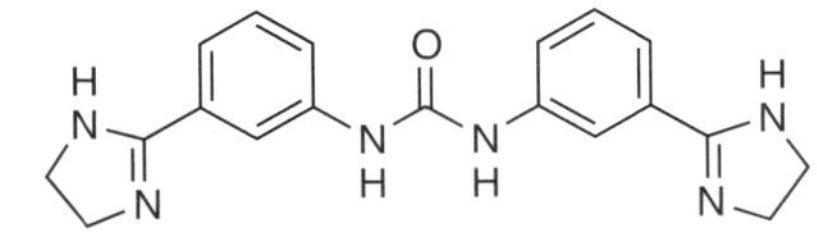 Compound-1
Acetyl Cholinesterase (Rollinger et al., 2004)	Alzheimer's disease	NPD	LigandScout	$IC_{50} = 168.6$ µM	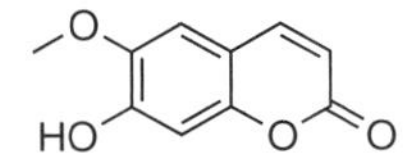 Scopoletin
ER (Zhao and Brinton, 2005)	Alzheimer's disease	Plant-based library	GOLD 2.0	$IC_{50} = 680$ nM	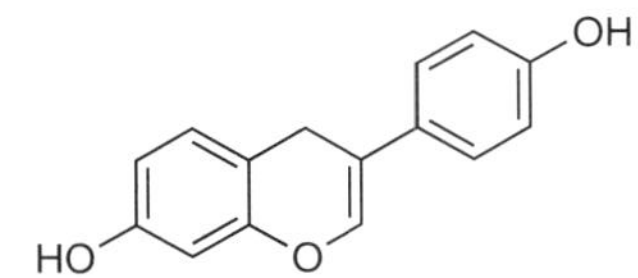 Compound-2
DPP-IV (Ward et al., 2005)	Diabetes	Commercially available compounds	GLIDE	% inhibition: 81.9% (30 µM)	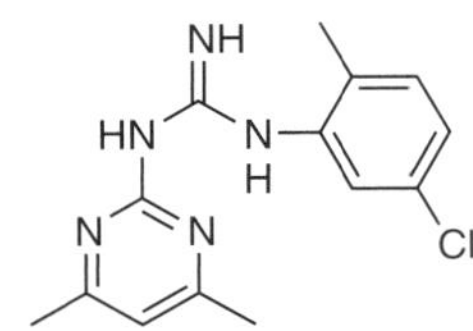 Compound-14
Human aldose reductase (Kraemer et al., 2004)	Diabetes	ACD	FlexX	$IC_{50} = 2.4 \pm 0.5$ µM	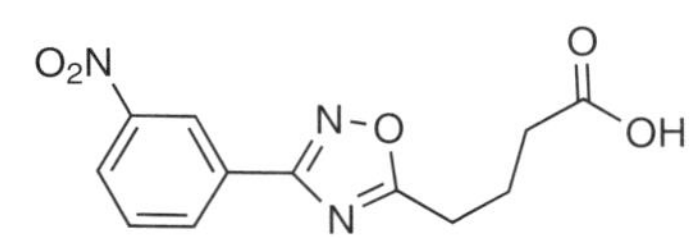 BTB02 809

(Continued)

TABLE 2.1 *(Continued)*

Target Protein	Disease Treatment Area	Small Moleule Library	VS Software Used	Potency in Binding Assay	Structure of a Representative Hit Compound
PTP1B (Doman et al., 2002)	Diabetes	ACD BioSpecs Maybridge	DOCK3.5	$IC_{50} = 4.1\ \mu M$	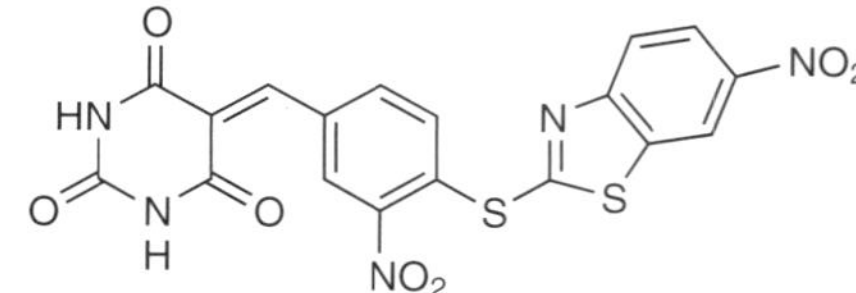 Compound-1
ADAM12 (Oh et al., 2004)	Cardiovascular disease	Virtual Library	PharmoMap/ PharmoScan	$IC_{50} = 16.7\ nM$	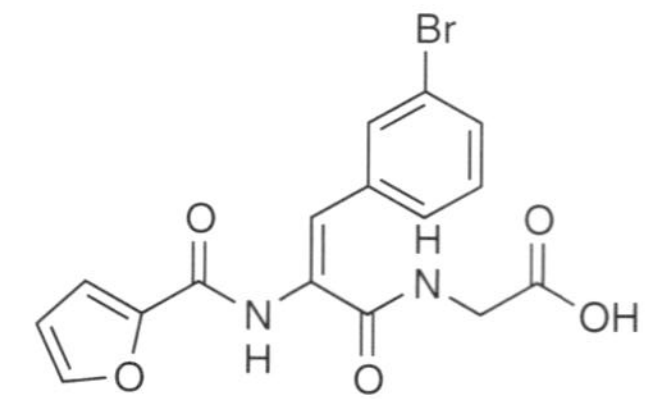 Compound-5
PDE-1 PDE-5 (Yamazaki et al., 2005)	Cardiovascular disease	SPECS	SPSS ver. 10.0	$IC_{50} = 1.9\ \mu M$ vs. PDE-1$IC_{50} =$ 0.7 μM vs. PDE-5	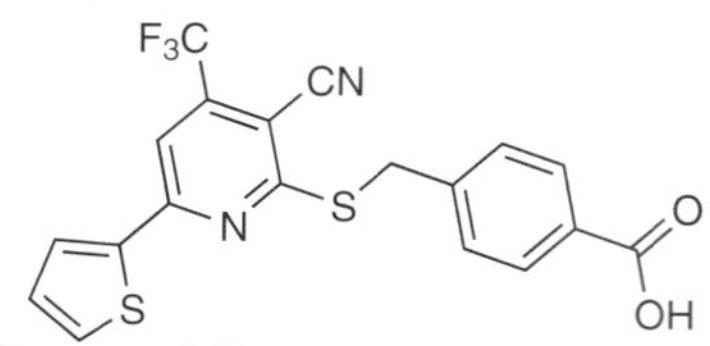 Compound-29
FT (Perola et al., 2000)	Cancer	ACD	EUDOC	$IC_{50} = 25\ \mu M$	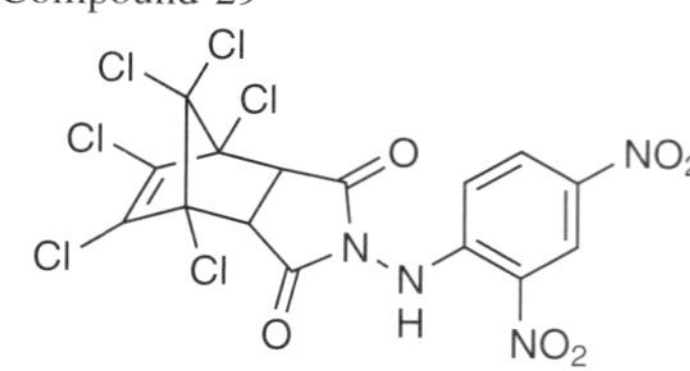 Compound-1

Stat3 (Song et al., 2005)	Cancer	NCI DB Merck Index Sigma-Aldrich Ryan Scientific	DOCK 4.0	ND	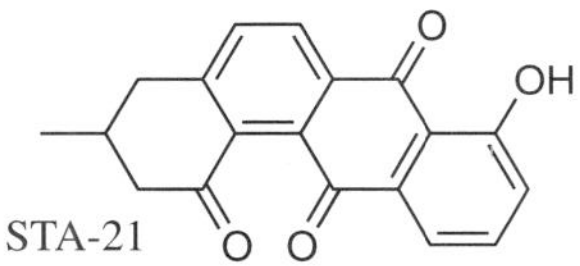 STA-21
CB 2 receptor (Salo, et al., 2005)	Alleviating pain inflammation	Maybridge	GOLD 4.0	$-\log EC_{50} = 5.3 \pm 0.2\ \mu M$	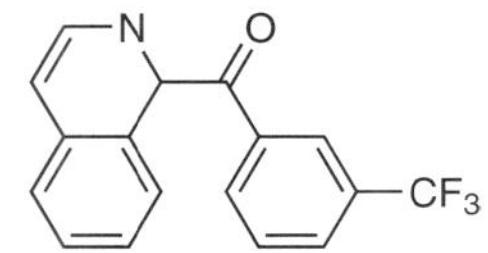 NRB 04079
NK 1 receptor (Evers and Klebe, 2004)	Inflammation immune response	Commercially available compounds	Selector Unity FlexX	$IC_{50} = 251$ nM	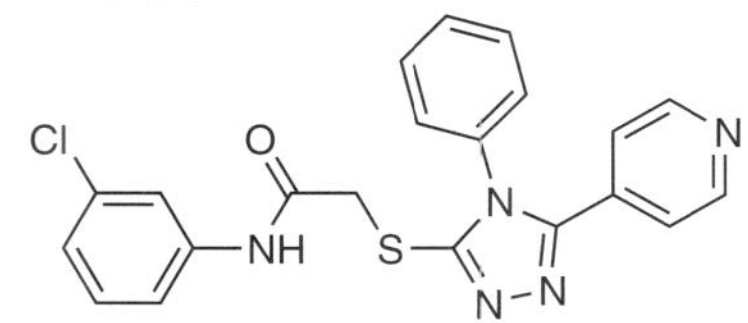 ASN-1377642
Alpha 1 A Adrenergic Receptor (Evers and Klabunde, 2005)	Cardiovascular disease	Aventis compound library	GOLD 2.0	$K_i = 1.4$ nM	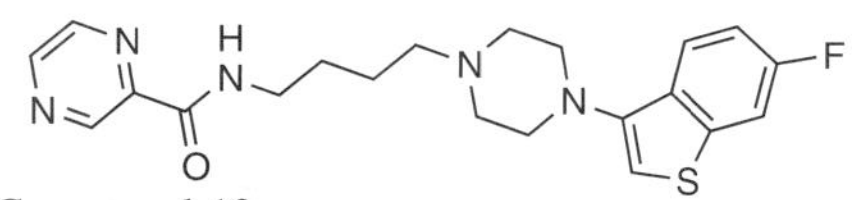 Compound-13
MCH 1 receptor (Clark et al., 2004)	Obesity	Commercially available compounds-	FlexS Unity FlexX	$IC_{50} = 55$ nM	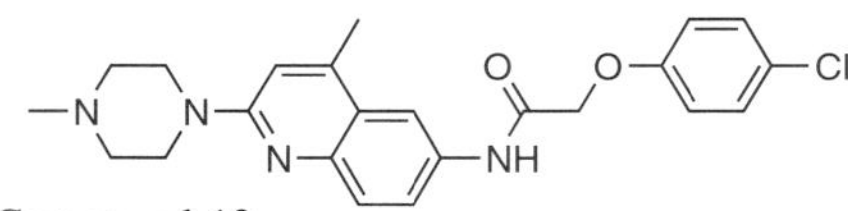 Compound-12

(Continued)

TABLE 2.1 *(Continued)*

Target Protein	Disease Treatment Area	Small Moleule Library	VS Software Used	Potency in Binding Assay	Structure of a Representative Hit Compound
Hsp90 (Barril et al., 2005)	Cancer	Commercially available compounds-	rDOCK	$IC_{50} = 600$ nM	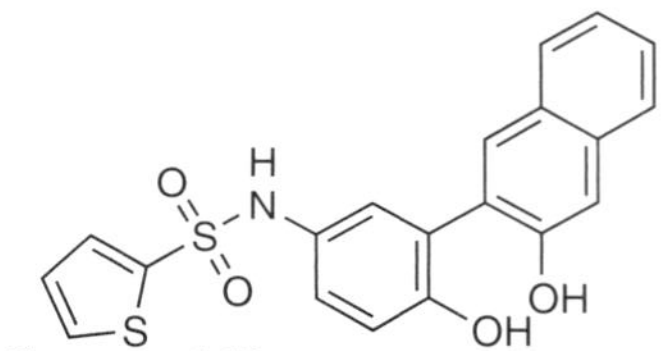 Compound-13
AICAR Transformylase (Li et al., 2004)	Cancer	NCI DB	AutoDock 2.4	$IC_{50} = 600$ nM	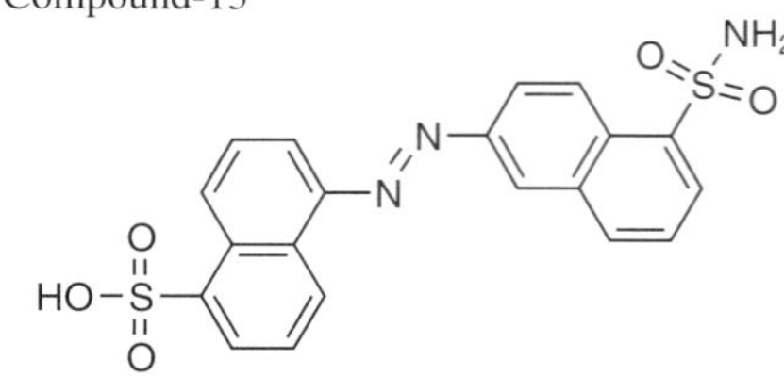 NSC-30171
Rac GTPase (Gao et al., 2004)	Cancer	NCI DB	FlexX	$IC_{50} = 50$ μM	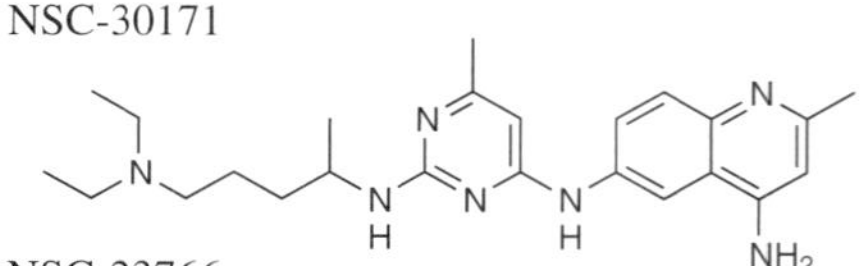 NSC-23766
Parasitic cysteine protease (Desai et al., 2004)	Parasitic disease	ChemBridge	GOLD	$IC_{50} = 1.0$ μM	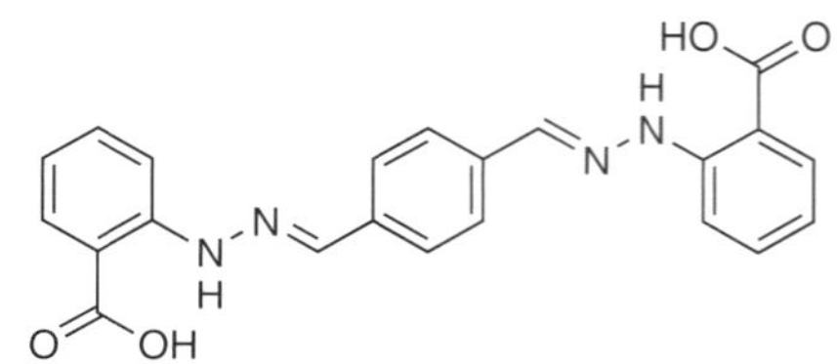 Compound-1

HRV coat protein (Steindl et al., 2005)	Infective disease	Maybridge	Cerius Software	$EC_{50} = 4.3\ \mu M$	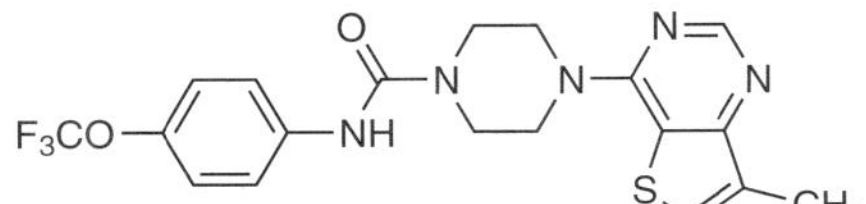 Compound-13
SARS-CoV protease (Liu et al., 2005)	Infective disease	NCI ACD MDDR	Dock 4.01	$K_i = 61\ \mu M$	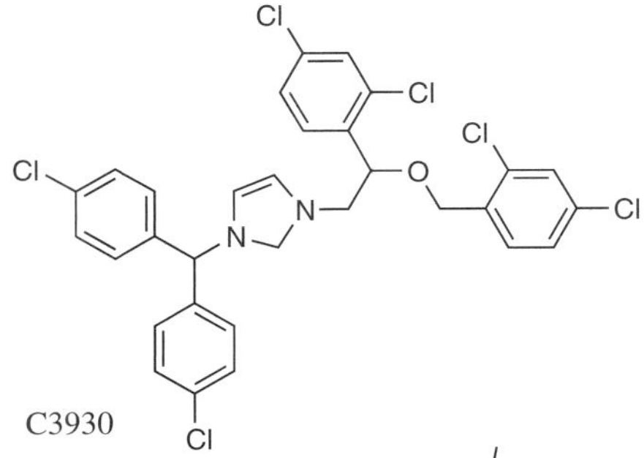 C3930
EGFR	Cancer	ChemBridge	ICM	$IC_{50} = 15.0\ \mu M$	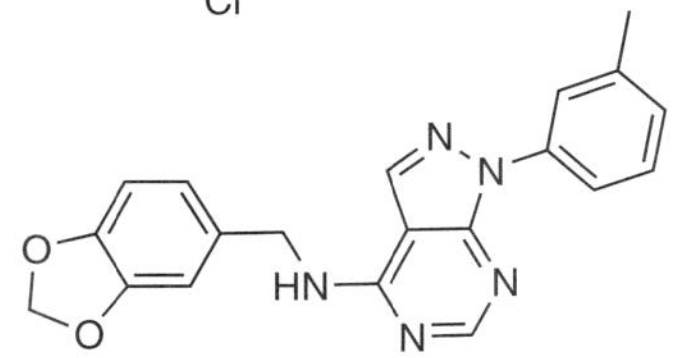 MSK-039
Chk1-kinase	Cancer	AstraZeneca	FlexX-Pharm	$IC_{50} = 450\ nM$	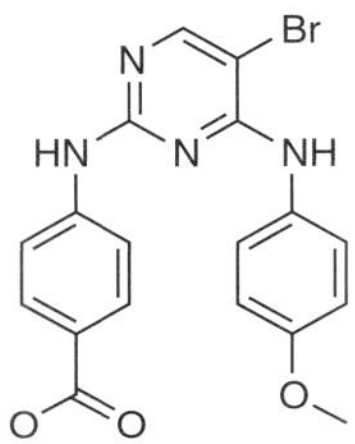 Pyrimidine hit

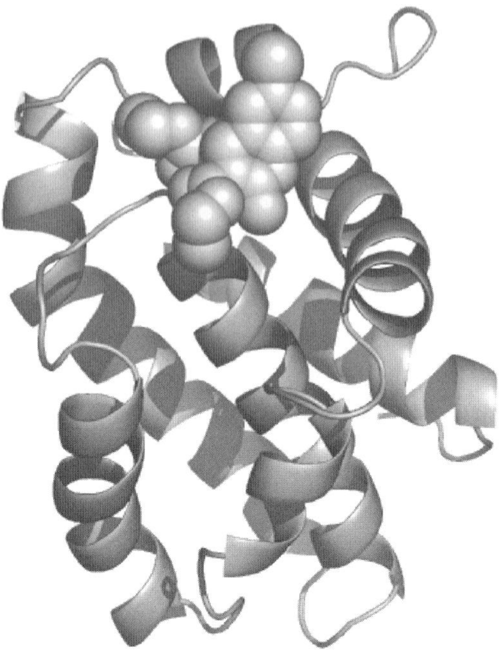

Figure 2.2 Structural model of HA14-1 (space-filling model) docked at the surface pocket of the Bcl-2 homology model (ribbon representation).

obtained from NCI and tested using an *in vitro* fluorescene polarization (FP) binding assay at an initial dose of 100 μM, and seven were found to have binding affinity (IC_{50} value) ranging from 1.6 to 14.0 μM). Compound 6 was the most potent, with an IC_{50} value of 4 μM in inhibition of cell growth using the 3-(4,5-dimethylthiazol-2-yl)-2,5-diphenyltetrazolium bromide assay. It was found that compound 6 induced apoptosis in cancer cells expressing high levels of Bcl-2. Furthermore, using NMR methods, compound 6 was shown to bind the BH3 binding site in Bcl-X_L, with a binding constant of 7 μM.

In a different study, Lugovskoy et al., (2002) outlined a novel approach based on a combination of structural and computational approaches to discover potent inhibitors, called BH3Is, of Bcl-2 family proteins. In this case, the authors used NMR chemical shift perturbation as an efficient tool for rapidly mapping the interaction interfaces and used direct NMR-derived constraints to restrict the conformational space for molecular modeling routines. The authors utilized the program TreeDock as a molecular modeling module. TreeDock uses the Lennard-Jones potential as the scoring function to obtain the protein–compound complexes based primarily on shape complementarity. The models of complexes were validated through an independent set of NMR restraints. This method was employed to analyze structure–activity relationships in the BH3I/Bcl-X_L complexes, and it was found that the free energies of the complexes calculated using the TreeDock routine correlated well with *in vitro* Bcl-X_L binding affinities of the compounds. The close homology between *in vitro* affinities of BH3Is and computed interaction energy allowed the authors to set up virtual screening of available online compound libraries. Ninety-three compounds were selected based on their similarity to BH3Is (more than 80%) and solubility (log *P* less than 6.0) from Chembridge (www.hit2lead.com) and Chemnavigator (www.chemnavigator.com) libraries. To incorporate ligand flexibility, nonfixed dihedral angles in the compounds were varied, and their backbones were superimposed with ones of the structurally closed BH3Is. Finally, the TreeDock module was used to calculate the interaction energy of these compounds with Bcl-X_L, keeping the same anchoring-docking atom pair as was used in the modeling of the structurally closed BH3I compound. From this screen, a novel analogue of BH3I-1, which was termed BH3I-1SCH_3 (structure shown in Table 2.1), was found, and the *in vitro* binding affinity of this compound to Bcl-X_L was tested ($IC_{50} = 8.5 \pm 0.6\,\mu M$). The results indicated that the compound binds Bcl-X_L with affinity comparable to that of BH3I-1.

2.2.2.2 *IAPs*

Current knowledge identifies the inhibitor of apoptosis proteins (IAPs) as a major control point in the execution of cell death (Deveraux and Reed, 1999). IAPs comprise a family of caspase-inhibiting proteins characterized by a shared conserved sequence region, termed the baculoviral IAP repeat (BIR) domain (Salvesen and Duckett, 2002). Eight endogenous IAPs are known in the human system, all of which inhibit apoptosis. So far, their main physiological roles seem to be (1) the establishment of a threshold under which caspases are kept at bay and (2) providing a pool of active caspases, which can rapidly execute death after release. Growing evidence also suggests the participation of IAP proteins in other cellular functions apart from inhibiting caspases, including protein degradation, cell-cycle control, and signal transduction.

The X-linked inhibitor of apoptosis (XIAP) is a promising new molecular target for the design of novel anticancer drugs aiming at overcoming apoptosis resistance of cancer cells to chemotherapeutic agents and radiation therapy. Recent studies demonstrated that the BIR3 domain of XIAP where caspase-9 and Smac proteins bind is an attractive site for designing small-molecule inhibitors of XIAP (Srinivasula et al., 2001). Using the DOCK program, the authors (Nikolovska-Coleska et al., 2004) have performed computational structure-based database searches of the TCM-3D database containing 8221 small organic molecules with diverse chemical structures isolated from traditional Chinese medicinal herbs to identify potential small-molecule inhibitors that bind to the XIAP BIR3 domain where caspase-9 and Smac bind. The high-resolution structure of the XIAP BIR3 domain in complex with the Smac protein was used to define a binding site for the database searches. The sum of the electrostatic and van der Waals interactions as calculated in the DOCK program was used as the ranking score. The top 1000 candidate small molecules with the best scores were rescored. After reranking, the top 200 compounds were considered as potential small-molecule inhibitors of XIAP. A total of 36 potential small-molecule inhibitors, primarily from the Developmental Therapeutics Program of the National Cancer Institute and some from commercial sources, were tested for their binding affinities to the XIAP BIR3 protein in the optimized FP-based biochemical binding assay. The authors have discovered five natural products from the TCM-3D that bind to the XIAP

BIR3 protein and directly compete with the SM7F peptide. Among these five inhibitors, embelin (structure shown in Table 2.1) is the most potent inhibitor, with an IC_{50} value of $4.1 \pm 1.1\ \mu M$ from three independent experiments in triplicates, which is slightly less potent than the natural 9-mer Smac peptide (IC_{50} value $= 2.8 \pm 0.7\ \mu M$). Embelin inhibits cell growth, induces apoptosis, and activates caspase-9 in prostate cancer cells with high levels of XIAP, but it has a minimal effect on normal prostate epithelial and fibroblast cells with low levels of XIAP. In stable XIAP-transfected Jurkat cells, embelin effectively overcomes the protective effect of XIAP to apoptosis, enhances etoposide-induced apoptosis, and has a minimal effect in Jurkat cells transfected with vector control. Taken together, the results showed that embelin is a fairly potent, nonpeptidic, cell-permeable, small-molecule inhibitor of XIAP and represents a promising lead compound for designing an entirely new class of anticancer agents that target the BIR3 domain of XIAP.

2.2.2.3 *Acetylcholinesterase*

A variety of neurological and neuromuscular disorders involve a diminution of cholinergic activity. Often the most effective treatments are ligands, which inhibit the breakdown of acetylcholine. Acetylcholinesterase inhibitors have been used clinically in the treatment of Alzheimer's disease. Alzheimer's disease is characterized by amyloid plaques and neurofibrillary tangles and is associated with a loss of cholinergic neurons. The loss of cholinergic neurons is progressive and results in profound memory disturbances and irreversible impairment of cognitive function. Recent efforts have been directed toward the development of novel strategies for the treatment of Alzheimer's disease. One strategy for the treatment of Alzheimer's patients has been the use of acetylcholinesterase inhibitors to increase the levels of acetylcholine in the synapse, thereby enhancing cholinergic activity in the affected brain regions. Novel inhibitors of acetylcholinesterase (AChE) have been discovered for the treatment of Alzheimer's disease using RBVS. Mizutani and Itai (2004) used the X-ray crystal structure of *Torpedo californica* AChE complexed with decamethonium (PDB code 1ACL). AChE has a deep cavity containing the catalytic triad (also called the "active site gorge"), and the residues in the active site gorge were defined as the ligand binding site. The software ADAM & EVE was used for the receptor-based virtual screening studies. This software uses the automated docking procedure and takes into account the flexibility of each small molecule in the database by exploring the conformational space fully and continuously. The authors screened commercially available databases (ACD and Maybridge) of 160,000 molecules. Promising ligand candidates were selected according to a number of user-defined parameters. The small-molecule database was filtered prior to docking, and compounds with molecular weight ≥ 300, number of atoms ≥ 40, number of heteroatoms ≥ 2, and ≤ 10, number of ring structures ≥ 2, and no carboxylate, phosphate, or sulfate functional groups were selected for docking. The criteria for the most stable docked model of each compound included intermolecular energy < 20.0 kcal/mol, intramolecular energy of each compound < 50.0 kcal/mol, and intermolecular hydrogen bonds ≥ 1. The intermolecular energy E_{inter} was used as a score function in the docking process and in selecting hit compounds, where E_{inter} was defined as

$$E_{\text{inter}} = E_{\text{vwd}} + E_{\text{elc}} - 2.5N_{\text{hb}}$$

where, E_{vwd} is the intermolecular van der Waals energy, E_{elc} is the electrostatic energy, and N_{hb} is the number of hydrogen bonds.

One hundred fourteen hit compounds were available and purchased for assay. The resulting hit compounds had different core structures, and a majority had structures that were quite different from the known AChE inhibitors. Of the 114 compounds tested, 35 showed inhibitory activities with IC_{50} values less than 100 μM. Thirteen of these compounds had IC_{50} between 0.5 and 10 μM. The most potent compound had an IC_{50} of 0.59 μM. The structure of a representative hit is shown in Table 2.1.

In another study of AChE inhibitors, Rollinger et al., (2004) generated a structure-based pharmacophore model utilizing an *in silico* filtering experiment for the discovery of promising candidates out of a 3D multiconformational database consisting of more than 110,000 natural products. Scopoletin and its glucoside scopolin emerged as potential AChE inhibitors by the virtual screening procedure. They showed moderate, but significant, dose-dependent and long-lasting inhibitory activities. In the *in vivo* experiments, scopoletin (structure shown in Table 2.1) and glucoside scopolin increased the extracellular acetylcholine (ACh) concentration in rat brain to about 170% and 300% compared to basal release, respectively. At the same concentration, the positive control galanthamine increased the ACh concentration to about the same level as scopoletin. These are the first *in vivo* results indicating an effect of coumarins on brain ACh.

2.2.2.4 *Estrogen Receptors*

Two nuclear estrogen receptors (ERs), ERα and ERβ, have been identified. There is roughly 60% homology in the ligand-binding domains between the human ERα and ERβ (Mosselman et al., 1996). ERβ is considered to be a pharmacological target to promote memory function and neuronal defense mechanisms against age-related neurodegeneration, such as Alzheimer's disease. Zhao and Brinton (2005) have used RBVS in this case to identify ERβ-selective compounds. It is a challenging task because there are only two residue differences in the active sites of ERα and ERβ. Structural and mutational studies have indicated that Met336 and Ile373 in ERβ replace the variant amino acids Leu384 and Met421 in the ligand-binding pocket of ERα, and these are the residues that need to be targeted to increase the selectivity of the ligand for one of the receptor subtypes. Using the crystal structure of ERβ LBD complexed with genistein and a plant-based natural library of 25,000 small molecules, the authors used an RBVS approach to identify ERβ-selective compounds. The receptor binding site was defined based on the position of genistein in the receptor and specified as all atoms within 10 Å of the center carbon of genistein. The program GOLD was used in the docking and scoring studies. Prior to database screening, 20 validation runs were performed on test complexes to figure out the GOLD parameter settings required to capture the reported binding mode of genistein. Using these parameters, the 3D natural source chemical database was docked into the ERβ active site and scored based on the GoldScore fitness function. Five hundred resulting top-scoring hits were filtered via visual screening. Based on visual analysis, 100 of these were redocked using Affinity. These 100 were selected based on their binding characteristics in the active site pocket. Finally, the compounds were subjected to a final round of screening in which the ones that passed the Lipinski's Rule of Five, and blood–brain barrier penetration properties were selected. Twelve representative hits from the database screening were assessed for their binding profiles to both ERs. Of these, five compounds showed binding selectivity to ERβ over ERα, and the most active compound (structure shown in Table 2.1) had an IC_{50} of 680 nM in ERβ binding assay, with over 100-fold binding selectivity to ERβ over ERα.

2.2.2.5 *Dipeptidyl Peptidase IV (DPP-IV)*

DPP-IV is a membrane-bound, homodimeric class II protein with a molecular weight of 110–150 kD per subunit. It is bound to the membrane by a transmembrane sequence of approximately 22 amino acids. The six cytosolic amino acids play no role in the binding functions. In addition to its protease function, DPP-IV has receptor properties and is an extracellular binding protein (Lambeir et al., 2003). DPP-IV has been regarded as one of the most promising therapeutic targets for the treatment of type 2 diabetes in recent years (Demutha et al., 2005). Inhibition of DPP-IV in normal and diabetic rats has been shown to prevent the degradation of GLP-1, leading to enhanced insulin secretion and improved glucose tolerance (Ahrén et al., 2002). DPP-IV is able to selectively cleave Xaa-Pro and Xaa-Ala dipeptides from the N-terminus of peptides and proteins (where "Xaa" is any amino acid).

Several DPP-IV crystal structures have recently been published, which allow structure-based drug design to be used in the search for new and different classes of DPP-IV inhibitors. Published DPP-IV crystal structures show a variety of bound ligands, including Val-Pyr, in the 1N1M (Rasmussen et al., 2003) structure. Other published structures including 1PGQ (Oefner et al., 2003), 1NU6, and 1NU8 (Thoma et al., 2003) have very similar conformations around the active site, affirming that this may be a suitable target for virtual screening. Ward et al. (2005) used a subset of 800,000 compounds for the starting point of the virtual screen. Filtering on physical properties and chemical filters reduced this number to around 500,000 compounds. Tautomer, protonation state, and stereogenic center enumeration using the in-house software and Corina increased the number of structures to around 750,000 in the single-conformer database. The number of structures in the multiconformer database is substantially higher, because up to 500 conformations from each compound are built. This multiconformational database was then processed against the two designed DPP-IV pharmacophores. Twenty thousand hits from each of the pharmacophores were elected based on the RMSD overlap of the compound with the pharmacophore and the overlap with the excluded volume of the active site. A single-conformer database of the 40,000 selected molecules was then docked into the DPP-IV crystal structure using GLIDE. Ten docked poses for each compound were postprocessed after which the top 8000 compounds were picked. Clustering and then visual inspection were used to select the final 4000 compounds for screening. These had activities ranging from 30% to 82% when tested at a concentration of 30 μM in an enzyme inhibition assay, and the most potent example (structure shown in Table 2.1) had an inhibition of 81.9% at 30 μM.

2.2.2.6 *Human Aldose Reductase*

Aldose reductase (AR, EC 1.1.1.21), a member of the aldo–keto reductase superfamily, is a monomeric enzyme that catalyzes the reduction of aldo sugars and a variety of aldehydes to their corresponding alcohols (Yabe-Nishimura, 1998). AR adopts a $(\beta/\alpha)_8$ TIM-barrel fold with the catalytic site located in the center of the barrel at the bottom of a deep hydrophobic cleft. The enzyme is NADPH-dependent, and the nicotinamide moiety of the cofactor is largely buried at the bottom of the barrel where it closes up the catalytic center (Borhani et al. 1992).

Inhibition of AR provides a therapeutic means for the prevention of complications associated with chronic diabetes. Kraemer et al. (2004) performed a virtual screen based on the high-resolution crystal structure of the inhibitor IDD594 in complex with human AR. AR operates on a large scale of structurally different substrates. To achieve this pronounced promiscuity, the enzyme can adapt rather flexibly to its substrates. Likewise, it has a similar adaptability for the binding of inhibitors. The authors applied a protocol of consecutive hierarchical filters to search the Available Chemicals Directory. In the first selection step, putative ligands were chosen that exhibit functional groups to anchor the anion-binding pocket of AR. Subsequently, a pharmacophore model based on the binding geometry of IDD594 and the mapping of the binding pocket in terms of putative "hot spots" of binding was applied as a second consecutive filter. In a third and final filtering step, the remaining candidate molecules were flexibly docked into the binding pocket of IDD594 with FlexX and ranked according to their estimated DrugScore values. Of the 206 compounds selected by this search and complemented by a cluster analysis and visual inspection, nine compounds were selected and subjected to biological testing. Of these, six compounds showed IC_{50} values in the micromolar range, with the best inhibitors BTB02809 (structure shown in Table 2.1) possessing an IC_{50} value of 2.4 ± 0.5 μM.

2.2.2.7 *Protein Tyrosine Phosphatase-1B (PTP-1B)*

Protein tyrosine phosphatase-1B (PTP-1B) is a tyrosine phosphatase that has been implicated as a key target for type 2 diabetes (Møller et al., 2000). This phosphatase hydrolyzes phosphotyrosines on the insulin receptor, thus deactivating it. Doman et al. (2002) defined the site targeted in the docking calculations by the positions of the two phosphotyrosine molecules observed in the complex with PTP-1B (Puius et al., 1997). The closed, ligand-bound conformation of the active site was used in the docking studies. Molecules were docked into the active site of PTP-1B in multiple conformations using the Northwestern University version of DOCK3.5. Two databases of molecules were used as follows: about 152,000 compounds from the ACD98.2 database and about 82,000 compounds from the BioSpecs 1999 and the Maybridge 1999 catalogs. Only molecules that had at least 17 and no more than 60 nonhydrogen atoms were docked. This amounted to 165,581 molecules docked in total. The atomic properties of these molecules, including van der Waals parameters and partial atomic charges, were calculated, and conformations were precalculated using SYBYL and stored in a flexibase. Compounds selected by the docking program were screened for their ability to inhibit the PTP-1B dephosphorylation of the insulin receptor peptide. The top-scoring 500 molecules from the flexible docking screen of the ACD database and the top-scoring 500 molecules from the flexible docking screen of the combined BioSpecs and Maybridge databases were considered for further evaluation. Of the 500 high-scoring ACD molecules, 118 were tested experimentally; 38 had IC_{50} values of 100 μM or better against PTP-1B. Additionally, 15 molecules from a preliminary rigid body docking screen of the ACD were also assayed; eight had IC_{50} values of 100 μM or lower (a combined hit rate of experimentally tested docking molecules of 34.6%). Of the 500 high-scoring BioSpecs/Maybridge molecules, 232 were tested experimentally; 81 had IC_{50} values of 100 μM or better against PTP-1B, a hit rate of 34.9%. Overall, 34.8% of the high-scoring docked molecules that were tested (127 of 365) had IC_{50} values of 100 μM or lower. The structure of a typical representative is outlined in Table 2.1. This representative had an IC_{50} of 4.1 μM for PTP-1B.

2.2.2.8 ADAMs

ADAMs (a disintegrin and metalloproteases) are type 1 transmembrane metalloproteases that are involved in cell–cell adhesion, surface proteolysis, and transmembrane signaling. Recently, ADAM12 was shown to play an important role in the development of cardiac hypertrophy by shedding heparin-binding epidermal growth factor (Asakura et al., 2002). Therefore, inhibition of ADAM12 could be a potent therapeutic strategy for cardiac hypertrophy and congestive heart failure. Some ADAM12 inhibitors have been reported to repress ADAM12 in both in vitro and *in vivo* experiments (Tokumaru et al., 2000). However, these inhibitors have broad spectra to MMPs. Oh et al. (2004) identified potent and selective ADAM12 inhibitors through structure-based virtual screening. The authors constructed the homology mode of ADAM12 based on the published structures of reference proteins that had over 30% sequence similarity to ADAM12. The reference proteins in this study were adamalysin (*Crotalus adamanteus*; PDB code 4AIG), acutolysin-C (*Agkistrodon acutus*; PDB code 1QUA), and acutolysin-C (*Agkistrodon acutus*; PDB code 1BUD). Modeling of ADAM12 was initially performed by using the HOMOLOGY module of INSIGHT2000, and it was further refined by using Discover (version 2.98) of INSIGHT2000. A pharmacophore map for the catalytic site of ADAM12 was generated by using PharmoMap/PharmoScan. Since ADAM12 is a Zn^{2+}-containing metalloenzyme, the Zn^{2+}-binding moieties are prerequisite for potent inhibitors against the enzymatic activity. Therefore, a focused virtual library, composed of 67,062 compounds containing a Zn^{2+}-binding moiety such as carboxyl, hydroxyl sulfur, or hydroxamate, was prepared. Virtual screening of this library for the ADAM12 pharmacophore using the PharmoMap/PharmoScan system resulted in 1217 compounds as initial hits, which were further refined by a consensus scoring function composed of Ligscore, PLP, JAIN, PMF, and LUDI. Applying a criterion of the top 30% of each score, 64 compounds were finally selected for biological testing. The best compound had an IC_{50} value of 16.7 nM in a cell-based assay for selective inhibition of ADAM12. The structure of this compound is shown in Table 2.1. To understand the mechanism underlying the highly selective inhibition of ADAM12 by the most active compound (called compound 5), the authors compared the docking modes of this active compound in the active sites of ADAM12 and TACE (also known as ADAM17). Although the compositions of amino acids in the active sites are highly conserved and the total binding energy is similar (−78 kcal/mol for the 5-ADAM12 complex and −53 kcal/mol for the 5-TACE complex), the binding modes of compound 5 in the active sites of ADAM12 and TACE were significantly different. Consistent with the proper docking of compound 5 in the pharmacophore of the Zn^{2+}-binding site of ADAM12, compound 5 fit into the active site of ADAM12 with the conserved Zn^{2+}-binding motif and hydrophobic interactions. However, compound 5 could not be docked in the active site of TACE because the hydrophobic pocket in this case was much smaller than that of ADAM12. Therefore, the hydrophobic interaction as well as the Zn^{2+}–carboxy group interaction might serve as an essential factor for the specific inhibition of ADAM12 by compound 5. The reason for this low nanomolar hit for virtual screening could be due to the combined use of the pharmacophore of the Zn^{2+} -binding site of ADAM12 and receptor-based docking to identify likely hits.

2.2.2.9 *Phosphodiesterase-1 and -5*

Phosphodiesterase types 1 (PDE-1) and 5 (PDE-5) are the major cGMP hydrolyzing enzymes in blood vessels and regulate the level of the mediator in concert with guanylyl cyclase, which catalyzes the synthesis of cGMP from GTP. PDE-1 and -5 have also been found in platelets. Therefore, inhibitors of PDE-1 and -5 are expected as therapeutics for cardiovascular diseases, such as hypertension, angina, cardiac failure, and obstructive arteriosclerosis (Wallis et al., 1999). Yamazaki et al. (2005) used classification and regression tree (CART) analysis to configure a prediction model for virtual screening. An optimum set of structural descriptors were selected as explanatory variables for CART analysis. A library of commercially available chemical compounds supplied by SPECS Inc. was screened for PDE-5 inhibitory activity by virtual screening. The authors used only 50,520 compounds that were not listed on the catalog of SPECS Inc. in September 1998 but were listed in October 1999 so that there would be a higher possibility that these compounds were actually in stock. In the first step of the virtual screening, the compounds with unfavorable physicochemical and pharmacokinetic properties were filtered out based on a modified Lipinski's Rule of Five; that is, molecular weight below 500, number of hydrogen bond acceptors below 10, number of hydrogen bond donors below 5, AlogP below 6, and number of rotatable bonds below 12. In the next step, the compounds with favorable biological activities were selected with the CART model for PDE-5 inhibition. SPSS ver.10.0 was used in both steps of the virtual screening. In the first step of virtual screening of a commercially available chemical library, 43,365 compounds were selected by a modified Lipinski's Rule of Five (85.8%). Among them, 1821 compounds were finally selected as PDE-5 inhibitor candidates (3.6%). Nineteen diverse and druglike compounds selected from the virtual screening hits were assayed for the inhibitions of PDE-1 and -5. Among them, seven compounds (37%) showed inhibitory activities against both PDE-1 and -5 below 10 μM as IC_{50}. The structure of a typical representative is shown in Table 2.1.

2.2.2.10 *Metalloenzymes*

Carbonic Anhydrase II, Matrix Metalloproteinase-3, Neutral Endopeptidase, Peptide Deformylase, and Xanthine Oxidase Metalloenzymes have therapeutic implications in many diseases such as cancer (Kruger and Figg, 2000; Bradshaw and Yi, 2002), arthritis (Beaudeux et al., 2004), glaucoma (Cvetkovic and Perry, 2003), and infectious diseases (Eswaramoorthy et al., 2002). However, these proteins are challenging targets for virtual screening because of the covalent-like interactions between the metal centers and their ligands and the large electrostatic potentials of the metals. This is because the molecular mechanics potential functions that underlie most modeling approaches have been parameterized to treat atoms as either interacting covalently or noncovalently, and metal–ligand interactions do not fit into either of these categories. Irwin et al. (2005) have explored the robustness of RBVS methods using classical scoring functions to identify potential ligands for several metalloenzymes. To do this, the authors have treated the metals in the enzymes like any other atoms in the enzyme, with van der Waals radii, potentials, and point charges. They have tested the protocol by docking a functionally annotated database of 95,000 compounds from the MDL Drug Data Report (MDDR) against the X-ray crystal structures of five metalloenzymes, such as carbonic anhydrase II, matrix metalloproteinase-3, neutral endopeptidase, peptide deformylase, and xanthine oxidase. The quality of the docking scheme was judged based on the enrichment of the annotated ligands among the top-scoring hits and reproduction of experimental geometries. In all five test cases, the annotated ligands within the MDDR were enriched at least 20 times over random. They found this approach to be reliable for zinc metalloenzymes, such as carbonic anhydrase II (where the highest-scoring docked geometry had an rms deviation of 0.4 Å from the crystal structure and 82-fold enrichment of hits in the top 0.1% of the database), matrix metalloproteinase-3 (where the highest-scoring docked geometry had an rms deviation of 0.7 Å from the crystal structure and 32-fold enrichment of hits in the top 0.1% of the database), and neutral endopeptidase (where the highest-scoring docked geometry had an rms deviation of 1.2 Å from the crystal structure and 189-fold enrichment of hits in the top 0.1% of the database). Docking against a nickel-containing enzyme, peptide deformylase (where the highest-scoring docked geometry had an rms deviation of 0.9 Å from the crystal structure and 47-fold enrichment of hits in the top 0.8% of the database), and against a molybdenum enzyme, xanthine oxidase (where the highest-scoring docked geometry had an rms deviation of 0.7 Å from the crystal structure and 36-fold enrichment of hits in the top 0.6% of the database), was reported as satisfactory. The authors applied this protocol to identify inhibitors of zinc-β-lactamase, an antibiotic resistance target from *Bacteroides fragilis*, for which there are no inhibitors in clinical use. The 33,000 fragment-like molecules used in this case were from the chemical library of the ZINC database (Irwin and Shoichet 2005). Out of the 50 top-ranked compounds from docking and scoring, 15 were purchased and evaluated. Of these, five compounds were found to have K_i values less than 120 μM, the best of which had a K_i of 2 μM. In the next application they identified 25 potential substrates of a Zn-dependent phosphotriesterase from *Pseudomonas diminuta* by screening 167,000 compounds in the ACD. Eight of the 25 hits were acquired, tested, and confirmed experimentally to be substrates. This study suggested that a simple, noncovalent scoring function may be used to identify inhibitors of at least some metalloenzymes.

Farnesyltransferase Farnesyltransferase (FT) catalyzes the transfer of the farnesyl group from the cosubstrate farnesyl pyrophosphate (FPP) to a cysteine residue in the C-terminal fragment of pro-Ras proteins. FT consists of an α subunit (48 kDa) and a β subunit (46 kDa). The two subunits form a large cationic residue-rich active site in the middle of the enzyme. The zinc divalent cation present in the active site is known to play a functional role in FT catalysis and to facilitate the binding of the substrates (Park et al., 1997). FT is a promising anticancer drug target (Gibbs and Oliff, 1997) because selective inhibitors of FT have been found to be able to inhibit tumor cell proliferation without substantially interfering with normal cell growth (Kohl et al., 1993; Prendergast et al., 1994; Sebti and Hamilton, 1997). From the standpoint of ligand docking, it is challenging to work with FT since the active site of FT is very large and contains a metal atom (Zn, in this case). Additionally, the force-field parameters of the zinc divalent cation required in docking studies has not been adequately represented in docking software programs. Perola et al. (2000) performed a virtual screening, containing 219,390 chemicals, of the ACD to identify prototypic inhibitors of zinc-bound FT. To achieve this, the authors developed a new docking program, EUDOC, and proposed an integrated, broad, conformation substitution-based (IBS) virtual-screening approach. Using the docking program and the IBS virtual-screening approach, 21 compounds were identified from the ACD database, four of which inhibited FT *in vitro* with IC_{50} values in the range of 25–100 μM. The most potent inhibitor (structure shown in Table 2.1) also inhibited FT in human lung cancer cells.

2.2.2.11 Stat3

Signal transducers and activators of transcription 3 (Stat3) are activated in response to cytokines and growth factors. The main domains of Stat3 protein include the tetramerization and leucine zipper at the N terminus, the DNA-binding domain, and the SH2 transactivation domain at the C-terminal end. The SH2 region is responsible for the binding of Stat3 to the tyrosine-phosphorylated receptors and for the dimerization that is necessary for DNA binding and gene expression (Zhong et al., 1994). Stat3 is activated by phosphorylation at tyrosine residue 705 (Tyr-705), which leads to dimer formation, nuclear translocation, recognition of Stat3-specific DNA binding elements, and activation of target gene transcription. The constitutive activation of Stat3 is frequently detected in breast carcinoma cell lines but not in normal breast epithelial cells. Blockade of Stat3 signaling inhibits cancer cell growth, indicating that Stat3 plays a role in the survival or growth of tumor cells. It has been reported that ~60% of breast tumors contain persistent activated Stat3. Thus, Stat3 represents an attractive target for the development of new anticancer therapy in the treatment of breast cancer.

To identify potential small-molecule compounds that can disrupt Stat3 dimerization, Song et al. (2005) used the X-ray crystal structure of Stat3β (PDB code 1BG1) and a docking and scoring strategy to screen a virtual library of 429,000 compounds collectively from NCI, Merck Index, Sigma-Aldrich, and Ryan Scientific compound databases. The docking program DOCK (version 4.0) was used to perform the virtual screening. The binding cavity on the Stat3 SH2 domain was the region targeted for docking. SYBYL software (version 6.9, Tripos Associates, St. Louis, MO) was used to assign the standard AMBER (a set of molecular mechanical force fields for the simulation of biomolecules) atomic partial charges on the Stat3β protein and the Gasteiger–Hückel atomic partial charges on each ligand molecule to be docked. Each compound with a molecular weight between 200 and 700 was docked into the targeted binding site in Stat3. The top 10% scored compounds from each database, as selected by DOCK, were extracted and combined together to provide a total of 35,000 candidate compounds. Based on the binding models of these compounds predicted by DOCK, the X-SCORE program (version 1.1) was applied to obtain an estimate of the binding affinities of these compounds for Stat3. The preselected 35,000 compounds were then reranked according to their binding affinities as estimated by X-SCORE. Of the best-scored 200 compounds selected by X-SCORE from the 35,000 compounds, physical samples of 100 compounds were obtained from NCI or purchased from Sigma-Aldrich or Ryan Scientific. One compound, termed STA-21 (structure shown in Table 2.1), was identified as the best match for this selection criteria. STA-21 inhibited Stat3 DNA binding activity, Stat3 dimerization, and Stat3-dependent luciferase activity at 20 μM. Moreover, STA-21 reduced the survival of breast carcinoma cells with constitutive Stat3 signaling but had minimal effect on the cells in which constitutive Stat3 signaling was absent at 20 μM or 30 μM.

2.2.2.12 *GPCRs*

Serotonin, Tachykinin NK1, Dopamine D2, and Chemokine CCR3 Receptors G-protein-coupled receptors (GPCRs) are implicated in many human diseases, such as inflammation, hypertension, pain, depression, obesity, depression, and anxiety. As a result, they are the targets for nearly 50% of all drug-discovery programs (Klabunde and Hessler, 2002). GPCRs are especially challenging targets because they are membrane-bound proteins, which make their experimental 3D structure determination difficult. Although ligand-based *in silico* approaches have been used to identify ligands for this family of proteins, nonetheless, structure-based drug discovery remains highly desirable for this family of proteins. Becker et al. (2004) have used *in silico* 3D models of several GPCRs and have used RBVS to identify novel, potent analogs targeting the serotonin receptor, tachykinin NK1 receptor, dopamine D2 receptor, and chemokine CCR3 receptor. In each case, the authors have constructed the PREDICT 3D model of the respective target protein from its amino acid sequence using the PREDICT algorithm and methodology (Shacham et al., 2001, 2004) and have used this model in RBVS studies. PREDICT is a *de novo* GPCR-modeling methodology that combines the properties of the protein sequence with those of its membrane environment without relying on any GPCR experimental structure. The algorithm searches through the receptor conformational space for the most stable 3D structure(s) of the transmembrane domain of the GPCR protein within the membrane environment. A binding pocket was defined in each case while docking and scoring were carried out using the program DOCK4.0 to screen more than 100,000 compounds for each target. This method was successful in identifying promising lead candidates for these GPCRs. For the serotonin 5-HT1A receptor, 78 hits were tested using *in vitro* assays. Of the 16 hits with K_i less than 5 μM, five were found to be novel compounds (new chemical entities—that is, compounds not previously described in literature), with the best compound having a potency of 1 nM. For the serotonin 5-HT4 receptor, 93 virtual hits were tested using *in vitro* binding assays, and 19 of those had K_i less than 5 μM. Of these, four had distinct chemical scaffolds, and the best novel hit had a K_i of 21 nM. In the case of the tachykinin NK1 receptor, there were 53 virtual hits that were tested using *in vitro* binding assays, and eight had K_i less than 5 μM. They were grouped into five distinct chemical scaffolds, and the best novel hit was a 56 nM compound. Finally, for the chemokine CCR3 receptor, of the 43 virtual hits identified by this method and tested using *in vitro* binding assays, five compounds had K_i less than 20 μM. The best novel hit had a K_i of 12 μM. This work has paved the way for the application of RBVS approaches for the GPCR family of proteins.

Human Cannabinoid CB2 Receptor The cannabinoid CB2 receptor is a GPCR located mainly in immune tissues. It is part of the endogenous cannabinoid system (ECS) that includes the endocannabinoids (e.g., 2-arachidonoylglycerol [2-AG]), their receptors (CB1 and CB2), metabolizing enzymes (fatty-acid amide hydrolase [FAAH] and monoglyceride lipase [MGL]), and a specific cellular uptake system. Since ECS is an attractive target for drug development, various cannabinergic ligands that act on one or more ECS target proteins have been developed during recent years (Di Marzo et al., 2004). The CB2 receptor ligands possess potential in alleviation of pain and inflammation as well as in treatment of chronic cough, gliomas, lymphomas, and osteoporosis. Many cannabinoid receptor ligands are nonselective in that they bind both to the central CB1 receptor as well as to the peripheral CB2 receptor. It is claimed that selective CB2 ligands should be devoid of the unwanted central nervous system (CNS) effects typical of Δ^9-tetrahydrocannabinol (Δ^9-THC), the major psychoactive component of *Cannabis sativa* L (Gaoni et al., 1964).

To identify novel selective CB2 lead compounds, Salo et al. (2005) constructed a comparative model of the CB2 receptor using the high-resolution bovine rhodopsin X-ray structure as a template. The HOMOLOGY module of InsightII (version 2000) modeling software was utilized to construct the CB2 receptor model, whereas structure optimization, visualization, and ligand structure construction were carried out using the SYBYL6.9 modeling package. Molecular dynamics (MD) simulations were performed with GROMACS v.3.1.4, and the stereochemical quality of the protein structures was checked with PROCHECK. GOLD2.0 was used for ligand docking, and both CScore and X-Score v. 1.1 were used for ranking the docking results. CoMFA, as implemented in SYBYL, was used for generating 3D-QSAR models. Lead-structure searches in the Maybridge molecular database were carried out with the UNITY4.4 module of SYBYL. Five different receptor conformers and three different database queries were used to virtually screen the Maybridge Database (~55,600 compounds) for the discovery of selective CB2 ligands. Both the docked CB2 ligand structures and the modeled receptor binding site were employed in building the 3D database queries. Query 1 (Q1) was based on the pharmacophoric points of the nonselective cannabinoid receptor agonist HU-210 and the multiple volume surface of six other CB2 agonists all in their docking conformations. The second query (Q2) was based on the binding cavity of the docked CB2 ligands, whereas the third query (Q3) was based on the combination of features taken from both the receptor and bound ligands. Finally, 86 hit molecules (0.15% of the whole database) were ordered from Maybridge to be tested for their *in vitro* activity at the CB2 receptor. In G-protein activation assays, 1-isoquinolyl[3-(trifluoromethyl)phenyl]methanone (NRB 04079) ($-\log EC_{50} = 5.3 \pm 0.2$) was found to act as a selective agonist at the human CB2 receptor. NRB 04079 (structure shown in Table 2.1), therefore, can serve as a lead structure for designing selective CB2 agonists that are useful in the management of pain.

Neurokinin-1 (NK1) Receptor The neurokinin-1 (NK1) receptor belongs to the family of GPCRs, which represents one of the most relevant target families in small-molecule drug design. Evers and Klebe (2004) describe the use of a homology model of the NK1 receptor based on the high-resolution X-ray structure of rhodopsin and the successful virtual screening based on this protein model. The NK1 receptor model was generated using the new MOBILE (modeling binding sites including ligand information explicitly) approach developed in the authors' laboratory. Starting with preliminary homology models, this software generates improved models of the protein binding pocket together with bound ligands. Ligand information is used as an integral part in the homology modeling process. For the construction of the NK1 receptor, antagonist CP-96345 was used to restrain the modeling. The quality of the obtained model was validated by probing its ability to accommodate additional known NK1 antagonists from structurally diverse classes. On the basis of the generated model and on the analysis of known NK1 antagonists, a pharmacophore model was deduced, which was subsequently used to guide the 2D and 3D database search with UNITY. As a following step, the remaining hits were docked into the modeled binding pocket of the NK1 receptor. Finally, seven compounds were selected for biochemical testing, from which one showed affinity in the submicromolar range.

Alpha1A Adrenergic Receptor The authors started with the constructed homology model of the alpha1A receptor based on the X-ray structure of bovine rhodopsin. The protein model was generated by applying ligand-supported homology modeling, using mutational and ligand SAR data to guide the protein modeling procedure. Evers and Klabunde (2005) performed a virtual screen of the company's compound collection to test how well this model is suited to identify alpha1A antagonists. They applied a hierarchical virtual screening procedure guided by 2D filters and 3D pharmacophore models. The ~23,000 filtered compounds were docked into the alpha1A homology model with GOLD and scored with PMF. From the top-ranked compounds, 80 diverse compounds were tested in a radioligand displacement assay. Thirty-seven compounds revealed K_i values better than 10 μM; the most active compound bound with 1.4 nM to the alpha1A receptor. These findings suggest that rhodopsin-based homology models used in conjunction with ligand-based pharmacophore models and 3D-QSAR modeling may aid in the identification of highly active compounds for the GPCR family of proteins.

Melanin-Concentrating Hormone-1 Receptor (MCH-1R) Melanin-concentrating hormone (MCH) has been known to be an appetite-stimulating peptide for a number of years. However, it is only recently that MCH has been discovered to be the natural ligand for a previously "orphan" G-protein-coupled receptor, now designated MCH-1R. This receptor has been shown to mediate the effects of MCH on appetite and body weight; consequently, drug-discovery programs have begun to exploit this information in the search for MCH-1R antagonists for the treatment of obesity. Clark et al. (2004) report the rapid discovery of multiple, structurally distinct series of MCH-1R antagonists using a variety of virtual screening techniques. The most potent of these compounds (12) demonstrated an IC_{50} value of 55 nM in the primary screen and exhibited antagonist properties in a functional cellular assay measuring Ca^{2+} release. More potent compounds were identified by follow-up searches around the initial hit.

2.2.2.13 Hsp90

Docking-based virtual screening was used to identify a new class of Hsp90 inhibitor (Barril et al., 2005) with submicromolar potency. The molecular chaperone Hsp90 is involved in the onset and progression of cancer (Maloney and Workman, 2002), thereby making it an attractive target for anticancer drug discovery. The N-terminal domain of this protein binds to ATP and has been reported to drive the chaperone activity of this protein. The authors used a four-step RBVS protocol (Barril et al., 2004) to identify ligands binding to this site of Hsp90. In the first step, the relevant binding pocket was selected from all the structurally available models for this enzyme. The open and helical form of the receptor was used in docking studies. The binding pocket for purine analogs with three bound water molecules was defined as the active site. In the second step, a virtual small-molecule library of 0.7 million compounds was constructed starting from 3.5 million commercially available compounds. These were selected on the basis of their druglikeness, absence of reactive groups, and vendor delivery timelines. Corina was used to generate the 3D structures of the ligand database. In the third step, rDock was used to dock the library, and the 9000 highest-scoring compounds were short-listed for further evaluation. In the fourth and final step, a postfiltering was done, and only those compounds that formed a hydrogen bond with one carboxylic oxygen of Asp93 and a second hydrogen bond with one of the interstitial water molecules, as found in all the reported cocrystal structures of Hsp90-inhibitor complexes, were selected. Of the top 1000 compounds selected for purchase, 719 were actually available and tested for their Hsp90 inhibition capability. Three of the identified hits showed greater than 50% inhibition in the primary assay, two of which showed submicromolar potency. However, all three compounds that showed good activity in fluorescent polarization assays exhibited poor cellular activity, suggesting that the compounds are not well absorbed by the cell. The authors went on to determine the X-ray crystal structures of the three active compounds complexed with Hsp90 to identify sites in the ligand that could be potentially modified to improve the cell-penetration ability of the respective hits (structure of a representative hit compound shown in Table 2.1).

2.2.2.14 AICAR Transformylase

AICAR transformylase (5-aminoimidazole-4-carboxamide ribonucleotide transformylase) has been proposed as a potential target for antineoplastic drug development. Li et al. (2004) carried out virtual screening of the AICAR transformylase active site against the NCI diversity set of 1990 compounds using AutoDock, and they also identified eight potential novel scaffolds with micromolar inhibition activity. Initially, 44 compounds were selected from the NCI diversity set for *in vitro* evaluation on the basis of computational simulation of free energy of binding to the AICAR Tfase active site. Prior to the assay, 10 of the compounds were found to be water insoluble and 18 precipitated in the assay buffer. Of the remaining 16 that were tested, eight proved to have micromolar activity in enzymatic inhibition assays. In the next step, the authors expanded their repertoire of active compounds by taking each of the five most active compounds and doing a similarity compound search on the complete NCI database of 213,628 compounds. They found 11 additional hits from the secondary round of similarity search and AutoDock screening. These 11 candidates also turned out to have micromolar potency in enzymatic inhibition assays. The best compound (structure shown in Table 2.1) in this screening turned out to have a measured IC_{50} of 600 nM in AICAR transformylase inhibition assays.

2.2.2.15 Rac GTPase

Rho family GTPases are molecular switches that control signaling pathways regulating cytoskeleton organization, gene expression, cell-cycle progression, cell motility, and other cellular processes (Etienne-Manneville and Hall, 2002). They can be activated through the interaction with the Dbl family guanine nucleotide exchange factors (GEFs) that catalyze their GTP/GDP exchange and connect them to the diverse stimuli from upstream cell surface receptors such as the GPCRs, growth factor receptors, cytokine receptors, and adhesion receptors (Zheng, 2001). Accumulating evidence has implicated Rho GTPases in many aspects of cancer development (Schmitz et al., 2000), because deregulated Rho GTPases have been discovered in many human tumors, including colon, breast, lung, myeloma, and head and neck squamous-cell carcinoma (Mira et al., 2000). Rho GTPases and the signal pathways regulated by them have thus been proposed as potential anticancer therapeutic targets (Erik Sahai, 2002).

Gao et al. (2004) have applied a structure-based virtual screening approach to the search for a Rac-GEF interaction-specific small-molecule inhibitor. In the 3D structure of the Rac1-Tiam1 complex, Trp^{56} of Rac1 is buried in a pocket formed by residues His^{1178}, $Ser1^{184}$, Glu^{1183}, and Ile^{1197} of Tiam1 and Lys^{5}, Val^{7}, Thr^{58}, and Ser^{71} of Rac1. To identify Rac1-specific inhibitors based on the structural features surrounding Trp^{56}, a putative inhibitor-binding pocket was created with residues of

Rac1 within 6.5 Å of Trp^{56} in the Rac1-Tiam1 monomer, including Lys^5, Val^7, Trp^{56}, and Ser^{71}. A 3D database search was performed to identify compounds whose conformations would fit into this pocket. The database was freely available from NCI, which includes the coordinates of >140,000 small chemical compounds. Taking the flexibility of the compounds into consideration during the screening process, the program UNITY, whose Directed Tweak algorithm allows a conformationally flexible 3D search, was applied. The small-molecule hits yielded by the UNITY program were next docked into the predicted binding pocket of Rac1 containing Trp^{56} by using the program FlexX, which allows flexible docking to protein-binding sites. By following the docking procedures, the compounds were ranked based on their predicted ability to bind to the binding pocket with the program CScore. CScore generates a relative, consensus score based on how well the individual scoring functions of the protein–ligand complex perform. Of the top 100 compounds that displayed high consensus scores, 58 compounds whose docking did not seem to involve residue Trp^{56} were crossed out by visual inspection. Taking into account the solubility and availability of the remaining compounds, 15 chemicals that show promising docking affinities by the scoring method were selected for further characterization. Among the compounds tested, NSC23766 (structure shown in Table 2.1) was the only one to significantly inhibit TrioN binding to Rac1. The inhibitory effect of NSC23766 appeared to be specific for the interaction between Rac1 and its GEF because it did not interfere with the Cdc42 binding to intersectin. It had a value of 50 μM of IC_{50}.

2.2.2.16 *Parasitic Cysteine Protease*

Cysteine proteases constitute an important class of enzymes that play vital roles in the life cycle of many parasites (Sajid and McKerrow, 2002). A possible strategy for combating parasitic infections is to inhibit cysteine proteases that are crucial to parasite metabolism and reproductive function. Papain-like cysteine proteases from *Leishmania donovani, Plasmodium falciparum* (falcipain-2 and falcipain-3), and *Trypanosoma cruzi* (cruzain) have been identified and functionally characterized (Sajid and McKerrow, 2002). Several studies have confirmed the efficacy of peptidyl cysteine protease inhibitors in arresting and killing parasites *in vivo* and in tissue culture models of parasite replication or cell invasion (Sajid and McKerrow, 2002). Vinyl sulfone-based peptides, such as cruzain and falcipains, are known to be irreversible inhibitors of cysteine proteases (Shenai et al., 2003) by forming covalent bonds with the thiolate of the catalytic cysteine. Although such irreversible inhibitors are quite potent with IC_{50} values in the nanomolar range, the poor selectivity for parasitic cysteine proteases over the human cysteine proteases remains a significant concern (Engel et al., 1998). Also, it is desirable to design reversible inhibitors to minimize the potential toxicity that can be observed with irreversible inhibitors.

Desai et al. (2004) constructed homology models of falcipain-2 and falcipain and compared them with crystal structures of cruzain. They found that the structure of the binding pocket appears to be highly conserved across the protozoal cysteine proteases. This suggests the possibility of developing common inhibitors of these enzymes to treat malaria, leishmaniasis, and trypanosomiasis. A relatively common binding pocket in each was defined to make sure that the common residues lining the active site were all included in the binding pocket definitions. Virtual screening was performed using the ChemBridge database (Express Pick, October 2001: ChemBridge Corporation, San Diego, CA) containing a library of about 241,000 compounds. The database was filtered to collect only druglike molecules for docking using the following protocol. Considering metals, ADME, and Lipinski's Rule of Five, the filtered database consisting of approximately 60,000 compounds was subjected to Concord to generate 3D coordinates and was subsequently used for docking. Docking was carried out in three consecutive stages employing three different settings in GOLD. Starting with about 60,000 molecules for the first stage of docking, a total of 5500 top-ranking molecules were selected for the next stage. Next, the top 1500 molecules selected based on the second stage of docking were screened against both falcipain-2 and falcipain-3. It is interesting to note that the top 100 molecules in the falcipain-3 run were among the top 200 molecules in the falcipain-2 run, with 78% of molecules in the top 100. Finally, a total of 24 druglike nonpeptide inhibitors were identified, of which 12 compounds showed dual activity against falcipain-2 and falcipain-3. Four compounds (structure of a representative hit shown in Table 2.1) were found to inhibit both the plasmodial cysteine proteases as well as the *L. donovani* cysteine protease. These leads can be optimized into broad spectrum antiprotozoal drugs.

2.2.2.17 Human Rhinovirus Coat Protein

The human rhinovirus (HRV) belongs to the family of picornaviruses and is the main cause for common colds and a variety of other respiratory illnesses, including otitis media and sinusitis, and for exacerbations of asthma and reactive airways disease. These illnesses still lack effective antiviral treatment. The viral capsid is a promising and intensively studied target for drug development. This protein shell encapsulates a single, positive RNA strand and consists of 60 copies of four different viral proteins. HRV coat protein inhibitors act as capsid-binding antiviral agents that block the uncoating of the viral particles and/or inhibit cell attachment (Hadfield et al., 1999). Their binding site is located within a hydrophobic pocket situated at the bottom of a depression, a so-called canyon, on the capsid surface. In the absence of an inhibitor, this pocket can be empty or occupied by a pocket factor, a lipid or fatty acid. Structural conservation in this region among the different serotypes permits the development of broad-spectrum anti-HRV agents (Kolatkar et al., 1999). Steindl et al. (2005) applied a combined approach of virtual screening techniques (pharmacophore model generation, docking, and ligand clustering in combination with a principal component analysis [PCA] performed on a set of molecular descriptors) to select a small set of possible new HRV coat protein inhibitors from a large commercially available database. The authors used the program Catalyst to generate pharmacophore models, also termed hypotheses. A pharmacophore model defines a special arrangement of chemical features that are shared by different ligands and are obviously responsible for binding and, as a consequence, for biological activity. Such a model can be used as a query tool in 3D database mining, thereby provoking an enrichment of active compounds in the obtained hit lists. Using the crystal structure of an inhibitor complexed with the HRV coat protein and the functional group definitions contained within the Catalyst software, a representation of the binding site was produced (structure-based pharmacophore model) and applied in screening the commercially available Maybridge database. The resulting hits were submitted to further investigation using a docking study as well as a PCA of a set of molecular descriptors, thus aiming at the final selection of a few compounds for biological testing. Application of a scoring function (SF) allows the estimation of binding affinities and therefore a ranking of the substances. PCA is a widely applied multivariate technique in drug discovery. It provides a reduction of data multidimensionality (i.e., the properties of the input molecules), gives an overview of the data, and detects trends, groupings, and outliers. In this study, PCA-based clustering was applied to estimate the similarity of the Maybridge hits with already known active molecules. The docking and scoring procedures as well as the PCA-based clustering were carried out within the Cerius software. The combination of pharmacophore model generation, docking, and PCA-based clustering presents a virtual screening and selection protocol aimed at the fast and reliable identification of potential new HRV coat protein inhibitors. Thereby, each step underwent thorough theoretical validation probing their performance on sets of known active and by all probability inactive compounds.

These combined efforts finally allowed the selection of six promising candidate structures that were tested in a biological assay exemplarily against one major receptor group virus serotype for *in vitro* anti-HRV activities. The ability of the selected compounds to inhibit HRV serotype 39 was evaluated. The biological results were determined by multiple-cycle CPE inhibition assay and expressed as EC_{50} values. Microscopic and spectrophotometric reading confirmed that all six of the Maybridge compounds inhibit viral growth on OH-I and WI-38 cells and show activities in the micromolar range. The conclusion can be drawn that at this level of activity, we could achieve a success rate of 100% in the selected hit list. Particularly, one of the six promising candidate shows an EC_{50} of 4.3 μM for OH-I cells. The structure of the candidate is shown in Table 2.1.

2.2.2.18 SARS-CoV Protease

Severe acute respiratory syndrome (SARS) is a respiratory illness that had a widespread dramatic outbreak in Asia, North America, and Europe in early 2003 (Lee et al., 2003). Evidence indicates that a previously unrecognized coronavirus exists, called SARS coronavirus, which is the leading hypothesis for the cause of SARS (Rota et al., 2003). It was known that the cleavage process of the SARS-CoV polyproteins by a special proteinase, the so-called SARS coronavirus 3C-like proteinase (CoV Mpro), is a key step for the replication of SARS-CoV (Fan et al., 2004). The SARS coronavirus 3C-like proteinase is considered as a potential drug design target for the treatment of SARS.

Liu et al. (2005) built a "flexible" 3D model for SARS 3C-like proteinase by using a homology modeling and multicanonical molecular dynamics method, and they used the model for virtual screening of chemical databases. A 3D homology model of SARS 3C-like proteinase based on the structure of transmissible-gastroenteritis-virus coronavirus 3C-like proteinase was built by using MODELLER6.0

program. Using the MPI Edition DOCK4.01 program, the authors screened several chemical databases including the National Cancer Institute Diversity Set (~230,000 compounds total), ACD-3D (Available Chemical Database, Release: ACD 3D 2002.2, ~280,000 compounds in total), and MDDR-3D (MDL Drug Database Report, Release: MDDR 3D 2002.2, 120,000 compounds in total). After DOCK procedures, strategies including pharmocophore modeling, consensus scoring, and "druglike" filtering were applied in order to accelerate the process and improve the success rate of virtual docking screening hit lists. A total of 126 molecules with more than 10 types of chemical structures were selected for further study. Forty compounds were purchased and tested by HPLC and colorimetric assay against SARS 3C-like proteinase. Three of them—including calmidazolium, a well-known antagonist of calmodulin—were found to inhibit the enzyme with an apparent K_i from 61 to 178 μM. These active compounds and their binding modes provide useful information for understanding the binding sites and for further selective drug design against SARS and other coronaviruses.

2.2.2.19 *Protein Kinase*

Epidermal Growth Factor Receptor (EGFR) EGFR is a member of the protein tyrosine kinase family and is frequently overexpressed in cancer cells, correlating with poor prognosis and survival (Normanno et al., 2005). EGFR therefore represents a rational target for the development of novel anticancer therapies. Inhibition of the EGFR tyrosine kinase activity by small molecules has proved to be effective for the treatment of cancer (Shawver et al., 2002). Cavasotto et al. (2006) used the X-ray crystal structure of EGFR (Stamos et al., 2002) bound to OSI-744 (PDB code 1M17) to screen a collection of 315,102 compounds of the ChemBridge Express Library. The conserved water molecule present within the binding site was included in the description of the binding site because of its observed strong contacts with the receptor. The ICM virtual-screening program (2005) was used in this study. The ICM-based VS procedure is based on fast docking of a flexible ligand to a grid representation of the receptor (Totrov and Abagyan, 2001) followed by scoring. The ICM score takes into account the ligand–receptor interaction energy, conformational strain energy of the ligand, conformational entropy loss, and desolvation effects. Self-docking of cocrystallized ligand OSI-744 showed a score of −35. The authors thus selected a cutoff score of −28 to take into account limitations stemming from the rigid representation of the receptor and inaccuracies in the scoring function. There is also strong evidence that kinase inhibitors form two hydrogen bonds to the hinge-region backbone atoms (corresponding in EGFR to carboxylic O in Q767, and carboxylic O and HN in M769). For this reason, 1246 compounds (from a total of 315,102) scoring better than −28 and forming at least two hydrogen bonds with the hinge-region backbone were clustered according to chemical similarity and visual inspection, resulting in a final list of 212 compounds. The 50 best-scoring molecules were evaluated for their effect on EGFR tyrosine kinase activity and their antiproliferative activity in cancer cells. The authors identified several compounds with antiproliferative effects on cancer cells. Amongst them, MSK-039 was discovered as a low-

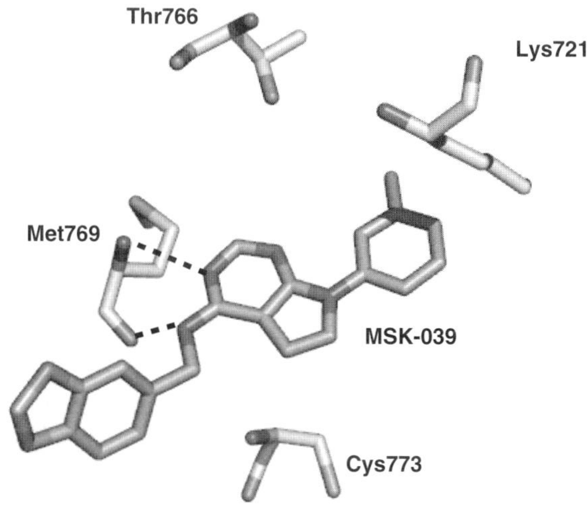

Figure 2.3 Structural model of MSK-039 docked at the EGFR kinase active site (only selected residues are shown). Dashed lines represent hydrogen bonds between ligand and active site residues. See color plates.

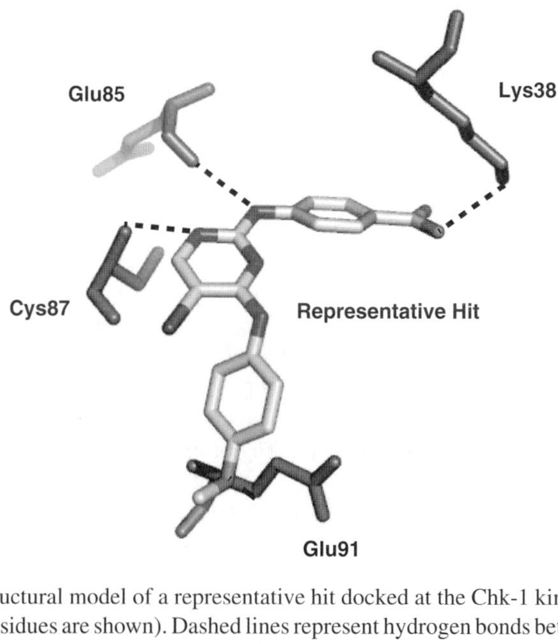

Figure 2.4 Structural model of a representative hit docked at the Chk-1 kinase active site (only selected residues are shown). Dashed lines represent hydrogen bonds between ligand and active site residues. See color plates.

micromolar inhibitor of EGFR tyrosine kinase activity. The predicted binding mode of MSK-039 is shown in Fig. 2.3, and its structure is shown in Table 2.1.

Checkpoint Kinase-1 (Chk-1 Kinase) In recent years, DNA damage checkpoint kinases have emerged as attractive targets for cancer therapy (Zhou and Elledge, 2000). The checkpoint kinase-1 (Chk-1 kinase) prevents cells with damaged DNA from entering mitosis by arresting them at the G2/M checkpoint. It has been suggested that inhibition of Chk-1 kinase would lead to abrogation of the G2/M checkpoint, allowing cells to enter a lethal mitosis with damaged DNA. Thus there is a great interest in developing inhibitors of Chk-1 kinase for use as potential sensitizing agents in combination with standard therapies that induce DNA damage. Lyne et al., (2004) carried out a virtual screen of the AstraZeneca compound collection for the identification of compounds likely to bind checkpoint kinase-1 (Chk-1 kinase), using a receptor-based strategy. In the first step, the compound collection was filtered by application of generic physical properties, followed by removal of compounds with undesirable chemical functionality. Subsequently, a 3D pharmacophore screen for compounds with kinase binding motifs was applied. A filtered database of approximately 200,000 compounds was then docked into the active site of Chk-1 kinase using the FlexX-Pharm (Hindle et al., 2002) program. For each compound that docked successfully into the binding site, up to 100 poses were saved. These poses were then postfiltered using a customized consensus scoring scheme for a kinase, followed by visual inspection of a selection of the docked compounds. This resulted in 103 compounds being ordered for testing in the project assay, and 36 of these (corresponding to four chemical classes) were found to inhibit the enzyme in a dose–response fashion with IC_{50} values ranging from 110 nM to 68 μM. The binding mode of the best hit from this study is shown in Fig. 2.4, and its structure is shown in Table 2.1.

2.3 CONCLUSIONS AND PROSPECTS

Several clear successes have been demonstrated over the past few years in the field of computer-aided virtual screening for lead identification using receptor-based, ligand-based, pharmacophore-based, or similarity-based approaches. Lead compounds have been found for a wide array of biological targets such as enzymes, ligand binding domains, and even small molecules disrupting protein–protein interactions. There is no general method to carry out these studies. Every system is unique, and what has led to these successes is an in-depth understanding of the system under investigation and fine-tuning of the virtual-screening scheme to achieve the desired result. In most cases, this was done by applying all available information to generate and validate the models, which included seeding the database to be screened with known actives and then defining the workflow or the screening protocol so as to identify the known ligands as high-ranking hits. RBVS requires database filtering, target preparation, docking, and postdocking analysis. Appropriate filters can significantly reduce the number of candidates for screening by eliminating structures with undesirable properties. Some of the key challenges in VS are appropriate treatment of ionization and tautomerization of ligand and protein residues, treatment of target/ligand flexibility, choice of force fields in docking calculations, solvation effects, dielectric constants, exploration of multiple binding modes, and, most importantly, the approximations in the scoring functions that can lead to false positives and miss true hits. In spite of the addressed limitations in this approach, it is still the best option available currently to explore a large chemical space in terms of cost-effectiveness, time commitment, and material as it allows access to a large number of possible ligands, most of which may be simply purchased and tested. Despite the challenges, this method has provided leads in several cases, and the most important feature of this approach is its ability to find novel chemical entities (NCE) for targets, which is very valuable from the medicinal chemistry perspective. With an increasing number of targets being identified by HTP genomics and proteomics efforts, virtual screening may provide an excellent complementary approach to HTP screening for valuable leads.

ACKNOWLEDGMENTS

We would like to thank members of the Huang laboratory and collaborators at Raylight Corporation and Chemokine Pharmaceutical Inc. for helpful discussion. This work was supported by grants from the National Institutes of Health, Raylight Corporation, and Chemokine Pharmaceutical Inc.

REFERENCES

MOE (2003). Montreal: Chemical Computing Group.

ICM (2005). La Jolla, CA: Molsoft LLC.

Ahrén, B., Simonsson, E., and Larsson, H. (2002). Islet hormone secretion under normal conditions in diabetes and in obesity. *Diabetes Care* **25**, 869–875.

Asakura, M., Kitakaze, M., et al. (2002). Cardiac hypertrophy is inhibited by antagonism of ADAM12 processing of HB-EGF: Metalloproteinase inhibitors as a new therapy. *Nat Med* **8**, 35.

Bajorath, J. (2002). Integration of virtual and high-throughput screening. *Nat Rev Drug Discov* **1**(11), 882–894.

Barril, X., Hubbard, R. E., and Morley, S. D. (2004). Virtual screening in structure-based drug discovery. *Mini Rev Med Chem* **4**(7), 779–791.

Barril, X., Brough, P., Drysdale, M., Hubbard, R. E., Massey, A., Surgenor, A., and Wright., L. (2005). Structure-based discovery of a new class of Hsp90 inhibitors. *Bioorg Med Chem Lett* **15**, 5187–5191.

Baxter, C. A., Murray, C. W., et al. (1998). Flexible docking using Tabu search and an empirical estimate of binding affinity. *Proteins.* **33**(3), 367–382.

Beaudeux, J. L., Giral, P., et al. (2004). Matrix metalloproteinases, inflammation and atherosclerosis: Therapeutic perspectives. *Clin Chem Lab Med.* **42**(2), 121–131.

Becker, O. M., Marantz, Y., et al. (2004). G protein-coupled receptors: *In silico* drug discovery in 3D. *Proc Natl Acad Sci U S A.* **101**(31), 11304–11309.

Blundell T. L., J. H., and Abell, C. (2002). High-throughput crystallography for lead discovery in drug design. *Nat Rev Drug Discov* **1**, 45–54.

Bohm, H. J. (1992). LUDI: Rule-based automatic design of new substituents for enzyme inhibitor leads. *J Comput Aided Mol Des* **6**(6), 593–606.

Bohm, H. J. (1994). The development of a simple empirical scoring function to estimate the binding constant for a protein–ligand complex of known three-dimensional structure. *J Comput Aided Mol Des.* **8**(3), 243–256.

Bohm, H. J. (1998). Prediction of binding constants of protein ligands: A fast method for the prioritization of hits obtained from *de novo* design or 3D database search programs. *J Comput Aided Mol Des* **12**(4), 309–23.

Borhani, D. W., Harters, T. M., and Petrash, J. M. (1992). The crystal structure of the aldose reductase NADPH binary complex. *J Biol Chem* **267**, 24841–24847.

Bradshaw, R. A., and Yi, E. (2002). Methionine aminopeptidases and angiogenesis. *Essays Biochem* **38**, 65–78.

Brooijmans, N., and Kuntz, I. D. (2003). Molecular recognition and docking algorithms. *Ann Rev Biophys Biomol Struct* **32**, 335–373.

Carlson, H. A., and McCammon, J. A. (2000). Accommodating protein flexibility in computational drug design. *Mol Pharmacol.* **57**(2), 213–218.

Cavasotto, C. N., Ortiz, M. A., et al. (2006). *In silico* identification of novel EGFR inhibitors with antiproliferative activity against cancer cells. *Bioorg Med Chem Lett.* **Jan 11**, Epub ahead of print.

Charifson, P. S., Corkery, J. J., Murcko, M. A., and Walters, W. P. (1999). Consensus scoring: A method for obtaining improved hit rates from docking databases of three-dimensional structures into proteins. *J Med Chem* **42**(25), 5100–5109.

Clark, D. E., Higgs, C., Wren, S. P., Dyke, H. J., Wong, M., Norman, D., Lockey, P. M., and Roach, A. G. (2004). A virtual screening approach to finding novel and potent antagonists at the melanin-concentrating hormone 1 receptor, *J Med Chem* **47**, 3962–3971.

Cosulich, S. C., Worrall, V., et al. (1997). Regulation of apoptosis by BH3 domains in a cell-free system. *Curr Biol* **7**(12), 913–920.

Cvetkovic, R. S., and Perry, C. M. (2003). Brinzolamide: A review of its use in the management of primary open-angle glaucoma and ocular hypertension. *Drugs Aging* **20**(12), 919–947.

Demutha, H. U., McIntoshb, C. H. S., and Pederson, R. A. (2005). Type 2 diabetes—Therapy with dipeptidyl peptidase IV inhibitors. *Biochim Biophys Acta* **1751**, 33–44.

Desai, P. V., Patny, A., Sabnis, Y., Tekwani, B., Gut, J., Rosenthal, P., Srivastava, A., and Avery, M. (2004). Identification of novel parasitic cysteine protease inhibitors using virtual screening.1. The ChemBridge database. *J Med Chem* **47**, 6609–6615.

Deveraux, Q. L., and Reed, J. C. (1999). IAP family proteins—Suppressors of apoptosis. *Genes Dev* **13**, 239–252.

Di Marzo, V., Bifulco, M., and De Petrocellis, L. (2004). The endocannabinoid system and its therapeutic exploitation. *Nat Rev Drug Discovery* **3**, 771–784.

Di Nola, A., Roccatano, D., et al. (1994). Molecular dynamics simulation of the docking of substrates to proteins. *Proteins* **19**(3), 174–182.

Doman, T. N., McGovern, S. L., Witherbee, B. J., Kasten, T. P., Kurumbail, R., Stallings, W. C., Connolly, D. T., and Shoichet, B. K. (2002). Molecular docking and high-throughput screening for novel inhibitors of protein tyrosine phosphatase-1B. *J Med Chem* **45**, 2213–2221.

Eldridge, M. D., Murray, C. W., Auton, T. R., Paolini, G. V., and Mee, R. P. (1997). Empirical scoring functions: I. The development of a fast empirical scoring function to estimate the binding affinity of ligands in receptor complexes. *J Comput Aided Mol Des.* **11**(5), 425–445.

Engel, J. C., Doyel, P. S., Hseih, I., and McKerrow, J. H. (1998). Cysteine protease inhibitors cure an experimental *Trypanosoma cruzi* infection. *J. Exp. Med.* **188**, 725–734.

Enyedy, I. J., Ling, Y., Nacro, K., Tomita, Y., Wu, X., Cao, Y., Guo, R., Li, B., Zhu, X., and Huang, Y., et al. (2001). Discovery of small-molecule inhibitors of Bcl-2 through structure-based computer screening. *J Med Chem* **44**, 4313–4324.

Erik Sahai, C. J. M. (2002). Rho–GTPases and Cancer. *Nat Rev Cancer* **2**(2), 133–142.

Eswaramoorthy, S., Kumaran, D., et al. (2002). A novel mechanism for *Clostridium botulinum* neurotoxin inhibition. *Biochemistry* **41**(31), 9795–9802.

Etienne-Manneville, S., and Hall, A. (2002). Rho GTPases in cell biology *Nature* **420**, 629–635.

Evers, A., and Klabunde, T. (2005). Structure-based drug discovery using GPCR homology modeling: Successful virtual screening for antagonists of the alpha1A adrenergic receptor. *J Med Chem* **48**, 1088–1097.

Evers, A., and Klebe, G. (2004). Successful virtual screening for a submicromolar antagonist of the neurokinin-1 receptor based on a ligand-supported homology model. *J Med Chem* **47**, 5381–5392.

Ewing, T. J., Makino, S., Skillman, A. G., and Kuntz, I. D. (2001). DOCK 4.0: Search strategies for automated molecular docking of flexible molecule databases. *J Comput Aided Mol Des* **15**(5), 411–428.

Fan, K., Wei, P., Feng, Q., Chen, S., Huang, C., Ma, L., Lai, B., Pei, J., Liu, Y., Chen, J., and Lai, L. (2004). Biosynthesis, purification and substrate specificity of SARS Coronavirus 3C-like proteinase. *J Biol Chem* **279**, 1637–1642.

Fische, E. (1894). Einfluss der Configuration auf die wirkung der Enzyme. *Ber Dt Chem Ges* **27**, 2985–2993.

Gao, Y., Dickerson, J. B., Guo, F., Zheng, J., and Zheng, Y. (2004). Rational design and characterization of a Rac GTPase-specific small molecule inhibitor. *Proc Natl Acad Sci USA* **101**(20), 7618–7623.

Gaoni, Y., and Mechoulam, R. (1964). Isolation, structure, and partial synthesis of an active constituent of hashish. *J Am Chem Soc.* **86**, 1646–1647.

Gibbs, J. B., and Oliff, A. (1997). The potential of farnesyltransferase inhibitors as cancer chemotherapeutics. *Annu Rev Pharmacol Toxicol* **37**, 143–166.

Gohlke, H., Hendlich, M., and Klebe, G. (2000). Knowledge-based scoring function to predict protein–ligand interactions. *J Mol Biol* **295**(2), 337–356.

Goodsell, D. S., Lauble, H., Stout, C. D., and Olson, A. J. (1993). Automated docking in crystallography: Analysis of the substrates of aconitase. *Proteins* **17**(1), 1–10.

Hadfield, A. T., Diana, G. D., and Rossmann, M. G. (1999). Analysis of three structurally related antiviral compounds in complex with human rhinovirus 16. *Proc Natl. Acad. Sci. U.S.A.* **96**, 14730–14735.

Halgren, T. A., Murphy, R. B., Friesner, R. A., Beard, H. S., Frye, L. L., Pollard, W. T., and Banks, J.L. (2004). Glide: A new approach for rapid, accurate docking and scoring. 2. Enrichment factors in database screening. *J Med Chem.* **47**(7), 1750–1759.

Halperin, I., Ma, B., Wolfson, H., and Nussinov, R. (2002). Principles of docking: An overview of search algorithms and a guide to scoring functions. *Proteins* **47**(4), 409–443.

Hart, T. N., and Read, R. J. (1992). A multiple-start Monte Carlo docking method. *Proteins.* **13**(3), 206–222.

Hindle, S., Rarey, M., et al. (2002). Flexible docking under pharmacophore type constraints. *J Comput Aided Mol Des* **16**(2), 129–149.

Irwin, J. J., Raushel, F. M., et al. (2005). Virtual screening against metalloenzymes for inhibitors and substrates. *Biochemistry* **44**(37), 12316–12328.

Irwin, J. J., and Shoichet, B. K. (2005). ZINC—A free database of commercially available compounds for virtual screening. *J Chem Inf Model* **45**(1), 177–182.

Jones, G., Willett, P., Glen, R. C., Leach, A. R., and Taylo, R. (1997). Development and validation of a genetic algorithm for flexible docking. *J Mol Biol* **267**(3), 727–748.

Kitchen, D. B., Decornez, H., et al. (2004). Docking and scoring in virtual screening for drug discovery: Methods and applications. *Nat Rev Drug Discov* **3**(11), 935–949.

Klabunde, T., and Hessler, G. (2002). Drug design strategies for targeting G-protein-coupled receptors. *Chembiochem* **3**, 928–944.

Kohl, N. E., Mosser, S. D., et al. (1993). Selective inhibition of ras-dependent transformation by a farnesyltransferase inhibitor. *Science* **260**, 1934–1937.

Kolatkar, P. R., Bella, J., Olson, N. H., Bator, C. M., and Baker, T. S., et al. (1999). Structural studies of two rhinovirus serotypes complexed with fragments of their cellular receptor. *EMBO J* **18**, 6249–6259.

Kraemer, O., Hazemann, I., Podjarny, A. D., and Klebe, G. (2004). Virtual screening for inhibitors of human aldose reductase. *Proteins: Struct Funct Bioinf* **55**, 814–823.

Kramer, B., Metz, G., Rarey, M., and Lengauer, T. (1999a). Ligand docking and screening with FlexX. *Med Chem Res* **9**, 463–478.

Kramer, B., Rarey, M., and Lengauer, T. (1999b). Evaluation of the FLEXX incremental construction algorithm for protein–ligand docking. *Proteins* **37**(2), 228–241.

Kruger, E. A., and Figg, W. D. (2000). TNP-470: An angiogenesis inhibitor in clinical development for cancer. *Expert Opin Investig Drugs* **9**(6), 1383–1396.

Lambeir, A. M., Durinx, C., Scharpe, S., and DeMeester, I. (2003). Dipeptidylpeptidase IV from bench to bedside: An update on structural properties, functions, and clinical aspects of the enzyme DPP IV. *Crit Rev Clin Lab Sci* **40**, 209–294.

Leach, A. R. (1994). Ligand docking to proteins with discrete side-chain flexibility. *J Mol Biol* **235**(1), 345–356.

Leach, A. R. (1996). *Molecular Modeling: Principles and Applications*. Harlow, UK: Addison-Wesley, Longman Limited.

Lee, N., Hui, D., Wu, A., Chan, P., Cameron, P., Joynt, G. M., Ahuja, A., Yung, M. Y., Leung, C. B., To, K. F., Lui, S. F., Szeto, C. C., Chung, S., and Sung, J. J. (2003). A major outbreak of severe acute respiratory syndrome in Hong Kong. *N Engl J Med* **348**, 1986.

Li, C., Xu, L., Wolan, D. W., Wilson, I. A., and Olson, A. J. (2004). Virtual screening of human 5-aminoimidazole-4-carboxamide ribonucleotide transformylase against the NCI diversity set by use of AutoDock to identify novel nonfolate inhibitors. *J Med Chem* **47**, 6681–6690.

Liu, Z., Huang, C., Fan, K., Wei, P., Chen, H., Liu, S., Pei,J., Shi, L., Li, B., Yang, K., Liu, Y., and Lai, L. (2005). Virtual screening of novel noncovalent inhibitors for SARS-CoV 3C-like proteinase. *J Chem Inf Model* **45**, 10–17.

Lugovskoy, A. A., Degterev, A. I., Fahmy, A. F., Zhou, P., Gross, J. D., Yuan, J., and Wagner, G. (2002). A novel approach for characterizing protein ligand complexes: Molecular basis for specificity of small-molecule Bcl-2 inhibitors. *J Am Chem Soc.* **124**, 1234–1240.

Lyne, P. D., Kenny, P. W., et al. (2004). Identification of compounds with nanomolar binding affinity for checkpoint kinase-1 using knowledge-based virtual screening. *J Med Chem* **47**(8), 1962–1968.

Maloney, A., and Workman, P. (2002). HSP90 as a new therapeutic target for cancer therapy: The story unfolds. *Expert Opin Biol Ther* **2**(1), 3–24.

McGann, M. R., Almond, H. R., Nicholls, A., Grant, J. A., and Brown, F. K. (2003). Gaussian docking functions. *Biopolymers* **68**(1), 76–90.

Meng, E. C., Shoichet, B. K., et al. (1992). Automated docking with grid-based energy evaluation. *J Comp Chem* **13**, 505–524.

Mira, J. P., Benard, V., Groffen, J., Sanders, L. C., and Knaus, U. G.. (2000). Endogenous, hyperactive Rac3 controls proliferation of breast cancer cells by a p21-activated kinase-dependent pathway. *Proc Natl Acad Sci USA* **97**(1), 185–189.

Mizutani, M. Y., and Itai, A. (2004). Efficient method for high-throughput virtual screening based on flexible docking: Discovery of novel acetylcholinesterase inhibitors. *J Med Chem* **47**, 4818–4828.

Møller, N. P., Iversen, L. F., Andersen, H. S., and McCormack, J. G. (2000). Protein tyrosine phosphatases (PTPs) as drug targets: Inhibitors of PTP-1B for the treatment of diabetes. *Curr Opin Drug Discovery* **3**, 527–540.

Mosselman, S., Polman, J., and Dijkema, R. (1996). ER beta: Identification and characterization of a novel human estrogen receptor. *FEBS Lett.* **392**, 49–53.

Muegge, I. (2000). A knowledge-based scoring function for protein–ligand interactions: Probing the reference state. *Perspect Drug Discov Des* **20**, 99–114.

Muegge, I. (2001). Effect of ligand volume correction on PMF scoring. *J Comput Chem* **22**(4), 418–425.

Muegge, I., and Martin, Y. C. (1999). A general and fast scoring function for protein-ligand interactions: A simplified potential approach. *J Med Chem* **42**(5), 791–804.

Muegge, I., and I. Enyedy (2004). Docking and scoring. In J. Tollenaere, H. De Winter, W. Langenaeker and P. Bultinck (Eds.), *Computational Medicinal Chemistry and Drug Discovery*. New York: Marcel Dekker, pp. 405–436.

Nikolovska-Coleska, Z., Xu, L., Hu, Z., Tomita, Y., Li, P., Roller, P. P., Wang, R., Fang, X., Guo, R., and Zhang, M., et al. (2004). Discovery of embelin as a cell-permeable, small-molecular weight inhibitor of XIAP through structure-based computational screening of a traditional herbal medicine three-dimensional structure database. *J Med Chem* **47**(10), 2430–2440.

Normanno, N., Bianco, C., et al. (2005). The ErbB receptors and their ligands in cancer: An overview. *Curr Drug Targets* **6**, 243–257.

Oefner, C., D-Arcy, A., MacSweeney, A., Pierau, S., Gardiner, R., and Dale, G. E. (2003). High-resolution structure of human apo dipeptidyl peptidase IV/CD26 and its complex with 1-[([2-[(5-iodopyridin-2-yl)amino]-ethyl]amino)-acetyl]-2-cyano-(*S*)-pyrrolidine. *Acta Crystallogr D Biol Crystallogr* **59**, 1206–1212.

Oh, M., Im, I., Lee, Y. J., Kim, Y. H., Yoon, J. H., Park, H. G, Higashiyama, S., Kima, Y. C., and Parka, W. J. (2004). Structure-based virtual screening and biological evaluation of potent and selective ADAM12 inhibitors. *Bioorg Med Chem Lett* **14**, 6071–6074.

Oshiro, C. M., Kuntz, I. D., et al. (1995). Flexible ligand docking using a genetic algorithm. *J Comput Aided Mol Des* **9**(2), 113–130.

Park, H. W., Boduluri, S. R., Moomaw, J. F., Casey, P. J., and Beese, L. S. (1997). Crystal structure of protein farnesyltransferase at 2.25 angstrom resolution. *Science* **275**, 1800–1804.

Perola, E., Xu, K., Kollmeyer, T. M., Kaufmann, S. H., Prendergast, F. G., and Pang, Y. P. (2000). Successful virtual screening of a chemical database for farnesyltransferase inhibitor leads. *J Med Chem* **43**, 401–408.

Prendergast, G. C., J. P. Davide, et al. (1994). Farnesyltransferase inhibition causes morphological reversion of ras-transformed cells by a complex mechanism that involves regulation of the actin cytoskeleton. *Mol Cell Biol* **14**, 4193–4202.

Puius, Y. A., Zhao, Y., Sullivan, M., Lawrence, D. S., Almo, S. C., and Zhang, Z. Y. (1997). Identification of a second aryl phosphate-binding site in protein-tyrosine phosphatase 1B: A paradigm for inhibitor design. *Proc Natl Acad Sci USA* **94**, 13420–13425.

Rarey, M., Kramer, B., Lengauer, T., and Klebe, G. (1996). A fast flexible docking method using an incremental construction algorithm. *J Mol Biol* **261**(3), 470–489.

Rasmussen, H. B., Branner, S., Wiberg, F. C., and Wagtmann, N. (2003). Different modes of dipeptidyl peptidase IV (CD26) inhibition by oligopeptides derived from the N-terminus of HIV-1 Tat indicate at least two inhibitor binding sites. *Nat Struct Biol* **10**, 19–25.

Rollinger, J. M., Hornick, A., Langer, S., and Prast, H. (2004). Acetylcholinesterase inhibitory activity of scopolin and scopoletin discovered by virtual screening of natural products. *J Med Chem* **47**, 6248–6254.

Rota, P. A., Oberste, M. S., Monroe, S. S., Nix, W. A., Campagnoli, R., Icenogle, J. P., Penaranda, S., Bankamp, B., Maher, K., Chen, M. H., Tong, S., Tamin, A., Lowe, L., Frace, M., DeRisi, J. L., Chen, Q., Wang, D., Erdman, D. D., Peret, T. C., Burns, C., Ksiazek, T. G., Rollin, P. E., Sanchez, A., Liffick, S., Holloway, B., Limor, J., McCaustland, K., Olsen-Rasmussen, M., Fouchier, R., Gunther, S., Osterhaus, A. D., Drosten, C., Pallansch, M. A., Anderson, L. J., and Bellini, W. J. (2003). Characterization of a novel Coronavirus associated with severe acute respiratory syndrome. *Science* **300**, 1394–1399.

Sajid, M., and McKerrow, J. (2002). Cysteine proteases of parasitic organisms. *Mol Biochem Parasitol* **120**, 1–21.

Salo, O. M. H., Raitio, K. H., Savinainen, J. R., Nevalainen, T., Lahtela-Kakkonen, M., Laitinen, J. T., Järvinen, T., and Poso, A. (2005). Virtual screening of novel CB2 ligands using a comparative model of the human cannabinoid CB2 receptor. *J Med Chem* **48**, 7166–7171.

Salvesen, G. S., and Duckett, C. S. (2002). IAP proteins: Blocking the road to death's door. *Nat Rev Mol Cell Biol* **3**, 401–410.

Sattler, M., Liang, H., Nettesheim, D., Meadows, R. P., Harlan, J. E, Eberstadt, M., Yoon, H. S, Shuker, S. B., Chang, B. S, and Minn, A. J., et al. (1997). Structure of Bcl-xL-Bak peptide complex: Recognition between regulators of apoptosis. *Science* **275**, 983–986.

Schmitz, A. A. P., Govek, E. E., Böttner, B., and Aelst, L. Van (2000). Rho GTPases: Signaling, migration, and invasion *Exp. Cell Res.* **261**(1), 1–12.

Sebti, S. M., and Hamilton, A. D. (1997). Inhibition of Ras prenylation: A novel approach to cancer chemotherapy. *Pharmacol Ther* **74**, 103–114.

Shacham, S., Topf, M., et al. (2001). Modeling the 3D structure of GPCRs from sequence. *Med Res Rev* **21**(5), 472–483.

Shacham, S., Marantz, Y., et al. (2004). PREDICT modeling and *in-silico* screening for G-protein coupled receptors. *Proteins* **57**(1), 51–86.

Shangary, S., and Johnson D. E. (2003). Recent advances in the development of anticancer agents targeting cell death inhibitors in the Bcl-2 protein family. *Leukemia* **17**, 1470–1481.

Shawver, L. K., Slamon, D., et al. (2002). Smart drugs: Tyrosine kinase inhibitors in cancer therapy. *Cancer Cell* **1**, 117–123.

Shenai, B. R., Lee, B. J., Alvarez-Hernandez, A., Chong, P. Y., Emal, C. D., Neitz, R. J., Roush, W. R., and Rosenthal, P. J. (2003). Structure–activity relationships for inhibition of cysteine protease activity and development of *Plasmodium falciparum* by peptidyl vinyl sulfones. *Antimicrob Agents Chemother* **47**, 154–160.

Song, H., Wang, R., Wang, S., and Lin, J. (2005). A low-molecular-weight compound discovered through virtual database screening inhibits Stat3 function in breast cancer cells. *Proc Natl Acad Sci USA* **102**(13), 4700–4705.

Srinivasula, S. M., Hegde, R., Saleh, A., Datta, P., Shiozaki, E., Chai, J., Lee, R. A., Robbins, P. D., Fernandes-Alnemri, T., Shi, Y., et al. (2001). A conserved XIAP-interaction motif in caspase-9 and Smac/DIABLO regulates caspase activity and apoptosis. *Nature* **410**, 112–116.

Stamos, J., Sliwkowski, M. X., et al. (2002). Structure of the epidermal growth factor receptor kinase domain alone and in complex with a 4-anilinoquinazoline inhibitor. *J Biol Chem* **2777**, 46265–46272.

Steindl, T. M., Crump, C. E., Hayden, F. G., and Langer, T. (2005). Pharmacophore modeling, docking, and principal component analysis based clustering: Combined computer-assisted approaches to identify new inhibitors of the human rhinovirus coat protein. *J Med Chem* **48**, 6250–6260.

Tao, P., and Lai, L. (2001). Protein ligand docking based on empirical method for binding affinity estimation. *J Comput Aided Mol Des* **15**(5), 429–446.

Thoma, R., Loffler, B., Stihle, M., Huber, W., Ruf, A., and Hennig, M. (2003). Crystal structure of human dipeptidyl Peptidase IV (DPPIV). *Structure* **1**, 947–959.

Tokumaru, S., Higashiyama, S., et al. (2000). Ectodomain shedding of epidermal growth factor receptor ligands is required for keratinocyte migration in cutaneous wound healing. *J Cell Biol* **151**, 209.

Totrov, M., and Abagyan, R. (1997). Flexible protein–ligand docking by global energy optimization in internal coordinates. *Proteins* **Suppl 1**, 215–220.

Totrov, M. and Abagyan, R. A. (2001). *Drug-Receptor Thermodynamics: Introduction and Experimental Applications.* New York: John Wiley & Sons.

Verdonk, M. L., Cole, J. C., Hartshorn, M. J., Murray, C. W., and Taylor, R. D. (2003). Improved protein–ligand docking using GOLD. *Proteins* **52**(4), 609–623.

Wallis, R. M., Corbin, J. D., Francis, S. H., and Ellis, P. (1999). Tissue distribution of phosphodiesterase families and the effects of sildenafil on tissue cyclic nucleotides, platelet function, and the contractile responses of trabeculae carneae and aortic rings *in vitro*. *Am J Cardiol* **83**, 3C.

Wang, J. L., Liu, D., Zhang, Z. J., Shan, S., Han, X., Srinivasula, S. M., Croce, C. M., Alnemri, E. S., and Huang, Z. (2000). Structure-based discovery of an organic compound that binds Bcl-2 protein and induces apoptosis of tumor cells. *Proc Natl Acad Sci USA* **97**, 7124–7129.

Wang, R., Lu, Y., and Wang, S. (2003). Comparative evaluation of 11 scoring functions for molecular docking. *J Med Chem* **46**(12), 2287–2303.

Ward, R. A., Perkins, T. D. J., and Stafford, J. (2005). Structure-based virtual screening for low molecular weight chemical starting points for dipeptidyl peptidase IV inhibitors. *J Med Chem* **48**, 6991–6996.

Westbrook, J., Feng, Z., Chen, L., Yang, H., and Berman, H. M. (2003). The Protein Data Bank and structural genomics. *Nucleic Acids Res* **31**, 489–491.

Yabe-Nishimura, C. (1998). Aldose reductase in glucose toxicity: A potential target for the prevention of diabetic complications. *Pharmacol Rev* **50**, 21–33.

Yamazaki K., K., N., Fujita, K., Sato, H., Asano, S., Dan, A., and Kanaoka, M. (2005). Identification of phosphodiesterase-1 and 5 dual inhibitors by a ligand-based virtual screening optimized for lead evolution. *Bioorg Med Chem Lett* **xxx**, xxxx.

Yin, X. M., Oltvai, Z. N., et al. (1994). BH1 and BH2 domains of Bcl-2 are required for inhibition of apoptosis and heterodimerization with Bax. *Nature* **369**, 272–273.

Zhao, L., and Brinton, R. D. (2005). Structure-based virtual screening for plant-based ERbeta-selective ligands as potential preventative therapy against age-related neurodegenerative diseases. *J Med Chem* **48**(10), 3463–3466.

Zheng, Y. (2001). Dbl family guanine nucleotide exchange factors. *Trends Biochem Sci* **26**(12), 724–732.

Zhong, Z., Z. Wen, et al. (1994). Stat3: a STAT family member activated by tyrosine phosphorylation in response to epidermal growth factor and interleukin-6. *Science* **264**, 95–98.

Zhou, B. B., and Elledge, S. J. (2000). The DNA damage response: Putting checkpoints in perspective. *Nature* **408**(6811): 433–439.

CHAPTER 3

VIRTUAL SCREENING IN DRUG DISCOVERY

*Malcolm J. McGregor, Zhaowen Luo, and Xuliang Jiang**

3.1 INTRODUCTION

The goal of screening small molecules for drug discovery is to deliver new hit compounds to medicinal chemists that can act as starting points for the development of drug candidates. Computational chemistry and molecular modeling provide tools that are commonly used to direct and increase the efficiency of laboratory screening by selecting or designing compounds to be tested (Bajorath, 2002; Jorgensen, 2004; Walters and Namchuk, 2003). This is driven by the fact that the number of compounds available for screening usually far exceeds the number that will actually go into the screen. Similar methods are then applied in the hit or lead optimization process; however, this review will mainly focus on the methodologies that are applied to computer-based, *in silico*, or "virtual" screening in the early stages of drug discovery.

The precise definition of a "hit" or "lead" molecule varies from one organization to another. At Serono, a compound from screening is regarded as a hit if it has favorable properties in the following respects: (i) potency in a biological or biochemical assay, (ii) novelty with regard to intellectual property, (iii) selectivity for the intended target, and (iv) tractability in terms of ease of synthesis. A lead has the properties of a hit plus a

*Corresponding author: xuliang.jiang@serono.com

Drug Discovery Research: New Frontiers in the Post-Genomic Era, Edited by Ziwei Huang

favorable ADME/Tox profile (absorption, distribution, metabolism, excretion, and toxicity) and proven *in vivo* activity in an animal model. Therefore, many factors need to be considered at the virtual screening stage to put together a compound set that is most likely to deliver multiple hit/lead candidates. Clearly, activity against the target of interest is an absolute requirement. Computational methods attempt to deliver the most active compounds possible, but usually a positive from high-throughput screening (HTS) in the 1–10 μM range is acceptable, with the compound then being optimized by medicinal chemistry. Computational methods can be applied to deliver compounds that are different from what is already known and provide the chemists with a range of compounds from different chemical classes, thereby giving the opportunity to accept or reject hits based on other considerations such as chemical tractability, intellectual property, and toxicity. Thus, a desirable property of a computational method is to be able to "jump" classes—that is, use the information derived from one class to correctly predict compounds from another chemical class. The results then also give a more useful structure–activity relationship (SAR) for further computational analysis. In a large-scale screening campaign it is also desirable to include a nontargeted set of diverse compounds. The advantage of screening a diversity set is that the approach does not make assumptions about must-have properties or substructures in potential actives and thus invites discovery of unexpected active chemical types or modes of action. There is no universal agreement as to how much computational effort should be expended into predicting other druglike properties at the screening stage, such as those associated with ADME/Tox, but most agree that at least a rough measure should be applied because many compounds, especially from sources such as combinatorial libraries and natural products, have been found to violate the most basic requirements for druglikeness. Filters for clearly undesirable compounds are also usually applied.

Virtual screening originated in the 1970s when compound database searches were introduced using two-dimensional (2D) structural fragments (Blake et al., 1977; Dromey, 1978; Feldmann et al., 1977; Feldmann and Heller, 1972; Saxberg et al., 1976). Subsequently, a wide variety of diverse methodologies have been introduced, and the field is still rapidly evolving. These methods can be usefully divided into two categories: (1) those that predict activity for a particular target and (2) those computations not related to a target. While it is not possible in this review to describe all the methods, the following two sections will describe some of the major approaches in each category.

3.2 TARGET-DEPENDENT METHODS

This category relies on having some knowledge of the biological target or compounds that bind to the target. Therefore, this category can be further subdivided into those that are ligand-based and those that are receptor (protein)-based. When a set of active ligands is available, the methods that can be applied are those of QSAR, substructure analysis, pharmacophore analysis, or simpler methods such as using a measure of molecular similarity to query compound libraries for potential active compounds. When a crystal structure of the protein target is available (or a homology model can be built based on a closely related protein), then docking methods can be employed.

3.2.1 Molecular Similarity Searching

Similar chemical structures tend to have similar biological activities (Fig. 3.1), although exceptions to this generalization are frequent, because it is often observed that a small change in chemical structure, sometimes involving a single atom or functional group, can have a profound affect on activity for a particular target (Martin et al., 2003; Matter, 1997; Schuffenhauer et al., 2003). Conversely, molecules with seemingly unrelated structures (at least when looked at on paper in a 2D representation) can exhibit similar activities (Bajorath, 2002). In spite of these shortcomings, the hit rate from similarity-based searches is higher than that from random screening (Martin et al., 2003).

Small molecules are usually represented (both on paper and in chemical databases) as atoms connected by bonds, and as such they are difficult to compare computationally in this form. To overcome this difficulty, several other representations have been devised to facilitate quantitative molecular comparisons. The most common representation is to encode chemical features in a fixed-length binary bitstring, referred to as a "molecular fingerprint." Examples include the MACCS keys (MDL Information Systems, Inc.), the UNITY fingerprint (Tripos, Inc.), and the Daylight fingerprint (Daylight Chemical Information Systems, Inc.). The length of these keys vary from hundreds to thousands or more, and each bit encodes the presence or absence of a chemical feature, which is usually an atom type, a functional group, or a connected fragment. The similarity of the fingerprints of two molecules is therefore a measure of overall chemical similarity. Several metrics have been proposed to compare bitstrings (Willett et al., 1998), each with its own advantages and disadvantages, but probably the most widely used is the Tanimoto coefficient, defined as $N_{ab}/(N_a + N_b - N_{ab})$, where N_a and N_b are the number of bits set in bitstrings a and b respectively, and N_{ab} is the number of bits set in both bitstrings (i.e., applying the AND operator). The value varies

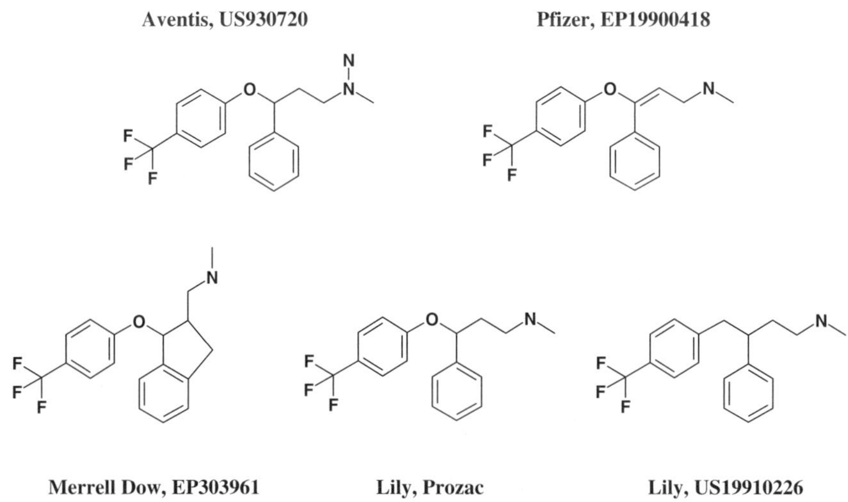

Figure 3.1 The chemical structures of a number of 5-HT reuptake inhibitors showing clear similarities in structure, illustrating the basis for the similarity and substructure methods described in the text. The common substructures are highlighted. See color plates.

between 0.0, where two bitstrings have no bits in common, and 1.0, where they are identical. This method discriminates well for most compounds in the molecular weight (MW) range of 200–600.

Many attempts have been conducted to improve the performance of this approach. Salim et al. (2003) showed that a combination of different similarity coefficients can improve similarity searching performance, although they failed to identity a single optimum combination for all activity classes. Another improvement for similarity searching is fingerprint scaling, whereby the contributions of each bit to the Tanimoto coefficient are scaled up if they also occur in a consensus fingerprint derived from a set of molecules active for a target. Xue et al. (2001, 2003, 2004) demonstrated that this method can improve the hit rate for most biological active classes. Some research groups have incorporated reduced molecular graphs into the approach, in which molecule structures are represented by topological graphs, where each node represents a substructure feature such as a ring or functional group. With this simplified structure representation, they found the method was more effective and could find more diverse compounds (Gillet et al., 2003; Harper et al., 2004).

Similarity searches using chemical fingerprints based on 2D structures will most likely pull out only analogs of known compounds. Although this is useful for generating SAR around a particular scaffold and producing "me too" compounds, the novelty of selected compounds is often an issue. Similarity searches based on fingerprints that incorporate three-dimensional (3D) molecular information are more able, in principle, to generalize to new scaffolds. McGregor and Muskal (1999) showed improved performance over 2D methods when the information of 3D pharmacophore fingerprints was used, but earlier studies found that 2D descriptors are equally or even more effective than 3D descriptors in selecting representative subsets of bioactive compounds (Brown and Martin, 1996; Matter, 1997).

The scope of fingerprints in similarity searches can also be extended beyond single targets to protein families, based on the observation that ligands for different targets in the same protein family are often seen to contain similar features or property ranges (Deng et al., 2004; Schuffenhauer et al., 2003). Therefore, the known active chemical structures for a whole target family can be grouped together and used in similarity searching; the resulting screening set then has an enlarged chemical space, and the same set can be screened repeatedly against targets of the same class.

3.2.2. Privileged Substructure Search

This approach is based on the observation that specific structural motifs in small molecule ligands often complement features in a target protein or protein family (Bondensgaard et al., 2004; Muller, 2003). These features may be common to the ligands of different members of a protein family (this is often the case with kinases). Sometimes a particular motif is a necessary requirement for optimal binding for a particular target—for example, the phenol group in estrogen receptor ligands (Anstead et al., 1997; Minutolo et al., 2003). By comparing structures of actives against inactives, these recurrent substructures can be identified, and a screening set is then selected by substructure search based on the identified fragments.

Many such features have been discovered by eye, either by examining sets of active compounds or by looking at ligand–receptor interactions in crystal structures. However, when data sets become larger and/or the features become more subtle, an automated procedure is desired. Examples of such procedures include an early artificial intelligence approach (Klopman, 1984), structural commonality analysis (Lewell et al., 1998), and a retrosynthetic analysis (Hajduk et al., 2000; Shemetulskis et al., 1996). Another example of such a method is discrete substructure analysis (DSA), developed by Serono Pharmaceutical Research Institute (Merlot et al., 2002, 2003). This method automatically enumerates all possible substructures up to a given size (typically 10–16 atoms) in a set of small molecules in 2D representation (the molecule connection table). Given a training set of actives and inactives, DSA calculates statistics that evaluate the occurrence of each fragment in the active set relative to the inactive set. Using the DSA Visualizer, the resulting statistics can be viewed interactively on a computer screen (Fig. 3.2). Each compound is displayed as a point on a 2D plot, and the compound structure can be displayed by clicking on the point. Thousands of fragments can be displayed and explored this way. Probably the most instructive statistics to plot are the percent occurrence in the active set against the Fisher log score—a measure of the significance of the occurrence of a fragment in the active set compared to the inactive set. Clearly, the most discriminating fragments are the ones that have a highly significant Fisher log score (i.e., enriched in the active

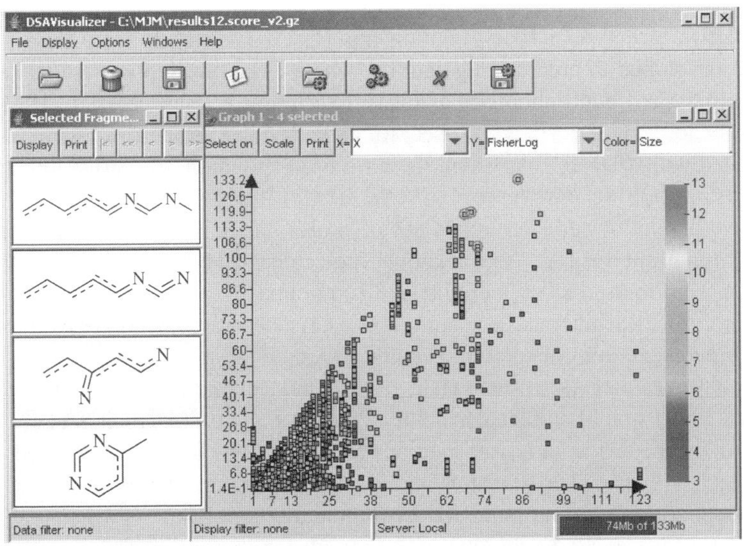

Figure 3.2 The DSA Visualizer for exploring substructure fragments in a set of compounds active for a particular target. This result was generated for a set of JNK kinase ligands. Each point is a fragment, which, when selected, is illustrated to the left. The horizontal axis is the number of compounds in the active set that contain the fragment. The vertical axis is the Fisher Log score—a measure of the significance of the occurrence of the fragment in the active set relative to an inactive set. The fragments are colored by size (number of atoms). The most significant fragments appear to the top right of the plot. The structures corresponding to the picked (highlighted) points are illustrated on the left. See color plates.

set) and also occur in a large percentage of actives. Fragments that are highly significant but of low occurrence are unlikely to pull out many actives in a test set (i.e., a prediction with many false negatives), whereas fragments that are of high occurrence but low significance are likely to pull out many inactives (false positives) in a test set (e.g., a benzene ring usually occurs as one of these fragments). There is still an element of subjectivity in selecting a small number of substructures to use to search a set of compounds for screening. Future work might include the ability to automatically generalize substructures by combining features of two or more of them, which presently is done manually.

The advantages of the DSA method are that it can rapidly be applied to large compound sets and many targets and that the results are in a chemically intuitive form. This method has been applied to the activity classes in the MDL drug data report (MDDR) and has been extended to predict toxicological outcomes. A disadvantage is a limited ability to cross chemical classes. Thus, if a substructure is considered to define a chemical class, then searching with this substructure cannot cross to another class. Also, the current method overlooks the potential requirement of multiple substructures instead of a single fragment in some targets. Nevertheless, the DSA method has been proven to be effective in the identification of novel hits and leads for several drug targets at Serono.

3.2.3. QSAR and Recursive Partitioning

Quantitative structure–activity relationship (QSAR) methods were some of the earliest developed to predict drug activity, where a relationship was found between drug activity and hydrophobicity as measured by log *P* (Hansch et al., 1986). The same methods have also been used to predict other properties, such as those associated with ADME/Tox, where the term used is quantitative structure–property relationships (QSPR) (Grover et al., 2000a, 2000b). The essence of all the methods is to derive a relationship between a set of molecular descriptors and a biological activity (or property) for a set of known molecules, giving a mathematical model that can then be used to predict untested compounds. A wide variety of molecular descriptors (Todeschini and Consonni, 2000) have been employed, derived from 2D and 3D representations of small molecules. Examples are topological state (Kier and Hall, 1989), E-state (Kier and Hall, 1990), binary fingerprints (McGregor and Muskal, 1999), and 3D molecular fields (Cramer et al., 1988; Cruciani et al., 2000).

An equally wide variety of methods of analysis has been used, derived mostly from the fields of applied statistics and machine learning. The most basic is linear regression, sometimes implemented as principal components regression (PCR) or partial least squares (PLS); the latter two methods are appropriate for dealing with the case in which there are more variables than compounds and/or when many variables are highly correlated, which often happens with molecular descriptors. These techniques work best when both the dependent and independent variables take on continuous values, are associated with small errors, and approximate to some typical kind of statistical distribution, usually Gaussian. However, it is not unusual to have molecular descriptors and activity data that are discontinuous, contain large errors, and/or do not belong to a typical distribution function. It is then more reasonable to classify compounds into categories such as active/inactive or high/medium/low. This would be the case when analyzing HTS results, which are usually of low quality. Examples of methods more suited to such data are discriminant function analysis, neural networks, genetic algorithms, and recursive partitioning (RP). As an example, we will describe in more detail the methodology of RP.

RP is a statistical method that constructs a decision tree to distinguish different classes of data (Chen et al., 1998) (Fig. 3.3). In its application in a virtual screening process, RP identifies relationships between specific chemical structure features (e.g., continuous descriptors, binary fingerprints, and/or substructure fragments) and biological activities. The types of biological activities can be categorical data (e.g., an activity class such as a "protein tyrosine kinase inhibitor") or numerical data (e.g., percentage inhibition data from HTS). The first step in the RP process is to generate descriptors or substructures for each compound in the collection with prior knowledge of biological activity. Numerical activities have to be converted into categories (e.g., good and bad or high, medium, and low). The RP algorithm evaluates each descriptor sequentially to determine the best descriptor at each node that separates actives from inactives, using a statistical test. As the RP decision tree grows, the whole data set is progressively split into smaller subsets until all compounds in the subsets are grouped into a homogeneous class or a parameter reaches its limit (e.g., the minimum number of compounds in each subset, or the maximum number of nodes). The descriptors associated with the detailed path by

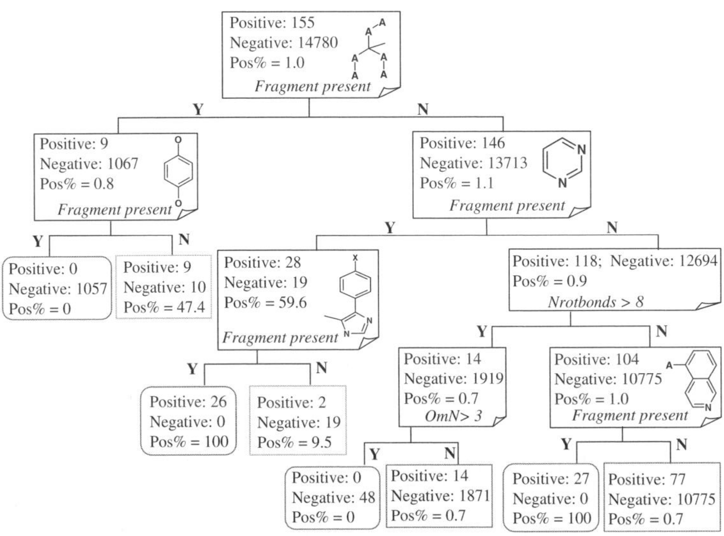

Figure 3.3 Example of a hypothetical recursive-partitioning tree. The nonterminal nodes are shown in black squares, and the terminal nodes are shown in colored squares. The reasons for the terminations are as follows: A homogeneous class is found (blue), the number of compounds is not sufficient (yellow; the minimum here is 25), and the maximum number of nodes is reached (red; the upper limit in this example is 15). The descriptors can be substructures or numerical properties (Nrotbonds is the number of rotatable bonds, and OmN is the number of oxygen atoms minus nitrogen atoms). The statistical test here is simply the percentage of positives (Pos%). See color plates.

which the molecules are split give an indication about their contribution to the activity or inactivity of the molecules in the subset. A major advantage of the RP approach, as compared to typical QSAR approaches, is that it does not assume a particular type of distribution in the activity data or a specific type of function that relates activity to the descriptors and therefore accommodates multiple binding modes of actives or multiple reasons for inactivity (Hawkins et al., 1997). This is important because the data from prior HTS or multiple sources (e.g., MDDR, literature, or patents) are usually not homogeneous or accurate enough to yield a reliable parametric regression equation. Like other QSAR approaches, cross-validation is an important quality-control step in the RP method to optimize the parameters, such as the minimum number of observations per subset and the maximum depth of a tree, before a final model is built for use in a virtual screen. One of the cross-validation methods is leave-N-out. This method splits data into N groups, builds a model from all but one group, and then predicts the activity of the left-out group. The process is repeated for each group, and the optimal parameter setting is then determined by comparing the overall quality of the predictions from all runs.

3.2.4 Pharmacophore Analysis

Another commonly used computational technique when a set of active ligands is known is that of pharmacophore analysis. In this context a pharmacophore is regarded as a set of atom (or atom-group) types (features) in a 3D spatial arrangement that represent the interactions made in common by a set of small molecule ligands with a protein receptor site (Langer and Wolber, 2004). The types of pharmacophore features used are based on the kinds of interactions seen in crystal structures of ligand–protein complexes. These usually consist of hydrogen bonding groups (H-bond donor and acceptor), groups with formal charges (positive/basic and negative/acidic), and hydrophobic centers. Other features may also be included. Often an aromatic group is used, and sometimes others, such as a combination H-bond donor/acceptor, are used. These are assigned to atoms or atom groups in a small molecule based on structural context. For example, a carboxylic acid is negatively charged at neutral pH, so this would be assigned the negative charge type; aliphatic amines are examples of a positively charged group; oxygen or nitrogen atoms with an attached hydrogen are H-bond donors, ones without an attached hydrogen are H-bond acceptors, and so on; and hydrophobic types are usually carbon atoms, or groups of them, removed from polar or charged groups. In this way, a small molecule is represented as a set of potential interactions with a protein, and the underlying chemical class or scaffold is unimportant in itself, only serving to present these kinds of functionality in 3D space. Thus one of the claimed advantages of this approach is that it can discover new chemical classes that bind to the same receptor.

All the methods start with a set of compounds that are known, or assumed, to bind to the same protein receptor site in a similar way (i.e., using a number of common pharmacophore interactions), even if they share little 2D chemical similarity. It is the aim of this method to calculate what these interactions are and output them as a set of pharmacophore points in a 3D spatial arrangement. This pharmacophore can then be used to computationally screen compounds of unknown activity on the assumption that ones that fit the pharmacophore are more likely to be active for that receptor.

Probably the best-known and most widely used program for pharmacophore analysis is Catalyst, available from Accelrys, Inc. (Barnum et al., 1996; Greene et al., 1994; Sprague, 1995). An algorithm for generating a set of 3D structures for a small molecule that optimally covers conformational space was a central part of the development of the methodology, as was an algorithm for deriving pharmacophores from a training set of molecules, termed "hypotheses" in Catalyst terminology. Several hypotheses are generated with relevant statistics, allowing the user some flexibility to choose which one(s) to use, based on other knowledge such as that derived from the crystal structure of the target. Input parameters can be tuned to control such things as the number of points to be generated. Molecule conformations with pharmacophore assignments for large numbers of compounds are stored in a Catalyst database that is then used in a search to fit each molecule to a hypothesis. Statistics are generated, and the results can be visualized. A recent development has been the ability to apply an excluded volume fit to a Catalyst search. This approach recognizes that ligand binding depends both on satisfying individual specific interactions and on overall shape complementarity with a binding site. Compounds too big to fit into the binding site, even though they contain the desired pharmacophore, are excluded.

Another more recent development is pharmacophore fingerprinting (Mason et al., 1999, 2001; McGregor and Muskal, 1999, 2000; Pickett et al., 1996). In this method, a basis set of three- or four-point pharmacophores is enumerated with a set of distance ranges. For example, the PharmPrint method (McGregor and Muskal, 1999, 2000) uses six pharmacophore types (H-bond donor/acceptor, positive/negative charge, hydrophobic, and aromatic) and six distance ranges (2–4.5, 4.5–7, 7–10, 10–14, 14–19, and 19–24 Å) (Fig. 3.4). When all possible combinations of three-point pharmacophores are enumerated, this gives a basis set of 10,549. Each pharmacophore is then either present or absent in a small-molecule 3D structure. Thus the fingerprint for the molecule can be represented as a binary bitstring. Fingerprints for different conformations can be either kept separate or combined. The pharmacophore fingerprints are then amenable to the same types of calculations (e.g., the Tanimoto coefficient) that are applied to bitstrings derived in other ways. QSAR models have been built using the pharmacophore fingerprint as a descriptor (McGregor and Muskal, 1999), and the mapping of biologically active space has been explored (McGregor and Muskal, 2000). It is also possible to calculate the fingerprint of a protein active site (Arnold et al., 2004; Mason et al., 1999). For example, if there is an H-bond donor on the protein in the binding site, then a ligand is favored if it contains an H-bond acceptor in a position that is a certain distance and direction relative to it. In this way, a fingerprint of the binding site can be built, and comparison to a small molecule fingerprint gives a measure of the goodness-of-fit in the site.

3.2.5 Docking

When the crystal structure of the target is available, there is an opportunity to employ docking methodologies, whereby in a computational simulation ligands are oriented

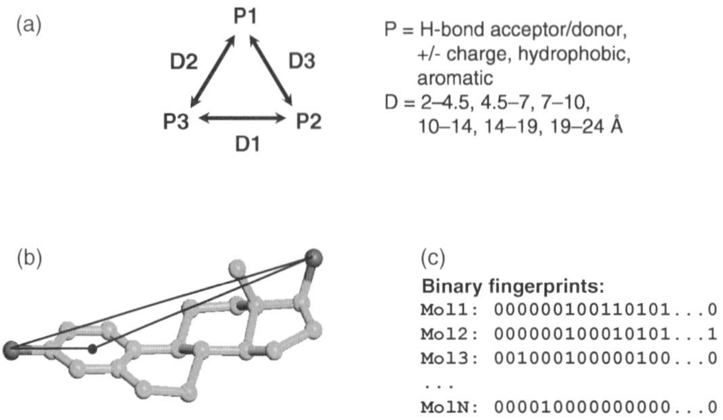

Figure 3.4 Pharmacophore fingerprinting: **(a)** The definition of a three-point pharmacophore, consisting of three pharmacophore points separated by three distance ranges; **(b)** a single three-point pharmacophore mapped to the estradiol molecule; and **(c)** the fingerprints that record the occurrence of every possible pharmacophore in a set of molecules.

in the active site of a protein and a measure of the goodness-of-fit is calculated, thereby explicitly implementing the key–lock principle of ligand–protein interactions (Fig. 3.5). Unlike ligand-based approaches, a target-based method, such as docking, does not depend on information from known ligands, so compounds selected by this method are not necessarily biased toward known chemotypes, providing an

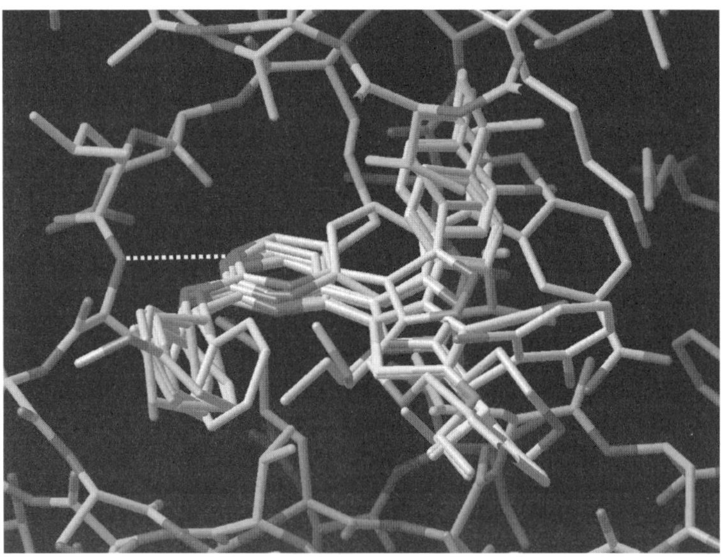

Figure 3.5 An example of molecular docking using the Glide program. This result was produced by docking a set of ligands to the crystal structure of JNK kinase (PDB code 1PMN). Several different ligands are overlapped, and common interactions can be seen to emerge (for example, the hydrogen bond to the hinge region, shown as a yellow dotted line), even though there is variation in the contacts made by the small-molecule ligands. See color plates.

opportunity to select or generate compounds that are truly novel. In addition, docking results can provide mechanistic explanations of ligand–receptor interactions at the atomic level, offering an excellent opportunity to then optimize hit or lead molecules based on a "rational design" strategy. In principle, one might expect docking to be the complete solution to the design of small-molecule ligands. In practice, the method is limited in several fundamental respects, but these limitations have not prevented its widespread use, both for selecting screening sets and for hit or lead optimization. Many commercial and academic software packages for docking are available, such as AUTODOCK (Morris et al., 2004), DOCK (Ewing et al., 2001; Kuntz et al., 1982), FlexX (Rarey et al., 1996), GOLD (Jones et al., 1997), Glide (Friesner et al., 2004; Halgren et al., 2004), and ICM (Cavasotto and Abagyan 2004; Stigler et al., 1999). Among them, Glide and GOLD are highly regarded in the pharmaceutical industry (Bissantz et al., 2000; Kontoyianni et al., 2004; Perola et al., 2004). A requirement for docking is the availability of a high-resolution crystal structure of the target of interest, preferably with bound ligand, although docking to good-quality homology models can still produce positive results (Diller and Li, 2003; Oshiro et al., 2004).

Most docking algorithms proceed in two steps: (1) the orientation of a particular conformation of a ligand in the protein binding site, often referred to as the ligand "pose"; and (2) the calculation of a score that correctly ranks different poses of the same molecule and the best pose of different molecules. Small molecules in a bound state in a protein active site have been often found to adopt a conformation different from that at the energy minimum of the unbound state, either in solution or *in vacuo* (Perola and Charifson, 2004). As a result, a docking algorithm needs to generate multiple conformations of the ligand to achieve the first step. It is, however, a challenge to sample the vast conformational space with reasonable accuracy and speed for a high number of rotatable bonds. Some programs, such as Glide, sacrifice speed for a more thorough search of conformational space, while others, such as FlexX, attempt to avoid the need to generate many conformations of the entire ligand by breaking it into fragments, orienting the fragments, and then joining them back together.

An area of critical importance is the scoring function, which calculates the interaction energy (and/or other terms contributing to the goodness-of-fit) between the ligand and receptor. Analysis of docking results from various studies indicate that it is largely the issue of inaccurate scoring functions that give rise to high false positives and/or false negatives for many targets (Bissantz et al., 2000; Kontoyianni et al., 2004; Stahl and Rarey, 2001). Most early implementations used scoring functions based on the nonbonded term of the molecular mechanics force field, incorporating a Lennard-Jones potential for van der Waal interactions and a Coulomb term for electrostatics and sometimes including a supplemental hydrogen bond energy term (Kitchen et al., 2004). Examples of force-field-based scoring functions are DOCK (Kuntz et al., 1982), Goldscore (Jones et al., 1997) and AutoDock (Morris et al., 2004). The drawback here is that the molecular mechanics force field only accounts for the potential energy of interaction between ligand and receptor, whereas the factor that governs the binding affinity is the difference in the Gibbs free energy of the whole system for bound and unbound states, which includes additional solvent and entropy effects. Accurate calculation of the latter terms is intractable in the context of high-throughput docking. This shortcoming has led to the addition of terms designed to take these effects into account and the development of empirical and knowledge-based scoring functions. The empirical scoring functions are usually the summations of various Gibbs energy changes parameterized by empirical values obtained from fitting the theoretical energy equations to experimental affinity data. LUDI (Bohm, 1994), FlexxScore (Rarey et al., 1996), ChemScore (Eldridge et al., 1997), and GlideScore (Friesner et al., 2004) belong to this type. The characteristics of knowledge-based scoring functions are that their potential energies are parameterized by the geometric distributions of atoms observed in known X-ray crystal structures: PMF (Muegge and Martin, 1999), SMoG (Dewitte et al., 1997), and DrugScore (Gohlke et al., 2000) are examples. These different scoring functions have been extensively evaluated (Perez and Ortiz, 2001; Stahl and Rarey, 2001; Terp et al., 2004; Wang and Wang, 2003), but no clear agreement has been reached as to which of the three types perform consistently better than the others. This situation has led some groups to implement a consensus scoring approach, in which the scores of several little-correlated scoring functions are combined, with reported enhancement of hit rates (Charifson et al., 1999; Jacobsson et al., 2003).

Most scoring functions have difficulty dealing with small steric clashes between ligand and receptor. A steric clash causes a large increase in interaction energy, which can lead to false-negative scores (Erickson et al., 2004). Steric clashes can be a result of errors in the conformation or pose of the ligand, or they can reflect a genuine incompatibility between ligand and receptor. Often steric clashes can be relieved by a change in the conformation of the receptor to accommodate the ligand, giving rise to the "induced fit" problem. If the change is major, involving movement of the main chain, then it is clearly unrealistic to expect current docking algorithms to deal with this problem. However, minor changes in side-chain conformation are very frequent, as seen when comparing crystal structures of the same protein with different ligands. Although soft docking techniques attempt to alleviate the problem of side-chain conformational changes (Jiang and Kim, 1991; Fernandez-Recio et al., 2002), it is difficult to devise a docking algorithm that can distinguish between the possible sources of steric clashes. An additional problem is that many ligands interact with their receptors through water-mediated hydrogen bonds, and it is also difficult for a docking algorithm to reproduce these kinds of interactions (Davis et al., 2003).

Some of these problems can be overcome with careful setup of the docking run by a user with knowledge of the specific target. For example, if it is known that there is a requirement for a bound water in the active site, this can be included as part of the receptor in the dock run. If it is seen that there is always a specific interaction or pharmacophore that occurs, this can be forced on the ligands in the dock run, either as part of the run if the software allows or in postprocessing of the results (Fradera et al., 2000; Good et al., 2003; Hindle et al., 2002; Verdonk et al., 2004). Kinase targets provide a good example, where there are known important interactions, such as a hydrogen bond to a main-chain NH group in the hinge region (Fig. 3.5) (Bossemeyer et al., 1993; Chuaqui et al., 2005; Toledo et al., 1999; Tong et al., 1997). If the ligands contain a privileged substructure known to bind in a certain place, then the docked poses can be filtered for RMS deviation to this substructure. Some commercially available docking packages, such as FlexX-Pharm (Hindle et al., 2002), GOLD, and

Glide, allow for the inclusion of these pharmacophore-type constraints in the docking stage. This greatly cuts down the search space of poses that needs to be covered. In an effort to overcome the induced fit problem, if multiple crystal structures are available, they can be used in parallel, and the final score can be either the best score (assuming this is the closest fit) or an average of all scores, possibly weighted by relevance of structure. Alternatively, a protein structure can be energy-minimized around an important ligand, if no crystal structure is available for that particular ligand, to bias the receptor toward ligands of that type. Moreover, scoring functions are known to scale poorly with some simple molecular descriptors, such as MW and the number of rotatable bonds. In our laboratory we have found value in normalizing docking scores for these kinds of descriptors in a postprocessing stage.

Despite the problems mentioned above, we anticipate that docking will continue to have a central role in computer-based drug discovery. Some of the problems are alleviated by increases in computer processing power and parallel computing, and the rapid increase in the number of experimentally determined crystal structures of proteins will extend this technology to a large proportion of pharmaceutical targets, such as G-protein-coupled receptors (GPCR) and ion channels, where docking technology currently depends on homology models with limited quality (Bissantz et al., 2003; Evers and Klabunde 2005; Zhorov et al., 2001).

3.3 TARGET-INDEPENDENT METHODS

The compounds selected from the second category of virtual screening methods are not dependent on the knowledge of any specific targets or their ligands, and this includes a wider set of considerations. One consideration is chemical diversity or coverage of biologically active space, the aim of which is to discover ligands from novel compound classes and maximize the chances of hitting a target for which there is little known information. Applying a measure of diversity also minimizes repetitive screening of the same or similar compounds. Another consideration is the calculation of ADME/Tox properties and the elimination of clearly undesirable compounds. The aim here is to screen molecules for properties that characterize a compound as druglike. These properties are independent of an individual target but may vary, depending on site of action or therapeutic area; for example, drugs acting on the central nervous system need to cross the blood–brain barrier and therefore need to have specialized permeability properties. Some of the methodologies may be common to the target-dependent and target-independent applications. For example, some of the same descriptors can be used in calculations of molecular similarity/diversity, ADME/Tox, and QSAR/QSPR.

3.3.1 Diversity Analysis

Screening a diversity set is relevant when there is no information available about a target structure or its ligands or when a compound collection is to be screened against many or unknown targets (such as in cell-based HTS); it also provides the opportunity to discover unexpected new chemotypes or modes of action. The concept of molecular diversity is based on the idea that by using calculated properties, small molecules can be meaningfully mapped to a property space, where compounds with similar biological activities are close in the space and compounds with different activities are far apart (Willett, 2000). Ideally, the activity for a target will be confined to a limited continuous region of this space. If this region is unknown (e.g., if there is no information about a target or the same compound set is to be screened against different targets), then the best chance of hitting the target is with a set that covers the space as comprehensively as possible. The property space can be produced by calculating a set of numerical molecular descriptors, in which case the dimensionality of the space is the number of descriptors used. The dimensionality can be reduced, if needed, by techniques such as principal components analysis (PCA), enabling visualization in 2D on paper or in 3D on an interactive computer screen. Alternatively, the molecular similarity methods described above can be used to give a measure of the distance between molecules in property space, and algorithms for diversity analysis can operate on these distances directly (Trepalin et al., 2002; Willett, 2000).

Diversity analysis involves selecting a subset of compounds that optimally covers the chemical space of an entire set, which can be a corporate collection, commercially available compounds, a combinatorial library, a virtual library, or a combination. Many algorithms have been devised for this purpose (Blake, 2004; McGregor and Pallai, 1997; Sauer and Schwarz, 2003a, 2003b; Trepalin et al., 2002; Young and Ge, 2004), and Fig. 3.6 illustrates three of them. Some algorithms aim to select a subset of compounds that are representative of the entire set, while others aim to cover the underlying property space as widely as possible, regardless of the distribution of the entire compound set in this space. Clustering is based on the observation that compounds are not usually evenly distributed in a property space but group together in certain regions, similar to galaxies in outer space. If clusters can be identified within which compounds have similar biological activities, then only a single representative (or a small number) of each need be chosen for screening, usually the calculated "centroid." Clustering has been applied to many kinds of

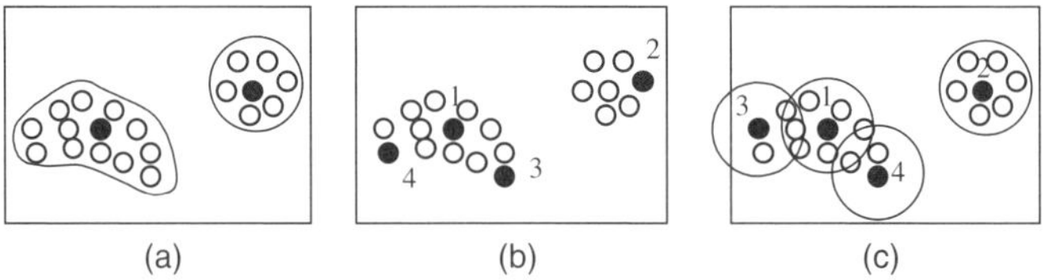

Figure 3.6 Schematic illustration of three different diversity methods. **(a)** *Clustering:* The compound closest to the centroid of each cluster is selected to represent the whole cluster. Some clustering algorithms have difficulty dealing with unevenly distributed clusters, and large clusters can be underrepresented. **(b)** *Maximum dissimilarity selection:* The first compound is randomly selected; in successive steps the chosen compound is the one with the greatest distance to its nearest neighbor in the chosen set. **(c)** *Maximum similarity elimination* (*or sphere exclusion*): A compound is randomly chosen, and then all compounds within a predefined similarity threshold are excluded from further analysis; the selection proceeds until no compounds are left. The latter two methods produce a more even sampling of property space but have a tendency to select outliers.

pattern-recognition problems, and there are well-known difficulties. For example, with small molecules, one problem is that clusters vary greatly in size, which gives rise to uneven sampling. Also, chemical space often gives rise to clusters with irregular shapes known to be difficult for most algorithms to deal with (Subhash, 1996). Another method, maximum dissimilarity selection (Snarey et al., 1997; Trepalin et al., 2002), proceeds by randomly choosing the first compound and then selecting the compound with the least similarity to it and adding it to the chosen set. Then, the next least similar compound to the existing set is then added until a certain number of compounds is reached. Maximum similarity elimination (also known as sphere exclusion) (Gobbi and Lee 2003; Taylor, 1995) proceeds by randomly selecting a compound and then excluding all the compounds that have a similarity higher than a given threshold from further analysis. This process continues until no compounds are left. This method ensures that no two selected compounds are more similar to each other than the threshold. Other methods aim to optimize a function of the chosen set. For example, for each compound in the chosen set, the MaxMin function (Snarey et al., 1997) is a summation of the distance to its nearest neighbor. A numerical algorithm needs to be applied to optimize this function.

It has been found that many diversity algorithms tend to be biased toward outliers in chemical space, and those unusual compounds are less likely to be drug-like. Thus, the search space should be limited to the regions that are populated by compounds with calculated favorable profiles. In the end, however, selecting a diversity set is a problem for which there is no objectively correct answer (Kubinyi, 1998), and it can be difficult to devise experiments in a practical setting to make an in-depth comparison of the methods. Several commercially available software packages for molecular modeling have modules designed for diversity analysis. Examples are DiverseSolution (DVS) from Tripos, Inc. and C2-Diversity from Accelrys, Inc. Both have methods for reducing a high-dimensional space to 3D for visualizing with computer graphics and provide a choice of algorithms for diverse subset selection.

3.3.2 ADME/Tox Filters

The majority of drug candidates fail in clinical trials due to ADME/Tox problems (*in vivo* absorption, distribution, metabolism, excretion, and toxicity) (FDA, 2004; Kennedy, 1997; Venkatesh and Lipper, 2000). These problems can be hard to resolve at an advanced stage of development, which has given rise to the desire to address these issues as early as possible in the drug-development cycle, either experimentally or computationally. However, *in vivo* ADME/Tox and pharmacokinetic outcomes are notoriously difficult to predict, even with *in vitro* experiments, due to the complex nature of the interaction of chemicals with biological systems. Due to the importance of the problem, much attention has been focused on it, and software is available for prediction of various ADME/Tox endpoints. However, many of these software packages are not suited for high-throughput treatment of compounds and are best used at the lead optimization stage. Additionally, many QSAR/QSPR models on ADME/Tox properties are found to be limited to certain chemical classes, and the goal of a global model that can be applied to all chemical classes proves elusive. Fortunately, many ADME/Tox endpoints are related to basic physicochemical properties of compounds, some of which are feasible to treat computationally at the screening stage. Examination of profiles of these properties for drug compounds show clear preferred ranges that can then be used to classify a new molecule as druglike or otherwise. The druglike structures are then assumed to be more likely to possess favorable ADME/Tox properties, without having to address individual ADME/Tox endpoints directly (Frimurer et al., 2000; Muegge et al., 2001; Sadowski and Kubinyi, 1998; Walters and Murcko, 2002; Xu and Stevenson, 2000). Some of these properties are the same molecular descriptors used in the target-dependent methods, such as QSAR. Some physiochemical properties, if unfavorable, can pose problems even at the screening stage; for example, poor solubility is known to compromise screening results as well as cell-based and *in vivo* measurements. The training sets for identifying druglike property profiles can be commercially available databases of bioactive compounds, such as the Comprehensive Medicinal Chemistry (CMC) or MDDR databases from MDL or the World Drug Index (WDI) database from Thomson Scientific.

A widely used virtual screening filter for compounds intended to be orally available drugs is the so-called "rule of five," derived from an analysis of property profiles of compounds that have survived preclinical stages and Phase I safety evaluation (Lipinski et al., 1997). Several computed properties were selected based on their clear relationship with oral bioavailability: MW, *c*log*P*, and counts of hydrogen bonding groups. The rule of five states that compounds are likely to have poor absorption or permeability properties if (1) there are more than five H-bond donors, (2) MW is over 500, (3) *c*log*P* is greater than 5, or (4) there are more than 10 H-bond acceptors. Compounds outside these ranges for two or more descriptors are predicted to be problematic, but the converse is not true; that is, compliance with the rule is no guarantee of good bioavailability. By comparing the properties of early-stage molecules with that of final drugs, Oprea et al. (2001) concluded that the properties of leadlike molecules often need to have more restrictive values: MW less than 450, *c*log*P* between −3.5 and 4.5, no more than four rings, no more than 10 nonterminal single bonds, no more than five H-bond donors, and no more than eight H-bond acceptors. The descriptors are easy to calculate and are often stored for every compound in a database, and compounds that violate the rule can be flagged. In addition to the use of the rule as a negative filter in compound selection, profiles of these kinds of descriptors can be examined for whole compound collections and adjustment made to the content to reflect the distributions found in druglike sets.

There may be an inconsistency between the aim of an *in vitro* HTS assay to generate potent compounds and the need for drug discovery to deliver bioactive molecules with oral availability. Most HTS assays are *in vitro* molecular assays. With potency as a major criterion, compounds that are larger, more hydrophobic, and less soluble tend to stand out at the HTS stage due to their propensity to be more potent. This can be compounded at the post-HTS optimization stage if medicinal chemists seek to improve *in vitro* activity by adding hydrophobic groups to compounds (Kubinyi, 2003; Lipinski et al., 1997). However, there is a clear relationship between, for example, high MW and poor intestinal and blood–brain barrier permeability (Navia and Chaturvedi, 1996; Pardridge, 1995). This highlights the need to address druglike properties early on in the drug-development pipeline.

Similarly, the *in vitro* nature of HTS technology does not favor compounds with optimum toxicity profiles. Consequently, toxicity problems are usually detected late in drug development. However, toxicity is a problem with many physiological outcomes, so there is no simple rule to filter toxic compounds. Many approaches have been explored, mostly aimed at identifying molecular substructures and/or descriptor patterns responsible for specific toxicity outcomes (Cronin et al., 2003; Enslein and Craig, 1982; Henry et al., 1982; Hodes, 1981; Mattioni et al., 2003; Pearl et al., 2001; Yuta and Jurs, 1981). Merlot et al. (2003) have extended their DSA method to analyze toxicological properties for large compound sets such as the MDL Toxicity database.

Other filters are also desired in order to enhance success rate in screening. Many false positives from an HTS program arise from compounds with the presence of a reactive functional group (such as an epoxide or an aldehyde) that is likely to interact covalently with proteins (Rishton, 1997, 2003) or "promiscuous inhibitors" that inhibit targets via aggregate forms (McGovern et al., 2002, 2003). Several computational approaches have been developed to exclude these types of "nuisance" compounds (Roche et al., 2002; Walters and Namchuk, 2003).

3.4 INTELLECTUAL PROPERTY

In the pharmaceutical industry, the novelty of compounds with regard to intellectual property (IP) is usually a crucial criterion in many aspects: hit/lead selection, compound library design, and evaluation of compound acquisition and maintenance. For a small number of molecules, the novelty evaluation is often performed manually against known data, for example, against the online information in the SciFinder database (Chemical Abstracts Service). This evaluation becomes harder for large numbers of compounds, but using the more efficient measures of molecular similarity discussed above it is possible to do a molecule-by-molecule comparison of a screening set against a large reference set to ensure a relatively high IP value for the screening compounds. Other methods, such as privileged substructure searches, can also be applied in a similar way. The reference sets are usually commercial databases installed in-house, such as the CMC, MDDR, and WDI. More specialized reference sets for individual target classes such as kinases and GPCRs are provided by Jubilant Biosys, Inc., Aureus Pharma, Inc., and Eidogen-Sertanty, Inc.

3.5 COMPOUND COLLECTION MAINTENANCE AND ACQUISITION

The success of a screening project, whether it is done virtually or experimentally, depends on the quality of the compound source. The physicochemical profile of hits in an organization tends to resemble that of the overall screened compounds, as observed by Lipinski at Pfizer, Inc. (Lipinski et al., 1997) and also by us at Serono. Many compounds in a corporate collection have been derived from hit to lead programs and are consequently close analogues and not diverse. This gives rise to the need to maintain a favorable druglike profile and diversity of chemotypes in a corporate collection, using the methods described previously. Additional logistics issues may be encountered that impact a computational selection of compounds for screening. For example, in most corporate collections compounds are stored on plates ready for screening. Therefore computational methods need to be applied at the plate level and not the compound level, which can complicate calculations. If the opportunity arises, selection can be made at the plate creation stage.

Other considerations also require computational support in order to have a high-quality compound collection. Corporate collections of compounds need to be replenished on a regular basis for a number of reasons: compound samples stored over a number of years deteriorate, especially in solution; certain compound classes may no longer hold the IP position that they once did; old compounds may fail newly implemented methods for assessing druglikeness; or old compound classes may no longer have relevance to current targets or therapeutic areas. Some organizations decide to simply discard all compounds older than a certain date. Companies looking to add to their collections, or smaller or newer companies screening for the first time, need to rely on either an outside source or a technology that can rapidly produce large numbers of compounds. There exist many vendors of compound sets for screening, and most will freely supply electronic files or databases of their libraries for computational analysis. Some supply focused sets for target types (e.g., kinase, GPCR) that have been computationally designed, or they work with customers on a contract basis to fulfill their requirements using their own proprietary design technologies. Therefore the same considerations with regard to druglikeness, IP, and diversity should be applied to purchased compounds. An additional consideration is complementarity to the existing collection, where the similarity methods described above can be applied to compare possible new compounds against the existing set.

3.6 PERSPECTIVE

The sections above have outlined the desirable properties that a compound screening set should possess and the methods available to achieve this goal. The methods are many and varied, so in case a reader seeking guidance is overwhelmed by possibilities, we would like to conclude by making some remarks about how screening sets are actually chosen at Serono. The overall scheme is summarized and illustrated in Fig. 3.7. The selection is driven by the nature and goals of the project and by the information already available. A final screening set is usually put together from a target set, a target-family set, and a diversity set chosen from our corporate collection. The compounds in the target set are selected based on any information that is available on a particular protein target, and the target-family set is based on knowledge of the protein structural or functional family of which the target is a member (e.g., protein kinases, or GPCR family A). It is highly desirable to have a known set of active ligands to create a target set, by either the similarity or DSA methods described above or a more sophisticated QSAR or pharmacophore model. If there are no known active compounds, then modeling can be based on the actives for the closest targets for which information is available. Each of these sets may be a combination of smaller

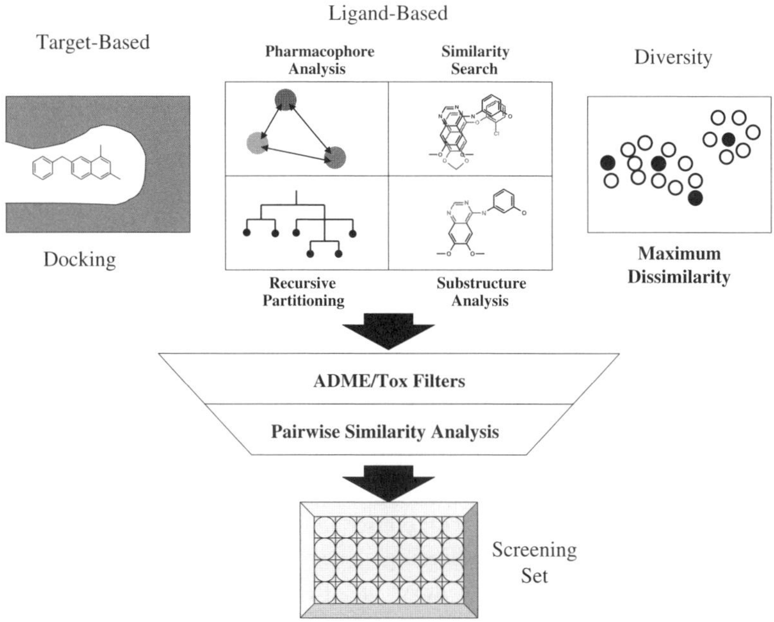

Figure 3.7 Illustration of a way that compounds chosen by different methods are combined to create a final screening set. The actual procedures can vary from one implementation to another. For example, the exclusion of non-druglike compounds is often performed before docking.

sets chosen by different methods. In addition, other sets may be added—for example, a newly acquired compound collection from a commercial vendor. To some extent, the methods used reflect the experience and preferences of the computational chemist on the project. At Serono we make extensive use of methods such as DSA, PharmPrint, and diversity analysis that have been developed in-house or by the authors elsewhere. In addition, we use a combination of commercial and in-house tools for fingerprinting of compounds in similarity and other methods. The different sets are then combined according to weights reflecting the amount of information we know about the target and the project team's desire to explore the possibility of discovering novel compound classes by serendipity. Because of the unpredictable nature of drug discovery, we rely on the combination of the sets and methods to avoid becoming too attached to one particular approach. Compounds with undesired properties in the combined set are excluded using the above-mentioned druglike and related filters. The remaining compounds are then subject to pairwise similarity analysis, and highly similar compounds are removed. The list may finally be subjected to inspection by eye (often by a medicinal chemist) before it is delivered to screening.

We have experienced mixed success with docking as a method for choosing screening compounds, where some small sets chosen by this method produced no positives in screening, an observation that is consistent with other findings; for example, docking does not always perform well in all target classes (Cummings et al., 2005). Consequently, we have not used high-throughput docking to a large extent in choosing screening sets, preferring to use more thorough (and therefore slower) docking methods at the post-HTS stage. However, we have found improvements in enrichment by implementing pharmacophore or substructure constraints in docking, either as part of the algorithm or in postprocessing. In addition, we have recently experimented with using methods such as DSA and PharmPrint as a filter to reduce the number of compounds before the docking stage. A more thorough docking search can then be implemented on a smaller set of compounds. This implements a "funnel" concept in which a large number of compounds are put through progressively more thorough methodology, reducing the number of compounds at each stage, to produce a small screening set highly enriched in actives. In a test case this produced positive results.

Despite the many unsolved problems, the appeal of docking remains. Because it is based on the key–lock principle of drug-target interaction, docking has the potential to be the ultimate solution to discovering a novel small-molecule binder to a protein. There have been many new docking methods developed recently, and much attention is being devoted to the subject, with consequent improvements in algorithms and results. In many cases, the correct binding modes of ligands to their targets have been successfully predicted, and some novel and highly active compounds have been identified (Desai et al., 2004; Li et al., 2004; Lyne et al., 2004). Structure-based design studies of molecules that eventually lead to clinical compounds have been well documented for their important roles in drug discovery (Davis et al., 2003).

ADME/Tox is then the other important issue involved in turning a small-molecule inhibitor into a drug candidate. There are many commercial software packages available for predicting a variety of endpoints (Cronin et al., 2003; Ekins, 2003). The accuracy of predicting some properties, such as metabolism and toxicity, is still an issue, at least at the high-throughput level. New methods as well as more training data are needed to improve the prediction quality. At present, it is only practical to calculate basic descriptors for large numbers of molecules, such as the rule-of-five descriptors, number of rotatable bonds, and polar surface area (Clark, 1999a,b; Ertl et al., 2000; Palm et al., 1996). The more time-consuming methods are only practical at the hit-to-lead stage.

Computational methods have become an indispensable tool in drug discovery. Accurate predictions of both high-affinity binding and a favorable ADME/Tox profile will move us closer to identifying drug candidates *in silico* prior to laboratory screening. With improved algorithms, we expect that docking and other methods will identify high-affinity binders with increasing frequency. In addition, improvements in ADME/Tox prediction accuracy and efficiency will enable successful prediction of the *in vivo* behavior of compounds on a large scale. A combination of these two technologies will be a powerful tool in the hands of the computational chemist. This progress will be aided by (1) the fact that information available from genomics and X-ray crystallography is growing at an unprecedented rate and (2) increases in computer power drive new kinds of computations that were intractable a few years before. While the prospects for performing a drug-discovery project completely *in silico* are still distant, we anticipate an increased role for computational methods in drug discovery and development.

ACKNOWLEDGMENTS

We would like to thank the members of the Design Technology Team at Serono for contributions to this work.

REFERENCES

Anstead, G. M., Carlson, K. E., and Katzenellenbogen, J. A. (1997). The estradiol pharmacophore: Ligand structure–estrogen receptor binding affinity relationships and a model for the receptor binding site. *Steroids* **62**, 268–303.

Arnold, J. R., Burdick, K. W., Pegg, S. C., Toba, S., Lamb, M. L., and Kuntz, I. D. (2004). SitePrint: Three-dimensional pharmacophore descriptors derived from protein binding sites for family based active site analysis, classification, and drug design. *J Chem Inf Comput Sci* **44**, 2190–2198.

Bajorath, J. (2002). Integration of virtual and high-throughput screening. *Nat Rev Drug Discov* **1**, 882–894.

Barnum, D., Greene, J., Smellie, A., and Sprague, P. (1996). Identification of common functional configurations among molecules. *J Chem Inf Comput Sci* **36**, 563–571.

Bissantz, C., Folkers, G., and Rognan, D. (2000). Protein-based virtual screening of chemical databases. 1. Evaluation of different docking/scoring combinations. *J Med Chem* **43**, 4759–4767.

Bissantz, C., Bernard, P., Hibert, M., and Rognan, D. (2003). Protein-based virtual screening of chemical databases. II. Are homology models of G-protein coupled receptors suitable targets? *Proteins* **50**, 5–25.

Blake, J. E., Famer, N. A., and Haines, R. C. (1977). An interactive computer graphics system for processing chemical structure diagrams. *J Chem Inf Comput Sci* **17**, 223–228.

Blake, J. F. (2004). Integrating cheminformatic analysis in combinatorial chemistry. *Curr Opin Chem Biol* **8**, 407–411.

Bohm, H. J. (1994). The development of a simple empirical scoring function to estimate the binding constant for a protein–ligand complex of known three-dimensional structure. *J Comput Aided Mol Des* **8**, 243–256.

Bondensgaard, K., Ankersen, M., Thogersen, H., Hansen, B. S., Wulff, B. S., and Bywater, R. P. (2004). Recognition of privileged structures by G-protein coupled receptors. *J Med Chem* **47**, 888–899.

Bossemeyer, D., Engh, R. A., Kinzel, V., Ponstingl, H., and Huber, R. (1993). Phosphotransferase and substrate binding mechanism of the cAMP-dependent protein kinase catalytic subunit from porcine heart as deduced from the 2.0 A structure of the complex with Mn^{2+} adenylyl imidodiphosphate and inhibitor peptide PKI(5–24). *EMBO J* **12**, 849–859.

Brown, R., and Martin, Y. (1996). Use of structure-activity data to compare structure-based clustering methods and descriptors for use in compound selection. *J Chem Inf Comput Sci* **36**, 572–584.

Cavasotto, C. N., and Abagyan, R. A. (2004). Protein flexibility in ligand docking and virtual screening to protein kinases. *J Mol Biol* **337**, 209–225.

Charifson, P. S., Corkery, J. J., Murcko, M. A., and Walters, W. P. (1999). Consensus scoring: A method for obtaining improved hit rates from docking databases of three-dimensional structures into proteins. *J Med Chem* **42**, 5100–5109.

Chen, X., Rusinko III, A., and Young, S. S. (1998). Recursive partitioning analysis of a large structure–activity data set using three-dimenstional descriptors. *J Chem Inf Comput Sci* **38**, 1054–1062.

Chuaqui, C., Deng, Z., and Singh, J. (2005). Interaction profiles of protein kinase-inhibitor complexes and their application to virtual screening. *J Med Chem* **48**, 121–133.

Clark, D. E. (1999a). Rapid calculation of polar molecular surface area and its application to the prediction of transport phenomena. 1. Prediction of intestinal absorption. *J Pharm Sci* **88**, 807–814.

Clark, D. E. (1999b). Rapid calculation of polar molecular surface area and its application to the prediction of transport phenomena. 2. Prediction of blood–brain barrier penetration. *J Pharm Sci* **88**, 815–821.

Cramer, R. D., III, Patterson, D. E., and Bunce, J. D. (1988). Comparative molecular field analysis (CoMFA). 1. Effect of shape on binding of steroids to carrier proteins. *J Am Chem Soc* **110**, 5959–5967.

Cronin, M. T., Jaworska, J. S., Walker, J. D., Comber, M. H., Watts, C. D., and Worth, A. P. (2003). Use of QSARs in international decision-making frameworks to predict health effects of chemical substances. *Environ Health Perspect* **111**, 1391–1401.

Cruciani, G., Pastor, M., and Guba, W. (2000). VolSurf: A new tool for the pharmacokinetic optimization of lead compounds. *Eur J Pharm Sci* **11** (Suppl 2), S29–S39.

Cummings, M. D., DesJarlais, R. L., Gibbs, A. C., Mohan, V., and Jaeger, E. P. (2005). Comparison of automated docking programs as virtual screening tools. *J Med Chem* **48**, 962–976.

Davis, A. M., Teague, S. J., and Kleywegt, G. J. (2003). Application and limitations of X-ray crystallographic data in structure-based ligand and drug design. *Angew Chem Int Ed Engl* **42**, 2718–2736.

Deng, Z., Chuaqui, C., and Singh, J. (2004). Structural interaction fingerprint (SIFt): A novel method for analyzing three-dimensional protein–ligand binding interactions. *J Med Chem* **47**, 337–344.

Desai, P. V., Patny, A., Sabnis, Y., Tekwani, B., Gut, J., Rosenthal, P., Srivastava, A., and Avery, M. (2004). Identification of novel parasitic cysteine protease inhibitors using virtual screening. 1. The ChemBridge database. *J Med Chem* **47**, 6609–6615.

Dewitte, R. S., Ishchenko, A. V., and Shakhnovich, F. I. (1997). SMoG: *De novo* design method on simple, fast, and accurate free energy estimates. 2. Case studies in molecular design. *J Am Chem Soc* **119**, 4608–4617.

Diller, D., and Li, R. (2003). Kinases, homology models, and high throughput docking. *J Med Chem* **46**, 4638–4647.

Dromey, R. G. (1978). A structural molecular formula for flexible and efficient substructure searching of large databases. *J Chem Inf Comput Sci* **18**, 163–168.

Ekins, S. (2003). *In silico* approaches to predicting drug metabolism, toxicology and beyond. *Biochem Soc Trans* **31**, 611–614.

Eldridge, M. D., Murray, C. W., Auton, T. R., Paolini, G. V., and Mee, R. P. (1997). Empirical scoring functions: I. The development of a fast empirical scoring function to estimate the binding affinity of ligands in receptor complexes. *J Comput Aided Mol Des* **11**, 425–445.

Enslein, K., and Craig, P. N. (1982). Carcinogenesis: a predictive structure–activity model. *J Toxicol Environ Health* **10**, 521–530.

Erickson, J. A., Jalaie, M., Robertson, D., Lewis, R., and Vieth, M. (2004). Lessons in molecular recognition: The effects of ligand and protein flexibility on molecular docking accuracy. *J Med Chem* **47**, 45–55.

Ertl, P., Rohde, B., and Selzer, P. (2000). Fast calculation of molecular polar surface area as a sum of fragment-based contributions and its application to the prediction of drug transport properties. *J Med Chem* **43**, 3714–3717.

Evers, A., and Klabunde, T. (2005). Structure-based drug discovery using GPCR homology modeling: Successful virtual screening for antagonists of the alpha1A adrenergic receptor. *J Med Chem* **48**, 1088–1097.

Ewing, T., Makino, S., Skillman, A., and Kuntz, I. (2001). DOCK 4.0: Search strategies for automated molecular docking of flexible molecule databases. *J Comput-Aided Mol Des* **15**, 411–428.

FDA. Challenge and Opportunity on the Critical Path to New Medical Products. 2004. U.S. Department of Health and Human Services, Food and Drug Administration.

Feldmann, R. J., and Heller, S. R. (1972). An application of interactive graphics—the nested retrieval of chemical structures. *J Chem Inf Comput Sci* **12**, 48–54.

Feldmann, R. J., Milne, G. W. A., Heller, S. R., Fein, A., Miller, J. A., and Koch, B. (1977). An interactive substructure search system. *J Chem Inf Comput Sci* **17**, 157–163.

Fernandez-Recio, J., Totrov, M., and Abagyan, R. (2002). Soft protein-protein docking in internal coordinates. *Protein Sci* **11**, 280–291.

Fradera, X., Knegtel, R. M., and Mestres, J. (2000). Similarity-driven flexible ligand docking. *Proteins* **40**, 623–636.

Friesner, R. A., Banks, J. L., Murphy, R. B., Halgren, T. A., Klicic, J. J., Mainz, D. T., Repasky, M. P., Knoll, E. H., Shelley, M., Perry, J. K., Shaw, D. E., Francis, P., and Shenkin, P. S. (2004). Glide: a new approach for rapid, accurate docking and scoring. 1. Method and assessment of docking accuracy. *J Med Chem* **47**, 1739–1749.

Frimurer, T. M., Bywater, R., Naerum, L., Norskov, L., and Brunak, S. (2000). Improving the odds in discriminating "drug-like" from "non drug-like" compounds. *J Chem Inf Comput Sci* **40**, 1315–1324.

Gillet, V. J., Willett, P., and Bradshaw, J. (2003). Similarity searching using reduced graphs. *J Chem Inf Comput Sci* **43**, 338–345.

Gobbi, A., and Lee, M. L. (2003). DISE: directed sphere exclusion. *J Chem Inf Comput Sci* **43**, 317–323.

Gohlke, H., Hendlich, M., and Klebe, G. (2000). Knowledge-based scoring function to predict protein–ligand interactions. *J Mol Biol* **295**, 337–356.

Good, A. C., Cheney, D. L., Sitkoff, D. F., Tokarski, J. S., Stouch, T. R., Bassolino, D. A., Krystek, S. R., Li, Y., Mason, J. S., and Perkins, T. D. (2003). Analysis and optimization of structure-based virtual screening protocols. 2. Examination of docked ligand orientation sampling methodology: Mapping a pharmacophore for success. *J Mol Graph Model* **22**, 31–40.

Greene, J., Kahn, S., Savoj, H., Sprague, P., and Teig, S. (1994). Chemical function queries for 3D database search. *J Chem Inf Comput Sci* **34**, 1297–1308.

Grover, I. I., Singh, I. I., and Bakshi, I. I. (2000a). Quantitative structure–property relationships in pharmaceutical research—Part 1. *Pharm Sci Technol Today* **3**, 28–35.

Grover, I. I., Singh, I. I., and Bakshi, I. I. (2000b). Quantitative structure–property relationships in pharmaceutical research—Part 2. *Pharm Sci Technol Today* **3**, 50–57.

Hajduk, P. J., Bures, M., Praestgaard, J., and Fesik, S. W. (2000). Privileged molecules for protein binding identified from NMR-based screening. *J Med Chem* **43**, 3443–3447.

Halgren, T. A., Murphy, R. B., Friesner, R. A., Beard, H. S., Frye, L. L., Pollard, W. T., and Banks, J. L. (2004). Glide: A new approach for rapid, accurate docking and scoring. 2. Enrichment factors in database screening. *J Med Chem* **47**, 1750–1759.

Hansch, C., Klein, T., McClarin, J., Langridge, R., and Cornell, N. W. (1986). A quantitative structure–activity relationship and molecular graphics analysis of hydrophobic effects in the interactions of inhibitors with alcohol dehydrogenase. *J Med Chem* **29**, 615–620.

Harper, G., Bravi, G. S., Pickett, S. D., Hussain, J., and Green, D. V. (2004). The reduced graph descriptor in virtual screening and data-driven clustering of high-throughput screening data. *J Chem Inf Comput Sci* **44**, 2145–2156.

Hawkins, D. M., Young, S. S., and Rusinko, A., III (1997). Analysis of a large structure–activity data set using recursive partitioning. *Quant Struct–Act Relat* **16**, 296–302.

Henry, D. R., Jurs, P. C., and Denny, W. A. (1982). Structure–antitumor activity relationships of 9-anilinoacridines using pattern recognition. *J Med Chem* **25**, 899–908.

Hindle, S. A., Rarey, M., Buning, C., and Lengaue, T. (2002). Flexible docking under pharmacophore type constraints. *J Comput Aided Mol Des* **16**, 129–149.

Hodes, L. (1981). Selection of molecular fragment features for structure–activity studies in antitumor screening. *J Chem Inf Comput Sci* **21**, 132–136.

Jacobsson, M., Liden, P., Stjernschantz, E., Bostrom, H., and Norinder, U. (2003). Improving structure-based virtual screening by multivariate analysis of scoring data. *J Med Chem* **46**, 5781–5789.

Jiang, F., and Kim, S. H. (1991). "Soft docking": Matching of molecular surface cubes. *J Mol Biol* **219**, 79–102.

Jones, G., Willett, P., Glen, R., Leach, A., and Taylor, R. (1997). Development and validation of a genetic algorithm to flexible docking. *J Mol Biol* **267**, 727–748.

Jorgensen, W. L. (2004). The many roles of computation in drug discovery. *Science* **303**, 1813–1818.

Kennedy, T. (1997). Managing the drug discovery/development interface. *Drug Discov Today* **2**, 436–444.

Kier, L. B., and Hall, L. H. (1989). Differential molecular connectivity in data-base fragment searching. *Pharm Res* **6**, 497–500.

Kier, L. B., and Hall, L. H. (1990). An electrotopological-state index for atoms in molecules. *Pharm Res* **7**, 801–807.

Kitchen, D. B., Decornez, H., Furr, J. R., and Bajorath, J. (2004). Docking and scoring in virtual screening for drug discovery: Methods and applications. *Nat Rev Drug Discov* **3**, 935–949.

Klopman, G. (1984). Artificial intelligence approach to structure–activity studies. Computer automated structure evaluation of biological activity of organic molecules. *J Am Chem Soc* **106**, 7315–7321.

Kontoyianni, M., McClellan, L., and Sokol, G. (2004). Evaluation of docking performance: Comparative data on docking algorithms. *J Med Chem* **47**, 558–565.

Kubinyi, H. (1998). Combinatorial and computational approaches in structure-based drug design. *Curr Opin Drug Discov Dev* **1**, 16–27.

Kubinyi, H. (2003). Drug research: Myths, hype and reality. *Nat Rev Drug Discov* **2**, 665–668.

Kuntz, I., Blaney, J., Oatley, S., Langridge, R., and Ferrin, T. (1982). A geometric approach to macromolecule–ligand interactions. *J Mol Biol* **161**, 269–288.

Langer, T. , and Wolber, G. (2004). Pharmacophore definition and 3D searches. *Drug Discov Today* **1**, 203–207.

Lewell, X. Q., Judd, D. B., Watson, S. P., and Hann, M. M. (1998). RECAP—Retrosynthetic combinatorial analysis procedure: A powerful new technique for identifying privileged molecular fragments with useful applications in combinatorial chemistry. *J Chem Inf Comput Sci* **38**, 511–522.

Li, C., Xu, L., Wolan, D. W., Wilson, I. A., and Olson, A. J. (2004). Virtual screening of human 5-aminoimidazole-4-carboxamide ribonucleotide transformylase against the NCI diversity set by use of AutoDock to identify novel nonfolate inhibitors. *J Med Chem* **47**, 6681–6690.

Lipinski, C. A., Lombardo, F., Dominy, B.W., and Fernandes, P. B. (1997). Experimental and computational approaches to estimate solubility and permeability in drug discovery and development settings. *Adv Drug Deliv Rev* **23**, 2–25.

Lyne, P. D., Kenny, P. W., Cosgrove, D. A., Deng, C., Zabludoff, S., Wendoloski, J. J., and Ashwell, S. (2004). Identification of compounds with nanomolar binding affinity for checkpoint kinase-1 using knowledge-based virtual screening. *J Med Chem* **47**, 1962–1968.

Martin, Y. C., Kofron, J., and Traphagen, L. M. (2003). Do structurally similar molecules have similar biological activity? *J Med Chem* **45**, 4350–4358.

Mason, J. S., Morize, I., Menard, P. R., Cheney, D. L., Hulme, C., and Labaudiniere, R. F. (1999). New 4-point pharmacophore method for molecular similarity and diversity applications: Overview of the method and applications, including a novel approach to the design of combinatorial libraries containing privileged substructures. *J Med Chem* **42**, 3251–3264.

Mason, J. S., Good, A. C., and Martin, E. J. (2001). 3-D pharmacophores in drug discovery. *Curr Pharm Des* **7**, 567–597.

Matter, H. (1997). Selecting optimally diverse compounds from structure databases: A validation study of two-dimensional and three-dimensional molecular descriptors. *J Med Chem* **40**, 1219–1229.

Mattioni, B. E., Kauffman, G. W., Jurs, P. C., Custer, L. L., Durham, S. K., and Pearl, G. M. (2003). Predicting the genotoxicity of secondary and aromatic amines using data subsetting to generate a model ensemble. *J Chem Inf Comput Sci* **43**, 949–963.

McGovern, S. L., Caselli, E., Grigorieff, N., and Shoichet, B. K. (2002). A common mechanism underlying promiscuous inhibitors from virtual and high-throughput screening. *J Med Chem* **45**, 1712–1722.

McGovern, S. L., Helfand, B. T., Feng, B., and Shoichet, B. K. (2003). A specific mechanism of nonspecific inhibition. *J Med Chem* **46**, 4265–4272.

McGregor, M. J., and Muskal, S. M. (1999). Pharmacophore fingerprinting. 1. Application to QSAR and focused library design. *J Chem Inf Comput Sci* **39**, 569–574.

McGregor, M. J., and Muskal, S. M. (2000). Pharmacophore fingerprinting. 2. Application to primary library design. *J Chem Inf Comput Sci* **40**, 117–125.

McGregor, M. J., and Pallai, P. V. (1997). Clustering of large databases of compounds: Using the MDL "keys" as structural descriptors. *J Chem Inf Comput Sci* **37**, 443–448.

Merlot, C., Domine, D., and Church, D. J. (2002). Fragment analysis in small molecule discovery. *Curr Opin Drug Discov Dev* **5**, 391–399.

Merlot, C., Domine, D., Cleva, C., and Church, D. J. (2003). Chemical substructures in drug discovery. *Drug Discov Today* **8**, 594–602.

Minutolo, F., Antonello, M., Bertini, S., Ortore, G., Placanica, G., Rapposelli, S., Sheng, S., Carlson, K. E., Katzenellenbogen, B. S., Katzenellenbogen, J. A., and Macchia, M. (2003). Novel estrogen receptor ligands based on an anthranylaldoxime structure: Role of the phenol-type pseudocycle in the binding process. *J Med Chem* **46**, 4032–4042.

Morris, G., Goodsell, D., Halliday, R., Huey, R., Hart, W., Belew, R., and Olson, A. (2004). Automated docking using a lamarckian genetic algorithm and an empirical binding free energy function. *J Comput Chem* **19**, 1639–1662.

Muegge, I., Heald, S. L., and Brittelli, D. (2001). Simple selection criteria for drug-like chemical matter. *J Med Chem* **44**, 1841–1846.

Muegge, I., and Martin, Y. C. (1999). A general and fast scoring function for protein–ligand interactions: A simplified potential approach. *J Med Chem* **42**, 791–804.

Muller, G. (2003). Medicinal chemistry of target family-directed masterkeys. *Drug Discov Today* **8**, 681–691.

Navia, M. A., and Chaturvedi, P. R. (1996). Design principles for orally bioavailable drugs. *Drug Discov Today* **1**, 179–189.

Oprea, T. I., Davis, A. M., Teague, S. J., and Leeson, P. D. (2001). Is there a difference between leads and drugs? A historical perspective. *J Chem Inf Comput Sci* **41**, 1308–1315.

Oshiro, C., Bradley, E., Eksterowicz, J., Evensen, E., Lamb, M., Lanctot, J., Putta, S., Stanton, R., and Grootenhuis, D. (2004). Performance of 3D-database molecular docking studies into homology models. *J Med Chem* **47**, 764–767.

Palm, K., Luthman, K., Ungell, A. L., Strandlund, G., and Artursson, P. (1996). Correlation of drug absorption with molecular surface properties. *J Pharm Sci* **85**, 32–39.

Pardridge, W. M. (1995). Transport of small molecules through the blood–brain barrier: Biology and methodology. *Adv Drug Deliv Rev* **15**, 5–36.

Pearl, G. M., Livingston-Carr, S., and Durham, S. K. (2001). Integration of computational analysis as a sentinel tool in toxicological assessments. *Curr Top Med Chem* **1**, 247–255.

Perez, C., and Ortiz, A. R. (2001). Evaluation of docking functions for protein–ligand docking. *J Med Chem* **44**, 3768–3785.

Perola, E., and Charifson, P. S. (2004). Conformational analysis of drug-like molecules bound to proteins: An extensive study of ligand reorganization upon binding. *J Med Chem* **47**, 2499–2510.

Perola, E., Walters, W. P., and Charifson, P. S. (2004). A detailed comparison of current docking and scoring methods on systems of pharmaceutical relevance. *Proteins* **56**, 235–249.

Pickett, S. D., Mason, J. S., and McLay, I. M. (1996). Diversity profiling and design using 3D pharmacophores: Pharmacophore-derived queries (PDQ). *J Chem Inf Comput Sci* **36**, 1214–1223.

Rarey, M., Kramer, B., Lengauer, T., and Klebe, G. (1996). A fast flexible docking method using an incremental construction algorithm. *J Mol Biol* **261**, 269–288.

Rishton, G. M. (1997). Reactive compounds and *in vitro* false positives in HTS. *Drug Discov Today* **2**, 382–384.

Rishton, G. M. (2003). Nonleadlikeness and leadlikeness in biochemical screening. *Drug Discov Today* **8**, 86–96.

Roche, O., Schneider, P., Zuegge, J., Guba, W., Kansy, M., Alanine, A., Bleicher, K., Danel, F., Gutknecht, E. M., Rogers-Evans, M., Neidhart, W., Stalder, H., Dillon, M., Sjogren, E., Fotouhi, N., Gillespie, P., Goodnow, R., Harris, W., Jones, P., Taniguchi, M., Tsujii, S., von der, S. W., Zimmermann, G., and Schneider, G. (2002). Development of a virtual screening method for identification of "frequent hitters" in compound libraries. *J Med Chem* **45**, 137–142.

Sadowski, J., and Kubinyi, H. (1998). A scoring scheme for discriminating between drugs and nondrugs. *J Med Chem* **41**, 3325–3329.

Salim, N., Holliday, J., and Willett, P. (2003). Combination of fingerprint-based similarity coefficients using data fusion. *J Chem Inf Comput Sci* **43**, 435–442.

Sauer, W. H., and Schwarz, M. K. (2003a). Molecular shape diversity of combinatorial libraries: A prerequisite for broad bioactivity. *J Chem Inf Comput Sci* **43**, 987–1003.

Sauer, W. H., and Schwarz, M. K. (2003b). Size doesn't matter: Scaffold diversity, shape diversity and biological activity of combinatorial libraries. *Chimia* **57**, 276–283.

Saxberg, B. E. H., Blom, D. S., and Kowalski, B. R. (1976). A flexible interactive graphics system for searching atom connectivity matrices. *J Chem Inf Comput Sci* **16**, 233–238.

Schuffenhauer, A., Floersheim, P., Acklin, P., and Jacoby, E. (2003). Similarity metrics for ligands reflecting the similarity of the target proteins. *J Chem Inf Comput Sci* **43**, 391–405.

Shemetulskis, N. E., Weininger, D., Blankley, C. J., Yang, J. J., and Humblet, C. (1996). Stigmata: An algorithm to determine structural commonalities in diverse datasets. *J Chem Inf Comput Sci* **36**, 862–871.

Snarey, M., Terrett, N. K., Willett, P., and Wilton, D. J. (1997). Comparison of algorithms for dissimilarity-based compound selection. *J Mol Graph Model* **15**, 372–385.

Sprague, P. (1995). Automated chemical hypothesis generation and database searching with catalyst. In B. V. K. Mueller (Ed.), *Perspectives in Drug Discovery and Design.* Leiden: ESCOM Science Publishers, pp.1–20.

Stahl, M., and Rarey, M. (2001). Detailed analysis of scoring functions for virtual screening. *J Med Chem* **44**, 1035–1042.

Stigler, R. D., Hoffmann, B., Abagyan, R., and Schneider-Mergener, J. (1999). Soft docking an L and a D peptide to an anticholera toxin antibody using internal coordinate mechanics. *Structure Fold Des* **7**, 663–670.

Subhash, S. (1996). Cluster analysis. In *Applied Multivariate Techniques*. New York: John Wiley & Sons.

Taylor, R., (1995). Simulation analysis of experimental design strategies for screening random compounds as potential new drugs and agrochemicals. *J Chem Inf Comput Sci* **35**, 59–67.

Terp, G., Johansen, B., Christensen, T., and Jørgensen, F. (2004). A new concept for multidimensional selection of ligand conformations (MultiSelect) and multidimensional scoring (MultiScore) of protein–ligand binding affinities. *J Med Chem* **44**, 2333–2343.

Todeschini, R., and Consonni, V. (2000). *Handbook of Molecular Descriptors*, Weinheim, Germany: Wiley-VCH.

Toledo, L. M., Lydon, N. B., and Elbaum, D. (1999). The structure-based design of ATP-site directed protein kinase inhibitors. *Curr Med Chem* **6**, 775–805.

Tong, L., Pav, S., White, D. M., Rogers, S., Crane, K. M., Cywin, C. L., Brown, M. L., and Pargellis, C. A. (1997). A highly specific inhibitor of human p38 MAP kinase binds in the ATP pocket. *Nat Struct Biol* **4**, 311–316.

Trepalin, S. V., Gerasimenko, V. A., Kozyukov, A. V., Savchuk, N. P., and Ivaschenko, A. A. (2002). New diversity calculations algorithms used for compound selection. *J Chem Inf Comput Sci* **42**, 249–258.

Venkatesh, S., and Lipper, R. A. (2000). Role of the development scientist in compound lead selection and optimization. *J Pharm Sci* **89**, 145–154.

Verdonk, M., Berdini, V., Hartshorn, M., Mooij, W., Murray, C., Taylor, M., and Watson, P. (2004). Virtual screening using protein-ligand docking: avoiding artificial enrichment. *J Chem Inf Comput Sci* **44**, 793–806.

Walters, W. P., and Murcko, M. A. (2002). Prediction of 'drug-likeness'. *Adv Drug Deliv Rev* **54**, 255–271.

Walters, W. P., and Namchuk, M. (2003). Designing screens: How to make your hits a hit. *Nat Rev Drug Discov* **2**, 259–266.

Wang, R., Lu, Y., and Wang, S. (2003). Comparative evaluation of 11 scoring functions for molecular docking. *J Med Chem* **46**, 2287–2003.

Willett, P. (2000). Chemoinformatics–Similarity and diversity in chemical libraries. *Curr Opin Biotech* **11**, 85–88.

Willett, P., Barnard, J. M., and Downs, G. M. (1998). Chemical similarity searching. *J Chem Inf Comput Sci* **38**, 983–996.

Xu, J., and Stevenson, J. (2000). Drug-like index: A new approach to measure drug-like compounds and their diversity. *J Chem Inf Comput Sci* **40**, 1177–1187.

Xue, L., Stahura, F. L., Godden, J. W., and Bajorath, J. (2001). Fingerprint scaling increases the probability of identifying molecules with similar activity in virtual screening calculations. *J Chem Inf Comput Sci* **41**, 746–753.

Xue, L., Godden, J. W., Stahura, F. L., and Bajorath, J. (2003). Profile scaling increases the similarity search performance of molecular fingerprints containing numerical descriptors and structural keys. *J Chem Inf Comput Sci* **43**, 1218–1225.

Xue, L., Stahura, F. L., and Bajorath, J. (2004). Similarity search profiling reveals effects of fingerprint scaling in virtual screening. *J Chem Inf Comput Sci* **44**, 2032–2039.

Young, S. S., and Ge, N. (2004). Design of diversity and focused combinatorial libraries in drug discovery. *Curr Opin Drug Discov Devel* **7**, 318–324.

Yuta, K., and Jurs, P. C. (1981). Computer-assisted structure-activity studies of chemical carcinogens. Aromatic amines. *J Med Chem* **24**, 241–251.

Zhorov, B. S., Folkman, E. V., and Ananthanarayanan, V. S. (2001). Homology model of dihydropyridine receptor: Implications for L-type Ca(2+) channel modulation by agonists and antagonists. *Arch Biochem Biophys* **393**, 22–41.

CHAPTER 4

COMPUTER-AIDED DRUG DESIGN

*Grace Shiahuy Chen and Ji-Wang Chern**

4.1 INTRODUCTION

Even though many routes for the discovery of new drugs exist, most clinically used drugs were discovered accidentally or by screening against the enzymatic products, chemicals, and natural products. In the traditional drug discovery process, a lead compound might be obtained out of around 80,000 compounds, and then lead optimization is carried out by numerous efforts in chemical synthesis to improve its activity or to reduce the toxicity. A structure–activity relationship (SAR) might be achieved after a variety of analogues are synthesized, and then SAR could be used as a guideline for future synthesis and optimization. Sulfonamide drugs, penicillin, and 5-fluorouracil (5-FU) were found in this way. Their action mechanisms were explored afterward. For example, many derivatives of sulfonamides were prepared to evaluate the antibacterial activity, and this exhaustive synthetic work led to the discovery of other treatments such as diuretics, drugs for leprosy, antimalarial agents, antigoatic agents, and antihypertensive drugs.

Advances in molecular biology and the Human Genome Project have revealed a superfluity of potential targets. The techniques of X-ray crystallography and nuclear magnetic resonance (NMR) enable the structural determination of many biological targets (enzymes or proteins). Then, pharmaceutical companies have invested heavily in lead discovery in target assessment technologies such as high-throughput screening

*Corresponding author: chern@jwc.mc.ntu.edu.tw

Drug Discovery Research: New Frontiers in the Post-Genomic Era, Edited by Ziwei Huang

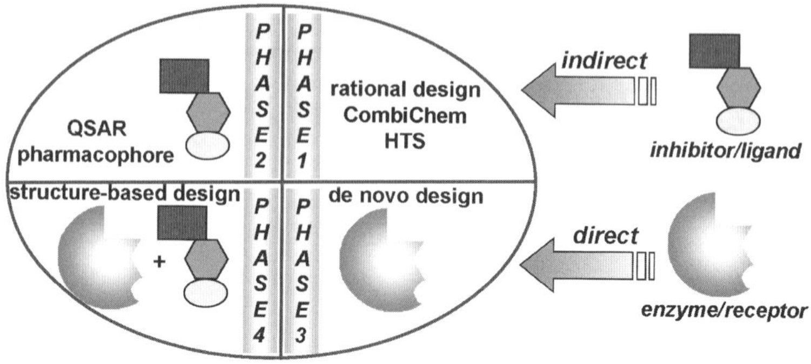

Figure 4.1 Strategies in drug design respective to known or unknown 3D structure of the target.

(HTS, Lutz et al., 1996) and combinatorial chemistry (CombiChem, Gollop et al., 1994; Gordon et al., 1994). Combinatorial synthesis makes large chemical libraries available, and HTS might afford more hits for finding potential lead compounds. However, problems of how to increase the probability of success in screening of vast chemical libraries still exist (Spencer, 1998; Bailey and Brown, 2001). In addition, screening large chemical databases remains an expensive and time-consuming process, especially for academic research. There are over 3000 structures deposited every year and more than 32,000 protein structures in total available today in the Protein Data Bank (PDB, Bernstein et al., 1977). Therefore, computational methods are the best alternatives to increase efficiency and reduce costs in the drug discovery process. In addition, three-dimensional (3D) molecular databases are available to implement the computational screening (Shadowski and Gasteiger, 1993; Manallack, 1996).

"Direct" and "indirect" are the two major strategies used in the current drug design process (Cohen et al., 1990). In relation to the availability of target (enzyme or protein) structures and ligands as shown in Fig. 4.1, drug discovery can be divided into four phases: no target 3D structure or ligands (phase 1), no target 3D structure but ligands available (phase 2), target 3D structure available but no ligands (phase 3), and both target 3D structure and ligands available (phase 4). In the early stage, no 3D target structures and little or no SAR are known, and therefore HTS and CombiChem are the two major approaches (phase 1). In the indirect strategy, the structural features and biological activity of inhibitors or ligands are taken into account for quantitative structure–activity relationship (QSAR) or pharmacophore-based design (phase 2). In the direct approach, the 3D features of the enzyme or receptor are directly considered for *de novo* design (phase 3) and structure-based design (phase 4). In this review, we will describe ligand- and structure-based design, employing computational techniques and their features.

4.2 LIGAND-BASED DRUG DESIGN

4.2.1 Quantitative Structure–Activity Relationship

When ligands or inhibitors are available, QSAR was the first method developed for virtual screening and has been advanced from a two-dimensional (2D) to a 3D model. Quantitative structure–activity relationship analysis uses statistical methods to study the correlation of biological activity, physicochemical properties, or toxicities to structural properties of candidate molecules. Computer modeling analyzes a series of chemical molecules possessing a diversity of binding affinities to a protein or an enzyme. A QSAR attempts to find consistent relationships between the variations in the values of molecular properties and the biological activity. In the classical or 2D QSAR techniques, traditional SAR methods such as the Hansch equation (Hansch and Fujita, 1964), Topliss tree (Topliss, 1972), and Craig plot (Craig, 1971) could be used to optimize the leads.

Modern 3D QSAR methods analyze the interaction fields around a molecule by calculating the interaction energy in a grid. This is repeated for all molecules in the series, with their alignment relative to each other being a crucial parameter. The interaction energies for each point in the grid are then subject to a QSAR across the set of molecules. The most well-known 3D QSAR technique is Comparative Molecular Field Analysis (CoMFA, http://www.tripos.com). The CoMFA approach calculates molecular properties such as steric, electronic, hydrogen bonding, and hydrophobic fields (Cramer et al., 1988).

4.2.2 Substructure Search

When a lead compound is known, the similarity or substructure search can be carried out to retrieve novel compounds, containing the main skeleton or group present in the original lead, from databases. Structural search is the first type of screening method broadly employed for screening chemical databases. Once important structural features or patterns are determined, they can be used to search databases for all molecules possessing the features. Then, these molecules can be verified in a biological assay. Recently, we discovered compound soluble blue as a hepatitis C virus (HCV) helicase inhibitor with an IC_{50} value of 40 μM (Fig. 4.2). Then, we used the 2D substructure search approach to find compounds based on the structure of

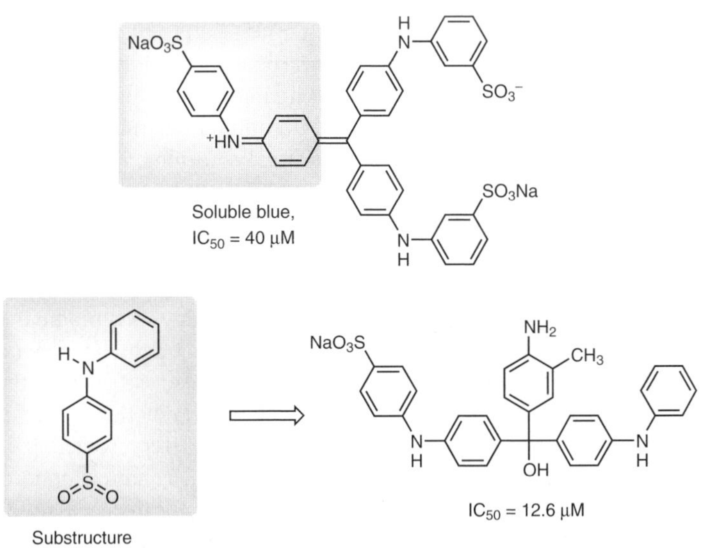

Figure 4.2 Example of HCV helicase inhibitor derived from substructure search.

soluble blue. The fragment was specified as a query substructure in which the polar groups SO_2 and NH might be involved in the H-bonding interactions with the enzyme. The substructure search was conducted using the ISIS/Base search program and the MDL database Available Chemicals Directory-Screening Compounds (ACD-SC, MDL Information Systems Inc., San Leandro, CA, USA, http://www.mdli,com). A more potent HCV helicase inhibitor with an IC_{50} value of 12.6 μM was identified (Chen et al., in preparation).

4.2.3 Pharmacophore-Based Design

The pharmacophore is an important concept in rational drug design. It represents a set of functional features in 3D geometry that can interact with a specific receptor to gain activity. In other words, a pharmacophore is the geometrical depiction of fragments or features believed to be responsible for the biological activity (Ghose and Wendoloski, 1998). In general, a pharmacophore is atom-based and defined as chemical properties such as acid, base, hydrophobe, or aromatic group. Sometimes, others such as plane and normal are included. When an unknown structure of a receptor or a membrane-associated protein is the target, the pharmacophore model becomes a useful tool for searching lead compounds since the structure-based virtual screening is not accessible. Pharmacophore-based virtual screening has been demonstrated as an efficient method and can reach hit rates of 1–20%.

Many approaches have been utilized in the generation of a pharmacophore. They can be divided into three categories: (1) similarity of a known agent against an enzyme or a ligand to a receptor: this method predicts the necessary shape of molecules in binding to a target protein through the shape-similarity of molecules; (2) prediction of bioactive molecules with different structures: this method analyzes a set of structurally different chemical molecules with diverse activity, finds the consistency and specificity, and determines the pharmacophore model of a protein or an enzyme; and (3) analysis of the target's active site: The application of the pharmacophore model can extend to the structure-based design. By analyzing the active site of a target, one can find a ligand possessing specific chemical features and suitable binding orientation. Combined pharmacophore-structure-based virtual screening would be a potential virtual screening approach but is more time-consuming. The generation of a pharmacophore can be achieved by DISOtech and RECEPTOR (Tripos, St. Louis, Missouri, USA; http://www.tripos.com) and CATALYST (Accelrys, San Diego, California, USA; http://www.accelrys.com). Once the pharmacophore is determined, it can be used to screen chemical libraries for compounds fitting to the pharmacophore. Through the biological assay to validate the model, a potential novel lead compound might be generated. For example, in CATALYST, pharmacophore models are also called hypotheses and consist of 3D geometrical arrangements of chemical functions. The chemical functions are the features related in 3D geometry and represented by a set of spheres as shown in Fig. 4.3 (Chen et al., 2001). In general, a pharmacophore consists of three to five features. The volume of the sphere stands for the accessible space of the chemical functions. For organic molecules, different structures could have the same chemical function and therefore result in the same biological activity.

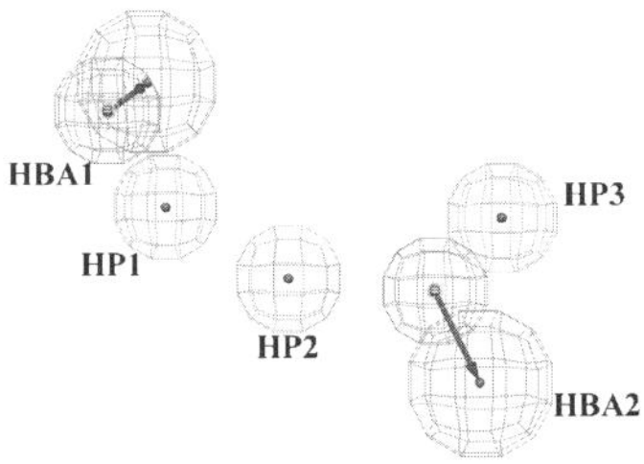

Distance constraints (Å) between any two features:

HBA1-HP1	2.2–4.2	HP1-HP2	3.2–5.2
HBA1-HP2	6.1–8.1	HP1-HBA2	7.2–9.2
HBA1-HBA2	9.8–11.8	HP1-HP3	8.9–10.9
HBA1-HP3	10.8–12.8	HP2-HBA2	3.3–5.3
HP2-HP3	5.6–7.6	HBA2-HP3	2.3–4.3

Figure 4.3 Pharmacophore of type II 5α-reductase inhibitors containing two hydrogen bond acceptors (HBAs) and three hydrophobic groups (HPs).

This approach is based on two fundamental assumptions: (1) a same binding mode for all compounds in the training set to a receptor and (2) the more binding interactions between compound and receptor, the higher the activity. The selection of the training set, a set of molecules, plays the essential role in determining the success of HypoGen. Compounds in a training set should contain a large variety of chemical structures and interact with the target via the same mechanism with a range of potency from highly active to inactive. There are two successive major steps in the generation of a pharmacophore. First, a conformational analysis is needed to determine all the possible low-energy conformations. The biological activity of a molecule is associated with a unique conformation, and it often is not the global minimum energy conformation (Ghose et al., 1999). The search of the conformational model is the major task in this approach (Bausen and Shands, 1996). Second, all the conformations in the training set are used to calculate their stereoelectronic properties, such as steric, electronic, lipophilic, and hydrogen bonding. Then, a pharmacophore is generated on the basis of the stereoelectronic properties of the training set. A 3D database search is used to identify compounds mapped to the pharmacophore. For example, HypoGen (Sutter et al., 2000; Kurogi and Güner, 2001) in CATALYST analyzes the relationship of biological activity and structures in the training set. Then, it automatically generates a chemical functions–based pharmacophore model. Following a schematic process as shown in Fig. 4.4, we generated a pharmacophore for 5α-reductase inhibitors, which was used to screen the NCI database (Milne et al., 1994) for novel nonsteroidal lead compounds. A series of isoflavone derivatives were identified as 5α-reductase inhibitors as shown in Fig. 4.5 (Chen et al., 2001).

Many successful lead identifications using pharmacophore searches have been reported, and Fig. 4.6 depicts five successful case studies. Novobiocin (**1**) is a known inhibitor of bacterial enzyme DNA gyrase, and a pharmacophore was obtained based on the structure of **1** (Böhm et al., 2000). Seven classes of novel DNA gyrase inhibitors were recognized, and among them a 3D guided optimization provided a highly potent DNA gyrase inhibitor, 3,4-disubstituted indazole (**2**), which was 10 times more potent than **1**. According to the X-ray structure of the thyroid hormone (**3**), a pharmacophore containing hydrophobic, hydrogen bond donor and acceptor features, and excluded volumes was defined (Greenidge et al., 1998). A new ligand (**4**) was subsequently shown to bind to THR-R with an IC_{50} value of 69 μM. A series of farnesyl protein transferase (FPT) inhibitors exemplified by Sch44342 (**5**) were analyzed by CATALYST to generate a pharmacophore model containing five features

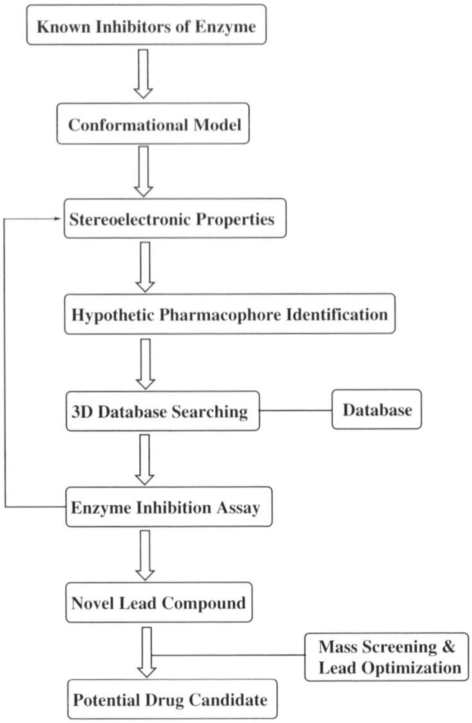

Figure 4.4 Schematic chart for pharmacophoric model by CATALYST.

of four hydrophobic and one hydrogen bond acceptor regions (Kaminski et al., 1997). A 3D search of the pharmacophore identified two peptides such as **6** and two dihydrobenzothiophenes such as **7** exhibiting *in vitro* FPT inhibitory activity with IC_{50} values below 5 μM. A Human Immunodeficiency Virus type 1 Integrase (HIV-1 IN) inhibitor (**8**) was taken as a lead to generate a pharmacophore, and after modification of the pharmacophore a 3D search led to highly potent inhibitors **9** and **10** (Nicklaus et al., 1997; Hong et al., 1997).

4.3 DE NOVO LIGAND DESIGN

With known 3D structure targets (enzymes or proteins), new inhibitors or ligands can be designed *de novo*. *De novo* design programs try to construct novel structures, using sets of predefined fragments, into an active site or onto a pharmacophore model. The combinatorial nature of *de novo* design leads to a large amount of chemical structures. GRID (Goodford, 1985) is the first *de novo* method. Then, *de novo* ligand design programs such as LEGEND (Nishibta and Itai, 1993), LeapFrog (http://www.tripos.com), LigBuilder (Wang et al., 2000), SPROUT (http://www.simbiosys.ca/sprout/index.html), CONCERTS (Pearlman and Murcko, 1996), PRO-LIGAND (Waszkowycz et al., 1994), TOPAS (Schneider et al., 2000), BUILDER (Roe and Kuntz, 1995), and LUDI (http://www.accelrys.com, Böhm, 1992a,b) were

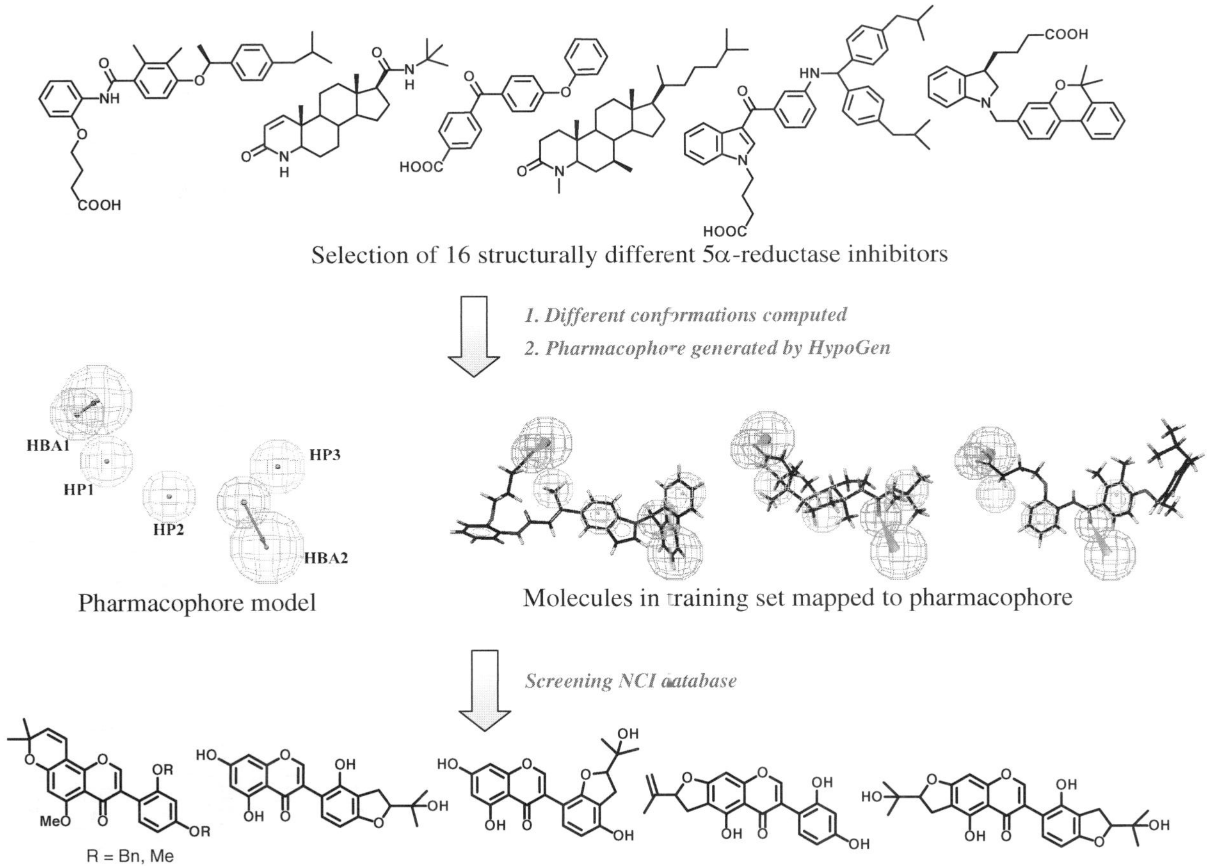

Figure 4.5 Isoflavone derivatives as 5α-reductase inhibitors from pharmacophore screening. See color plates.

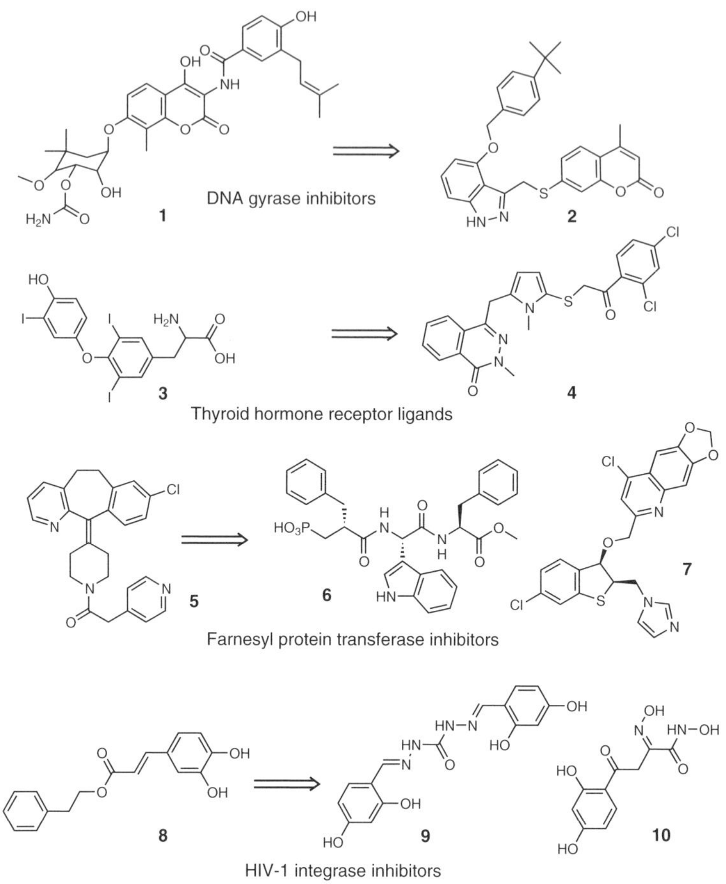

Figure 4.6 Lead identification by pharmacophore approach.

developed to build up structures according to the active site of the receptor. *De novo* design generates functional groups in suitable forms and suitable conformations to interact with the protein surface. More importantly, *de novo* design programs can also construct structures including novel skeletons. They show how suitable small fragments can be positioned into clefts of protein structures. For example, Ludi has two major processes: (1) It can be used to design inhibitors *de novo*, and (2) it can be used to suggest modifications for a known ligand. When run in link mode, Ludi fits fragments in a way that simultaneously exploits available hydrogen bonding and hydrophobic sites on the receptor and suggests how fragments might be bonded to the existing ligand. The processes of Ludi *de novo* design are the following: (1) On the basis of the 3D structure of the receptor, the program computes the points of interaction that will be fulfilled by a ligand within the protein's active site. The interaction points are hydrogen bondings and hydrophobic sites. Four directional interaction sites are defined for portions of the binding site: lipophilic–aliphatic, lipophilic–aromatic, hydrogen donor, and hydrogen acceptor. (2) Fragments from a library of fragments are aligned with these interaction sites and are scored with respect to their degree of overlap. (3)

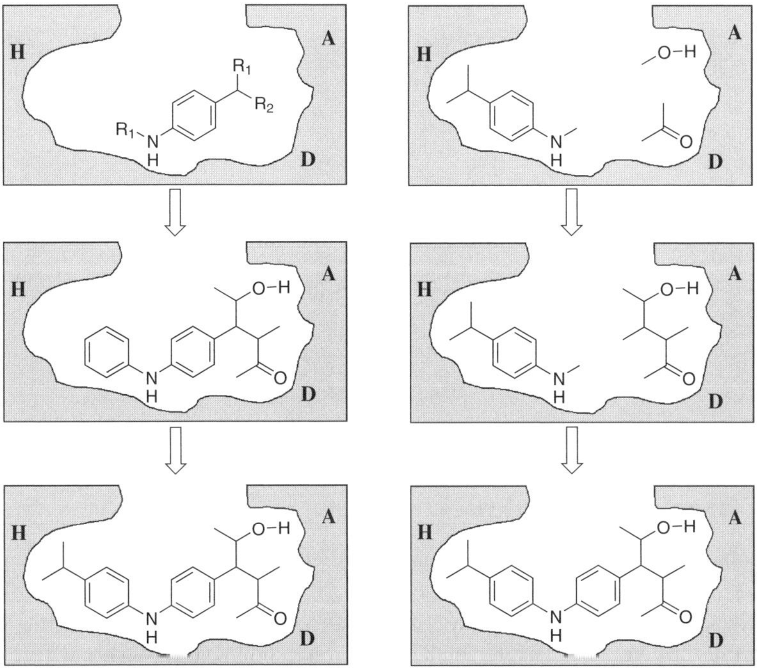

(a) growth strategy (b) linkage strategy

Figure 4.7 Growth and linkage strategies in *de novo* ligand design.

Fragments are then connected in a straightforward way (the two closest hydrogens between fragments are removed and a linker of the appropriate length is attached to each terminal atom in the fragments). (4) Conformational flexibility can be accounted for by including multiple conformations of a fragment in the library. Most *de novo* programs contain two strategies of ligand generation: (a) growth (Liu et al., 1999; Rotstein and Murcko, 1993) and (b) linkage (Böhm, 1996) as shown in Fig. 4.7.

De novo ligand design led to a novel series of factor Xa (f Xa) inhibitors incorporating an amidino 6,5-fused bicyclic moiety such as (**11**, SE170, Fig. 4.8) with a K_i value of 0.3 nM and 350-fold selective for fXa over trypsin (Han et al., 2000). A second example is the ligand against human cannabinoid receptor 1 (CB1). The TOPAS program was used to design a library of ligands and led to a very potent ligand (**12**) with a K_i value of 0.3 μM. A potent CDK4 inhibitor (**13**) with nanomolar inhibitory activity was developed by LEGEND.

4.4 STRUCTURE-BASED DESIGN

When the 3D structure of the target enzyme or protein is known, the lead identification and optimization can be efficiently processed with the aid of molecular modeling. It is especially beneficial to have the 3D structure of a protein–ligand complex so that the active or binding site can be explored.

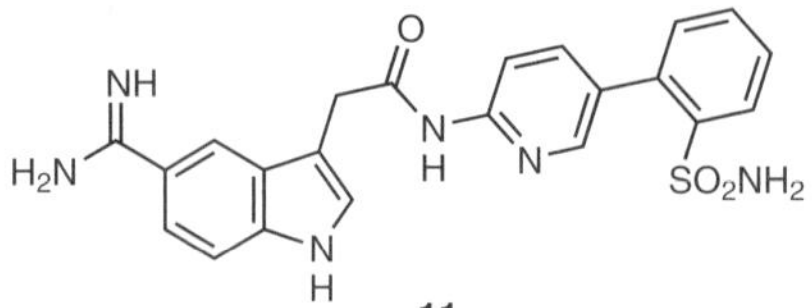

11

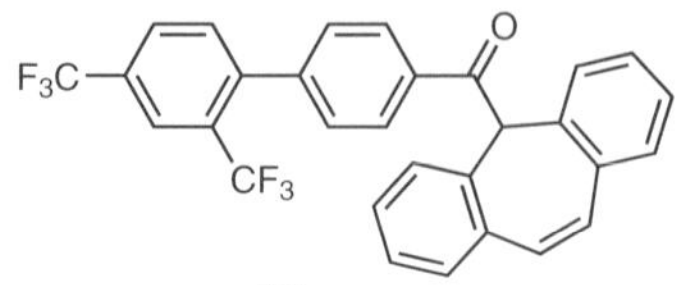

12

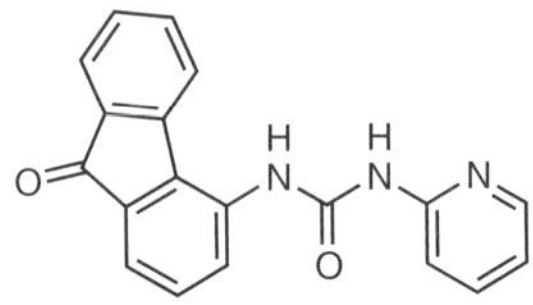

13

Figure 4.8 Leads identified by *de novo* ligand design.

4.4.1 Docking Techniques

Molecular docking techniques dock small molecules into the protein binding site. Success of the docking depends on the 3D target protein structure with or without a ligand bound. The 3D structures have to be detailed at atomic resolution. With the vast 3D protein structures available in the PDB, one should keep in mind that small changes in structure significantly influence the results of docking (Muegge, 1999), and crystal structures employed in docking should have resolution below 2.5 Å. Ligand–receptor docking has become the choice to accelerate structure-based drug design and has been successfully used in the discovery of drug candidates. Docking procedures try to identify correct poses of ligands in the binding site of a protein and to predict the binding affinity between the ligand and the protein. Comprehensive reviews on the docking techniques have been appeared in the literature (Boyd and Lipkowitz, 2001). Most methods available now treat the ligands flexibly and the proteins rigidly. Van Leeuwen has compiled some docking algorithms (http://www.bio.vu.nl/nvtb/Docking.html). Table 4.1 depicts some docking methods used for single molecule docking, and Table 4.2 shows the tools used only for single molecule docking. Among them, DOCK, FlexX (Rarey et al., 1997), and GOLD (Jones et al. 1995) are frequently used platforms in the docking approach.

4.4.1.1 Program DOCK

The computer program DOCK (Meng et al., 1992; Kuntz et al., 1994) is the most widely used computational docking method and provides a rapid method by using the 3D structure of a target receptor to identify putative ligands among commercially available compounds. When a ligand fits to a receptor well, it may bind to the receptor with high affinity, and it has the potential to be an inhibitor for a biological protein receptor. DOCK discovers how a ligand and a receptor fit together. DOCK has been utilized to discover leads in a number of diverse protein systems such as HIV-1 protease (Rutenber et al., 1993), thymidylate synthase (Shoichet et al., 1993), influenza hemagglutinin (Bodian et al., 1993), purine salvage enzymes in parasitic

TABLE 4.1 Software Available for Single Molecule Docking and Database Screening

Software	Docking Method(s)	Scoring Method(s)	Web Site
GLIDE	Rigid docking	Empirical scoring, force-field scoring	www.schrodinger.com
DOCK	Flexible docking	Force-field scoring, chemical scoring, contact scoring	www.cmpharm.ucsf.edu/kuntz/dock.html
FlexX	Flexible docking	Empirical scoring	cartan.gmd.de/FlexX
LigandFit	Monte Carlo	LIGSCORE, PLP, PMF, LUDI	www.accelrys.com
FRED	Rigid docking	Chemscore, PLP, Gaussian shape scoring, SCreenScore	www.eyesopen.com/fred.html
DockVision	Monte Carlo, genetic algorithm	Force-field scoring	www.dockvision.com
GOLD	Genetic algorithm	Van der Waals and hydrogen bond potentials	www.ccdc.cam.ac.uk/prods/gold

protozoa (Somoza et al., 1998), and macrophage migration inhibitory factor (Orita et al., 2001). Several successful examples in structure-based lead generation have been reported. This structure-based computational method, using the DOCK suite to screen a large database of small molecules, was shown to be a powerful tool in identifying drug leads.

In DOCK, spheres are created to map and fill the molecular surface of an active site. A set of spheres represents the ligand. In general, the docking process includes four primary tasks: (1) creating a negative image of the active site represented by a set of overlapping spheres whose centers are putative locations for ligand atoms, (2) matching sphere centers to atoms of small molecules from a database, (3) scoring and ranking the docked orientations of compounds on the basis of their complementarities to the protein surface, and (4) reviewing the docked orientations of the best-scoring compounds. The DOCK algorithm (Lightstone et al., 2000) was illustrated in Fig. 4.9.

DOCK is used to locate favorable binding orientations for a ligand molecule and a receptor molecule. The program evaluates the ligand orientations by a contact (shape-fitting) score, a forced field interaction energy, and a user-defined chemical

TABLE 4.2 Software Available Only for Single-Molecule Docking

Software	Docking Method	Scoring Method(s)	Web Site
Liaison	Flexible docking	Free–energy scoring	www.schrodinger.com
AutoDock	Genetic algorithm	Force-field scoring, free-energy scoring	www.scripps.edu/pub/olson-web/doc/autodock/index.html
Affinity	Flexible docking	Molecular mechanics, force–field scoring	www.accelrys.com

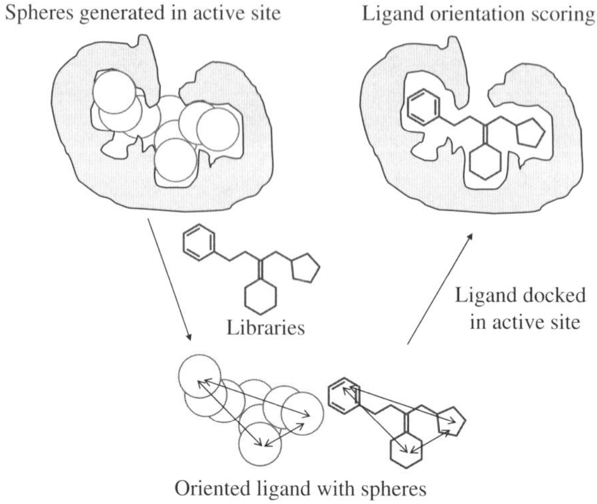

Figure 4.9 DOCK algorithm.

scoring scheme. The binding energy is taken to be the sum of the van der Waals attractive, van der Waals dispersive, and Coulombic electrostatic energies. The best-scoring orientation of each molecule is saved. The molecules in the list of best-scoring compounds may have the potential as new ligands for the receptor. The docking procedure is divided into four general stages: ligand preparation, site characterization, scoring grid calculation, and docking itself, as shown in Fig. 4.10 (http://www.cmpharm.ucsf.edu/kuntz/dock.html).

4.4.1.2 *Rigid Docking*

Early versions of the DOCK program (Kuntz et al., 1982) performed rigid-body docking based on the distance-compatible match searches. Later, chemical properties are taken to match spheres (Shoichet et al., 1993). In DOCK 4.0, different scoring functions (Meng et al,. 1992, 1994; Gschwend and Kuntz, 1996) and solvation (Shoichet et al., 1999; Zou et al., 1999) are available. Rigid-body docking, also called orientation searching, is a process that considers the orientations of ligands. The conformations of ligands are kept. First, compounds are docked using rigid-body docking to search for favorable binding orientations based on the interaction force field scoring that included van der Waals and electrostatic terms. The number of orientations can be defined. Since this process is less time-consuming, it can be used to screen large-scale databases. This method can rank a larger amount of compounds that might bind to the receptors. This procedure can narrow down the size of databases to be in-house databases, which then can be used for flexible docking. Program FlexX employs a pose clustering technique (Rarey et al., 1996a,b).

4.4.1.3 *Flexible Ligand Docking*

In general, the biologically active conformations of ligands are not the conformations with the lowest energy (global minimum). Therefore, flexibility is considered for

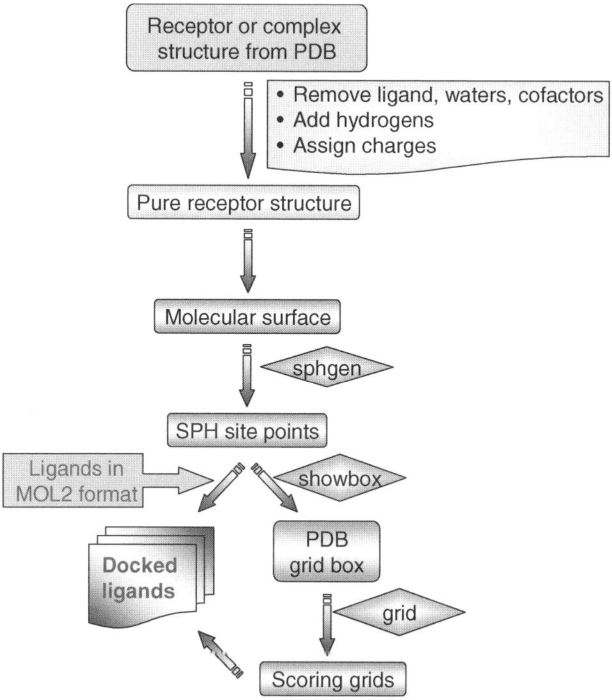

Figure 4.10 Modified docking task flowchart.

ligands. The conformation of a flexible molecule is searched or relaxed in flexible ligand docking. Normally, it is approached by fragmentation, ensembles, or genetic algorithms. For example, in DOCK, only the torsion angles are modified, and the bond lengths and angles are not changed. Only the interactions between segments are considered. There are two processing stages involved in this procedure: identification of rigid segments and conformation search. A flexible molecule is considered as a collection of rigid segments. Each segment contains the largest set of adjacent atoms separated by nonrotatable bonds. Segments are separated by rotatable bonds. This procedure can predict the binding more accurately, but it is much more time- and memory-consuming. Therefore, it will be employed on smaller and in-house databases.

4.4.2 Scoring Function

The binding affinity of a noncovalent ligand–protein complex is governed by the change in enthalpy and entropy of the whole system. In fact, a docked complex involving ligands and proteins actually is associated with the changes of free energies. However, the binding of free energies is difficult to estimate. The docking result is estimated by the binding affinity, and the scoring functions range from computer-intensive theoretical calculations to simple empirical methods (Murcko and Murcko, 1995; Tame, 1999). Since docking programs need to screen large databases

14

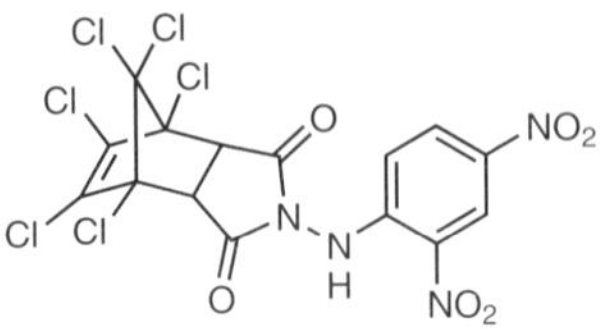

15

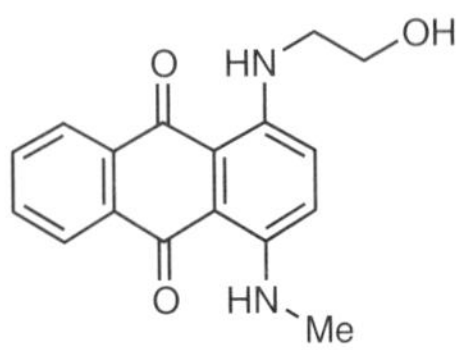

16

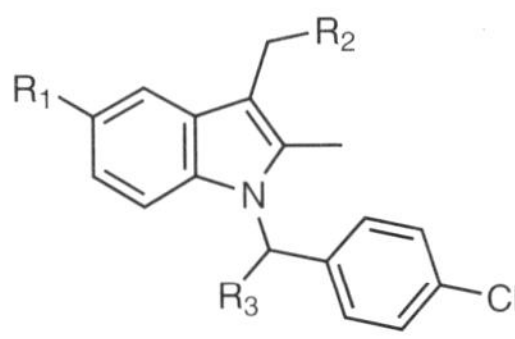

17

Figure 4.11 Successful cases from docking.

efficiently, in practical terms the scoring functions favor speed over theoretical accuracy to get qualitative scores. As a result, most scoring functions estimate the binding affinity according to hydrophobic, surface area, hydrogen bonds, and solvent accessible area. Scoring functions describing protein–ligand interactions play a key role in identifying the correct binding geometry of a protein–ligand complex. Researchers have invested a great deal of effort to achieve accurate predictions of binding affinities for ranking different complexes (Guida, 1994; Colman, 1994; Kuntz, 1992; Lybrand, 1995; Rosenfeld et al., 1995; Lengauer and Rarey, 1996).

In general, scoring functions are divided into four categories: (1) force-field-derived scoring functions considering nonbonded interactions (Jones et al., 1997; Weiner et al., 1984, 1986) or solvations (Zhang et al., 1997; Shoichet et al., 1999); (2) regression-based scoring functions (Horton and Lewis, 1992; Böhm, 1998; Eldridge et al., 1997; Head et al., 1996); (3) knowledge-based scoring functions (Gohlke et al., 2000; Muegge and Martin, 1999; Verkhivker et al., 1995); and (4) other scoring functions including chemical scores, contact scores, or shape scores (Shoichet et al., 1992; Nicklaus et al., 1995; Stouch and Jurs, 1986; Wang et al., 1998; Flower, 1998). FK506 binding protein (FKBP) ligand **14** (Burkhard et al., 1999), farnesyl transferase inhibitor **15** (Perola et al., 2000), HIV-1 RNA TAR inhibitor **16** (Filikov et al., 2000), and Aldose reductase inhibitors **17** (Iwata et al., 2001) are successful cases from docking (Fig. 4.11).

4.5 CONCLUSION

Pharmacophoric models and *de novo* ligand design provide potential leads that would possess novel chemical structures, and virtual screening of 3D pharmaceutical or chemical libraries would give bioactive compounds commercially available or prepared feasibly according to known methods. As shown in Fig. 11.12, ligand- and receptor-based virtual screening can be used in combination with each other and can

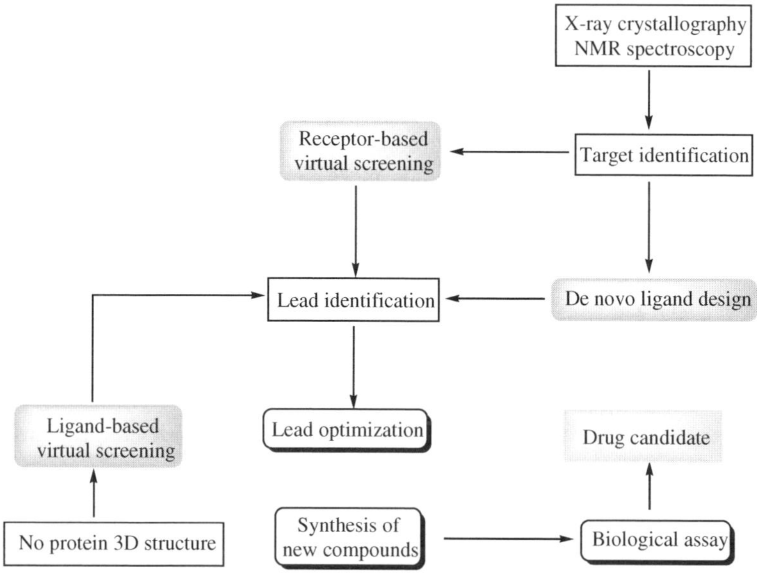

Figure 4.12 Schematic illustration of virtual screening.

be broadly employed in the discovery of bioactive compounds. Indisputably, one needs to come back to bench work for synthesis and biological assay, even though *in silico* drug design and virtual screening can accelerate the process of drug discovery. Currently, molecular modeling including ligand- and structure-based drug designs play significant roles in drug discovery. Especially, the advances in high-speed computing platforms have made multiprocessor workstations and PC clusters affordable to the majority of researchers. With the improvement of computing methods and scoring functions, virtual screening is becoming important and may be indispensable in the drug discovery.

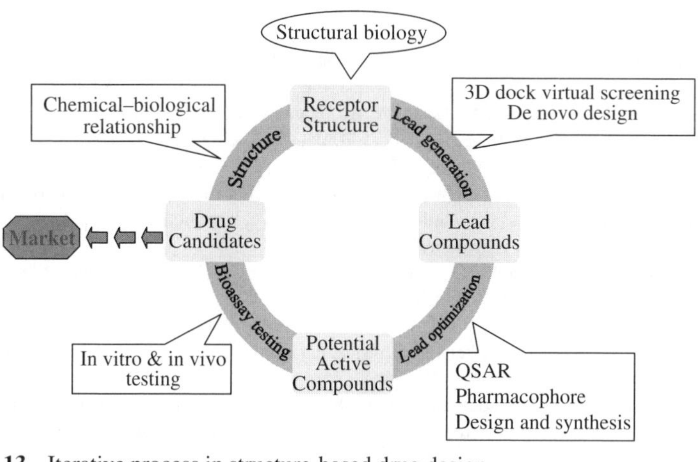

Figure 4.13 Iterative process in structure-based drug design.

As a whole, the available 3D structures of proteins or protein–ligand complexes enable lead generation by means of molecular modeling to increase efficiency. With the help of QSAR, pharmacophore, and synthesis, new secondary leads can be designed. In cooperation with experimental biological assays and with further rounds of an iterative process (Fig. 4.13), the concept of structure-based drug design would provide potent and specific compounds more productively in the drug discovery process.

REFERENCES

Bailey, D., and Brown, D. (2001). High-throughput chemistry and structure-based design: Survival of the smartest. *Drug Disc Today* **6**, 57–59.

Bausen, D. D., and Shands, E. F. B. (1996). Systematic search strategies in conformational analysis. *Drug Disc Today* **1**, 429–437.

Bernstein, F. C., Koetzle, T. F., Williams, J. B., Meyer, E. F., Brice, M. D., Rodgers, J. R., Kennard, O., Shimanouchi, T., and Tasumi, M. (1977). The Protein Data Bank: A computer-based archival file for macromolecular structures. *J Mol Biol* **112**, 535–542.

Bodian, D. L., Yamasaki, R. B., Buswell, R. L., Stearns, J. F., White, J. M., and Kuntz, I. D. (1993). Inhibition of the fusion-inducing conformational change of influenza hemagglutinin by benzoquinones and hydroquinones. *Biochemistry* **32**, 2967–2978.

Böhm, H. J. (1992a). The computer program LUDI: A new method for the de novo design of enzyme inhibitors. *J Comput Aided Mol Des* **6**, 61–78.

Böhm, H. J. (1992b). LUDI: Rule-based automatic design of new substituents for enzyme inhibitor leads. *J Comput Aided Mol Des* **6**, 593–606.

Böhm, H. J. (1996). Computational tools for structure-based ligand design. *Prog Biophys Mol Biol* **66**, 197–210.

Böhm, H. J. (1998). Prediction of binding constants of protein ligands—A fast method for the prioritization of hits obtained from *de novo* design or 3D database search programs. *J Comput Aided Mol Des* **12**, 309–323.

Böhm, H. J., Boehringer, M., Bur, D., Gmuender, H., Huber, W., Klaus, W., Kostrewa, D., Kuehne, H., Luebbers, T., Meunier-Keller, N., and Mueller, F. (2000). Novel inhibitors of DNA gyrase: 3D structure based biased needle screening, hit validation by biophysical methods, and 3D guided optimization. A promising alternative to random screening. *J Med Chem* **43**, 2664–2674.

D. B. Boyd and K. B. Lipkowitz (Eds.) (2001). *Reviews in Computational Chemistry*, Vol. 17. New York: Wiley-VCH.

Burkhard, P., Hommel, U., M. Sanner, M., and Walkinshaw, M. D. (1999). The discovery of steroids and other novel FKBP inhibitors using a molecular docking program. *J Mol Biol* **287**, 853–858.

Chen, G.. S., Chang, C.-S., Kan, W. M., Chang, C.-L., Wang, K. C., and Chern, J.-W. (2001). Novel lead generation through hypothetical pharmacophore 3D database searching: Discovery of isoflavonoids as nonsteroidal inhibitors of rat 5α-reductase. *J Med Chem* **44**, 3759–3763.

Chen, C.-S., Chiou, C.-T., Chen, G. S., Hwang, L.-H., Liaw, S.-H., Chen, D.-S., Chern, J.-W. Discovery of triphenylmethane derivatives as novel inhibitors of hepatitis C virus helicase through virtual screening and substructure searching. In preparation.

Cohen, N. C., Blaney, J. M., Humblet, C., Gund, P., and Barry, D. C. (1990). Molecular modeling software and methods for medicinal chemistry. *J Med Chem* **33**, 883–894.

Colman, P. M. (1994). Structure-based drug design. *Curr Opin Struct Biol* **4**, 868–874.

Craig, P. N. (1971). Interdependence between physical parameters and selection of substituent groups for correlation studies. *J Med Chem* **14**, 680–684.

Cramer, R. D. I., Patterson, D. E., and Bunce, J. D. (1988). Comparative molecular field analysis (CoMFA). 1. Effect of shape on binding of steroids to carrier proteins. *J Am Chem Soc* **110**, 5959–5967.

Eldridge, M. D., Murray, C. W., Auton, T. R., Paolini, G. V., and Mee, R. P. (1997). Empirical scoring functions: 1. The development of a fast empirical scoring function to estimate the binding affinity of ligands in receptor complexes. *J Comput Aided Mol Des* **11**, 425–455.

Filikov, A. V., Mohan, V., Vickers, T. A., Griffey, R. H., Cook, P. D., Abagyan, R. A., and James, T. L. (2000). Identification of ligands for RNA targets via structure-based virtual screening: HIV-1 TAR. *J Comput Aided Mol Des* **14**, 593–610.

Flower, D. R. (1998). SERF: A program for accessible surface area calculations. *J Mol Graph Model* **15**, 238–244.

Ghose, A. K., Crippen, G. M., Revankar, G. R., Mckernan, P. A., Smee, D. F., and Robins, R. K. (1999). Analysis of the *in vitro* antiviral activity of certain ribonucleosides against Parainfluenza virus using a novel computer aided receptor modeling procedure. *J Med Chem* **32**, 746–756.

Ghose, A. K., and Wendoloski, J. J. (1998). In perspective in drug discovery and design. *Kluwer/Escom*, **9/10/11**, p. 253.

Gohlke, H., Hendlich, M., and Klebe, G. (2000). Knowledge-based scoring function to predict protein-ligand interactions. *J Mol Biol* **295**, 337–356.

Gollop, M. A., Barrett, R. W., Dower, W. J., Fodor, S. P. A., and Gordon A. M. (1994). Applications of combinatorial technologies to drug discovery. 1. Background and peptide combinational libraries. *J Med Chem* **37**, 1233–1251.

Goodford, P. J. (1985). A computational procedure for determining energetically favorable binding sites on biologically important macromolecules. *J Med Chem* **28**, 849–857.

Gordon, E. M., Barrett, R. W., Dower, W. J., Fodor, S. P. A., and Gallop, M. (1994). Applications of combinatorial technologies to drug discovery. 2. Combinatorial organic synthesis, library screening strategies, and future directions. *J Med Chem* **37**, 1385–1401.

Greenidge, P. A., Carlsson, B., Bladh, L. G., and Gillner, M. (1998). Pharmacophores incorporating numerous excluded volumes defined by X-ray crystallographic structure in three-dimensional database searching: application to the thyroid hormone receptor. *J Med Chem* **41**, 2503–2512.

Gschwend, D. A., and Kuntz, I. D. (1996). Orientational sampling and rigid-body minimization in molecular docking revisited: On-the-fly optimization and degeneracy removal. *J Comput Aided Mol Des* **10**, 123–132.

Guida, W. C. (1994). Software for structure-based drug design. *Curr Opin Struct Biol* **4**, 777–781.

Han, Q., Dominguez, C., Stouten, P. F. W., Park, J. M., Duffy, D. E., Galemmo, R. A., Jr., Rossi, K. A., Alexander, R. S., Smallwood, A. M., Wong, P. C., Wright, M. M., Luettgen, J. M., Knabb, R. M., and Wexler, R. R. (2000). Design, synthesis, and biological evaluation of potent and selective amidino bicyclic factor Xa inhibitors. *J Med Chem* **43**, 4398–4415.

Hansch, C., and Fujita, T. (1964). ρ–σ–π analysis. A method for the correlation of biological activity and chemical structure. *J Am Chem Soc* **86**, 1616–1626.

Head, R. D., Smythe, M. L., Oprea, T. L., Waller C. L., Green, S. M., and Marshall, G. M. (1996). VALIDATE: A new method for the receptor-based prediction of binding affinities of novel ligands. *J Am Chem Soc* **118**, 3959–3969.

Hong, H., Neamati, N., Wang, S., Nicklaus, M. C., Mazumder, A., Zhao, H., Burke, T. R. J., Pommier, Y., and Milne, G. W. A. (1997). Discovery of HIV-1 integrase inhibitors by pharmacophore searching. *J Med Chem* **40**, 930–936.

Horton, N., and Lewis, M. (1992). Calculation of the free energy of association for protein complexes. *Protein Sci* **1**, 169–181.

Iwata, Y., Arisawa, M., Hamada, R., Kita, Y., Mizutani, M. Y., Tomioka, N., Itai, A., and Miyamoto, S. (2001). Discovery of novel aldose reductase inhibitors using a protein structure-based approach: 3D-database search followed by design and synthesis. *J Med Chem* **44**, 1718–1728.

Jones, G., Willett, P., Glen, C., and Leach, A. R. (1995). A genetic algorithm for flexible molecular overlay and pharmacophore elucidation. *J Comput Aided Mol Des* **9**, 532–549.

Jones, G., Willett, P., Glen, R. C., and Leach, A. R. (1997). Development and validation of a genetic algorithm for flexible docking. *J Mol Biol* **267**, 727–748.

Kaminski, J. J., Rane, D. F., Snow, M. E., Weber, L., Rothofsky, M. L., Anderson, S. D., and Lin, S. L. (1997). Identification of novel farnesyl protein transferase inhibitors using three-dimensional database searching methods. *J Med Chem* **40**, 4103–4112.

Kuntz, I. D., Blaney, J. M., Oatley, S. J., and Langridge, R. L. (1982). A geometric approach to macromolecule-ligand interactions. *J Mol Biol* **161**, 269–288.

Kuntz, I. D. (1992). Structure-based strategies for drug design and discovery. *Science* **257**, 1078–1082.

Kuntz, I. D., Meng, E. C., and Shoichet, B. K. (1994). Structure-based molecular design. *Acc Chem Res* **27**, 117–123.

Kurogi, Y., and Güner, O. (2001). Pharmacophore modeling and three-dimensional database searching for drug design using catalyst. *Curr Med Chem* **8**, 1035–1055.

Lengauer, T., and Rarey, M. (1996). Computational methods for biomolecular docking. *Curr Opin Struct Biol* **6**, 402–406.

Lightstone, F. C., Prieto, M., Singh, A. K., Piqueras, M. C., Whittal, R. M., Balhorn, R., and Roe, D. (2000). Identification of novel small molecule ligands that bind to tentanus toxin. *Chem Res Toxicol* **13**, 356–362.

Liu, H., Duan, Z., Luo, Q., and Shi, Y. (1999). Structure-based ligand design by dynamically assembling molecular building blocks at binding site. *Proteins* **36**, 462–470.

Lutz, M. W., Menius, J., Choi, T. D., Laskody, R. G., Domanico, P. L., Goetz, A. S., and Saussy, D. L. (1996). Experimental design for high throughput screening. *Drug Disc Today* **1**, 277–286.

Lybrand, T. P. (1995). Ligand–protein docking and rational drug design. *Curr Opin Struct Biol* **5**, 224–228.

Manallack, D. T. (1996) Getting that hit: 3D database searching in drug discovery. *Drug Disc Today* **1**, 231–238.

Meng, E. C., Shoichet, B. K., and Kuntz, I. D. (1992). Automated docking with grid-based energy evaluation. *J Comput Chem* **13**, 505–524.

Meng, E. C., Kuntz, I. D., Abraham, D. J., and Kellogg, G. E. (1994). Evaluating docked complexes with the hint exponential function and empirical atomic hydrophobicities. *J. Compt Aided Mol Des* **8**, 299–306.

Milne, G. W. A., Nicklaus, M. C., Driscoll, J. S., Wang, S., and Zaharevitz, D. W. (1994). The NCI drug information system 3D database. *J Chem Inf Comput Sci* **34**, 1219–1224.

Muegge, I. (1999). The effect of small changes in protein structure on predicted binding modes of known inhibitors of influenza virus neuraminidase: PMF-scoring in DOCK4. *Med Chem Res* **9**, 490–500.

Muegge, I., and Martin, Y. C. (1999). A general and fast scoring function for protein-ligand interactions: A simplified potential approach. *J Med Chem* **42**, 791–804.

Murcko, A., and Murcko, M. A. (1995). Computational methods to predict binding free energy in ligand–receptor complexes. *J Med Chem* **38**, 4953–4967.

Nicklaus, M. C., Wang, S., Driscoll, J. S., and Milne, G. W. A. (1995). Conformational changes of small molecules binding to proteins. *Bioorg Med Chem* **3**, 411–428.

Nicklaus, M. C., Neamati, N., Hong, H., Mazumder, A., Sunder, S., Chen, J., Milne, G. W., and Pommier, Y. (1997). HIV-1 integrase pharmacophore: Discovery of inhibitors through three-dimensional database searching. *J Med Chem* **40**, 920–929.

Nishibta, Y., and Itai, A. (1993). Confirmation of usefulness of a structure construction program based on three-dimensional receptor structure for rational lead generation. *J Med Chem* **36**, 2921–2928.

Oprea, T. I., and Marshall, G. R. (1998). Receptor-based prediction of binding affinities. *Perspect Drug Disc Des* **9/10/11**, 63–84.

Orita, M., Yamamoto, S., Katayama, N., Aoki, M., Takayama, K., Yamagiwa, Y., Seki, N., Suzuki, H., Kurihara, H., Sakashita, H., Takeuchi, M., Fujita, S., Yamada, T., and Tanaka, A. (2001). Coumarin and chromen-4-one analogues as tautomerase inhibitors of macrophage migration inhibitory factor: Discovery and X-ray crystallography. *J Med Chem* **44**, 540–547.

Pearlman, D. A., and Murcko, M. A. (1996). CONCERTS: Dynamic connection of fragments as an approach to *de novo* ligand design. *J Med Chem* **39**, 1651–1663.

Perola, E., Xu, K., Kollmeyer, T. M., Kaufmann, S. H., Prendergast, F. G., and Pang, Y.-P. (2000). Successful virtual screening of a chemical database for farnesyltransferase inhibitor leads. *J Med Chem* **43**, 401–408.

Rarey, M., Kramer, B., Lengauer, T., and Klebe, G. (1996a). A fast flexible docking method using an incremental construction algorithm. *J Mol Biol* **261**, 470–489.

Rarey, M., Wefing, S., and Lengauer T. (1996b). Placement of medium-sized molecular fragments into active sites of proteins. *J Comput Aided Mol Des* **10**, 41–54.

Rarey, M., Kramer, B., and Lengauer, T. (1997). Multiple automatic base selection: Protein–ligand docking based on incremental construction without manual intervention. *J Comput Aided Mol Des* **11**, 369–384.

Roe, D. C., and Kuntz, I. D. (1995). BUILDER v.2: Improving the chemistry of a *de novo* design strategy. *J Comput Aided Mol Des* **9**, 269–282.

Rosenfeld, R., Vajda, S., and Delisi, C. (1995). Flexible docking and design. *Annu Rev Biophys Biomol Struct* **24**, 677–700.

Rotstein, S. H., and Murcko, M. A. (1993). GroupBuild: A fragment-based method for de novo drug design. *J Med Chem* **36**, 1700–1710.

Rutenber, E., Fauman, E. B., Keenan, R. J., Fong, S., Furth, P. S., Ortiz de Montellano, P. R., Meng, E., Kuntz, I. D., DeCamp, D. L., Salto, R., Rose, J. R., Craik, C. S., and Stroud, R. M. (1993). Structure of a nonpeptide inhibitor complexed with HIV-1 protease—Developing a cycle of structure-based drug design. *J Biol Chem* **268**, 15343–15346.

Sadowski, J., and Gasteiger, J. (1993). From atoms and bonds to three-dimensional atomic coordinates: Automatic model builders. *Chem Rev* **93**, 2567–2581.

Schneider, G., Lee, M. L., Stahl, M., and Schneider, P. (2000). *De novo* design of molecular architectures by evolutionary assembly of drug-derived building blocks. *J Comput Aided Mol Des* **14**, 487–494.

Shoichet, B. K., Bodian, D. L., and Kuntz, I. D. (1992). Molecular docking using shape descriptors. *J Comput Chem* **13**, 380–397.

Shoichet, B. K., Stroud, R. M., Santi, D. V., Kuntz, I. D., and Perry, K. M. (1993). Structure-based discovery of inhibitors of thymidylate synthase. *Science* **259**, 1445–1450.

Shoichet, B. K., Leach, A. R., and Kuntz, I. D. (1999). Ligand solvation in molecular docking. *Proteins Struct Funct Genet* **34**, 4–16.

Somoza, J. R., Skillman, A. G., Munagala, N. R., Oshiro, C. M., Knegtel, R. M., A., Mpoke, S., Fletterick, R. J., Kuntz, I. D., and Wang, C. C. (1998). Rational design of novel antimicrobials: blocking purine salvage in a parasitic protozoan. *Biochemistry* **37**, 5344–5348.

Spencer, R. W. (1998). High-throughput screening of historic collections: Observations on file size, biological targets, and file diversity. *Biotech Bioeng* **61**, 61–67.

Stouch, T. R., and Jurs, P. C. (1986). A simple method for the representation, quantification, and comparison of the vaolumes and shapes of chemical compounds. *J Chem Inf Comput Sci* **26**, 4–12.

Sutter, J., Güner, O., Hoffmann, R., Li, H., and Waldman, M. (2000). Pharmacophore perception, development, and use in drug design. In International University Line, La Jolla, CA, USA.

Tame, J. R. H. (1999). Scoring functions: A view from the bench. *J Comput Aided Mol Des* **13**, 99–108.

Topliss, J. G. (1972). Utilization of operational schemes for analog synthesis in drug design. *J Med Chem* **15**, 1006–1011.

Verkhivker, G., Appelt, K., Freer, S. T., and Villafranca, J. E. (1995). Empirical free energy calculations of ligand–protein crystallographic complexes. 1. Knowledge-based ligand–protein interaction potentials applied to the prediction of human immunodeficiency virus 1 protease binding affinity. *Protein Eng* **8**, 677–691.

Wang, R. X., Liu, L., Lai, L. H., and Tang, Y. Q. (1998). SCORE: A new empirical method for estimating the binding affinity of a protein–ligand complex. *J Mol Model* **4**, 379–394.

Wang, R., Gao, Y., and Lai, L. (2000). LigBuilder: A multi-purpose program for structure-based drug design. *J Mol Model* **6**, 498–516.

Waszkowycz, B., Clark, D. E., Frenkel, D., Li, J., Murray, C. W., Robson, B., and Westhead, D.R. (1994). PRO_LIGAND: An approach to *de novo* molecular design. 2. Design of novel molecules from molecular field analysis (MFA) models and pharmacophores. *J Med Chem* **37**, 3994–4002.

Weiner, S. J., Kollman, P. A., Case, D. A., Singh, U. C., Ghio, C., Alagona, G., Profeta, S., Jr., and Weiner, P. (1984). A new force field for molecular mechanical simulation of nucleic acids and proteins. *J Am Chem Soc* **106**, 765–784.

Weiner, S. J., Kollman, P. A., Nguyen, D. T., and Case, D. A. (1986). An all atom force field for simulations of proteins and nucleic acids. *J Comput Chem* **7**, 230–252.

Zhang et al. (1997).

Zou, X. Q., Sun, Y. X., and Kuntz, I. D. (1999). Inclusion of salvation in ligand binding free energy calculations using the generalized-born model. *J Am Chem Soc* **121**, 8033–8043.

CHAPTER 5

FOCUSED LIBRARY DESIGN BASED ON HIT AND TARGET STRUCTURES: METHOD AND APPLICATION IN DRUG DISCOVERY

*Weiliang Zhu, Jian Li, Zhen Gong, Hong Liu, and Hualiang Jiang**

5.1 INTRODUCTION

The chemical space was estimated to have chemical structures in the order of magnitude of up to 10^{100} (Walters et al., 1998). This inconceivably large number gives hope that every therapeutic target would have an appropriate chemical as its ligand and that there might exist several different molecules exhibiting isofunctional activity (Schneider, 2002). Therefore, the task of drug discovery is to find such molecules that specifically bind to a certain therapeutic target relevant to a disease. There are many approaches developed for the task, one of which is target structure-based drug discovery (Lyne, 2002), which is now an important technique in cases where the three-dimensional (3D) structure of a therapeutic target is available. Numerous excellent reviews have been published on this approach (Lyne, 2002;

*Corresponding author: hljiang@mail.shcnc.ac.cn

Drug Discovery Research: New Frontiers in the Post-Genomic Era, Edited by Ziwei Huang

Jansen and Martin, 2004; Alvarez, 2004; Kitchen et al., 2004; Beavers and Chen, 2004; Erlanson, 2004). Computationally, there are at least three different target structure-based methodologies established for identifying or creating chemicals as potential drug leads, namely, docking-based virtual screening, *de novo* drug design, and combinatorial library design.

5.1.1 Docking-Based Virtual Screening

Virtual screening can be defined as a process of reducing a library containing a great number of compounds to a limited number of potentially promising compounds for a specific therapeutic target by means of computational methods. The screening consists of positioning in a target a series of small molecules contained in a chemical database ("docking") followed by scoring. Thus, molecular recognition is the fundamental basis for virtual screening (Alvarez, 2004). This action mechanism was firstly proposed by Emil Fischer in 1894 as the "lock-and-key" principle whereby steric and electrostatic complementarity between proteins and their ligands drive the complexation process (Fischer, 1894). Docking-based virtual screening algorithms attempt to generate and identify the most complementary match between a ligand from a ligand database and its macromolecular target (Brooijmans et al., 2003; Glen and Allen, 2003; Halperin et al., 2002). Scoring functions are used in the approach for ranking both the different ligands and the different orientations of the same ligand. There have been many publications describing the identification of novel ligands by docking-based virtual screening methods (Alvarez, 2004). The most remarkable characteristic of this approach is that the potential ligands discovered are usually structurally diverse.

When high-performance computing is utilized, the virtual screening of compound libraries is very time-efficient and cost-effective relative to real high-throughput-screening (HTS) (Lengauer et al., 2004). Therefore, it is a promising tool as a preliminary step toward discovering drug hits or leads. Furthermore, it also provides the interaction details between a target and its ligands for further lead optimization. Thus, docking-based virtual screening has become part of today's lead discovery arsenal for the identification of more diverse hits and has been established as a complement to real HTS. However, as the screening is performed in compound databases, the discovered potential leads are limited within the available compounds in the databases. In practical terms, there are true ligands missed due to algorithmic limitations in conformational sampling (Brooijmans et al., 2003) and imperfect scoring functions (Wang et al., 2003), no matter which docking programs are employed. On the other hand, the user-defined parameters and selected algorithms throughout the docking process might result in different results by different users for a same project.

5.1.2 *De novo* Drug Design

De novo drug design does not start from a database of complete molecules but aims at building a complete molecule from molecular bricks ("building blocks") to chemically fill the binding sites of a target molecule (Rotstein and Murcko, 1993). The complete chemical entries could be constructed through linking the "building blocks" together, or by growing from an "embryo" molecule with the guidance of evaluation of binding affinity. The "building blocks" could be either atoms or fragments (functional groups or small molecules). However, using atoms as "building blocks" is thought to be inefficient; therefore, this approach is seldom used nowadays. In the fragment-linking approach, the binding site is mapped to identify the possible anchor points for functional groups. These groups are then linked together and form a complete molecule. In the sequential-growing approach, the molecule grows in the binding site controlled by an appropriate search algorithm that evaluates each growing possibility with a scoring function.

Different from docking-based virtual screening, fragment-based *de novo* drug design can perform sampling in whole compound space, obtaining novel structures that are not limited in available databases. However, the quality of a growing step strongly depends on the previous steps. Any step chemically growing incorrectly would lead to an unacceptable result. For the fragment-linking approach, there is the problem of how to choose linkers to connect fragments together as complete structures. The most remarkable drawback of this approach might be the synthetic accessibility of the designed structures.

5.1.3 Combinatorial Library Design

Generally speaking, combinatorial chemistry is a collection of methods that allow the simultaneous chemical synthesis of large numbers of compounds using a variety of starting materials (Tropsha and Zheng, 2002). The primary benefit that combinatorial chemistry brings to drug discovery is speed and the large size of the designed libraries. For instance, one billion compounds can be synthesized from a molecular scaffold with three different substitution positions if each of the positions has 1000 different substituents. Therefore, combinatorial library provides a large number of structures for docking-based virtual screening. Actually, it is one of the most exciting developments in medicinal chemistry in the last decade (Gallop et al., 1994; Gordon et al., 1994; Terrett et al., 1995). Coupled with automation technologies and HTS, it offers great potential for discovering structurally novel drug leads. However, many products in the huge library are redundant. In addition, it was found that the large numbers of compounds synthesized did not result in a remarkable increase in drug candidates, although the number of compounds synthesized and screened has increased by several orders of magnitude (Oprea, 2002). It also does not make sense to validate and assay billions or millions of compounds.

5.2 FOCUSED LIBRARY BASED ON HIT AND TARGET STRUCTURES

5.2.1 Strategy

Because success in new drug development is directly tied to identifying drugable molecules, one of the biggest challenges is how to efficiently discover the molecules with novel structures, high affinities to therapeutic targets, ideal ADMET profiles, and good manufacturing features. Different approaches for receptor structure-based drug discovery have different characteristics as discussed above. An effective way of discovering druglike molecules can be undertaken by using the advantages of different approaches. Moreover, 3D structural information and properties of a studied therapeutic target must be focused for specificity in discovering drug leads. Here we introduce (1) a focused library design method based on hit and target structures and (2) its application (Chen et al., 2005).

The focused library is built with fragments and geared toward one particular molecular target, which is actually a representative sample of the full chemical space. Starting with the structures of hits and a therapeutic target, the overall skeleton of potential ligands is schematically split into several fractions, or "fragments," based on the interaction mechanism and the physicochemical properties of the binding site. An individual library, called a fragment library, is constructed for each fragment, taking into account the binding features of the fragments to the binding site. Finally, focused libraries on the studied target are constructed with the judgments from structural diversity, druglikeness, ADMET profiles, and binding affinities (Chen et al., 2005).

Some stochastic methods for exploring the whole chemical space are applied, of which methods a genetic algorithm (GA) is generally used because of its high efficiency in searching large combinatorial spaces (Gillet et al., 2002; Weber et al., 1998; Sheridan and Kearsley, 1995). There is no guarantee that the globally best solution will be found. However, good results are usually found much more quickly than a purely random search or a systematic search (Holland, 1975). A software package called LD1.0 was developed based on the approach as shown in Fig. 5.1 (Chen et al., 2005). The critical challenges are, first, to select sets of fragments that have the best potential to be parts of new drug leads for a given target and, second, to set up proper criteria for product judgment, including structural diversity, druglikeness, ADMET, and binding affinities.

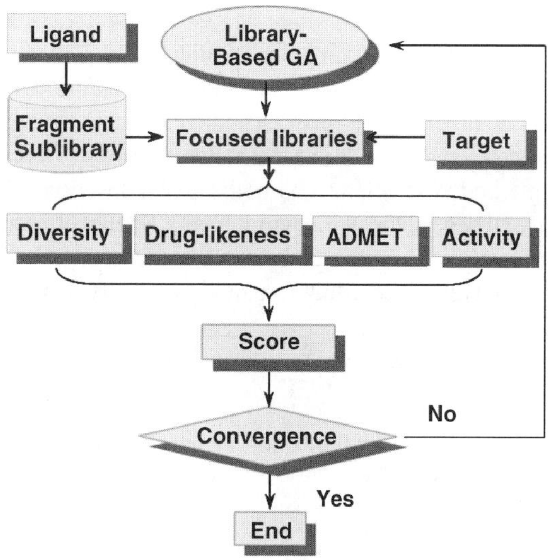

Figure 5.1 Flowchart of the software LD1.0 for focused library design (Chen et al., 2005).

5.2.2 Fragments or Building Blocks

The quality of fragment libraries has a critical effect on the final focused library. There are at least three ways to construct the fragment libraries (Chen et al., 2005).

1. ***From Known Drugs or Inhibitors***: Extracting fragments from known drugs or ligands (inhibitors or activators) of the studied target is an effective approach for collecting building blocks. A specific fragment library should be constructed based on available structural information of its ligands and the binding mechanism with the target.
2. ***From Existing Drug Fragment Libraries***: Some fragment libraries, such as the fragment library in the Ludi module of the InsightII software, are often used for *de novo* drug design (InsightII 2000.1, 2002). The LD1.0 software has a default fragment database for those targets with little information about their ligands.
3. ***From Inhibitors of Similar Targets***: Homology proteins usually share similar structural features and characteristics, especially at the binding site or active site. Therefore, the ligands for different targets belonging to the same family should share some common fragments in their inhibitors. Thus, the fragment libraries for a target could be constructed by referring to the structures of the ligands of its homology proteins.

5.2.3 Diversity of the Focused Library

Initially the focus in combinatorial library design was on selecting diverse sets of compounds on the assumption that maximizing diversity would result in a broad coverage of bioactivity space and hence would maximize the chances of finding drug leads (Jorgensen and Pedersen, 2001). Suggestions and assumptions on how to assess diversity have been studied during the last decade (Willett and Winterman, 1986; Alexander, 2000; Brown and Martin, 1996; Matter and Potter, 1999; Flower, 1998b; Ashton et al., 1996; Warr, 1997; Pearlman and Smith, 1998; Gillet et al., 1999). It was reported that two-dimensional (2D) structural descriptions are fast and accurate for calculating molecular diversity (Flower, 1998b; Brown, 1997). The descriptors employed in LD1.0 are a set of the 39 best structural descriptors proposed by Flower (1998a). Distance method is selected to normalize each descriptor. Then, a Euclidean space is described by all descriptors with their weights. So the difference (or similarity), d_{ij}, of two molecules can be represented by the distance between the two molecular points in the space, calculated according to equation (5.1) (Chen et al., 2005),

$$d_{ij} = \sqrt{\sum_k (\hat{x}_k^i - \hat{x}_k^j)^2} \tag{5.1}$$

where $\hat{x}_k^j$ is the normalized value of the kth descriptor for the jth molecule, which can be calculated using equation (5.2),

$$\hat{x}_k^j = w_k \left(\frac{x_k^j - \bar{x}_k}{\sigma_k} \right) \tag{5.2}$$

TABLE 5.1 Molecular Libraries for Testing the Modified Descriptor Set in Evaluating Molecular Diversity

Molecular Library	No. 1	No. 2	No. 3	No. 4
Molecule number from library No. 1	23	0	12	18
Molecule number from library No. 2	0	23	11	5
$D_{k,Nor}$	0.0000	1.0000	0.7726	0.3882

where x_k^j is the value of the kth descriptor for the jth molecule, $\bar{x}_k$ is the mean value of the descriptor k, σ_k is variance difference of the descriptor k, and w_k is weight of the descriptor k. Then, a library's diversity, D_k, can be calculated using equation (5.3),

$$D_k = \frac{1}{n(n-1)} \sum_{i=1}^{n} \sum_{j=1}^{j<i} d_{ij} \tag{5.3}$$

where n is the number of molecules in the library. So D_k is a sum of the distances between any two molecules in a library. The final scores of all libraries in a certain GA generation are normalized using equation (5.4),

$$D_{k,Nor} = \frac{D_k - D_{\min}}{D_{\max} - D_{\min}} \tag{5.4}$$

where $D_{k,Nor}$ is the final score of the kth focused library. $D_{\max}$ and $D_{\min}$ are the maximum and minimum D_k values of a specific generation, respectively. Thus, the highest score of library diversity is 1 and the lowest is 0.

The reasonability of the descriptor set for calculating molecular diversity was verified by two libraries, No. 1 and No. 2, which were constructed from two sets of molecules randomly selected from the MACCS-II Drug Data Report (MDDR) with molecular weight less than 350. The library No. 1 has 23 molecules that are composed of only three elements, C, H, and O. The library No. 2, also randomly selected from MDDR, has 23 molecules as well, but without any restraint in their structures. Therefore, library No. 1 should possess lower diversity than library No. 2 (Chen et al., 2005). Thus, $D_{k,Nor}$ for libraries No. 1 and No. 2 should equal to 0.0 and 1.0, respectively. Meanwhile, two more libraries, No. 3 and No. 4, were constructed by mixing some molecules from library No. 1 with some molecules from library No. 2 (Table 5.1). Indeed, the calculation result using the molecular diversity descriptor set shows that the library diversity correlates with the number of the molecules from library No. 2, demonstrating that the structural diversity descriptor set is reasonably good (Table 5.1).

5.2.4 Druglikeness (Zheng et al., 2005)

Druglikeness is another key factor that should be considered during library design. It has been estimated that ~40% of compounds fail to be developed into drugs due to their poor pharmacokinetic properties (Stanton et al., 2000). Therefore, initial strategies toward this goal should be involved in use of computational filters to

TABLE 5.2 Definitions of the Newly Developed Descriptors (Zheng et al., 2005)

Descriptor	Definition
UNSATP	UNSAT/BDUH
NO_C3	(N + O)/C3
C3	Number of sp^3 hybridized C atoms
N	Number nitrogen atoms
O	Number of oxygen atoms
BDUH	Number of bonds that don't contain H and X atoms
UNSAT	RNG + BD2 + 2*BD3 + (BDAR + 1)/2

remove compounds deemed to be chemically unsuitable for drug development. This approach has already been applied by several research groups (Gallop et al., 1994; Gillet et al., 1999; Lewell et al., 1998). In 1997, a set of assumptions about necessary features for a "good" drug candidate was suggested and embodied in a so-called "Rule of Five" by Lipinski et al. (1997). Then, other scientists developed new methods to improve prediction performance of the "Rule of Five" (Galvez et al., 2001; Ajay et al.,1998; Sadowski and Kubinyi, 1998; Ghose and Crippen, 1987; Xu and Stevenson, 2000). However, most of the methods have limitations in virtual screening, such as being time-consuming and performing poorly. Therefore, how to discriminate a druglike compound from a non-druglike one is still a great challenge in library design. Considering the advantages and disadvantages of the available computational methods of predicting druglikeness (SSKEYS; Viswanadhan et al., 1989; Gillet et al., 1998; The SPRESI database; Wagener et al., 2000; Muegge et al., 2001), a new chemistry space filter was developed and embedded in the software package LD1.0 (Chen et al., 2005; Zheng et al., 2005).

Two new descriptors, UNSATP and NO_C3, which are ratios of the molecular physicochemical properties, are defined in Table 5.2. The new descriptors are molecular size or molecular weight (MW) and database independent. Table 5.3 shows the initial values and distributions for the new descriptors.

To further improve the descriptors' performance and robustness, the UNSATP is combined with NO_C3 to create a new chemistry space filter (Table 5.3).

TABLE 5.3 Initial Values and Distributions for the New Descriptors and the New Chemistry Space Filter in Different Libraries (Chen et al., 2005)

	Initial Value	ACD[a]	MDDR	CMC[b]
UNSATP	$0.15 < \text{UNSATP} \leq 0.40$	31.76%	57.17%	62 .73%
NO_C3	$0.2 < \text{NO\_C3} \leq 1.2$	39.35%	68.09%	63.38%
Chemistry Space filter	$0 \leq \text{UNSATP} \leq 0.43$ $0.10 \leq \text{NO\_C3} \leq 1.8$	39.09%	69.05%	70.39%

[a]The Available Chemicals Directory (ACD).
[b]CMC, comprehensive medicinal chemistry.

Considering that ACD is not a pure non-druglike database, the final value for the new descriptors was reset and optimized to get maximum capability of the chemistry space filter in differentiating druglike and non-druglike compound databases (Table 5.3). The result indicates that this chemistry space filter can distinguish the MDDR and CMC more efficiently than individual NO_C3 or UNSATP. Therefore, this filter is suitable as a chemistry space filter to discriminate druglikeness and non-druglikeness (Zheng et al., 2005).

As the new filter is atomic type-based, it shows a favorable computational speed. The evaluation of drug-likeness can be completed within 2 hours for the MDDR database (~80,000 compounds) and within 4 hours for the ACD database (~300,000 compounds) on a computer with a CPU of 733MHz and RAM of 256M.

5.2.5 Activity Assessment

Molecular activity could be defined by its binding affinity to target protein. Nowadays, molecular docking is the most commonly used method to evaluate binding strength of a ligand to its target(s). One of the most popular molecular docking programs is DOCK4.0 (Kuntz et al., 1982; Oshiro and Kuntz, 1995; Ewing and Kuntz, 1997). The best energy score from DOCK 4.0 was used in LD1.0 for assessing molecular bioactivity. The score could be very different in different systems, suggesting that the ordinary methods for normalization cannot be used in the case of the activity score. To solve this problem, the sigmoid function, equation (5.5), was introduced into LD1.0 for dealing with the normalization,

$$y = \frac{1 - e^{ax}}{1 + e^{ax}} \tag{5.5}$$

where a is a constant, x is the binding energy from the output file of DOCK 4.0, and y is the normalized activity score; for the binding energy in the output file of DOCK4.0 is always no greater than 0.00 kcal/mol, and y is always between 0 and 1. Therefore, sigmoid function is a good choice for the normalization of the bioactivity score. Figure 5.2 demonstrates the plot of sigmoid function against binding energy when a is

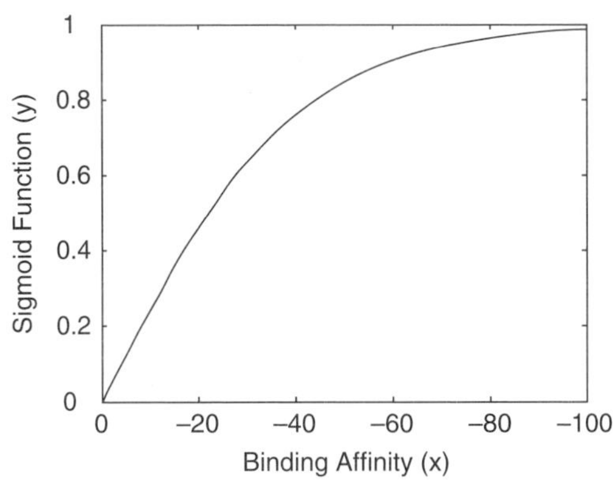

Figure 5.2 The plot of sigmoid function (y) against binding score from DOCK 4.0.

equal to 0.05. If the value of x is between 0.00 and -80.00 kcal/mol, the normalized activity score, y, is between 0.0000 and 0.9640 (Fig. 5.2).

5.2.6 Test Cases for LD1.0

To verify the reliability of the software package LD1.0, two tests were carried out on building COX-2-focused and PPARγ-focused libraries (Chen et al., 2005).

5.2.6.1 Focused Library Design for COX-2 Inhibitors

Cyclooxygenase (COX) is a key enzyme associated with arachidonic acid (AA) metabolism. COX has two isoforms, namely, COX-1 and COX-2 (Marnett, 2000; Garavito and Dewitt, 1999; O'Banion, 1999). It is believed that the inhibition of COX-1 causes the side effects seen with nonsteroidal anti-inflammatory drugs (NSAIDs). Therefore, the selective inhibitors of COX-2 would constitute a novel approach to the treatment of inflammation with fewer side effects (Masferrer et al., 1994). The LD1.0 software was used to build a COX-2-focused library to see whether the optimized focused library contains the marketed and published COX-2 selective drugs or inhibitors.

Fragment Libraries SC58635, an FDA-approved drug for anti-inflammation, is one of the NSAIDs (Graul et al., 1997), which could be divided into three parts as shown in Fig. 5.3 based on its interaction characteristic with the binding site of COX-2. The head of SC58635 is part A, *p*-aminosulfo-phenyl; the tail is part C, *p*-methyl-phenyl; and the body is part B, 3-trifluoro-1*H*-pyrazol-1-yl, which acts as a linkage between parts A and C. Three fragment libraries were constructed for parts A, B, and C with 12, 16, and 4 fragments, respectively.

Focused Library These fragments from different parts could react with each other to produce 768 compounds. Starting from these fragments and based on the

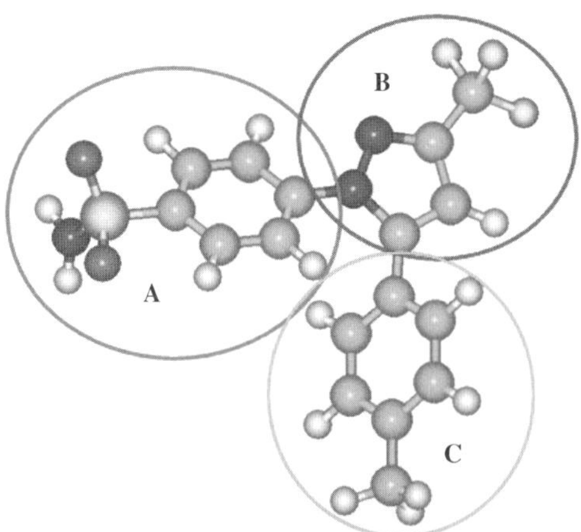

Figure 5.3 Structure of SC58635 and fragment division for COX-2 inhibitors. See color plates.

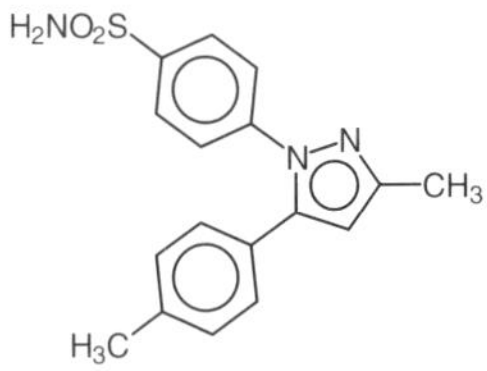

M1 (SC58635)
pIC50 = 7.39

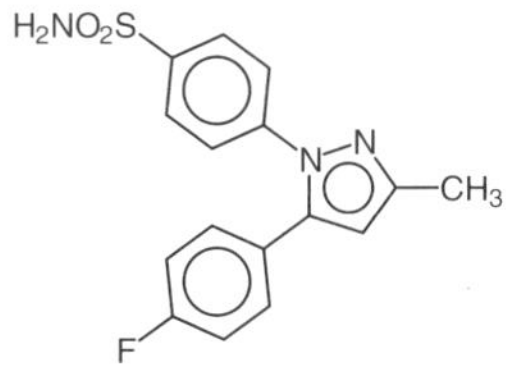

M2
pIC50 = 7.39

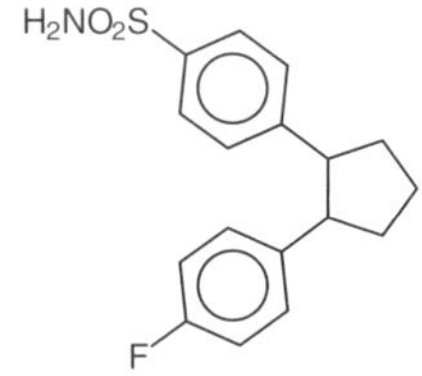

M3
pIC50 = 8.2

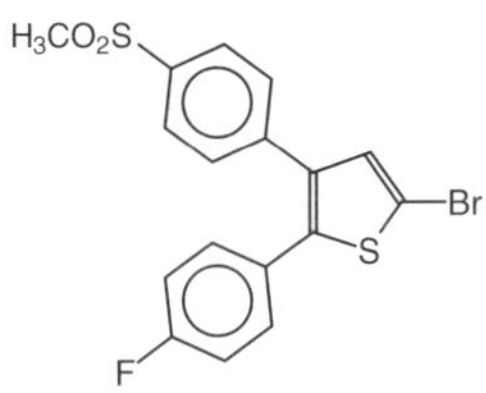

M4 (DuP697)
pIC50 = 8.0

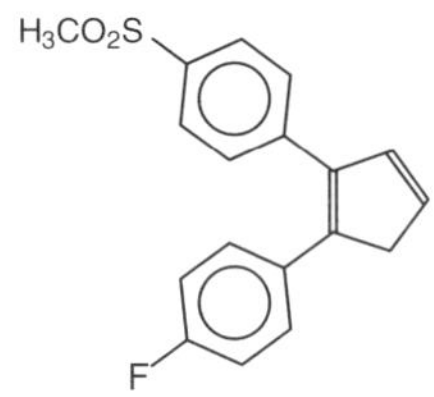

M5
pIC50 = 7.6

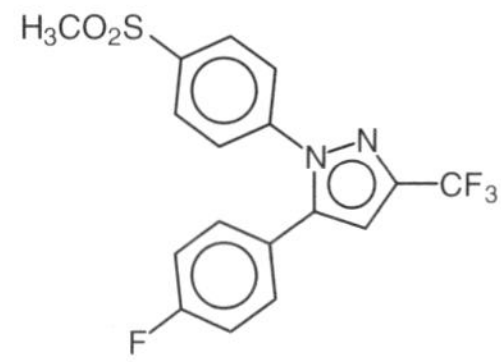

M6(SC58125)
pIC50 = 7.3

Figure 5.4 Successfully reproduced inhibitor structures in COX-2-focused library by LD1.0.

X-ray crystallographic structure of COX-2 (PDB entry 6COX), a COX-2-focused library with a population of $3 \times 3 \times 3$ was built. The program terminated normally after running for 28 generations. It took one CPU on an SGI R12000 Origin3800 computer 5.42 hours. Based on the optimized fragments, a focused library containing 27 compounds has been constructed. A preliminary literature survey showed that six of them are the reported COX-2-selective inhibitors (Fig. 5.4) (Marot et al., 2000; Huff et al., 1995). For comparison, a pure molecular docking was performed using DOCK4.0 for the 768 structures, costing more than 40.4 hours. The docking result shows that none of the six experimentally tested COX-2 inhibitors is found via the pure docking. Accordingly, LD1.0 should be a fast and reliable tool for building up a focused library against a specific target structure.

5.2.6.2 *Focused Library Design for PPARγ Agonist*

The peroxisome proliferator-activated receptors (PPARs) form a subfamily of the nuclear receptor superfamily. Three isoforms, encoded by separate genes, have been identified: PPARγ, PPARα, and PPARδ (also named PPARβ). PPARγ is an important target for the treatment of type II diabetes, and its agonists are therefore expected to be novel drugs for the disease (Michalik and Wahli, 1999; Kliewer et al., 1992). Indeed, thiazolidinediones (TZDs)—for example, Rosiglitazone and Pioglitazone (glitazones), approved by FDA as drugs for type II diabetes—are high-affinity PPARγ agonists. To further verify LD1.0, a test on designing a focused library of PPARγ agonists was performed based on the X-ray crystallographic structure of PPARγ (PDB entry 2PRG) by using LD1.0 (Chen et al., 2005).

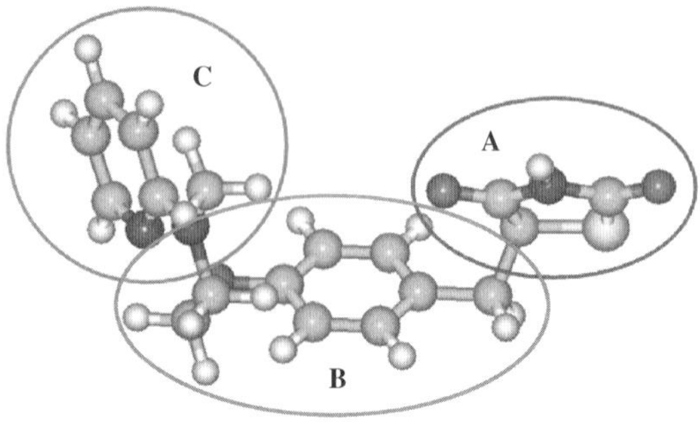

Figure 5.5 Common structural characteristic of PPARγ agonists and possibly structural fragment division. See color plates.

Fragment Library It was noticed that most of the PPARγ agonists can be divided into three parts as shown in Fig. 5.5. Part A is a hydrophilic head, part C is a hydrophobic tail, and part B is a linker between parts A and C. After analyzing a large number of PPAR agonists and referring to other known drug fragment libraries, three fragment libraries were constructed for parts A, B, and C with 118, 88, and 98 fragments, respectively. These fragments could react with each other to give about 10^6 structures. Taking into account the feasibility of fast bioassay, a PPARγ-focused library was designed with a population of $10 \times 10 \times 10$ by 3×10 building blocks.

Focused Library The calculation conditions and parameters were set the same as those for COX-2-library construction. After running 434 generations, the program LD1.0 terminated normally, and the optimized fragments for parts A, B, and C were reported. Indeed, the parts A and B of TZDs were found among the optimized fragments for parts A and B. However, the tail of TZDs was not found in the optimized building blocks for part C. This is in agreement with the high structural flexibility of the entrance of the PPARγ binding site. On the other hand, it also hints that there might be some compounds in the library that are more active than the usual TZDs. Indeed, a bioassay experiment on some of the compounds showed that some compounds in the designed focused library are more powerful as agonists than those FDA-approved TZD drugs (unpublished data). This testing example demonstrates again that, for a focused library based on hit and target structures, the LD1.0 software as well as the newly modified descriptor sets for druglikeness, structural diversity, and bioactivity can optimize the building fragments, reduce the library size, and increase the hit rate.

5.3 APPLICATION: DISCOVERING NOVEL CHEMICAL INHIBITORS OF CYCLOPHILIN A

Cyclophilin A (hCypA) is one of the most important members of the 15 known human cyclophilins, which play a critical role in a variety of biological processes, including enhancing the rate of folding/unfolding of proteins via its peptidyl-prolyl isomerase (PPIase) activity (Li et al., 2006a,b; Handschumacher et al., 1984; Galat, 2003;

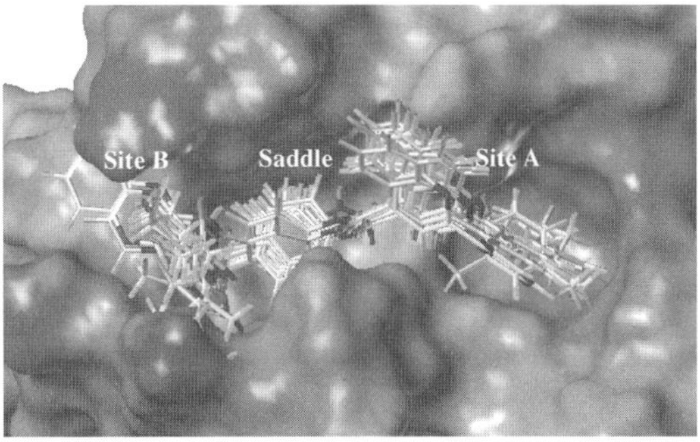

Figure 5.6 Three-dimensional structure of hCypA with designed inhibitors displayed. The CypA surface was colored by electrostatic potential. See color plates.

Dornan et al., 2003), binding to the HIV-1 Gag polyprotein for facilitating viral replication (Luban et al., 1993). Therefore, CypA inhibitors are of therapeutic significance. However, inhibitors of CypA are mainly derived from the nature sources (Handschumacher et al., 1984; Siekierka et al., 1989; Clane et al., 1989; Sedrani et al., 2003) and peptide analogues (Li et al., 2000), which are all large molecules. In order to discover potent small chemical ligands for CypA, a CypA-focused library was constructed by using LD1.0 (Chen et al., 2005).

5.3.1 Focused Library Design

Thirty-five binders of CypA with binding affinities under submicromolar or micro molar level were discovered by docking-based virtual screening targeting the 3D structure of CypA (PDB entry 1NMK), and four of them showed high CypA PPIase inhibition activities with IC_{50}s of 2.5-6.2 μM (Li et al., 2006a,b). According to the binding models of these binders, the skeleton of potential CypA inhibitors is divided into three parts: Part A interacting with the small pocket of CypA (site A), part B is located in the large pocket (site B), and part L interacting with the "saddle" is a linker between A and B (Fig. 5.6). Five fragments for part A, 17 fragments for part B, and three linkers for part L are selected. These fragments from different parts could react with each other to generate 255 molecules. Starting from these fragments and the X-ray structure of the hCypA, the LD1.0 software with default parameters was employed to build an optimized CypA-focused library with a population of $4 \times 3 \times 12$ (Li et al., 2006).

5.3.2 Chemical Synthesis and Biological Assay

With the guidance of the designed hCypA-focused library, 16 of the 144 designed compounds in the library were synthesized. Surface plasmon resonance (SPR) measurements were used for determining the binding affinity of the 16 molecules to CypA, which reveal that the compounds can bind to CypA *in vitro* at the micromolar range ($K_D = 0.076$–41.0 μM, Table 5.4). Thus, the hit rate of the focused library approach is ~100% in regard to SPR binding. The PPIase activity assessment

TABLE 5.4 Chemical Structures, SPR Binding Affinities, and PPIase Inhibitory Activities of the 5 Compounds Selected from the Focused Library (Li et al., 2006a,b)

Compound	Structure	K_D^a (μM)	IC_{50} (μM)[b]
4a	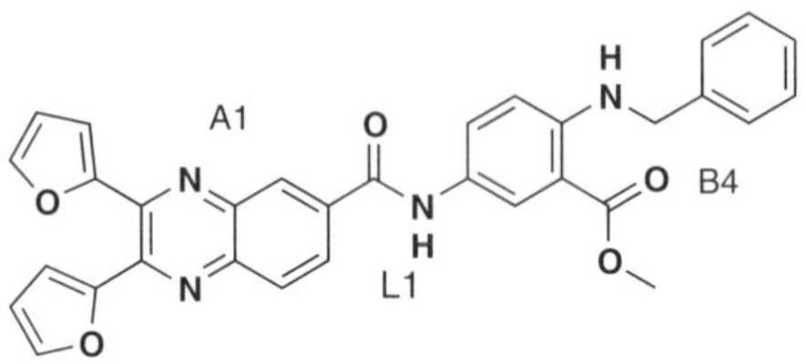	8.19	0.25
4c	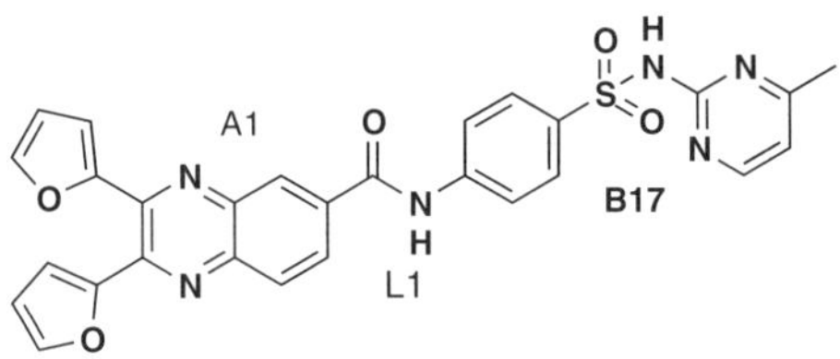	41.0	6.43
4e	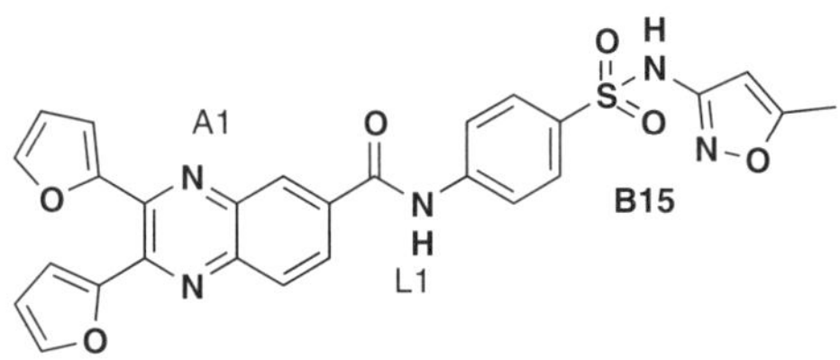	38.7	2.86
4f	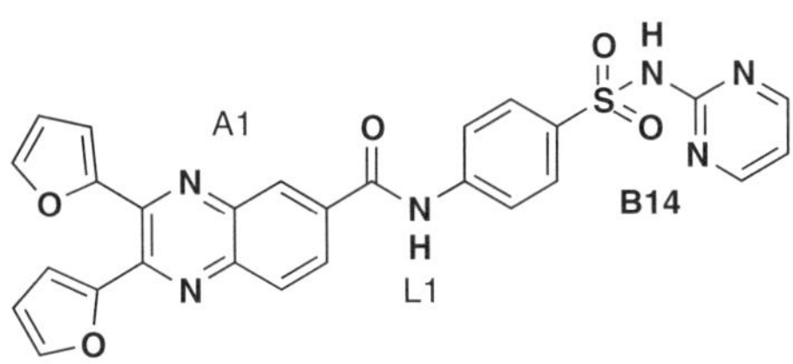	22.8	0.96
4g	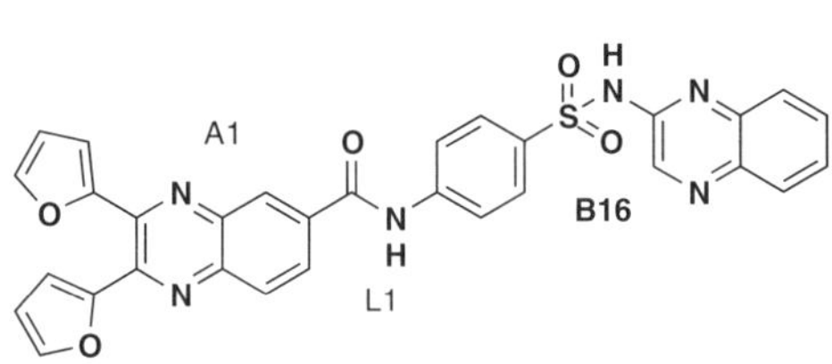	35.3	0.78

[a]Data obtained from SPR-based binding affinity assay. K_D represents the dissociation constant.
[b]Data are means of three independent experiments.

shows that five compounds (i.e., 4a, 4c, and 4e–g in Table 5.4) have the IC_{50} values of 0.25, 6.43, 2.86, 0.96, and 0.78 μM, respectively, demonstrating that the hit rate with the enzymatic inhibitory activity at the micromolar level is as high as ~31.25%. Remarkably, both the binding affinities and inhibitory activities of the compounds from focused library design increase on an average ~2 times that of the compounds discovered previously by using virtual screening. This demonstrates the effectiveness of the focused library design approach LD1.0 (Li et al., 2006a,b).

5.4 SUMMARY

It could be concluded that the focused library based on hit and target structures is an approach with high efficiency in discovering potent novel drug leads with high specificity against a therapeutic target. The software package LD1.0 with calculation methods integrated for druglikeness, structural diversity, bioactivity, and favorable ADMET properties seems to be powerful in practical terms for new drug development.

ACKNOWLEDGMENTS

This work was supported by grants from the State Key Program of Basic Research of China (2002CB512802, 2004CB518901) and the Key Program of Science and Technology (2005BA711A04).

REFERENCES

Ajay, A., Walters, W. P., and Murcko, M. A. (1998). Can we learn to distinguish between "drug-like" and "nondrug-like" molecules? *J Med Chem* **41**, 3314–3324.

Alexander, G. (2000). Molecular dataset diversity indices and their applications to comparison of chemical databases and QSAR analysis. *J Chem Inf Comput Sci* **40**, 414–425.

Alvarez, J. C. (2004). High-throughput docking as a source of novel drug leads. *Curr Opin Chem Biol* **8**, 365–370.

Ashton, M. J., Jaye, M. C., and Mason, J. S. (1996). New perspectives in lead generation II: Evaluating molecular diversity. *Drug Discov Today* **1**, 71–78.

Beavers, M. P., and Chen, X. (2004). Structure-based combinatorial library design: Methodologies and applications. *J Mol Graph Model* **20**, 463–468.

Brooijmans, N., and Kuntz, I. D. (2003). Molecular recognition and docking algorithms. *Annu Rev Biophys Biomol Struct* **32**, 335–373.

Brown, R. D. (1997). Descriptors for diversity analysis. *Perspect Drug Discov Des* **7–8**, 31–49.

Brown, R. D., and Martin, Y. C. (1996). Use of structure–activity data to compare structure-based clustering methods and descriptors for use in compound selection. *J Chem Inf Comput Sci* **36**, 572–584.

Chen, G., Zheng, S., Luo, X., Shen, J., Zhu, W., Liu, H., Gui, C., Zhang, J., Zheng, M., Puah, C. M., Chen, K., and Jiang, H. (2005). Focused combinatorial library design based on structural diversity, druglikeness and binding affinity score. *J Comb Chem* **7**, 398–406.

Clane, R. Y., Lim, S., Samaan, A., Collier, D. S. T., Pollard, S. G., White, D. J. G., and Thiru, S. (1989). Rapamycin for immunosuppression in organ allografting. *Lancet North Am Ed* **334**, 227–227.

Dornan, J., Taylor, P., and Walkinshaw, M. D. (2003). Structures of immunophilins and their ligand complexes. *Curr Top Med Chem* **3**, 1392–1409.

Erlanson, D. A., McDowell, R. S., and O'Brien, T. (2004). Fragment-based drug discovery. *J Med Chem* **47**, 3463–3482.

Ewing, T. J. A., and Kuntz, I. D. (1997). Critical evaluation of search algorithms for automated molecular docking and database screening. *J Comput Chem* **18**, 1175–1189.

Fischer, E. (1894). Einfluss der Configuration auf die wirkung der Enzyme. *Ber Dt Chem Ges* **27**, 2985–2993.

Flower, D. R. (1998a). DISSIM: A program for the analysis of chemical diversity. *J Mol Graph Model* **15**, 239–253.

Flower, D. R. (1998b). On the properties of bit string-based measures of chemical similarity. *J Chem Inf Comput Sci* **38**, 379–386.

Galat, A. (2003). Peptidylprolyl cis/trans isomerases (immunophilins): Biological diversity–targets–functions. *Curr Top Med Chem* **3**, 1315–1347.

Gallop, M. A., Barrett, R. W., Dower, W. J., Fodor, S. P. A., and Gordon, E. M. (1994). Applications of combinatorial technologies to drug discovery. 1. Background and peptide combinatorial libraries. *J Med Chem* **37**, 1233–1251.

Galvez, J., Julian-Ortiz, J. V., and Garcia-Domenech, R. (2001). General topological patterns of known drugs. *J Mol Graph Model* **20**, 84–94.

Garavito, R. M., and Dewitt, D. L. (1999). The cyclooxygenase isoforms: Structural insights into the conversion of arachidonic acid to prostaglandins. *Biochim Biophys Acta* **1441**, 278–287.

Ghose, A., & Crippen, G. (1987). Atomic physicochemical parameters for three-dimensional-structure-directed quantitative structure–activity relationships. 2. Modeling dispersive and hydrophobic interactions. *J Chem Inf Comput Sci* **27**, 21–35.

Gillet, V. J., Willett, P., and Bradshaw, J. (1998). Identification of biological activity profiles using substructural analysis and genetic algorithms. *J Chem Inf Comput Sci* **38**, 165–179.

Gillet, V. J., Willett, P., Bradshaw, J., Green, and D. V. S. (1999). Selecting combinatorial libraries to optimize diversity and physical properties. *J Chem Inf Comput Sci* **39**, 169–177.

Gillet, V. J., Khatib, W., Willett, P., Fleming, P. J., and Green, D. S. (2002). Combinatorial library design using a multiobjective genetic algorithm. *J Chem Inf Comput Sci* **42**, 375–385.

Glen, R. C., and Allen, S. C. (2003). Ligand–protein docking: Cancer research at the interface between biology and chemistry. *Curr Med Chem* **10**, 763–777.

Gordon, E. M., Barrett, R. W., Dower, W. J., Fodor, S. P. A., and Gallop, M. A. (1994). Applications of combinatorial technologies to drug discovery. 2. Combinatorial organic synthesis, library screening strategies, and future directions. *J Med Chem* **37**, 1385–1401.

Graul, A., Martel, A. M., and Castaner J. (1997). Celecoxi: Anti-inflammatory, cyclooxygenase-2 inhibitor. *Drugs Future* **22**, 711–714.

Halperin, I., Ma, B., Wolfson, H., and Nussinov, R. (2002). Principles of docking: An overview of search algorithms and a guide to scoring functions. *Proteins* **47**, 409–443.

Handschumacher, R. E., Harding, M. W., Rice, J., Drugge, R. J., and Speicher, D. W. (1984). Cyclophilin: A specific cytosolic binding protein for cyclosporin A. *Science* **226**, 544–547.

Holland, J. H. (1975). *Adaptation in Natural and Artificial Systems.* Ann Arbor: University of Michigan Press.

Huff, R., Collins, P., Kramer, S., Seibert, K., Koboldt, C., Gregory, S., and Isakson, P. (1995). A structural feature of *N*-[2-(cyclohexyloxy)-4-nitrophenyl] methanesulfonamide (NS-398) that governs its selectivity and affinity for cyclooxygenase 2 (COX2). *Inflamm Res* **44**(Suppl. 2), S145–S146.

InsightII 2000.1 (2002) *Accelrys Inc.*; San Diego, CA, USA.

Jansen, J. M., and Martin, E. J. (2004). Target-biased scoring approaches and expert systems in structure-based virtual screening. *Curr Opin Chem Biol* **8**, 359–364.

Jorgensen, A. M., and Pedersen, J. T. (2001). Structural diversity of small molecule libraries. *J Chem Inf Comput Sci* **41**, 338–345.

Kitchen, D. B., Decornez, H., Furr, J. R., and Bajorath, J. (2004). Docking and scoring in virtual screening for drug discovering: Methods and application. *Nat Rev Drug Discov* **3**, 935–949.

Kliewer, S. A., Umesono, K., Noonan, D. J., Heyman, R. A., and Evans, R. M. (1992). Convergence of 9-cis retinoic acid and peroxisome proliferator signalling pathways through heterodimer formation of their receptors. *Nature* **358**, 771–774.

Kuntz, I. D., Blaney, J. M., Oatley, S. J., Langridge, R., and Ferrin, T. E. (1982). A geometric approach to macromolecular-ligand interactions. *J Mol Biol* **161**, 269–288.

Lengauer, T., Lemmen, C., Rarey, M., and Zimmermann, M. (2004). Novel technologies for virtual screening. *Drug Discov Today* **9**, 27–34.

Lewell, X. Q., Judd, D. B., Watson, S. P., and Hann, M. M. (1998). RECAP—retrosynthetic combinatorial analysis procedure: A powerful new technique for identifying privileged molecular fragments with useful applications in combinatorial chemistry. *J Chem Inf Comput Sci* **38**, 511–522.

Li, J., Chen, J., Gui, C., Zhang, L., Qin, Y., Xu, Q., Zhang J., Liu, H., Shen, X., and Jiang, H. (2006a). Discovering novel chemical inhibitors of human cyclophilin A: Virtual screening, synthesis and bioassay. *Bioorg Med. Chem* **14**, 2209–2224.

Li, J., Zhang, J., Chen, J., Luo, X., Zhu, W., Shen, J., Liu, H., Shen, X., and Jiang, H. (2006b). Strategy for discovering chemical inhibitors of human cyclophilin A: Focused library design, virtual screening, chemical synthesis and bioassay. *J Combin Chem* **8**, 326–337.

Li, Q., Moutiez, M., Charbonnier, J. B., Vaudry, K., Menez, A., Quemeneur, E., and Dugave, C. (2000). Design of a Gag pentapeptide analogue that binds human cyclophilin A more efficiently than the entire capsid protein: New insights for the development of novel anti-HIV-1 drugs. *J Med Chem* **43**, 1770–1779.

Lipinski, C. A., Lombardo, F., Dominy, B. W., and Feeney, P. J. (1997). Experimental and computational approaches to estimate solubility and permeability in drug discovery and development settings. *Adv Drug Deliv Rev* **23**, 3–25.

Luban, J., Bossolt, K. L., Franke, E. K., Kalpana, G. V., and Goff, S. P. (1993). Human immunodeficiency virus type 1 gag protein binds to cyclophilins A and B. *Cell* **73**, 1067–1078.

Lyne, P. D. (2002). Structure-based virtual screening: An overview. *Drug Discov Today* **7**, 1047–1055.

Marnett, L. (2000). Cyclooxygenase mechanisms. *Curr Opin Chem Biol* **4**, 545–552.

Marnett, L. J., and Kalgutkar, A. S. (1999). Cyclooxygenase 2 inhibitors: Discovery, selectivity and the future. *Tips* **20**, 465–469.

Marot, C., Chavatte, P., and Lesieur, D. (2000). Comparative molecular field analysis of selective cyclooxygenase-2 (cox-2) inhibitors. *Quant Struct–Act Rel* **19**, 127–134.

Masferrer, J. L., Zweifel, B. S., Manning, P. T., Hauser, S. D., Leahy, K. M., Smith, W. G., Isakson, P. C., and Seibert, K. (1994). Selective inhibition of inducible cyclooxygenase 2 *in vivo* is antiinflammatory and nonulcerogenic. *Proc Natl Acad Sci USA* **91**, 3228–3232.

Matter, H., and Potter, T. (1999). Comparing 3D pharmacophore triplets and 2D fingerprints for selecting diverse compound subsets. *J Chem Inf Comput Sci* **39**, 1211–1225.

Michalik, L., and Wahli, W. (1999). Peroxisome proliferator-activated receptors: Three isotypes for a multitude of functions. *Curr Opin Biotechnol* **10**, 564–570.

Muegge, I., Heald, S. L., and Brittelli, D. (2001). Simple selection criteria for drug-like chemical matter. *J Med Chem* **44**, 1841–1846.

O'Banion, M. K. (1999). Cyclooxygenase-2: molecular biology, pharmacology and neurobiology. *Crit Rev Neurobiol* **13**, 45–82.

Oprea T. I. (2002). Chemical space navigation in lead discovery. *Curr Opin Chem Biol* **6**, 384–389.

Oshiro, C. M., and Kuntz, I. D. (1995). Flexible ligand docking using a genetic algorithm. *J Comput Aided Mol Des* **9**, 113–130.

Pearlman, R. S., and Smith, K. M. (1998). Novel software tools for chemical diversity. *Perspect Drug Discov Des* **9–11**, 339–353.

Rotstein, S. H., and Murcko, M. A. (1993). GroupBuild: A fragment-based method for *de novo* drug design. *J Med Chem* **36**, 1700–1710.

Sadowski, J., and Kubinyi, H. (1998). A scoring scheme for discriminating between drugs and nondrugs. *J Med Chem* **41**, 3325–3329.

Schneider, G. (2002). Trends in virtual combinatorial library design. *Curr Med Chem* **9**, 2095–2101.

Sedrani, R., Kallen, J., Martin Cabrejas, L. M., Papageorgiou, C. D., Senia, F., Rohrbach, S., Wagner, D., Thai, B., Jutzi Eme, A. M., France, J., Oberer, L., Rihs, G., Zenke, G., and Wagner, J. (2003). Sanglifehrin-cyclophilin interaction: degradation work, synthetic macrocyclic analogues, x-ray crystal structure, and binding data. *J Am Chem Soc* **125**, 3849–3859.

Sheridan, R. P., and Kearsley, S. K. (1995). Using a genetic algorithm to suggest combinatorial libraries. *J Chem Inf Comput Sci* **35**, 310–320.

Siekierka, J. J., Hung, S. H., Poe, M., Lin, C. S., and Sigal, N. H. (1989). A cytosolic binding protein for the immunosuppressant FK506 has peptidyl-prolyl isomerase activity but is distinct from cyclophilin. *Nature* **341**, 755–757.

SSKEYS, MDL Information Systems Inc., San Leandro, CA.

Stanton, R. V., Mount, J., and Miller, J. L. (2000). Combinatorial library design: Maximizing model-fitting compounds within matrix synthesis constraints. *J Chem Inf Comput Sci* **40**, 701–705.

Terrett, N. K., Gardner, M., Gordon, D. W., Kobylecki, R. J., and Steele, J. (1995). Combinatorial synthesis: The design of compound libraries and their application to drug discovery. *Tetrahedron* **51**, 8135–8173.

The SPRESI database is produced by the All-Union Institute of Scientific and Technical Information of the Academy of Science of the USSR (VINITI) in Moscow and by the Central Information Processing for Chemistry (ZIC) in Berlin. This database consists of data extracted from 1000 journals and patents, books, and other sources from 1975 to 1990. SPRESI is distributed by Daylight Chemical Information Systems, Inc., Mission Viejo, CA.

Tropsha, A., and Zheng, W. (2002). Rational principles of compound selection for combinatorial library design. *Comb Chem High Throughput Screen* **5**, 111–123.

Viswanadhan, V. N., Ghose, A. K., Revankar, G. R., and Robins, R. K. (1989). Atomic physicochemical parameters for three dimensional structure directed quantitative structure–activity relationships. 4. Additional parameters for hydrophobic and dispersive interactions and their application for an automated superposition of certain naturally occurring nucleoside antibiotics. *J Chem Inf Comput Sci* **29**, 163–172.

Wagener, M., and van Geerestein, V. J. (2000). Potential drugs and nondrugs: Prediction and identification of important structural features. *J Chem Inf Comput Sci* **40**, 280–292.

Walters, W. P., Stahl, M. T., and Murcko, M. A. (1998). Virtual screening—an overview. *Drug Discov Today* **3**, 160–178.

Wang, R., Lu, Y., and Wang, S. (2003). Comparative evaluation of 11 scoring functions for molecular docking. *J Med Chem* **46**, 2287–2303.

Warr, W. A. (1997). Combinatorial chemistry and molecular diversity. An overview. *J Chem Inf Comput Sci* **37**, 134–140.

Weber, L. (1998). Evolutionary combinatorial chemistry: Application of genetic algorithms. *Drug Discov Today* **3**, 379–385.

Willett, P., and Winterman, V. (1986). A comparison of some measures for the determination of inter-molecular structural similarity. *Quant Struct–Act Rel* **5**, 18–25.

Xu, J., and Stevenson, J. (2000). Drug-like index: A new approach to measure drug-like compounds and their diversity. *J Chem Inf Comput Sci* **40**, 1177–1187.

Zheng, S., Luo, X., Chen, G., Zhu, W., Shen, J., Chen, K., and Jiang, H. (2005). A new rapid and effective chemistry space filter in recognizing a druglike database. *J Chem Inf Mod* **45**, 856–862.

PART II

CHEMICAL AND SYNTHETIC APPROACHES IN DRUG DISCOVERY

CHAPTER 6

BEYOND NATURAL PRODUCTS: SYNTHETIC ANALOGUES OF BRYOSTATIN 1

Paul A. Wender, Jeremy L. Baryza, Michael K. Hilinski, Joshua C. Horan, Cindy Kan, and Vishal A. Verma*

6.1 INTRODUCTION

Historically, humans have looked to nature as a source of medicinal remedies. In the more recent past, many biologically active natural products were clinically advanced without structural modification because they were too difficult to modify or too little was known about how modifications would improve performance. While such unmodified natural products have occasionally succeeded as drugs (e.g., paclitaxel), it is clear that many natural leads have failed because they are not "designed" for therapeutic use. The increasing effectiveness of synthesis coupled with our growing understanding of therapeutic pathways—advanced by the sequencing of the human genome—is changing this situation. It has now become possible to examine a natural lead with an eye toward synthesis

*Corresponding author: wenderp@stanford.edu

Drug Discovery Research: New Frontiers in the Post-Genomic Era, Edited by Ziwei Huang

and biological function and design a potentially superior clinical candidate that can be synthesized in a practical fashion (function-oriented synthesis) (Wender et al., 2004). Thus, while many natural products will continue to be beyond the reach of practical synthesis, insightful design could allow synthetic access to targets that possess similar, if not superior, activity. The focus of this chapter is to provide an overview on how one such natural product, bryostatin 1, provided the starting point for the design of new leads that exhibit activity superior to the natural product. We report on the background of this effort, the evolution of our function-oriented design strategy, and studies that provide an important approach for evaluating the functional activity of these analogues.

In 1968, extracts from bugula neritina, a marine bryozoan, were found to have potent anticancer activity (Pettit et al., 1970). This activity was ultimately traced to bryostatin 1, whose complex structure was finally determined in 1982 (Fig. 6.1) (Pettit et al., 1982). Subsequently, 18 other naturally occurring bryostatins were identified and characterized. These structures differ primarily at the C7 and C20 positions (Mutter and Wills, 2000). Originally collected in the Gulf of Mexico, bryostatins have since been isolated from bryozoa collected from the Pacific coast of California, the Gulf of California, and the Gulf of Japan. There is also evidence that the bryostatins are produced by some Atlantic bryozoa (McGovern and Hellberg 2003). It has been postulated that the bryostatins are natively produced by symbiotic bacteria in order to protect host larvae from predation (Lopanik et al., 2004).

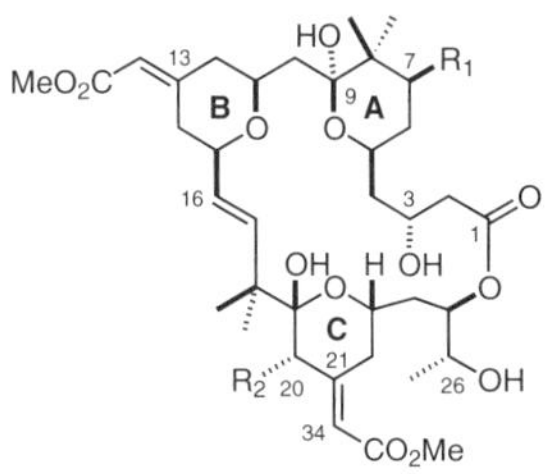

Bryostatin	R_1	R_2	K_i for PKC (nM)
1	O_2CCH_3	O_2C	1.4
2	OH	O_2C	5.9
4	$O_2CC(CH_3)_3$	$O_2CCH_2CH_2CH_3$	1.3
5	$O_2CC(CH_3)_3$	O_2CCH_3	1.0
6	$O_2CCH_2CH_2CH_3$	O_2CCH_3	1.2
7	O_2CCH_3	O_2CCH_3	0.8
8	$O_2CCH_2CH_2CH_3$	$O_2CCH_2CH_2CH_3$	1.7
9	O_2CCH_3	$O_2CCH_2CH_2CH_3$	1.3
10	$O_2CC(CH_3)_3$	H	1.6
11	O_2CCH_3	H	4.8
12	$O_2CCH_2CH_2CH_3$	O_2C	
13	$O_2CCH_2CH_2CH_3$	H	
14	$O_2CC(CH_3)_3$	OH	
15	O_2CCH_3	O_2C OH	
18	$O_2CC(CH_3)_3$	H	4.8

(Note: Z-C21-C34 enoate)

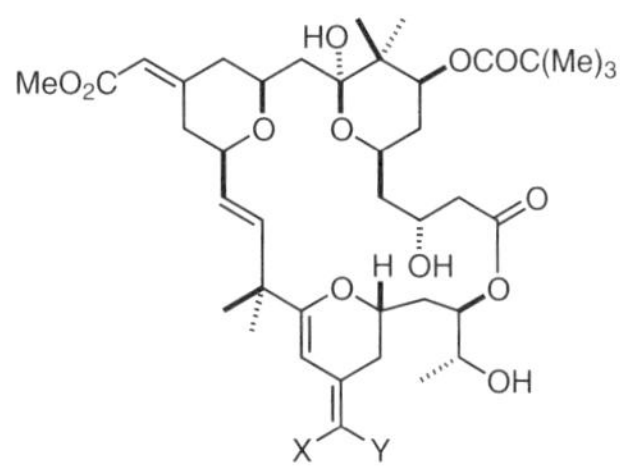

Bryostatin	X	Y	K_i for PKC (nM)
16	H	CO_2Me	118
17	CO_2Me	H	188

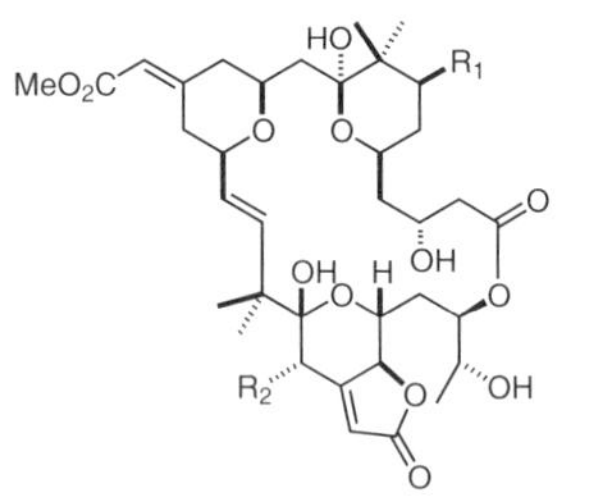

Bryostatin	R_1	R_2	K_i for PKC (nM)
3	OAc	$O_2C(CH)_4nPr$	2·8
20	OAc	H	

Figure 6.1 The natural bryostatin family.

TABLE 6.1 ***In vivo*** **and** ***in vitro*** **Potency of the Bryostatins Against Murine P388 Lymphocytic Leukemia**

Bryostatin	GI_{50} (P388 cells) (μg/mL)	Life Extension *in vivo* (Dose)
1	0.89	52–96% (10–70 μg/kg)
2	N/A	62% (46 μg/kg)
4	10^{-3}–10^{-4}	62% (46 μg/kg)
5	1.3×10^{-3}–2.6×10^{-4}	88% (185 μg/kg)
6	3.0×10^{-3}	82% (185 μg/kg)
7	2.6×10^{-5}	77% (92 μg/kg)
10	7.6×10^{-4}	64% (10 μg/kg)
11	1.8×10^{-5}	64% (92.5 μg/kg)
12	1.4×10^{-2}	47–68% (30–50 μg/kg)
13	5.4×10^{-3}	NA[a]
16	9.3×10^{-3}	NA
17	1.9×10^{-2}	NA
18	3.3×10^{-3}	NA

[a]NA, not available.

Bryostatin 1 exhibits activity against murine P388 lymphocytic leukemia both *in vitro* and *in vivo* (Table 6.1). While bryostatin 1 is only moderately potent *in vitro*, *in vivo* dosing of bryostatin 1 results in significant life extension with a remarkably low dose of only 10–70 μg/kg. The reason for the discrepancy between the *in vitro* and *in vivo* potency is not understood but might involve other cellular systems and their resulting interactions that are not seen with a single cell line *in vitro*.

Bryostatin 1 is currently in Phase I and II clinical trials (descriptions of NIH supported trials can be found at http://www.clinicaltrials.gov). As of this writing, the NIH database shows that 14 clinical trials involving bryostatin are underway, and five new trials are recruiting patients. Although bryostatin has not excelled as a single agent therapeutic, it has been shown to increase the potency and activity of other antineoplastic agents (Mutter and Wills, 2000). Examples include agents that act upon microtubules, such as dolastatin, and auristatin (Mohammad et al., 1998b), the microtubule depolymerizer vincristine (Al-Katib et al., 1998), and the microtubule stabilizer paclitaxel (Wang et al., 1998). Bryostatin also synergizes with DNA replication inhibitors such as arabinosylcytosine (ara-C) (Wang et al., 2002) and 2-chlorodeoxyadenosine (2-CdA) (Beck et al., 2004), DNA cross-linking agents such as cisplatin (Basu and Lazo, 1992), and the estrogen receptor agonist tamoxifen (McGown et al., 1998). In a landmark clinical trial, treatment of cancer of the esophagus with a combination of bryostatin 1 and paclitaxel resulted in a 60% response rate as opposed to the 17% response seen with paclitaxel alone (Marsa, 2002). It has also been observed that the order of dosing is critical in obtaining a synergistic effect, which might lead to new insights about the mechanism of action. In recent clinical trials, bryostatin has performed synergistically with ara-C (Cragg et al., 2002) and vincristine (Dowlati et al., 2003).

Since synergy is observed with a variety of agents whose mechanisms of action differ, it is thought that bryostatin might predispose cells toward apoptosis, resulting in an increased response to various therapeutic strategies (Farrow et al., 2002; Wall et al., 1999; Mohammad et al., 1998a). Mechanistic studies have shown that bryostatin increases the amount of the proapoptotic protein Bax relative to the antiapoptotic protein Bcl-2. This imbalance results in cytochrome c release, which induces the onset of apoptosis. Bryostatin has also been shown to promote the release of tumor necrosis factor alpha (TNF-α), which can induce apoptosis through an extrinsic pathway (Wang et al., 2003; Cartee et al., 2003).

Another way in which bryostatin potentiates the activity of anticancer agents is in its ability to reverse multidrug resistance (MDR) (Elgie et al., 1998). In HeLa cells, bryostatin 1 is able to inhibit the action of mutant *p*-glycoprotein (P-gp), an efflux pump responsible for resistance to many cancer therapeutics (Spitaler et al., 1998). An additional study showed reduced expression of P-gp in cells treated with bryostatin (Al-Katib et al., 1998). Since P-gp is overexpressed in resistant cells, it should be possible to increase selectivity in targeting malignant cells.

Bryostatin is also able to differentiate cancer cells into new potentially treatable cell types. Resistance to differentiation is overcome with bryostatin in acute myeloid leukemia cells, allowing differentiation into dendritic-like leukemia cells (Matsui et al., 2005; Roddie et al., 2002).

A drawback to most anticancer therapies is the suppression of immune system activity. In contrast, bryostatin stimulates the immune system *in vitro* and *in vivo* (Rabah et al., 2001; Curiel et al., 2001; Oz et al., 2000; Scheid et al., 1994). In various studies it has been shown that bryostatin promotes T-cell production, antigen presentation (Kudinov et al., 2003), dendritic cell activation, and differentiation (Do et al., 2004). Bryostatin has also been effective in adoptive immunotherapy, wherein antigens are generated *ex vivo* (Parviz et al., 2003; Bear et al., 2001).

The recommended dose of bryostatin from Phase I clinical trials is nothing short of spectacular. Bryostatin is effective when dosed at 25 $\mu g/m^2$ per week or approximately 50 μg per week in an average adult. In comparison, antineoplastic agents that require milligram doses per day are considered potent. For example, an 8-week treatment cycle would require 2.6 g of paclitaxel, whereas a similar treatment would require only 1.1 mg of bryostatin. The dose-limiting toxicity of bryostatin 1 is myalgia (muscle pain); consequently, higher amounts of bryostatin for use as a therapeutic have not been explored.

While the mode of action of the bryostatins is still under investigation, bryostatin's activity is thought to be mediated through binding to protein kinase C (PKC; $K_i = 1.4$ nM) (Hennings et al., 1987; Wender et al., 1998b), an enzyme family that plays a critical role in cell signaling. PKC is involved in many cellular processes, such as apoptosis, cell-cycle regulation, vesicle transport, gene expression, cellular motility, ion channel regulation, neuronal growth, and cellular differentiation (Dempsey et al., 2000). Abnormal PKC signaling has been identified in many cancer types (Gavrielides et al., 2004; Lahn et al., 2004; Mackay and Twelves, 2003; da Rocha et al., 2002; Caponigro et al., 1997). PKC is also a target for the potent tumor-promoting phorbol esters as well as endogenous diacylglycerols (DAGs) (Gomez-Fernandez et al., 2004; Silinsky and Searl, 2003; de Vries et al., 1988). While bryostatins induce a subset of phorbol responses, including activation and subsequent downregulation of PKC, they antagonize other responses, notably tumor promotion (Kraft et al., 1986; Berkow and Kraft, 1985; Ramsdell et al., 1986; Hennings et al., 1987).

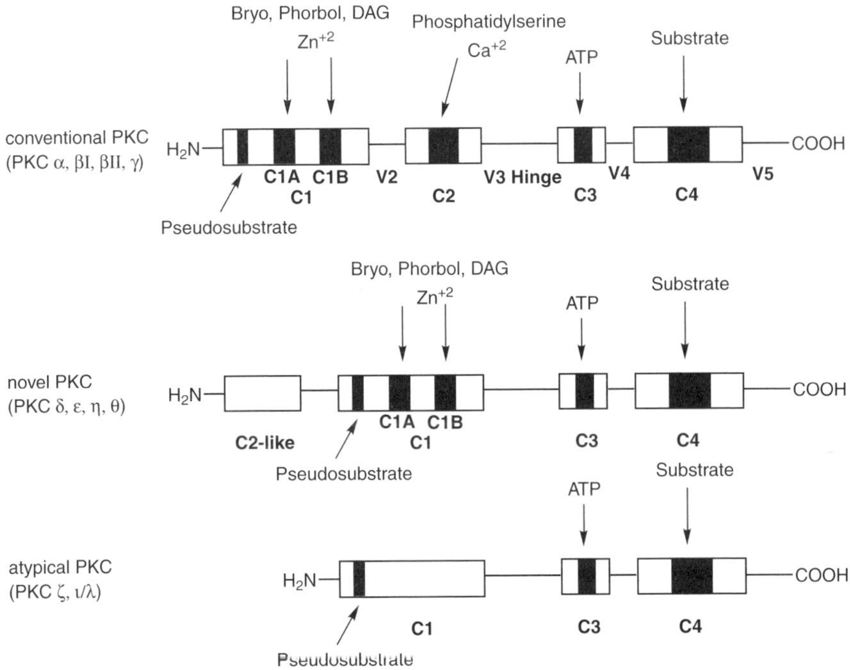

Figure 6.2 The conserved subunits of conventional, novel, and atypical PKCs.

PKC isozymes can be grouped into three classes: the calcium-dependent conventional isozymes (α, βI, βII, γ), the calcium-independent novel isozymes (δ, ε, η, θ), and atypical isozymes (ξ, ι/λ). Only the conventional and novel families bind DAG, the bryostatins, and phorbol esters at the C1 domain (Fig. 6.2). This binding results in the modulation of PKC activity and distinguishes bryostatin from many other kinase targeting agents that bind at the ATP binding site or the catalytic site. Ligands that bind to the C1 domain offer a selectivity advantage over ligands for the ATP binding domain of kinases. All kinases (~500) (Manning et al., 2002) have an ATP binding site, whereas approximately 50 proteins possess a C1 domain.

Although bryostatin holds great promise as a uniquely potent therapeutic agent, its limited supply presents a currently insurmountable obstacle to its widespread therapeutic use. The bryostatins have been isolated from only one source, *Bugula neritina*, in extremely low yields. In the most substantial effort, which has supplied all the material used in clinical trials, 18 g of bryostatin 1 was isolated from 14 tons of sponge, a yield of only $1.4 \times 10^{-4}\%$ (Schaufelberger et al., 1991). While aquaculture has been explored as a method to overcome difficulties in collection, it does not circumvent the issue of yield (Mendola, 2003). Efforts are underway to identify the genetic mechanisms responsible for bryostatin production with the goal of transplanting such a mechanism to another host for overexpression and subsequent harvesting of bryostatin (Hildebrand et al., 2004). These efforts are important for accessing bryostatin and its natural derivatives. However, it is abundantly clear that bryostatin is *not* produced for human therapy. Like many natural products, it is an important therapeutic lead but not an optimized therapeutic agent.

The complexity of the bryostatins has challenged synthetic chemists since the structure of bryostatin 1 was first reported in 1982. Over two decades of work has resulted in the total syntheses of bryostatins 2, 3, and 7 by Evans, Yamamura, and Masamune respectively (Evans et al., 1999; Ohmori et al., 2000; Kageyama et al., 1990). While impressive in content, each synthesis requires over 70 steps and is consequently unlikely to be a cost-effective source. Of course, the main justification for these efforts is to access bryostatin, or a functional equivalent, to pursue its therapeutic value. The identification of functionally equivalent or superior analogues of bryostatin that can be produced cost-effectively represents a solution to the low abundance of the natural product.

6.2 BRYOSTATIN ANALOGUE DESIGN

A major intellectual challenge and a key to solving the bryostatin supply and clinical performance problems is to identify those functionalities in bryostatin that are responsible for its activity and to design a molecule with these features that could be synthesized in a practical fashion. Along with bryostatin, several other natural products are known to bind competitively to the C1 domains of PKC. These natural products include the phorbol esters, the endogenous activator DAG, gnidimacrin, aplysiatoxin, ingenol, and teleocidin B-4 (Fig. 6.3).

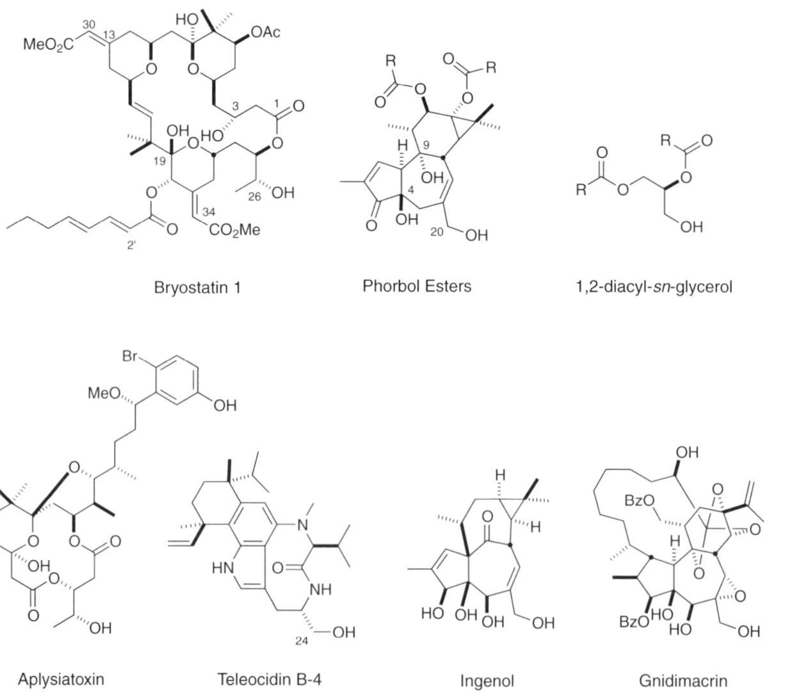

Figure 6.3 Ligands that bind to the C1 regulatory domains of PKC.

Because these ligands all bind to PKC, it was supposed that they might share part or all of a common pharmacophore, a similar three-dimensional orientation of a subset of atoms in each molecule. While this is not a requirement, it provides a working hypothesis that allows for a rational approach to the design of PKC activators and inhibitors. Comparison of the interatomic distances of all possible heteroatomic triads of phorbol and other PKC ligands revealed that the best triad overlay corresponded to the C4, C9, and C20 oxygens of the phorbol esters (Wender et al., 1986). The second-best correlation was with the C3, C9, and C20 triad. The acyl chains, also shown to be required for activity, were spatially oriented in a similar fashion as the lipophilic moieties of the other species. These findings are in agreement with known phorbol structure–activity relationships that include a long acyl chain at C12 or C13, a C20 alcohol, and a C3 ketone and/or C4 hydroxyl group being required for activity (Hecker, 1978).

When the structures employed in elucidating the phorbol pharmacophore were compared to the heteroatomic triads in bryostatin 1, it was found that the C1, C19, and C26 heteroatomic triad of bryostatin had the best overlay with the C4, C9, and C20 heteroatoms of phorbol (RMSD = 0.16 Å) (Fig. 6.4) (Wender et al., 1998b).

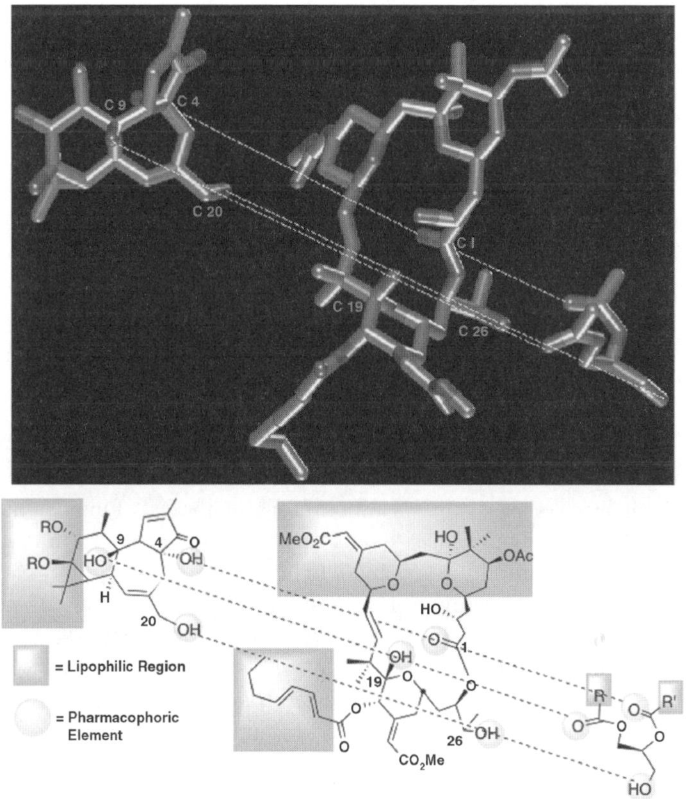

Figure 6.4 The preferred pharmacophoric correlation of 1,2-diacyl-*sn*-glycerol, the phorbol esters, and bryostatin 1. See color plates.

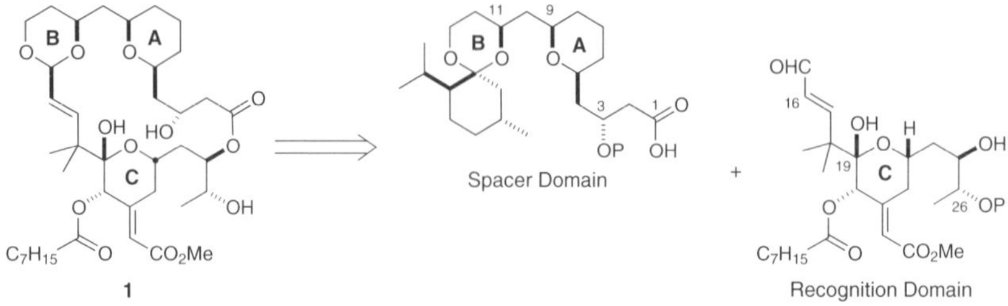

Scheme 6.1 Retrosynthetic analysis of analogue **1**.

Moreover, when the correlated atoms were compared, their lone pair electrons or attached hydrogen atoms assumed a similar orientation. Structure–activity relationships established using the naturally occurring bryostatins substantiated this hypothesis, as well as additional information concerning which elements of the molecular architecture are essential for potency. These studies revealed that acylation of C26 resulted in loss of potency (Wender et al., 1998b) and suggests that the C3 hydroxyl is involved in a transannular hydrogen bonding network that holds the macrocycle in an active conformation (Pettit et al., 1982, 1991; Kamano et al., 1996). Also, while hydrogenation of the C13–C30, C2–C3, and C4–C5 olefins did not affect potency, hydrogenation of the C21–C34 olefin resulted in a significant potency loss (Pettit et al., 1992).

Using the pharmacophore hypothesis, it became possible to propose structurally simplified analogues of bryostatin that should have similar activity. Given that alterations to the A- and B-rings of bryostatin had little effect on potency (Pettit et al., 1991; Kamano et al., 1996), it was proposed that these rings could be simplified while maintaining the rigidity of the macrocycle. In addition, saturation of the C20 side-chain was proposed based on the previously discussed hydrogenation of the natural product. Finally, introduction of an additional B-ring oxygen was anticipated to simplify the synthesis and make it convergent. The ability to simultaneously design for function and select for ease of synthesis was key to advancing this effort. Structure **1** emerged from this analysis (Scheme 6.1) (Wender et al., 1998b). Its synthesis was planned to arise from the assembly and coupling of two fragments, the spacer and recognition domains, followed by a proposed macro-transacetalization process. Completion of the synthesis allowed for a major test of the pharmacophore hypothesis. Of great significance to this approach, compound **1** binds to PKC ($K_i = 3.4$ nM) with an affinity comparable to bryostatin 1.

The synthesis of analogue **1** allowed for the preparation of additional compounds (**2**–**6**, Fig. 6.5) to probe the structural requirements for analog binding. It is noteworthy that the C3 hydroxyl plays a significant role in PKC binding. NMR studies suggest that it forms a bifurcated transannular hydrogen-bond array, presumably preorganizing the macrocycle for binding. Deletion of the C3 hydroxyl or inversion of the C3 stereochemistry reduced affinity by two orders of magnitude. Acylation or oxidation of the C26 alcohol resulted in a complete loss of PKC binding. Deletion of the macrocycle also resulted in total abrogation of potency (Wender et al., 1999).

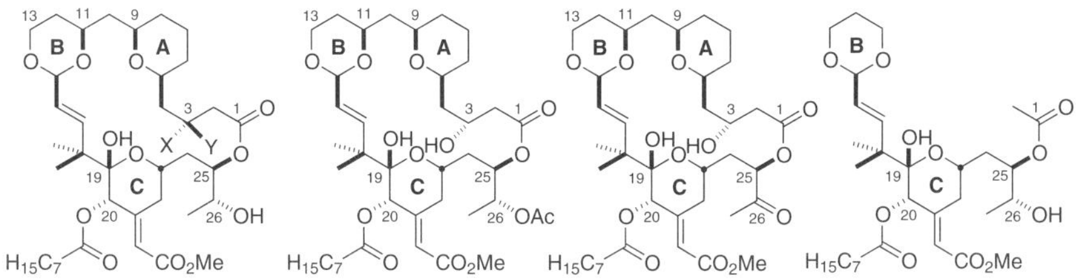

1 X = OH, Y = H
2 X = H, Y = OH
3 X = Y = H

4 **5** **6**

Compound	K_i
1	3.4 nM
2	285 nM
3	297 nM
4	>10 mM
5	>500 nM
6	>10 mM

Figure 6.5 Structure-binding studies in accord with the pharmacophore hypothesis.

6.3 THE DESIGN AND SYNTHESIS OF A C26 DES-METHYL ANALOGUE

When compared to other PKC ligands, the hydroxyl at C26 in bryostatin corresponds to the hydroxyl groups at C20 in phorbol, C3 in DAG, and C24 in teleocidin (Fig. 6.3). However, these counterparts to the C26 hydroxyl are primary rather than secondary alcohols. An analogue possessing a C26 primary alcohol was anticipated to retain biological activity while reducing molecular complexity by removal of a stereocenter. Molecular modeling of the proposed C26 des-methyl analogue (**15**, Scheme 6.2) suggested a global minimum that closely resembled the crystal structures of the bryostatins.

Analogue **15** was prepared from an advanced intermediate in the synthesis of analogue **1** (Scheme 6.2) (Lippa, 1999). Starting from intermediate **7**, hydrogenation followed by oxidative cleavage and olefination of the resultant aldehyde afforded terminal olefin **9**. This three-step sequence successfully deleted the C26 methyl group. Following conversion of the C17 silyl ether to aldehyde **11**, the C25–C26 diol was reintroduced via Sharpless asymmetric dihydroxylation. Protection as the bis-TES ether provided intermediate **12**. After a four-step homologation of the C17 aldehyde to enal **13**, treatment with aqueous HF simultaneously cleaved the silyl ethers while hydrolyzing the methyl ketal. The C26 primary hydroxyl group was selectively protected as the TBS ether. The completed recognition domain **14** was coupled to the completed spacer domain, **21** (*vide infra*), via a Yamaguchi esterification. Finally, treatment with HF·pyridine effected a macrotransacetalization to close the macrocycle and set the C15 stereocenter while simultaneously cleaving the two silyl ethers to provide the completed analogue **15**.

In agreement with prediction, the C26 des-methyl analogue displays *picomolar* affinity for PKC with a K_i of 250 pM. When tested *in vitro* against various human

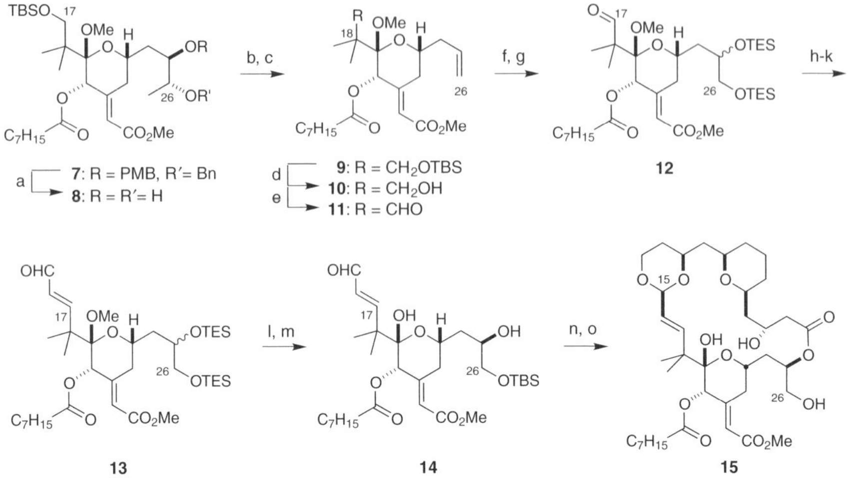

Scheme 6.2 Degradative approach to a C26 des-methyl analogue. (a) H_2, $Pd(OH)_2$/C, EtOAc, rt, 67%; (b) $Pb(OAc)_4$, Et_3N, PhH, 0°C; (c) $Cp_2Ti(\mu\text{-}Cl)(\mu\text{-}CH_2)AlMe_2$, THF, 0°C, 56% over two steps; (d) HF · pyridine, THF, rt; (e) Dess–Martin periodinane, CH_2Cl_2, rt, 60% over two steps; (f) $(DHQD)_2AQN$, $K_2OsO_2(OH)_4$, K_2CO_3, $K_3Fe(CN)_6$, *t*-BuOH/H_2O 1:1, rt, 70%, $\beta{:}\alpha = 2.2{:}1$; (g) TESCl, pyridine, CH_2Cl_2, rt, 100%; (h) allyl-BEt_2, Et_2O, −10°C; (i) Ac_2O, DMAP, CH_2Cl_2, rt, 97% over two steps; (j) OsO_4 (cat.), NMO, THF, H_2O, rt; (k) i. $Pb(OAc)_4$, Et_3N, PhMe, rt; ii. DBU, rt, 73% over 2 steps; (l) aq. HF, CH_3CN, H_2O, rt; (m) TBSCl, imidazole, CH_2Cl_2, DMF, rt, 57% β 25% α; (n) 2,4,6,-trichlorobenzoyl chloride, Et_3N, **21**, PhMe, rt, then **14**, DMAP, PhMe, rt, 79%; (o) HF · pyridine, THF, rt, 73%.

cancer cell lines, **15** displayed greater potency than bryostatin 1 in 24 of 35 cases. In some cell lines, such as MOLT-4 and NCI-H460, **15** is three orders of magnitude more potent than bryostatin 1 at inhibiting cell growth (see Section 6.6).

The exceptional potency of **15** prompted the development of a more efficient synthesis to address the supply problem and provide access to potentially superior clinical candidates (Wender et al., 2002). Building on the existing synthesis of the C1–C13 spacer domain, **21**, the synthesis began with the ozonolysis of **16**, which, after *in situ* reduction, produced pentane-1,3,5-triol (Scheme 6.3). Desymmetrization of the resulting triol via acetal formation with (−)-menthone (Harada et al., 1993; Moune et al., 1997) and subsequent oxidation yielded aldehyde **17**. A hetero Diels–Alder cycloaddition between **17** and Danishefsky's diene using Jacobsen's tridentate Cr(III) catalyst (Joly and Jacobsen, 2002) provided pyranone **18** with exceptional selectivity. Luche reduction, vinyl ether formation, and subsequent Claisen rearrangement provided the unsaturated aldehyde **19**. Hydrogenation of **19** followed by asymmetric allylation of the aldehyde and silylation of the resultant alcohol afforded silyl ether **20**. Oxidative cleavage of the terminal olefin provided spacer domain **21**. The overall conversion of **16** to **21** required 11 steps and resulted in an overall yield of 11%.

A shorter, more selective synthesis of the spacer domain was subsequently developed to provide **30**. In this approach, the (−)-menthone cyclic ketal was replaced with an acetone-derived ketal, and the C3 TBS ether was replaced with a TBDPS ether (Scheme 6.4) (Wender et al., 2003b). The synthesis of **30** began with the

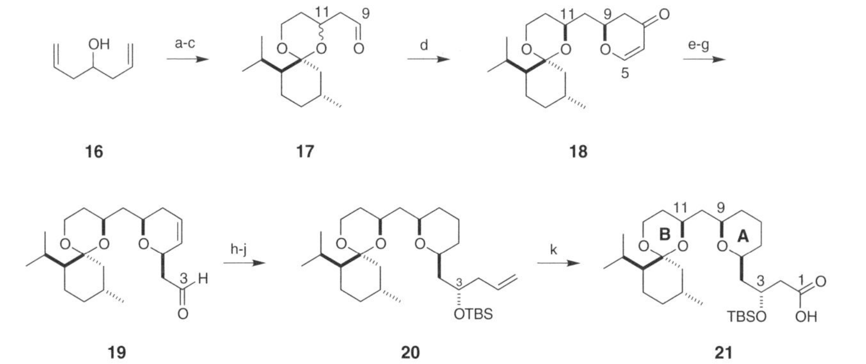

Scheme 6.3 An efficient synthesis of spacer domain **21**. (a) i. O_3, MeOH, $-78°C$, ii. $NaBH_4$, $-78°C \rightarrow$ rt, 90%; (b) (−)-menthone, *p*TsOH · H_2O, $CH(OEt)_3$, Et_2O, rt, 71%; (c) $(COCl)_2$, DMSO, Et_3N, CH_2Cl_2, $-78°C$, 87% (β:α = 1.6:1); (d) i. Danishefsky's diene, Jacobsen's tridentate Cr(III) catalyst, 4 Å MS, acetone, rt, ii. TFA, rt, 88%; (e) $NaBH_4$, $CeCl_3 \cdot 7H_2O$, MeOH, $-40°C$, 92%; (f) $Hg(OAc)_2$, isobutylvinyl ether, rt; (g) decane, 150°C, 83% over two steps; (h) H_2, $Pd(OH)_2/C$, EtOAc, rt, 85%; (i) (−)-$(Ipc)_2BOMe$, allyl-MgBr, CH_2Cl_2, $-78°C \rightarrow$ rt; (j) TBSCl, imidazole, THF, rt, 69% over two steps; (k) $KMnO_4$, $NaIO_4$, 1:1 *t*-BuOH:pH 7 buffer, rt, 84%.

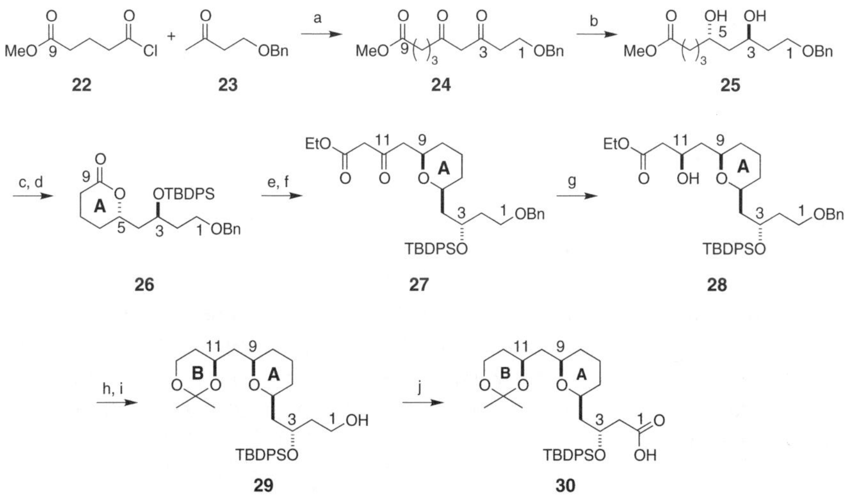

Scheme 6.4 Selective synthesis of spacer domain **30**. (a) i. LDA, THF, **23**, $-78°C$, ii. **22**, 68%; (b) Ru-(*S*)-BINAPCl$_2$, MeOH, H_2 (95 atm), 30°C, 92%; (c) silica, PhMe, reflux, 95%; (d) TBDPSCl, imidazole, DMF, rt, 85%; (e) ethylacetoacetate, LDA, THF, $-78°C$; (f) Et_3SiH, TFA, $-30°C$, 70% over two steps; (g) Ru-(*R*)-BINAPCl$_2$, EtOH, H_2 (78 atm), rt, 91%; (h) i. H_2 (1 atm), $Pd(OH)_2/C$, Et_2O, rt, ii. $LiBH_4$, rt, 96%; (i) i. 2,2-dimethoxypropane, *p*TsOH, DMF, rt, ii. silica, CH_2Cl_2, 93%; (j) TEMPO, NaOCl, $NaClO_2$, CH_3CN, 50°C, 92%.

acylation of 4-benzyloxy-2-butanone **23** with acid chloride **22** to provide diketone **24**. A Noyori asymmetric hydrogenation was employed to set the C3 and C5 stereocenters in excellent yield and in greater than 95:5 enantioselectivity to provide diol **25** (Kitamura et al., 1988; Noyori et al., 1987). The secondary alcohols were then differentiated through lactonization of the C5 alcohol with the proximate ester. Protection of the remaining alcohol gave lactone **26**. The remainder of the carbon backbone (C10–C13) of the spacer domain was then introduced by addition of the dianion of ethyl acetoacetate to lactone **26**. The resulting lactol was reduced to set the C9 stereocenter, and a second asymmetric reduction converted β-ketoester **27** to alcohol **28**. The C1 benzyl ether was cleaved and the ethyl ester reduced in one operation to give the corresponding triol. The 1,3-diol was protected as acetonide **29** and the primary alcohol oxidized to the carboxylic acid (Zhao et al., 1999), yielding spacer domain **30** in 10 steps and 25% overall yield.

Coordinated with the synthesis of the spacer domain, the synthesis of the C15–C26 recognition domain **14** began with the monoprotection and oxidation of inexpensive diol **31** to generate aldehyde **32** (Scheme 6.5). Reaction of **32** with the Grignard reagent derived from 4-chloro-1-butanol (Cahiez et al., 1978) followed by oxidation and asymmetric Keck allylation (Keck and Krishnamurthy, 1998; Yu et al., 1997) provided homoallylic alcohol **33** and set the C23 stereocenter. Dehydrative cyclization of **33** followed by epoxidation and *in situ* methanolysis yielded a mixture of diastereomers, the major of which was oxidized to ketone **34**. Conversion of ketone **34** to enoate **35** was readily accomplished in one step. A highly diastereoselective reduction of **35** followed by esterification afforded ester **37**. Deprotection of **37** followed by oxidation gave aldehyde **38**. In contrast to our original four-step sequence, the transformation of aldehyde **38** to enal **39** was achieved in only one step using a vinyl zincate reagent. Sharpless asymmetric dihydroxylation followed by hydrolysis of the C19 ketal and selective protection of the C26 alcohol provided recognition domain **14** in 17 steps and 3% overall yield.

Synthesis of analogue **15** was completed by coupling **14** and **21** using the Yamaguchi esterification protocol followed by macrotransacetalization (Scheme 6.2). This transacetalization procedure closed the macrocycle, set the C15 stereocenter under thermodynamic control, and allowed for global deprotection, providing analogue **15** in 19 steps (longest linear sequence) and 2% yield. The spacer domain **30** was also coupled to fragment **14** using a PyBroP-mediated esterification to provide the corresponding ester, which was treated with HF·pyridine to again close the macrocycle via a transacetalization and to cleave the protecting groups. Both syntheses provide novel strategies for the practical and scalable synthesis of **15**. Additionally, the convergent nature of these routes is ideal for the rapid and efficient synthesis of related analogues.

6.4 BRYOSTATIN ANALOGUES WITH MODIFIED SPACER DOMAINS

While investigating conditions to diversify the completed analogs, it was found that treatment of **15** with dimethyldioxirane (DMDO) resulted in selective oxygenation at

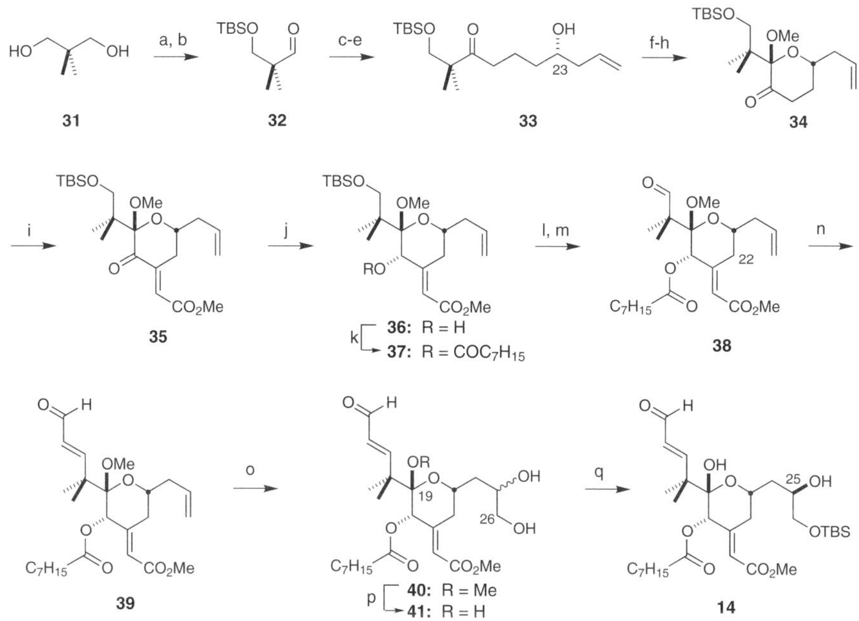

Scheme 6.5 An efficient synthesis of recognition domain **14**. (a) NaH, TBSCl, THF, rt; (b) $SO_3 \cdot$pyridine, Et_3N, DMSO, CH_2Cl_2, rt; (c) i. 4-chloro-1-butanol, MeMgCl, THF, $-78°C \rightarrow$ rt, ii. Mg, reflux, iii. **32**, $-78°C$; (d) $(COCl)_2$, DMSO, CH_2Cl_2, Et_3N, $-78°C$, 54% over four steps; (e) 10 mol% (*R*)-BINOL, 4 Å MS, 5 mol% Ti(O*i*-Pr)$_4$, $B(OMe)_3$, allyl-$SnBu_3$, CH_2Cl_2, rt, 77%; (f) cat. *p*TsOH$\cdot H_2O$, 4 Å MS, MePh, rt, 85%; (g) MMPP, $NaHCO_3$, 2:1 CH_2Cl_2:MeOH, 0°C, 78% (dr = 4:1); (h) 10 mol% TPAP, NMO, 4 Å MS, 6:1 CH_2Cl_2:CH_3CN, $0°C \rightarrow$ rt, 78%; (i) K_2CO_3, $OHCCO_2Me$, MeOH, rt, 72%; (j) $NaBH_4$, $CeCl_3 \cdot 7H_2O$, MeOH, $-30°C$; (k) $C_7H_{15}CO_2H$, DIC, DMAP, CH_2Cl_2, rt, 93% over two steps; (l) $3HF \cdot Et_3N$, THF, rt; (m) Dess–Martin periodinane, $NaHCO_3$, CH_2Cl_2, $0°C \rightarrow$ rt, 87% over two steps; (n) i. (*Z*)-1-bromo-2-ethoxyethene, *t*-BuLi, Me_2Zn, **38**, Et_2O, $-78°C$, ii. 1M HCl, $-78°C \rightarrow$ rt, 90%; (o) $(DHQD)_2PYR$, $K_2OsO_2(OH)_4$, $K_3Fe(CN)_6$, K_2CO_3, 1:1 *t*-BuOH:H_2O, 0°C, 71%, $\beta:\alpha = 2.5:1$; (p) *p*TsOH$\cdot H_2O$, MeCN, H_2O; (q) TBSCl, imidazole, CH_2Cl_2, rt, 46% over two steps for C25 β-isomer.

C9 (**42**, Scheme 6.6) (Wender et al., 2005b). This highly selective C–H insertion reaction effectively reinstalled the C9 hemiketal present in the natural product without affecting other sensitive functionality in the molecule such as the C16–C17 olefin, the C15 and C19 acetals, or the C26 alcohol. Similar reactivity was observed when **1** was subjected to the same reaction conditions. However, partial oxidation of the C26 alcohol to the corresponding ketones **5** and **44** was observed with this substrate. This novel transformation produced new bryostatin analogues in a single step and represents one of the most complex intermolecular C–H activations reported to date.

According to the proposed pharmacophore, the hemiketal functionality at C9 is not essential for PKC affinity (Wender et al., 1998b). This hypothesis is supported by the observation that analogues **1** and **15**, which lack a C9–OH group, bind PKC as well or better than bryostatin 1. Analogues **42** and **43,** along with their nonhydroxylated variants, **1** and **15,** allowed the effect of the C9 hemiketal on affinity for PKC to

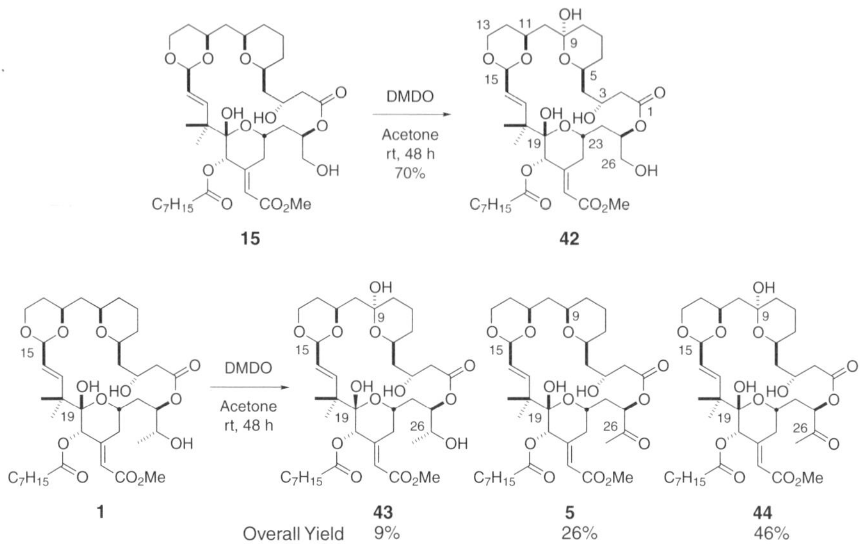

Scheme 6.6 Functionalization at the C9 position via selective C–H insertion.

be examined. The binding constants for new analogues **42** and **43** are 2 nM and 4 nM, respectively. Consistent with the proposed pharmacophore, the potencies of these compounds are similar to those found for **1** and **15**. These data serve as a negative control further supporting the pharmacophore hypothesis.

The potent binding affinities of analogues **1** and **15** corroborate the hypothesis that the top portion serves as a lipophilic spacer domain and holds the recognition elements in the proper orientation for binding to PKC. These agents demonstrate that it is possible to dramatically reduce complexity in the spacer domain without adversely affecting binding affinity. These structural simplifications provide a significant increase in the synthetic accessibility of the analogues compared to the natural product.

While the existing spacer domains **21** and **30** do not contain much of the functionality present in the natural product, they still possess four stereocenters and require 10 or 11 steps to synthesize. Consequently, further simplifications of the spacer domain were explored. The A-ring tetrahydropyran, devoid of pharmacophoric elements, was targeted for further structural modification.

The greatest simplification of the A-ring is the removal of carbons C6 to C8, simplifying the structure dramatically by removal of two stereocenters. However, this approach was expected to increase the conformational flexibility of the spacer domain. In order to confer a greater level of conformational rigidity, a sterically demanding substituent was introduced at the C9 position in place of the A-ring. These design concepts were realized in the synthesis of four new simplified spacer domains (Fig. 6.6).

The synthesis of these spacer domains began with oxidation of alcohol **49** to aldehyde **17β** (Scheme 6.7) (Wender et al., 1998a, 2004, Wender 2005a). Addition of the desired nucleophile followed by installation of the correct stereochemistry gave

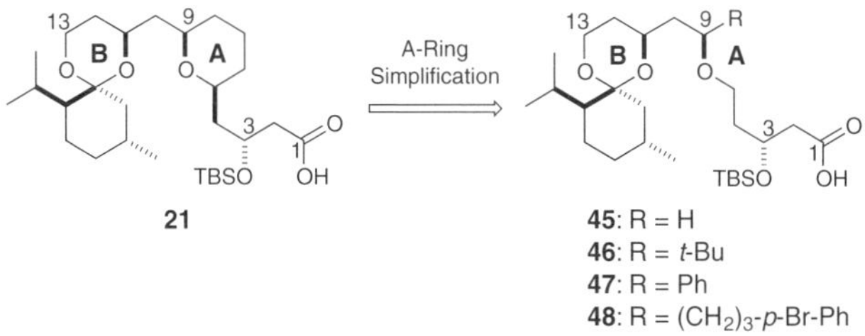

Figure 6.6 Simplified A-ring spacer domains.

alcohols **51** and **53**. Preparation of intermediate **55** started with an asymmetric Brown's allylation of **17β** followed by TBS protection to give silyl ether **54**. Cross metathesis with 4-bromostyrene and subsequent reduction of the olefin produced silyl ether **55**.

After desilylation of **55**, each of the secondary alcohols was carried independently through a parallel synthetic sequence to complete the individual spacer domains (Scheme 6.8). Alkylation with allyl bromide gave terminal olefins **56–59**. Hydroboration followed by oxidation gave aldehydes **60–63**. Asymmetric allylation and subsequent TBS protection allowed for isolation of the silyl ethers **64**, **65**, **66**, and **68**. Oxidative cleavage of the terminal olefins gave the completed spacer domains **45–48**.

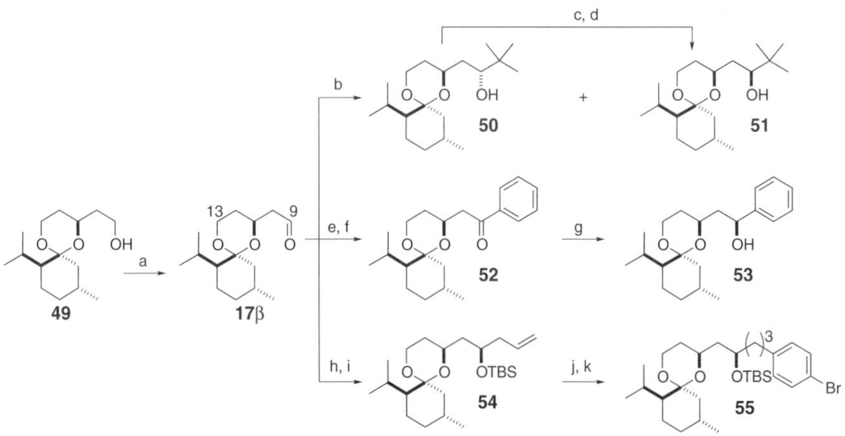

Scheme 6.7 The synthesis of intermediates **51**, **53**, and **55**. (a) Dess–Martin periodinane, CH_2Cl_2, rt, 90%; (b) *t*-BuLi, Et_2O, −78°C, 58% (1:1 mixture of **50** and **51**); (c) TPAP, NMO, 4 Å MS, CH_2Cl_2, rt; (d) $NaBH_4$, $CeCl_3 \cdot 7H_2O$, MeOH, −78°C → rt, 87% (over two steps; 9:1 of **51** and **50**); (e) PhMgBr, CH_2Cl_2, −78°C → rt; (f) Dess–Martin periodinane, CH_2Cl_2, rt, 84% (over two steps); (g) $NaBH_4$, $CeCl_3 \cdot 7H_2O$, MeOH, −78°C → rt, 78%; (h) (−)-$(Ipc)_2$BOMe, allylMgBr, Et_2O, −78°C → rt, 94%; (i) TBSCl, imidazole, DMF, rt, 91%; (j) Grubbs' second generation catalyst, 4-bromostyrene, 1,2-DCE, 40°C, 59%; (k) 5% Rh/Al_2O_3, H_2 (1 atm), EtOH, rt, 73%.

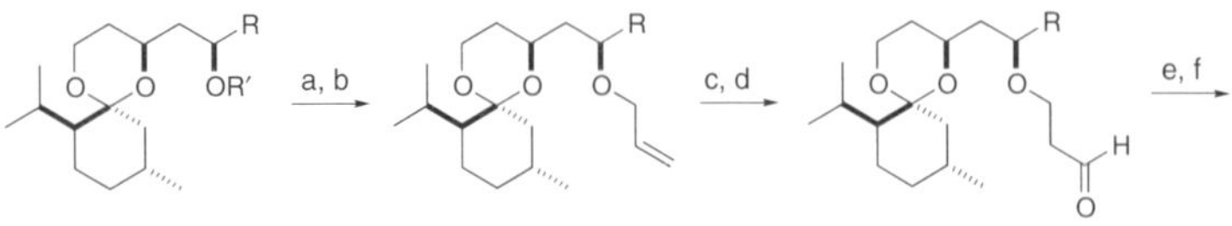

49: R = R′ = H
51: R = *t*-Bu; R′ = H
53: R = Ph; R′ = H
55: R = $(CH_2)_3$-*p*-Br-Ph
R′ = TBS

56: R = H: 96%
57: R = *t*-Bu: 88% BORSM
58: R = Ph
59: R = $(CH_2)_3$-*p*-Br-Ph:
90% (2 steps)

60: R = H: 79% (over 2 steps)
61: R = *t*-Bu: 78% (2 steps)
62: R = Ph: 67% (3 steps)
63: R = $(CH_2)_3$-*p*-Br-Ph:
67% (2 steps)

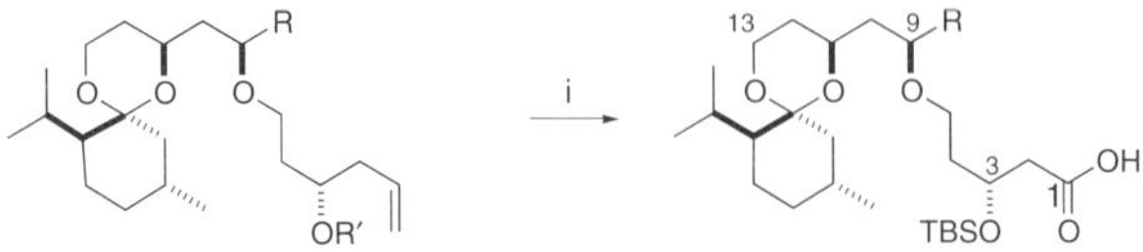

64: R = H, R′ = TBS: 53% (over 2 steps)
65: R = *t*-Bu, R′ = TBS: 67% (2 steps)
g **66**: R = Ph, R′ = TBS: (diast. mixt.)
h **67**: R = Ph, R′ = H: 32% (3 steps)
66: R = Ph, R′ = TBS: 90%
68: R = $(CH_2)_3$-*p*-Br-Ph, R′ = TBS:
78% (2 steps)

45: R = H: 76%
46: R = *t*-Bu: 79%
47: R = Ph: 70%
48: R = $(CH_2)_3$-*p*-Br-Ph: 75%

Scheme 6.8 Completion of spacer domains **45–48**. (a, **55** only) TBAF, THF, rt; (b) *t*-BuOK, allyl bromide, THF, rt; (c) i. 9-BBN, THF, rt, ii. NaOH, H_2O_2; (d) Dess–Martin periodinane, CH_2Cl_2, rt; (e) (−)-$(Ipc)_2$BOMe, allylMgBr, CH_2Cl_2, −78°C → rt; (f) TBSCl, imidazole, CH_2Cl_2 or DMF, rt; (g) TBAF, THF, rt; (h) TBSCl, imidazole, CH_2Cl_2, rt; (i) $NaIO_4$, $KMnO_4$, 1:1 *t*-BuOH:pH 7 buffer, rt.

A second strategy toward simplified spacer domains focused on synthetic accessibility and structural simplification. Given that the syntheses of spacer domains **21** and **45–48** were linear sequences of 10–16 steps, it would be beneficial to develop a route where two or more complex pieces could be coupled together to complete the molecule. It was envisioned that replacing the ether linkage in the A-ring-modified spacer domains with an ester linkage would increase the convergency of the synthesis (Scheme 6.9) (Wender and Lippa, 2000).

The preparation of this family of spacer domains began with synthesis of secondary alcohol **71**. Prenylation of **17β** gave a mixture of alcohols, which was recycled through an oxidation–reduction sequence to give **72β**. Reductive ozonolysis

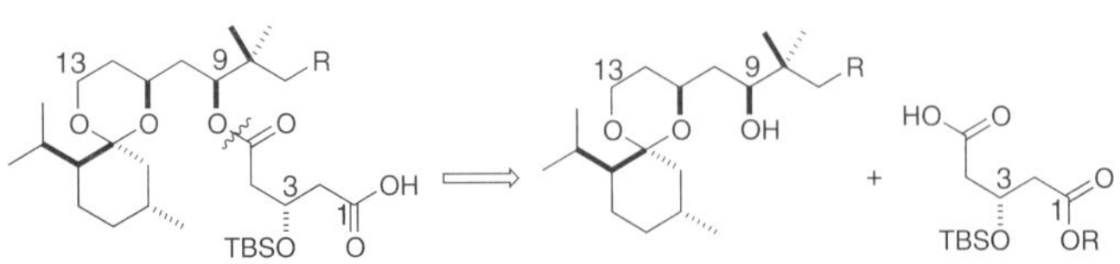

69: R = $OCOCH_2Cl$
70: R = H

71: R = $OCOCH_2Cl$
51: R = H

Scheme 6.9 The design of ester-linked A-ring spacer domains.

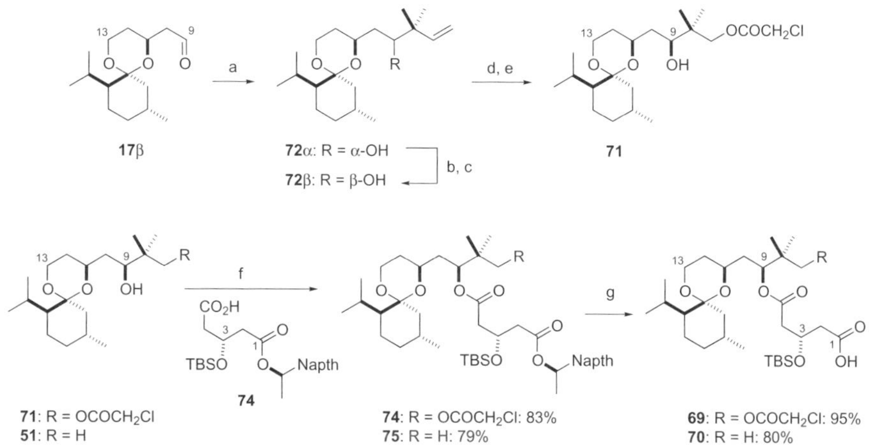

Scheme 6.10 Synthetic route to spacer domains **69** and **70**. (a) Prenyl bromide, In^0, DMF, 84% (3:1, **72α**:**72β**); (b) Dess–Martin periodinane, CH_2Cl_2, rt; (c) $NaBH_4$, $CeCl_3 \cdot 7H_2O$, MeOH, 0°C, 71% (over two steps; 3:1, **72β**:**72α**); (d) i. O_3, CH_2Cl_2, MeOH, −78°C, ii. $NaBH_4$, 0°C; (e) $(ClCH_2CO)_2O$, pyridine, CH_2Cl_2, 0°C, 63% (over two steps; based on recovered diol); (f) 2,4,6-trichlorobenzoyl chloride, DMAP, Et_3N, PhMe; (g) H_2 (1 atm), $Pd(OH)_2/C$, EtOAc.

of the terminal olefin followed immediately by esterification with chloroacetic anhydride yielded the selectively protected derivative **71**. The chloroacetate group was chosen as a temporary protecting group that could be later substituted with a variety of functionally diverse groups. The secondary alcohols **71** and **51** were then independently taken through the remaining two steps of the sequence. Yamaguchi esterification with known acid **73** followed by cleavage of the chiral auxiliary afforded the completed spacer domains **69** and **70** (Scheme 6.10).

The six A-ring-modified spacer domains reviewed in this chapter were synthesized over the course of several years. A consequence of this was that some of these spacer domains were coupled to the C26 methyl recognition domain **76**, used to construct analogue **1**, while those synthesized at a later time were coupled to the C26 des-methyl recognition domain **14** to complete the final analogues.

Of the spacer domains with the ether linkage, only **45** and **46** were coupled to **76** (Scheme 6.11) (Wender et al., 1998a). The coupling of these domains was accomplished using Yamaguchi's esterification protocol. The C3 silyl group was removed and the macrocyclic acetal was closed under acidic conditions. Hydrogenolysis of the C26 benzyl ether gave the completed bryostatin analogues **79** and **80**.

Spacer domains **69** and **70**, with the ester linkage, were also coupled with recognition domain **76** in a similar manner (Scheme 6.12). The chloroacetate of **84** was then removed by treatment with excess thiourea.

The C26 benzyl ethers were deprotected using the standard conditions to provide three new analogues: **86**, **87**, and **88** (Scheme 6.13).

Intermediate **85** was further diversified before removing the benzyl group to provide access to various esters in the A-ring region of the molecule (Scheme 6.14). In

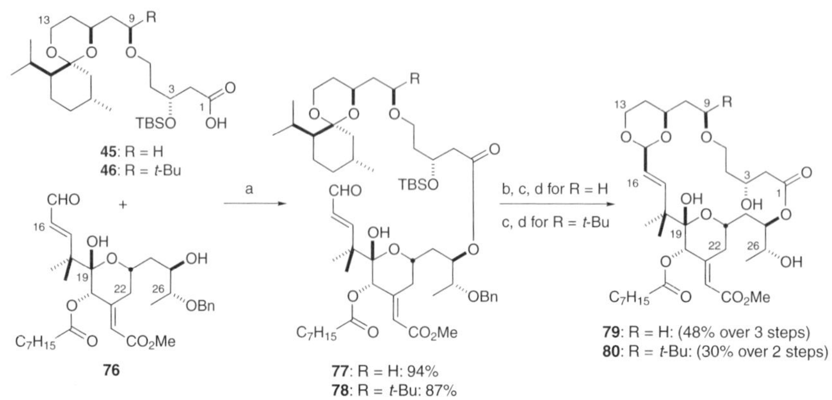

Scheme 6.11 Completion of C26 methyl analogues **79** and **80**. (a) 2,4,6-Trichlorobenzoylchloride, Et_3N, DMAP, CH_2Cl_2; (b) HF·pyridine, THF; (c) Amberlyst-15 resin, CH_2Cl_2; (d) $Pd(OH)_2/C$, H_2 (1 atm), EtOAc.

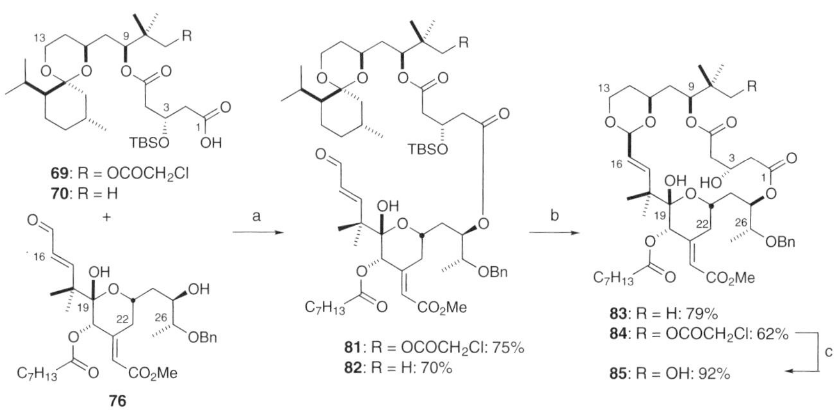

Scheme 6.12 Coupling and closure steps for ester-linked analogues. (a) 2,4,6-Trichlorobenzoyl chloride, Et_3N, DMAP, MePh; (b) excess HF·pyridine, THF; (c) excess thiourea, THF, 4 days.

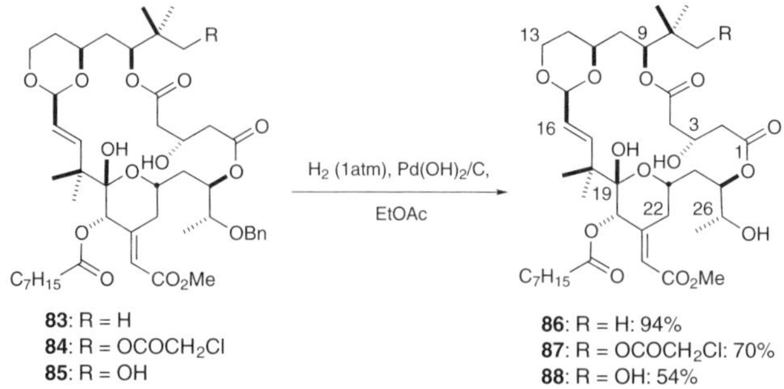

Scheme 6.13 Completion of C26 methyl analogues **86**, **87**, and **88**.

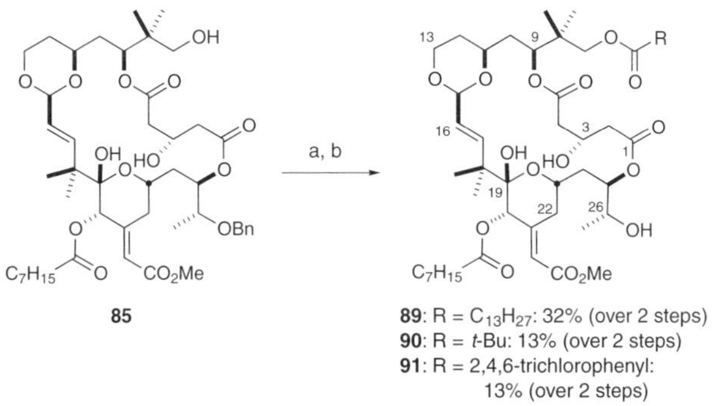

Scheme 6.14 Generation of three new analogues through diversification of intermediate **85**. (a) $C_{13}H_{27}CO_2H$ or *t*-$BuCO_2H$, 2,4,6-trichlorobenzoyl chloride, Et_3N, DMAP, MePh; (b) H_2 (1 atm), $Pd(OH)_2/C$, EtOAc.

this manner the myristate (**89**) and pivalate (**90**) analogues were generated. The 2,4,6-trichlorophenyl ester analogue **91** was generated as a byproduct in the second reaction.

All compounds exhibit high affinity for PKC (Fig. 6.7). Gratifyingly, most of the compounds show only slightly diminished activity compared with the parent compound **1** (3.4 nM) or the natural product (1.4 nM).

Attempts to couple the ester-based spacer domain **70** to the picolog recognition domain **14** to generate the des-methyl version of analogue **86** were hampered by decomposition of the material during purification. However, no stability issues were

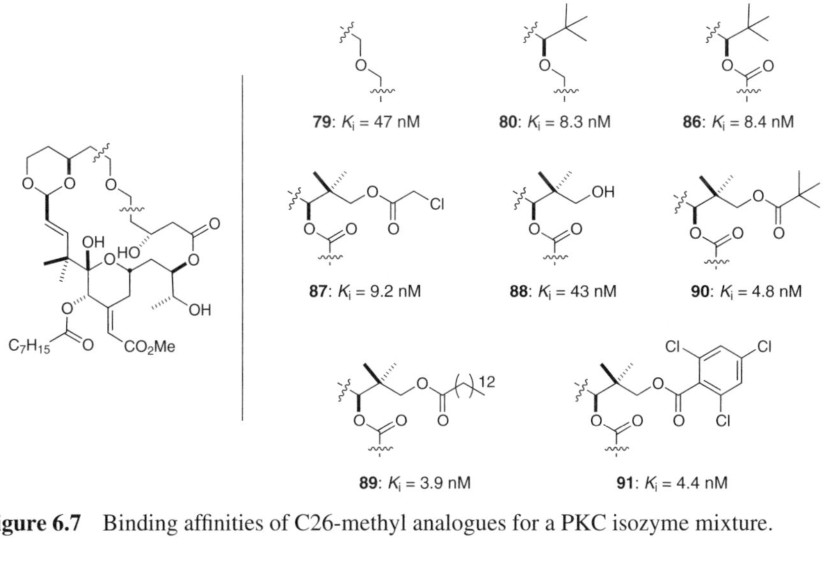

Figure 6.7 Binding affinities of C26-methyl analogues for a PKC isozyme mixture.

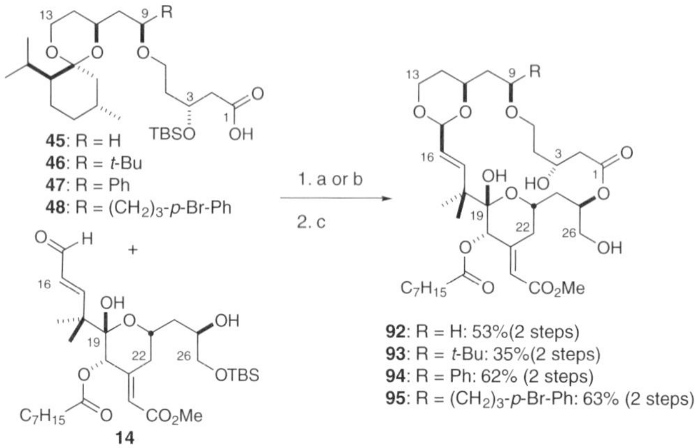

Scheme 6.15 Completion of A-ring modified analogues **92–95**. (a for **45**) 2,4,6-trichlorobenzoyl chloride, Et_3N, DMAP, MePh, rt; (b for **46–48**) PyBroP, $(i\text{-Pr})_2EtN$, DMAP, CH_2Cl_2, rt; (c) HF·pyridine, THF, 0°C → rt.

observed when spacer domains **45–48** were used in place of **70**. The spacer domains were coupled to the recognition domain and the macrocycle closed in the manner described previously to generate four new C26 des-methyl bryostatin analogues (Scheme 6.15).

All analogues exhibit single-digit nanomolar affinity for PKC (Fig. 6.8). While these A-ring modified analogues are roughly an order of magnitude less potent than the parent compound **15** in this assay, the activities are still very high. It is

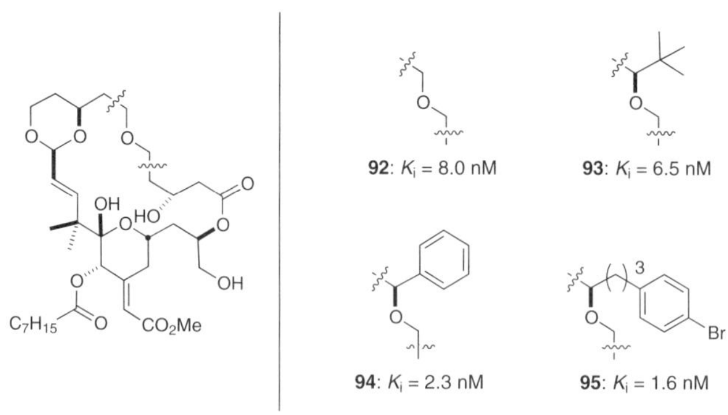

Figure 6.8 Binding affinity of C26 des-methyl analogues for PKC isozyme mixture.

interesting to note that removal of the C26 methyl group from analogue **1** resulted in a 10-fold increase in binding affinity. Similarly, removing the C26 methyl from compound **79** resulted in a six-fold increase in potency. However, removal of the same methyl group from compound **80** only resulted in a negligible change in binding affinity.

6.5 BRYOSTATIN ANALOGUES WITH MODIFIED RECOGNITION DOMAINS

The fundamental understanding that the key pharmacophoric substructures of bryostatin are contained in the recognition domain has led to a focus on spacer domain modifications while leaving most of the original recognition domain of bryostatin 1 intact, in order to tune potency and function. However, early structure–activity relationship studies of the natural products and closely related derivatives (Wender et al., 1998b; Pettit et al., 1992) suggest strategies for the design of recognition domain analogues that incorporate more significant changes. Comparison of the PKC binding affinities of bryostatins 1, 2, 4–11, and 18 (Fig. 6.1) shows that variations in the C20 ester have only a modest effect on potency. A derivative in which both the C13 and C21 enoates were hydrogenated shows greatly reduced potency as compared to a derivative in which only the C13 enoate was hydrogenated, suggesting that unsaturation in the C-ring is important for PKC affinity. Bryostatins lacking the C19 hydroxyl group and the C20 side chain (bryostatins 16 and 17) show decreased potency. These initial comparisons, while crucial to early directions in analogue design, do not address a number of issues such as the role of the C21 substituent and the range of tolerated C20 side chains. A hypothesis for the necessity of the C21 enoate for potency is that it serves as a conformational constraint on the C-ring. Consequently, a synthetically simplified recognition domain should either retain unsaturation at C21 or contain an alternative conformational constraint. In analogue **104**, a C20–C21 carbonate (Scheme 6.16) is such an alternative constraint.

The synthesis of the desired analogue started from pyranone **96** (Wender et al., 2003a; Koehler, 1999), which was diastereoselectively converted into diol **97** (Scheme 6.16). Treatment of **97** with triphosgene generated the carbonate **98**. Oxidative deprotection of the PMB groups followed by allylic oxidation provided enal **100**. Hydrolysis of the mixed ketal provided recognition domain **101**. Coupling to spacer domain **21** followed by TBS removal, macrotransacetalization, and hydrogenation of the benzyl group provided completed analogue **104**. This analogue shows substantially reduced affinity for PKC ($K_i > 10\,\mu\text{M}$), suggesting that unsaturation might be required for potency.

Subsequent studies on analogues with simplified recognition domains focused on preserving unsaturation at C21 while modifying the C34 ester functionality. Analogues without the ester moiety could prove to be more stable *in vivo*. Furthermore, it was envisioned that modifications at C34 could allow for rapid diversification at a late stage. With these issues in mind, analogues **105** (incorporating an aryl group) and **106** (incorporating a benzyl ester that could be further elaborated) were targeted (Fig. 6.9) (Pattabiramen, 2004).

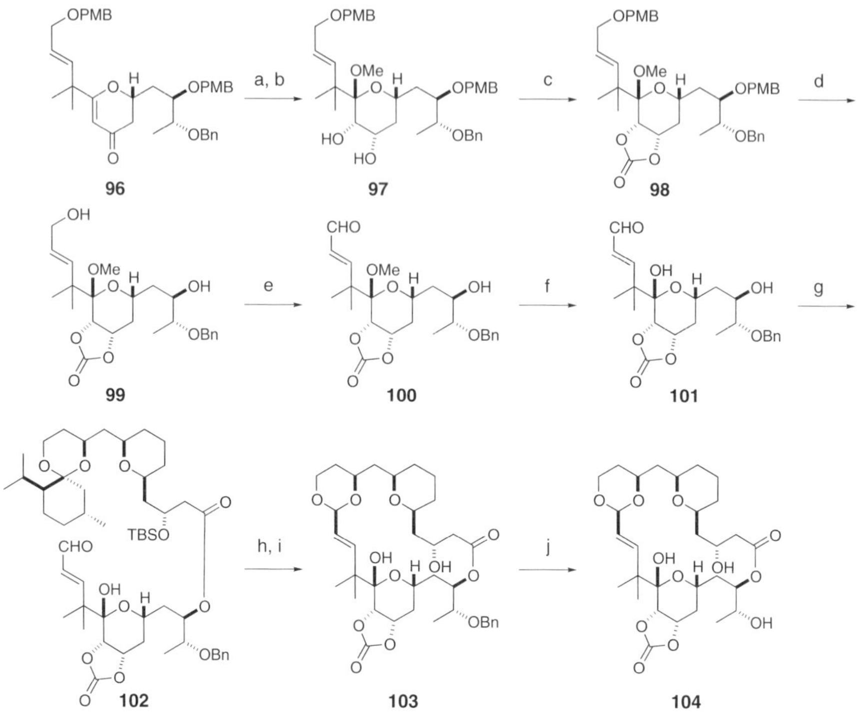

Scheme 6.16 Synthesis of a recognition domain analogue incorporating a C20–C21 carbonate. (a) $NaBH_4$, $CeCl_3 \cdot 7H_2O$, MeOH, −20°C; (b) *m*-CPBA, $NaHCO_3$, MeOH, 0°C, 71% over two steps; (c) triphosgene, pyridine, CH_2Cl_2, 93%; (d) DDQ, CH_2Cl_2/H_2O, 66%; (e) MnO_2, CH_2Cl_2, 80%; (f) 48% aq. HF, CH_3CN/H_2O (9:1), 45°C; (g) 2,4,6-trichlorobenzoyl chloride, Et_3N, PhMe, then **21**, DMAP, 62%; (h) HF · pyridine, pyridine, THF; (i) Amberlyst-15, 4 Å MS, CH_2Cl_2, 62% over two steps; (j) H_2 (1 atm), $Pd(OH)_2/C$, EtOAc, 88%.

Analogue **105** was synthesized starting from ketone **34** (Scheme 6.17), an advanced intermediate in the synthesis of **15**. Sharpless asymmetric dihydroxylation followed by protection as an acetonide provided **108**. Condensation with benzaldehyde provided enone **109**, which was reduced and esterified to provide **110**. Elaboration to the completed recognition domain **113** using conditions developed for the synthesis of **15** proceeded smoothly. Coupling of **113** with spacer domain **21** following the standard conditions provided the completed analogue **105**. This new analogue has a binding affinity of 100 nM.

Synthesis of analogue **106** began from ketone **34** (Scheme 6.18). Condensation with glyoxylate followed by esterification with benzyl chloroformate gave enone **114**. Using conditions developed for the synthesis of **15**, **114** was elaborated to recognition domain **120**. Coupling of recognition domain **120** with spacer domain **21** followed by closure of the macrocycle provided the completed analogue **106**. This new analogue exhibits a PKC affinity of 120 nM.

These analogues demonstrate the importance of the C21 enoate. Removal (**104**) or modification (**105** and **106**) of this functionality results in a significant loss of PKC

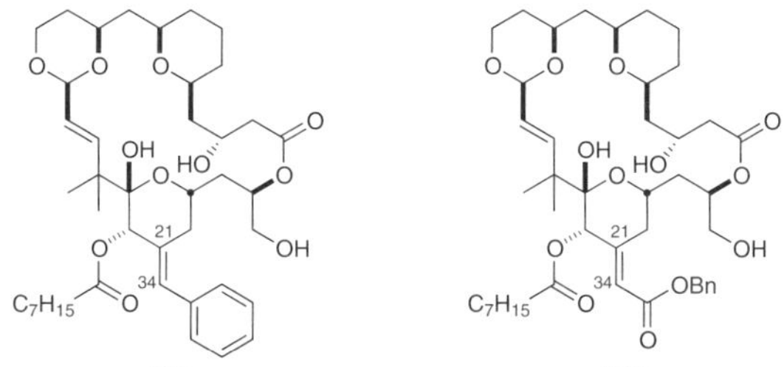

Figure 6.9 Benzylidene and benzoate analogues.

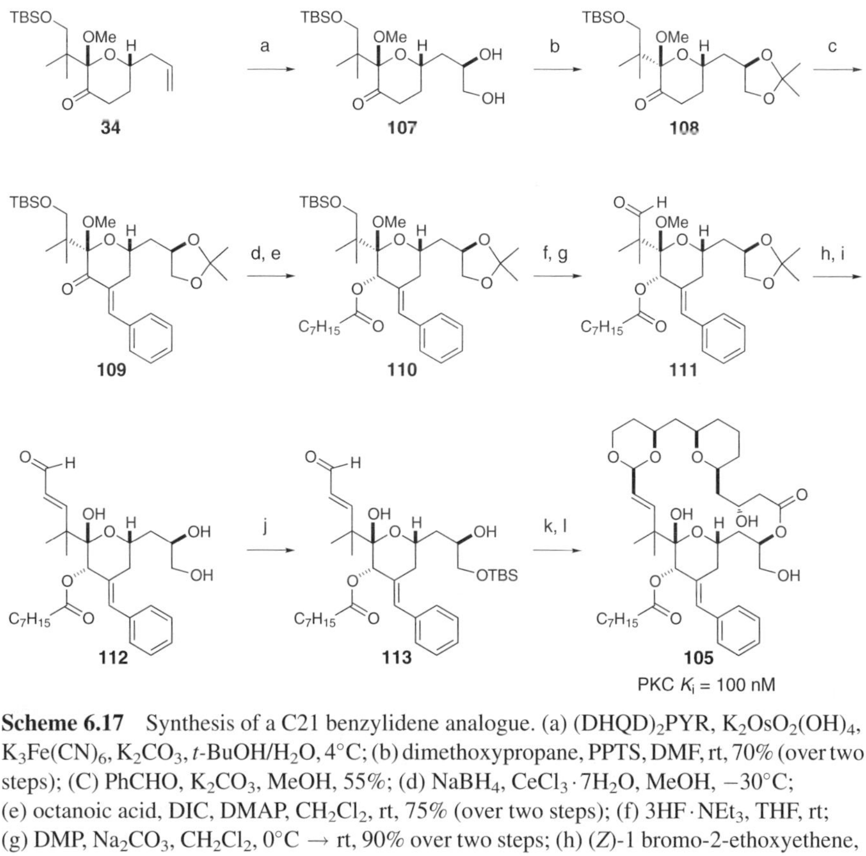

Scheme 6.17 Synthesis of a C21 benzylidene analogue. (a) $(DHQD)_2PYR$, $K_2OsO_2(OH)_4$, $K_3Fe(CN)_6$, K_2CO_3, *t*-BuOH/H_2O, 4°C; (b) dimethoxypropane, PPTS, DMF, rt, 70% (over two steps); (C) PhCHO, K_2CO_3, MeOH, 55%; (d) $NaBH_4$, $CeCl_3 \cdot 7H_2O$, MeOH, −30°C; (e) octanoic acid, DIC, DMAP, CH_2Cl_2, rt, 75% (over two steps); (f) 3HF · NEt_3, THF, rt; (g) DMP, Na_2CO_3, CH_2Cl_2, 0°C → rt, 90% over two steps; (h) (*Z*)-1 bromo-2-ethoxyethene, *t*-BuLi, Me_2Zn, Et_2O, −78°C; (i) *p*TsOH, MeCN/H_2O, rt, 55% (over two steps); (j) TBSCl, imidazole, CH_2Cl_2, rt, 85%; (k) **21**, PyBroP, DIEA, DMAP, CH_2Cl_2; (l) HF · pyridine, THF, rt, 80% over two steps.

affinity. While bulky C34 substituents are detrimental, smaller substituents in this region might be better tolerated.

The structure–activity analysis of the natural bryostatins (Fig. 6.1) and their semisynthetic analogues suggests that changes could be made to the C20 substituent without sacrificing efficacy *in vivo*. Additionally, computer modeling and pharmacophore analysis of bryostatin and synthetic analogues suggest that the C20 substituent might correspond to the *sn*-1 substituent of DAG and the C12 substituent of the phorbol esters. These substituents have been shown to affect the selectivity of PKC activation for both DAG and the phorbol esters (Madani et al., 2001; Wada et al., 2002; Wang et al., 1999). A systematic investigation of C20 substituents would enhance the ability to tune the biological activity of analogues.

To explore possible correlations between the carbon content of the C20 ester and affinity to PKC, functionalization of the C20 position was studied (Wender and Hinkle, 2000). Following protection of enal **121** as a dimethyl acetal, the octanoate ester was hydrolyzed to the corresponding alcohol (Scheme 6.19). Acylation followed by PMB deprotection and cleavage of both acetals provided recognition domain **123**. Coupling to spacer domain **21** followed by silyl deprotection, macrotransacetalization,

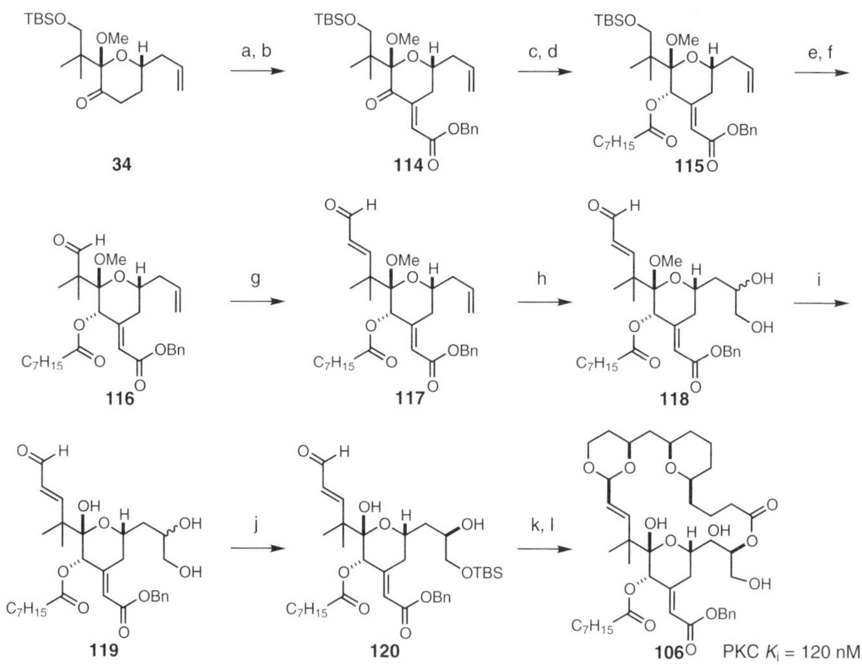

Scheme 6.18 Synthesis of a C21 benzoate analogue. (a) Glyoxylic acid, NaOH, THF/H_2O, rt; (b) benzyl chloroformate, DMAP, CH_2Cl_2, rt, 82% over two steps; (c) $NaBH_4$, $CeCl_3 \cdot 7H_2O$, MeOH, −30°C; (d) octanoic acid, DIC, DMAP, CH_2Cl_2, rt, 69% over two steps; (e) $3HF \cdot NEt_3$, THF; (f) DMP, Na_2CO_3, CH_2Cl_2, rt, 70% over two steps; (g) (*Z*)-1-bromo-2-ethoxyethene, *t*-BuLi, Me_2Zn, Et_2O, −78°C to rt, 63%; (h) $(DHQD)_2PYR$, $K_2OsO_2(OH)_4$, $K_3Fe(CN)_6$, K_2CO_3, *t*-BuOH/H_2O, 4°C, 61%; (i) *p*TsOH, MeCN/H_2O, rt, 64%; (j) TBSCl, imidazole, CH_2Cl_2, rt, 76%; (k) **21**, PyBroP, DIEA, DMAP, CH_2Cl_2, rt; (l) HF · pyridine, THF, rt, 80% over two steps.

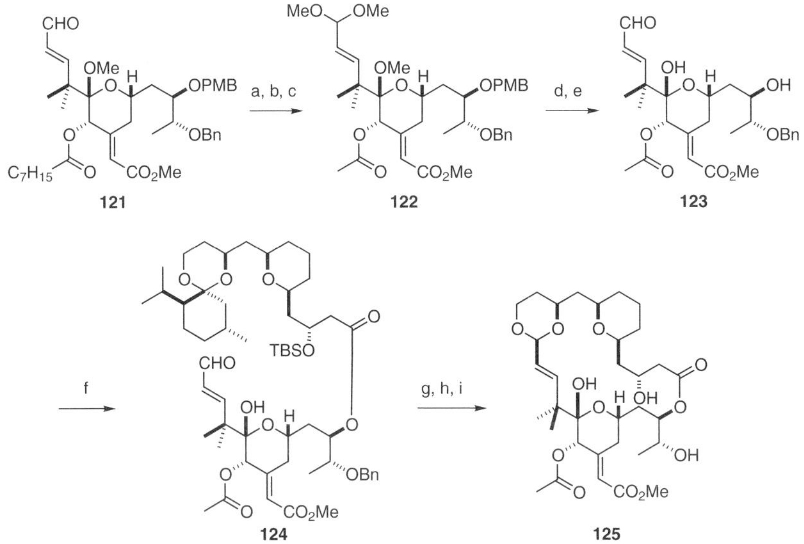

Scheme 6.19 Synthesis of C20 acetate analogue **125.** (a) $CH(OMe)_3$, PPTS, MeOH, rt; (b) K_2CO_3, MeOH, rt; (c) Ac_2O, DMAP, CH_2Cl_2, rt, 82% for three steps; (d) DDQ, CH_2Cl_2/H_2O, rt, 91%; (e) aq. HF, CH_3CN, rt, 89%; (f) 2,4,6-trichlorobenzoyl chloride, Et_3N, DMAP, then **21**, CH_2Cl_2, rt, 63%; (g) HF · pyridine, CH_3CN, rt, 82%; (h) amberlyst-15 resin, CH_2Cl_2, rt, 83%; (i) $Pd(OH)_2/C$, H_2 (1 atm), EtOAc, 93%.

and hydrogenation of the C26 benzyl protecting group provided analogue **125**, containing a C20 acetate.

Analogues **126** and **127** (Fig. 6.10) were prepared using a similar sequence. They have affinities for PKC similar to that of **1**. Within this series, it is evident that there is a positive correlation between carbon content of the C20 ester and PKC binding affinity. However, replacing the octanoate ester with a benzoate ester does not affect binding affinity, suggesting that diverse structural types might be tolerated.

The success of these efforts prompted the further investigation of modifications to the C20 position (Wender and Baryza, 2005). To efficiently explore variations at the

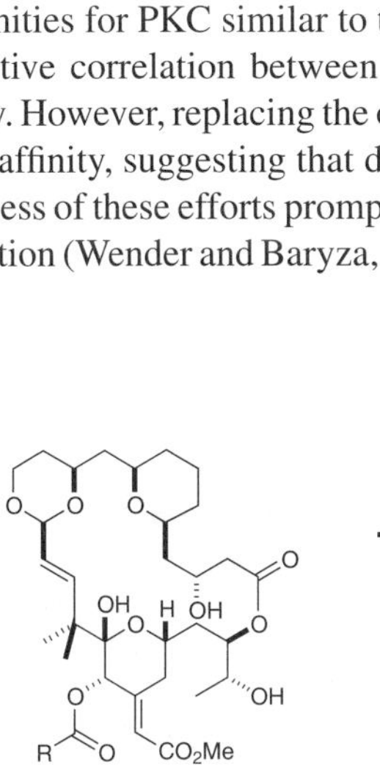

Compound	R	K_i (nM)
1	C_7H_{15}	3.4
125	CH_3	232
126	$C_{13}H_{27}$	1.5
127	Ph	7

Figure 6.10 PKC binding constants for first-generation C20 ester analogues.

C20 position, a functional handle was incorporated into the molecule that could be selectively diversified as late as the final analogue. An aniline, which could be revealed by selective reduction of an aryl nitro group, was chosen as the diversification handle (**134**, Scheme 6.20). The 3-nitro-benzoate was introduced by

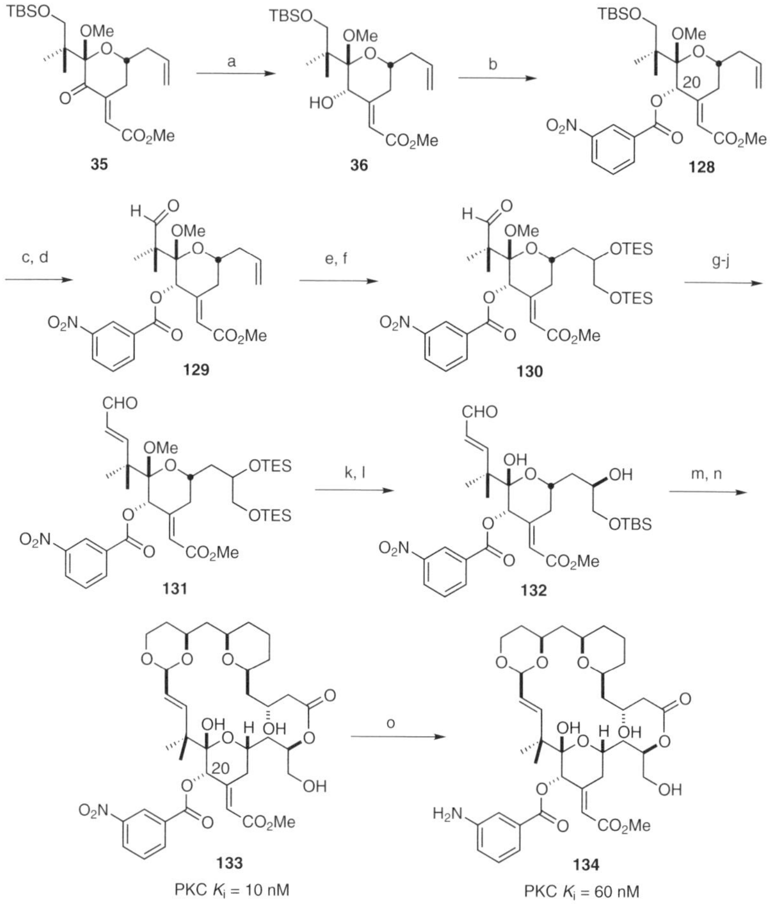

Scheme 6.20 Preparation of aryl amine for late-stage diversification. (a) $NaBH_4$, $CeCl_3 \cdot 7H_2O$, MeOH, −30°C; (b) 3-$(NO_2)C_6H_4CO_2H$, DIC, DMAP, CH_2Cl_2, rt, 75% over two steps; (c) $3HF \cdot NEt_3$, CH_2Cl_2, rt, 58%; (d) DMP, $NaHCO_3$, CH_2Cl_2, rt, 86%; (e) $K_2OsO_2(OH)_4$, $(DHQD)_2PYR$, $K_3Fe(CN)_6$, K_2CO_3, H_2O, *t*-BuOH, 4°C, 67% (2:1 β:α); (f) Et_3SiCl, imidazole, CH_2Cl_2, rt, 98%; (g) Et_2BOMe, allylMgCl, Et_2O, rt; (h) Ac_2O, DMAP, CH_2Cl_2, rt, 67% over two steps; (i) OsO_4, NMO, H_2O, THF, rt; (j) i. $Pb(OAc)_4$, PhMe, ii. DBU, rt, 60% over two steps; (k) *p*TsOH · H_2O, H_2O, MeCN, rt; (l) TBSCl, imidazole, CH_2Cl_2, rt, 43% over two steps; (m) **21**, PyBroP, *i*-Pr_2NEt, DMAP, CH_2Cl_2, rt, 75%; (n) HF · pyridine, THF, rt, 80%; (o) H_2 (1 atm), 10% Pd/C, EtOAc, 99%.

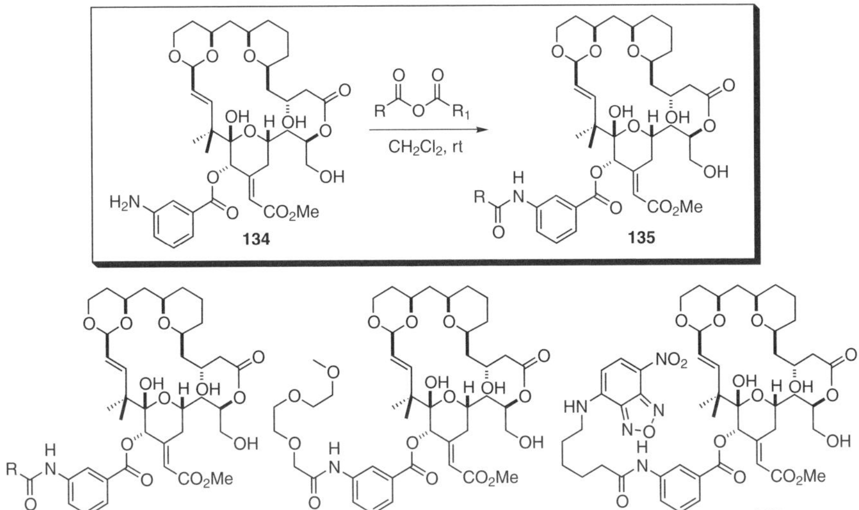

135
R = CH_3 PKC K_i = 91 nM
R = C_7H_{15} PKC K_i = 12 nM
R = Ph PKC K_i = 18 nM

136
PKC K_i = 77 nM

137
PKC K_i = 21 nM

Scheme 6.21 PKC binding constants for C20 aryl amide analogues.

esterification to provide **128**. This compound was elaborated to the corresponding recognition domain through a sequence similar to that used for the synthesis of **15**. Coupling with spacer domain **21** and subsequent one-step deprotection/macrotransacetalization gave nitroaryl analogue **133**, which was selectively hydrogenated to aniline **134**.

Reaction of aniline **134** with a variety of anhydrides resulted in exclusive acylation of the aniline moiety. Using this strategy, an initial set of analogues was prepared (Scheme 6.21). All these analogues have low nanomolar affinity for PKC. The general trend for increased potency with increasing carbon count seems to hold, while many of the more polar groups are well-tolerated.

6.6 BIOLOGICAL STUDIES ON ANALOGUES OF BRYOSTATIN

It is important that designed analogues of bryostatin possess not only the affinity, but also the function of the natural product. Determination of binding affinity is an important first screen to eliminate compounds that are unlikely to act on the target of interest. However, binding potency in a cell-free environment does not necessarily correlate to cellular or *in vivo* activity.

An important biological activity of bryostatin is its anticancer activity; therefore, analogues **1** and **15** were sent to the NCI for testing against the 60 cell-line panel used for profiling new anticancer agents. In many cell lines, both analogue **1** and

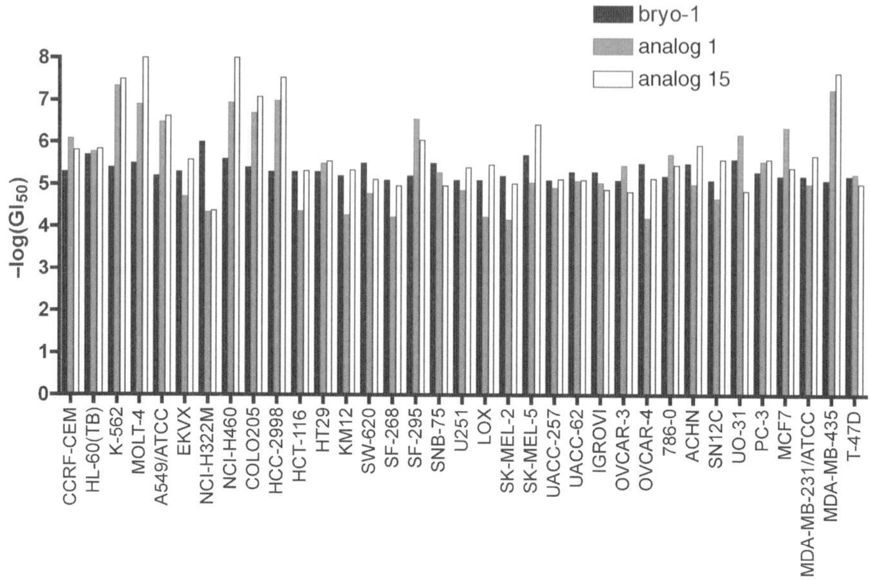

Figure 6.11 Growth inhibitory activity of bryostatin 1 and analogues **1** and **15** against select cancer cell lines.

analogue **15** outperformed bryostatin 1 at inhibiting the growth of cancer cells, in some cases by up to three orders of magnitude (Fig. 6.11) (Wender et al., 2002, Wender et al., 2004).

Bryostatin 1 is known to activate PKC, an activity likely to be responsible for at least some of the therapeutic effects it displays. Other C1 domain containing protein families are also likely to be involved (Kazanietz, 2002; Brose et al., 2004), but the activation and translocation of PKC to cellular membranes is a specific cellular effect of bryostatin treatment. Therefore, a PKC translocation assay was implemented as a way to measure potency and activity of bryostatin analogues in living cells.

Activation of PKC is tightly coupled to the location of the protein. Inactive PKC is kept in various intracellular locations. Activating cofactors increase the membrane affinity of the kinase and induce its translocation to cellular membranes. Membrane association drives the release of an autoinhibitory sequence from the kinase active site, which turns on signaling by the kinase. This union of translocation and activation has led to the widespread use of translocation as a surrogate measure for PKC activation.

For the measurement of PKC translocation, plasmids encoding for a fusion protein of PKC and green fluorescent protein (GFP) are expressed in living cells (Fig. 6.12). By treating these cells with bryostatin or bryostatin analogues and observing the localization of the fluorescence with confocal microscopy, the response of the protein can be observed in real time. Analysis of the images allows for quantification of the rate and degree of response as well as the final localization of the protein. As PKC translocates to the membrane, fluorescence in

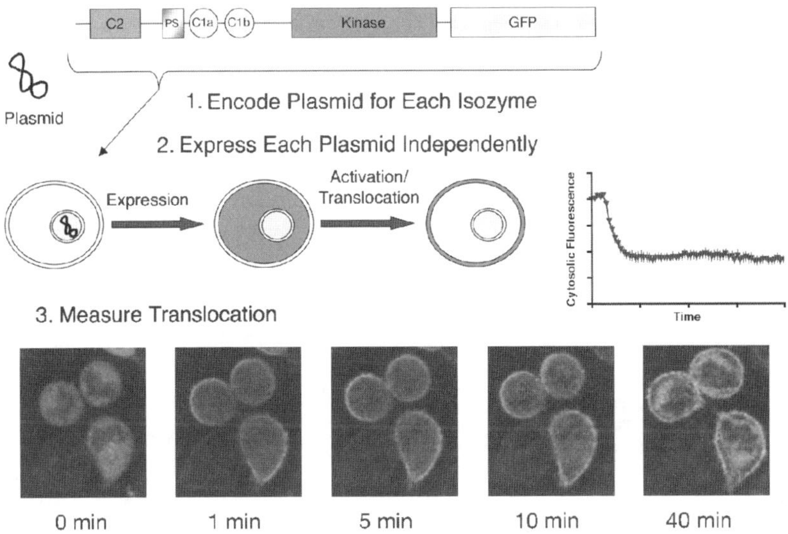

Figure 6.12 The PKC-GFP assay.

the cytosol decreases, providing plots of the type shown in Fig. 6.12. The use of isozyme specific constructs allows for the response of individual PKC isozymes to be measured (Baryza et al., 2004).

Comparison of the response of PKCδ to bryostatin and to analogues **1** and **15** confirmed that the analogues are more potent than the natural product at inducing translocation of this isozyme. While bryostatin 1 failed to induce translocation of PKCδ at doses below 5 nM, analogue **1** still possessed activity at 1 nM, and analogue **15** maintained the ability to induce translocation even at 0.10 nM concentration (Fig. 6.13).

The ability of the analogues to induce translocation of PKCδ contrasts with their binding affinities for PKC. While this is not unexpected, the findings have led to an increased interest in testing a larger set of compounds against a wider range of PKC isozymes.

Over the last several years, aided by the sequencing of the human genome, other families of proteins containing C1 domains have been identified, including RasGRP (Caloca et al., 2003), the chimaerins (Hall et al., 2005), the Munc proteins (Betz et al., 1998), diacylglycerol kinase (DGK) (van Blitterswijk and Houssa, 2000), and protein kinase D (PKD) (Rozengurt et al., 2005). Where tested, bryostatin has been effective at activating these proteins as well (Lorenzo et al., 2000; Rhee et al., 2002; Shindo et al., 2003; Matthews et al., 1997). It is likely that some of the effects previously ascribed to PKC are in fact caused by these proteins.

Activation of RasGRP appears to be responsible for some of the immune system effects previously attributed to PKC. RasGRP is a protein family that is activated by translocation to the plasma membrane, where it activates Ras. RasGRP1 appears to play a critical role in T-cell receptor signaling. When analogues **1** and **15** were tested in assays measuring RasGRP1 activation, the compounds displayed the

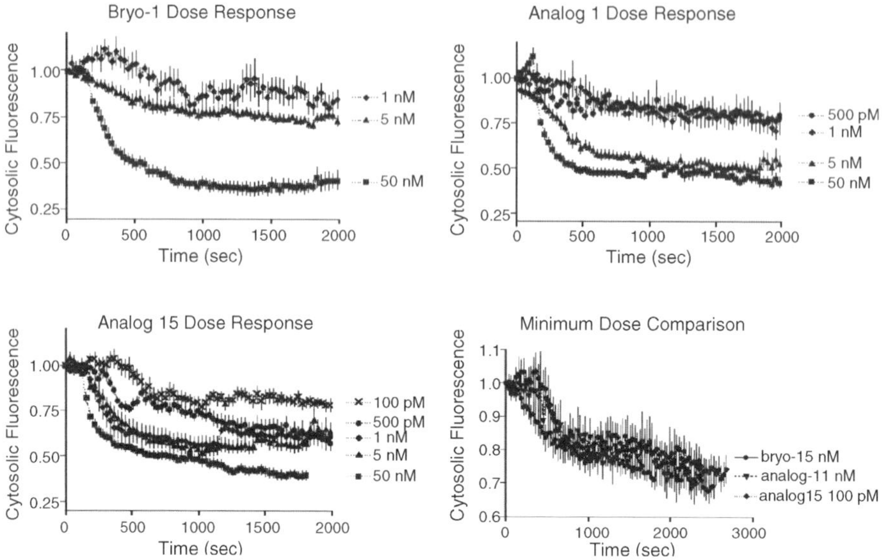

Figure 6.13 Translocation of PKCδ by bryostatin, analogue **1** and analogue **15**.

same activity as bryostatin. The analogues, as well as bryostatin, induced RasGRP1 activation, (as measured by both RasGRP1 translocation and ERK phosphorylation), CD69 expression (a marker for T-cell activation), and thymocyte aggregation (Stone et al., 2004).

Bryostatin and the phorbol esters both have immunostimulatory properties. *in vivo*, immune system activation is possibly responsible for some of the anticancer activity of bryostatin. In a study of immune cell activation, it was shown that phorbol-12-myristate-13-acetate (PMA) increased the immunogenicity of chronic lymphocytic leukemia (CLL) cells taken from treated and untreated patients. In the same study, analogue **15** was tested and found to display the same effects on CLL cells as PMA, increasing the clinical potential of this finding. This could lead to immunotherapy-based treatments for CLL and other hematopoetic cancers based on bryostatin analogues (Hammond et al., 2005).

The phorbol esters have been widely used in biochemical studies to measure the consequences of activating C1 domain containing proteins. Due to their tumor-promoting effects, there is great reluctance to use them in therapeutic applications, although there have been some rare instances of testing the therapeutic effects of phorbol treatment in humans. With the exception of their effects on tumor growth, bryostatin and the phorbol esters share many biological effects (Sako et al., 1987). Bryostatin has been approved for therapeutic testing in humans, so in cases where the phorbol treatment induces a potentially beneficial effect in biochemical studies, it is natural to switch to bryostatin for *in vivo* testing and development. The limited availability and cost of bryostatin effectively prohibit its use in place of phorbol esters. Bryostatin analogues are therefore attractive as agents in these applications.

6.7 CONCLUSION AND OUTLOOK

The established and growing links between PKC signaling and human diseases suggest that the need for and therapeutic potential of agents that act on PKC will grow. The most widely used agents, the phorbol esters, are tainted by their tumor-promoting effects. Bryostatin has shown anticancer activity, but the low isolation yield and structural complexity of the natural product effectively prohibit its widespread use. Therefore, functional analogues of bryostatin, which possess bryostatin-like activity but can be prepared in reliable quantities and tuned for function, are compounds with enormous potential for the treatment of human disease. More generally, these studies collectively illustrate the emerging role of rational drug design coupled to synthesis in the generation and advancement of new therapeutic leads. Whereas traditionally the focus of synthesis has been on making natural products themselves, the growing realization that few natural products are optimal for human use and the creative challenge of designing more effective but simpler leads collectively create new and highly significant opportunities for synthesis. While there is great satisfaction and value in achieving the synthesis of a known natural product, there is often even greater satisfaction and impact in the design and synthesis of previously unknown agents, especially as in the case of the bryologs, when these agents have activities and promise beyond natural products.

ACKNOWLEDGMENTS

Financial support for aspects of the work described herein was provided by grants from the National Institutes of Health (CA 31845) and GPC Biotech. This work is a product of many exceptionally talented, creative, and dedicated investigators whose names appear as coauthors and in the references provided.

REFERENCES

Al-Katib, A. M., Smith, M. R., Kamanda, W. S., Pettit, G. R., Hamdan, M., Mohamed, A. N., Chelladurai, B., and Mohammad, R. M. (1998). Bryostatin 1 down-regulates mdr1 and potentiates vincristine cytotoxicity in diffuse large cell lymphoma xenografts. *Clin Cancer Res* **4**, 1305–1314.

Baryza, J. L., Brenner, S. E., Craske, M. L., Meyer, T., and Wender, P. A. (2004). Simplified analogues of bryostatin with anticancer activity display greater potency for translocation of PKCdelta-GFP. *Chem Biol* **11**, 1261–1267.

Basu, A., and Lazo, J. S. (1992). Sensitization of human cervical-carcinoma cells to *cis*-diamminedichloroplatinum(II) by bryostatin-1. *Cancer Res* **52**, 3119–3124.

Bear, H. D., Roberts, J., Cornell, D., Tombes, M. B., and Kyle, B. (2001). Adoptive immunotherapy of cancer with pharmacologically activated lymph node lymphocytes: A pilot clinical trial. *Cancer Immunol Immunother* **50**, 269–274.

Beck, F. W. J., Eilender, D. S., Dandashi, M. H., Siddiq, F., Snell, D. C., Godmere, M. A., Al-Katib, A. M., and Mohammad, R. M. (2004). Factors affecting bryostatin 1-enhanced 2-CdA cytotoxicity in resistant B-cell chronic lymphocytic leukemia. *Int J Mol Med* **14**, 113–119.

Berkow, R. L., and Kraft, A. S. (1985). Bryostatin, a non-phorbol macrocyclic lactone, activates intact human polymorphonuclear leukocytes and binds to the phorbol ester receptor. *Biochem Biophys Res Commun* **131**, 1109–1116.

Betz, A., Ashery, U., Rickmann, M., Augustin, I., Neher, E., Sudhof, T. C., Rettig, J., and Brose, N. (1998). Munc13-1 is a presynaptic phorbol ester receptor that enhances neurotransmitter release. *Neuron* **21**, 123–136.

Brose, N., Betz, A., and Wegmeyer, H. (2004). Divergent and convergent signaling by the diacylglycerol second messenger pathway in mammals. *Curr Opin Neurobiol* **14**, 328–340.

Cahiez, G., Alexakis, A., and Normant, J. F. (1978). Organomagnesium derivatives of terminal alcohols. *Tetrahedron Lett* **33**, 3013–3014.

Caloca, M. J., Wang, H. B., and Kazanietz, M. G. (2003). Characterization of the Rac-GAP (Rac-GTPase-activating protein) activity of β2-chimaerin, a 'non-protein kinase C' phorbol ester receptor. *Biochem J* **375**, 313–321.

Caponigro, F., French, R. C., and Kaye, S. B. (1997). Protein kinase C: A worthwhile target for anticancer drugs? *Anti-Cancer Drugs* **8**, 26–33.

Cartee, L., Maggio, S. C., Smith, R., Sankala, H. M., Dent, P., and Grant, S. (2003). Protein kinase C-dependent activation of the tumor necrosis factor receptor-mediated extrinsic cell death pathway underlies enhanced apoptosis in human myeloid leukemia cells exposed to bryostatin 1 and flavopiridol. *Mol Cancer Ther* **2**, 83–93.

Cragg, L. H., Andreeff, M., Feldman, E., Roberts, J., Murgo, A., Winning, M., Tombes, M. B., Roboz, G., Kramer, L., and Grant, S. (2002). Phase I trial and correlative laboratory studies of bryostatin I (NSC 339555) and high-dose 1-B-D-arabinofuranosylcytosine in patients with refractory acute leukemia. *Clin Cancer Res* **8**, 2123–2133.

Curiel, R. E., Garcia, C. S., Farooq, L., Aguero, M. F., and Espinoza-Delgado, I. (2001). Bryostatin-1 and IL-2 synergize to induce IFN-gamma expression in human peripheral blood T cells: Implications for cancer immunotherapy. *J Immunol* **167**, 4828–4837.

da Rocha, A. B., Mans, D. R., Regner, A., and Schwartsmann, G. (2002). Targeting protein kinase C: New therapeutic opportunities against high-grade malignant gliomas? *Oncologist* **7**, 17–33.

de Vries, D. J., Herald, C. L., Pettit, G. R., and Blumberg, P. M. (1988). Demonstration of sub-nanomolar affinity of bryostatin 1 for the phorbol ester receptor in rat brain. *Biochem Pharmacol* **37**, 4069–4073.

Dempsey, E. C., Newton, A. C., Mochly-Rosen, D., Fields, A. P., Reyland, M. E., Insel, P. A., and Messing, R. O. (2000). Protein kinase C isozymes and the regulation of diverse cell responses. *Am J Physiol Lung Cell Mol Physiol* **279**, L429–L438.

Do, Y., Hegde, V. L., Nagarkatti, P. S., and Nagarkatti, M. (2004). Bryostatin-1 enhances the maturation and antigen-presenting ability of murine and human dendritic cells. *Cancer Res* **64**, 6756–6765.

Dowlati, A., Lazarus, H. M., Hartman, P., Jacobberger, J. W., Whitacre, C., Gerson, S. L., Ksenich, P., Cooper, B. W., Frisa, P. S., Gottlieb, M., Murgo, A. J., and Remick, S. C. (2003). Phase I and correlative study of combination bryostatin 1 and vincristine in relapsed B-cell malignancies. *Clin Cancer Res* **9**, 5929–5935.

Elgie, A. W., Sargent, J. M., Alton, P., Peters, G. J., Noordhuis, P., Williamson, C. J., and Taylor, C. G. (1998). Modulation of resistance to ara-C by bryostatin in fresh blast cells from patients with AML. *Leuk Res* **22**, 373–378.

Evans, D. A., Carter, P. H., Carreira, E. M., Charette, A. B., Prunet, J. A., and Lautens, M. (1999). Total synthesis of bryostatin 2. *J Am Chem Soc* **121**, 7540–7552.

Farrow, B., Thomas, R. P., Wang, X. F., and Evers, B. M. (2002). Activation of conventional PKC isoforms increases expression of the pro-apoptotic protein Bad and TRAIL receptors. *Int J Gastrointest Cancer* **32**, 63–72.

Gavrielides, M. V., Frijhoff, A. F., Conti, C. J., and Kazanietz, M. G. (2004). Protein kinase C and prostate carcinogenesis: targeting the cell cycle and apoptotic mechanisms. *Curr Drug Targets* **5**, 431–443.

Gomez-Fernandez, J. C., Torrecillas, A., and Corbalan-Garcia, S. (2004). Diacylglycerols as activators of protein kinase C. *Mol Membr Biol* **21**, 339–349.

Hall, C., Lim, L., and Leung, T. (2005). C1, see them all. *Trends Biochem Sci* **30**, 169–171.

Hammond, C., Shi, Y., Mena, J., Tomic, J., Cervi, D., He, L., Millar, A. E., Debenedette, M., Schuh, A. C., Baryza, J. L., Wender, P. A., Radvanyi, L., and Spaner, D. E. (2005). Effect of serum and antioxidants on the immunogenicity of protein kinase C-activated chronic lymphocytic leukemia cells. *J Immunother* **28**, 28–39.

Harada, T., Inoue, A., Wada, I., Uchimura, J. J., Tanaka, S., and Oku, A. (1993). Stereoselective acetalization of 1,3-alkanediols controlled by intramolecular van-der-Waals attractive interactions and its application to an enantiodifferentiating transformation of sigma-symmetrical 1,3,5-pentanetriols. *J Am Chem Soc* **115**, 7665–7674.

Hecker, E. (1978). In T. J. Slaga, A. Sivak, and R. K. Boutwell (Eds.), *Carcinogenesis—A Comprehensive Survey*, Vol. 2, p. 11.

Hennings, H., Blumberg, P. M., Pettit, G. R., Herald, C. L., Shores, R., and Yuspa, S. H. (1987). Bryostatin 1, an activator of protein kinase C, inhibits tumor promotion by phorbol esters in SENCAR mouse skin. *Carcinogenesis* **8**, 1343–1346.

Hildebrand, M., Waggoner, L. E., Liu, H. B., Sudek, S., Allen, S., Anderson, C., Sherman, D. H., and Haygood, M. (2004). bryA: An unusual modular polyketide synthase gene from the uncultivated bacterial symbiont of the marine bryozoan Bugula neritina. *Chem Biol* **11**, 1543–1552.

Joly, G. D., and Jacobsen, E. N. (2002). Catalyst-controlled diastereoselective hetero-Diels–Alder reactions. *Org Lett* **4**, 1795–1798.

Kageyama, M., Tamura, T., Nantz, M. H., Roberts, J. C., Somfai, P., Whritenour, D. C., and Masamune, S. (1990). Synthesis of bryostatin-7. *J Am Chem Soc* **112**, 7407–7408.

Kamano, Y., Zhang, H. P., Morita, H., Itokawa, H., Shirota, O., Pettit, G. R., Herald, D. L., and Herald, C. L. (1996). Conformational analysis of a marine antineoplastic macrolide, bryostatin 10. *Tetrahedron* **52**, 2369–2376.

Kazanietz, M. G. (2002). Novel "nonkinase" phorbol ester receptors: The C1 domain connection. *Mol Pharmacol* **61**, 759–767.

Keck, G. E., and Krishnamurthy, D. (1998). Catalytic asymmetric allylation reactions: (*S*)-1-(phenylmethoxy)-4-penten-2-ol. *Org Synth* **75**, 12–18.

Kitamura, M., Ohkuma, T., Inoue, S., Sayo, N., Kumobayashi, H., Akutagawa, S., Ohta, T., Takaya, H., and Noyori, R. (1988). Homogeneous asymmetric hydrogenation of functionalized ketones. *J Am Chem Soc* **110**, 629–631.

Koehler, M. F. T. (1999). Protein kinase C activators: Structure, function and synthesis. Ph.D. Dissertation, Stanford University.

Kraft, A. S., Smith, J. B., and Berkow, R. L. (1986). Bryostatin, an activator of the calcium phospholipid-dependent protein kinase, blocks phorbol ester-induced differentiation of human promyelocytic leukemia cells HL-60. *Proc Natl Acad Sci USA* **83**, 1334–1338.

Kudinov, Y., Wiseman, C. L., and Kharazi, A. I. (2003). Phorbol myristate acetate and bryostatin 1 rescue IFN-gamma inducibility of MHC class II molecules in LS1034 colorectal carcinoma cell line. *Cancer Cell Int.* **3**, 4–19.

Lahn, M., Kohler, G., Sundell, K., Su, C., Li, S., Paterson, B. M., and Bumol, T. F. (2004). Protein kinase C alpha expression in breast and ovarian cancer. *Oncology* **67**, 1–10.

Lippa, B. (1999). The design, synthesis and biological evaluation of the first totally synthetic bryostatin analogues. Ph.D. Dissertation, Stanford University.

Lopanik, N., Lindquist, N., and Targett, N. (2004). Potent cytotoxins produced by a microbial symbiont protect host larvae from predation. *Oecologia* **139**, 131–139.

Lorenzo, P. S., Beheshti, M., Pettit, G. R., Stone, J. C., and Blumberg, P. M. (2000). The guanine nucleotide exchange factor RasGRP is a high-affinity target for diacylglycerol and phorbol esters. *Mol Pharmacol* **57**, 840–846.

Mackay, H. J., and Twelves, C. J. (2003). Protein kinase C: A target for anticancer drugs? *Endocr Relat Cancer* **10**, 389–396.

Madani, S., Hichami, A., Legrand, A., Belleville, J., and Khan, N. A. (2001). Implication of acyl chain of diacylglycerols in activation of different isoforms of protein kinase C. *FASEB J* **15**, 2595–2601.

Manning, G., Whyte, D. B., Martinez, R., Hunter, T., and Sudarsanam, S. (2002). The protein kinase complement of the human genome. *Science* **298**, 1912–19334.

Marsa, L. (2002). *The Los Angeles Times* February 11, p. S1.

Matsui, W., Smith, B. D., Vala, M., Beal, N., Huff, C. A., Diehl, L. F., and Jones, R. J. (2005). Requirement for myeloid growth factors in the differentiation of acute promyelocytic leukaemia. *Br J Haematol* **128**, 853–862.

Matthews, S. A., Pettit, G. R., and Rozengurt, E. (1997). Bryostatin 1 induces biphasic activation of protein kinase D in intact cells. *J Biol Chem* **272**, 20245–20250.

McGovern, T. M., and Hellberg, M. E. (2003). Cryptic species, cryptic endosymbionts, and geographical variation in chemical defences in the bryozoan *Bugula neritina*. *Mol Ecol* **12**, 1207–1215.

McGown, A. T., Jayson, G., Pettit, G. R., Haran, M. S., Ward, T. H., and Crowther, D. (1998). Bryostatin 1-tamoxifen combinations show synergistic effects on the inhibition of growth of P388 cells *in vitro*. *Br J Cancer* **77**, 216–220.

Mendola, D. (2003). Aquaculture of three phyla of marine invertebrates to yield bioactive metabolites: Process developments and economics. *Biomol Eng* **20**, 441–458.

Mohammad, R. M., Beck, F. W. J., Katato, K., Hamdy, N., Wall, N., and AlKatib, A. (1998a). Potentiation of 2-chlorodeoxyadenosine activity by bryostatin 1 in the resistant chronic lymphocytic leukemia cell line (WSU-CLL): Association with increased ratios of dCK/5'-NT and Bax/Bcl-2. *Biol Chem* **379**, 1253–1261.

Mohammad, R. M., Varterasian, M. L., Almatchy, V. P., Hannoudi, G. N., Pettit, G. R., and AlKatib, A. (1998b). Successful treatment of human chronic lymphocytic leukemia xenografts with combination biological agents auristatin PE and bryostatin 1. *Clin Cancer Res* **4**, 1337–1343.

Moune, S., Niel, G., Busquet, M., Eggleston, I., and Jouin, P. (1997). Total synthesis of dolatrienoic acid: A subunit of dolastatin 14. *J Org Chem* **62**, 3332–3339.

Mutter, R., and Wills, M. (2000). Chemistry and clinical biology of the bryostatins. *Bioorg Med Chem* **8**, 1841–1860.

Noyori, R., Ohkuma, T., Kitamura, M., Takaya, H., Sayo, N., Kumobayashi, H., and Akutagawa, S. (1987). Asymmetric hydrogenation of beta-keto carboxylic esters—A practical, purely chemical access to beta-hydroxy esters in high enantiomeric purity. *J Am Chem Soc* **109**, 5856–5858.

Ohmori, K., Ogawa, Y., Obitsu, T., Ishikawa, Y., Nishiyama, S., and Yamamura, S. (2000). Total synthesis of bryostatin. *Angew Chem, Int Ed* **39**, 2290–2294.

Oz, H. S., Hughes, W. T., Rehg, J. E., and Thomas, E. K. (2000). Effect of CD40 ligand and other immunomodulators on *Pneumocystis carinii* infection in rat model. *Microb Pathog* **29**, 187–190.

Parviz, M., Chin, C. S., Graham, L. J., Miller, C., Lee, C., George, K., and Bear, H. D. (2003). Successful adoptive immunotherapy with vaccine-sensitized T cells, despite no effect with vaccination alone in a weakly immunogenic tumor model. *Cancer Immunol Immunother* **52**, 739–750.

Pattabiramen, K. (2004). The design, synthesis, and evaluation of guanidine-based molecular transporters. Ph.D. Dissertation, Stanford University.

Pettit, G. R., Day, J. F., Hartwell, J. L., and Wood, H. B. (1970). Antineoplastic components of marine animals. *Nature* **227**, 962–963.

Pettit, G. R., Herald, C. L., Doubek, D. L., Herald, D. L., Arnold, E., and Clardy, J. (1982). Antineoplastic agents 86. Isolation and structure of bryostatin 1. *J Am Chem Soc* **104**, 6846–6848.

Pettit, G. R., Herald, D. L., Gao, F., Sengupta, D., and Herald, C. L. (1991). Antineoplastic agents 200. Absolute-configuration of the bryostatins. *J Org Chem* **56**, 1337–1340.

Pettit, G. R., Sengupta, D., Blumberg, P. M., Lewin, N. E., Schmidt, J. M., and Kraft, A. S. (1992). Structural modifications of bryostatin-2. *Anti-Cancer Drug Des* **7**, 101–113.

Rabah, D., Grant, S., Ma, C., and Conrad, D. H. (2001). Bryostatin-1 specifically inhibits *in vitro* IgE synthesis. *J Immunol* **167**, 4910–4918.

Ramsdell, J. S., Pettit, G. R., and Tashjian, A. H., Jr. (1986). Three activators of protein kinase C, bryostatins, dioleins, and phorbol esters, show differing specificities of action on GH4 pituitary cells. *J Biol Chem* **261**, 17073–17080.

Rhee, J. S., Betz, A., Pyott, S., Reim, K., Varoqueaux, F., Augustin, I., Hesse, D., Sudhof, T. C., Takahashi, M., Rosenmund, C., and Brose, N. (2002). β-phorbol ester- and diacylglycerol-induced augmentation of transmitter release is mediated by Munc13s and not by PKCs. *Cell* **108**, 121–133.

Roddie, P. H., Horton, Y., and Turner, M. L. (2002). Primary acute myeloid leukaemia blasts resistant to cytokine-induced differentiation to dendritic-like leukaemia cells can be forced to differentiate by the addition of bryostatin-1. *Leukemia* **16**, 84–93.

Rozengurt, E., Rey, O., and Waldron, R. T. (2005). Protein kinase D signaling. *J Biol Chem* **280**, 13205–13208.

Sako, T., Yuspa, S. H., Herald, C. L., Pettit, G. R., and Blumberg, P. M. (1987). Partial parallelism and partial blockade by bryostatin-1 of effects of phorbol ester tumor promoters on primary mouse epidermal-cells. *Cancer Res* **47**, 5445–5450.

Schaufelberger, D. E., Koleck, M. P., Beutler, J. A., Vatakis, A. M., Alvarado, A. B., Andrews, P., Marzo, L. V., Muschik, G. M., Roach, J., Ross, J. T., Lebherz, W. B., Reeves, M. P., Eberwein, R. M., Rodgers, L. L., Testerman, R. P., Snader, K. M., and Forenza, S. (1991). The large-scale isolation of bryostatin-1 from *Bugula neritina* following current good manufacturing practices. *J Nat Prod* **54**, 1265–1270.

Scheid, C., Prendiville, J., Jayson, G., Crowther, D., Fox, B., Pettit, G. R., and Stern, P. L. (1994). Immunomodulation in patients receiving intravenous Bryostatin 1 in a phase I clinical study: Comparison with effects of Bryostatin 1 on lymphocyte function *in vitro*. *Cancer Immunol Immunother* **39**, 223–230.

Shindo, M., Irie, K., Masuda, A., Ohigashi, H., Shirai, Y., Miyasaka, K., and Saito, N. (2003). Synthesis and phorbol ester binding of the cysteine-rich domains of diacylglycerol kinase (DGK) isozymes. DGKγ and DGKβ are new targets of tumor-promoting phorbol esters. *J Biol Chem* **278**, 18448–18454.

Silinsky, E. M., and Searl, T. J. (2003). Phorbol esters and neurotransmitter release: more than just protein kinase C? *Br J Pharmacol* **138**, 1191–1201.

Spitaler, M., Utz, I., Hilbe, W., Hofmann, J., and Grunicke, H. H. (1998). PKC-independent modulation of multidrug resistance in cells with mutant (V185) but not wild-type (G185) P-glycoprotein by bryosratin 1. *Biochem Pharmacol* **56**, 861–869.

Stone, J. C., Stang, S. L., Zheng, Y., Dower, N. A., Brenner, S. E., Baryza, J. L., and Wender, P. A. (2004). Synthetic bryostatin analogues activate the RasGRP1 signaling pathway. *J Med Chem* **47**, 6638–6644.

van Blitterswijk, W. J., and Houssa, B. (2000). Properties and functions of diacylglycerol kinases. *Cell Signal* **12**, 595–605.

Wada, R., Suto, Y., Kanai, M., and Shibasaki, M. (2002). Dramatic switching of protein kinase C agonist/antagonist activity by modifying the 12-ester side chain of phorbol esters. *J Am Chem Soc* **124**, 10658–10659.

Wall, N. R., Mohammad, R. M., and AlKatib, A. M. (1999). Bax:Bcl-2 ratio modulation by bryostatin 1 and novel antitubulin agents is important for susceptibility to drug induced apoptosis in the human early pre-B acute lymphoblastic leukemia cell line, Reh. *Leuk Res* **23**, 881–888.

Wang, Q. M. J., Bhattacharyya, D., Garfield, S., Nacro, K., Marquez, V. E., and Blumberg, P. M. (1999). Differential localization of protein kinase C delta by phorbol esters and related compounds using a fusion protein with green fluorescent protein. *J Biol Chem* **274**, 37233–37239.

Wang, S. J., Guo, C. Y., Castillo, A., Dent, P., and Grant, S. (1998). Effect of bryostatin 1 on taxol-induced apoptosis and cytotoxicity in human leukemia cells (U937). *Biochem Pharmacol* **56**, 635–644.

Wang, S. J., Wang, Z. L., Dent, P., and Grant, S. (2003). Induction of tumor necrosis factor by bryostatin 1 is involved in synergistic interactions with paclitaxel in human myeloid leukemia cells. *Blood* **101**, 3648–3657.

Wang, Z. L., Wang, S. J., Dai, Y., and Grant, S. (2002). Bryostatin 1 increases 1-beta-D-arabinofuranosylcytosine-induced cytochrome c release and apoptosis in human leukemia cells ectopically expressing Bcl-xL. *J Pharmacol Exp Ther* **301**, 568–577.

Wender, P. A., and Baryza, J. L. (2005). Identification of a tunable site in bryostatin analogues: C20 bryologs through late stage diversification. *Org Lett* **7**, 1177–1180.

Wender, P. A., and Hinkle, K. W. (2000). Synthesis and biological evaluation of a new class of bryostatin analogues: The role of the C20 substituent in protein kinase C binding. *Tetrahedron Lett* **41**, 6725–6729.

Wender, P. A., and Lippa, B. (2000). Synthesis and biological evaluation of bryostatin analogues: The role of the A-ring. *Tetrahedron Lett* **41**, 1007–1011.

Wender, P. A., Koehler, K. F., Sharkey, N. A., Dellaquila, M. L., and Blumberg, P. M. (1986). Analysis of the phorbol ester pharmacophore on protein kinase C as a guide to the rational design of new classes of analogues. *Proc Natl Acad Sci USA* **83**, 4214–4218.

Wender, P. A., DeBrabander, J., Harran, P. G., Hinkle, K. W., Lippa, B., and Pettit, G. R. (1998a). Synthesis and biological evaluation of fully synthetic bryostatin analogues. *Tetrahedron Lett* **39**, 8625–8628.

Wender, P. A., DeBrabander, J., Harran, P. G., Jimenez, J. M., Koehler, M. F. T., Lippa, B., Park, C. M., Siedenbiedel, C., and Pettit, G. R. (1998b). The design, computer modeling, solution structure, and biological evaluation of synthetic analogues of bryostatin 1. *Proc Natl Acad Sci USA* **95**, 6624–6629.

Wender, P. A., Hinkle, K. W., Koehler, M. F. T., and Lippa, B. (1999). The rational design of potential chemotherapeutic agents: Synthesis of bryostatin analogues. *Med Res Rev* **19**, 388–407.

Wender, P. A., Baryza, J. L., Bennett, C. E., Bi, C., Brenner, S. E., Clarke, M. O., Horan, J. C., Kan, C., Lacote, E., Lippa, B., Nell, P. G., and Turner, T. M. (2002). The practical synthesis of a novel and highly potent analogue of bryostatin. *J Am Chem Soc* **124**, 13648–13649.

Wender, P. A., Koehler, M. F., and Sendzik, M. (2003a). A new synthetic approach to the C ring of known as well as novel bryostatin analogues. *Org Lett* **5**, 4549–4552.

Wender, P. A., Mayweg, A. V. W., and VanDeusen, C. L. (2003b). A concise, selective synthesis of the polyketide spacer domain of a potent bryostatin analogue. *Org Lett* **5**, 277–279.

Wender, P. A., Baryza, J. L., Brenner, S. E., Clarke, M. O., Craske, M. L., Horan, J. C., and Meyer, T. (2004). Function oriented synthesis: The design, synthesis, PKC binding and translocation activity of a new bryostatin analog. *Curr Drug Discov Tech* **1**, 1–11.

Wender, P. A., Clarke, M. O., and Horan, J. C. (2005a). Role of the A-ring of bryostatin analogues in PKC binding: Synthesis and initial biological evaluation of new A-ring-modified bryologs. *Org Lett* **7**, 1995–1998.

Wender, P. A., Hilinski, M. K., and Mayweg, A. V. (2005b). Late-stage intermolecular CH activation for lead diversification: A highly chemoselective oxyfunctionalization of the C-9 position of potent bryostatin analogues. *Org Lett* **7**, 79–82.

Yu, C. M., Choi, H. S., Yoon, S. K., and Jung, W. H. (1997). Effects of subjoin Lewis acid on the catalytic asymmetric allylic transfer reactions of aldehydes promoted by BINOL-Ti(IV) complex. *Synlett*, 889–890.

Zhao, M. Z., Li, J., Mano, E., Song, Z. G., Tschaen, D. M., Grabowski, E. J. J., and Reider, P. J. (1999). Oxidation of primary alcohols to carboxylic acids with sodium chlorite catalyzed by TEMPO and bleach. *J Org Chem* **64**, 2564–2566.

CHAPTER 7

THE PROMISES AND PITFALLS OF SMALL-MOLECULE INHIBITION OF POLY(ADP-RIBOSE) GLYCOHYDROLASE (PARG)

*Amanda C. Nottbohm and Paul J. Hergenrother**

7.1 INTRODUCTION TO PARP AND PARG

Driven by DNA damage, the poly(ADP-ribose) polymerases (PARPs) utilize NAD^+ to create poly(ADP-ribose) (PAR) on the glutamic acid residues of proteins, drastically altering the overall charge and size of the modified protein and ultimately initiating DNA repair (Bürkle, 2001). PAR biopolymers consist of up to 200 ADP-ribose units, and polymer levels may increase more than 100-fold in minutes (Meyer-Ficca et al., 2004). This poly(ADP-ribosyl)ation of proteins is transient, because the enzyme poly(ADP-ribose)glycohydrolase (PARG) operates in both an endo- and exoglycosidic fashion to break down PAR into ADP-ribose monomers (Davidovic et al., 2001). If more severe DNA damage occurs, PARP-1 becomes cleaved by caspases into 89- and

*Corresponding author: hergenro@uiuc.edu

Drug Discovery Research: New Frontiers in the Post-Genomic Era, Edited by Ziwei Huang

24-kDa subunits (Scovassi and Poirier, 1999; Lazenik et al., 1994), which separates the DNA binding domain from the automodification and catalytic domains of PARP, thus inactivating the enzyme. Most likely this PARP-1 cleavage plays a role in preserving cellular energy for apoptosis by preventing futile cycles of DNA repair (Putt et al., 2004). Finally, in cases of extreme DNA damage, PARP-1 is overactivated, which depletes the cell's valuable energy resources, thus leading to death by necrosis (Pieper et al., 1999). The cellular activity and processing of PARP-1 in response to these three levels of DNA damage is depicted in Fig. 7.1. While PAR has been shown to play a role in DNA repair and replication (Dantzer et al., 2000), it has also been implicated in control of chromosome migration during cell division (Shall and de Murcia, 2000), activation of transcription after DNA damage (Hassa and Hottiger, 1999), activation of the proteasome (Ulrich et al., 1999), and regulation of telomere maintenance (Smith et al., 1998) and is required for spindle assembly and function (Chang et al., 2004). Failure to dynamically regulate PAR metabolism causes increased cytotoxic sensitivity and leads to early embryonic lethality (Koh et al., 2004).

7.1.1 Potential Applications of PARP and PARG Inhibitors

Due to these key roles of PAR in the cell, the PARP isozymes and PARG are potential drug targets. Overactivation of PARP caused by severe DNA damage quickly leads to the depletion of NAD^+, and this loss of energy inside the cell usually culminates in cell death through necrosis (Pieper et al., 1999). In cases of ischemic injury, when oxygen deprivation alone drastically decreases cellular energy output, overstimulation of PARP can also cause a drop of NAD^+ and ATP to less than 20% of the normal level (Berger, 1985). Inhibition of either PARP or PARG in such cases could potentially maintain cellular energy at acceptable levels, thus decreasing necrotic cell death. Conversely, when utilized in the context of tumor cells, PARP or PARG inhibitors might enhance the cytotoxicity of cancer therapies already in use (Li et al., 2001). It is presumed that the addition of such an inhibitor prevents the recovery of tumor cells from lethal DNA damage after radiation or other cancer therapies. The recently illuminated requirement for PAR during spindle assembly (Chang et al., 2004) also hints at the possible utility of spindle-associated PARP and PARG as cancer drug targets. Finally, both PARP and PARG are envisioned as potential targets for neuroprotection. Just as inhibition of these enzymes could alleviate damage in cases of ischemic injury, inhibition of PARP and PARG may potentially reduce neuronal death (Ying et al., 2001). Additionally, although it is unknown if abnormal accumulation of PAR is found in humans, a number of neurodegenerative disorders such as Alzheimer's and Parkinson's diseases are associated with the accrual of protein aggregates, and study of PAR accumulation in cases of PARG inhibition might be useful in understanding these conditions (Hanai et al., 2004). Clearly, PAR's pivotal role in the cell indicates that both PARP and PARG inhibitors have considerable medicinal potential versus a variety of disease states (Fig. 7.1).

7.1.2 PARP Inhibitors and Their Drawbacks

Although they have limited cellular uptake and low potency and specificity, the most commonly used PARP inhibitors are nicotinamide, the endogenous inhibitor of

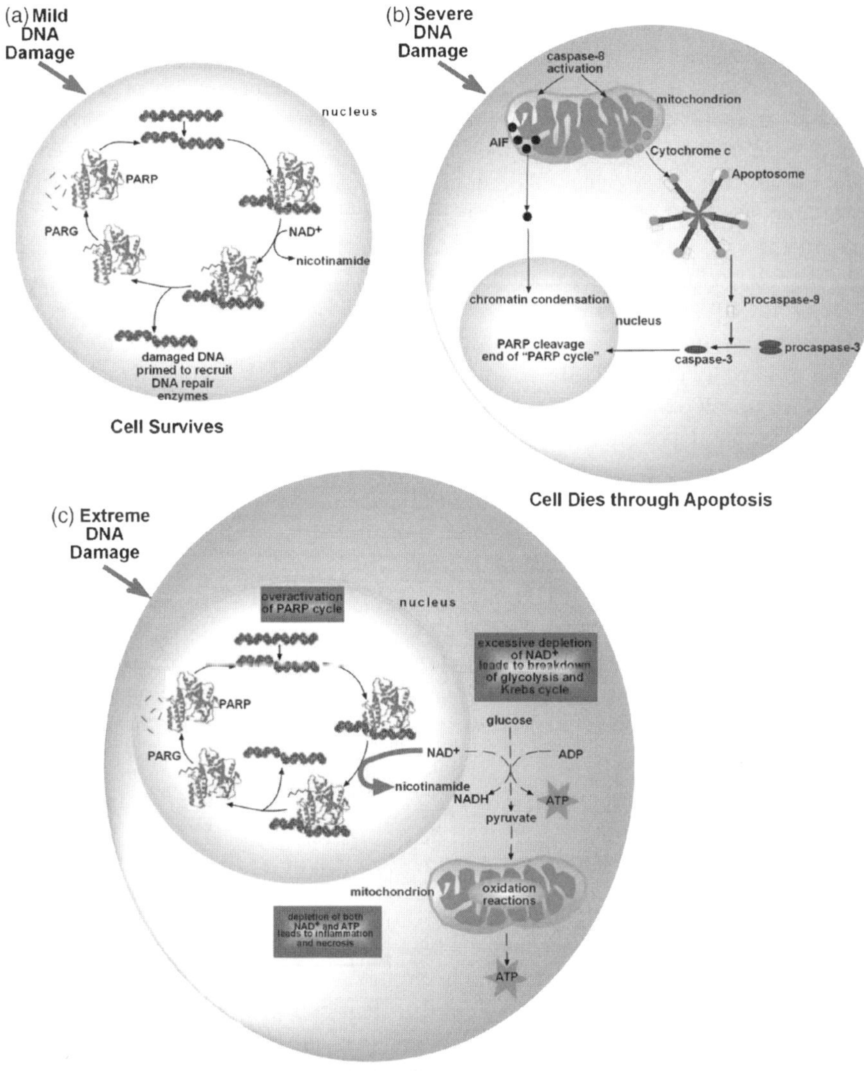

Figure 7.1 (a) In response to mild DNA damage, PARP-1 binds to DNA and catalyzes the formation of poly(ADP-ribose) polymers onto protein acceptors including itself. Because of electrostatic repulsion between the damaged DNA and the newly formed polymer, PARP-1 releases the DNA, which is then primed to recruit DNA repair enzymes such as αXRCC1 and DNA ligase. PARP-1 remains in an inactive state until the enzyme PARG degrades the poly(ADP-ribose) polymer, which allows it to once again bind damaged DNA. (**b**) Upon more severe DNA damage, the cell initiates the apoptotic cascade. Release of cytochrome c ultimately leads to cleavage of PARP-1, thus preventing unnecessary cycles of DNA repair to save cellular energy for apoptosis. (**c**) Extreme DNA damage causes PARP-1 overactivation, which results in a loss of cellular energy. With NAD^+ severely depleted, the cell cannot make ATP, and this leads to inflammation and death through necrosis. See color plates.

PARP, and 3-aminobenzamide (Virag and Szabo, 2002; Wilson et al., 2004; Contoni et al., 1987; Szabo et al., 1998). Many other PARP inhibitory compounds fall into two classes: monoaryl amides and bi-, tri-, or tetracyclic lactams, which often contain either an aromatic carboxamide or a carbamoyl group attached to a polyaromatic heterocyclic skeleton (Virag and Szabo, 2002). To date, the PARP family is thought to comprise at least 18 isozymes (Pion et al., 2003), and due to the highly conserved NAD^+ binding site of these PARPs, it is likely that isozyme-specific inhibitors will be difficult to develop (Smith, 2001). Although some preliminary studies have been completed that compare the inhibitory effects of phenanthridinones on PARP-1 versus PARP-2 (Perkins et al., 2001), the issue of isozyme specificity has not been fully examined. While it cannot be denied that PARP inhibitors have been shown to have a beneficial effect in many *in vivo* studies (Virag and Szabo, 2002; Tentori et al., 2002; Southan and Szabo, 2003), before they can be utilized in the clinic, some important safety aspects must first be considered. Although PARP-deficient mice do not show an increased level of spontaneous tumors, they do display an increased number of chemically induced tumors in comparison to wild-type ones (Tsutsumi et al., 2001), and even the most well-documented PARP inhibitors display undesired metabolic side effects by knocking out the normal physiologic functions of PAR (Lu et al., 2003; Milam and Cleaver, 1984). Long-term studies have not yet been completed, but due to PARP's active roles in DNA repair and preservation of genomic integrity (Dantzer et al., 2000; Shall and de Murcia, 2000; Hassa and Hottiger, 1999; Smith et al., 1998; Chang et al., 2004), extended PARP inhibition could increase mutation rates and ultimately lead to cancer (Virag and Szabo, 2002).

7.1.3 PARG Inhibition: An Attractive Alternative to PARP Inhibitors

Obstruction of DNA repair mechanisms by PARP inhibition might be avoided by inhibiting PARG instead of PARP. It has been suggested that while the concentration of PARG in the cell is 13–50 times less than that of PARP (it has been reported that the number of PARG molecules present per cell may be as low as 2000), inhibitors of PARG should be just as useful as any previously demonstrated PARP inhibitors; in fact, the specific catalytic activity of PARG is 50- to 70-fold higher than that of PARP (Hatakeyama et al., 1986; Rapizzi et al., 2004). While PARP inhibition prevents the formation of PAR and consequently disrupts its normal metabolic functions, PARG inhibition indirectly slows PARP activity by locking PARP in the fully automodified (with PAR) and inactive state. PARG inhibition could possibly allow PAR to maintain its normal function within the cell (Lu et al., 2003; Milam and Cleaver, 1984). Finally, in stark contrast to the burgeoning PARP family of enzymes, there has only been one PARG identified to date (Pion et al., 2003), and thus its inhibition would not be complicated by the need for isozyme specificity. By searching for direct inhibitors of PARG enzymatic activity, it is possible that potent, specific, and pharmacologically acceptable drugs will be discovered.

7.2 PROPERTIES OF PARG

As mentioned earlier, there are at least 18 proposed members of the PARP family (Pion et al., 2003), but only one PARG has been identified to date, and its sequence is highly conserved among mammals (Davidovic et al., 2001). Three PARG isoforms are known: the 110-kDa nuclear form, and the more abundant 103- and 99-kDa cytoplasmic forms (Meyer-Ficca et al., 2004; Bonicalzi et al., 2003). Homology searches have indicated that the PARG sequence contains both a nuclear localization signal (NLS) (Lin et al., 1997) and a leucine-rich nuclear export signal (NES) (Shimokawa et al., 1999), suggesting that PARG shuttles between the cytoplasm and nucleus to maintain tight control of PAR catabolism (Fig. 7.2) (Bonicalzi et al., 2003). Interestingly, PARG is a substrate for caspase-3, but the functional implications of this cleavage are unclear to date (Affar et al., 2001). Although it is stable between pH 5 and 9, PARG's optimum pH for enzymatic activity has been found to be 7.0 to 7.5. The enzyme exhibits substantial thermostability, and its activity increases up to temperatures of 50°C; thereafter, PARG activity decreases sharply (Hatakeyama et al., 1986).

7.2.1 Biochemistry of PARG

As highlighted earlier, although PARG has a low cellular concentration, it is constitutively active and has a high specific activity (5400 turnovers per minute)

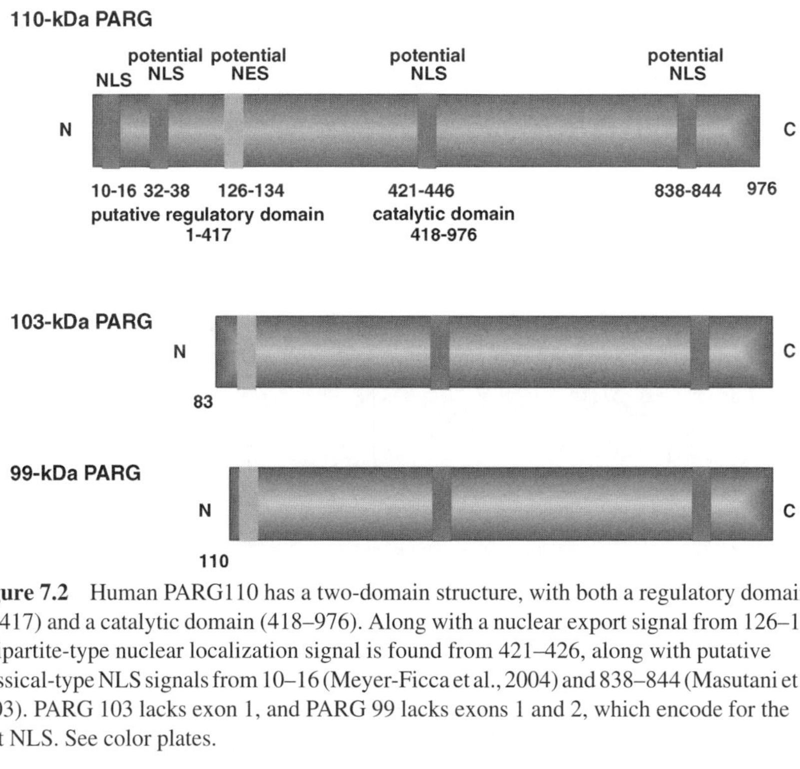

Figure 7.2 Human PARG110 has a two-domain structure, with both a regulatory domain (1–417) and a catalytic domain (418–976). Along with a nuclear export signal from 126–134, a bipartite-type nuclear localization signal is found from 421–426, along with putative classical-type NLS signals from 10–16 (Meyer-Ficca et al., 2004) and 838–844 (Masutani et al., 2003). PARG 103 lacks exon 1, and PARG 99 lacks exons 1 and 2, which encode for the first NLS. See color plates.

(Hatakeyama et al., 1986); as such, the *in vivo* half-life of PAR in DNA-damaged cells is only ~1 minute (Alvarez-Gonzalez and Althaus, 1989). The nuclear PAR concentration is approximately one to two times the K_M of PARG (0.3 μM) (Davidovic et al., 2001; Hatakeyama et al., 1986), and it has been found that the rates of PAR hydrolysis are a function of both concentration and structure (Brochu et al., 1994a). Long-chain polymers are catabolized more quickly, leaving behind 20- to 35mers,

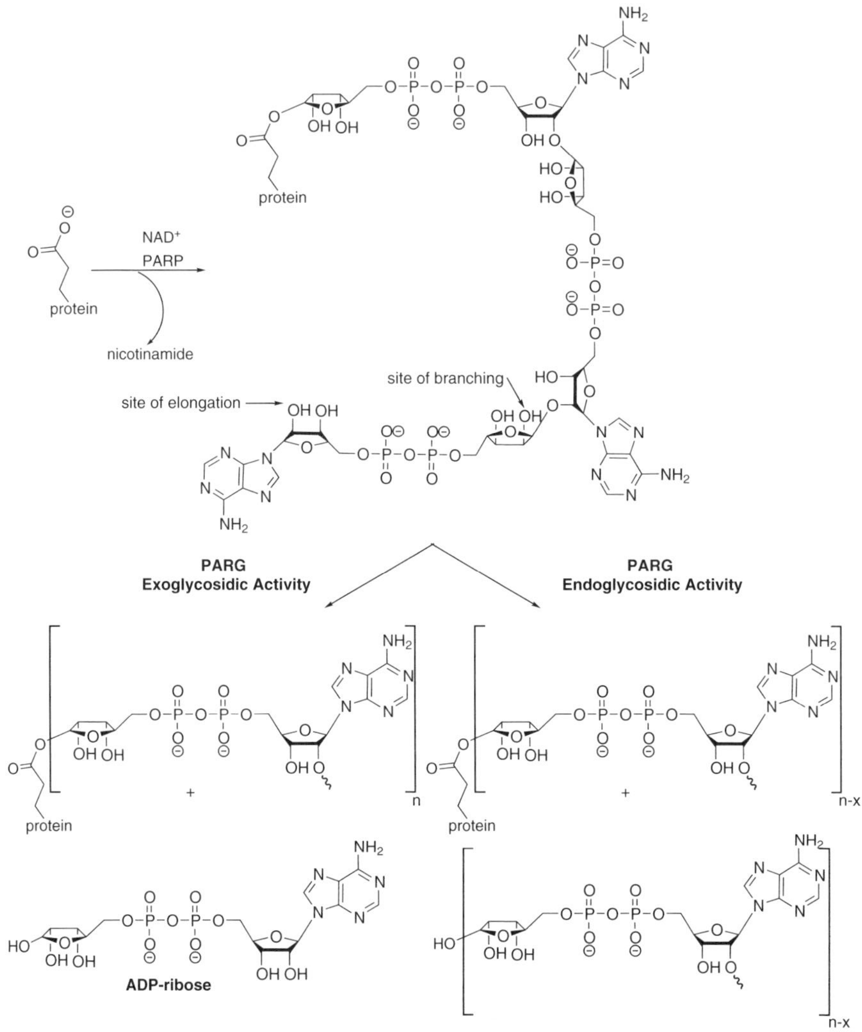

Figure 7.3 PARP catalyzes the formation of poly(ADP-ribose) polymers (from NAD^+) on glutamic acid residues of protein substrates. This process can occur from either the 2′ position (elongation) or 2″ position (branching). PARG catabolizes the ADP-ribose polymers in both an exo- and endo-glycosidic fashion. ADP-ribose monomers are directly produced from exoglycosidic activity, whereas the endoglycosidic mode of action gives smaller ADP-ribose polymers that can be converted into ADP-ribose through further processing.

which are poorer substrates for PARG (Lin et al., 1997; Burzio et al., 1976). In early stages of PAR catabolism, the enzyme degrades larger polymers in both an endo- and exoglycosidic fashion (Fig. 7.3), leaving behind both ADP-ribose monomers and smaller polymers, usually less than half the size of the original polymer. Conversely, smaller PAR are degraded at a rate that is 20-fold slower (for larger PAR the K_M is approximately 100-fold lower than that of small polymers), suggesting that PARG differentially controls the levels of large and small PAR (Hatakeyama et al., 1986). Interestingly, it is known that binding of photolabeled adenosine diphosphate (hydroxymethyl)pyrrolidinediol (ADP-HPD), an ADP-ribose analogue inhibitor of PARG, competes with the binding of short PAR (5- to 15mer) but not with the longer, highly branched polymer (Ramsinghani et al., 1998). For PARG, substrate identification and binding appears to be a complex event, with possible multiple attachment points or subsites for polymer recognition. It also must be noted that these striking differences in substrate processing may be attributed to higher-order structures of large PAR; those polymers with a chain length of more than nine ADP-ribose units exhibit a secondary structure, while smaller polymers do not (Minaga and Kun, 1983).

7.2.2 Structure of PARG

No crystal structure for PARG has yet been reported. However, from studies of the catalytic fragment of recombinant bovine PARG, it is believed that the active site consists of at least some residues in the Leu-771 to Arg-801 region. Photoderivatization studies with a known PARG inhibitor indicate Tyr-796 (Y796), which is a highly conserved residue in PARGs across a wide range of organisms, as a possible amino acid important for initial binding of PAR. Site-directed mutagenesis to replace Y796 with an alanine residue (Y796A) leads to an eight-fold decrease in catalytic efficiency (K_{cat}/K_M), a difference attributed to the ring-stacking ability of the Y796 residue with the adenine of the PAR substrate (Koh et al., 2003a,b). From homology studies it is known that Gly-262 and Gly-263 in *Arabidopsis* PARG are found in all mammalian PARGs, and a third Gly-264 is highly conserved (Panda et al., 2002). When Gly-262 is replaced with a glutamic acid residue to create the G262E mutant PARG enzyme, *Arabidopsis* seedlings accumulate much higher levels of PAR as compared to wild-type. These data, together with the high degree of amino acid conservation in this region of the PARG sequence, suggest that this GGG motif may play an important role in the regulation of PARG activity (Panda et al., 2002).

7.2.3 Genetic Studies with PARG

Mutation in the PARG gene, *tej*, in plants has been demonstrated to influence the transcription of genes involved in circadian rhythm regulation (Panda et al., 2002). Using a cycling bioluminescence reporter phenotype, *tej* mutants with abnormal circadian oscillations were discovered in *Arabidopsis thaliana*. While the effects of this recessive *tej* mutation are independent of light input, *tej* plants exhibit a period length 2 hours longer than the wild-type and always flower earlier under both long- and short-day conditions. Interestingly, PARP inhibitors have been shown to rescue the mutant *tej* phenotype. By slowing or stopping the rate of PAR synthesis, the known PARP inhibitor 3-aminobenzamide causes *tej* mutants to exhibit a wild-type circadian rhythm, thus indicating that poly(ADP-ribosyl)ation establishes a period length of the *Arabidopsis* circadian oscillator (Panda et al., 2002).

To date few studies have been completed with PARG knockout organisms. Loss of the catalytic domain of PARG causes lethality under normal conditions for development (25°C) in the larval stages of *Drosophila melanogaster*, but if the incubation temperature is increased to 29°C, approximately 25% of the PARG knockouts survive to the adult stage (Hanai et al., 2004). These organisms lacking PARG are associated with excessive accumulation of PAR in the central nervous system (CNS), progressive neurodegeneration, reduced locomotor activity, and shortened lifespan. Accumulation of PAR in the CNS could be explained by any of the following: PAR cannot be diluted in the CNS because neuronal cells in the CNS do not divide; PARP is possibly more active in the brain, and therefore polymer builds up more quickly there (Menegazzi et al., 1991); or the other enzymes responsible for degrading PAR are less active in the brain than in other organs of *D. melanogaster.* Whatever the reason for the excessive amounts of PAR found in the CNS of these PARG knockouts, these studies clearly suggest that PAR metabolism is a necessary part of normal neuronal function.

Through gene targeting in embryonic stem cells and in mice, deletion of the 110-kDa PARG protein, which is normally found in the nucleus (Meyer-Ficca et al., 2004), has been thoroughly studied. While no full-length PARG mRNA is detected in the PARG knockout embryonic fibroblasts, a 60-kDa truncated version of PARG remains present (Cortes et al., 2004). Mice deficient in the full-length PARG do not show any obvious phenotype, are fertile, and do not exhibit excessive buildup of PAR in tissue as might first be expected. This is attributed to the fact that the 60-kDa PARG remains unaffected in such cases and can still be observed by Western blot analysis. On the other hand, specific PARG activity is significantly lower in cells obtained from 110-kDa PARG-deficient animals, and they are hypersensitive to DNA alkylating agents, radiation, lipopolysaccharide-induced septic shock, and streptozotocin-induced diabetes (Cortes et al., 2004).

Most recently, *in vivo* studies have indicated that catabolism of PAR by PARG is necessary for genomic stability in higher-order eukaryotes (Koh et al., 2004). In contrast to the above study in which PARG activity remained in the 110-kDa PARG knockout mice due to the presence of alternatively spliced isoforms, complete PARG knockouts are lethal. While PARG $^{+/-}$ mice are viable and fertile, homozygous null mice (PARG $^{-/-}$) arrest before gastrulation at embryonic day 3.5. Under conditions of PARP inhibition (treatment with the PARP inhibitor benzamide), it is possible to culture PARG $^{-/-}$ embryonic trophoblast stem (TS) cells, although removal of the PARP inhibitor causes abnormal morphology, and after 3 days approximately 60% of the PARG $^{-/-}$ TS cells undergo apoptosis (Koh et al., 2004). Cells remaining after removal of the benzamide PARP inhibitor also exhibit slow growth rates, PAR accumulation, and increased sensitivity to the cytotoxic agent MNNG. Due to the lethality of the knockouts, it has been difficult to study PARG via traditional genetic approaches.

7.3 PARG PURIFICATION AND ASSAYS

Although provocative data indicate PARG inhibitors could be useful for a range of maladies, the lack of detailed structural knowledge has hampered their development. As a result, substrate analogues and compounds identified through random screening are the only known PARG inhibitors. However, recent advances in both PARG production and high-throughput assays should enable the discovery of a next generation of PARG inhibitors. Described in this section are the methods used to obtain the PARG enzyme, assay protocols, and a comprehensive look at known PARG inhibitors.

7.3.1 PARG Purification and Recombinant Protein Expression

Due to the very low content of PARG in various tissues and its instability during purification, difficulties have been encountered in obtaining pure quantities of this enzyme. Since its initial discovery in 1971 (Miwa and Sugimura, 19971), PARG has been purified and characterized from a wide range of cell types, which has allowed for full enzymatic characterization. Initial attempts at obtaining PARG focused on its extensive purification from calf thymus and other tissues, a process that can involve up to six purification steps and produces low yields of purified enzyme (Hatakeyama et al., 1986; Minaga and Kun, 1983; Brochu et al., 1994b; Thomassin et al., 1990; Shah et al., 1995). Additionally, although PARG is known to have a molecular weight of about 110 kDa (Lin et al., 1997; Winstall et al., 1999), during purification it is easily proteolyzed to 59- and 74-kDa fragments (Affar et al., 2001), which originally caused uncertainty as to whether different types of PARG protein exist in the cell (Maruta et al., 1991). However, full-length PARG was purified in 1997 (Lin et al., 1997) utilizing recombinant expression from the 4.1-kb PARG cDNA in *Escherichia coli* to produce the 110-kDa protein, thus eliminating the controversy surrounding PARG's size.

7.3.2 PARG Enzymatic Activity Assays

Many assays for measuring PARG enzymatic activity are limited by the fact that they are neither convenient nor high-throughput. The most common technique used to evaluate PARG activity first involves the preparation of ^{32}P-labeled PAR using PARP and ^{32}P-labeled NAD$^+$. Then, following purification and reaction of the radiolabeled PAR with PARG, either a thin-layer chromatography (TLC) technique or high-resolution polyacrylamide gel electrophoresis is utilized to separate the catabolized ADP-ribose from PAR, and radioactivity is measured (Ménard and Poirier, 1987; Pachecho-Rodriguez and Alvarez-Gonzalez, 1999). Although both these radiometric assays have been helpful in determination of IC_{50} values for PARG inhibitors (Koh et al., 2003a; Masutani et al., 2002; Slama et al., 1995) and in studies of PARG kinetics (Koh et al., 2003a; Maruta et al., 1997), neither is particularly useful in screening large libraries of compounds in a high-throughput manner. Additionally, PARG works in an endo- and exoglycosidic fashion, and these TLC-based methods are only able to quantitate monomeric ADP-ribose; the direct products of endoglycosidic cleavage cannot be detected (Putt and Hergenrother, 2004).

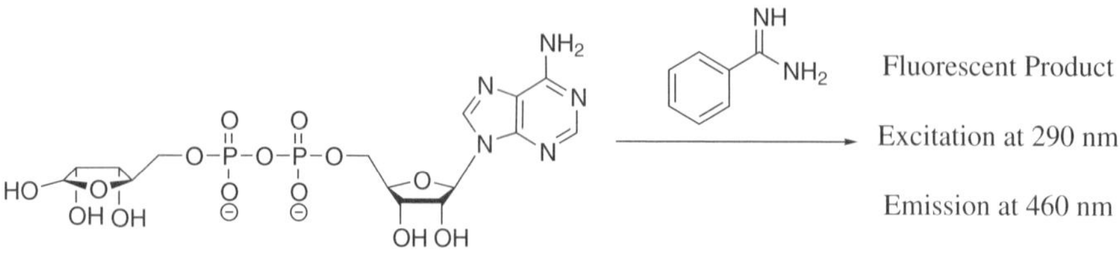

Figure 7.4 Conversion of ADP-ribose to a fluorescent dye with benzamidine.

More recently, a nonradiometric assay involving the conversion of the ADP-ribose product of PARG activity into a fluorescent molecule has been developed. Taking advantage of the fact that reducing sugars are known to react with aromatic amidines to give fluorescent products (Kai et al., 1985; Coquet et al., 1991; Kakita et al., 2002), this assay involves the reaction of benzamidine with ADP-ribose (a reducing sugar) to create a fluorophore (Fig. 7.4) (Putt and Hergenrother, 2004). A simple fluorescence reading can be used to measure the amount of free ADP-ribose hemiacetal, thus quantifying PARG activity. One limitation of this assay is that it cannot be used in cases in which potential PARG inhibitors contain aldehydes or reducing sugars. Unlike the aforementioned PARG assays, this fluorescence-based method is capable of detecting both exo- and endoglycosidic cleavage of PAR, and this method is highly sensitive. Additionally, development of an assay based on fluorescence rather than radioactivity greatly reduces both cost and time expenditures and allows compounds to be screened in a high-throughput manner.

7.4 PARG INHIBITORS

From a drug-development perspective, the perfect PARG inhibitors would be highly potent and specific compounds of modest size (molecular weight <500). Although the current classes of inhibitors do not necessarily fit into all the categories of an ideal pharmaceutical, they have certainly offered insights into the structure and function of PARG. These compounds can be broadly grouped into the three categories of DNA intercalators, tannins, and substrate analogues. Below, all classes of PARG inhibitors along with their *in vivo* effects will be discussed.

7.4.1 DNA Intercalators

Of the few classes of PARG inhibitors known, the ones that potentially offer the best cell permeability are polyaromatic compounds typically thought of as DNA intercalators, depicted in Fig. 7.5. These compounds were initially evaluated as a consequence of their known DNA binding properties and the known role of PARG in the DNA damage and repair process.

Using the ^{32}P-PAR assay, the K_I for ethacridine was estimated at <7 μM, the K_I for Tilorone was ~ 7 μM, and the K_I for profavine was 36 μM, while another well-known DNA intercalator, ethidium bromide, is not inhibitory (Tavassoli et al., 1985).

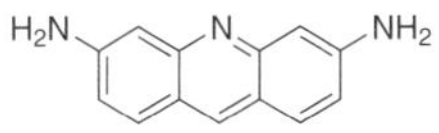

Proflavine
K_i 36 μM

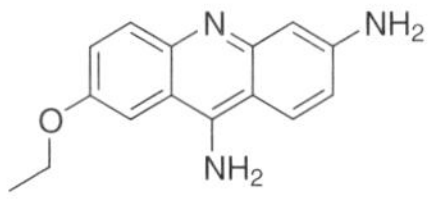

Ethacridine
$K_i < 7$ μM

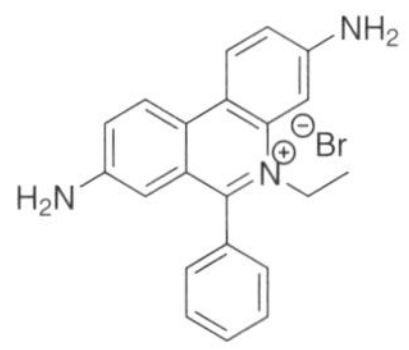

Ethidium bromide
not inhibitory

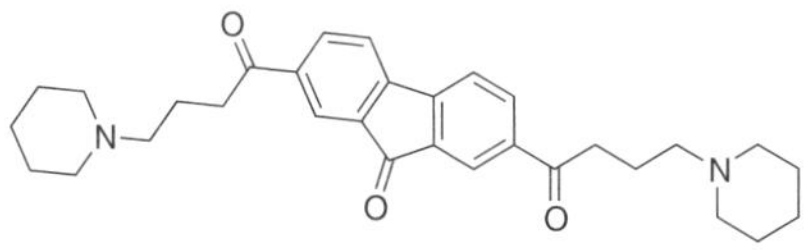

Tilorone
K_i 7.3 μM

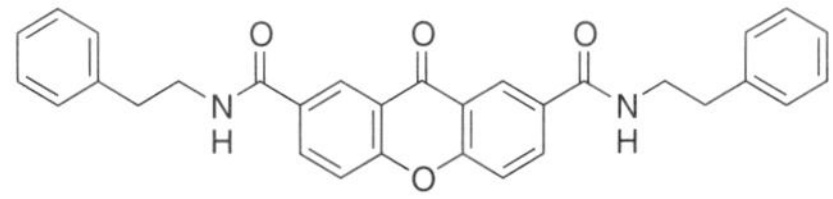

2,7-Bis[N-(3-phenyl-propyl)carbamoyl]-9-Oxo-9H-fluorene
(GPI 16552)
IC_{50} of 1.7 μM

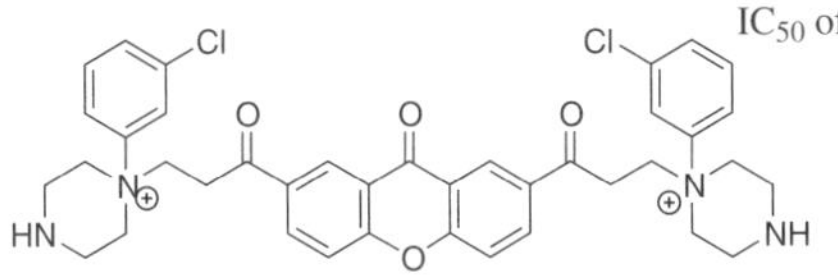

2,7-Bis-{3-[1-(3-chloro-phenyl)-piperazine-1-yl]-propionyl}-xanthen-9-one
IC_{50} of 1.1 μM

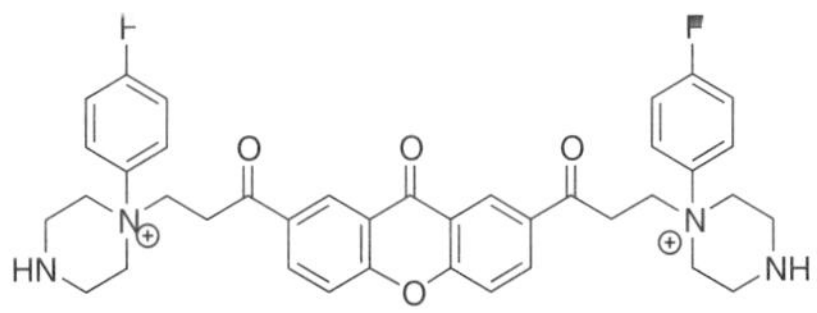

2,7-Bis-{3-[1-(4-fluoro-phenyl)-piperazine-1-yl]-propionyl}-xanthen-9-one
IC_{50} of 2.9 μM

Figure 7.5 Intercalator and intercalator-like inhibitors of PARG and their IC_{50} values.

These were all relative to a K_M for the PAR substrate that was calculated as 1.7 μM in this particular assay system. In general, these inhibitors appear to give competitive inhibition, and there is some evidence that this inhibition may be due to the compounds forming a complex with the PAR polymer, thus blocking substrate binding to the enzyme; this seems to be particularly true for ethacridine (Tavassoli et al., 1985; Bernardi et al., 1997). Further support of this hypothesis is found in the fact that ethacridine and other intercalators not only inhibit the degradation of PAR by PARG, but also prevent degradation of the polymer by snake venom phosphodiesterase and hinder ethanol/acetate precipitation of PAR (Tavassoli et al., 1985). Recent work has focused on the design and synthesis of symmetrically disubstituted aromatic compounds that exhibit PARG-inhibitory properties (Li et al., 2001). Representative members of two of the most potent classes, 2,7-substituted fluorenes and xanthen-9-ones, which are structurally very similar to the DNA intercalator Tilorone, are also depicted in Fig. 7.5. In conclusion, while these compounds are modestly potent inhibitors of PARG and have been used in several *in vivo* studies (see below), questions about their selectivity for PARG inhibition versus DNA binding are likely to doom their use as therapeutic agents or chemical tools to extensively investigate PARG biology.

7.4.2 Tannins

A survey of the literature shows that the naturally occurring polymethoxy-phenolic compounds called tannins are by far the most well-studied PARG inhibitors (Fig. 7.6). Initially extracted from green tea leaves and pinecone fractions, it has been found that the oligomeric ellagitannins such as nobotanin B, E, and K are the most potent tannin inhibitors of PARG, with *in vitro* IC_{50} values ranging from 0.44 μM for nobotanin K to

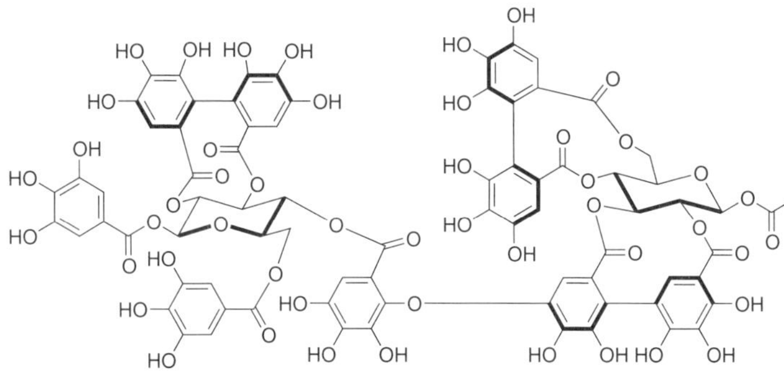

Nobotanin B
IC_{50} 4.8 μM

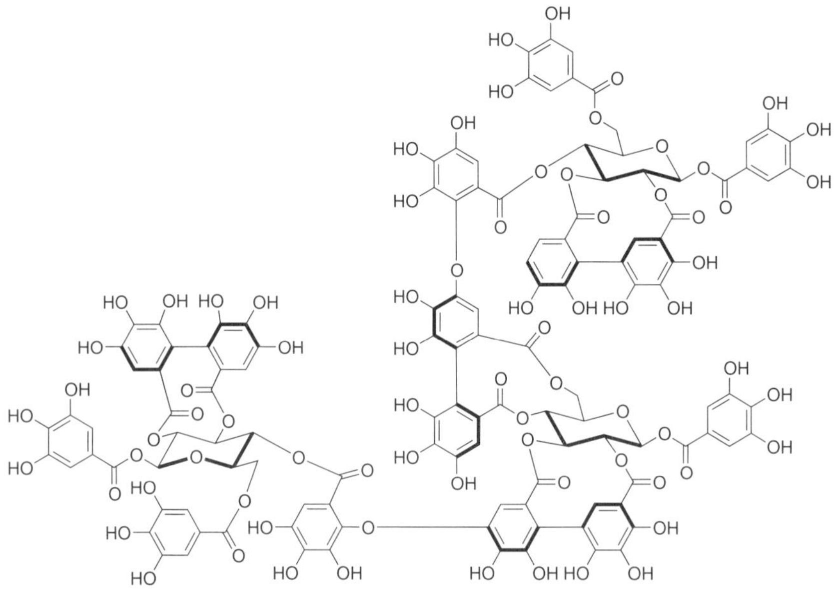

Nobotanin E
IC_{50} 1.4 μM

Figure 7.6 Tannins and their representative IC_{50} values.

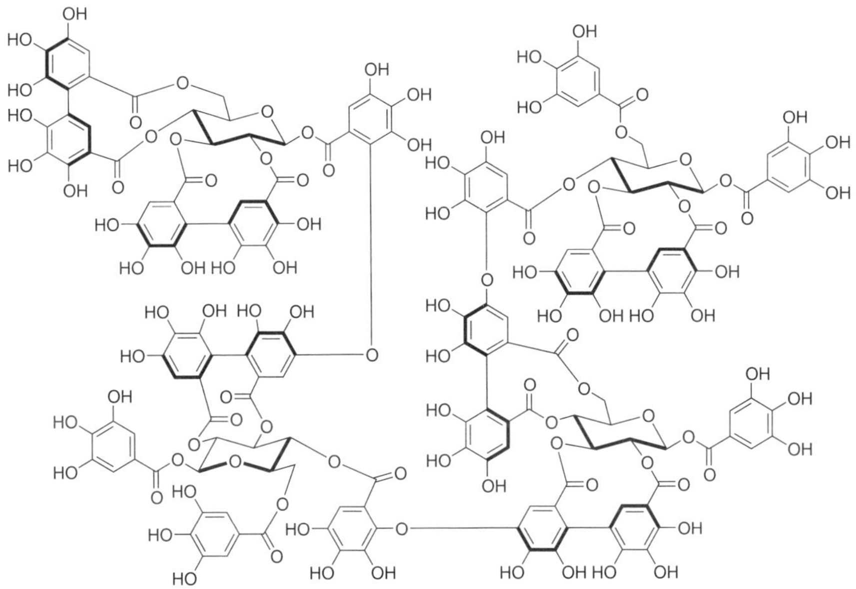

Nobotanin K
IC_{50} 0.44 μM

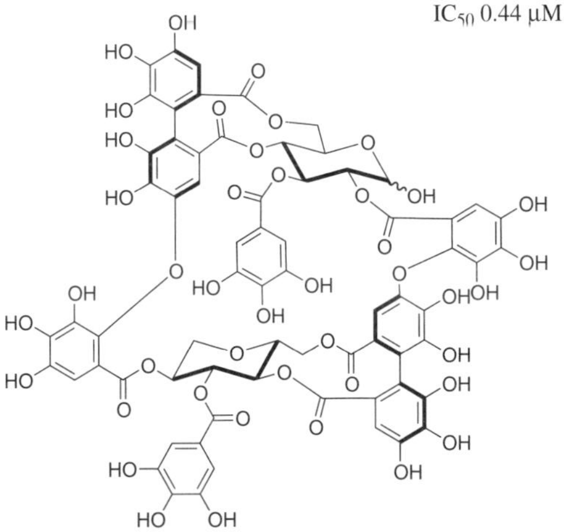

Oenothein B
IC_{50} 0.44 μM

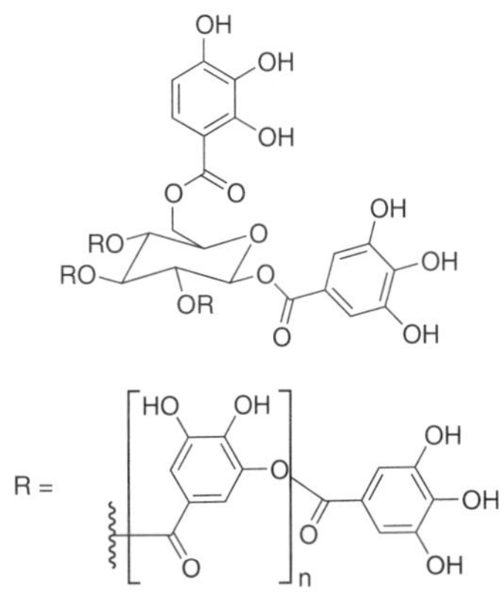

Gallotannins
IC_{50} 18-33 μM

Figure 7.6 (*Continued*)

4.8 μM for nobotanin B (Tsai et al., 1992). In comparison, the circular dimeric ellagitannin Oenothein B (Oen B) inhibits with an IC_{50} of 3.8 μM (Maruta et al., 1997). Inhibitory activity of oligomeric ellagitannins increases with the number of monomeric residues (dimeric < trimeric < tetrameric), with condensed tannins having little effect on PARG activity even at very high concentrations, possibly indicating that differences in inhibitory effects are related to variations in secondary structures of the tannins (Aoki et al., 1993). Only hydrolyzable tannins (gallotannins and ellagitannins) were able to inhibit PARG. While tannins of the trimeric and tetrameric variety exhibit mixed-type inhibition, nobotanin B, (a dimer) has been found to competitively inhibit PARG with respect to PAR (Tsai et al., 1992).

Several studies with cell extracts have confirmed the inhibitory properties of these phenolic phytochemicals against PARG. When 200 μM tannin was added to HeLa nuclear extracts, a significant 40-fold increase of the PAR level was observed, and it was confirmed that this increase was indeed due to inhibition of the catalytic activity of PARG (Keil et al., 2004).

7.4.3 Substrate Analogues

As expected, substrate analogues of PAR have been synthesized and are shown to be both potent and specific inhibitors of PARG (Fig. 7.7). Though ADP-ribose monomer

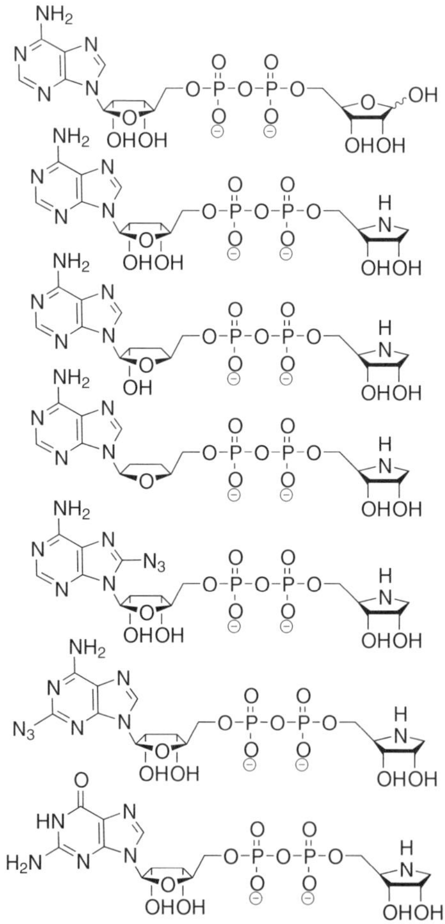

Entry	Name	IC_{50} (μM)
1	ADP-ribose	120
2	APD-HPD	0.12
3	APD-HPM	3.07
4	ADP-HP	19.2
5	8-N_3-APD-HPD	0.39
6	2-N_3-APD-HPD	290
7	APD-GPD	>1000

Figure 7.7 Substrate analogue PARG inhibitors and their comparative IC_{50} values.

units inhibit PARG with an IC_{50} of 120 μM, ADP-HPD, an NH-analogue of ADP-ribose, has been shown to inhibit the action of PARG with a 1000-fold lower IC_{50} value (entry 2) (Slama et al., 1995). Structure–activity analysis utilizing ADP-HPD analogues containing various alterations to the purine base has revealed the importance of the adenine moiety for optimal inhibition. It was discovered that substitution at the 2-position of adenine negatively affects PARG inhibition (entry 7), while substitution at the 8-position has no adverse effects (entry 5). Both the pyrrolidine *cis*-hydroxyls of ADP-HPD are necessary for inhibitor binding; removal of one increases the IC_{50} value to 3.07 μM (entry 3), and complete elimination of these hydroxyls results in a 160-fold increase in IC_{50} value as compared to ADP-HPD (entry 4) (Koh et al., 2003a).

7.4.4 Other Inhibitors

Additionally, through random screening it has been found that some cyclic peptides are capable of inhibiting PARG activity. One cyclic hexadepsipeptide (known as PD 124,966) isolated from the fermentation products of actinomycetes was found to have an IC_{50} value of 40 μg/ml, while pargamicin, which is purified from the fermentation broth of *Amicolatopsis* exhibits an IC_{50} value of 28 μg/ml (Masutani et al., 2002). Interestingly, linear peptides containing the same piperazic acid residues as PD 124,966 and pargamicin appear to have no effect on PARG activity.

7.5 IN VIVO STUDIES

7.5.1 Studies with DNA Intercalators

The second-generation intercalator-like PARG inhibitors are ideal for *in vivo* studies due to their low molecular weights and good cell permeability properties. One member of the Tilorone family of PARG inhibitors, GPI 16552 or 2,7-bis [*N*-(3-phenyl-propyl)carbamoyl]-9-oxo-9*H*-fluorene, has exhibited neuroprotective properties in a rat model of focal cerebral ischemia. Both a 30-minute pre-ischemia treatment or a 1- or 2-hour post-ischemia treatment with GPI 16552 reduces the total infarct volume by 47–53%, with greater protection observed in the cortical areas (43–59%) than in the subcortical areas (28–40%) of the brain (Lu et al., 2003). Furthermore, application of the novel PARG inhibitor GPI 18214 substantially reduces the septic-shock-like syndrome caused by zymosan in mice (Genovese et al., 2004). Zymosan is a nonbacterial, nonendotoxic agent that produces liver, intestine, lung, and kidney failure in test animals. In mice treated with 40 mg/kg of GPI 18214, which has an IC_{50} value of 3.0 μM, all signs of pancreatic, renal, and liver damage were abolished, and lung injury was significantly reduced after induction of zymosan (Genovese et al., 2004). The protective effects of GPI 18214 correlate with reduced neutrophil and PMN lysosome infiltration into the intestine and lung, which have been known to augment tissue damage through inflammatory effects (Hassa and Hottiger, 2002; Lefer and Lefer,

1993). It is hypothesized that these PARG inhibitors might reduce infarct volumes not only by stopping the vicious cycle of NAD^+ depletion, but also by their anti-inflammatory effects, as described below (Lu et al., 2003; Rapizzi et al., 2004; Genovese et al., 2004).

7.5.2 Studies with Tannins

The *in vivo* effects of tannins have been thoroughly studied. While the more potent PARG inhibitors nobotanin E (trimer) and K (tetramer) have little inhibitory effect on the catabolism of PAR in intact 34I cells, it has been found that nobotanin B causes concentration-dependent inhibition of PAR catabolism (Tsai et al., 1992). The inability of nobotanin E and K to inhibit PARG is due to the poor penetration of these compounds into plasma membranes. When cell extracts incubated with various tannins were analyzed by HPLC, it was found that only nobotanin B could be taken up by the cells in a significant level (Tsai et al., 1992), thus hinting at the major downfall of this high-molecular-weight class of compounds.

PARG enzymatic activity also appears to directly aid the process of DNA damage repair (Oei and Ziegler, 2000). Using the specific competitive inhibitor Oen B, the role of PAR catabolism by PARG in synchronized HeLa S3 cells at G1 phase has been studied. Because the process of excision repair synthesis of DNA requires large amounts of energy, it has been hypothesized that the energy produced during PAR catabolism by PARG is directly involved in this DNA repair. In fact, when ^{3}H-labeled thymidine 5′-triphosphate (^{3}H-dTTP) was added to poly(ADP-ribosyl)ated nuclei from G1-sychronized HeLa cells that were subsequently treated with the DNA-damaging agent MNNG, ^{3}H-dTMP incorporation into the DNA began after a short 3-minute lag (Maruta et al., 1997). On the other hand, in the presence of Oen B very little incorporation of ^{3}H-dTMP was observed, suggesting that energy in the form of ATP produced by PAR catabolism serves as a direct energy source for repair synthesis. Furthermore, in cases of moderate DNA damage, such as that caused by NO donors such as spermine nonoate (SNO), addition of the PARG inhibitor gallotannin greatly decreases the ability of rat germinal cells to recover (Di Meglio et al., 2004). In this case, the important role of PARG in protection from DNA damage might be twofold. First, PARP-1 seeks out and repairs DNA strand breaks while suppressing transcription to prevent the expression of damaged genes (Masson et al., 1998; Ziegler and Oei, 1999). If PARP-1 is inhibited by its own auto-poly(ADP-ribosyl)ation (Lautier et al., 1993; D'Amours et al., 1999; Zahradka and Ebisuzaki, 1982), PARG works to restore PARP to its original active status; in short, PARG inhibition ultimately prevents PARP's DNA repair mechanism from functioning properly. Additionally, as explained above, by metabolizing PAR into ADP-ribose monomers and thus adding to the NAD^+ energy stores, PARG possibly supplies energy needed for the DNA ligation step in base excision repair (Oei and Ziegler, 2000).

On the other hand, in cases of severe DNA damage, PARG inhibition can reduce the amount of cell death observed. A significant decrease in H_2O_2-induced cell death is observed when cultured murine astrocytes are preincubated with as little as 100 nM gallotannin (Ying and Swanson, 2000). Amazingly, gallotannin is 10-fold more potent than the PARP inhibitor benzamide in preventing oxidative cell death. Later studies have shown that nobotanin B, a PARG inhibitor isolated from the plant *Tibouchina semidecandra*, also offers neuroprotection against both oxidative (induced by both H_2O_2 and *N*-methyl-D-aspartate) and excitotoxic (induced by MNNG) cell death (Ying et al., 2001). Attempts to discover the mechanism of tannin cytoprotection led to the discovery that while gallotannin increases the amount of PAR accumulation in astrocyte cultures, the known PARP inhibitor benzamide actually decreases protein poly(ADP-ribosyl)ation (Ying et al., 2001). This indicates that PARG inhibitors do not necessarily inhibit PARP-1 directly but instead slow the turnover of PAR and therefore stop the vicious cycle of NAD^+ consumption that ultimately leads to cell death through necrosis. While the above explanation of PARG inhibitors involving prevention of an energy crisis in the cell seems to be the most accepted, other mechanisms could possibly contribute to the known protective effects of PARG inhibition in cases of severe DNA damage. Since PARP-1 is inhibited by its own automodification by PAR (D'Amours et al., 1999; Zahradka and Ebisuzaki, 1982), PARG inhibitors could work to indirectly knock out PARP by preventing catabolism of the inhibitory polymers. Another enzyme, Ca^{2+}/Mg^{2+}-dependent endonuclease, which is responsible for DNA fragmentation, is also inhibited by poly(ADP-ribosyl)ation and could be indirectly inhibited by PARG in a similar fashion (Tanaka et al., 1984; Yakovlev et al., 2000). Although the exact mechanism of action is not totally clear, PARG inhibition by tannin-related drugs has potential as a neuroprotective tool.

In vivo it has been found that PAR metabolism, which is controlled by the concerted action of PARP and PARG, plays a role in the transfer of metabolites, proteins, and drugs across cellular membranes through ATP-binding cassette (ABC) transporters (Dumitriu et al., 2004). UVB irradiation causes oxidative stress in cells due to the generation of peroxynitrite (PN) and inhibits the activity of ABC transporters (Deliconstantinos et al., 1996). Application of either 3,4-dihydro-5(4-(1-piperindinyl)butoxy)-1(2H)-isoquinoline (DPQ), a PARP inhibitor, or the known PARG inhibitor gallotannin restores function of ABC transporters following irradiation, most likely because ADP-ribose itself actually serves as an ABC transport inhibitor (Dumitriu et al., 2004). Multiple-drug resistance of tumor cells has often been linked to activity of ABC transporters (Cole et al., 1992), and this discovery highlights yet another potential application of PARG to cancer therapy.

Remarkably, PARG has also been linked to the regulation of transcription and consequently the expression of proinflammatory genes. When a PARG inhibitor such as gallotannin is incubated with macrophages, PAR accumulates, which in turn causes expression of inducible nitric-oxide synthase (iNOS) and cyclooxygenase-2 (COX-2), but not interleukin-1β or tumor necrosis factor-α (Rapizzi et al., 2004). This discovery is further supported by the fact that silencing of PARG with small interfering RNA (siRNA) prevents gallotannin-mediated expression of iNOS and COX-2 (Rapizzi et al., 2004). At this point, it is necessary to mention that chromatin condensation, which is one of the markers of gene down-regulation, has been shown to be tied to PARG and its associated PAR catabolism (de Murica et al 1986). Conversely, inhibition of PARG and the consequent buildup of

PAR is associated with chromatin loosening (de Murcia et al., 1988), which allows for gene transcription. Taken together, these data indicate that PARG inhibition might increase immune cell activation and therefore have major impacts on inflammation.

Through the use of the tannins as putatively selective inhibitors of PARG, the neuroprotective effects of PARG inhibition have been demonstrated. However, recently it has been suggested that while gallotannin may act as a PARG inhibitor, this compound is promiscuous and ineffective *in vivo* (Falsig et al., 2004). Initial questions concerning the specificity of gallotannin for PARG first arose upon close examination of its properties as provided by its manufacturer, namely that gallotannin is known to form insoluble complexes with proteins and acts as a fixative/stain. In models of H_2O_2- and MNNG-mediated cytotoxicity, it was found that while gallotannin prevents cell death in oxidative models as expected, in the studies involving nonoxidative DNA damage this polyphenol has no effect. Additionally, two other polyphenols, quercetin and catechin, were shown to alleviate cell death in the same model at concentrations similar to gallotannin, although they have never been reported as PARG inhibitors. One would also expect that PARG inhibition by gallotannin in living cells would enhance PAR production, but it has actually been found to reduce PAR formation, an effect attributed to its reactive oxygen species scavenging ability (Falsig et al., 2004). The strong H_2O_2-scavenging properties of tannins at low concentrations have been well-documented (Yokozawa et al., 1998), and the major component of gallotannin, tetra-galloyl-glucose, contains 13 phenolic hydroxyl groups, and 12 are in ortho positions and primed to react with hydroxyl radicals. While gallotannin does act as a PARG inhibitor in cell-free assays, it also has antioxidant properties and can protect cells from oxidative stress without PARG inhibition. Therefore, because gallotannin is not ideal for the evaluation of PARG in cellular death models, all results obtained previously using this compound must be carefully reviewed (Falsig et al., 2004).

7.5.3 Studies with Substrate Analogues

Along with an array of microtubules and their associated proteins, which act to align and separate chromosomes, it has been found that PAR is a nonproteinaceous, nonchromosomal component of the spindle necessary for bipolar spindle assembly and function (Chang et al., 2004). Enriched at both the spindle poles and kinetochores, cellular PAR concentration markedly increases during metaphase and anaphase–telophase due to heightened PARP activity (Bakondi et al., 2002), which suggests that PARG itself might be important in regulating spindle disassembly/assembly as well. After addition of 100 μg/ml of PARG to *Xenopus* egg extract spindles, a rapid breakdown of spindle structure is observed. Because of this hydrolytic activity of PARG, which can indeed be inhibited by the addition of ADP–HPD, microtubules were observed to turn outward from the spindle center, causing disconnection of the two half-spindles. Along with PARP and PAR, PARG is found to colocalize with spindle microtubules in BSC1 and HeLa cells, once again suggesting that it actively works to regulate PAR levels in the spindle (Chang et al., 2004).

7.6 PROSPECTUS

As highlighted above, three major classes of PARG inhibitors exist to date: ADP-ribose analogues, DNA intercalators, and tannins. While the last two classes of compounds show considerable potency, neither one is ideal because each class has high toxicity, high molecular weight, and an inability to act as specific inhibitors. Because of their highly charged nature, ADP-ribose analogues are not cell-permeable and are therefore unsuitable for use as inhibitors in a biological context. Additionally, such substrate analogues closely resemble biologically relevant molecules, which may ultimately be a source of unwanted side effects. Despite the lack of optimal inhibitors, it is clear that PARG is an attractive target for drug discovery and development, and it seems as if the scope of potential inhibitors is only beginning to be understood. In consideration of the aforementioned problems, other classes of novel and potent PARG inhibitors are currently being sought. Armed with the new methods to rapidly screen for PARG inhibition, it is possible that the next potent inhibitor might be a member of a combinatorial library or small molecule collection. As new and selective inhibitors of PARG are developed, it is certain that these tools will be used to elucidate and define new and exciting roles for cellular poly(ADP-ribosyl)ation.

REFERENCES

Affar, E. B., Germain, M., Winstall, E., Vodenicharov, M., Shah, R. G., Salvesen, G. S., and Poirer, G. G. (2001). Caspase-3-mediated processing of poly(ADP-ribose) glycohydrolase during apoptosis. *J Biol Chem* **276**, 2935–2942.

Alvarez-Gonzalez, R., and Althaus, F. R. (1989). Poly(ADP-ribose) catabolism in mammalian cells exposed to DNA-damaging agents. *Mutat Res* **218**, 67–74.

Aoki, K., Nishimura, K., Abe, H., Maruta, H., Sakagarni, H., Hatano, T., Okuda, T., Yoshida, T., Tsai, Y.-J., Uchiumi, F., and Tanuma, S-I. (1993). Novel inhibitors of poly(ADP-ribose) glycohydrolase. *Biochim Biophys Acta* **1158**, 251–256.

Bakondi, E., Bai, P., Szabó, E., Hunyadi, J., Gergely, P., Szabó, C., and Virág, L. (2002). Detection of poly(ADP-ribose) polymerase activation in oxidatively stressed cells and tissues using biotinylated NAD substrate. *J Histochem Cytochem* **50**, 91–98.

Berger, N. A. (1985). Poly(ADP-ribose) in the cellular response to DNA damage. *Radiat Res* 4–15.

Bernardi, R., Rossi, L., Poirer, G. G., and Scovassi, A. I. (1997). Analysis of poly(ADP-ribose) glycohydrolase activity in nuclear extracts from mammalian cells. *Biochim Biophys Acta* **1338**, 60–68.

Bonicalzi, M. E., Vodenicharov, M., Coulombe, M., Gagne, J. P., and Poirier, G. G. (2003). Alteration of poly(ADP-ribose) glycohydrolase nucleocytoplasmic shuttling characteristics upon cleavage by apoptotic proteases. *Biol Cell* **95**, 635–644.

Brochu, G., Duchaine, C., Thibeault, L., Lagueux, J., Shah, G. M., and Poirier, G. G. (1994a). Mode of action of poly(ADP-ribose) glycohydrolase. *Biochim Biophys Acta—Gene Structure & Expression* **1219**, 342–350.

Brochu, G., Shah, G. M., and Poirier, G. G. (1994b). Purification of poly(adp-ribose) glycohydrolase and detection of its isoforms by a zymogram following one- or two-dimensional electrophoresis. *Anal Biochem* **218**, 265–272.

Bürkle, A. (2001). PARP-1: A regulator of genomic stability linked with mammalian longevity. *Chembiochem* **2**, 725–728.

Burzio, L. O., Riquelme, P. T., Ohtsuka, E., and Koide, S. S. (1976). Evidence for two variants of poly(ADP-ribose) glycohydrolase in rat testis. *Arch Biochem Biophys* **173**, 306–331.

Cantoni, O., Sestili, P., Spadoni, G., Balsamini, C., Cucchiarini, L., and Cattabeni, F. (1987). Analogues of benzamide containing a sulfur atom as poly(ADP-ribose) transferase inhibitors. *Biochem Int* **15**, 329–337.

Chang, P., Jacobson, M. K., and Mitchison, T. J. (2004). Poly(ADP-ribose) is required for spindle assembly and structure. *Nature* **432**, 645–649.

Cole, S. P., Bhardwaj, G., Gerlach, J. H., Mackie, J. E., Grant, C. E., Almquist, K. C., Stewart, A. J., Kurz, E. U., Duncan, A. M., and Deeley, R. G. (1992). Overexpression of a transporter gene in a multi-drug resistant human lung cancer cell line. *Science* **258**, 1650–1654.

Coquet, A., Veuthey, J. L., and Haerdi, W. (1991). Selective post-column fluorigenic reation with benzamidine for trace level dection of reducing saccharides in liquid chromatography. *J Chromatography* **552**, 255–263.

Cortes, U., Tong, W. M., Coyle, D. L., Meyer-Ficca, M. L., Meyer, R. G., Petrilli, V., Herceg, Z., Jacobson, E. L., Jacobson, M. K., and Wang, Z. Q. (2004). Depletion of the 110-kilodalton isoform of poly(ADP-ribose) glycohydrolase increases sensitivity to genotoxic and endotoxic stress in mice. *Mol Cell Biol* **24**, 7163–7178.

D'Amours, D., Desnoyers, S., D'Silva, I., and Poirier, G. G. (1999). *Biochem J* **342**, 249–268.

Dantzer, F., de La Rubia, G., Menissier-De Murcia, J., Hostomsky, Z., de Murcia, G., and Schreiber, V. (2000). Base excision repair is impaired in mammalian cells lacking poly(ADP-ribose) polymerase-1. *Biochemistry* **39**, 7559–7569.

Davidovic, L., Vodenicharov, M., Affar, E. B., and Poirier, G. G. (2001). Importance of poly (ADP-ribose) glycohydrolase in the control of poly(ADP-ribose) metabolism. *Exp Cell Res* **268**, 7–13.

Deliconstantinos, G., Villiotou, V., and Stavrides, J. C. (1996). Increase of particulate nitric oxide synthase activity and peroxynitrite synthesis in UVB-irradiated keratinocyte membranes. *Biochem J* **320**, 997–1003.

de Murcia, G., Huletsky, A., Lamarre, D., Gaudreau, A., Poujet, J., Daune, M., and Poirier, G. G. (1986). Modulation of chromatin superstructure induced by poly(ADP-ribose) synthesis and degredation. *Proc Natl Acad Sci USA* **261**, 7011–7017.

de Murcia, G., Huletsky, A., and Poirier, G. G. (1988). Modulation of chromatin structure by poly(ADP-ribosyl)ation. *Biochem Cell Biol* **66**, 626–635.

Di Meglio, S., Tramontano, F., Cimmino, G., Jones, R., and Quesada, P. (2004). Dual role for poly(ADP-ribose)polymerase-1 and-2 and poly(ADP-ribose)glycohydrolase as DNA-repair and pro-apoptotic factors in rat germinal cells exposed to nitric oxide donors. *Biochim. Biophys. Acta - Mol Cell Res* **1692**, 35–44.

Dumitriu, I. E., Voll, R. E., Kolowos, W., Gaipl, U. S., Heyder, P., Kalden, J. R., and Herrmann, M. (2004). UV irradiation inhibits ABC transporters via generation of ADP-ribose by concerted action of poly(ADP-ribose) polymerase-1 and glycohydrolase. *Cell Death Differ* **11**, 314–320.

Falsig, J., Christiansen, S. H., Feuerhahn, S., Bürkle, A., Oei, S. L., Keil, C., and Leist, M. (2004). Poly(ADP-ribose) glycohydrolase as a target for neuroprotective intervention: Assessment of currently available pharmacological tools. *Eur J of Pharmacol* **497**, 7–16.

Genovese, T., Di Paola, R., Catalano, P., Li, J. H., Xu, W. Z., Massuda, E., Caputi, A. P., Zhang, J., and Cuzzocrea, S. (2004). Treatment with a novel poly (ADP-ribose) glycohydrolase inhibitor reduces development of septic shock-like syndrome induced by zymosan in mice. *Crit Care Med.* **32**, 1365–1374.

Hanai, S., Kanai, M., Ohashi, S., Okamoto, K., Yamada, M., Takahashi, H., and Miwa, M. (2004). Loss of poly(ADP-ribose) glycohydrolase causes progressive neurodegeneration in *Drosophila melanogaster. Proc Natl Acad Sci USA* **101**, 82–86.

Hassa, P. O., and Hottiger, M. O. (1999). A role of poly (ADP-ribose) polymerase in NF-κB transcriptional activation. *Biol Chem* **380**, 953–959.

Hassa, P. O., and Hottiger, M. O. (2002). The functional role of poly(ADP-ribose)polymerase 1 as novel coactivator of NF-kappaB in inflammatory disorders. *Cell Mol Life Sci* **59**, 1534–1553

Hatakeyama, K., Nemoto, Y., Ueda, K., and Hayaishi, O. (1986). Purification and characterization of poly(ADP-ribose) glycohydrolase. Different modes of action on large and small poly(ADP-ribose). *J Biol Chem* **261**, 14902–14911.

Kai, M., Tamura, K., Yamaguchi, M., and Ohkura, Y. (1985). Aromatic amidines as fluorogenic reagents for reducing carbohydrates. *Anal Sci* **1**, 255–263.

Kakita, H., Kamishima, H., Komiya, K., and Kato, Y. (2002). Simultaneous analysis of monosaccharides and oligosaccharides by high-performance liquid chromotography with post column fluorescence derivatization. *J Chromatography* **961**, 77–82.

Keil, C., Petermann, E., and Oei, S. L. (2004). Tannins elevate the level of poly(ADP-ribose) in HeLa cell extracts. *Arch Biochem Biophys* **425**, 115–121.

Koh, D. W., Patel, C. N., Ramsinghani, S., Slama, J. T., Oliveira, M. A., and Jacobson, M. K. (2003a). Identification of an inhibitor binding site of poly(ADP-ribose) glycohydrolase. *Biochemistry* **42**, 4855–4863.

Koh, D. W., Coyle, D. L., Mehta, N., Ramsinghani, S., Kim, H., Slama, J. T., and Jacobson, M. K. (2003b). SAR analysis of adenosine diphosphate (hydroxymethyl)pyrrolidinediol inhibition of poly(ADP-ribose) glycohydrolase. *J Med Chem* **46**, 4322–4332.

Koh, D. W., Lawler, A. M., Poitras, M. F., Sasaki, M., Wattler, S., Nehls, M. C., Stöger, T., Poirer, G. G., Dawson, V. L., and Dawson, T. M. (2004). Failure to degrade poly(ADP-ribose) causes increased sensitivity to cytotoxicity and early embryonic lethality. *Proc Natl Acad Sci USA* **101**, 17699–17704.

Lautier, D., Laguexu, J., Thibodeau, J., Menard, L., and Poirier, G. G. (1993). Molecular and biochemical features of polyADP(ribose) metabolism. *Mol Cell Biochem* **122**, 171–193.

Lazenik, Y. A., Kaufmann, S. H., Desnoyers, S., Poirier, G. G., and Earnshaw, W. V. (1994). Cleavage of poly(ADP-ribose) polymerase by a proteinase with properties like ICE. *Nature* **371**, 346–347.

Lefer, A. M., and Lefer, D. J. (1993). Pharmacology of the endothelium in ischemia–reperfusion and circulatory shock. *Annu Rev Pharmacol Toxicol* **33**, 71–90.

Li, J.-H., Ferraris, D. V., Kletzly, P. W., Li, W., Wang, E. Y., Xing, A. D., Xu, W., and Zhang, J. (2001). Symmetrically disubstituted aromatic compounds and pharmaceutical compositions for inhibiting poly (ADP-ribose) glycohydrolase, and methods for their use. US Patent 6,635,786.

Lin, W. S., Ame, J. C., Aboulela, N., Jacobson, E. L., and Jacobson, M. K. (1997). Isolation and characterization of the cDNA encoding bovine poly(adp-ribose) glycohydrolase. *J Biol Chem* **272**, 11895–11901.

Lu, X. C. M., Massuda, E., Lin, Q., Li, W. X., Li, J. H., and Zhang, J. (2003). Post-treatment with a novel PARG inhibitor reduces infarct in cerebral ischemia in the rat. *Brain Res* **978**, 99–103.

Maruta, H., Matsumura, N., and Tanuma, S. (1997). Role of (ADP-ribose) catabolism in DNA repair. *Biochem Biophys Res Comm* **236**, 265–269.

Maruta, H., Inageda, K., Aoki, T., Nishina, H., and Tanuma, S. (1991). Characterization of two forms of poly(ADP-ribose) glycohydrolase in guinea pig liver. *Biochemistry* **30**, 5907–5912.

Masson, M., Niedergang, V., Schreiber, S., Muller, J., Menissier-de Murcia, G., and de Murcia, G. (1998). XRCC1 is specifically associated with poly(ADP-ribsoe) polymerase and negatively regulates its activity following DNA damage. *Mol Cell Biol* **18**, 3563–3571.

Masutani, M., Shimokawa, T., Igarashi, M., Hamada, M., Shibata, A., Oami, S., Nozaki, T., Nakagama, H., Sugimura, T., Takeuchi, T., and Hori, M. (2002). Inhibition of poly(ADP-ribose) glycohydrolase activity by cyclic peptide antibiotics containing piperazic acid residues. *Proc Japan Acad Ser B*, **78**, 15–17.

Masutani, M., Nakagama, H., and Sugimura, T. (2003). Poly(ADP-ribose) and carcinogenesis. *Genes, Chromosomes & Cancer* **38**, 339–348.

Ménard, L., and Poirier, G. G. (1987). Rapid assay of poly(ADP-ribose)glycohydrolase. *Biochem Cell Biol* **65**, 668–673.

Menegazzi, M., Grassi-Zucconi, G., Carcerero De Prati, A., Ogura, T., Poltronieri, P., Nyunoya, H., Shiratori-Nyunoya, Y., Miwa, M., and Suzuki, H. (1991). Differential expression of poly(ADP-ribose) polymerase and DNA polymerase beta in rat tissues. *Exp Cell Res* **197**, 66–94.

Meyer-Ficca, M. L., Meyer, R. G., Coyle, D. L., Jacobson, E. L., and Jacobson, M. K. (2004). Human poly(ADP-ribose) glycohydrolase is expressed in alternative splice variants yielding isoforms that localize to different cell compartments. *Exp Cell Res* **297**, 521–532.

Milam, K. M., and Cleaver, J. E. (1984). Inhibitors of poly(adenosine diphosphate-ribose) synthesis: effect on other metabolic processes. *Science* **223**, 589–591.

Minaga, T., and Kun, E. (1983). Spectral analysis of the conformation of polyadenosine diphosphoribose. Evidence indicating secondary structure. *J Biol Chem* **258**, 725–730.

Miwa, M., and Sugimura, T. (1971). Splitting of the ribose–ribose linkage of poly(adenosine diphosphate-ribose) by calf thymus extract. *J Biol Chem* **246**, 6362–6364.

Oei, S. L., and Ziegler, M. (2000). ATP for the DNA ligation step in base excision repair is generated from poly(ADP-ribose). *J Biol Chem* **275**, 23234–23239.

Pacheco-Rodriguez, G., and Alvarez-Gonzalez, R. (1999). Measurement of poly(ADP-ribose) glycohydrolase activity by high resolution polyacrylamide gel electrophoresis: Specific inhibition by histones and nuclear matrix proteins. *Mol Cell Biochem* **193**, 13–18.

Panda, S., Poirer, G. G., and Kay, S. A. (2002). tej defines a role for poly(ADP-ribosyl)ation in establishing period length of the *Arabidopsis* circadian oscillator. *Dev Cell* **3**, 51–61.

Perkins, E., Sun, D., Nguyen, A., Tulac, S., Francesco, M., Tavana, H., Nguyen, H., Tugendreich, S., Barthmaier, P., Couto, J., Yeh, E., Thode, S.; Jarnagin, K., Jain, A., Morgans, D., and Melese, T. (2001). Novel inhibitors of poly(ADP-ribose) polymerase/PARP1 and PARP2 identified using a cell-based screen in yeast. *Cancer Res* **61**, 4175–4183.

Pieper, A. A., Verma, A., Zhang, J., and Snyder, S. H. (1999). Poly (ADP-ribose) polymerase, nitric oxide and cell death. *Trends Pharmacol Sci* **20**, 171–181.

Pion, E., Bombarda, E., Stiegler, P., Ullmann, G. M., Mely, Y., de Murcia, G., and Gerard, D. (2003). Poly(ADP-ribose) polymerase-1 dimerizes at a 5′ recessed DNA end *in vitro*: A fluorescence study. *Biochemistry* **43**, 12409–12417.

Putt, K. S., and Hergenrother, P. J. (2004). A nonradiometric, high-throughput assay for poly(ADP-ribose) glycohydrolase (PARG): Application to inhibitor identification and evaluation. *Anal Biochem* **333**, 256–264.

Putt, K. S., Beilman, G. J., and Hergenrother, P. J. (2004). Direct quantitation of poly(ADP-ribose) polymerase (PARP) activity as a means to distinguish necrotic and apoptotic death in cell and tissue samples. *Chembiochem* **5**, 1–3.

Ramsinghani, S., Koh, D. W., Amé, J.-C., Strohm, M., Jacobson, M. K., and Slama, J. T. (1998). Syntheses of photoactive analogues of adenosine diphosphate (hydroxymethyl)pyrrolidinediol and photoaffinity labeling of poly(ADP-ribose) glycohydrolase. *Biochemistry* **37**, 7801–7812.

Rapizzi, E., Fossati, S., Moroni, F., and Chiarugi, A. (2004). Inhibition of poly(ADP-ribose) glycohydrolase by gallotannin selectively up-regulates expression of proinflammatory genes. *Mol Pharm* **66**, 890–898.

Scovassi, A. I., and Poirier, G. G. (1999). Poly(ADP-ribosylation) and apoptosis. *Mol Cell Biochem* **199**, 125–137.

Shah, G. M., Poirer, D., Duchaine, C., Brochu, G., Desnoyers, S., Lagueux, J., Verreault, A., Hoflack, J. C., Kirkland, J. B., and Poirer, G. G. (1995). Methods for biochemical study of poly(ADP-ribose) metabolism *in vitro* and *in vivo*. *Anal Biochem* **227**, 1–13.

Shall, S., and de Murcia, G. (2000). Poly(ADP-ribose) polymerase-1: What have we learned from the deficient mouse model? *Mutat Res* **460**, 1–15.

Shimokawa, T., Masutani, M., Nagasawa, S., Nozaki, T., Ikota, N., Aoki, Y., Nakagama, H., and Sugimura, T. (1999). Isolation and Cloning of rat poly(ADP-ribose) glycohydrolase: presence of a potential nuclear export signal conserved in mammalian orthologs. *J Biochem* **126**, 748–755.

Slama, J. T., Aboul-Ela, N., and Jacobson, M. K. (1995). Specific inhibition of poly (ADP-ribose) glycohydrolase by adenosine diphosphate (hydroxymethyl) pyrrolidinediol. *J Med Chem* **38**, 4332–4336.

Smith, S. (2001). The world according to PARP. *Trends Biochem Sci* **26**, 174–179.

Smith, S., Giriat, I., Schmitt, A., and de Lange, T. (1998). Tankyrase, a poly(ADP-ribose) polymerase at human telomeres. *Science* **282**, 1484–1487.

Southan, G. J., and Szabo, C. (2003). Poly(ADP-ribose) polymerase inhibitors. *Curr Med Chem* **10**, 321–340.

Szabo, C., Virag, L., Cuzzocrea, S., Scott, G. S., Hake, P., O'Connor, M. P., Zingarelli, B., Salzman, A., and Kun, E. (1998). Protection against peroxynitrite-induced fibroblast injury and arthritis development by inhibition of poly(ADP-ribose) synthase. *Proc Natl Acad Sci USA* **95**, 3867–3872.

Tanaka, Y., Yoshihara, K., Itaya, A., Kamiya, T., and Koide, S. S. (1984). Mechanism of the inhibition of Ca^{2+}, Mg^{2+}-dependent endonuclease of bull seminal plasma induced by ADP-ribosylation. *J Biol Chem* **259**, 6579–6585.

Tavassoli, M., Tavassoli, M. H., and Shall, S. (1985). Effect of DNA intercalators on poly(ADP-ribose) glycohydrolase activity. *Biochim Biophys Acta* **827**, 228–234.

Tentori, L., Portarena, I., and Graziani, G. (2002). Potential clinical applications of poly(ADP-ribose) polymerase inhibitors. *Pharmacol Res* **45**, 73–85.

Thomassin, H., Jacobson, M. K., Guay, J., Verreault, A., Aboul-ela, N., Menard, L., and Poirier, G. G. (1990). An affinity matrix for the purification of poly(ADP-ribose) glycohydrolase. *Nucleic Acids Res* **18**, 4691–4694.

Tsai, Y.-J., Aoki, T., Maruta, H., Abe, H., Sakagami, H., Hatano, H., Okuda, T., and Tanurna, S. (1992). Mouse mammary tumor virus gene expression is suppressed by oligomeric ellagitannins, novel inhibitors of poly(ADP-ribose) glycohydrolase. *J Biol Chem* **267**, 14436–14442.

Tsutsumi, M., Masutani, M., Nozaki, T., Kusuoka, O., Tsujiuchi, T., Nakagama, H., Suzuki, H,. Konishi, Y., and Sugimura, T. (2001). Increased susceptibility of poly(ADP-ribose) polymerase-1 knockout mice to nitrosamine carcinogenicity. *Carcinogenesis* **22**, 1–3.

Ullrich, O., Reinheckel, T., Sitte, N., Hass, R., Grune, T., and Davies, K. J. (1999). Poly-ADP ribose polymerase activates nuclear proteasome to degrade oxidatively damaged histones. *Proc Natl Acad Sci USA* **96**, 6223–6228.

Virag, L., and Szabo, C. (2002). The therapeutic potential of poly(ADP-ribose) polymerase inhibitors. *Pharmacol Rev* **54**, 375–429.

Wilson, G. L., Patton, N. J., McCord, J. M., Mullins, D. W., and Mossman, B. T. (1984). Mechanisms of streptozotocin- and alloxan-induced damage in rat B cells. *Diabetologia* **27**, 587–591.

Winstall, E., Affar, E. B., Shah, R., Bourassa, S., Scovassi, A. I., and Poirier, G. G. (1999). Poly(ADP-ribose) glycohydrolase is present and active in mammalian cells as a 110-kDa protein. *Exp Cell Res* **246**, 395–398.

Yakovlev, A. G., Wang, G., Stoica, B. A., Boulares, H. A., Spoonde, A. Y., Yoshihara, K., and Smulson, M. E. (2000). A Role of the Ca^{2+}/Mg^{2+}-dependent endonuclease in apoptosis and its inhibition by poly(ADP-ribose) polymerase. *J Biol Chem* **275**, 21302–21308.

Ying, W. H., and Swanson, R. A. (2000). The poly(ADP-ribose) glycohydrolase inhibitor gallotannin blocks oxidative astrocyte death. *Neuroreport* **11**, 1385–1388.

Ying, W. H., Sevigny, M. B., Chen, Y. M., and Swanson, R. A. (2001) Poly(ADP-ribose) glycohydrolase mediates oxidative and excitotoxic neuronal death. *Proc Natl Acad Sci USA* **98**, 12227–12232.

Yokozawa, T., Chen, C. P., Dong, E., Tanaka, T., Nonaka, G. I., and Nishioka, I. (1998). Study on the inhibitory effect of tannins and flavonoids against the 1,1-diphenyl-2 picrylhydrazyl radical. *Biochem Pharmacol* **56**, 213–222.

Zahradka, P., and Ebisuzaki, K. (1982). A shuttle mechanism for DNA–protein interactions. The regulation of poly(ADP-ribose) polymerase. *Eur J Biochem* **127**, 579–585.

Ziegler, M., and Oei, S. L. (1999). Poly(ADP-ribosyl)ation stimulates DNA repair and silence transcription. *BioEssays* **23**, 543–547.

CHAPTER 8

CYCLIC ADP-RIBOSE ANALOGUES WITH MINIMAL STRUCTURE: SYNTHESIS AND CALCIUM-RELEASE ACTIVITY

Lihe Zhang and Andreas H. Guse*

8.1 INTRODUCTION

Ca^{2+} signaling is one of the most important intracellular signal transduction pathways in diverse cells. The Ca^{2+} mobilization is crucial for a wide variety of cell functions—for example, muscle contraction; secretion of neurotransmitters, hormones, and enzymes; fertilization of oocytes; and lymphocyte activation and proliferation. Local Ca^{2+} signals are mainly due to release from internal stores, and these calcium-release events are controlled by various second messengers and regulated by Ca^{2+} channel-associated proteins. Some small molecules—D-myo-inositol 1,4,5-triphosphate (IP_3) (Streb et al., 1983), cyclic adenosine diphosphate ribose (cADPR) (Lee et al., 1989), and nicotinic acid adenine dinucleotide phosphate (NAADP) (Lee and Aarhus, 1995)—are Ca^{2+} mobilizing messengers that act on specific intracellular receptors.

*Corresponding author: zdszlh@mail.bjmu.edu.cn

Drug Discovery Research: New Frontiers in the Post-Genomic Era, Edited by Ziwei Huang

The different isoforms of ryanodine receptors (RyRs) (Galeone et al., 1991) and inositol triphosphate receptors (IP_3R) (Joseph et al., 1995) are involved in both local and global Ca^{2+} responses.

cADPR (Fig. 8.1), a metabolite of NAD^+, was discovered in 1987 and is a signaling molecule to regulate calcium mobilization via RyR from intracellular stores in a wide variety of biological systems such as sea urchin eggs, pancreatic beta cells, smooth and cardiac muscles, T-lymphocytes, and cerebella neurons (Lee, 2002). The structure of cADPR was determined by X-ray crystallographic analysis in 1994 (Lee et al., 1994). Three genes encoding mammalian RyR1, RyR2, and RyR3 proteins show a high degree of homology except in three divergent regions (Rossi and Sorrentino, 2000; Conklin et al., 1999; Takeshima et al., 1998; Bhat and Ma, 1999). Each protein consists of about 5000 amino acids. A large number of key proteins are involved in specifically shaping an RyR-dependent Ca^{2+} signal. FK 506-binding protein (FKBP), protein kinase A, calmodulin-dependent protein kinase, protein phosphatases 1 and 2A, calsequestrin, sorcin, triadin, and junctin are RyR-associated proteins that contribute to the endoplasmic/sarcoplasmic reticulum Ca^{2+} signaling system (Bers, 2004). Calmodulin also modulates RyR channels from both skeletal and cardiac muscles (Mackrill, 1999). For example, it was found that the Ca^{2+} release channels in skeletal muscle comprise four 565-kD type 1 RyR subunits and four molecules of the 12-kD protein FKBP12. FKBP12 stabilizes the RyR1 complex and enables the four subunits to open and close coordinately (Marx et al., 1998). In pancreatic β cells and in smooth muscle binding of ligands to FKBP, either the naturally occurring ligand cADPR or the drug FK 506 induces a conformational

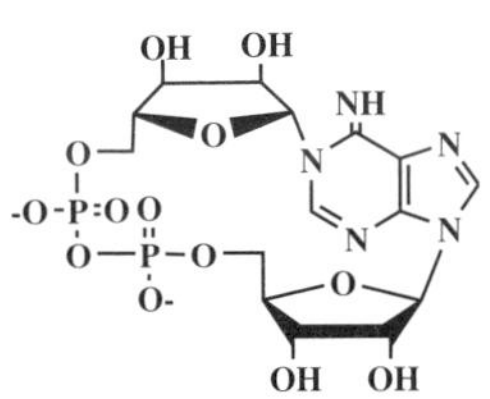

cADPR

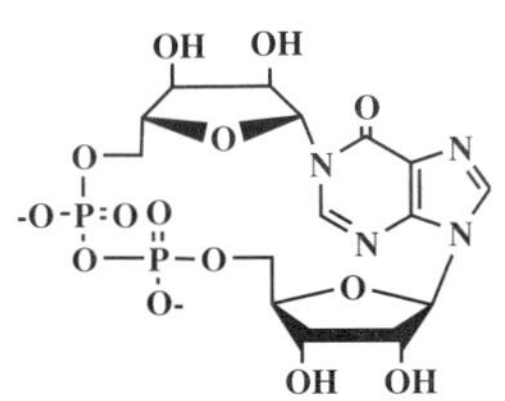

cIDPR

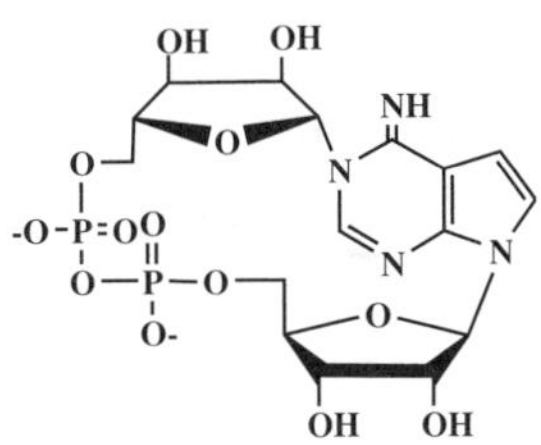

7-deaza-cADPR

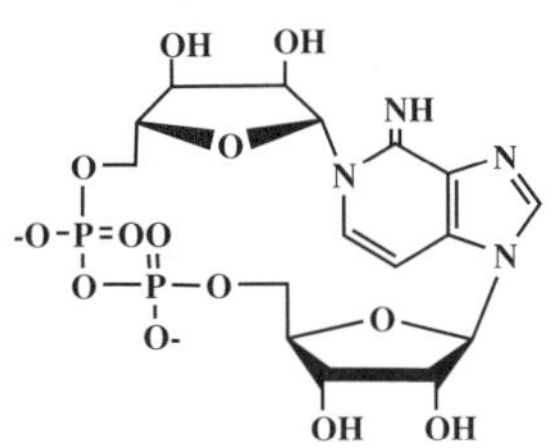

3-deaza-cADPR

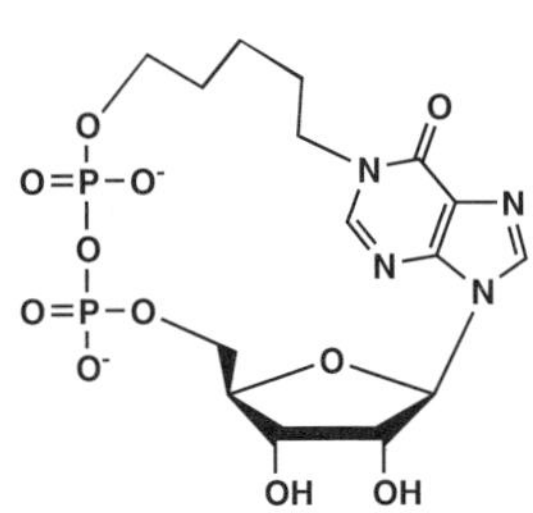

N^1-pentyl-cIDPcR

Figure 8.1 The structures of cADPR, cIDPR, 3-deaza-cADPR, 7-deaza-cADPR, N^1-pentyl-cIDPcR, and cADPR triphosphate mimic.

change of FKBP resulting in its release from RyR, which in turn resulted in the opening of the RyR Ca^{2+} channel (Noguchi et al., 1997). However, it is unclear whether this is the general mechanism of action of cADPR or whether cADPR elicits calcium release by direct binding to RyR. A significant approach to exploring the molecular mechanism of calcium release is to investigate the structure–activity relationship of cADPR and cADPR analogues with their calcium-mobilizing activities. Agonists or antagonists of cADPR that resulted from such studies are valuable tools to study cellular signal transduction and are also useful as lead structures for drug discovery.

Since the discovery of cyclic ADP-ribose as a messenger for agonist-induced Ca^{2+} release, more than 100 papers have been published describing a role for this novel metabolite in regulating calcium mobilization in different biological systems. Some reviews have described the syntheses and properties of cADPR analogues, including the modifications on the adenine moiety, the southern and northern ribose, and the pyrophosphate groups (Guse, 2004; Potter and Walseth, 2004; Shuto and Matsuda, 2004). However, due to the difference in pharmacological properties between the various modified cADPR analogues and the different cell models employed, our view of the structure-activity relationship is still ambiguous. According to the biosynthesis of cADPR from β-NAD^+ (Scheme 8.1), a variety of cADPR analogues were obtained by the enzymatic and chemo-enzymatic approach using nicotinamide dinucleotide-type precursors and ADP-ribosyl cyclase for the synthesis of the cyclic pyrophosphate. cADPR analogues with a substitution at the 8-position of the adenine ring (including 8- Br, NH_2, CH_3 and MeO) (Walseth and Lee, 1993; Reyes et al., 1999; Guse et al., 1997) were found to be antagonists of cADPR (Fig. 8.1). 8-Br-cADPR, although not as potent as 8-NH_2-cADPR, has been shown to be cell-permeant and can be used as a valuable tool. 7-Deaza-8-Br-cADPR is a cell-permeant antagonist and is about twice as effective as 8-Br-cADPR (Guse et al., 1999). 7-Deaza-cADPR was a partial agonist in sea urchin egg homogenates and

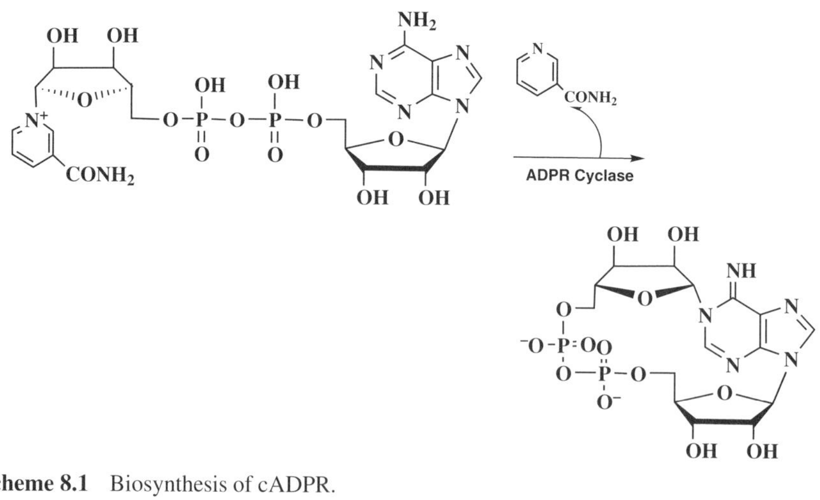

Scheme 8.1 Biosynthesis of cADPR.

3-deaza-cADPR was about 70-fold more potent in sea urchin egg homogenates, as compared to cADPR (Bailey et al., 1997; Sethi et al., 1997; Wong et al., 1999) (Fig. 8.1). It seems that either the 3-deaza- or the 7-deaza-purine ring in cADPR could retain the conjugated plane shape and be recognized by the receptor for cADPR. In the case of the N^1-coupled cyclic inosine diphosphate ribose mimics (cIDPR 2) (Fig. 8.1), different functional groups at the 8-position resulted in completely different biological activities; for example, 8-Br-cIDPR acts as a membrane-permeant agonist of Ca^{2+} signaling (Guse et al., 1999). However, due to the limitation of substrate-specificity of the ADP-ribosyl cyclase, the enzymatic and chemo-enzymatic method cannot be used for the construction of a larger variety of mimics modified in the northern or southern ribose moiety of cADPR. To build up a more detailed structure–activity profile for the calcium-releasing activity of cADPR, a method for the total syntheses of cADPR and its mimics is urgently needed.

8.2 INOSINE-BASED ANALOGUES OF cADPR WITH MODIFICATIONS IN THE NORTHERN RIBOSE MOIETY

Matsuda's group developed a total synthetic approach to construct cADPR analogues (Shuto et al., 1998). They found an efficient intramolecular cyclization for the construction of cyclopyrophosphate (Scheme 8.2). Their strategy was to introduce a bulky group into the 8-position of purine nucleoside; this maneuver can restrict the conformation of diphosphate intermediate in a syn-form in which the two phosphate moieties are close to each other and facilitate intramolecular condensation in the presence of EDC. Matsuda and co-workers also reported that the molecular cyclization was completed by I2/3A MS as a promoter in pyridine in quantitative yield. In this way, they synthesized cIDP-carbocyclic-ribose (Fukuoka et al., 2000) (cIDPcR) and cyclic ADP-carbocyclic-ribose (Shuto et al., 2001) (cADPcR). cADPcR, in which the 4′-oxo in the northern ribose moiety of cADPR was substituted by a methylene group caused a significant release of Ca^{2+} in sea urchin eggs; in the same system, cADPcR was 3–4 times more potent than cADPR while the 8-substituted analogues were less efficacious.

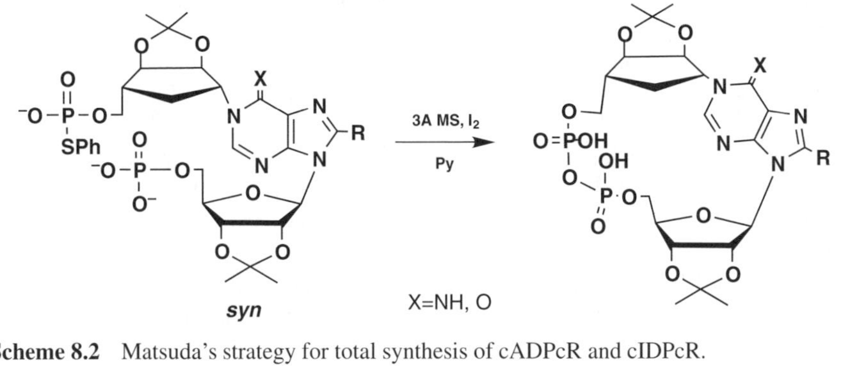

Scheme 8.2 Matsuda's strategy for total synthesis of cADPcR and cIDPcR.

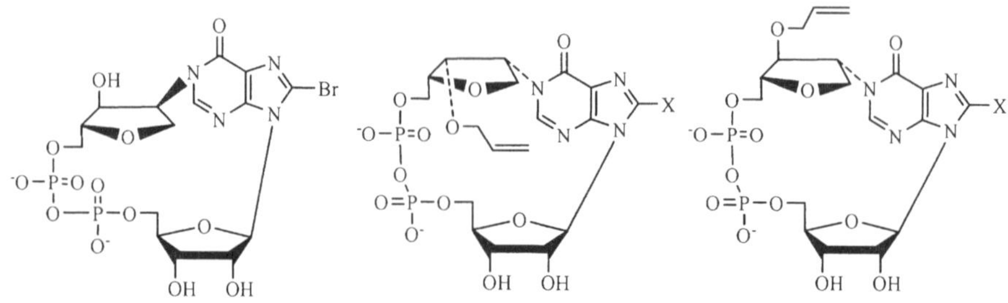

3

4a X = H
4b X = Br

5a X = H
5b X = Br

Figure 8.2 The structures of N^1-glucosyl-cIDPR mimics 3, 4a, 4b, 5a, and 5b.

In contrast, in mammalian cells, cADPcR acted much more weakly as compared to the sea urchin system (Guse et al., 2002). However, it was found that cIDPcR showed only an insignificant Ca^{2+}-mobilizing effect ($EC_{50} > 10^6$ μM), and 8-Br-cIDPcR was almost inactive in sea urchin egg homogenates (Shuto and Matsuda, 2004). Using a similar synthetic approach, an N^1-pentyl substituted cIDPcR analogue was reported, but the biological data have not appeared yet (Galeone et al., 2002a). A novel class of stable cIDPR mimics (3–5) (Fig. 8.2) was synthesized, in which the N^1-ribosyl moiety is replaced by different configurational glycosyl moieties and the N^1-glycosyl linkage is shifted from the N^1–$C^{1'}$ to the N^1–$C^{2''}$ position of furanose to improve the stabilization (Huang et al., 2002a).

The key steps for the synthesis of N^1-glycosyl cIDPR analogues involve substitution at the N^1 position of inosine and intramolecular cyclophosphorylation of the corresponding bisphosphate (Scheme 8.3). 1,4-anhydro-2-*O*-triflyl-3,5-*O*-benzylidene-L-xylitol,5-MMTr-3-*O*-allyl-2-0-triflyl-1,4-anhydro-D-xylitol and 5-*O*-benzoyl-3-*O*-allyl-2-)-triflyl-1,4-anhydro-D-ribitol were synthesized from the corresponding sugar (Scheme 8.3); the active triflates reacted with protected 8-bromo-inosine in the presence of K_2CO_3 and 18-crown-6 by an S_N2 substitution to give the key intermediate, N^1-glycosyl substituted inosine derivative, accompanied by a litter of O^6-substituted derivative. The intramolecular cyclization was performed by adding a solution of the cyclic precursor slowly over 20 hr, using a syringe pump, to a large excess of promoter I2/3A MS in pyridine at room temperature. The cyclic products (3, 4a, 4b, 5a, 5b) of triethylammonium salts were obtained, and some debrominated cyclic products (4a, 5a) were also found in addition to the normal products during the cyclization (Scheme 8.3).

Rat brain microsomes were one of the first mammalian cell preparations in which cADPR and IP_3 were shown to trigger Ca^{2+} release independently of each other. The five novel cyclic nucleotide analogues (3, 4a, 4b, 5a, 5b) were tested for their abilities to release calcium from rat brain microsomes. In addition, Ca^{2+} mobilizing effects of extracellularly added compounds were analyzed in intact HeLa cells.

cADPR or the synthetic samples (3, 4a, 4b, 5a, 5b) induced Ca^{2+} release from brain microsomes with a threshold concentration of less than 15 μM under these

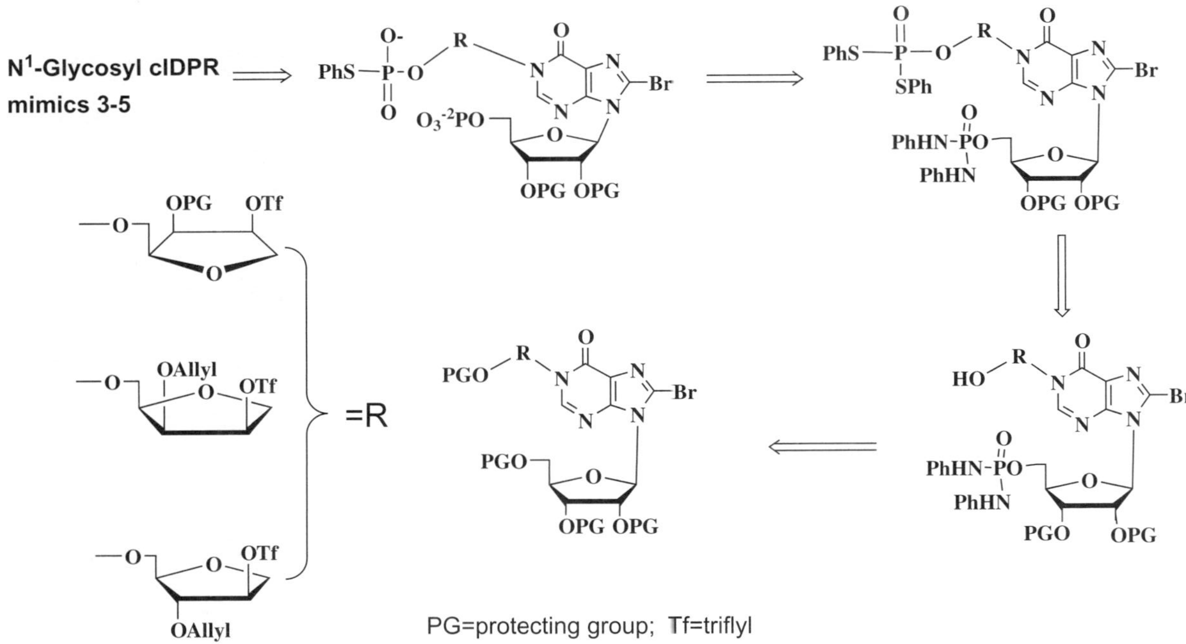

PG=protecting group; Tf=triflyl

Scheme 8.3 Retrosynthesis of N^1-glycosyl c-IDPR.

experimental conditions. Ca^{2+} was soon resequestered into the microsomes (< 400 sec). It is interesting to find that all samples except compound 4b induced the calcium release from brain microsomes under these conditions. The order of potency is 5a > 4a > cADPR > 5b > 3. It is evident that the 8-nonsubstituted analogues (4a, 5a) have more potent calcium-modulating activity than the corresponding 8-bromo cIDPR mimics (4b, 5b). Compound 4b was not able to induce the calcium release from brain microsomes even at the concentration of 60 μM. In addition to their activity in broken cell preparations, these cIDPR mimics modified in the N^1-ribosyl moiety were also active in intact HeLa cells. The natural compound cADPR (200 μM) itself exhibited no Ca^{2+} releasing activity, presumably because it cannot cross the cell membrane. However, samples 3, 4b, and 5b showed significant calcium release under the same conditions. The 8-unsubstituted analogue 5a only elicited calcium release at concentrations as high as 400 μM, and 4a exhibited no calcium release activity. Taken together, these data suggest that the modification in the northern ribose moiety of cADPR could retain the calcium release activity and also increase the membrane permeability of cADPR. It seems that the N^1-ribosyl moiety and the 6-amino group in cADPR are not critical structural factors for

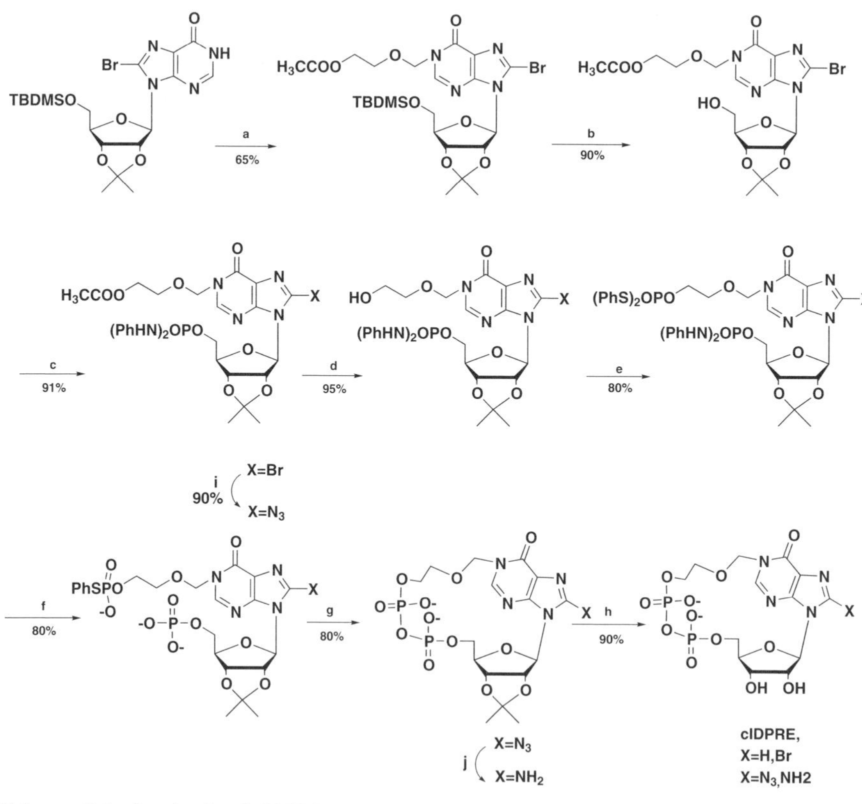

Scheme 8.4 Synthesis of cIDPRE.

biological activity. The data indicated that further modifications in the northern ribose might also work—for example, a mimic of cADPR with complete replacement of the northern ribose with an ether strand (cIDPRE) (Scheme 8.4).

An N^1 substitution was carried out regioselectively on the protected inosine with 2-chloromethoxyethyl acetate in the presence of excess DBU to afford N^1-substituted inosine in 69% yield (Huang et al., 2002b). After deprotection and phosphorylation, the cyclic precursor was obtained. The intramolecular cyclization of bisphosphate intermediate was completed by using Matsuda's strategy. Compared to the yields of the cyclization in the syntheses of cIDPR and its analogues, a more flexible N^1-ether strand of cIDPRE results in a lower yield for the synthesis. A series of 8-substituted cIDPRE were also reported (Gu et al., 2004). The pharmacological activities of cIDPRE and 8-Br-, 8-N_3-, 8-NH_2-, and 8-Cl-cIDPRE were analyzed in intact and permeabilized human Jurkat T-lymphocytes (Figs. 8.3 and 8.4). The results indicate that cIDPRE permeates the plasma membrane, releases Ca^{2+} from an intracellular, cADPR-sensitive Ca^{2+} store, and subsequently initiates Ca^{2+} release-activated Ca^{2+} entry. The Ca^{2+}-releasing activity of cIDPRE resulted in a lower efficacy of cIDPRE as compared to cADPR (Fig. 8.4d). Analogues of cIDPRE modified at C8 of inosine had strikingly differential effects: while 8-N_3-cIDPRE and 8-NH_2-cIDPRE were similarly effective in their agonistic activity as compared to cIDPRE, the

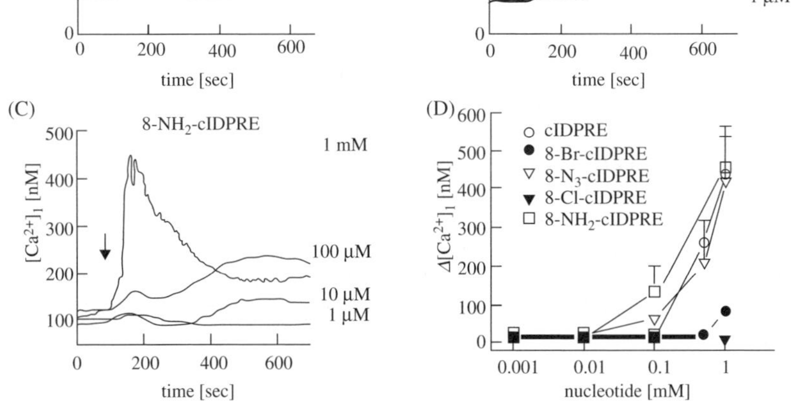

Figure 8.3 Effect of the cIDPRE and 8-substituted cIDPRE analogues on Ca^{2+} signaling in T cells $[Ca^{2+}]_i$ was analyzed in Fura2-loaded Jurkat T cells. (**A–C**) Characteristic tracings in response to cIDPRE and two of its analogues. (**D**) Concentration–response relationship of cIDPRE and its analogues.

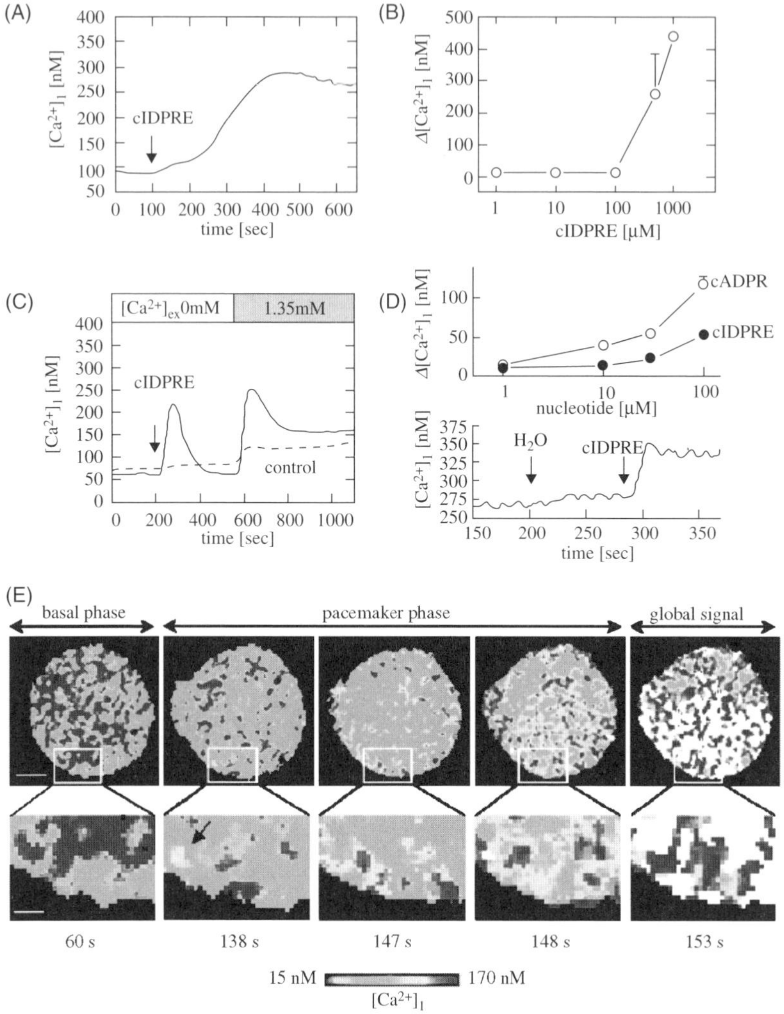

Figure 8.4 Effect of the cADPR mimic cIDPRE on Ca^{2+} signaling in T cells. (**A**) cIDPRE was added as indicated (final concentration 0.5 mM). (**B**) Concentration–response data are given as mean ± SEM ($n = 3$–5). (**C**) cIDPRE was added as indicated. Then, $CaCl_2$ was readded. The dashed line shows a control experiment in which buffer was added instead of cIDPRE. (**D**) Cells were permeabilized. *Lower panel*: Vehicle (H_2O) and cIDPRE (100 μM) were added as indicated. *Upper panel*: Data are mean ± SEM ($n \geq 3$; error bars are partially not visible because the values are smaller than the symbol size). (**E**) Characteristic confocal ratiometric pseudocolor images of a single Jurkat cell and magnifications of a defined subcellular region are shown. See color plates.

halogenated derivatives 8-Br- and 8-Cl-cIDPRE did not significantly elevate $[Ca^{2+}]$. Comparison of the calcium-release activities of different cIDPRE derivatives revealed a remarkable structure–activity relationship: While any functional group larger than the natural H-atom at C8 in cADPR turned the molecule into a cADPR antagonist—for example, 8-NH_2-cADPR, 8-Br-cADPR, 8-CH_3O-cADPR, 8-CH_3-cADPR—in the case of the cyclic IDP-ribose mimics, different functional groups at the 8-position resulted in completely different biological activities.

Thus, cIDPRE, 8-NH_2-, and N_3-cIDPRE are novel and important agonist tools similarly suitable for studying the structure–activity relationship of cADPR and for use as membrane-permeant cADPR mimics in intact cells. As an example subcellular Ca^{2+} signals can be induced by cIDPRE, as demonstrated by confocal Ca^{2+} imaging at the single cell level (Fig. 8.4e). A role for cADPR-mediated amplification of subcellular calcium signals and spatiotemporal signal propagation has been analyzed using the membrane-permeant cIDPRE (Kunerth et al., 2004).

8.3 cIDP-DE: A MIMIC OF cADPR WITH MODIFICATIONS IN THE NUCLEOBASE AND NORTHERN AND SOUTHERN RIBOSE MOIETIES

Concerning modifications on the southern ribose moiety of cADPR, it was found that the 3′-hydroxyl group was essential for calcium release in the sea urchin egg system, whereas the 2′-hydroxyl was not. *O*-Methylation of the 3′-hydroxyl group generated an antagonist (Ashamu et al., 1997) (Fig. 8.5). The stable cyclic aristeromycin diphosphoribose (c-Aris DPR) contains a replacement of the southern ribose with a carbocyclic five-membered ring and is an agonist slightly more potent as compared to cADPR in the T-lymphocyte system (Bailey et al., 1996) (Fig. 8.5). The Mayol group reported the synthesis of a cIDPcR analogue in which a pentyl group was included instead of the southern ribose, but no biological data were described (Galeone et al., 2002b) (Fig. 8.5). Comparison of the calcium-release activities of cIDPRE and cADPR revealed that the greater flexibility of the whole molecule introduced by replacement of the northern ribose by the ether strand either still allows fitting into the binding site or allows the functional groups to support a conformation much more like cADPR. Because of the same structure of the northern and southern ribose occurring

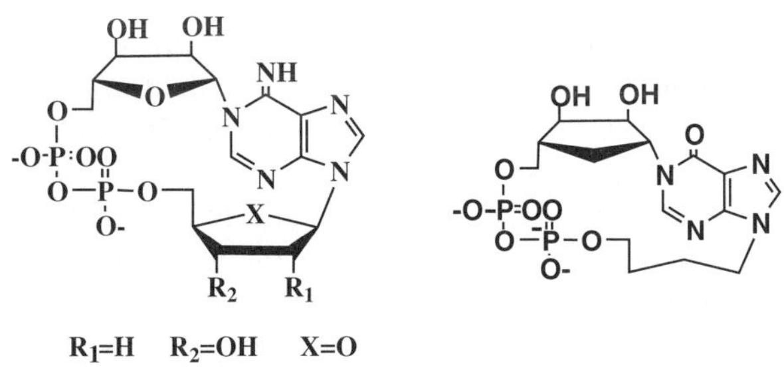

R_1=H R_2=OH X=O
R_1=OH R_2=H X=O
R_1=H R_2=OMe X=O
R_1=R_2=OH X=CH_2

Figure 8.5 The structures of 2′-*O*- and 3′-*O*-methyl cADPR, c-Aris-DPR, and N^9-pentyl-cIDPcR.

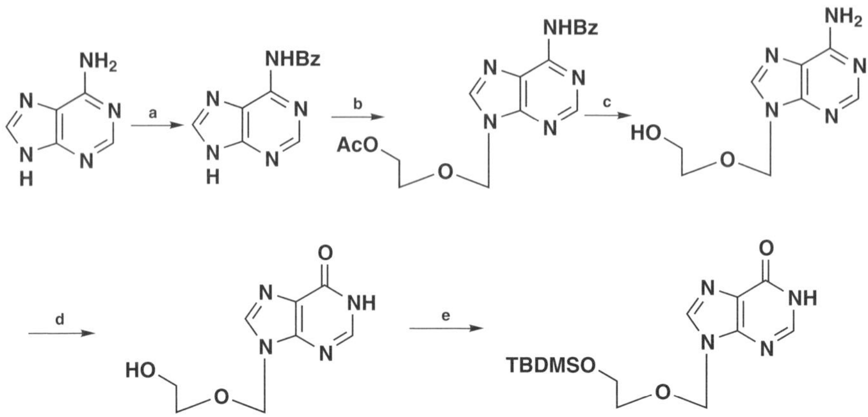

Scheme 8.5 Synthesis of N^9-substituted hypoxanthine.

in cADPR, it appeared interesting to synthesize the cIDPR analogue with replacement of both northern and southern riboses using two ether strands. Synthesis and biological evaluation of the respective compound N^1-[(phosphoryl-O-ethoxy)-methyl]-N^9-[(phosphoryl-O-ethoxy)-methyl]-hypoxantine-cyclicpyrophosphate (cIDP-DE) was reported recently (Guse et al., 2005). The key steps for the synthesis of cIDP-DE include the synthesis of the N^9-substituted hypoxanthine (Scheme 8.5) and the intramolecular cyclization of the bisphosphate intermediate (Scheme 8.6). It was very interesting to find that the intramolecular cyclization of bisphosphate intermediate was completed in 61% yield, although more flexible N^1 and N^9 ether strands probably resulted in a larger spatial distance of the two phosphate groups (Scheme 8.6). Interestingly, cIDP-DE displayed moderate fluorescence (λ_{ex} 278 nm, λ_{em} 355 nm) in water at room temperature. However, no fluorescence was observed in cIDPR or cIDPRE. The reason for these spectroscopic properties might be the more flexible N^1 and N^9 ether strands in cIDP-DE, which allow an easier π-electronic transition in the purine ring. The fluorescent character of cIDP-DE might be used for the mechanistic studies of cADPR-mediated calcium release.

Interestingly, the evaluation of the biological activity of cIDP-DE indicates that it is a membrane-permeant analogue of cADPR with agonistic properties. In saponin-permeabilized Jurkat T cells, cIDP-DE released Ca^{2+} from intracellular stores but was much weaker as compared to cADPR. Regardless of this weak direct effect, a pronounced activity of cIDP-DE on Ca^{2+} signaling was observed in intact Jurkat T cells. The Ca^{2+}-mobilizing activity of cIDP-DE was also observed in Ca^{2+} imaging experiments. The biological activity of cIDP-DE in Jurkat T cells was quite similar to cIDPRE in terms of (i) Ca^{2+} release activity in permeabilized cells and (ii) induction of both Ca^{2+} release and Ca^{2+} entry in a Ca^{2+}-free/Ca^{2+}-reintroduction protocol.

The Ca^{2+}-mobilizing activity of cIDP-DE was also confirmed in cardiac myocytes freshly prepared from mouse hearts. In the myocytes, addition of cIDP-DE induced global Ca^{2+} signaling with a very slow onset, obviously due to a slower uptake into the cell as compared to the Jurkat T cells. However, spatiotemporal

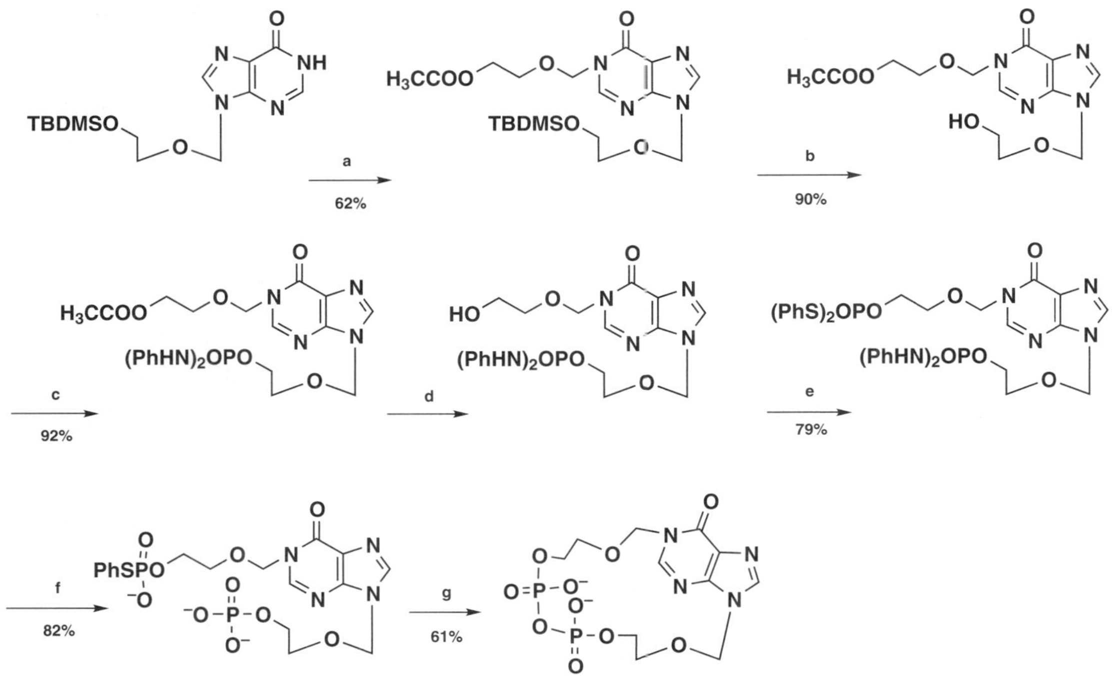

Scheme 8.6 Synthesis of cIDP-DE

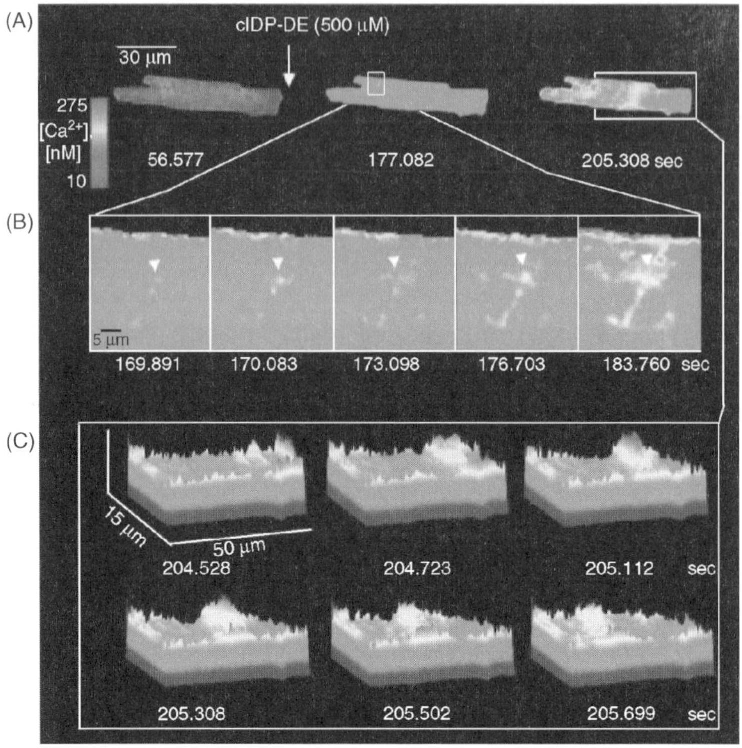

Figure 8.6 cIDP-DE-mediated local Ca^{2+}signaling and global Ca^{2+} waves in intact mouse cardiac myocytes. (**A**) Pseudocolor images of the whole cell before (56.677 sec) and after addition of cIDP-DE (177.082 and 205.308 sec). (**B**). A region of interest marked on the 177.082-sec image is displayed together with additional time points at larger magnification. The arrowhead points to a small "hot spot" region with oscillating $[Ca^{2+}]$ and ongoing recruitment of neighboring Ca^{2+} microdomains. (**C**) The part of the myocyte marked in the 205.308-sec image is shown at different time points as a surface plot to visualize the development and movement of a global Ca^{2+} wave starting from the right edge of the cell. See color plates.

Ca^{2+} signaling pattern including local spots with oscillating $[Ca^{2+}]_i$ and global Ca^{2+} waves were observed upon cIDP-DE addition (Fig. 8.6).

Taken together, this similarity of cIDP-DE and cIDPRE, especially in induction of Ca^{2+} signaling in intact T cells, indicates that the further replacement of the southern ribose by the second ether bridge in cIDP-DE did not significantly influence its biological activity. Compounds of the type cIDP-DE or cIDPRE may thus be used as membrane-permeant agonists of the cADPR-/Ca^{2+}-signaling system in intact cells.

8.4 MODEL FOR THE INTERACTION OF cADPR WITH ITS RECEPTOR

The structure of cADPR in solution was investigated using NMR by Sekine and co-workers (Wada et al., 1995). They reported a predominant C2′-*endo* conformation for the southern ribose unit and a flat northern ribose conformation in the cADPR structure in solution. These structural characteristics in solution are very similar to those of the X-ray crystallographic analysis (Lee et al., 1994). However, in the structure of cIDP-DE, two more flexible ether strands are included instead of both the northern and southern riboses of cADPR. Nevertheless, cIDP-DE remains biologically active for calcium release in T-lymphocytes and cardiac myocytes. Although the structure of the receptor for cADPR is not clear so far, the data reported above may indicate some clues of the structural requirements for the design of future cADPR agonists.

To model the structure of cIDP-DE, its conformation was minimized by quantum mechanics calculation (Fig. 8.7). The different ways of superimpositions

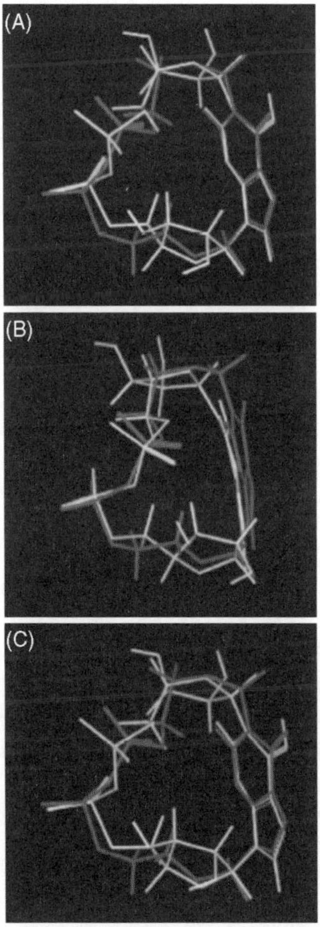

Figure 8.7 Superimposition of minimized conformations of cADPR and cIDP-DE. The conformations of cADPR (yellow) and cIDP-DE (pink) were minimized using the density functional theory (DFT) method. (**A**) Superimposition based on nucleobases; (**B**) superimposition based on diphosphates; (**C**) superimposition based on whole molecules. See color plates.

of the energetically minimized conformations of cIDP-DE (pink color) and cADPR (yellow color) clearly show that both compounds are characterized by an almost identical three-dimensional structure (Fig. 8.7). Despite the replacement of the northern and southern ribose of cADPR by two flexible ether strands, the distance and orientation between nucleobase and diphospho-bridge were maintained and the whole conformation was still almost unchanged, as visualized by superimposition of either the nucleobase (Fig. 8.7A) or the diphospho-bridge (Fig. 8.7B). The optimal superimposition of the two whole molecules in addition suggests that the two ether groups in cIDP-DE nicely replace the corresponding C1–O–C4 units of the two riboses (Fig. 8.7C). Because of the flexible ether strands used as linkers in cIDP-DE, it may well be that the binding pocket of the cADPR receptor helps induce the correct "binding conformation" during the process of ligand binding, a process known in enzymology as "induced fit."

In summary, recent data from our own and other laboratories indicate the following crucial elements for the structure–activity relationship of cADPR: (1) 7-Deaza-cADPR and 3-deaza-cADPR are agonists of cADPR in the sea urchin egg system. It seems that the 3-deaza- or 7-deaza-purine ring in cADPR would retain the conjugated plane shape and be recognized by the sea urchin egg cADPR receptor. (2) cADPR analogues with substitutions at the 8-position of the adenine ring (including 8-substituted 7-deaza-cADPR) were found to be antagonists of cADPR in both sea urchin eggs and many high eukaryotic cell types. However, in the case of the N^1-coupled cyclic inosine pyrophosphate-ribose mimics (N^1-cIDPR), 8-Br-cIDPR acts as a membrane permeant agonist of cADPR-induced Ca^{2+} signaling. (3) Despite the more radical structure alterations in cIDPRE and cIDP-DE, these analogues are agonists of cADPR. The N^1- and N^9-ribosyl moieties and the 6-amino group in cADPR obviously are not critical structural factors for biological activity in mammalian cells. (4) Modification of cIDPRE at C8 of inosine had strikingly different effects, with 8-N_3-cIDPRE and 8-NH_2-cIDPRE being effective agonists and 8-Br- and 8-Cl-cIDPRE without biological activity. (5) With modification in the pyrophosphate moiety of cADPR using three phosphate groups in line, the triphosphate mimic (Fig. 8.1) was a highly effective agonist in bovine brain microsomes (Zhang et al., 1996).

Therefore, the proposed model for the interaction of cADPR with its receptor protein would be as follows: (1) The binding pocket may specifically recognize the purine base or other aromatic rings with similar shape as purine on the one side and the di- or triphosphate bridge or multinegative groups on the other side of the ligand. (2) The two parts of the binding pocket located close to the northern and southern ribose moieties may be sufficiently variable, for example, one electronegative atom, such as oxygen, and two spacers of the right length appear to be sufficient for binding of the ligand and for weak biological activity. (3) The binding pocket interacting with the 8-position in purine moiety may be an active trigger whose function—agonist or antagonist—depends on the structure of adenine or hypoxanthine in the binding molecule. However, as long as the binding protein for cADPR has not been identified, this remains speculation.

ACKNOWLEDGMENTS

We wish to thank all former and current members of our groups who have contributed to this project. The project is supported by National Natural Science Foundation of China (to LHZ), The Ministry of Science and Technology of China (to LHZ), the Deutsche Forschungsgemeinschaft (to AHG), the Hertie-Foundation (to AHG), the Wellcome Trust (to AHG), and the Deutsche Akademische Austauschdienst/Chinese Scholarship Council (to LHZ and AHG).

REFERENCES

Ashamu, G. A., Sethi, J. K., Galione, A., and Potter, B. V. L. (1997). *Biochemistry* **36**, 9509–9517.

Bailey, V. C., Fortt, S. M., Summerhill, R. J., Galeone, A., and Potter, B. V. L. (1996). *FABS Lett* **379**, 227–230.

Bailey, V. C., Sethi, J. K., Fortt, S. M., Galeone, A., and Potter, B. V. L. (1997). *Chem Biol* **4**, 51.

Bers, D. M. (2004). *J Mol Cell Cardiol* **37**, 417–429.

Bhat, M. B., and Ma, J. (1999). *J Biol Chem* **277**, 8597–8601.

Conklin, N. W., Barone, V., Sorrentino, V., and Coronado, R. (1999). *Biophys J* **77**, 1394–1403.

Fukuoka, M., Shuto, S., Shirato, M., Sumita, Y., Veno, Y., and Matsuda (2000). *J Org Chem* **65**, 5238–5248.

Galeone, A., Lee, H. C., and Busa, W. B. (1991). *Science* **253**, 1143.

Galeone, A., Mayol, L., Oliviero, G., Piccialli, G., and Varra, M. (2002a). *Eur J Org Chem* **24**, 4234–4238.

Galeone, A., Mayol, L., Oliviero, G., Piccialli, G., and Varra, M. (2002b). *Tetrahedron* **58**, 363–368.

Gu, X. F., Yang, Z. J., Zhang, L. R., Kunerth, S., Fliegert, R., Weber, K., Guse, A. H., and Zhang, L. H. (2004). *J Med Chem* **47**, 5674–5682.

Guse, A. H., da Silva, C. P., Weber, K., Armah, C. N., Ashamu, G. A., Schulze, C., Potter, B. V. L., Mayr, G. W., and Hilz, H. (1997). *Eur J Biochem* **245**, 411.

Guse, A. H., da Silva, C. P., Berg, I., Skapenko, A. L., Weber, K., Heyer, P., Hohenegger, M., Ashamu, G. A., Schulze-Koops, H., Potter, B. V. L., and Mayer, G. W. (1999). *Nature* **398**, 70–73.

Guse, A. H., Cakir-Kiefer, C., Fukoka, M., Shuto, S., Weber, K., Bailey,V. C., Matsuda, A., Mayer, G. W., Oppenheimer, N., Schuber, F., and Potter, B. V. L. (2002). *Biochemistry* **41**, 6744–6751.

Guse, A. H. (2004). Regulation of calcium signaling by the second messenger (cADPR). *Curr Mol Med* **4**, 239–248.

Guse, A. H., Gu, X. F., Zhang, L. R., Weber, K. Krämer, E.,Yang, Z. J., Jin, H. W., Li, Q., Carrier, L., and Zhang, L. H. (2005). *J Biol Chem.* in press.

Huang, L. J., Zhao, Y. Y., Yuan, L., Min, J. M., and Zhang, L. H. (2002a). *J Med Chem* **45**, 5340–5352.

Huang, L. J., Zhao, Y. Y., Yuan, L., Min, J. M., and Zhang, L. H. (2002b). *Bioorg Med Chem Lett* **12**, 887–889.

Joseph, S. K., Lin, C., Pierson, S., Thomas, A. P., and Maranto, A. R. (1995). *J Biol Chem* **270**, 23310–23316.

Kunerth, S., Langhorst, M. F., Schwarzmann, N., Gu, X. F., Huang,L. J., Yang, Z. J., Zhang, L. R., Mills, S. J., Zhang, L. H., Potter, B. V. L., and Guse, A. H. (2004). *J Cell Sci* **117**, 2141–2149.

Lee, H. C. (2002). *Cyclic ADP-Ribose and NAADP, Structures, Metabolism and Functions.* Dordrecht: Kluwer Academic Publishers.

Lee, H. C., and Aarhus, R. (1995). *J Biol Chem* **270**, 2152–2157.

Lee, H. C., Walseth, T. F., Bratt, G. T., Hayes, R. N., and Clapper, D. L. (1989). *J Biol Chem* **264**, 1608–1615.

Lee, H. C., Aarhus, R., and Levitt, D. (1994). *Nat Struct Biol* **1**, 143–144.

Mackrill, J. J. (1999). *Biochem J* **337**, 345–361.

Marx, S. O., Ondrias, K., and Marks, A. R. (1998). *Science* **281**, 818–821.

Noguchi, N., Takasawa, S., Nata, K., Tohgo,A., Kato, I., Ikehata, F., Yonakura, H., and Okamoto, H. (1997). *J Biol Chem* **272**, 3133.

Potter, B. V. L., and Walseth, T. F. (2004). *Curr Mol Med* **4**, 303–311.

Reyes-Harde, M., Empson, R., Potter, B. V. L., Galeone, A., and Stanton, P. K. (1999). *Proc Natl Acad Sci USA* **96**, 4061.

Rossi, D., and Sorrentino, V. (2000). *Cell Calcium* **32**, 307–319.

Sethi, J. K., Empson, R. M., Bailey, V. C., Potter, B. V. L., and Galeone, A. (1997). *J Biol Chem* **272**, 16358.

Shuto, S., and Matsuda, A. (2004). *Curr Med Chem* **11**, 827–845.

Shuto, S., Shirato, M., Sumita, Y., Ueno, Y., and Matsuda, A. (1998). *J Org Chem* **63**, 1986–1994.

Shuto, S., Fukuoka, M., Manikowasky, A., Ueno, Y., Nakano, T., Kuroda, R., Kuroda, H., and Matsuda, A. (2001). *J Am Chem Soc* **123**(36), 8750–8759.

Streb, H., Irvine, R. F., Berridge, M. J., and Schulz, I. (1983). *Nature*, **306**, 67–69.

Takeshima, H., Komazaki, S., Hirose, K., Nishi, M., Noda, T., and Lino, M. (1998). *EMBO J.* **17**, 3309–3316.

Wada, T., Inageda, K., Aritomo, K., Tokita, K., Nishina, H., Takahashi, K., Katada, T., and Sekine, M. (1995). *Nucleosides Nucleotides* **14**, 1301–1341.

Walseth, T. F., and Lee, H. C. (1993). *Biochim Biophys Acta* **1178**, 235.

Wong, L., Aarhus, R., Lee, H. C., and Walseth, T. F. (1999). *Biochim Biophys Acta* **1472**, 555.

Zhang, F. J., Yamada, S., Gu, Q. M., and Shi, C. J. (1996). *Bioorg Med Chem Lett* **6**, 1203.

CHAPTER 9

EVALUATION OF NEUROIMMUNOPHILIN LIGANDS

*Junhai Xiao, Aihua Nie, and Song Li**

9.1 INTRODUCTION

Neurodegenerative disease—including diseases such as Huntington's disease (HD), Parkinson's disease (PD), and Alzheimer's disease (AD)—is caused by chronic degeneration of the nervous system. HD is a terminal progressive neurodegenerative disorder. HD patients mainly develop chorea (uncontrollable jerking movements) and psychiatric dysfunction that generally includes intense irritability and the loss of emotional control. PD is also a terminal progressive neurodegenerative disorder whereby patients experience a debilitating loss of movement functionality that includes muscle rigidity, bradykinesia (slowness of movement), and resting tremor. AD is a form of dementia that invariably leads to complete loss of all cognitive abilities and premature death. The mechanisms of these diseases are still unclear, so there are no specific and effective strategies to treat them. Currently, drugs used to treat HD, PD, and AD mainly include neuroprotectants, nerve growth factor, and other small organic molecules such as neuroimmunophilin ligands.

*Corresponding author: lis@nic.bmi.ac.cn

Drug Discovery Research: New Frontiers in the Post-Genomic Era, Edited by Ziwei Huang

9.1.1 Neuroprotectants

There are three main mechanisms of neuronal cell death that may act separately or cooperatively to cause neurodegeneration: metabolic compromise, excitotoxicity, and oxidative stress (Alexi et al., 2000): "Metabolic compromise of neurons is caused by stroke, asphyxiation, hypoglycemia, and certain respiratory poisons." These neurotoxic poisons are mainly mitochondrial poisons and cause a depletion of ATP, which leads to preferential neurodegeneration in the basal ganglia. Excitotoxicity occurs due to a dysfunction of excitatory amino acid neurotransmisson—usually a stimulation of glutamate receptors that becomes pathological. The toxic effects of excitatory amino acids are mainly due to the activation of the *N*-methyl-D-aspartate (NMDA) glutamate receptor, which leads to an influx of Ca^{2+} and then toxic overloading (Greene and Greenamyre, 1996). Per Alexi et al., "Oxidative stress is due to the actions of highly reactive free radicals such as the ROSs superoxide anion and hydroxyl radical, and the RNS peroxynitrite." These free radicals can lead to the depletion of ATP (Szabo et al., 1996) and induce neuronal cell death (Morel et al., 1999). Alexi et al. add: "The lethal triplet of metabolic compromise, excitotoxicity, and oxidative stress may also act cooperatively in causing neuronal cell death. For example, metabolic impairment may elicit secondary excitotoxicity." Metabolic compromise also may cause oxidative stress by inducing the production of free radicals both from the electron transport chain and due to the burden of increased intracellular Ca^{2+} on mitochondria function. The oxidative stress may also cause metabolic impairment and initiate excitotoxic pathways. The byproduct HNE that yields from oxidative stress-induced lipid peroxidation can impair glucose transport, which can lead to energetic failure and render neurons more sensitive to excitotoxicity (Pedersen et al., 1999). Thus, all the materials inhibiting or eliminating those mechanisms may have the potential to treat neurodegenerative diseases. These materials that can protect neurons from damage but not promote the regeneration of nerves are termed neuroprotectants.

A characteristic of neurodegenerative disease is formation of β-amyloid. On one side, β-amyloid can activate the Ca^{2+} channel of cell which results in the accumulation of intracellular Ca^{2+} (Hölscher, 1998). This intracellular Ca^{2+} then induces further activation of Ca^{2+}-dependent proteases and free-radical generation to lesion neurons. The glutamate mediate system also participates in this procedure. Thus, glutamate receptor antagonists, Ca^{2+} channel inhibitors, and inhibitors of related enzymes all have the potential to treat neurodegenerative disease. On the other side, the accumulation of β-amyloid will trigger abnormal protein phosphorylation reactions responsible for dysfunction and eventual death of neurons in the brain (Jin and Saitoh, 1995). The depositing of β-amyloid is modified by advanced glycation end products (AGEs) (Durany et al., 1999), which implies that the inhibitors of AGEs may also have the potential to treat neurodegenerative disease. The ability of glutamic acid to kill neurons (excitotoxicity) seems to be mediated, in most cases, by an interaction with NMDA receptors, leading to an uncontrollable rise in intracellular calcium concentrations and thence cell lysis and death (Doble, 1995). Since glutamaterigic neurodegeneration appears to better explain the physiological and anatomical hallmarks of neurodegeneration, the focus of research has shifted toward

NMDA receptor inhibitors, such as riluzole (Arendt et al., 1998) and memantine (Parsons et al., 1999). Besides the NMDA receptor family, the AMPA/kainate receptor family is also involved in neurodegeneration and may also have the potential for neurodegenerative disease therapy. Antioxidants are another type of neuroprotectant. Antioxidants can clean the highly reactive free radicals such as the ROSs superoxide anion, hydroxyl radical, and the RNS peroxynitrite, which cause oxidative stress. There are other materials involved in oxidative stress, such as the products of inflammatory reactions—that is, prostaglandins (PGs: PGE1 and PGA1), cytokines, and complement proteins. Thus, the combination of multiple antioxidants and NSAIDs may be more beneficial in the prevention of neurodegenerative disease. Drugs that act mainly on the acetylcholine neurotransmitter system—such as the acetylcholine esterase inhibitors tacrine, velnacrine, or Aricept—are also used to treat neurodegenerative disease, especially AD, because it was believed that the main cause for AD is a degeneration of basal brain nuclei that use acetylcholine as a neurotransmitter.

Considerable efforts have been made to find more effective neuroprotectants, and there must be more progress in this active research field. All the known neuroprotectants can only relieve symptoms or slow the neurodegeneration of neurons, but they cannot stop the progress of neurodegeneration or repair the damaged neurons. Thus, using neuroprotectants cannot completely cure neurodegenerative disease and so is not an ideal therapeutic method.

9.1.2 Nerve Growth Factor (NGF)

Nerve growth factor is an endogenetic factor that can stimulate the growth of neurons and the extension of nerve fibers. NGF is a macromolecule including Zn^{2+} and two proteins that consist of three subunits: α unit, β unit, and γ unit. NGF is unstable in acidic and basic environments and disassociates when $pH < 5$ or $pH > 8$. Only the β unit has a biofunction that consists of 118 residues. The net charge on its surface is electropositive since positive groups are greater than negative groups. Whether *in vivo* or *in vitro*, the neurons cannot survive and grow without the existence of NGF. There are several cell types that can excrete NGF, including fibroblasts, smooth muscle cells, skeletal muscle cells, glial cells, neurilemma cell, and so on. These cells mainly come from mesenchyme and are related to the normal function of neuraxons. There are specific receptors on the membranes of neuraxons that can recognize and complex with NGF and then transport into cells through a positive transport procedure. The endogenetic nerve growth factor acts on sensory and sympathetic neurons of the peripheral nervous system and on basal forebrain and striatal cholinergic neurons of the central nervous system. Evidence showed that NGF effectively attenuates lesion-induced cholinergic deficits and cognitive impairments in animal models. In addition, a clinical study with chronic NGF treatment in an AD patient showed promise. Olson et al. showed that NGF treatment increased blood flow, 11C nicotine uptake, and 11C binding in the cerebral cortex. In addition, NGF infusions normalized EEG patterns and improved performance in word-recognition tests (Lapchak, 1993). Besides NGF, other neurotrophic factors such as activity-dependent neurotrophic factor (ADNF) and glial cell line-derived neurotrophic factor (GDNF) all show notable activity in neruoprotection and neuroregeneration (Bohn, 1999). A short peptide sequence derived from ADNF (a peptidergic active site) protected neurons from death associated with a broad range of toxins, including those related to Alzheimer's disease, the human immunodeficiency virus, excitotoxicity, and electrical blockade. This peptide is proposed as a lead compound for drug development against neurodegeneration (Gozes and Brenneman, 1996). Although using endogenetic factor to stimulate the regeneration of nerves should be the best therapeutic method for neurodegeneration disease, the common shortcomings of biopolymers such as having low bioavailability, degrading easily, passing the brain–blood barrier (BBB) with difficulty, and being difficult to take orally reduce their clinical application in neurodegenerative disease therapy. Studies addressing this problem are applying viral vector-mediated transfer of the NGF, GDNF, and ADNF genes to the CNS in order to deliver biosynthesized neurotrophic factors to a specific location in a chronic manner. Recent studies suggest that these GDNF gene-therapy approaches are effective in rat models of Parkinson's disease (Fink et al., 1997). However, the effectiveness of gene therapy in humans still needs to be proven.

9.1.3 Neuroimmunophilin Ligands

The ideal way to treat neurodegenerative disease would be to use NGF-like organic compounds that can be taken orally, are stable *in vivo*, and pass through the BBB easily. The ligands of neuroimmunophilin are the best choice. Neuroimmunophilin is primarilly a type of immunophilin that complexes with immunosuppressants such as FK506, CsA, and rapamycin (RAPA) and produces immunosuppressive responses. These proteins are also known as peptidylprolyl *cis–trans* isomerases (PPIases) since they can catalyze the interconversion of *cis* and *trans* rotamers of amide bonds adjacent to proline residues in peptidic substrates (Fischer et al., 1984). The enzymes complex with CsA termed cyclophilin (CyP), and the enzymes complex with FK506 or RAPA termed FKBPs (FK506 bind proteins).

FKBPs are a family of proteins that exist extensively in nature. The three-dimensional structure of this family is highly conserved. The sequence homology is more than 50% among different species. A large number of immunophilins belonging to the FKBP family have been discovered in the past several years. Currently, there are about 20 known FKBPs in nature. FKBP immunophilins known to be present in humans include FKBP12, FKBP12.6, FKBP13, FKBP25, FKBP37, FKBP51, and FKBP52 (by convention, members of the FKBP family are named by appending to the prefix "FKBP" the apparent molecular weight in kilodaltons). In addition to the activity of PPIases, FKBPS are also involved in some other biological processes, such as translation (FKBP25, FKBP37), protein secretion of endoplast (FKBP13), single mediation of EGFR on the cell membrane (FKBP12), the influx of Ca^{2+} intracellular (FKBP12), and involvement in the formation of steroid receptor complexes (Smith et al., 1993). Functions of the other members of the FKBP family are still unknown. FKBP12 is the most well-known member of the FKPBs. It complexes FK506 and further inhibits the activity of Calcineurin phosphatase, which results in the interruption of proliferation of T-lymphocyte (Liu et al., 1991). FKBP51 acts the same way as FKBP12 (Baughman et al., 1995). The first three-dimensional structure of the

FKBPs is the complex of FKBP12 and FK506 (Van Duyne et al., 1991; Lepne et al., 1992). The binding site consists of residues Tyr26, Phe36, Asp37, Arg42, Phe46, Glu54, Val55, Ile56, Trp59, Tyr82, His87, Ile91, and Phe99. All these residues make a hydrophobic cave and complex with FK506 through hydrophobic and electronic interaction.

Steiner et al. (1992) showed the high density of FKBP mRNA in brain tissue and a close homology between the structure of NINA A, a protein in the neural retina of *Drosophila*, and cyclophilin. This implies a physiological role for the immunophilins in the nervous system. The results of Dawson et al. (1993) showed that FK506 enhanced phosphorylation of nitric oxide synthase and protected against glutamate neurotoxicity. The neuroprotective effect of FK506 presumably involves the inhibition of calcineurin, preventing the dephosphorylation of nitric oxide synthase and its subsequent activation. GAP-43, a growth-associated protein, was also found to be a physiologically important substrate for calcineurin. Also, calcineurin and protein kinase C may regulate the levels of free calmodulin available in neurons (Liu and Storm, 1989). In addition, regeneration of sciatic and facial nerves is always accompanied by the increase of mRNA of GAP-43. All these results imply a physiological role for the FKBPs in the regeneration of the nervous system. Under the direction of this hypothesis, a small organic compound with neuroregenerative activity was found in FKBP ligands. The findings of Lyons et al. (1994) that FK506 enhances neurite outgrowth in PC12 rat pheochromocytoma cells and in rat sensory ganglia by increasing sensitivity to nerve growth factor initiate a new era of finding small organic neuroregeneration compounds.

9.2 NATURAL NEUROIMMUNOPHILIN LIGANDS

Being well-known targets of immunosuppressant for many years, many neuroimmunophilin ligands have been found before FKBPs became the targets of neuroregeneration. Among them are the well-known natural neuroimmunophilin ligands FK506, rapamycin, and cyclosporine A.

9.2.1 FK506

FK506 (Fig. 9.1A) is a 23-member ring macrolide lactone that was discovered in 1984 by scientists at Fujisawa Pharmaceutical Company in the fermentation broth of the filamentous bacterium, *Streptomyces tsukubaenisis* (Kino et al., 1987), launched in 1993 for the treatment of immunosuppression in organic transplantation (Starzl et al., 1990). It acts by binding to its receptors, FK506-binding proteins (FKBPs), which in turn can bind to and regulate a Ca^{2+}-dependent phosphatase, calcineurin; it also inhibits the activity of calcineurin and inhibits interleukin-2 production in activated T cells. Lyons et al. (1994) found that FK506 was significantly effective in neuroregeneration *in vitro*. It promotes neurite outgrowth both in PC12 cells and in sensory neuronal cultures of dorsal root ganglia. It can also stimulate functional regrowth of damaged sciatic, cortical cholinergic, dopamine, and 5-HT neurons (Snyder et al., 1998). The search results of other researchers showed FK506 can improve H_2O_2-induced cell damage by increasing glutathione levels in NG108-15 cells (Tanaka et al., 2001), increase the rate of axonal regeneration in rat sciatic nerve (Gold et al., 1995), increase neuronal expression of GAP-43, improve functional recovery after spinal cord injury in rats (Madsen et al., 1998), and enhance the survival of cultured and grafted rat embryonic dopamine neurons (Gastilho et al., 2000). In an in vivo model of focal cerebral ischemia, Sharkey and Butcher (1994) found that FK506 exhibited a powerful neuroprotective activity. Administered up to 60 min post-occlusion, the minimum effective neuroprotective dose is comparable with the immunosuppressant dose in humans, suggesting that FK506 may have clinical potential for the treatment of stroke. Kitamura et al. (1994) found that in the MPTP-induced dopamine depletion in the striatum of young C57BL/6 mice model, pretreatment with FK-506 significantly protected MPTP-induced DA depletion in the striatum.

9.2.2 Rapamycin

Rapamycin (Fig. 9.1B) is another kind of immunosuppressant that can bind to FKBPs. The complex of rapamycin and FKBP12 is more stable than that of FK506, and rapamycin can inhibit the neuroprotective activity of FK506. Recent research results showed that rapamycin may be the one of the only immunosuppressants that can stimulate the regeneration of nerves (Parker et al., 2000). In the study, rapamycin markedly increased neurite outgrowth in PC12 cells in the presence of a low concentration of nerve growth factor, while FK506 (0.1–1000 nM) had little effect on neurite outgrowth in PC12 cells in either the presence or absence of nerve growth factor. The authors implied that this result is due to the ability of rapamycin to inhibit cell cycle progression. The results of Costantini and Isaacson (2000) showed that rapamycin can antagonize the elongation effects of DA neurites of FK506 and cyclosporine A.

9.2.3 Cyclosporine A

Cyclosporine A (CsA, Fig. 9.1C) is a potent immunosuppressive drug that has widespread clinical uses in organ transplantation and the treatment of autoimmune disorders. Cyclosporine A is a cyclic peptide that consists of 11 amino acids. It complexes with cyclophilin and leads to immunosuppression. Research results show that the cyclophilin level in neuronal areas is higher than the level in T-cells (Lad et al., 1991), and this initiates the study of its potential function on the nervous system. Uchion et al. (1995) found that cyclosporine A can dramatically ameliorate CA1 hippocampal damage following transient forebrain ischemia in the rat. Other cyclosporine A-mediated neuroprotection includes attenuation of mitochondrial swelling, stabilization of the permeability transition pore, and neurofilament compaction (Herdegan et al., 2000). In animal models of nerve grafts, cyclosporine A has been proved to increase the survival of grafted rat embryonic dopamine neurons (Gastilho et al., 2000). In addition, cyclosporine A (15 mg/kg) can also inhibit xenograft rejection in rats and enhanced tissue survival *in vitro* (Wennberg et al., 2001). In a study that investigates whether cyclosporine A is able to facilitate axonal

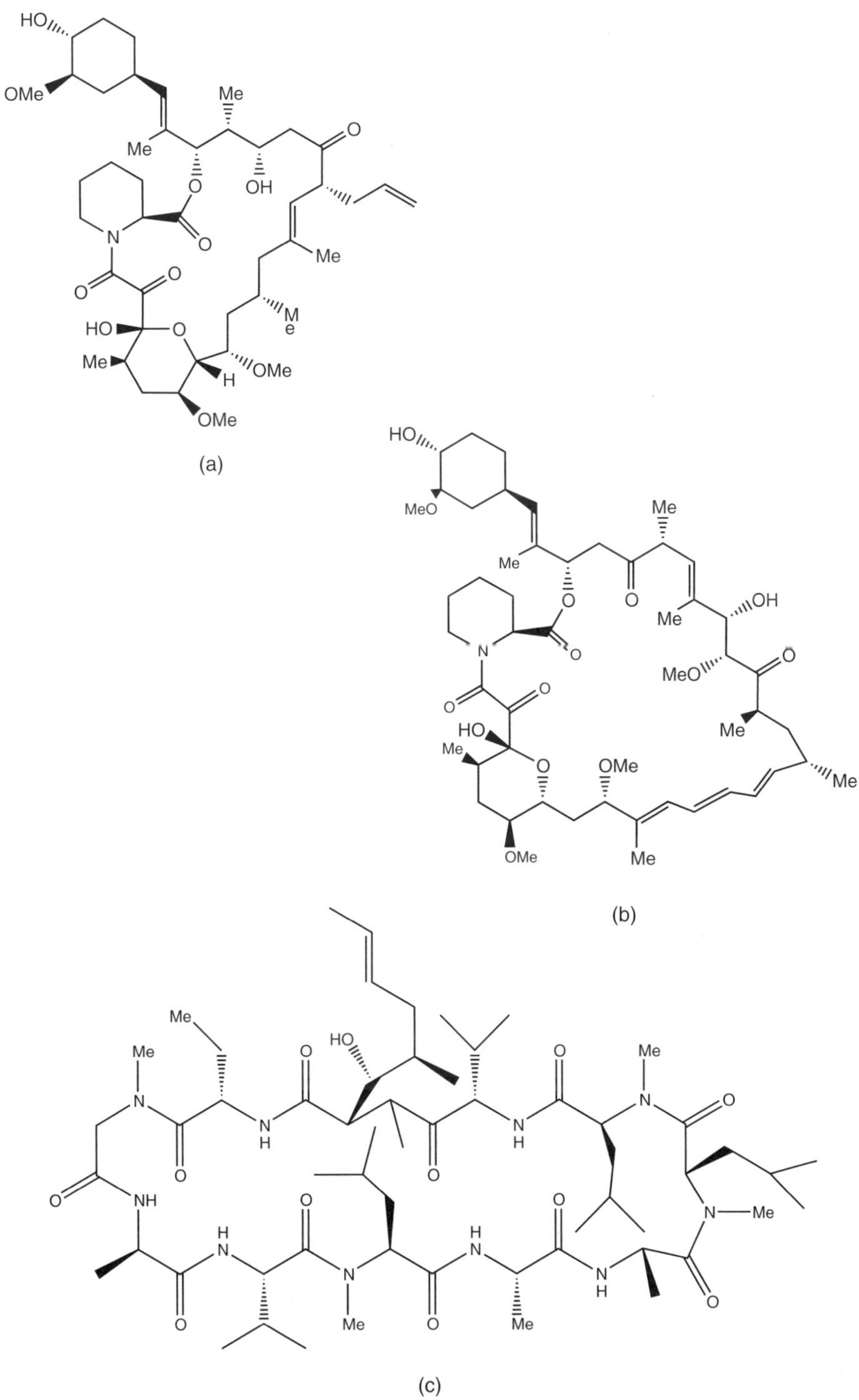

Figure 9.1 Structure of FK506 (**a**), rapamycin (**b**), and cyclosporine A (**c**).

regeneration in rats, strong evidence of morphological axonal regeneration was observed 15 days after surgery in all cyclosporine A-treated animals (Palladini et al., 1996). The function of enhancing neurite branching or elongation is also proved (Costantini and Isaacson, 2000).

In general, FK506 has significant potential for neuroprotection, neuroregeneration, and passing the brain–blood barrier easily. Cyclosporine A also has neuroprotective and neuroregenerative activity but has to pass the brain–blood barrier. Rapmycin has no neuroprotective activity but has significant neuroregenerative potential. Thus, among the natural immunophilin ligands, FK506 should be the best candidate for treatment of neurodegenerative disease. However, immunosuppressive responses limit its clinical application.

The above-mentioned neuroregenerative and neuroprotective properties of FK506 and other immunophilin ligands suggests an exciting new therapeutic approach for the treatment of neurological disorders. Promising drug candidates should be small-molecule FKBP ligands with oral bioavalibility, which are devoid of immunosuppressive activity and are able to cross the blood–brain barrier. Many efforts have been made in this area, and hundreds of novel nonimmunosuppressant neuroimmunophilin ligands have been found.

9.3 NEUROREGENERATIVE MECHANISM OF NEUROIMMUNOPHILIN LIGANDS

The neuroregenerative mechanism of neuroimmunophilin ligands is poorly known. The original mechanism is based on the calcineurin and Ca^{2+} channel. Dawson et al. (1993) assumed that the complex of FKBP12 and FK506 inhibits the function of calcineurin and in turn inhibits the activity of nitric oxidate synthase (NOS), which makes FK506 produce the neuroprotective activity in focal ischemia. However, the evidence that nonimmunosuppressive analogues of FK506 that bind FKBP12 but can't complex with calcineurin also exhibit neuroprotective activity denied the calcineurin-involved neuroregenerative mechanism.

Since FKBP12 is the primary target of many nonimmunosuppressive ligands, the relevant mechanism may involve this protein. This mechanism has been well accepted for several years. To examine if FKBP12 is required for neurotrophic effects of immunophilin ligands, Guo et al. (2001) cultured dopaminergic neurons from FKBP12 knockout mice and found that FK506 still protects dopaminergic neurons against MPP+ toxicity. Gold et al. also found that FK506 maintains its neurotrophic activity in primary hippocampal cell cultures from FKBP-12 knockout mice (Gold et al., 1999). These two independent studies proved that without the participation of FKBP12, immunophilin ligands can also exhibit neuroregenerative activity. Recently, a novel nonimmunosuppressive analogue of FK506 named V-13661 exhibited the same neuroregenerative activity. However, this compound can't bind to FKBP12. This result uncovered that the neuroregenerative activity of immunophilin ligands is not related to FKBP12 (Costantini et al., 2001). All the above results show there may be another mechanism for neuroregenerative effects of immunophilin ligands.

The results of Gold showed that FKBP52 may be a key protein in the mechanism of neuroregenerative effects of FKBPs ligands (Gold et al. 1999). FKBP52 is a member of the FKBP family, which consists of 459 amino acid residues. It contains three function domains, named D1, D2, and D3, respectively. Each domain is connected with a short hydrophilic amino acid sequence. The sequence homology of D1 (A1-A149, FKBP52-I) and FKBP12 is about 55%. This domain can bind to FK506 and be of the same PPIase activity as FKBP12. The D1 domain can also bind to ATP and GTP, and it phosphorylates under the catalytic of tyrosine protein kinase II. Also, FKBP52 is a component of a subclass of steroid receptor complexes (Tai et al., 1992). The binding of FKBP52 and FK506 is weak, and the complex cannot inhibit the effect of calcineurin. This is the main difference between FKBP12 and FKBP52. Since it participates in the steroid receptor complex, people presumed that the neuroregenerative effects of immunophilin ligands may relate to the steroid receptor/hsp-90/FKBP52 complex.

Two related research results support this presumption: The first is molybdate, which acts in contrast to steroids (Raaka et al., 1985) and can inhibit the neuroregenerative effects of FK506; the other is that the geldanamycin, which inhibits the complex of FKBP52/p23/Hsp-90 and has neuroregenerative activity (Paratt and Toft, 1997). These results imply that FK506 may complex with FKBP52, which in turn disassociated the steroid receptor complex, and the disassociated components may further act as neuroregeneratants. The findings of Ishimoto et al. (1998) that Hsp-90 can promote the regeneration of telencephalic neurons and spinal neurons proved the rationality of this presumption. But how it stimulates the growth of nerves is still unclear. Gold et al. (1997) found that injecting FK506 10 mg/kg daily could significantly increase the mRNA level of GAP-43, a protein related to the growth of neurons.

Based on this evidence, Gold presumed a more rational mechanism as shown in Fig. 9.2 (Gold et al., 1999): "Binding of steroid hormones (steroid) to the steroid receptor complex leads to dissociation of the complex; this enables the ligand binding component to translocate to the nucleus, where it binds to the steroid response element (SRE) on steroid hormone-response genes. In addition, steroid hormones, by disrupting the complex, act like geldanamycin and FK506 to activate hsp-90. This activation leads to a conformational change that dissociates p23, one proposed mediator of nerve regeneration. Hsp-90 may also stimulate mitogen-activated protein kinase/extracellular signal-regulated kinase-2 (MAP kinase/ERK2) pathways, providing a potential cross-talk with signal transduction pathways for neurotrophic factors (e.g., NGF). Downstream effectors ultimately mediating nerve regeneration include c-jun and GAP-43 (which contains an AP-1 site in its promotor region for activation by c-Jun homodimers), both of which show a markedly increased expression during nerve regeneration that is further augmented by FK506. In contrast, molybdate prevents dissociation of the complex and inhibits the ability of all ligands to increase neurite outgrowth (nerve regeneration)."

Based on the FKBP52-mediated hypothesis, Baulieu and co-workers presumed that the neurotrophic activity of immunophilin ligands may relate to the mediation of phytanoly-COA-α-hydroxylase (PAHX), a particular and specific target of FKBP52 (Chambraud et al., 1999). PAHX is a peroxisomal enzyme. Per the study: "Inactivation of this enzyme is responsible for Refsum disease in humans. The protein also

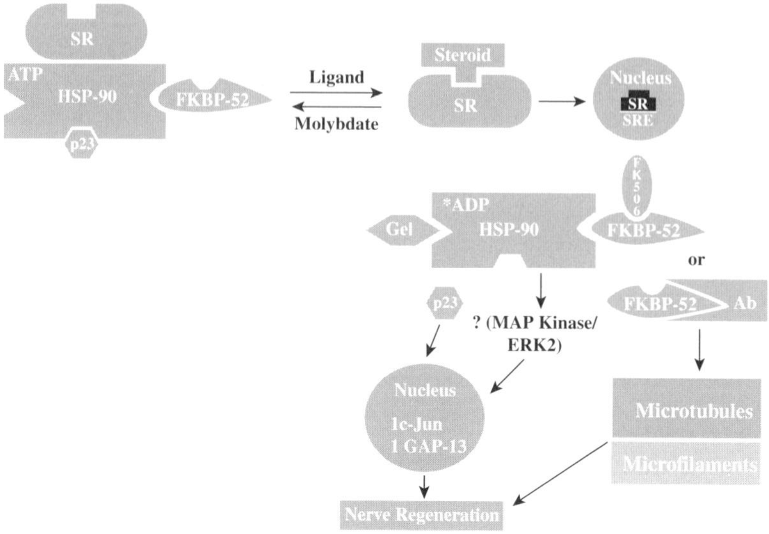

Figure 9.2 Presumed neuroregeneration mechanism of a neuroimmunophilin ligand (Gold et al., 1999).

corresponds to the mouse protein LN1, which could be involved in the progress of lupus nephritis." The results showed that "PAHX has the physical capacity to interact with the FKBP12-like domain of FKBP52, but not with FKBP12...." Also, "... the specific association of PAHX and FKBP52 is maintained in the presence of FK506. Although the function of PAHX/FKBP52 complex is still unclear, this observation suggests that PAHX is a serous candidate for studying the cellular signaling pathway(s) involving FKBP52 in the presence of immunosuppressants."

9.4 NONIMMUNOSUPPRESSANT NEUROIMMUNOPHILIN LIGANDS

As mentioned above, the best way to treat neurodegenerative disease is by using small organic molecular NGF-like compounds. Neuroimmunophilin ligands seem to be the best choice at present. However, the unwanted immunosuppressive response limits their clinical application. It invokes the research of neuroimmunophilin ligands, which satisfies the demand of neurodisorder disease treatment without immunosuppressive activity.

Design and synthesis of low-molecular-mass, monofunctional FKBP ligands first provided compounds with structural modification of the FKBP binding domain of FK506 and the immunosuppressive mechanisms of FK506. FK506 is a bifunctional molecule whose structure is divided into two distinct parts: the part that binds to FKBP12, termed binding domain, and the part that binds to calcineurin, termed effective domain (Fig. 9.3a). FK506 complexes with FKBP12, and the complex further binds to calcineurin. The immunosuppressive response is produced by this trinity (Dumont et al., 1990). Based on the immunosuppressive mechanisms of

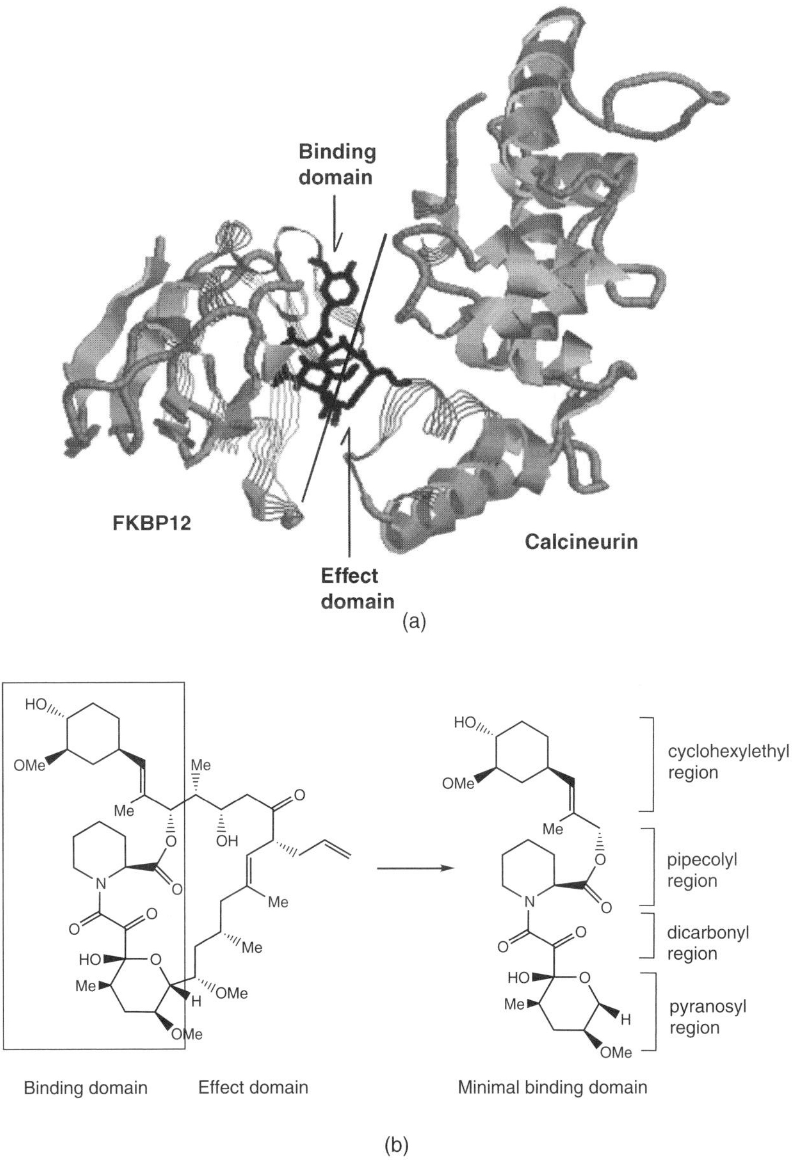

Figure 9.3 (**a**) Binding domain and effect domain of FK506. FKBP12 and calcineurin are displayed in cartoon model. FK506 is displayed in black stick model. (**b**) In general the minimal binding domain can be divided into four regions: cyclohexylethyl, pipecolyl, dicarbonyl, and pyranosyl.

FK506, scientists assumed that blocking the binding of the FK506-FKBP12 complex and calcineurin may remove the immunosuppressive activity of neuroimmunophilin ligands. According to this suggestion, Steiner et al. (1997a) designed and synthesized a nonimmunosuppressive analogue of FK506, that is, L-685818. The structural difference between L-685818 and FK506 is evidenced by the ethyl substitution of

C-21 and the hydroxyl substitution of C-18. This variation blocks the binding of calcineurin at the effective domain of FK506 and leads to the loss of immunosuppressive response and neurotoxicity of FK506 (Dumont et al., 1992). It also exhibits the same neuroregenerative activity as FK506. This find is of great significance to research in low-molecular-weight neuroregenerative agents because it gets rid of the primary therapeutic obstacle of neuroimmunophilin ligands.

9.4.1 Classical Compounds of Nonimmunosuppressant Neuroimmunophilin Ligands

Following the discovery of L-685818, many analogues of FK506 were designed and synthesized. Among those exciting compounds, the classical candidates are L-685818, V-10367, GPI-1046, and VX-853.

L-685818 (Fig. 9.4a) is the first immunophilin ligand without immnosuppressive response as mentioned above. Compared to FK506, the only structural difference

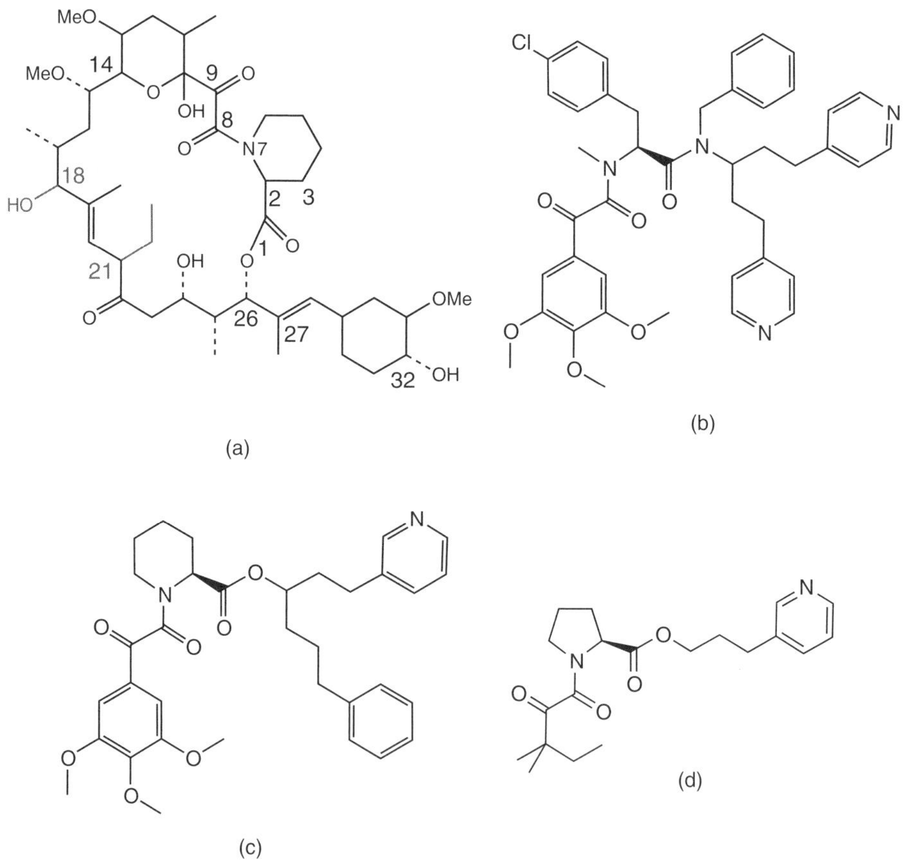

Figure 9.4 (a) L-685818, (b) VX-853, (c) V-10367, and (d) GPI-1046.

of L-685818 is the additional hydroxy group in C-18 and the ethyl substitution of C-21 propyl. It cannot bind to calcineurin and has no immunosuppressive response. Research results showed that it exhibited the same effect in sciatic nerve regeneration and function recovery as FK506 does. VX-853 (Fig. 9.4b) is a novel compound developed by Vertex with neuroregenerative and neuroprotective responses. It has finished its phase II trial on the treatment of nerve disorder caused by diabetes (Herdegan et al., 2000).

V-10,367 (Fig. 9.4c) was developed by Armistead et al. (1995), who adopted structure-based design strategy to discover FK506 analogues. It highly inhibits the PPIase activity of FKBP12 with $K_i = 0.5$ nM. The complex of V-10,367 and FKBP12 cannot bind to calcineurin, so there is no immunosuppressive activity. *In vitro*, V-10,367 has been shown to potentiate neurite outgrowth induced by submaximal concentrations of NGF from both PC12 cells and the human neuroblastoma cell line SH-SY5Y. It can also accelerate the function recovering of rat sciatic nerves *in vivo*. Per Costantini et al. (1998), V-10,367... "increased the number of neurites extended by tyrosine hydroxylase positive (TH+) DA neurons in embryonic day 14 primary DA neuronal cultures. In contrast, the immunosuppressive immunophilin ligand FK506 increased the length of TH+ neurites. After oral administration in MPTP-treated mice, V-10,367 completely protected against MPTP-induced loss of striatal TH+ axonal density, while FK506 did not."

GPI 1046 (Fig. 9.4d) is the most famous nonimmunosuppressive analogue of FK506. It exhibits highly neuroregenerative activity. Per Steiner et al. (1997b): "*In vitro*,... GPI-1046 elicited neurite outgrowth from sensory neuronal cultures with picomolar potency with maximal effects comparable to nerve growth factor. *In vivo*, GPI-1046 stimulated the regeneration of lesioned sciatic nerve axons and myelin levels. In the central nervous system, GPI-1046 promoted protection and/or sprouting of serotonin-containing nerve fibers in somatosensory cortex following parachloroamphetamine treatment." Steiner et al. (1997b) add: "GPI-1046 also induced regenerative sprouting from spared nigrostriatal dopaminergic neurons following 1-methyl-4-phenyl-1,2,3,6-tetrahydropyridine toxicity in mice or 6-hydroxydopamine (6-OHDA) toxicity in rats. The rotational abnormality in 6-OHDA treated rats was alleviated by GPI-1046." However, some reports showed that in the MPTP lesioning model of monkeys, GPI-1046 did not exhibit neuroregenerative activity (Eberling et al., 2002).

9.4.2 Structure-Based Design of Nonimmunosuppressant Neuroimmunophilin Ligands

Discovery of nonimmunosuppressant neuroimmunophilin ligands is mainly based on the three-dimensional structure of FKBP12 and the interaction of FK506 and FKBP12. FKBP12 is a member of the FKBP family. It consists of 107 amino residues and complexes tightly with FK506. The overall structure of FKBP12 is a five-stranded antiparallel β-sheet wrapping with a right-handed twist around a short α-helix. FK506 and rapamycin bind in a hydrophobic cavity between the α-helix and the interior wall of the β-sheet that is lined with a number of highly conserved, aromatic residues (Fig. 9.5a). The entrance of the cavity consists of aromatic residues 39–46, 50–55, and

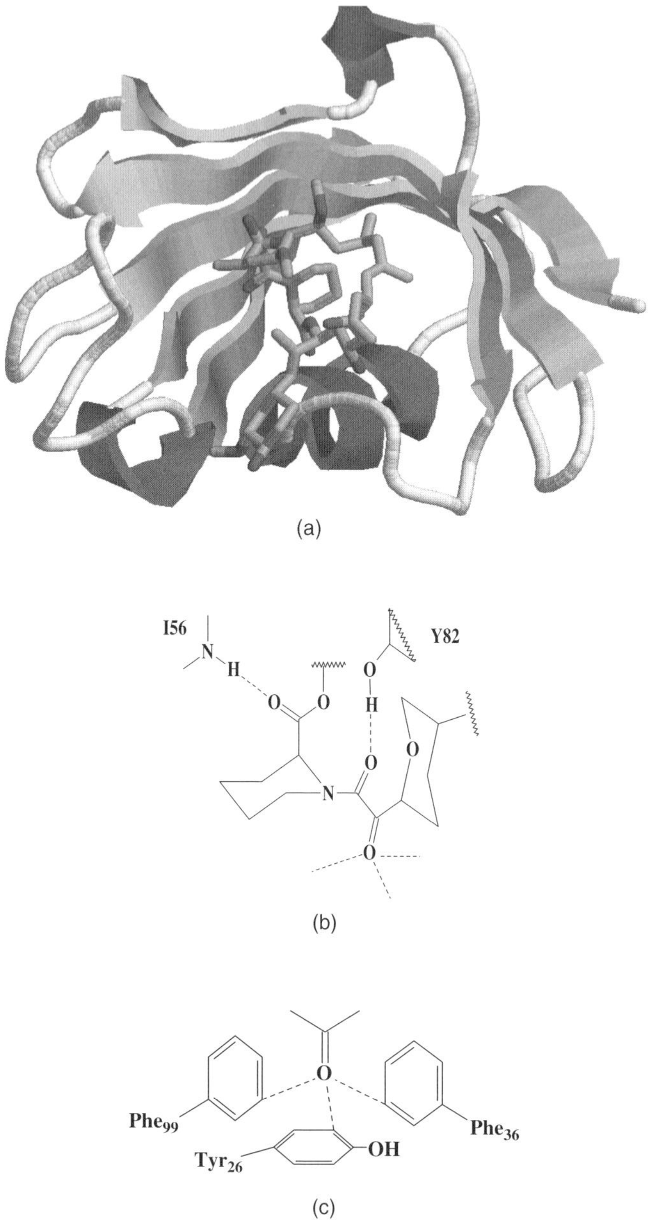

Figure 9.5 (**a**) Complex structure of FKBP12-FK506; the protein is displayed in cartoon model, and FK506 is shown in stick model. (**b**) Key interaction between FK506 and FKBP12. I56 and Y82 make two essential hydrogen bonds with FK506. (**c**) The ϵ-H of Tyr26, Phe36, and Phe99 build the specific carbonyl oxygen binding site of FK506. See color plates.

82–85. The wall of the cavity is formed by side chains of residues Tyr20, Phe46, Phe99, and Val55–Ile56. The indole ring of Trp59 makes the bottom of the cavity (Van Duyne et al., 1993).

The major interactions of FKBP12 and FK506 are hydrogen bond and hydrophobic interactions. In the complex of FK506 and FKBP12 the pipecolinyl ring binds deeply in the cavity defined by Trp59 and the side chains of Tyr26, Phe46, Val55, Ile-56, and Phe99, whereas the α-dicarbonyl amide is hydrogen bonded to the Ile-56 NH group and the Tyr-82 OH group. The pyranose ring is buried in the hydrophobic pocket formed by Phe36, Asp37, Tyr82, His87, Ile90, and Ile91, and the cyclohexyl ester chain is engaged in hydrophobic interactions within a shallow groove on the surface of the FKBPs. Besides the hydrogen bond of the α-dicarbonyl amide of FK506, there are other three hydrogen bonds between FK506 and FKBP12. The NH group of Ile56 is hydrogen bonded to oxygen of the carbonyl of C-1. The carbonyl group of Glu54 makes a hydrogen bond with the OH group of C-24. Also, the carbonyl group of Gln53 makes a hydrogen bond with the OH group of C-24 through a water molecule. There is another special interaction between oxygen of carbonyl of C-9 and residues of FKBP12. In Tyr26, Phe36, and Phe99 the aromatic ϵ-H makes a cavity that interacts with oxygen of the carbonyl of C-9, which makes the complex more stable (Fig. 9.5b).

According to the key interactions between FK506 and FKBP12, the structure of FK506 can be divided into four regions: the pipecolyl region, the dicarbonyl region, the pyranosyl region, and the cyclohexylethyl region, as shown in Fig. 9.3b (Holt et al., 1994). According to the variation of each group, many novel nonimmunosuppressive neuroimmunophilin ligands were designed and synthesized.

The first simple acyclic analogues of FK506 with high affinity to FKBP12 were reported by SmithKline Beecham. Based on the four-regions model, they designed and synthesized many analogues of FK506 with a generic structure as shown in Fig. 9.6 (Holt et al., 1994). The compound, incorporating an R1 phenyl group, an R2 phenethyl group, and an R3 (dimethylethyl)methyl group, is a very high-affinity FKBP12 ligand. The crystal structure for this compound bound to FKBP12 shows that the pipecolinyl ring and dicarbonyl groups bind identically as they do in FK506. The R3 (dimethylethyl)methyl group fills the same site that the pyran ring of FK506 occupies. The phenethyl R2 group makes hydrophobic surface contacts in the site that the cyclohexyl group of FK506 occupies. The residue Phe46 makes VDW contacts with R1 phenyl group as well as VDW contacts with the ethyl group of the R3 group (Fig. 9.7a). While solved in different space groups, the structure of the protein is

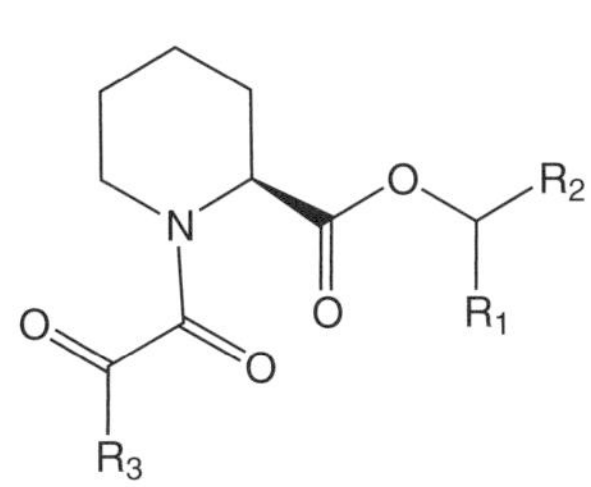

Figure 9.6 Generic structure of Holt's FK506 analogues.

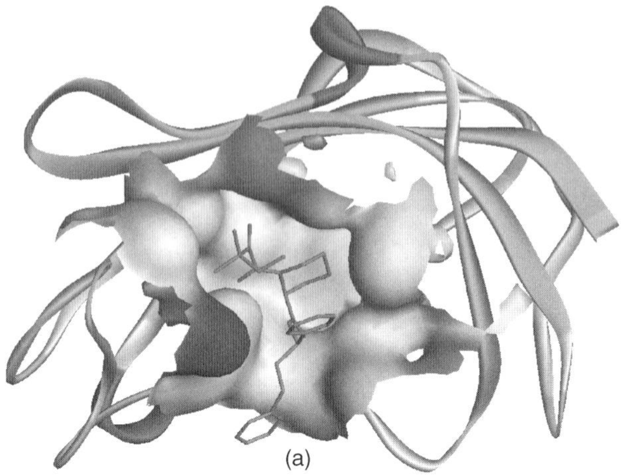

(a)

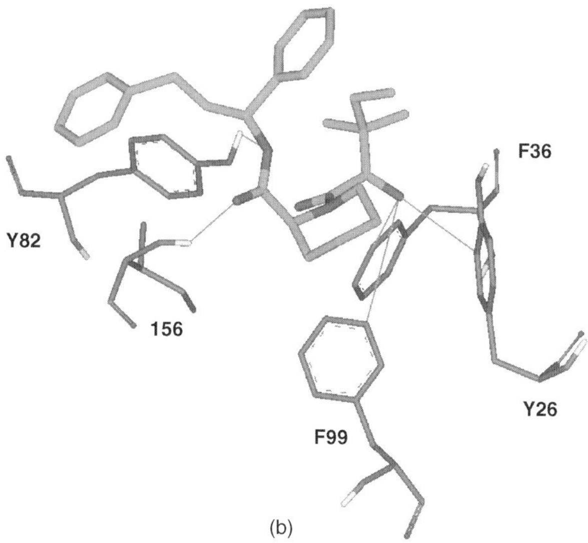

(b)

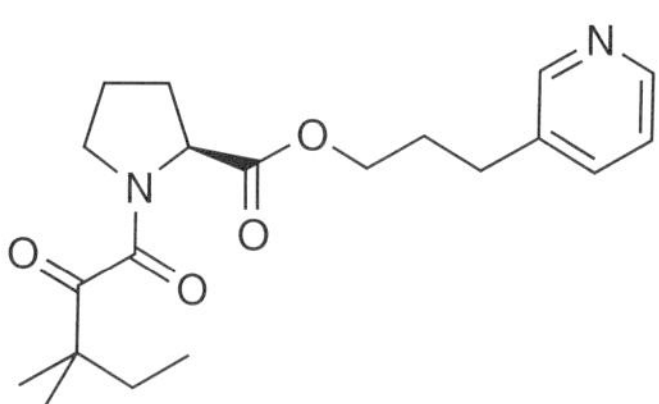

(d)

nearly identical between the complexes of FK506 and SmithKline Beecham's compound (Fig. 9.7b).

A slight conformation change in the side chain of Phe46 of FKBP12 takes place to maximize VDW contacts with the ligand. The side chains of Asp37 are in slightly different conformations in the two complexes. In the FK506 complex the side chains of Asp37 approach the ligand more closely to make a hydrogen bond with the hemiketal hydroxyl group. The flexible region of the protein between Ala84 and His94 is very similar in both complexes. The crystal structure means that the simplified structure of FK506 binding domain can interact with FKBP12, as FK506 does (Holt et al., 1993).

Computational methods are commonly applied for lead identification and optimization. Although there are only a few examples of completely new chemical entities discovered by such approaches, many scientists still believe in its power and adopt it to find novel neuroimmunophilin ligands. Ludi is a suggestive program for the *de novo* design of ligands for proteins. Babine et al. (1995) applied this approach and found a new type of ligand of FKBP12. Using the crystal structure of the FKBP12-FK506 complex as a starting point, Ludi suggested complementary fragments for the hydrophobic pipecolinic acid binding site of FKBP12. Per Babine et al. (1995): "Among the many fragments suggested, adamantane appeared to fill this site well and seemed well suited for further elaboration. Examination of the docked ligand for the formation of potential hydrogen bonds suggested the removal of a methylene bridge with the introduction of a carbonyl group. This should allow for a hydrogen bond to the backbone amide NH of Ile-56." Among the suggestions of potential groups to fill the FK506 pyran binding site of FKBP12, Babine et al. found that aromatic groups appended through a linker to the bridgehead position of the new ring system seemed be the best choice. Babine et al. add: "Finally, Ludi suggested a hydroxyl group in the meta position of an aromatic ring tethered by a methylene linker [Fig. 9.8]. This hydroxyl group has the potential to make favorable electrostatic contact with Asp-37. Thus, compounds of the generic structure 1 were *de novo* designed . . ." Among the designed compounds, the most effective one was compound A. A 2.4-Å crystal structure was obtained for the complex of compound A and FKBP12. The complex showed that the bicyclic ring system makes VDW contact with the hydrophobic residues Val55, Ile56, Trp59, Phe46, Tyr82, and Phe99. The carbonyl group makes an H-bond to the backbone NH of Ile56. The sulfur atom fills a small hydrophobic, electropositive cavity formed by Tyr26, Phe36, and Phe99. The pyridine ring makes VDW contact with His87, Ile91, Tyr82, and Phe99.

In another paper, Burkhard et al. (1999) utilized the molecular docking computer program SANDOCK to screen small molecule three-dimensional databases for

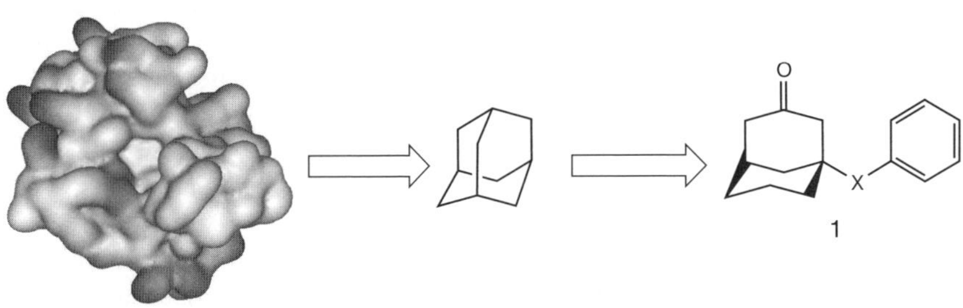

1

Figure 9.8 Ludi-based ligand design pipeline. See color plates.

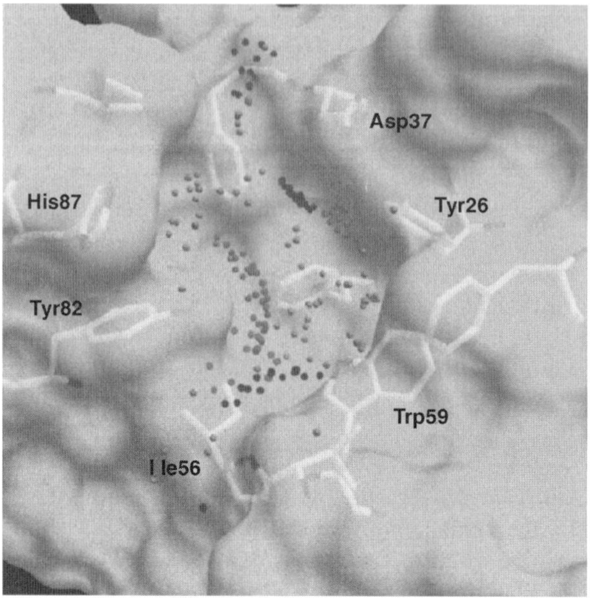

Figure 9.9 Site points in FKBP. The dots that were used during the SANDOCK runs are color-coded according to their chemical properties: hydrophobic site points (gray), hydrogen bond acceptor site points (red), hydrogen bond donor site points (blue), and hydrogen bond donor/acceptor site points (magenta). The molecular surface of the protein was calculated by the program MSMS (Sanner et al., 1996) and visualized with the graphics program DINO (Philippsen, 1998; http://www.bioz.unibas.ch/ _xray/dino). See color plates.

FKBP ligands. Based on the site point of FKBP (Fig. 9.9), many potential compounds were selected. As confirmed by fluorescence quenching tests, several compounds with micromolar K_d values were identified. Interestingly, several steroids were characterized to be novel leads for FKBP12 ligands—for example, 5β Pregnan 3.20 dion ($K_d = 7\ \mu M$). The biological significance of these FKBP-steroid interactions remains to be elucidated. However, so far there is no evidence that the binding of steroids to FKBPs accounts for neuronal actions of steroids.

Our laboratory has researched neuroimmunophilin ligands since 1997. With the integration of virtual library, virtual screening, and combinatorial chemistry, we designed, built, and screened several different generic structure libraries, each consisting of thousands of compounds. Compound 308 exhibits remarkable activity in many *in vitro* and *in vivo* neuroregeneration models and was chosen as the prime candidate for further evaluation. The co-crystal structure of compound 308 and FKBP12 was obtained and solved the key interaction between them (Sun et al., 2003). The complex shows that the 1,4-thiazine and sulfanilamide groups bind identically as the pipecolinyl ring and dicarbonyl groups do in FK506. The *p*-methyl-phenyl group fills the same site that the pyran ring of FK506 occupies. The 3-carbonyl group of the 1,4-thiazine group makes a hydrogen bond to the backbone amide NH of Ile56. The carbonyl of the carboxyl group makes another

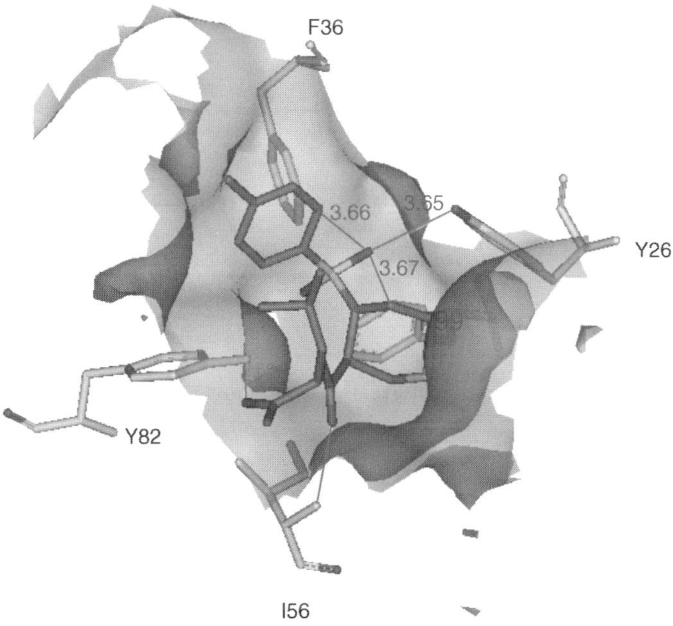

Figure 9.10 Complex of compound 308 and FKBP12. The protein is only displayed using the surface of the active site (orange); compound 308 is shown in stick model. The hydrogen bond between the ligand and Y82 and I56 still exists. The oxygen of sulfonyl group acts as the carbonyl oxygen of α-ketone in the specific ϵ-hydrogen site of Y26, F36, and F99. See color plates.

hydrogen bond to the hydroxyl group of Tyr82. The 1,4-thiazine group and *p*-methylphenyl group make VDW contact with the hydrophobic residues Tyr26, Phe36, Phe46, Val55, Trp59, His87, and Ile90 (Fig. 9.10). At present, this compound is going to phase I clinical trial for therapy of stroke.

9.4.3 SAR of Nonimmunosuppressant Neuroimmunophilin Ligands

Structure–activity relationship (SAR) studies were performed for a large number of simple, analogous molecules (Fig. 9.11) in terms of their ability to inhibit FKBP12 (Armistead et al., 1995) and were reviewed recently (Hamilton and Steiner, 1997, 1998). As mentioned above, the binding domain of FK506 is consisted of four regions: the pipecolyl region, the dicarbonyl region, the pyranosyl region, and the cyclohexylethyl region. It is suggested that the minimal binding requires the central pipecolic acid ring with the ester group (Holt et al., 1994). SAR study results show that there are three key factors of neuroregenerative activity of this minimal binding structure. They are the presence of ester bond, the chirality of ester bond, and the presence of pipecolic ring. The disappearance of the ester group will lead to the decrease of activity. Changing the chirality of the ester bond from S-type to R-type will cause the activity to decrease in two degrees, while open-chain compounds have

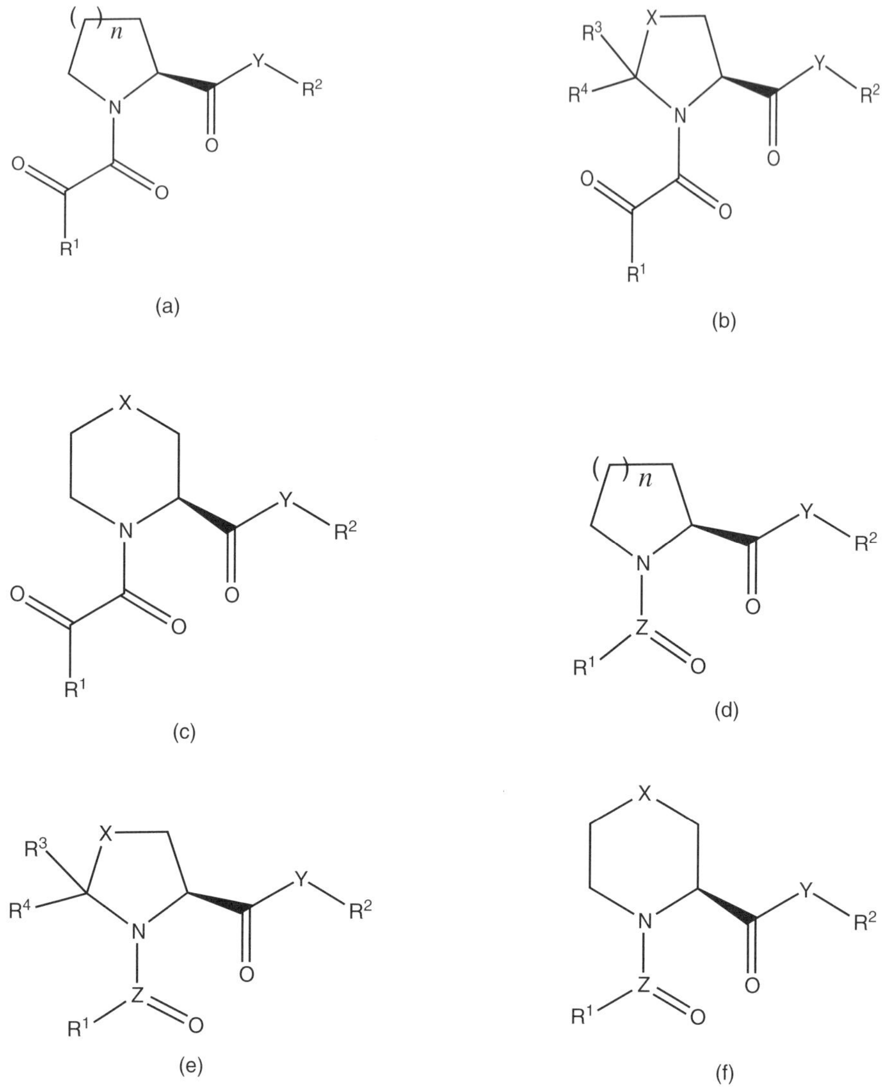

Figure 9.11 Generic structures of nonimmunosuppressive neuroimmunophilin ligands ($n = 1, 2$; X = O, S; Y = O, S, NH; Z = C, SO).

approximately 10-fold weaker affinity for the enzyme as compared to FK506 and rapamycin (Teague and Stocks, 1993; Teague et al., 1994). These findings imply that the binding of the ester bond and the surface of FKBP12 needs a stable conformation to support it, which makes the binding more complete. The pipecolic acid, pyrrolidine, pyridine, and other 5 or 6 heterocyclic ring just satisfied with this requested conformation.

The α-dicarbonyl amide group was identified as being essential for enzyme inhibition because derivatives obtained by replacement of either or both of the carbonyl groups corresponding to positions C-8 and C-9 of FK506 such as peptides ($K_i = \geq 1\,\mu M$) (Hauske et al., 1992), sulfonamides ($K_i = \geq 160\,nM$) (Holt et al., 1994; Duffi, 1992) and ureas ($K_i = \geq 120\,nM$) (Dragovich et al., 1996) were substantially less potent. The Abbot group examined the effect of modifications to the ketone carbonyl groups in simple acyclic compounds (Wang et al., 1994). Their results were consistent with the conclusion that ketone is the most potent group.

Although the binding properties for some of these α-dicarbonyl amides have been critically discussed, the variation of pyranose ring and cyclohexylethyl moiety of the FK506 binding domain is also studied extensively. The SmithKline Beecham results showed that substitution of the pyranosyl group with a (dimethylethyl)methyl group resulted in a threefold increase in affinity for FKBP12 (Luengo et al., 1994). The Vertex group has reported that a 3,4,5-trimethoxy phenyl group is a useful pyran replacement in acyclic FKBP12 ligands (Armistead et al., 1995). Substitution of cyclohexylethyl moiety with either a phenyl group or a phenethyl group resulted in an 11-fold and 18-fold increase in FKBP12 affinity, respectively. Incorporating both phenyl groups and phenethyl groups is a very-high-affinity FKBP12 ligand (7 nm). Among these compounds, high-affinity FKBP12 ligands were bulky hydrophobic alkyl groups such as 1,1-dimethyl-propyl or (3,4,5-trimethoxyl)phenyl as substituents for the pyranose ring region of FK506, as well as simple alkyl or alkyl aryl esters, instead of the lead cyclohexylethyl moiety. This confirmed the conclusion that these moieties mainly make hydrophobic VDW contact with FKBP12.

9.5 CONCLUSIONS AND FUTURE PROSPECTS

Because of the complicated mechanisms of neruodegenerative disease, there are still few available clinical therapeutants for these diseases' treatment. All available drugs for the treatment of frequently occurring neurodegenerative disorders such as Alzheimer's and Parkinson's diseases only alleviate symptoms and delay the neuronal atrophy by compensating or increasing impairments of the neurotransmitter metabolism (Brodaty, 1999; Grunblatt et al., 2000). Growth factors such as BDNF, GDNF, NGF, and NT3 have been shown to promote nerve regrowth *in vivo* (Apfel et al., 1994; McMahon and Priestley, 1995; Gash et al., 1996) and seem be the perfect therapeutants for nervous disorders. Most of them are currently being evaluated in animal models of neurodegeneration. But the therapeutic utility of these proteins is severely limited due to their lack of oral bioavailability and inability to cross the blood–brain barrier, necessitating their delivery directly to the site of injury. Abnormal neuronal sprouting has also been observed following administration of protein growth factors (Voolf et al., 1996). Clinical potential of other available or presently examined neuroregenerative therapeutics, such as gangliosides (Ekstrom and Tomlinson, 1990) or the dihydropyridine Ca^{2+} antagonist nimodipine, is hampered by the observed adverse side effects (Candelise and Ciccone, 2000; Ahmed et al., 2000). As an alternative approach, neural transplantation of competent neuronal cells or tissues is considered to hold promise as a future therapeutic tool to treat progressive and irreversible neural disorders (Dunnett, 1999). These neural transplantation techniques, however, are still in an early developmental stage, and many of the adherent problems such as viability and function of the graft remain to be solved. New therapeutic approaches are required to treat neural diseases.

The neuroimmunophilin ligands have been demonstrated to promote regrowth of damaged facial and sciatic nerves in the peripheral nervous system and the regeneration of damaged dopaminergic, serotonergic, and cholinergic neurons in the central nervous system. This broad scope of action suggests therapeutic utility in a variety of neurodegenerative disorders such as Parkinson's disease, Alzheimer's disease, diabetic and peripheral neuropathies, and spinal cord injuries. The studies surveyed above document the remarkable ability of FKBP ligands to cause functional regeneration of a variety of damaged neuronal pathways. In terms of specificity of action, bioavailability, and stability, these compounds might be essentially advantageous compared to peptidic growth factors, being presently evaluated, such as NGF, BDNF, NT3, and GDNF. Unlike the growth factors, which exert neurotrophic activities on overlapping but limited populations of CNS neurons, FKBP ligands do not cause aberrant sprouting of healthy neuronal processes *in vivo* (Steiner et al., 1997b).

Many unanswered questions remain regarding the actions of these compounds. The mechanism whereby the nerve regenerative effects of FK506 and small FKBP ligands are exerted is still unknown. FKBP12 is present in high concentrations in neurons, and many compounds produce neurotrophic effects *in vitro* in concentrations at which less than 0.1% of the enzyme would be inhibited. This implies that there may be other FKBPs, present in lower concentrations in nerve cells, that mediate the actions of these compounds. The ability of neuroimmunophilin ligands to produce potent neurotrophic effects *in vitro* and *in vivo* at low concentrations, along with the lack of correlation of these effects with inhibition of rotamase activity, suggests another possibility that interaction of the compounds with FKBP or an FKBP-like protein results in formation of an activated complex, leading to a gain of function for the FKBP. All these hypotheses need more evidence to prove.

Although there are many things such as targets and mechanism that need to be elucidated, the intensive interest and effort devoted to exploring the therapeutic utility of neuroimmunophilin ligands is continuing. It is likely that this will be a major new area of medicinal chemistry and neuroscience research for years to come.

ACKNOWLEDGMENTS

We would like to acknowledge Dr. Wang Lili for their contribution during the preparation of the manuscript, especially for discussion of ligand efficiency and potential neuroregenerative mechanism. We are also grateful to A. B. for administrative assistance and for overall scientific input and abbreviation preparation.

ABBREVIATIONS

ATP	Adenosine triphosphate
5-HT	5-Hydroxytryptamine
6-OHDA	6-hydroxydopamine
AD	Alzheimer's disease
ADNF	activity-dependent neurotrophic factor
AGEs	advanced glycation end products
AMPA	alpha-amino-3-hydroxy-5-methyl-4-isoxazolepropionic acid
AP-1	activator protein 1
BBB	brain–blood barrier
BDNF	brain-derived neurotrophic factor
CA1	hippocampal domain CA1
CNS	central nervous system
CsA	Cyclosporine A
DA	dopamine
EGFR	epidermal growth factor receptor
FK506	an immunosuppressant named Tacrolimus
FKBP12	FK506 binding protein with molecular weight 12 kDa
FKBP12.6	FK506 binding protein with molecular weight 12.6 kDa
FKBP13	FK506 binding protein with molecular weight 13 kDa
FKBP25	FK506 binding protein with molecular weight 25 kDa
FKBP37	FK506 binding protein with molecular weight 37 kDa
FKBP51	FK506 binding protein with molecular weight 51 kDa
FKBP52	FK506 binding protein with molecular weight 52 kDa
FKBPs	fK506 binding protein family
GAP-43	growth-associated protein 43
GDNF	glial cell-line-derived neurotrophic factor
GTP	guanosine triphosphate
HD	Huntington's disease
HNE	human neutrophil elastase
Hsp-90	heat shock protein 90
LN1	lupus nephritis protein
MAP kinase/ERK2	a 42-kDa serine/threonine protein kinase
MPP	*N*-methyl-4-phenylpyridinium
MPTP	1-methyl-4-phenyl-1,2,3,6-tetrahydropyridine
NGF	nerve growth factor
NINA A	a protein in the neural retina of *Drosophila*
NMDA	*N*-methyl-D-aspartic acid
NOS	nitric oxidate synthase
NSAIDs	nonsteroidal anti-inflammatory drugs
PAHX	phytanoly-COA-α-hydroxylase
PC12	rat adrenal pheochromocytoma cell line
PD	Parkinson's disease
PGA1	prostaglandin A1
PGE1	prostaglandin E1
PGs	prostaglandins
PPIases	peptidylprolyl *cis–trans* isomerases

RAPA	rapamycin
RNS	reactive nitrogen species
ROSs	reactive oxygen species
SAR	structure and activity relationship
SRE	steroid response element
TH+	tyrosine hydroxylase positive
VDW	van Der Waals

REFERENCES

Ahmed, N., Nasman, P., and Wahlgren, N. G. (2000). Effect of intravenous nimodipine on blood pressure and outcome after acute stroke. *Stroke* **31**, 1250–1255.

Alexi, T., Borlongan, C. V., Faull, R. L., Williams, C. E., Clark, R. G., Gluckman, P. D., and Hughes, P. E. (2000). Neuroprotective strategies for basal ganglia degeneration: Parkinson's and Huntington's diseases. *Prog Neurobiol* **60**, 409–470.

Apfel, S. C., Arezzo, J. C., Brownlee, M., Federoff, H., and Kessler, J. A. (1994). Nerve growth factor administration protects against experimental diabetic sensory neuropathy. *Brain Res* **634**, 7–12.

Arendt, T., Bruckner, M., Gertz, H., and Marcova, L. (1998). Cortical distribution of neurofibrillary tangles in Alzheimer's disease matches the pattern of neurons that retain their capacity of plastic remodeling in the adult brain. *Neuroscience* **83**, 991–1002.

Armistead, D. M., Badia, M. C., Deininger, D. D., Duffy, J. P., Saunders, J. O., Tung, R. D., Thomson, J. A., DeCenzo, M. T., Futer, O., Livingston, D. J., Murcko, M. A., Yamashita, M. M., and Navia M.A. (1995). Design, synthesis and structure of non-macrocyclic inhibitors of FKBP12, the major binding protein for the immunosuppressant FK506. *Acta Crystallogr* **D51**, 522–528.

Babine, R. E., Bleckman, T. M., Kissinger, C. R., Showalter, R., Pelletier, L. A., Lewis, C., Tucker, K., Moomaw, E., Parge, H. E., and Villafranca, J. E. (1995). Design, synthesis and x-ray crystallographic studies of novel FKBP12 ligands. *Bioorg Med Chem Lett* **5**, 1719–1724.

Baughman. G., Wiederrecht, G. J., Campbell, N. F., Martin, M. M., and Bourgeois, S. (1995). FKBP51, a novel T-cell-specific immunophilin capable of calcineurin inhibition. *Mol Cell Biol* **15**, 4395–4402.

Bohn, M. C. (1999). A commentary on glial cell line-derived neurotrophic factor (GDNF). From a glial secreted molecule to gene therapy. *Biochem Pharmacol* **57**, 135–142.

Brodaty, H. (1999). Realistic expectations for the management of Alzheimer's disease. *Eur Neuropsychopharmacol* **9**, S43–S52.

Burkhard, P., Hommel, U., Sanner, M., and Walkinshaw, M. D. (1999). The discovery of steroids and other novel FKBP inhibitors using a molecular docking program. *J Mol Biol* **287**, 853–858.

Candelise, L., and Ciccone, A. (2000). Gangliosides for acute ischaemic stroke. *Cochrane Database of Systematic Rev* **2**, CD000094.

Chambraud, B., Radanyi, C., Camonis, J. H., Rajkowski, K., Schumacher, M., and Baulieu, E. E. (1999). Immunophilins, Refsum disease, and lupus nephritis: The peroxisomal enzyme phytanoyl-COA-α-hydroxylase is a new FKBP-associated protein. *Proc Natl Acad Sci USA* **96**, 2104–2109.

Costantini, L. C., and Isacson, O. (2000). Immunophilin ligands and GDNF enhance neurite branching or elongation from developing dopamine neurons in culture. *Exp Neurol* **164**, 60–70.

Costantini, L. C., Chaturvedi, P., Armistead, D. M., McCaffrey P. G.., Deacon, T. W., and Isacson O. (1998). A novel immunophilin ligand: Distinct branching effects on dopaminergic neurons in culture and neurotrophic actions after oral administration in an animal model of Parkinson's disease. *Neurobiol Dis* **5**, 97–106.

Costantini, L. C., Cole, D., Chaturvedi, P., and Isacson, O. (2001). Immunophilin ligands can prevent progressive dopaminergic degeneration in animal models of Parkinson's disease. *Eur J Neurosci* **13**, 1085–1092.

Dawson, T. M., Steiner, J. P., Dawson, V. L., Dinerman, J. L., Uhl, G. R., and Snyder, S. H. (1993). Immunosuppressant FK506 enhances phosphorylation of nitric oxide synthase and protects against glutamate neurotoxicity. *Proc Natl Acad Sci USA* **90**, 9808–9812.

Doble, A. (1995). Excitatory amino acid receptors and neurodegeneration, *Therapie* **50**, 319–337.

Dragovich, P. S., Barker, J. E., French, J., Imbacuan, M., Kalish, V. J., Kissinger, C. R., Knighton, D. R., Lewis, C. T., Moomaw, E. W., Parge, H. E., Pelletier. L. A., Prins, T. J., Showalter, R. E., Tatlock, J. H., Tucker, K. D., and Villafrance, J. E. (1996). Structure-based design of novel urea-containing FKBP12 inhibitors. *J Med Chem* **39**, 1872–1884.

Duffi, J. P. (1992). Novel immunosuppressive compounds. International Patent Application WO92/213313.

Dumont, F. J., Melino, M. R., Staruch, M. J., Koprak, S. L., Fischer, P. A., and Sigal, N. H. (1990). The immunosuppressive macrolides FK-506 and rapamycin act as reciprocal antagonists in murine T cells. *J Immunol* **144**, 1418–1424.

Dumont, F. J., Staruch, M. J., Koprak, S. L., Siekierka, J. J., Lin, C. S., Harrison, R., Sewell, T., Kindt, V. M., Beattie, T. R., and Wyvratt, M. (1992). The immunosuppressive and toxic effects of FK506 are mechanistically related: Pharmacology of a novel antagonist of FK506 and rapamycin. *J Exp Med* **176**, 751–760.

Dunnett, S. B. (1999). Repair of the damaged brain. *Neuropathol Appl Neurobiol* **25**, 351–362.

Durany, N., Munch, G., Michel, T., and Riederer P. (1999). Investigations on oxidative stress and therapeutical implications in dementia. *Eur Arch Psychiatry Clin Neurosci* **249** *Suppl 3*, 68–73.

Eberling, J. L., Pivirotto, P., Bringas, J., Steiner, J. P., Kordower, J. H., Chu, Y., Emborg, M. E., and Bankiewicz, K. S. (2002), The immunophilin ligand GPI-1046 does not have neuroregenerative effects in MPTP-treated monkeys. *Exp Neurol* **178**, 236–242.

Ekstrom, P. A., and Tomlinson, D. R. (1990). Impaired nerve regeneration in streptozotocin-diabetic rats is improved by treatment with gangliosides. *Exp Neurol* **109**, 200–203.

Fink, D. J., Poliani, P. L., Oligino, T., Krisky, D. M., Goins, W. F., and Glorioso, J. C. (1997). Development of an HSV-based vector for the treatment of Parkinson's disease. *Exp Neurol* **144**, 103–121.

Fischer, G., Bang, H., and Mech, C. (1984). Detection of enzyme catalysis for cis-trans isomerization of peptide bonds using proline containing peptides as substrates. *Biomed Biochim Acta* **43**, 1101–1112.

Gash, D. M., Zhang, Z., Ovadia, A., Cass, W. A., Yi, A., Simmerman, L., Russell, D., Martin, D., Lapchak, P. A., Collins, F., Hoffer, B. J., and Gerhardt, G. A. (1996). Functional recovery in parkinsonian monkeys treated with GDNF. *Nature* **380**, 252–255.

Gastilho, R. F., Hansson, O., and Brundin, P. (2000). FK506 and cyclosporin A enhance the survival of cultured and grafted rat embryonic dopamine neurons. *Exp Neurol* **164**, 94–101.

Gold, B. G., Katoh, K., and Storm-Dickerson, T. (1995). The immunosuppressant FK506 increases the rate of axonal regeneration in rat sciatic nerve. *J Neurosci* **15**, 7509–7516.

Gold, B. G., Michelle, Z. P., Wang, M. S., Chaturvedi, P., and Armistead D. M. (1997). A nonimmunosuppressant FKBP12 ligand increases nerve regeneration. *Exp Neurol* **147**, 269–278.

Gold, B. G., Densmore, V., Shou, W., Matzuk, M. M., and Gordon, H. S. (1999). Immunophilin FK506-binding protein 52 (not FK506-binding protein 12) mediates the neurotrophic action of FK506. *J Pharmacol Exp Therapeutics* **289**, 1202–1210.

Gozes, I., and Brenneman, D. E. (1996). Activity-dependent neurotrophic factor (ADNF). An extracellular neuroprotective chaperonin? *J Mol Neurosci* **7**, 235–244.

Greene, J. G., and Greenamyre, J. T. (1996). Bioenergetics and glutamate excitotoxicity. *Prog Neurobiol* **48**, 613–634.

Grunblatt, E., Mandel, S., and Youdim, M. B. (2000). Neuroprotective strategies in Parkinson's disease using the models of 6-hyroxydopamine and MPTP. *Ann NY Acad Sci* **899**, 262–273.

Guo, X., Dawson, V. L., and Dawson, T. M. (2001). Neuroimmunophilin ligands exert neuroregeneration and neuroprotection in midbrain dopaminergic neurons. *Eur J Neurosci* **13**, 1683–1693.

Hamilton, G. S., and Steiner J. P. (1997). Neuroimmunophilin ligands as novel therapeutics for the treatment of degenerative disorders of the nervous system. *Curr Pharm Des* **3**, 405–428.

Hamilton, G. S., and Steiner, J. P. (1998). Immunophilins: Beyond immunosuppression. *J Med Chem* **41**, 5119–5143.

Hauske, J. R., Dorff, P., Julin, S., DiBrino, J., Spencer, R., and Williams R. (1992). Design and synthesis of novel FKBP inhibitors. *J Med Chem* **35**, 4284–4296.

Herdegan, T., Fischer, G., and Gold, B. G. (2000). Immunophilin ligands as a novel treatment for neurological disorders. *Trends Pharmacol Sci* **21**, 3–5.

Holt, D. A., Luengo, J. I., Yamashita, D. S., Oh, H. J., Konialian, A. L., Yen, H. K., Rozamus, L. W., Brandt, M., Bossard, M. J., Levy, M. A., Eggleston, D. S., Liang, J., Schultz, L. W., Stout, T. J., and Clardy,

J. (1993). Design, synthesis, and kinetic evaluation of high-affinity FKBP ligands and the x-ray crystal structures of their complexes with FKBP12. *J Am Chem Soc* **115**, 9925–9938.

Holt, D. A., Konialian-Beck, A. L., Oh, H. J., Yen, H. K., Rozamus, L. W., Krog, A. J., Erhard, K. F., Ortiz, E., Levy, M. A., Brandt, M., Bossard, M. J., and Luengo, J. I. (1994). Structure–activity studies of synthetic FKBP ligands as peptidyl-prolyl isomerase inhibitors. *Bioorg Med Chem Lett* **4**, 315–320.

Hölscher, C. (1998). Possible causes of Alzheimer's disease: Amyloid fragments, free radicals, and calcium homeostasis. *Neurobiol Dis* **5**, 129–141.

Ishimoto, T., Kamei, A., Koyanagi, S., Nishide, N., Uyeda, A., Kasai, M., and Taguchi, T. (1998). HSP90 has neurite-promoting activity in vitro for telencephalic and spinal neurons of chick embryos. *Biochem Biophys Res Commun* **253**, 283–287.

Jin, L. W., and Saitoh, T. (1995). Changes in protein kinases in brain aging and Alzheimer's disease. Implications for drug therapy. *Drugs Aging* **6**, 136–149.

Kino, T., Hatanaka, H., Hashimoto, M., Nishiyama, M., Goto, T., Okuhara, M., Kohsaka, M., Aoki, H., and Imanaka, H. (1987). FK506, a novel immunosuppressant isolated from a Streptomyces. I. Fermentation, isolation, and physicochemical and biological characteristics. *J Antibiot* **40**, 1249–1255.

Kitamura, Y., Itano, Y., Kubo, T., and Nomura, Y. (1994). Suppressive effect of FK506, a novel immunosuppressant, against MPTP-induced dopamine depletion in the striatum of young C57BL/6 mice. *J Neuroimmunol* **50**, 221–224.

Lad, R. P., Smith, M. A., and Hilt, D. C. (1991). Molecular cloning and regional distribution of rat brain cyclophilin. *Mol Brain Res* **9**, 239–244.

Lapchak, P. A. (1993). Nerve growth factor pharmacology: Application to the treatment of cholinergic neurodegeneration in Alzheimer's disease. *Exp Neurol* **124**, 16–20.

Lepne, C. A., Thomson, J. A., and Moore, J. M. (1992). Solution structure of FK506 bound to FKBP-12. *FEBS Lett* **302**, 89–96.

Liu, J., Farmer, J. D., Jr., Lane, W. S., Friedman, J., Weissman, I., and Schreiber S. L. (1991). Calcineurin is a common target of cyclophilin-cyclosporin A and FKBP-FK506 complexes. *Cell* **66**, 807–815.

Liu, Y. C., and Storm, D. R. (1989). Dephosphorylation of neuromodulin by calcineurin. *J Biol Chem* **264**, 12800–12804.

Luengo et al. (1994).

Lyons, W. E., George, E. B., Dawson, T. M., Steiner, J. P., and Snyder, S.H. (1994). Immunosuppressant FK506 promotes neurite outgrowth in cultures of PC12 cells and sensory ganglia. *Proc Natl Acad Sci USA* **91**, 3191–3195.

Madsen, J. R., MacDonald, P., Irwin, N., Goldgerg, D. E., Yao, G. L., Meiri, K. F., Rimm, I. J., Stieg, P. E., and Benowitz, L. I. (1998). Tacrolimus (FK506) increases neuronal expression of GAP-43 and improves functional recovery after spinal cord injury in rats. *Exp Neurol* **154**, 673–683.

McMahon, S. B., and Priestley, J. V. (1995). Peripheral neuropathies and neurotrophic factors: Animal models and clinical perspectives. *Curr Opin Neurobiol* **5**, 616–624.

Morel, P., Tallineau, C., Pontcharraud, R., Piriou, A., and Hughet, F. (1999). Effects of 4-hydroxynonenal, a lipid peroxidation product, on dopamine transport and Na+/K+-ATPase in rat striatal synaptosomes. *Neurochem Int* **33**, 531–540.

Palladini, G., Caronti, B., Pozzessere, G., Teichner, A., Buttarelli, F. R., Morselli, E., Valle, E., Venturini, G., Fortuna, A., and Pontieri, F. E. (1996). Treatment with cyclosporine A promotes axonal regeneration in rats submitted to transverse section of the spinal cord. II. Recovery of function. *J Hirnforsch* **37**, 145–153.

Paratt, W. B., and Toft, D. O. (1997). Steroid receptor interactions with heat shock protein and immunophilin chaperones. *Endo Rev* **18**, 306–360.

Parker, E. M., Monopoli, A., Ongini, E., Lozza, G., and Babij, C. M. (2000). Rapamycin, but not FK506 and GPI-1046, increases neurite outgrowth in PC12 cells by inhibiting cell cycle progression *Neuropharmacology* **39**, 1913–1919.

Parsons, C. G., Danysz, W., and Quack, G. (1999). Memantine is a clinically well tolerated *N*-methyl-D-aspartate (NMDA) receptor antagonist–a review of preclinical data. *Neuropharmacology* **38**, 735–767.

Pedersen, W. A., Cashman, N. R., and Mattson, M. P. (1999). The lipid peroxidation product 4-hydroxynanenal impairs glutamate and glucose transport and choline acetyltransferase activity in NSC-19 motor neuron cells. *Exp Neurol* **115**, 1–10.

Raaka, B. M., Finnerty, M., Sun, E., and Samuels, H. H. (1985). Effects of molybdate on steroid receptors in intact GH_1 cells. Evidence for dissociation of an intracellular 10 S receptor oligomer prior to nuclear accumulation. *J Biol Chem* **260**, 14009–14015.

Sharkey, J., and Butcher, S. P. (1994). Immunophilins mediate the neuroprotective effects of FK506 in focal cerebral ischaemia. *Nature* **371**, 336–339.

Smith, D. F., Albers, M. W., Schreiber, S. L., Leach, K. L., and Deibel M. R. (1993). FKBP54, a novel FK506-binding protein in avian progesterone receptor complexes and HeLa extracts. *J Biol Chem* **268**, 24270–24273.

Snyder, S. H., Sabatini, D. M., Lai, M. M., Steiner, J. P., Hamilton, G. S., and Suzdak, P. D. (1998). Neural actions of immunophilin ligands. *Trends Pharmacol Sci* **19**, 21–26.

Starzl, T. E., Fung, J., Jordan, M., Shapiro, R., Tzakis, A., McCauley, J., Johnston, J., Iwaki, Y., Jain, A., and Alessiani, M. (1990). Kidney transplantation under FK506. *JAMA* **264**, 63–67.

Steiner, J. P., Dawson, T. M., Fotuhi, M., Glatt, C. E., Snowman, A. M., Cohen, N., and Snyder, S. H. (1992). High brain densities of the immunophilin FKBP colocalized with calcineurin. *Nature* **358**, 584–587.

Steiner, J. P., Connolly, M. A., Valentine, H. L., Hamilton, G. S., Dawson, T. M., Hester, L., and Snyder, S. H. (1997a). Neurotrophic actions of nonimmunosuppressive analogues of immunosuppressive drugs FK506, rapamycin and cyclosporin A. *Nat Med* **3**, 421–428.

Steiner, J. P., Hamilton, G. S., Ross, D. T., Valentine, H. L., Guo, H., Connolly, M. A., Liang, S., Ramsey, C., Li, J.-H. J., Huang, W., Howorth, P., Soni, R., Fuller, M., Sauer, H., Nowotnik, A. C., and Suzdak, P. D. (1997b). Neurotrophic immunophilin ligands stimulate structural and functional recovery in neurodegenerative animal models. *Proc Natl Acad Sci USA* **94**, 2019–2024.

Sun, F., Li, P. G., Ding, Y., Wang, L. W., Bartlam, M., Shu, C., Shen, B. F., Jiang, H. L., Li, S., and Rao, Z. H. (2003). Design and structure-based study of new potential FKBP12 inhibitors, *J Biophy* **85**, 3194–3201.

Szabo, C., Zingarelli, B., O'Connor, M., and Salzman, A. L. (1996). DNA strand breakage, activation of poly (ADP-ribose) synthetase, and cellular energy depletion are involved in the cytotoxicity of macrophages and smooth muscle cells exposed to peroxynitrite. *Proc Natl Acad Sci USA* **93**, 1753–1758.

Tai, P.-K. K., Albers, M. W., Chang, H., Faber, L. E., and Schreiber, S. L. (1992). Association of a 59-kilodalton immunophilin with the glucocorticoid receptor complex. *Science* **256**, 1315–1318.

Tanaka, K., Fujita, N., Yoshioka, M., and Ogawa, N. (2001). Immunosuppressive and non-immunosuppressive immunophilin ligands improve H_2O_2-induced cell damage by increasing glutathione levels in NG108-15 cells. *Brain Res* **889**, 225–228.

Teague, S. J., Cooper, M. E., Donald, D. K., and Furber, M. (1994). Synthesis and study of a non macrocyclic FK506 derivative. *Bioorg Med Chem Lett* **4**, 1581–1584.

Teague, S. J., and Stocks, M. J. (1993). The affinity of the excised binding domain of FK506 for the immunophilin FKBP12. *Bioorg Med Chem Lett* **3**, 1947–1950.

Uchino, H., Elmér, E., Uchino, K., Lindvall, O., and Siesjo, B. K. (1995). Cyclosporin A dramatically ameliorates CA1 hippocampal damage following transient forebrain ischaemia in the rat. *Acta Physiol Scand* **155**, 469–471.

Van Duyne, G. D., Standaert, R. F., and Karplus, P. A. (1991). Atomic structure of FKBP-FK506, an immunophilin-immunosuppressant complex. *Science* **252**, 839–842.

Van Duyne, G. D., Standaert, R. F., Karplus, P. A., Schreiber, S. L., and Clardy, J. (1993). Atomic structures of the human immunophilin FKBP-12 complexes with FK506 and rapamycin. *J Mol Biol* **229**, 105–124.

Voolf, C. J., Ma, Q. P., Allchorne, A., and Poole, S. (1996). Peripheral cell types contributing to the hyperalgesic action of nerve growth factor in inflammation. *J Neurosci* **16**, 2716–2723.

Wang, G. T., Lane, B., Fesik, S. W., Petros, A., Luly, J., and Krafft, G. A (1994). Synthesis and FKBP binding of small molecule mimics of the tricarbonyl region of FK506. *Bioorg Med Chem Lett* **4**, 1161–1166.

Wennberg, L., Czech, K. A., Larsson, L. C., Mirza, B., Bennet, W., Song, Z., and Widner, H. (2001). Effects of immunosuppressive treatment on host responses against intracerebral porcine neural tissue xenografts in rats. *Transplantation* **71**, 1797–1806.

CHAPTER 10

CHIRAL DRUGS AND THE ASSOCIATED ASYMMETRIC SYNTHESIS

*Jiange Zhang and Guoqiang Lin**

10.1 INTRODUCTION

Since Arthur Cushny (Challener, 2001a) established 80 years ago the relevance of chirality to the pharmaceutical industry by showing that one enantiomer of hyoscyamine possessed greater pharmacological activity than the other, drug chirality has become more and more important in both pharmaceutical industries and the academic world over the past three decades. Drug chirality is now a major theme in the design, discovery, development, launching, and marketing of new drugs with the development of stereochemistry, pharmacology, and other subjects. The importance of drug chirality is not only a major challenge to the minds of practicing scientists, but also a highly fertile field for the development of technologies for the production of high-value pharmaceuticals. New developments in technologies for isolating, preparing, and purifying chiral materials have greatly increased the opportunities for utilizing enantiopure compounds in commercial applications. Novel techniques for classical resolution, new methodologies for developing selective enzymes for biocatalysis, advances in the application of microorganisms for chemical production, and continued progress in the area of asymmetric synthesis have all contributed to the growth of this field.

*Corresponding author: lingq@mail.sioc.ac.cn

Drug Discovery Research: New Frontiers in the Post-Genomic Era, Edited by Ziwei Huang

10.2 SIGNIFICANCE OF CHIRALITY

10.2.1 Terminology in Stereochemistry

10.2.1.1 Superimposability

The word "chiral" is derived from the Greek word "cheir," meaning hand, or "handedness" in a general sense. The left and right hands are mirror images of each other, and no matter how the two are arranged, one cannot be placed directly over the top of the other in the exact same orientation. Chiral objects are commonly known as a person's right and left hands, snail shells, and clockwise- or counterclockwise-threaded screws. Simply speaking, this is a unique character of nature. Accordingly, a chiral molecule is a molecule that is not superimposable on its mirror image, whereas an achiral compound has a superimposable mirror image (Müller and Kohler, 2004). Nowadays, chiral molecules have ever-increasing importance in the pharmaceutical area in both academy and industry settings. Chiral drugs refer to those molecules that belong to chiral molecules where the chirality exists. It is not surprising that the synthesis of chiral compounds has become an important subject for research. At the moment, most new drugs and those under development consist of a single optically active isomer.

10.2.1.2 Stereoisomerism and Enantiomers

Chirality is a fundamental property of many three-dimensional objects—in other words, a phenomenon caused by a process called stereoisomerism. Stereoisomers are compounds that have the same atoms connected in the same order but differ from each other in the way that the atoms are oriented in space. Chiral molecules that behave like the image and mirror image of each other and are not superimposable are called enantiomers. To take a simple example, lactic acid can be obtained in two forms, **1** and **2**, which are clearly related as mirror images to each other (Fig. 10.1).

10.2.1.3 Absolute Configuration

It is very important to define the absolute configuration of a chiral molecule in order to understand its function in a biosystem. Many biological activities are exclusive to one specific absolute configuration. Without a good understanding of absolute configuration

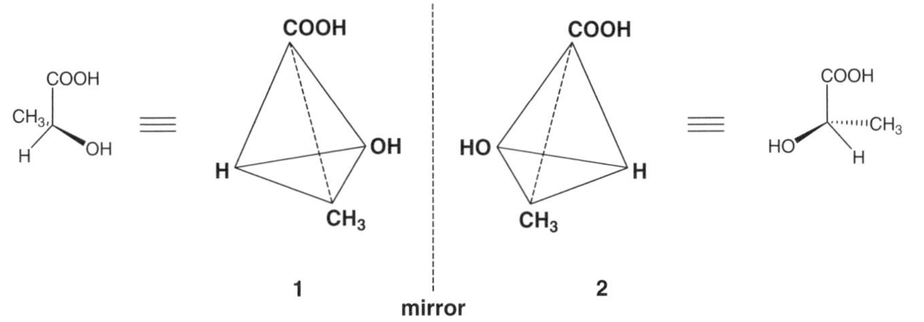

Figure 10.1 Mirror images of lactic acid.

of a molecule, we often cannot understand its chemical and biological behavior. Under normal conditions, the two enantiomers of a chiral compound will have identical chemical and physical properties such as the same boiling/melting point and the same solubility in normal achiral solvent (Eliel and Wilen, 1994). Also, their chemical reaction will be identical under achiral conditions. Enantiomers do differ from each other in the direction in which they rotate the plane-polarized light, and this is called optical activity or optical rotation. Thus, **2**, which rotates the plane-polarized light in a clockwise direction, is denoted as (+)-lactic acid by a plus sign (+) prior to the name of the compound, while the enantiomer (**1**), which has an equal but opposite rotation under the same conditions, is denoted as (−)-lactic acid by a minus sign (−) prior to the name of the compound. However, under chiral conditions, these enantiomers may behave in a very different way. For example, the physical property or chemical reactivity may change significantly under chiral conditions. Determination of absolute configuration of a chiral center is accomplished by assigning the spatial orientation of a molecule and correlating this orientation to the negative or positive sign of the rotation of polarized light by the substance under a given condition (Lin et al., 2001).

10.2.1.4 Diastereoisomers

Many molecules contain more than one chiral center. If a compound has two chiral carbon atoms, there are a total of four possible stereoisomers. Two of the isomers will be mirror images of each other, and therefore enantiomers. The other two isomers are called diastereoisomers—that is, stereoisomers that are not enantiomers (Challener, 2001a). The relationship for 2-chloro-3-hydroxybutane is exemplified by Fig. 10.2. In such conditions, there exist possibly two pairs of enantiomers and two pairs of diastereoisomers, for four individual compounds.

10.2.1.5 Determining Enantiomer Composition (ee)

Measuring the enantiomer composition is very important in the chemistry of chiral drugs, because people working in this area need this information to evaluate the biological data to make sure they are viable. The enantiomer composition of the sample is described by the enantiomer excess (*ee*), which describes the excess of one enantiomer over the other, where (*R*) and (*S*) are the composition of *R* and *S* enantiomers, respectively. In a similar manner, the diastereoisomer composition of a sample can be described by the diastereoisomer excess (*de*), which refers to the

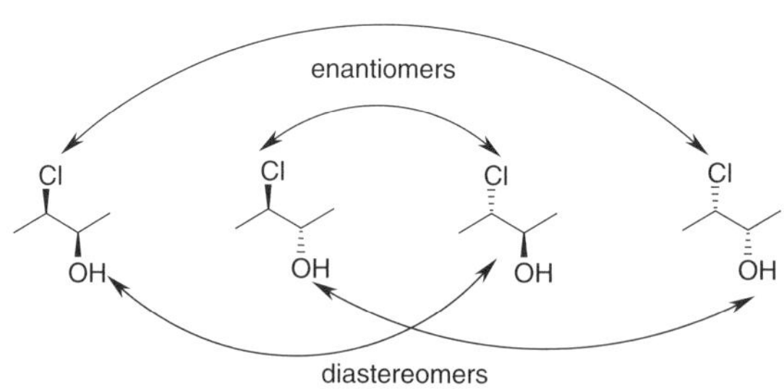

Figure 10.2 Enantiomers and diastereomers.

excess of one diastereoisomer over the other. A variety of methods for determination of *ee* are available (see Eliel and Wilen, 1994).

10.2.2 Chirality in Drug Development

10.2.2.1 Chirality in Living Systems

Louis Pasteur, over 100 years ago, might have been the first scientist to propose the idea of chemical "handedness." His initial theories arose from his study of racemic tartaric acid, which is composed of visibly discernible crystalline forms of enantiomeric D- and L-tartaric acid. From this, the concepts of asymmetry and chirality have been arisen (Hillier and Reider, 2002).

One of the main features of the living world is its chirality, because most of the biological macromolecules of living systems occur in nature in one enantiomer form only (Mislow, 1999). An intrinsic property of the "building blocks of life," such as the amino acid and carbohydrate building blocks of proteins, carbohydrates, and nucleic acids, is that they are all composed of chiral molecules (Maier et al., 2001). Moreover, they are homo-chiral: The 21 essential amino acids are all L-enantiomers, whereas most carbohydrates have the D-configuration (Bonner, 2000). The structures of the L- and D-configurations of alanines **3** and **4** are shown in Fig. 10.3.

10.2.2.2 Stereoselective Actions of Chiral Drugs

Almost 80 years ago Arthur Cushny (Challener, 2001a) first found a difference in pharmacological action between optical isomers, or enantiomers, of a chiral molecule. He showed that the natural, levorotatory alkaloid hyoscyamine was twice as potent as atropine, which is racemic, in antagonizing cholinergic stimuli. A few years later, Easson and Stedman postulated a three-point interaction between a drug and its receptor to explain stereoselectivity in drug action (Waldeck, 2003).

Metabolizing enzymes, protein-binding sites, and/or drug receptor sites have three-dimensional structures; therefore, if the drug molecule is chiral, thus forming optically active isomers, this interaction is usually stereoselective. This means that with the chiral human enzymes and cell surface receptors, the two enantiomers of a racemic drug may be absorbed, activated, or degraded in very different ways, both

$$ee = \left[\frac{(S) - (R)}{(S) + (R)}\right] \times 100\%$$

L-(+)- alanine **3** D-(-)- alanine **4**

Figure 10.3 D- and L-configurations of alanine.

in vivo and *in vitro*. The two enantiomers may have unequal degrees or different kinds of activity. One may be therapeutically effective, while the other may be ineffective or even toxic. This is illustrated by the following examples.

Propafenone (PPF) (**5**) is an anti-arrhythmic drug that is clinically used as a racemic mixture. The enantiomers have almost the same activity in their sodium channel-blocking activity, but (*S*)-PPF is 100 times more potent at the β-adrenergic receptors (Pires de Abreu et al., 1999). Amphetamine (**6**) and its analogues are substances that have a stimulating effect on the central nervous system (CNS). D-Amphetamine is 3–4 times more effective in central nervous system stimulation than its isomer, whereas the latter is slightly more potent in its cardiovascular action (Varesio and Veuthey, 1995). β-Adrenergic block agents exhibit stereoselective actions at β-adrenoceptors, whereby (−)-(*S*)-propranolol (**7**) is about 40 times more potent than the enantiomer. Nevertheless, for local anesthetic and antiarrhythmic activities, the enantiomers are essentially equipotent (Triggle, 1997). Ephedrine and its alkaloids are used for the treatment of asthma, nasal congestion, and obesity. There are two chiral centers in the molecule, and therefore there exist four possible isomers: (+)-ephedrine (**8a**), (−)-ephedrine (**8b**), (+)-pseudoephedrine (**9a**), and (−)-pseudoephedrine (**9b**). (−)-Ephedrine and (−)-pseudoephedrine are isomers occurring in nature. The ephedrine isomers exhibit direct and indirect effects on both α- and β-adrenergic receptors (ARs). (−)-Ephedrine (**8b**) has been shown to be the most potent of the four isomers with various behavioral measures (Young and Glennon, 1998). In addition to sympathomimetic effects, it has been reported that ephedrines can exert CNS dopaminergic effects similar to those of the amphetamines. Systemic administration of (−)-ephedrine in rats increased extracellular levels of dopamine (DA) in the nucleus accumbens, decreased feeding, and increased locomotor activity (Wellman et al., 1998).

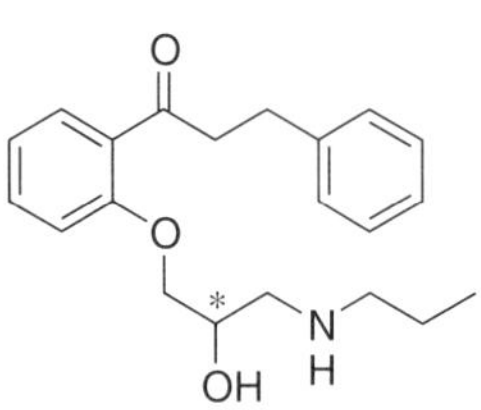

5

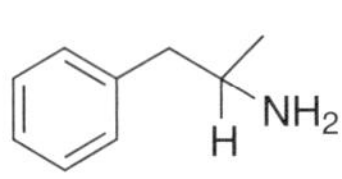

6

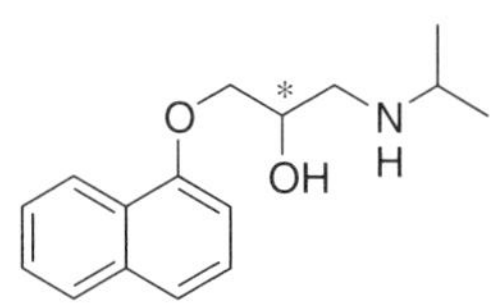

7

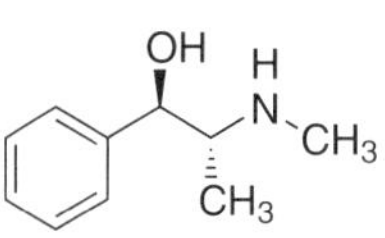

(+)- Ephedrine

8a

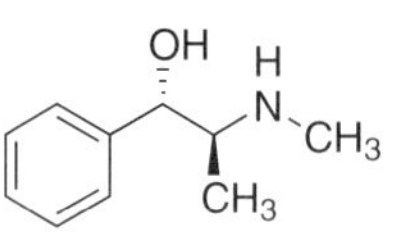

(−)- Ephedrine

8b

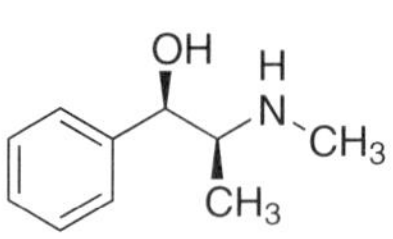

(+)-Pseudoephedrine

9a

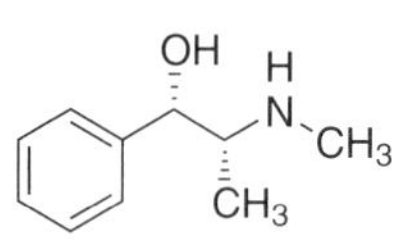

(−)- Pseudoephedrine

9b

In order to further investigate the mechanism of action for the ephedrine isomers *in vivo*, the direct effects of the four-ephedrine isomers on human β_1-, β_2-, and β_3-AR subtypes expressed in CHO cells were examined by Vansal and Feller. The

TABLE 10.1 Cyclic AMP Effects of Ephedrine Isomers on Human β-Adrenoceptor Subtypes Expressed in CHO Cells

	(β_1-Adrenoceptors[a]			(β_2-Adrenoceptors[a]			(β_3-Adrenoceptors[a]		
Compound	$pK_{act}(a)$	I.A. (%)	R.P.	$pK_{act}(a)$	I.A. (%)	R.P.	$pK_{act}(a)$	I.A. (%)	R.P.
(−)-Isoproterenol	8.13 ± 0.10(10)	100	1	9.46 ± 0.09(16)	100	1	7.74 ± 0.07(16)	100	1
(−)-Ephedrine	4.14 ± 0.07(6)	66	9,773	ND[b] (4)	21	ND	ND (4)	22	ND
(+)-Ephedrine	6.26 ± 0.12(6)	68	74	6.48 ± 0.24(4)	78	955	4.35 ± 0.35(4)	31	2,454
(+)-Pseudoephedrine	3.51 ± 0.05(6)	53	41,688	5.00 ± 0.25(4)	47	28,843	ND (4)	7	ND
(−)-Pseudoephedrine	2.95 ± 0.19(6)	53	151,358	5.15 ± 0.16(4)	50	20,420	ND (4)	17	ND

[a]Data for I.A. (intrinsic activity) are expressed as percentage responses to that of (−)-isoproterenol (3×10^{-5} M = 100%). pK_{act} values are expressed as means ± SEM.R.P. = relative potency, calculated as EC_{50} of drug/ EC_{50} of (−)-isoproterenol.

[b]ND, not determined due to low intrinsic activity.

Source: Data were taken from Vansal and Feller (1999).

results are listed in Table 10.1. Their studies demonstrated that the potencies of the four ephedrine isomers on the human β-AR subtypes are different. (−)-Ephedrine was the most potent of the four ephedrine isomers on all three human β-AR subtypes and was the only ephedrine isomer to possess some agonist activity on the human β_3-AR (Vansal and Feller, 1999).

Today, enantioselectivity is well accepted in molecular and atomic detail. Through site-directed mutagenesis, the amino acids responsible for the stereoselective recognition of ligand molecules have been identified in a number of cases, in particular for the β-adrenergic receptor, a prototypic G-protein-coupled receptor. Novel ligands with one or more chiral centers and complicated stereochemistry continue to be developed as potential medicines, despite reservations put forward by regulatory agencies (Soudijn et al., 2004).

The HIV-1 proteinase has been recognized as an attractive target for antiviral therapy. The structure of the naturally occurring L-enzyme, the HIV-1 proteinase with 99 amino acids, had been determined earlier (Miller et al., 1989). Later, the D-enantiomer was synthesized by chemical methods (Milton et al., 1992). Each enantiomer is equally active on hexapeptide substrate analogues; however, only the D-enzyme only being active on D-peptides and the L-enzyme on L-peptides (Lamzin, 1995). Therefore, HIV-l protease in its D- and L-forms exhibits opposing chiral substrate selectivity (Milton, 1992). Crixivan (**10**) is a protease inhibitor for AIDS, a single enantiomer in pure form. The X-ray structure at 1.9-Å resolution showed a stick model of Crixivan (**10**) bound to the active site of HIV-1 protease (Zurer, 2005), as shown in Fig. 10.4.

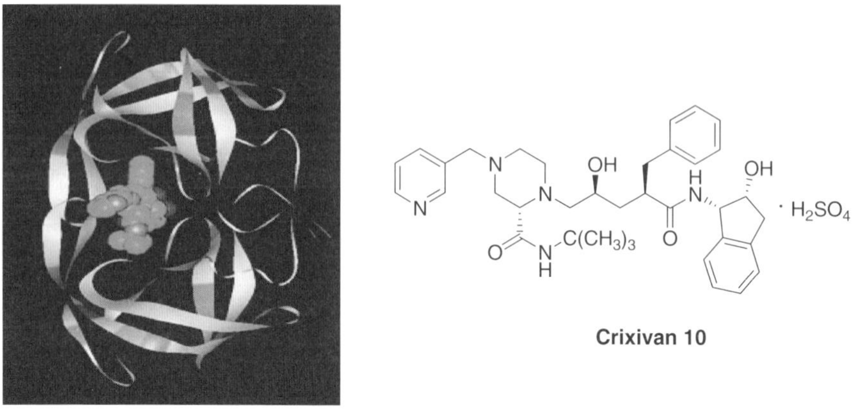

Crixivan 10

Figure 10.4 Crixivan bound to HIV protease (taken from www.biotech100.com). See color plates.

10.2.3 Pharmacological Implications of Chiral Drugs

Molecular chirality is a fundamental consideration in drug discovery (Zhang et al., 2005). In the asymmetric environment of receptors and enzymes, two enantiomers of chiral drugs may form different spatial relationships; therefore, there may be significant differences in their pharmacodynamic activity and their pharmacokinetic properties.

10.2.3.1 Pharmacodynamic Activity

Differences in the potency of stereoisomers may be related to differences in their affinity or intrinsic activity at receptor sites (Rentsch, 2002). The effects of differently acting stereoisomeric drugs can be categorized basically as follows: (a) Two enantiomers may have nearly identical qualitative and quantitative pharmacological activity. For example, carvedilol (**11**) is a newer agent that is marketed as the racemate for the treatment of hypertension and congestive heart failure (Tenero et al., 2000). Carvedilol is a nonselective β- and α-adrenergic receptor blocking agent, and the nonselective β-blocking activity resides mainly in the (*S*)-carvedilol, while the α-blocking effect is shared by (*R*)- and (*S*)-enantiomers (Bartsch et al., 1990). Sotalol (**12**) is a chiral β-adrenergic blocking drug used in therapy as a racemic mixture. The (*R*)-enantiomer possesses the majority of the β-blocking activity; however, the (*R*)- and (*S*)-enantiomers of sotalol share an equivalent degree of III antiarrhythmic potency (Mehvar and Brocks, 2001). (b) The two enantiomers of a chiral drug may have similar modes of action but may differ in their affinity to a receptor or an enzyme, resulting in different reaction rates. Under these conditions, enantiomers may have opposite or different effects—for example, isoprenaline (**13**), where the (−)-enantiomer is a β_1-adrenoceptor agonist and the (+)-enantiomer has antagonistic effects (Lopez-Vidriero et al., 1985). Picenadol (**14**) is a unique opioid mixed agonist–antagonist analgesic currently under clinical evaluation. The (+)-isomer of picenadol is a potent opiate agonist, whereas the (−)-isomer is an opioid antagonist (Zimmerman et al., 1985). (c) All the pharmacological activity may reside in one enantiomer; in this case the other enantiomer may be regarded as an impurity.

The impurity may be inactive or have desirable or undesirable activity. The maleate of (*S*)-(−)-Timolol (**15**) is a potent and nonselective β-adrenergic blocking agent. In comparison with timolol, the maleate of (*R*)-(+)-enantiomer (**16**) is 4 times less potent in reducing intraocular pressure in human subjects, 49 times less potent as a β_1- and β_2-adrenoceptor or antagonist in animal subjects, 13 times less potent as a bronchoconstrictor in normal subjects, and considerably less effective in reducing the heart rate of exercising human subjects (Hanna and Lau-Cam, 1995). Ofloxacin (**17**) is a fluoroquinolone compound being able to be concentrated in phagocytic cells (García et al., 2000). (*S*)-(−)-Ofloxacin and (*R*)-(−)-ofloxacin are optically active isomers of racemic ofloxacin. However, (*S*)-(−)-ofloxacin is considered to be the more active isomer and is 8–128 times more potent than (*R*)-(−)-ofloxacin with the different bacterial strains *in vitro* (Zeng et al., 1999). As a single enantiomer, ofloxacin has been used in a clinical medicine called levofloxacin. (d) The two enantiomers of a chiral drug may act on different receptors or enzymes, resulting in different pharmacological activity. The (*R*)- and (*S*)-isomers of verapamil (**18**) have different magnitudes of effect on hemodynamics and the atrioventricular conduction system. (*S*)-Isomer is 6–10 times more potent than (*R*)-isomer in regard to the Ca^{2+} channel blocking effect (Sakuta and Okamoto, 1994). However, both enantiomers are equally effective in increasing cellular accumulation of anticancer drugs. Verapamil may be a useful candidate drug for the treatment of multidrug resistance in cancer patients (Höll et al., 1992). (e) There are many chiral drugs for which one or both enantiomers have the desired effect and only one enantiomer causes unwanted side effects. A well-known disaster is the contergan tragedy. Contergan was a sedative that contained racemic thalidomide (**19**). Both enantiomers had the desired therapeutic effects, whereas only the (*S*)-enantiomer had teratogenic effects and caused severe malformations of human babies (Man et al., 2003). Racemic dropropizine (**20**) has long been used in human therapy as an antitussive agent. Recent studies have revealed that (*S*)-dropropizine possesses the same antitussive activity as the racemic mixture but has much lower selective activity on the CNS (Salunkhe and Nair, 2001). Therefore, particular clinical significance is attached to drugs of which one enantiomer may contribute to the side effects or toxic effects. (f) In contrast to the above-mentioned enantiomers, the inactive enantiomer might antagonize the side effects of the active isomer. In such cases, an enantiomerically pure compound is not preferred. Tramadol (**21**) is a centrally acting analgesic that is not associated with the classical side effects of opiate drugs, such as respiratory depression, constipation, or sedation (Sacerdote et al., 1999). This compound is manufactured as an equal ratio of two enantiomers, each one displaying different affinities for various types of receptors. The (+)-enantiomer is a selective agonist for μ receptors with preferential inhibition to serotonin reuptake and enhances serotonin efflux in the brain, whereas the (−)-enantiomer mainly inhibits noradrenaline reuptake. The incidence of side effects, particularly opioid-mediated ones, was higher with (+)-enantiomer than with (±)-tramadol or (−)-enantiomer. In regard to efficacy and side effects, the racemate of tramadol is superior to the administration of either enantiomer alone for the treatment of severe postoperative pain (Rojas-Corrales et al., 1998). The (−)-enantiomer of dobutamine (**22**), which possesses mainly α_1-adrenoceptor agonist activity, is in contrast to (+)-dobutamine, which possesses predominantly β_1- and β_2-adrenoceptor agonist activity. The effects of (±)-dobutamine on cardiac performance are mediated, which is due mainly to an increase in stroke volume, with heart rate being only minimally affected (Ruffolo and Messick, 1985).

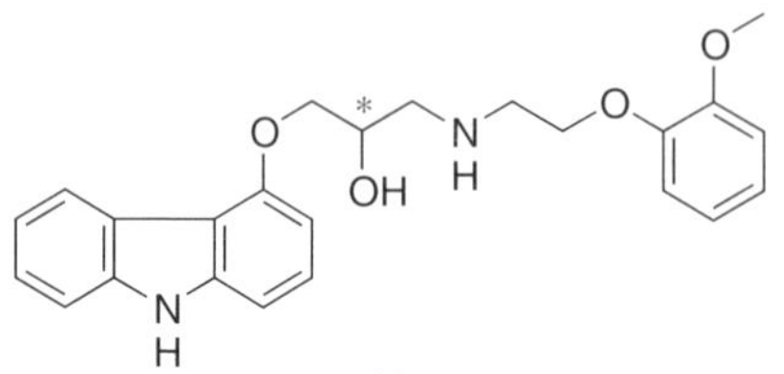

11

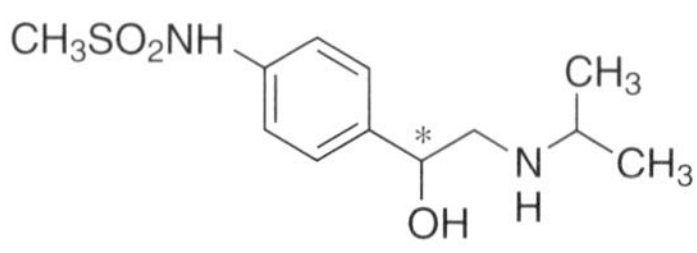

12

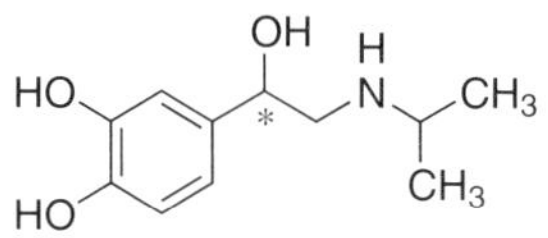

13

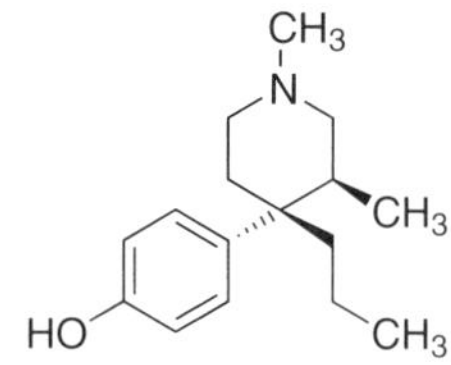

14

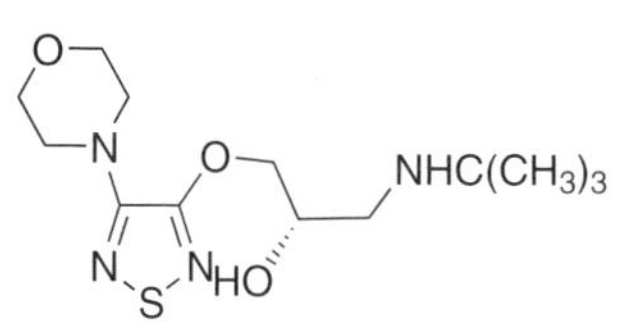

15

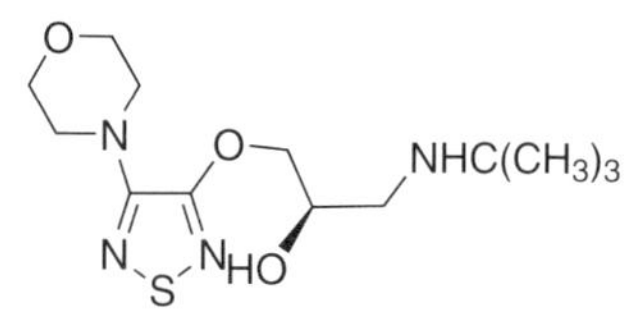

16

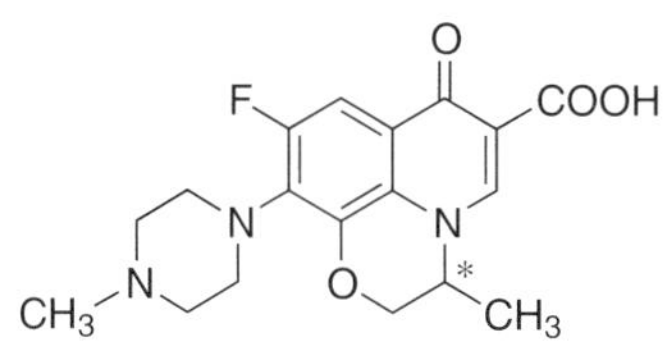

17

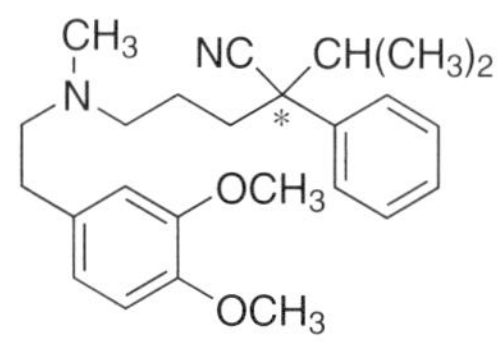

18

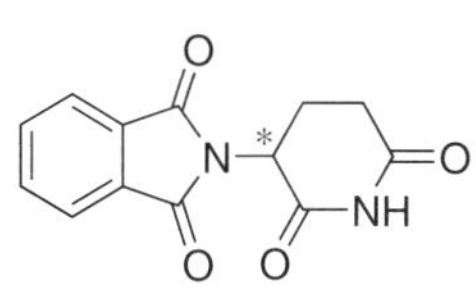

19

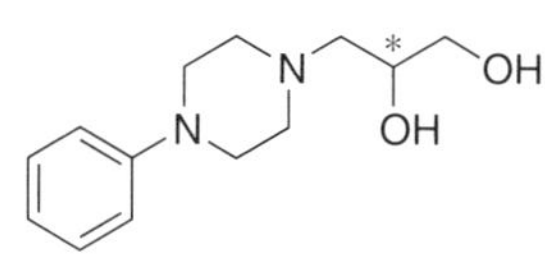

20

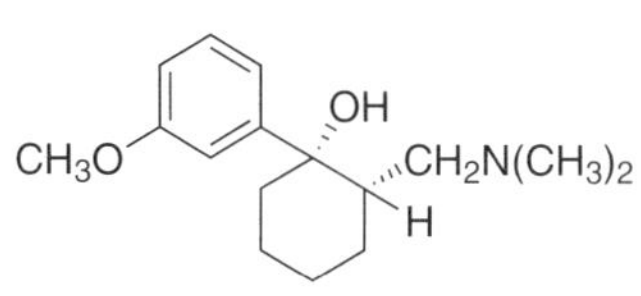

21

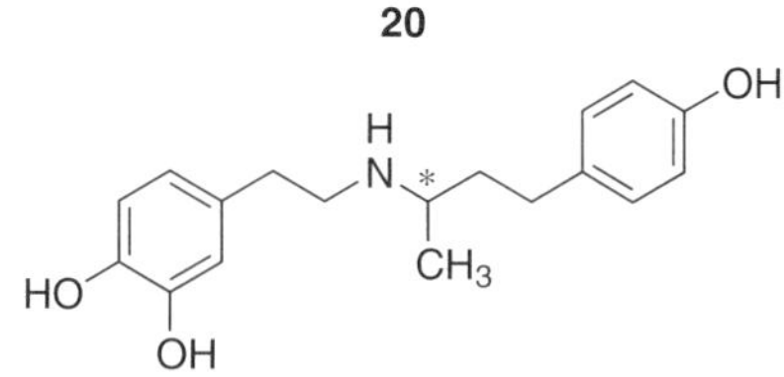

22

10.2.3.2 Pharmacokinetic Properties

The administration of racemic mixtures often results in far more complex pharmacokinetics than is the case when single enantiomers are used. Different enantiomers may have different plasma concentration time profiles as a result of chiral discrimination in pharmacokinetic processes. These differences may be further influenced by other factors such as the route of administration, the age and sex of the subject, disease states, and genetic polymorphism in cytochrome P450 (CYP) isoenzymes involved in drug metabolism (Lane and Baker, 1999). Thus, it is necessary to measure the stereoselectivity of pharmacokinetic processes (absorption, protein binding, metabolism, transport, and excretion) and to determine their contribution to the observed overall stereoselectivity of drug action.

Methadone (**23**) was introduced to manage opioid dependence in 1965. Methadone's therapeutic benefits reside in the (*R*)-enantiomer. One reason is that methadone's (*R*)-enantiomer shows 10-fold higher affinity for μ and κ opioid receptors and up to 50 times the anti-nociceptive activity in clinical studies and animal models compared with the (*S*)-enantiomer. Another important reason is that methadone's enantiomers show markedly different pharmacokinetics. (*R*)-Enantiomer shows a significantly greater unbound fraction and total renal clearance than (*S*)-enantiomer. This reflects (*S*)-enantiomer's higher affinity for the main plasma-binding protein. These stereoisomeric differences contribute to the wide variations in some of methadone's pharmacokinetic parameters between patients (Foster et al., 2000; Baumann et al., 2002).

The potent histamine H_1-receptor antagonist cetirizine (Zyrtec) is a racemic mixture of (*R*)-levocetirizine (**24**) (now available under the trademark Xyzal) and (*S*)-dextrocetirizine (**25**). In binding assays, levocetirizine has demonstrated a two-fold higher affinity for the human H_1-receptor compared to cetirizine and an approximately 30-fold higher affinity than dextrocetirizine (Gillard et al., 2002). However, the pharmacokinetic parameters indicate that levocetirizine is rapidly and extensively absorbed and poorly metabolized, compared after administration alone or in the racemate. Its apparent volume of distribution is smaller than that of dextrocetirizine ($0.41\,L\,kg^{-1}$ vs. $0.60\,L\,kg^{-1}$). Moreover, the nonrenal (mostly hepatic) clearance of levocetirizine is also significantly lower than that of dextrocetirizine ($11.8\,mL\,min^{-1}$ vs. $29.2\,mL\,min^{-1}$). All evidence available indicates that levocetirizine is intrinsically more active and more efficacious than dextrocetirizine and for a longer duration (Tillementa et al., 2003).

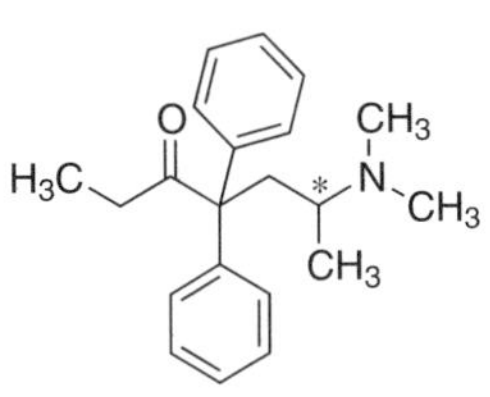

23

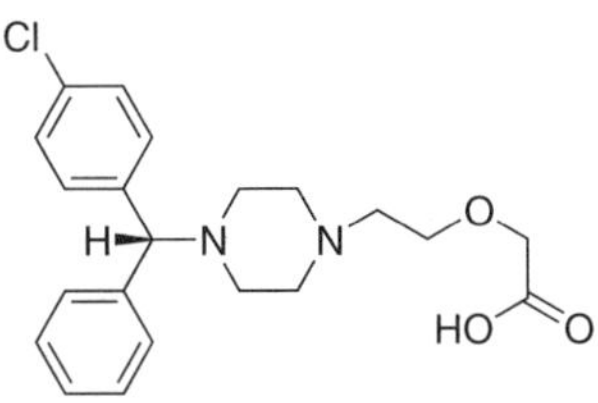

(*R*)-levocetirizine **24**

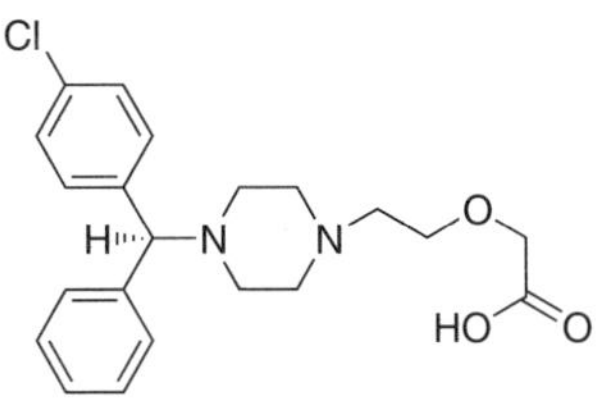

(*S*)-dextrocetirizine **25**

The main metabolites of trimipramine (**26**) are desmethyl-trimipramine (**27**), 2-hydroxy-trimipramine (**28**), and 2-hydroxy-desmethyl-trimipramine (**29**) (as shown in Fig. 10.5). However, trimipramine appears to show stereoselective metabolism with

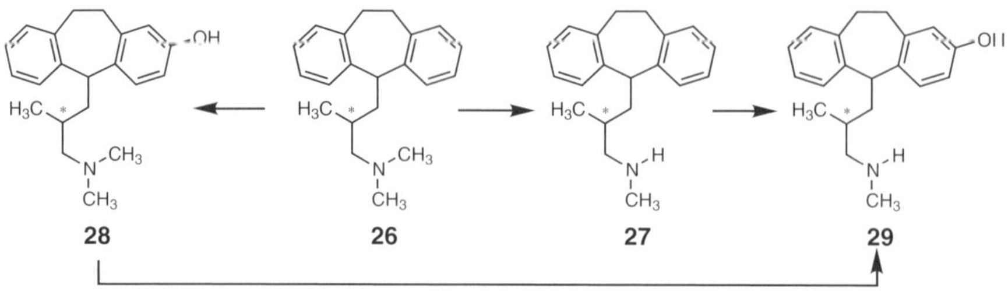

Figure 10.5 Metabolism of trimipramine.

preferential *N*-demethylation of D-trimipramine and preferential hydroxylation of L--trimipramine, mediated by CYP2-D6. CYP2C19 appears to be involved in demethylation and favors the D-enantiomer, while CYP3A4 and CYP3A5 seem to metabolize L-trimipramine to a currently undetermined metabolite. Direct investigations of the implications of this stereoselective differences are needed (Eap et al., 2000).

10.2.3.3 Drivers for Chiral Drugs Market

For about the past two decades, the pharmacopoeia was dominated by racemates, but since the emergence of new technologies in the 1980s that allowed the preparation of pure enantiomers in significant quantities, the awareness and interest in the stereochemistry of drug action has increased (Eichelbaum and Gross, 1996; Caldwell, 1999).

Drug chirality became a major theme in the design, discovery, development, launching, and marketing of new drugs after the importance of stereoselective pharmacodynamics and pharmacokinetics was widely recognized (Agranat et al., 2002). It further became evident that the use of stereochemically pure drugs could be advantageous, because they would be expected to reduce the total dose given, simplify the dose–response relationship, remove a source of intersubject variability, and minimize toxicity due to the inactive stereoisomer and enhanced therapeutic window (Caldwell, 1999). These factors have led to an increasing preference for single enantiomers in both industry and regulatory authorities.

10.2.3.4 Market Information

Economic interests are obvious and essential driving forces in the spectacular development of new chiral substances and technological improvements (Maier et al., 2001). Today, approximately 80% of chiral intermediates and related products go into the pharmaceuticals market, and it is expected that, in the future, pharmaceuticals will remain the key driver for chiral chemistry development. In 1999, worldwide single-isomer drug sales reached \$115 billion, up 16% from \$99 billion in 1998, according to Richard L. DiCicco, president of the consulting firm Technology Catalysts International Corp (TCI). Single-isomer drugs accounted for 30% of the \$335 billion total drug sales worldwide in 1998, edging up to 32% of the \$360 billion market in 1999 (Stinson, 2000). These data are shown in Table 10.2. Worldwide sales of chiral drugs in single-enantiomer dosage forms continued

TABLE 10.2 Chiral Drug Sales Hurtle Past $100 Billion and Show No Sign of Slowing

	Global Sales in $ (Millions)				Global Sales in $ (Millions)		
	1998	1999	2003		1998	1999	2003
Cardiovascular	21,906	24,805	26,012	Respiratory	4,305	5,087	8,795
Antibiotics/antifungals	19,756	20,907	23,265	Gastrointestinal	1,718	2,998	5,355
Hormones/endocrinology	12,297	13,760	17,345	Ophthalmic	1,482	1,794	2,070
Cancer	8,006	9,420	13,360	Dermatological	1,124	1,270	1,540
Central nervous system	7,027	8,592	13,720	Analgesics	842	1,045	1,135
Hematology	6,730	8,580	11,445	Vaccines	568	676	1,100
Antiviral	6,131	7,540	13,446	Other	7,947	8,527	7,425

TOTAL	**1998**	**1999**	**2003**
$ (Millions)	**$99,389**	**$115,001**	**$146,013**

Source: Data were taken from Stinson (2000).

growing at a more than 13% annual rate to $133 billion in 2000, according to TCI in Falls Church, Virginia. At a future growth rate estimated by TCI, the figure could hit $200 billion in 2008 (Stinson, 2001). According to the market research firm Freedonia Group (Rouhi, 2002), demand for chiral raw materials, intermediates, and active ingredients would grow by 9.4% annually between 2000 and 2005.

An overall look at the 20-year period 1983–2002 (Bristol, 1984–2003) indicates that single enantiomers surpassed achirals, whereas racemates represented the minority category at 23% of worldwide approved drugs. However, even in the 1980s, single enantiomers were not a minor component of approved drugs and of approved chiral drugs. In subsequent years the percentage of achiral drugs decreased gradually from 43% to 34% of all approved drugs. The achirals:chirals ratios were 43:57 (1983–1986), 36:64 (1987–1990), 34:66 (1991–1994), 41:59 (1995–1998), and 34:66 (1999–2002).

The distribution of FDA-approved NMEs (new molecular entities) according to chirality character covers only 12 years (1991–2002). The distribution of FDA-approved drugs for the whole 12-year period bears some similarity to the worldwide distribution, in that 44% were single enantiomers, 42% were achiral drugs, and a minority (14%) were racemic drugs. The distribution of the 15 FDA-approved drugs in the period of January–August 2003 was 64% from single enantiomers, 14% from racemates, and 22% from achirals (Caner, 2004). All the data are shown in Table 10.3.

The chiral technology market is also evolving due to regulatory changes and the demand for single isomers. The U.S. market for chiral technology reached $1.2 billion in 2003, comprising over 40% of the total worldwide market. Rising at an average annual growth rate of 8.8%, the U.S. market is expected to exceed $1.8 billion in 2008.

The chiral industry may be divided into two main categories: the manufacturing of chiral compounds and the analysis of chiral compounds. Chiral

TABLE 10.3 Annual Distribution of Worldwide and FDA-Approved Drugs (NMEs) According to Chirality Character in the Period 1983–2002

Year	Racemates[a] World-wide	Racemates[a] FDA	Single Enantiomers World-wide	Single Enantiomers FDA	Achiral World-wide	Achiral FDA	Year	Racemates[a] World-wide	Racemates[a] FDA	Single Enantiomers World-wide	Single Enantiomers FDA	Achiral World-wide	Achiral FDA
	(%)	(%)	(%)	(%)	(%)	(%)		(%)	(%)	(%)	(%)	(%)	(%)
1983	37	NA[b]	26	NA	37	NA	1993	16	17	45	29	39	54
1984	28	NA	26	NA	46	NA	1994	38	5	38	57	24	38
1985	38	NA	22	NA	40	NA	1995	21	37	46	30	33	33
1986	26	NA	26	NA	48	NA	1996	9	14	41	43	50	43
1987	18	NA	49	NA	33	NA	1997	24	8	30	38	46	54
1988	26	NA	39	NA	35	NA	1998	15	9	50	41	35	50
1989	29	NA	26	NA	45	NA	1999	13	4	52	46	35	50
1990	33	NA	35	NA	32	NA	2000	9	19	62	37	29	44
1991	20	9	40	65	40	26	2001	0	0	68	60	32	40
1992	21	33	44	42	35	25	2002	6	0	55	53	39	47

[a]Including diastereomeric mixtures.
[b]NA, not applicable.

Source: Data were taken from Bristol (1984–2003) and Caner et al. (2004).

manufacturing will continue to dominate the market as it rises from revenues of over \$1 billion in 2003 to the estimated over \$1.6 billion in 2008. The manufacturing category is further subdivided into synthesis and separation. The chiral analysis figures include the market for chiral chromatography, chiral polarimetry, chiral NMR (nuclear magnetic resonance) shift, and others. Total sales of reagents in the analysis of chiral products reached \$164.6 million in 2003 and are expected to grow at an AAGR of 6.6% to \$226.4 million by 2008 (see Fig. 10.6).

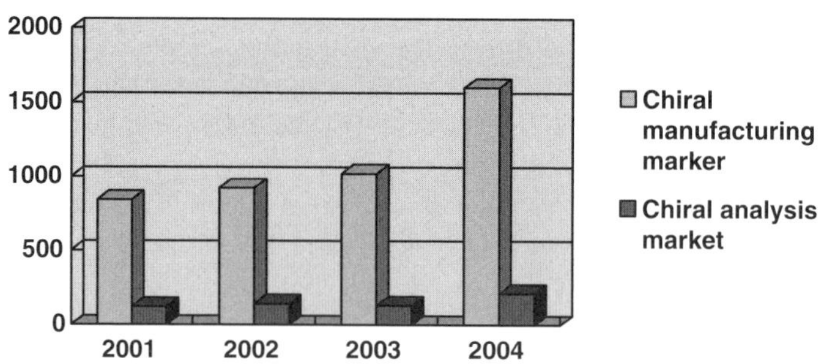

Figure 10.6 Sales of chiral technology products in the United States, 2001–2003 and 2008 (estimated) (taken from www.bccresearch.com).

10.2.3.5 Regulatory Issues

Regulatory changes have a significant effect on the production of chiral chemicals, particularly those prepared for use in formulated pharmaceutical and agrochemical products. Regulatory control of chiral drugs began in the United States with the publication in 1992 of formal guidelines on the development of chiral drugs in a document entitled *Policy Statement for the Development of New Stereoisomeric Drugs* (FDA, 1992) and was followed in the European Union (EU) in 1994 by *Investigation of Chiral Active Substances*. The governing bodies have issued specific rules pertaining to the development of stereoisomeric drugs. These guidelines require that applicants must recognize the occurrence of chirality in new drugs, attempt to separate the stereoisomers, assess the contribution of the various stereoisomers to the activity of interest, and then make a rational selection of the stereoisomeric form that is proposed for marketing. These common features are the basis for the consideration of this topic under the auspices of the International Conference on Harmonisation (Caldwell, 1996).

10.3 SOURCES OF CHIRAL DRUGS

There are three main methodologies for preparing chiral molecules in homo- or enriched enantiomeric forms: The first technique involves isolation of natural abundant products or semisynthetic drugs derived from natural products, the second involves biotransformation of inexpensive, available feedstocks, and the third involves a chemical control method using optically active compounds (obtained from one of the first two methods) or prochiral starting materials. The chemical control methods of the preparation of enantiomers can be characterized in detail as shown in Fig. 10.7.

A survey by Frost and Sullivan estimated that in 2002, 55% of chiral products worldwide were generated by traditional chiral pool and separation, 35% by chemocatalysis, and 10% by biocatalysis.

10.3.1 Natural Products

Natural products or semisynthetic drugs derived from natural resources present fertile preparation of enantiopure compounds by various chemical methods and diversified field for discovering therapeutic of single enantiomers (Rouhi, 2003).

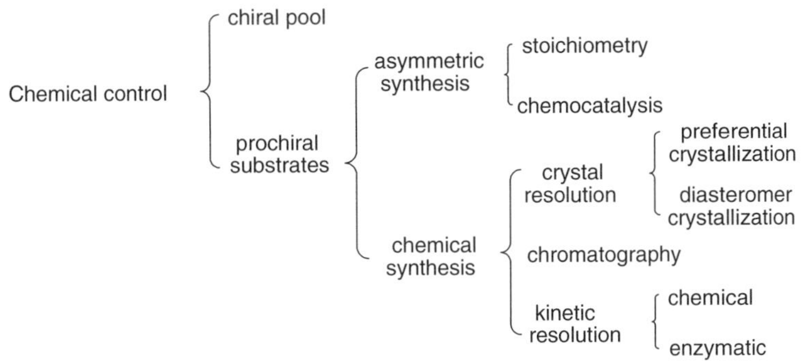

Figure 10.7 Chemical control methods of the preparation of enantiomers.

For example, (−)-quinine is an antimalarial (**30**), whereas its quasienantiomer, (+)-quinidine (**31**), is an antiarrhythmic compound. Similarly, the (−)-enantiomer of the predominant natural (+)-gossypol (**32**) has male-contraceptive and anticancer activities.

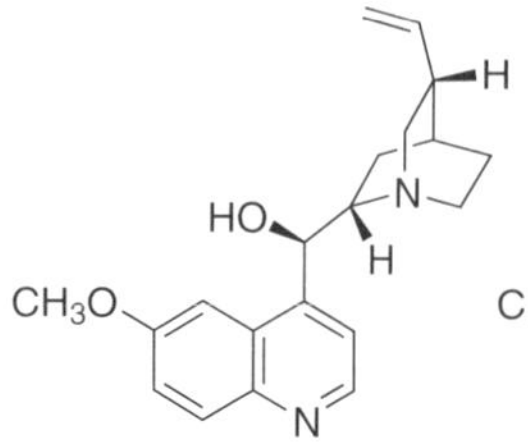

30

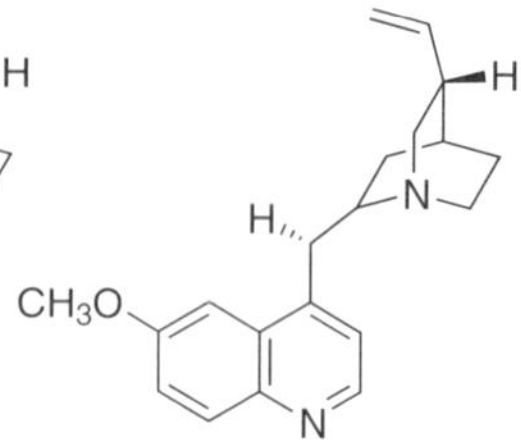

31

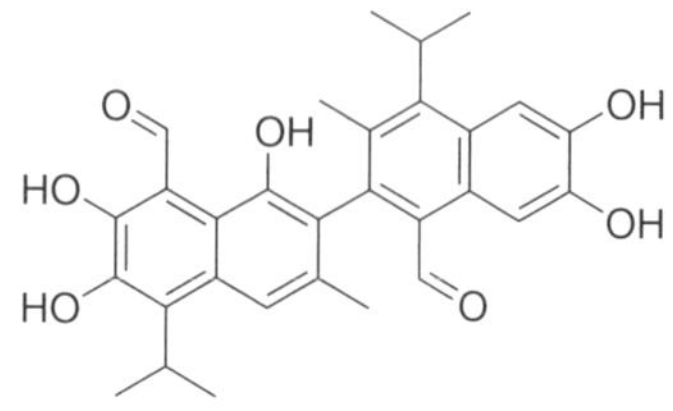

32

Isolation of natural products can be achieved in several ways. For example, many terpenes are extracted from the various plants in which they are produced. Steam distillation at atmospheric pressure is often used to extract the terpene molecules from plants. Essential oils are analyzed by gas chromatography in order to estimate the concentration of active molecule that is removed from the mixture by preparative low-pressure liquid chromatography and a polar solvent as an elutant. The chemical and optical purities of the monoterpenes are dependent upon the specific plant source, the location where it was grown, and the processing conditions. Table 10.4 lists different terpenes and plant sources they are derived from (Martin et al., 2004).

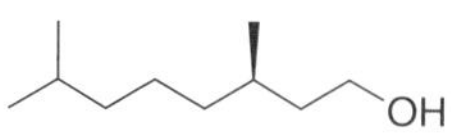

33

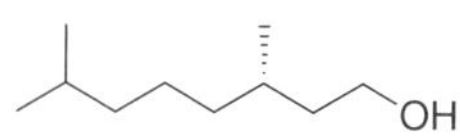

34

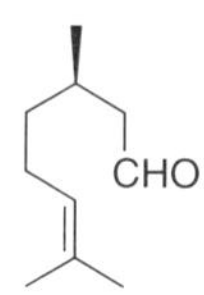

35

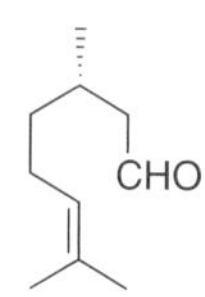

36

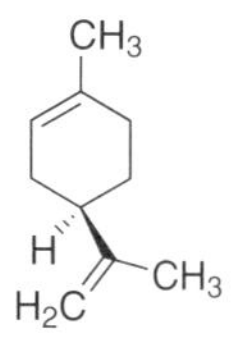

37

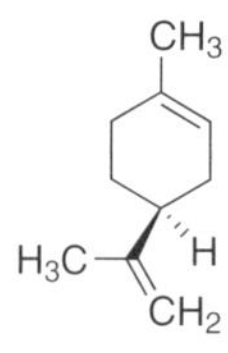

38

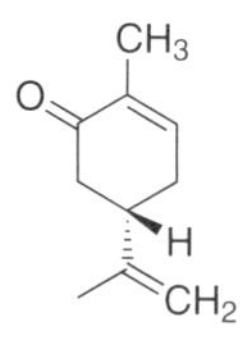

39

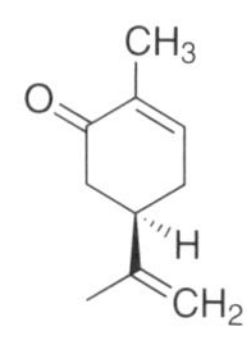

40

TABLE 10.4 Monoterpenes and Their Plant Sources

Terpene		Source
(*R*)-(+)-Citronellol	33	*Citronella*, *Eucalyptus citriodora*
(S)-(-)-Citronellol	34	*Geranium*, rose
(*R*)-(+)-Citronellal	35	*Citronella*
(*S*–)-(−)-Citronellal	36	*Eucalyptus citriodora*
(*R*)-(+)-Limonene	37	Orange, lemon, grapefruit, tangerine, bergamot, caraway, dill
(*S*–)-(−)-Limonene	38	Commercial
(*S*)-(+)-Carvone	39	Caraway, dill
(*R*–)-(−)-Carvone	40	*Mentha dulcis*, *Mentha spicata* (spearmint)

10.3.2 Biotransformation (Biocatalysis)

The pharmaceutical industries are doing everything they can to develop a mass-synthesis technique that yields single enantiomer compounds, the drug intermediates, or active pharmaceutical ingredients. While enantiomerically selective organic synthesis is the traditional approach, using enzymes and enzyme-containing microorganisms to biocatalyze a reaction is becoming increasingly important. There has been an increasing awareness of the enormous potential of microorganisms and enzymes for the transformation of synthetic chemicals with high chemoselectivity, regioselectivity, stereoselectivity, and specificity (Patel, 2002). Large numbers of microorganisms or enzymes have been employed for the production of optically active compounds by means of kinetic resolution or stereospecific chemical transformations (Rasool et al., 2005).

10.3.2.1 Kinetic Resolution—Enzymatic

Resolutions generally employ isolated enzymes. They are pretty basic technologies but can be very effective (McCoy, 2001). Naproxen (**41**) is one of the early chiral molecules developed and used as a single enantiomer. *Trichosporon* sp. (TSL), a newly found strain isolated from a locally fermented cottage cheese, has been found to be highly stereoselective in the resolution of (*S*)-(+)-naproxen (*ee* >99%, $E \sim 500$) from the corresponding racemic methyl ester. The viability of the resolution process has been further improved by the development of a simple racemization process for the enriched (*R*)-(−)-ester (Koul et al., 2003) (Scheme 10.1).

Secnidazole (**42c**) showed an anti-amoebic potency 10 times higher than that of metronidazole (**42a**) and ornidazole (**42b**). The latest developments of those drugs reveal that ornidazole (**42b**) is still a subject of research for use as antifertility agents in male animals. In order to further reveal the relationship between configuration and bioactivities of these two chiral drugs, the necessity of gaining the optically pure enantiomers of ornidazole (**42b**) and secnidazole (**42c**) is quite obvious. Our laboratory attempted to acquire enantiomerically pure ornidazole (**42b**) and secnidazole (**42c**) via a lipase-catalyzed resolution; accordingly, ornidazole (**42b**) and secnidazole (**42c**) were esterified with vinylacetate employing different lipases (Scheme 10.2), and the results are listed in Table 10.5. Among the available lipases tested, AKL exhibited the best result for the kinetic resolution of both (**42b**) and (**42c**), given the high values and isolated yields (Tian et al., 2003).

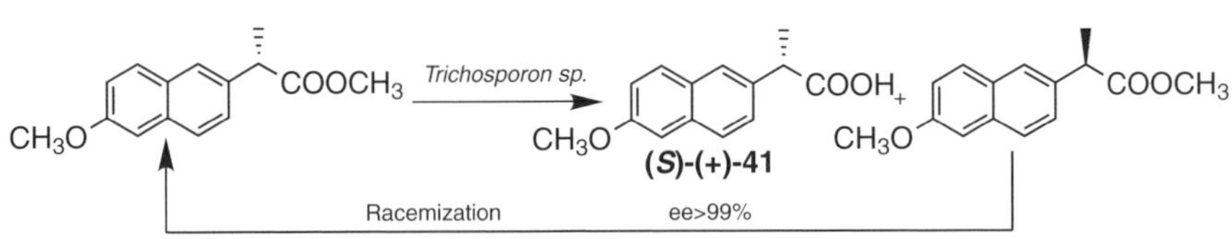

Scheme 10.1

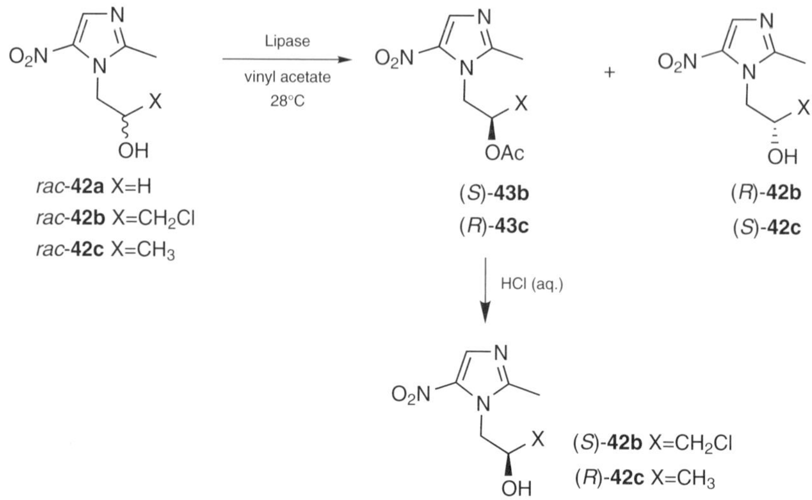

Scheme 10.2

10.3.2.2 Chemo-enzymatic Synthesis

Fermentation and enzyme-based catalysis are starting to challenge traditional synthetic methods of producing optically active pharmaceutical intermediates. As these biocatalytic methods gain ground, a few organic-chemistry-based intermediate manufacturers have climbed aboard, but not many open seats remain for the rest (McCoy, 1999). Historically, the synthesis of single isomers of complicated molecules has been expensive and time-consuming, and biocatalysis was not employed because of many factors restricting its use. Improvements in genetic engineering, immobilization, and stabilization technologies have enabled the introduction of stable enzymes that meet the highly specific needs required for chemical manufacturing, such as solvent stability and the ability to function at different temperature and pH ranges (Rogers, 1999).

TABLE 10.5 Enzymatic Esterfication of Ornidazole (42b) and Secnidazole (42c)

Entry	Compound	Lipase	Time (d)	Conversion (%)	Product	Yield (%)[a]	*ee* (%)
1	*rac*-**42b**	PSL	20	14	—	—	—
2	*rac*-**42b**	AKL	18	44	(−)-(*S*)-**42b**/ (+)-(*R*)-**42b**	43.0/ 52.2	95(>99)[b]/94(>99)[b]
3	*rac*-**42b**	CAL	10	5	—	—	—
4	*rac*-**42c**	AKL	10	44	(−)-(*R*)-**42c**/ (+)-(**S**)-**42c**	37.5/ 51.8	94(>99)[b,c]/90(>99)[b,c]

[a]Isolated yield.

[b]Enantiomeric excess (*ee*) after crystallization.

[c]Determined by chiral HPLC analysis of secnidazole acetate.

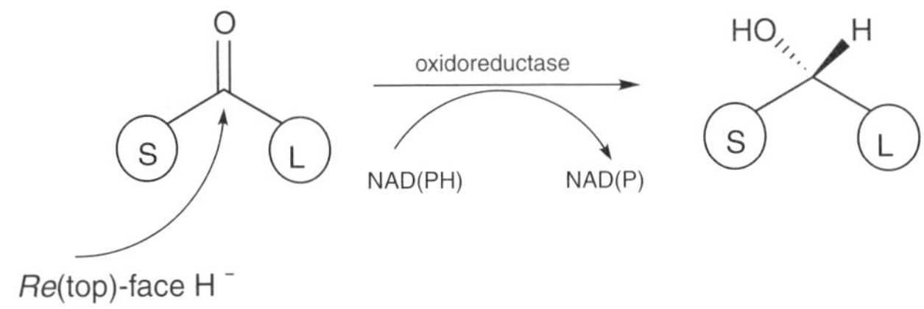

Scheme 10.3 Reduction following *anti*-Prelog's rule.

As part of our continuing work toward utilizing enzymes and enzyme-containing microorganisms to biocatalyze organic synthesis, biotransformation has become a powerful tool to provide chiral synthons of drugs or pharmaceutical intermediates, which are otherwise difficult to obtain by conventional chemical methods.

The Asymmetric Carbonyl Reduction Mediated by Geotrichum sp. Baker's yeast (*saccharomyces cerevisiae*) is well known for its ability to convert various classes of compounds, especially carbonyl-containing substrates, into the corresponding optically active alcohols, usually with (*S*)-configuration. For the synthesis of various kinds of natural products and physiologically active compounds, it is desirable to find a microorganism that is capable of performing reduction to provide products with the opposite enantioselectivity, a complement to baker's yeast. For this purpose, we carried out a number of screenings on soil samples and found a strain of fungus, *Geotrichum* sp., called *G*. 38, which shows the desired properties (Chen, 2003) when the carbonyl group is reduced to alcohol with *anti*-Prelog stereogenic outcome (Scheme 10.3).

Synthesis of (−)-Pyrenophorin α, β-Unsaturated ketone (**44**) could be converted to the major product (**45**) with >98% *ee* and 86% chemical yield by *G*. 38. **45** is a precursor of 7-(*S*)-hydroxy-4,4-ethylenedioxy-2-octenoic acid (**46**), which in turn is a building block for the synthesis of (−)-pyrenophorin (**47**) and its analogues, a kind of naturally occurring antifungal macrodiolide (Scheme 10.4).

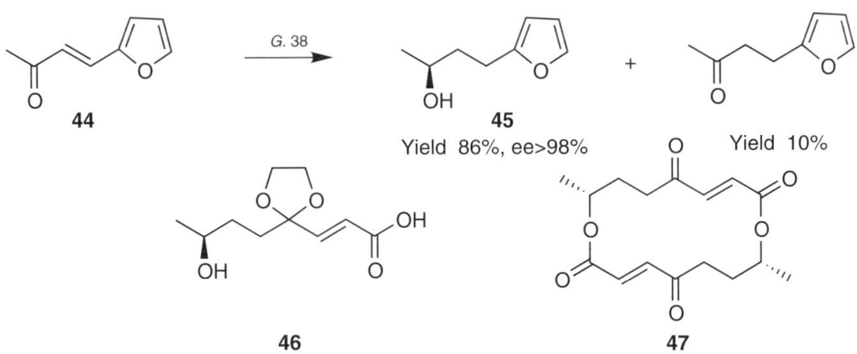

Scheme 10.4

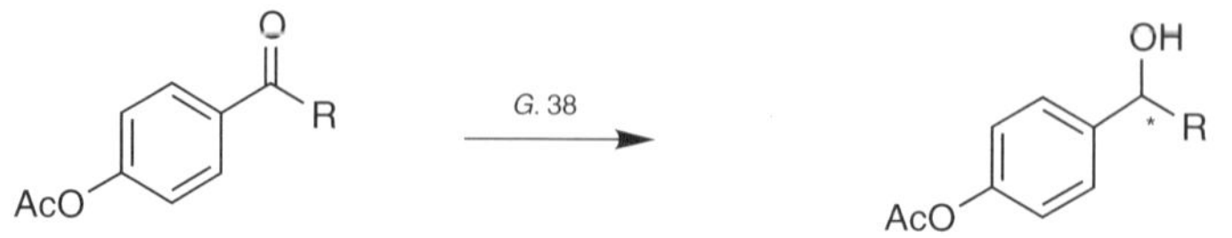

48 R=CH$_2$Cl

49 R=CO$_2$Et

50 R=CH$_2$Cl, (*S*) —alcohol (yield 90%, *ee* 95%)

51 R=CO$_2$Et, (R) — alcohol (yield 93%, *ee* 99%)

Scheme 10.5

Synthesis of (*R*) - and (*S*)-Denopamine (*R*)-(−)-Denopamine is the active form of denopamine, a new selective β_1-agonist that is important for the treatment of congestive heart failure without promotion of increased myocardial oxygen consumption or heart rate (Inamasu et al., 1987). There are few examples where enantioselective synthesis of denopamine has been achieved by using the CBS (boron-oxazolidin reaction) enantioselective catalytic reduction process (Corey and Link, 1991). The demand for finding more-convenient and higher-efficiency methods of synthesis remains. The microbial transformation of ketones with *G*.38 and its application to the syntheses of the key intermediates of (*R*)-denopamine were performed (Gu et al., 1993, 1995). Ketones **48** and **49** were incubated with *G*.38 to produce the corresponding alcohols **50** and **51** with (*S*)- and (*R*)-configurations, respectively (Scheme 10.5). It is necessary for protecting the phenolic hydroxyl group of ketones to obtain a high chemical and optical yield. In the case of unprotected ketones, poor *ee*s (30% for R = CH$_2$Cl, 39% for R = CO$_2$Et) were observed. Coupling of **50** with 3,4-dimethoxyphenyl-ethylamine afforded (*S*)-denopamine (**52**). On the other hand, **51** reacted with 3,4-dimethoxyphenyl-ethylamine and was then reduced by diborane, giving the (*R*)-enantiomer **53** (Scheme 10.6).

Synthesis of (*R*)- and (*S*)-Fluoxetine Fluoxetine is an important antidepressant drug for the treatment of unipolar mental depression. The enantiomers of fluoxetine exhibit their selective 5-hydroxy-tryptamine (5-HT) uptake inhibition with about equal potencies (Piperaki et al., 1995). However, many differences of pharmacological properties, as well as that of NE, have been reported in the literature

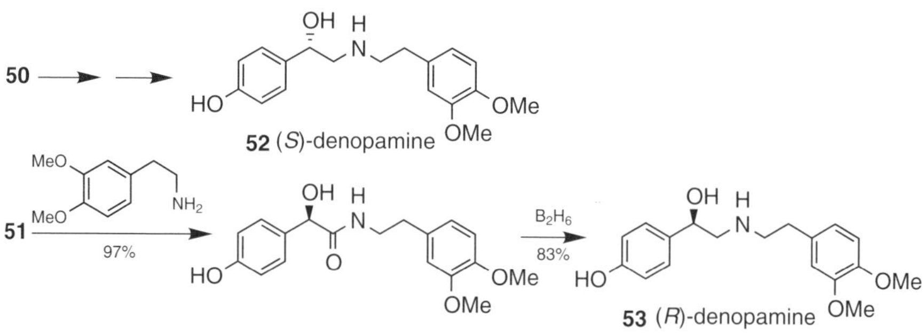

Scheme 10.6

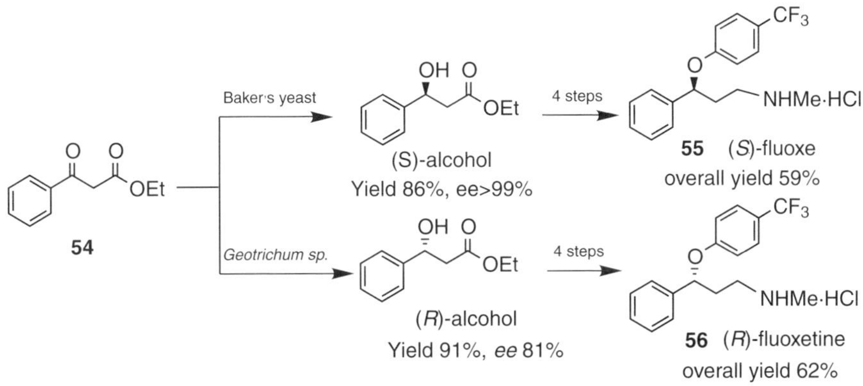

Scheme 10.7

regarding their necessitating the enantioselective synthesis of these compounds. Chemical enantiomeric routes to fluoxetine have been described by Sharpless (Gao and Sharpless, 1988), using selective reduction of 2,3-epoxycinnamoyl alcohol with Red-Al, and by Corey and Reichard, (1989), comprising the CBS enantiomeric catalytic reduction process, respectively. With the aid of microorganisms, we have accomplished the synthesis of both enantiomers of fluoxetine in good yields with high optical purities (Gu et al., 1992). Ethyl benzoyl acetate **54** was fermented with baker's yeast or *Geotrichum* sp. to give (*S*)- and (*R*)-alcohol, respectively, which provided the target compounds (*S*)-fluoxetine (**55**) and (*R*)-fluoxetine (**56**) through four steps of transformation as shown in Scheme 10.7. In the sense of overall yield, these processes provide the highest yield, to the best of our knowledge.

Asymmetric Hydrocyanation Mediated by (R)-Oxynitrilase or (S)-Oxynitrilase The (*R*)-oxynitrilase (HNL, hydroxynitrile lyase, EC.4.1.2.10), which catalyzes the reversible condensation of hydrogen cyanide with aldehydes, is a useful and promising enzyme for biotransformation. The resulting optically active cyanohydrins are expedient starting materials for the preparation of several important classes of compounds such as α-hydroxyketones, α-hydroxyacids, and β-aminoalcohols as well as amino-nitriles and aziridines. It has so far been reported that the (*R*)-oxynitrilase catalyzed reactions were carried out with purified enzymes or crude enzyme preparations from almonds, flax, plums, cherries, apricots, and so on (Lin et al., 1999). In general, micro-aqueous buffer media were chosen as favorable reaction conditions (Scheme 10.8).

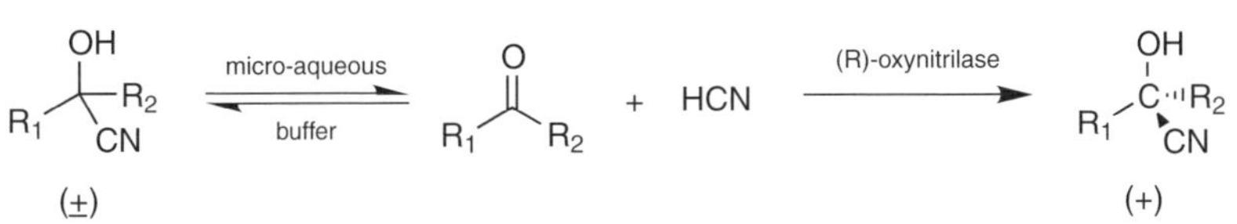

Scheme 10.8

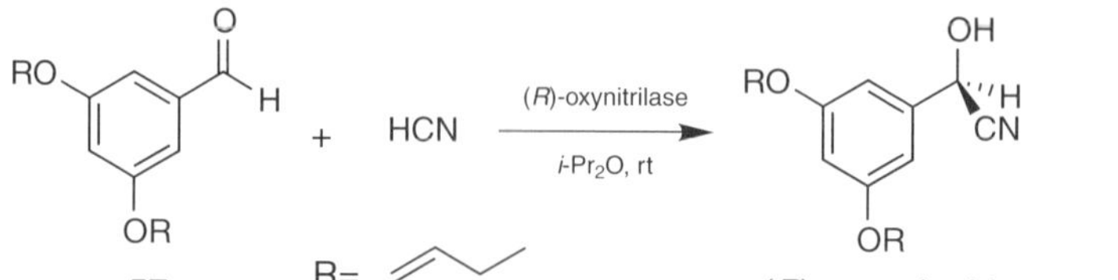

57 (R)-cyanohydrins 58

Scheme 10.9

We (Han et al., 1998) have disclosed that it was not a problem over a wide range of temperatures when the reactions were carried out under the micro-aqueous conditions. In that case the organic phase served as a big reservoir of substrates and products, where the catalytic enzyme power that retained essential water spread homogeneously and acted as a highly effective mode of transfer. Higher yields and enantiomeric selectivities were achieved as the nonenzymatic addition and decomposition were almost suppressed. To avoid using highly toxic potassium or sodium cyanide, acetone cyanohydrin was used as a cyano donor. In the last few years, a representative example of the utility of this reaction in the synthesis of pharmaceuticals was the preparation of adrenergic bronchodilators (*R*)-terbutaline (Effenberger and Jäger, 1997). With regard to the synthesis of (*R*)-terbutaline (**63**), Effenberger and his co-workers found that the protected benzaldehydes (**57**) are the most suitable substrates. The corresponding (*R*)-cyanohydrins (**58**) can be obtained with high conversion and excellent optical yields (>98% *ee*) (Scheme 10.9).

Stereoselective syntheses of (*R*)-terbutaline (**63**) started from the *O*-bisallyl protected cyanohydrin (*R*)-**58** via a Ritter *N*-tertiary butylation to provide the amide (*R*)-**60**. Hydrogenation of (*R*)-**60** gave the amino alcohol (*R*)-**61**. After deprotection of the hydroxyl functions, (*R*)-terbutaline hydrochloride (*R*)-**63**·HCl was obtained in an overall yield of 44% with >98% *ee* (Scheme 10.10).

Asymmetric Oxidations The direct conversion of a carbon–hydrogen bond to a carbon–hydroxyl bond with defined regio- and stereospecificity has given the organic chemist the ability to produce a wide range of compounds with the use of a

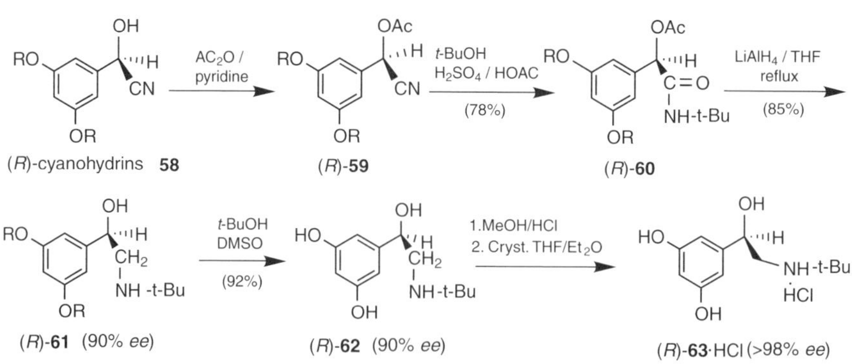

Scheme 10.10

microorganism. The first application of this technique was for the synthesis of the anti-inflammatory corticosteroids and followed the discovery in 1950 by Murray and Peterson that the fungus *Rhizopus arrhizus* (isolated from the air in Kalamazoo, Michigan) was able to convert progesterone (**64**) into its 11α-hydroxy derivative (**65**) in high isolable yield. Since that time, microbial hydroxylation has been extensively applied in the production of steroids on a commercial scale (Holland, 1982).

Microbial hydroxylation of progesterone (**64**) in the culture of *Acremonium strictum* PTCC 5282 produced two hydroxylated pregnene-like steroids (Faramarzi et al., 2003) (**66** and **67**). Steroid 15α-hydroxylation of 13β-ethyl-4-gonene-3,17-dione (**68**) is an industrially important intermediate of the synthesis of Gestoden, a widely used contraceptive drug. Jekkel et al. (1998) reported that **68** could be hydroxylated to produce the 15α-hydroxylated product **69** by the enzyme of *Fusarium nivale* (VJ-63 strain) in a high yield.

Omapatrilat (**70**) is an antihypertensive drug, which acts by inhibiting angiotensin-converting enzyme (ACE) and neutral endopeptidase (Patel, 2001b). A novel L-lysine-ε-aminotransferase has been isolated and then cloned, overexpressed in *Escherichia coli*, and produced on large-scale fermentation by using the selective enrichment technique. The isolated enzyme is capable of oxidizing the ε-amino group of L-lysine, which can be used in the oxidation of the dipeptide **71** to the aldehyde intermediate **72**. Through the conversion of **72** to **73**, a key chiral intermediate for the synthesis of omapatrilat (**70**) was obtained. The aminotransferase reaction required α-ketoglutarate as the amine acceptor. Glutamate formed during this reaction was recycled back to α-ketoglutarate by glutamate oxidase (GOX) from *Streptomyces noursei* SC 6007. The conversion of **71** to **73** proceeded with an overall yield of 65–70% (Zaks, 2001) (Scheme 10.11).

Enantioselective Hydrolysis and Esterification Esterases (EC 3.1.1.x) represent a diverse group of hydrolases catalyzing the cleavage and formation of ester bonds and are widely distributed in animals, plants, and microorganisms. Besides lipases, a considerable number of microbial carboxyl esterases have also been discovered and overexpressed (Bornscheuer, 2002). For example, the *S*-monoester **74** is a key intermediate for the synthesis of β_3-receptor agonists. The enantioselective enzymatic hydrolysis of diester **75** to the desired acid ester **74** by pig liver esterase has been demonstrated. The reaction yields and *ee* of **74** were dependent upon the solvent used. High *ee* values (>91%) were obtained with methanol, ethanol, and toluene as a cosolvent; ethanol gave the highest reaction yield (96.7%) and *ee* (96%) (Patel et al., 1998).

Chiral tertiary benzylic centers are present in the family of aromatase inhibitory drugs such as (*R*)-(+)-amino-glutethimide (**76**), which was originally developed as an anticonvulsant. The first enzymatic attempt to construct an asymmetric quaternary carbon of (*R*)-(+)-**76** has been undertaken by Fadel and Garcia-Argote. (*R*)-(+)- and (*S*)-(−)-aminoglutethimide was synthesized in 97% *ee via* enzymatic hydrolysis (PLE acetonic powder) of malonate derivative (*R*)-(+)-**77** followed by chemical modification (Fadel and Garcia-Argote, 1996) (Scheme 10.12).

Another enantioselective synthesis of (*R*)-(+)-**76** via lipase-catalyzed kinetic resolution of (±)-4-cyano-4-phenyl-1-hexanol (**78**) as appropriate precursors possessing

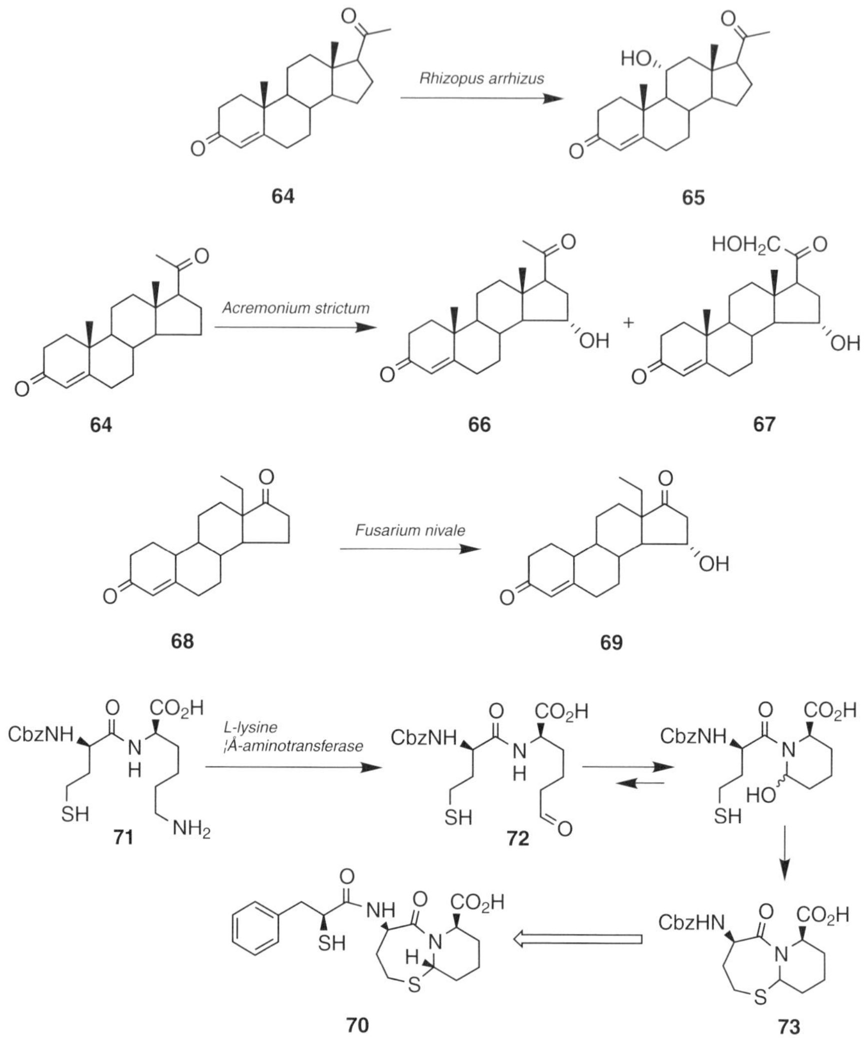

Scheme 10.11

an asymmetric quaternary carbon has been reported. Enzymatic transesterification of primary alcohol (±)-**78** using *Pseudomonas cepacia* (Amano PS, PCL) provided the enantiopure alcohol (*R*)-(−)-**78** with 99% *ee* at conversion of 86%, while that of (±)-**78** using *Pseudomonas fluorescens* (Amano AK, LAK) provided the (*S*)-(+)-**78** with 96% *ee* at conversion of 86%. Chemical transformation of substrate (*R*)-(−)-**78** gave (*R*)-(+)-aminoglutethimide (**76**) in enantioselectively high yield (Im, 2003) (Scheme 10.13).

Two *Nocardioides* strains containing the novel hydrolases C-13 taxolase and C-10 deacetylase, capable of cleaving the corresponding sidechain esters from taxol, have been isolated. Interestingly, in addition to the hydrolysis reaction, the intracellular C-10 deacetylase was found to catalyze the acylation of 10-deacetyl baccatine

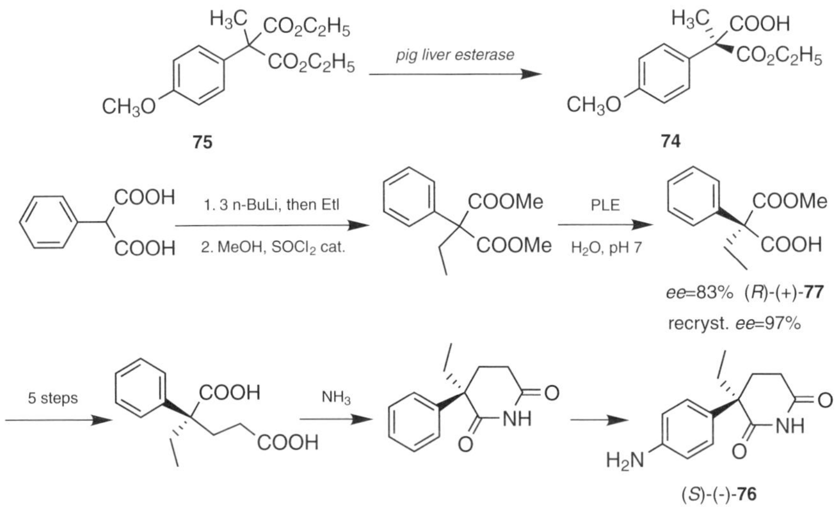

Scheme 10.12

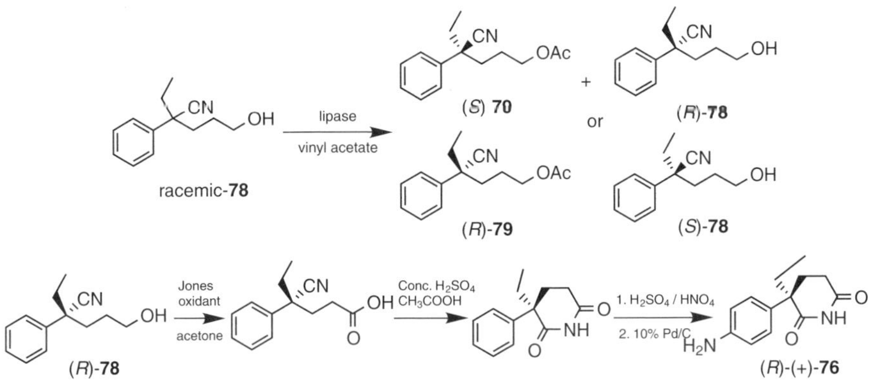

Scheme 10.13

(**80**) to baccatine III (**81**), an intermediate in the semisynthetic synthesis of taxol (**82**). The process proceeded in 51% yield and required no protection of the 7-hydroxyl group because of the high regioselectivity of the hydrolase (Patel et al., 2000) (Scheme 10.14).

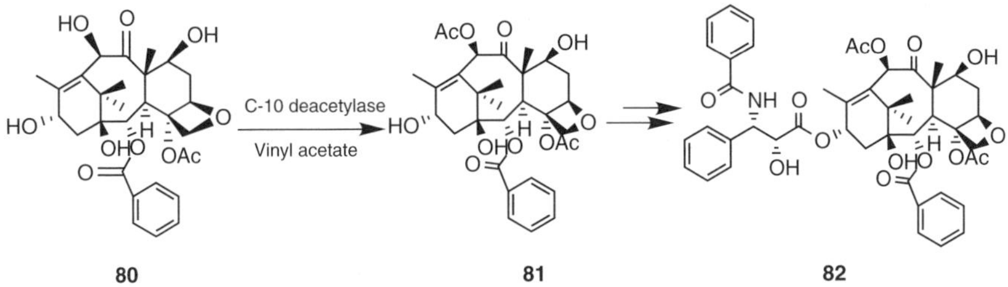

Scheme 10.14

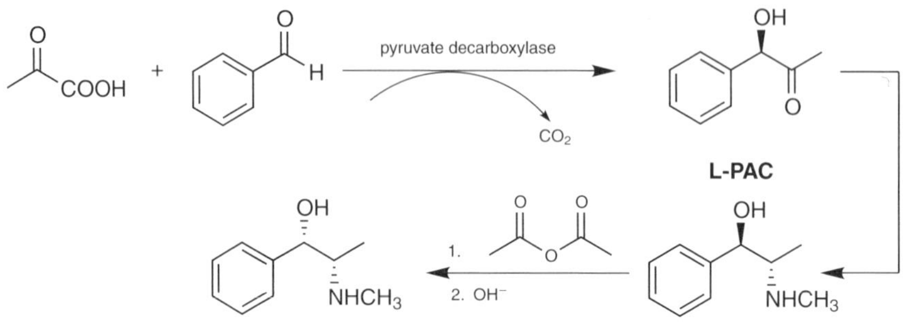

D-pseudoephedrine **9b**

L-ephedrine **8a**

Scheme 10.15

Other Chemo-enzymatic Syntheses The use of different classes of enzymes for the catalysis of many different types of chemical reactions has enabled a wide variety of chiral drugs to be generated. These include numerous hydrolytic enzymes and oxidoreductases. With the development of biocatalysts and fermentation technologies, a variety of enzymes and microbial cultures are regarded as tools for organic synthesis, such as aminotransferases, aldolases, decarboxylases, monooxygenases, and dioxygenases (Patel, 2001a).

L-Ephedrine (**8a**) hydrochloride, used as a decongestant and anti-asthmatic, is obtained from the extracts of a medicinal plant belonging to the genus *Ephedra*. In the pharmaceutical industry, L-phenylacetylcarbinol (L-PAC) is the commercial precursor for the production of L-ephedrine (**8a**) and D-pseudoephedrine (**9b**). L-PAC is produced from benzaldehyde and pyruvate catalyzed by the enzyme pyruvate decarboxylase (PDC), which is known to exist in certain yeast strains, especially the genera *Succharomyces* and *Candida* (Leksawasdi et al., 2005; Tripathi et al., 1997) (Scheme 10.15).

For enzymatic synthesis of eugenyl α-glucoside (**83**) as a promising pro-drug for a hair restorer and a derivative of spices in one step, selective α-glucosylation of eugenol (**84**) was carried out using the α-glucosyl transfer enzyme of *Xanthomonas campestris* WU-9701. The maximum molar conversion yield based on the amount of eugenol supplied reached 52% (Sato, 2003) (Scheme 10.16).

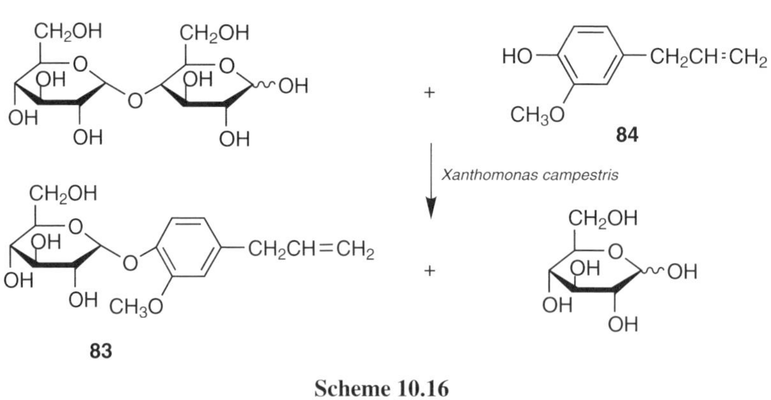

Scheme 10.16

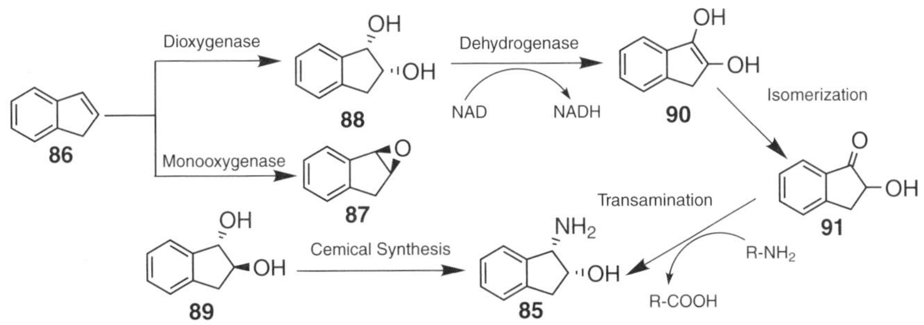

Scheme 10.17

Crixivan (**10**) is an orally active HIV protease inhibitor for the treatment of AIDS. It contains five asymmetric centers allowing 32 possible stereoisomers, only one of which, in enantiomeric purity in the structure of **10**, is the active ingredient. Production of the key intermediate *cis*-amino indanol (**85**) is technically demanding for synthesis of Crixivan (**10**). Its chemical synthesis proceeds through epoxidation of indene to indan oxide at *ee* 87%, followed by two steps to afford **85** in greater than 99% *ee*. Two alternative processes currently exist: biotransformations of indene (**86**) either through the transformation of indandiol (**88**) at >95% *ee* followed by its conversion to **85**, or through from indan oxide (**87**) at high enantio-purity followed by its conversion to **85** (Buckland, 1999). In both, *cis*-(1*S*, 2*R*)-indandiol (**88**) or *trans*-(1*R*, 2*R*)-indandiol (**89**) are potential precursors to **85**. Enrichment and isolation of microbial cultures yielded two *Rhodococcus* species—strains B 264-1 (MB 5655) and I-24 (MA 7205)—enabling the biotransformation of indene (**86**) to *cis*-(1*S*, 2*R*) indandiol (**88**) and *trans*-(1*R*, 2*R*) indandiol (**89**), respectively. Isolated MB 5655 was found to have a toluene dioxygenase, whereas MA 7205 was found to harbor toluene and naphthalene dioxygenases as well as a naphthalene monooxygenase, which catalyzes the above biotransformation (Patel et al., 2000) (Scheme 10.17).

10.3.3 Chemical Control

10.3.3.1 Separation of Enantiomers of Racemate of Chiral Drugs

If the chiral pool is not, or cannot, be employed, resolution is thought to be the most frequently used alternative industrial method for obtaining single enantiomers, although it is so far difficult to estimate exactly what the balance is between the use of chiral pool, asymmetric synthesis, and resolution (Crosby, 1997).

Despite the disadvantage that over half the amount of the racemate has to be discarded or transformed to a racemate again for reuse via chemical sequences, our rough impression based on current business trends is that some 30–50% of the single enantiomers required are obtained by classical resolution procedures (Rouhi, 2004). Although a number of stereoselective syntheses have been described and applied to the production of single enantiomeric substances, relatively few are selected for large-scale preparations, particularly at the early stages of development of new drugs. Obviously, it is urgent for researchers to provide some amounts of pure enantiomers for primary pharmacological tests before a manufacturing route (protocol) can be

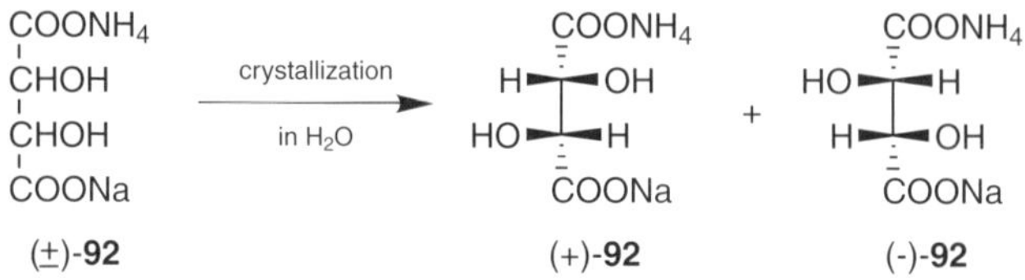

Scheme 10.18

selected. At these early stages, the development of an asymmetric synthesis would be both expensive and time-consuming, and thus techniques as simple as possible for the separation of enantiomers have an interesting potential (Maier et al., 2001).

Preferential Crystallization Spontaneous resolution is a phenomenon in which a particular racemate (racemic mixture or conglomerate) undergoes separate crystallization of both of (+) and (−)-enantiomers (antipodes) from solution. This was discovered by Pasteur in 1848 in the aqueous crystallization of ammonium sodium (±)-tartarate (**92**). As shown in Scheme 10.18, 50% each of (+)-**92** and (−)-**92** enantiomers are theoretically formed (Matsuura and Koshima, 2005).

Although the conglomerate formation is valuable for optical resolution, the frequency of the conglomerate occurrence has been estimated to be in the range of only 5–10% of all racemates. In order to find the controlling factors of the conglomerate crystal formation, extensive research has been carried out. The molten state of 1,1-binaphthyl and the mechanically stirred crystallization of $NaClO_3$ from its aqueous solution are two typical chiral breaking examples of crystallization by "external influences or measures" to affect the resolution of racemates (Kitoha et al., 2005). In another example, the spontaneous resolution of racemic **93** with $Cd(ClO_4)_2 \cdot 6H_2O$ in the presence of 2-butanol under solvothermal conditions favors the formation of crystal **94**, while a similar reaction in the presence of ethanol only favors the formation of crystal **95** (Scheme 10.19). The crystal structural determination shows that both **94** and **95** crystallize in chiral enantiomorphous space groups (Ye et al., 2005).

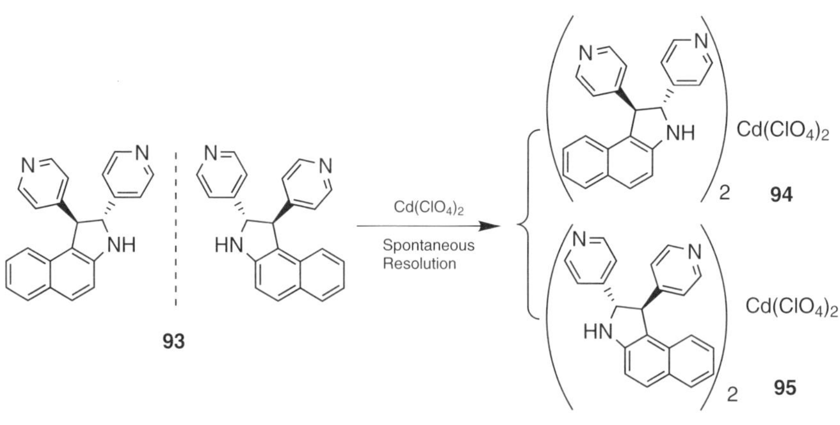

Scheme 10.19

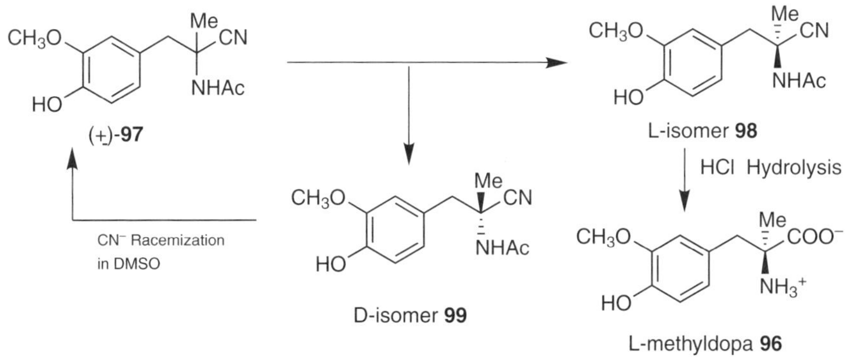

Scheme 10.20

Because of practical difficulties in discriminating one enantiomeric crystal from another by the visual observation of their crystal morphology, the so-called preferential crystallization is carried out. The preferential crystallization is one of the practical techniques for obtaining enantiomerically pure compounds, where the seeding of a small amount of the enantiomer in its supersaturated solution can preferentially crystallize one of the enantiomers. Resolution via preferential crystallization is known to be practical and cost-effective in that it requires no chiral reagent or enzyme and can be applied to industrial-scale preparations (Jacques et al., 1981). The Merck Company has a long history in the preparation of L-α-methyldopa (**96**) via a continuous fluidized-bed crystallization resolution-racemization process. The intermediate (±)-**97** for the synthesis of L-α-methyldopa possesses the properties of a conglomerate, which can be separated by preferential crystallization. The resulting L-isomer (**98**) was hydrolyzed to L-α-methyldopa with concentrated HCl. The undesired D-isomer (**99**) was racemized with NaCN in DMSO and returned for the resolution (Scheme 10.20). Using this process, Merck produced millions of kilograms of L-α-methyldopa over the lifetime of the product (Grabowski, 2005).

The antibiotic chloramphenicol has been used widely for the treatment of numerous microbial infections for nearly 50 years. Only the D-(−)-threo-isomer (**100**) of four chloramphenicol stereoisomers has antibacterial activity. The stereochemistry of chloramphenicol is another example of industrial-scale preparation via preferential crystallization. The threo-racemate of a synthetic intermediate is separated from the corresponding erythro-racemate at some stage in the synthetic sequence, and the desired D-(−)-threo isomer chloramphenicol is then obtained via resolution of the fully elaborated bases **101** and **102** into its enantiomers using the D-threo base, **101** as crystal seed, followed by introduction of the dichloroacetyl group (Liu and Fang, 1997).

Diastereomer Crystallization Preferential crystallization is a cheap and efficient resolution process, but the major limitation is that the racemic mixture must crystallize as a conglomerate, and the resolution results are composed of simultaneous nucleation and growth steps (Beilles et al., 2001). However, approximately 10% of all compounds that exist as racemates can be classified as

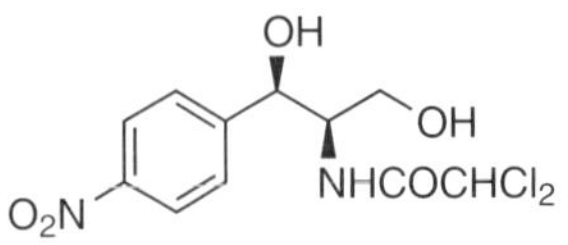

100

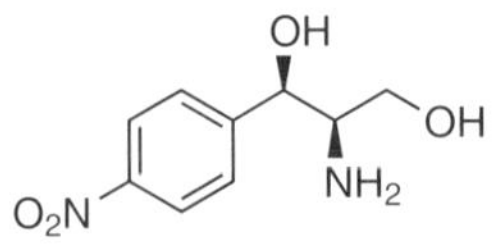

101

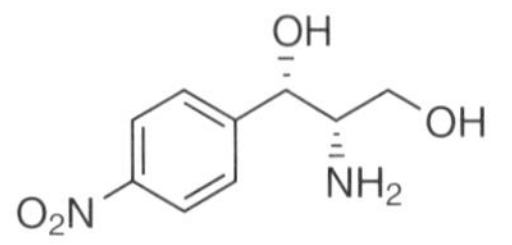

102

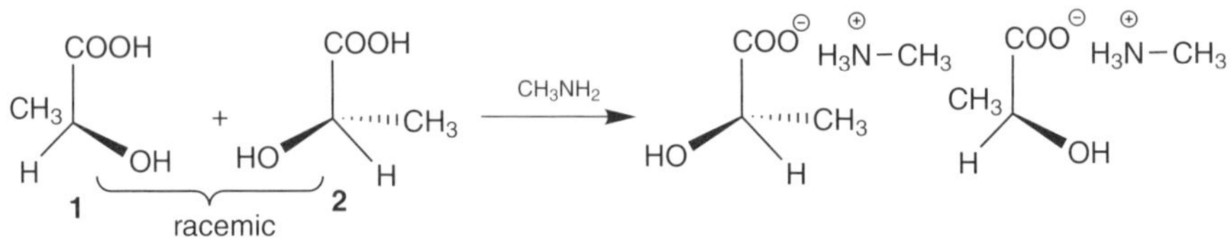

Scheme 10.21

conglomerate while the rest of them, ~90%, cannot be separated by preferential crystallization. However, these racemic compounds can be separated by diastereomer crystallization. In this method, a target racemate is mixed with a chiral reagent (known as the resolving agent) to give a mixture of diastereomers, which is repetitively recrystallized from an appropriate solvent to give the enantiomerically pure diastereomer. After removal of the resolving agent, the target enantiopure isomer can be finally isolated. Most commonly, the racemic mixture is a carboxylic acid, and the chiral reagent is an amine (or vice versa, a racemic mixture of an amine and a chiral carboxylic acid). An example of this process is illustrated in the resolution of (+)-lactic acid using ethylamine (Scheme 10.21).

The first amino acids were resolved by Emil Fischer in 1899 using the naturally occurring alkaloids strychnine (**103**) and brucine (**104**) to separate the racemic *N*-benzoyl derivatives of alanine, glutamic acid, and aspartic acid (Brittain, 1990). Although quite a lot of racemic compounds have been separated by diastereomer crystallization, it is still difficult to predict appropriate conditions for the successful resolution of a given target racemate. The practical and economical factors associated with the resolution of compounds by diastereomer crystallization must be considered, such as the choice of resolving agent, solvent composition, racemization of the unwanted isomer, and recovery of the chiral auxiliary (Challener, 2001b).

Resolving Agents Although the process of crystallization is affected by numerous factors, such as cooling rate, stirring, trace contaminants, and solvent, the crystal structures of substances obtained during optical-resolution advancements in the use of computer-aided molecular modeling have provided some assistance to professionals in the rational design and selection of resolving agents (Kinbara, 2005). The criteria for good resolving agents are as follows: (a) Diastereomeric compounds crystallize well; (b) formation of diastereomeric salts occurs readily; (c) diastereomeric salts can be separated easily; (d) inexpensive; (e) available in optically active pure form; (f) chiral center is as close as possible to the functional group; (g) responsible for salt formation; (h) available as both enantiomers; (i) have tight rigid structures; (j) strong acids/base give best results; (k) molecular weight as low as possible; and (l) easily

TABLE 10.6 Drugs Prepared via Classical Resolution

Chiral Drugs	Resolving Agent	Activity
Ampicinin	D-Camphosulfonic acid	Antibiotic
Ethambutol	L-(+)-Tartartic acid	Tuberculostatic
Chloramphenicol	D-Camphosulfonic acid	Antibiotic
Propoxyphene	D-Camphosulfonic acid	Analgesics
Dexbrompheniramine	D-Phenylsuccinic acid	Antihistamine
Fosfomycin	R-(+)-Phenethylamine	Antibiotic
Thiamphenicol	D-(−)-Tartartic acid	Anti-infective
Naproxen	Cinchonidine	Anti-inflammatory
Diltiazem	R-(+)-Phenethylamine	Calcium antagonist

Source: Bayley and Vaidya (1992).

recovered (Sheldon, 1993). Many types of pharmaceuticals have been prepared utilizing resolving agents by formation of diastereomeric salts. A list of a few selected drugs and the associated resolving agents is shown in Table 10.6.

The chiral pool is one key source for resolving agents. There are also several commercially available synthetic resolving agents. Figures 10.8 and 10.9 demonstrate examples of common chiral acids (Sakai et al., 2003) and bases used as agents.

Designed New Resolving Agents Most of the popular basic resolving agents such as brucine, ephedrine, and morphine are naturally originating, highly toxic amines. Furthermore, only one enantiomer of each is commonly available.

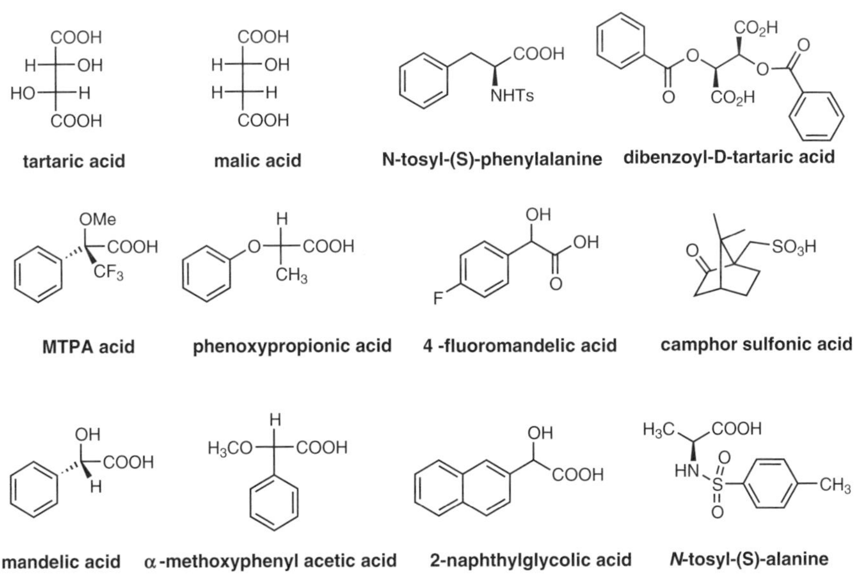

Figure 10.8 Examples of common chiral acids used as agents.

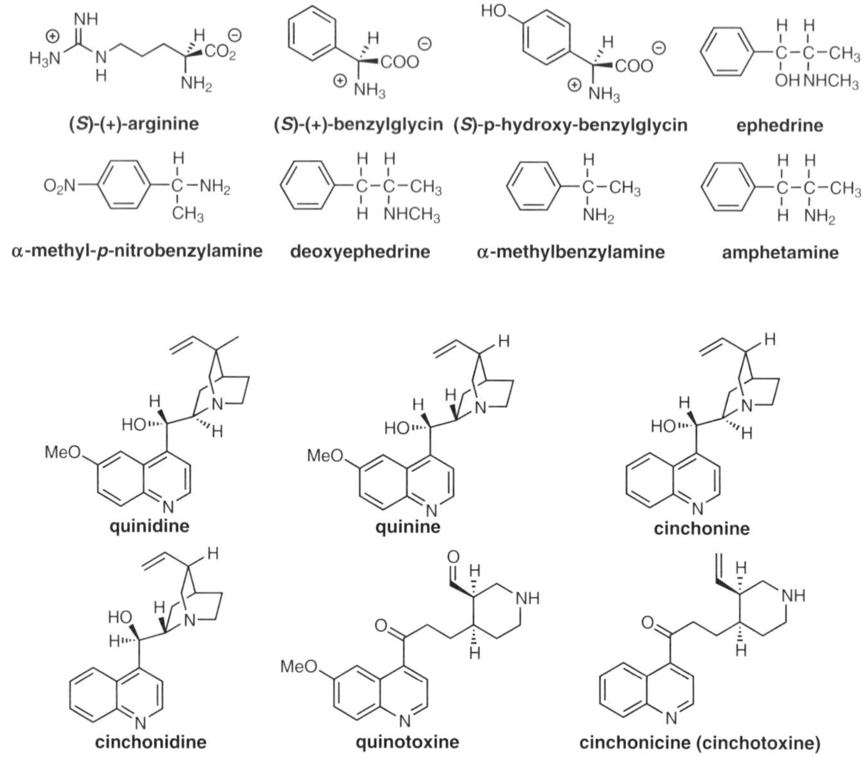

Figure 10.9 Examples of common chiral bases used as agents.

Therefore, the development of novel basic resolving agents, both of whose enantiomeric forms are readily accessible, is a subject of practical importance. A new chiral acid, the hydrogen phthalate of isopropylidene glycerol (**105**), has been shown to resolve a wide range of 1-arylalkylamines with high efficiency and with the same stereochemical outcome (Pallavicini et al., 2000).

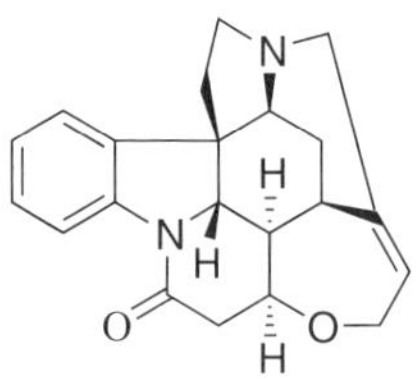
103

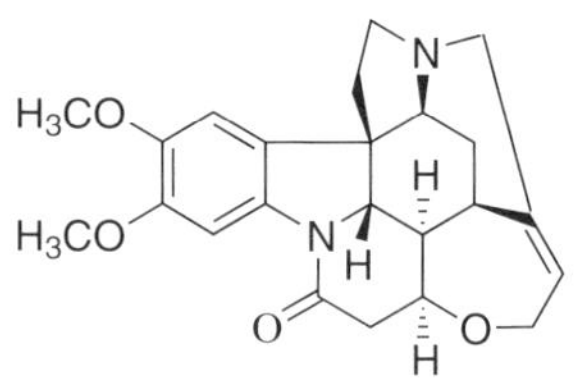
104

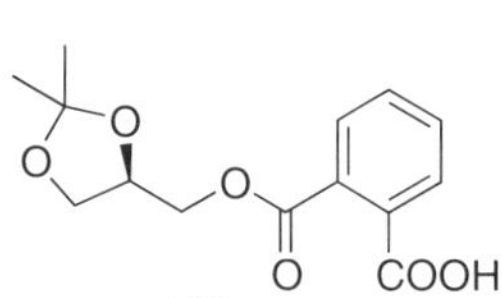
105

Systematic studies on the resolution of 2-arylalkanoic acids (**106**) by enantiopure amino alcohols, expecting that the hydroxy group of the amino alcohol would play the same role as that of 1-arylglycolic acids in the formation of a hydrogen-bond sheet in the less-soluble diastereomeric salts, have been done by the group of Kinbara et al. They found that (1*S*, 3*S*)-**109** is directly hydrogen-bonded to the adjacent hydrogen-bond column to form the supramolecular hydrogen-bond sheet. In addition, water molecules assisted stabilization of the supramolecular sheet externally. Hence, the skeleton of (1*S*, 3*S*)-**109** is the most preferable for formation of a supramolecular sheet, and further design of basic resolving agents should be possible based on this skeleton (Brittain, 1990).

Diastereomer Interconversion A perceptible fact is that crystallization of diastereomeric salts can occur in conjunction with racemization of the unwanted isomer. Asymmetric transformation of diastereomeric salts, also known as diastereomer interconversion, can yield a theoretical 100% of one isomer. 2-Acyl-3, 6-dihydroxycyclo-hex-2-enones are isolated from the insects that exhibit kairomonal activity. The optically pure isomers (−)-(*S*)- and (+)-(*R*)-2-acetyl-3,6-dihydroxy cyclohex-2-enone (**110**) can be obtained through an enantioconvergent synthesis, starting from the racemic α-ketol (**111**) via diastereomeric *N*-tosyl-(*S*)-proline esters (**112**) by a repeated 4-dimethylaminopyridine (DMAP)-catalyzed equilibration and crystallization of the undesired diastereomer. At last, the enantiomeric **110** can be given by hydrolysis of the diastereomerically pure ester **112** (Zaitsev and Mikhal'chuk, 2001) (as shown in Scheme 10.22).

Kinetic Resolution—Chemical Kinetic resolution can be achieved with chemical or enzymatic methods to promote selective reaction of one enantiomer over the other, giving a mixture of enantio-enriched starting material or product. The desired component is then isolated. Kinetic resolutions with enzymatic method have been described in Section 10.3.2.1. An interesting example of a parallel kinetic resolution involving asymmetric ketone reduction was reported by Kishi in the context of batrachotoxin synthesis. Batrachotoxins (**113**) are extremely potent neurotoxins that act as selective and irreversible Na^+-channel activators (Kurosu and Kishi, 1998).

As shown in Scheme 10.23, essentially perfect selectivity was achieved in the reduction of model steroidal ketones using the commercial CBS catalyst, and the epimeric alcohol products were separated as the corresponding acetates by

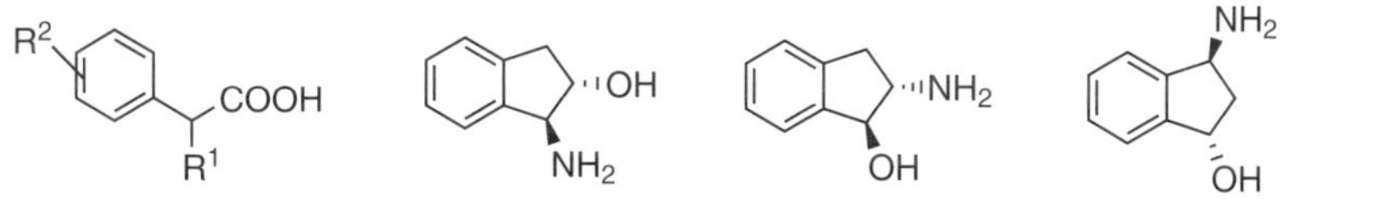

106 **(1S, 2S)-107** **(1S, 2S)-108** **(1S, 3S)-109**

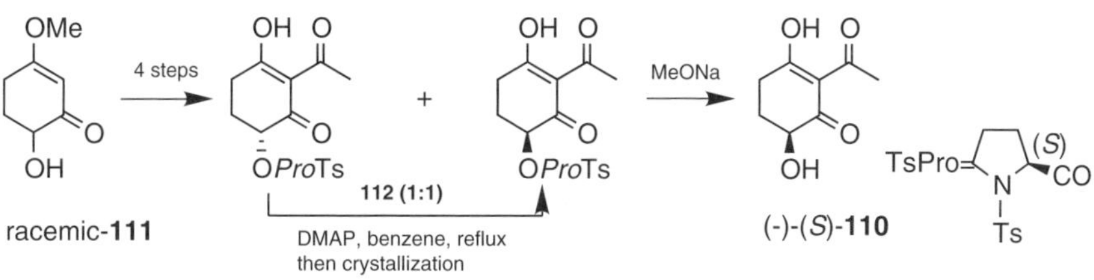

Scheme 10.22

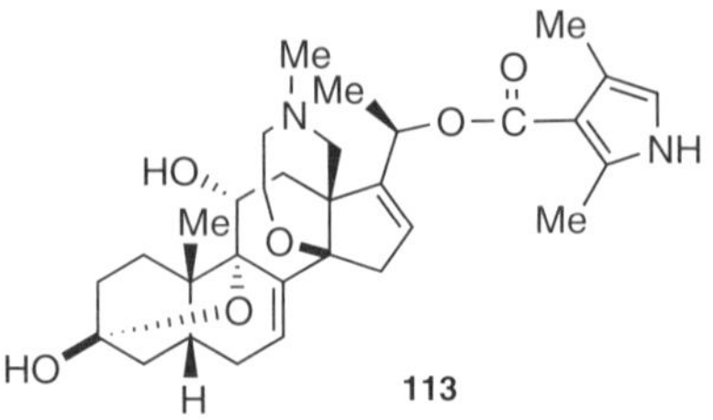

113

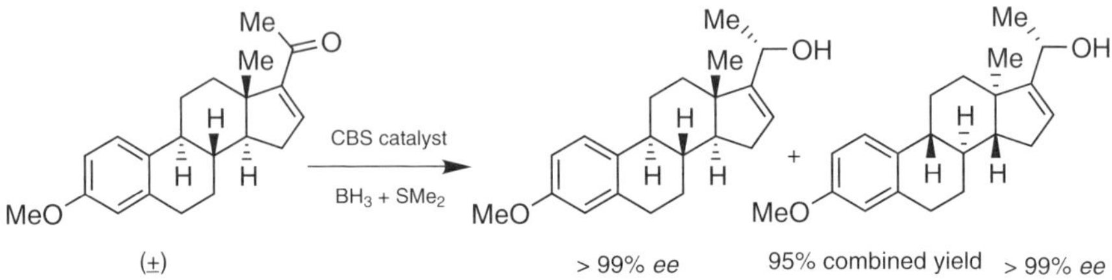

Scheme 10.23

chromatography. Limiting the reaction to 60% conversion led to kinetic resolution with recovered ketone in 40% isolated yield and 93% *ee* ($k_{rel} = 27$).

Sulfoxides possess the same stereochemical behavior as other chiral compounds that exist between the substituents of the stereogenic sulfur atom: a lone electron pair, an oxygen, and two different carbon ligands, which are able to differentiate the diastereotopic faces of a proximal or even remote reaction center (Carreno, 1995). Many chiral sulfoxides exhibit interesting biological activities, such as omeprazole (**114**), used as protonpump inhibitors to treat acid-related diseases, and OPC-29030 (**115**), used as a platelet adhesion inhibitor (Jia et al., 2004). Through kinetic resolution or, perhaps better, enantioselective sulfide oxidation coupled to kinetic resolution, enantiopure sulfoxides can be obtained in high *ee*.

The kinetic resolution of racemic sulfoxides alone with the Ti (IV)/BINOL/H_2O system has been shown to be an effective method for the preparation of enantiomerically pure sulfoxides, albeit with moderate isolated yields in most cases. The most promising system developed thus far is one in which alkyl aryl sulfides undergo enantioselective oxidation with aqueous TBHP as a stoichiometric oxidant in the presence of the $Ti(OiPr)_4$/BINOL catalyst. The same catalyst system can be applied to the kinetic resolution of racemic substrates as well as to provide enantiopure sulfoxides in moderate yield (Scheme 10.24) (Komatsu et al. 1992, 1993).

10.3.3.2 Asymmetric Synthesis and Its Application

In addition to chiral intermediates, enantioslective technologies (i.e., asymmetric synthesis) will also be on display at CPhI where the pharmaceutical industries have done a close collaboration with academicians. Innovative drug companies have always faced competition from one another as each strives to bring out the next member of a certain class of compounds for treatment of a disease. Asymmetric reactions have been applied

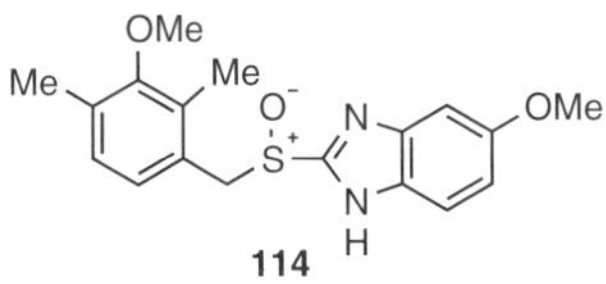

114

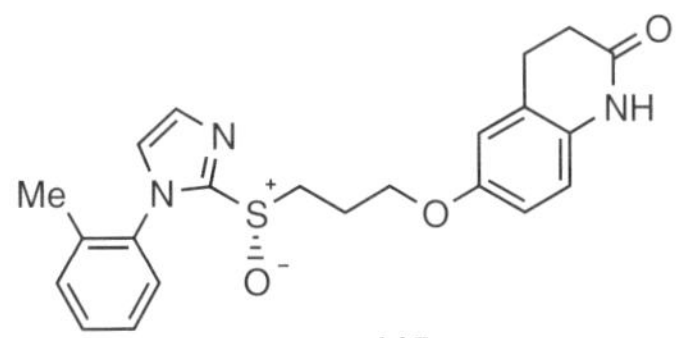

115

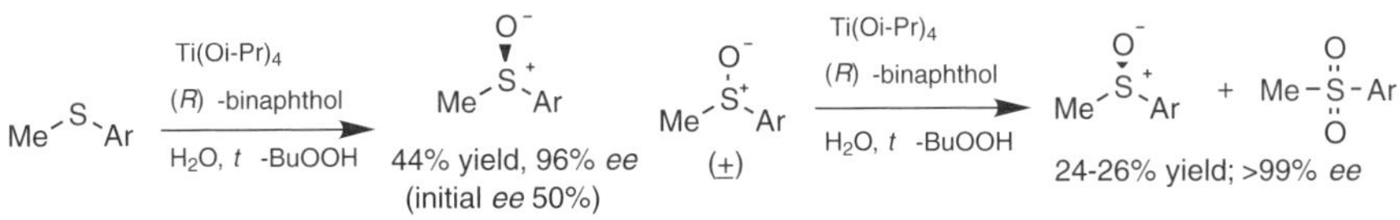

Scheme 10.24

in several industrial processes, such as the asymmetric synthesis of L-DOPA (**116**), a drug for the treatment of Parkinson's disease, via Rh(DIPAMP)-catalyzed hydrogenation of the enamide (Knowles, 1986), as shown in Scheme 10.25.

The industrial synthesis of (−)-menthol (**117**) and (+)-citronellal (**118**) through asymmetric isomerization of an allylic amine and asymmetric hydrogenation reaction (Akutagawa, 1997) are shown in Scheme 10.26.

Currently, the side chain (**119**) of efficient semisynthesis of taxol and decetaxol (Deng and Jacobsen, 1992) can also be synthesized via a highly enantioselective epoxidation reaction catalyzed by the readily available (salen)Mn(III) complex (**120**), as shown in Scheme 10.27.

Asymmetric Hydrogenation Reaction Hydrogen is the simplest molecule and is fully available in abundance at very low cost; therefore, asymmetric hydrogenation is a core technology in research and industry. Since Kang's discovery of a practical synthesis of an air-stable ferrocenyl bis(phosphine) and its application in the rhodium(I)-catalyzed enantioselective hydrogenation of dehydroamino acid derivatives, several ferrocenyl phosphines have found industrial applications in the synthesis of chiral

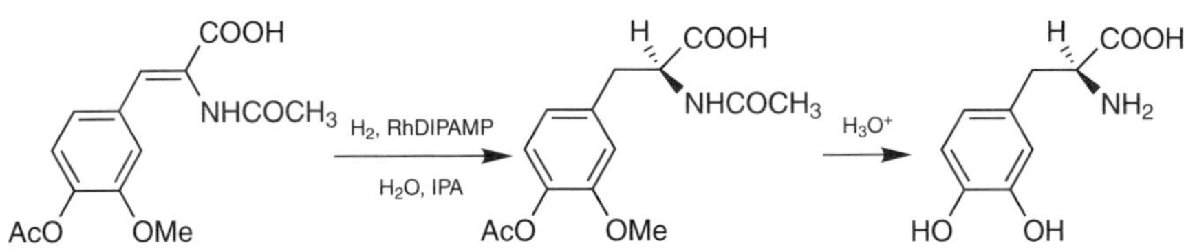

116

Scheme 10.25

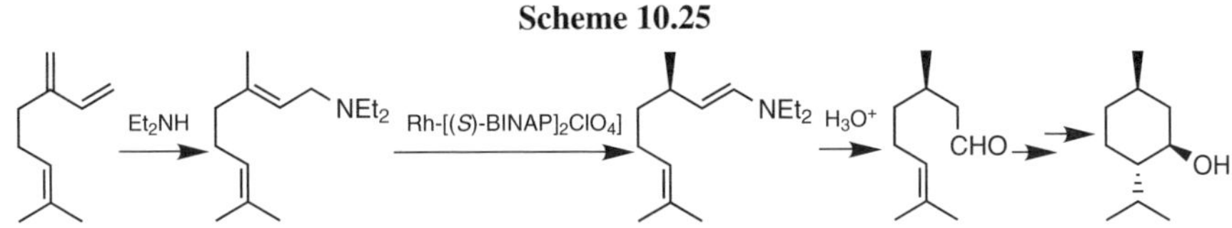

99% *ee* 118 117

Scheme 10.26

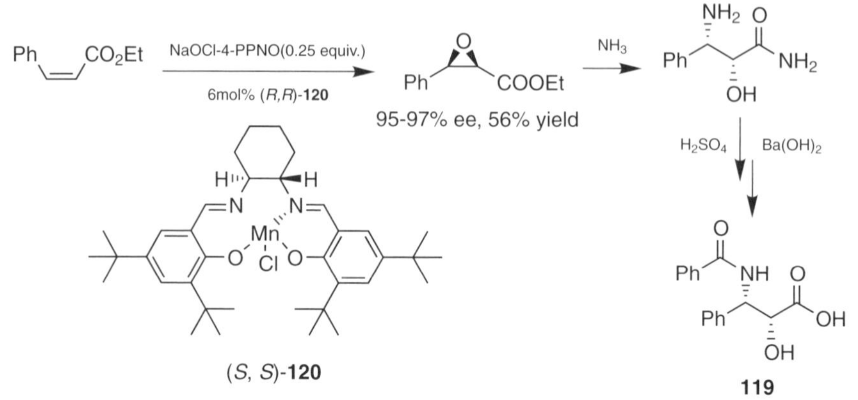

Scheme 10.27

pharmaceuticals and agrochemicals. A process has been developed by Ciba-Geigy (now Novartis) for the production of an herbicide, (*S*)-metolachlor (**121**), which has been synthesized by chloroacetylation of the corresponding amine obtained by the Ir-catalyzed enantioselective hydrogenation of the requisite imine (Blaser and Spindler, 1997), as shown in Scheme 10.28. Similar technologies involving an Rh catalyst-containing ligand (**123**) were developed by Lonza Fine Chemicals in partnership with Ciba-Geigy for the production of the vitamin (+)-biotin (**124**) (Imwinkelried, 1997) (Scheme 10.29).

The antibacterial agent levofloxacin (**125**) can be prepared by utilizing catalytic asymmetric hydrogenation of the cyclic imine (**126**) with the iridium (I) complex of (2*S*, 4*S*)-BPPM in the presence of bismuth (III) iodide to get the key intermediate (**127**), which can be easily converted to levofloxacin (**125**) in six steps, with high enantiometric purity (90% *ee*) and a chemical yield of 96% (Satoh, 1998) (Scheme 10.30).

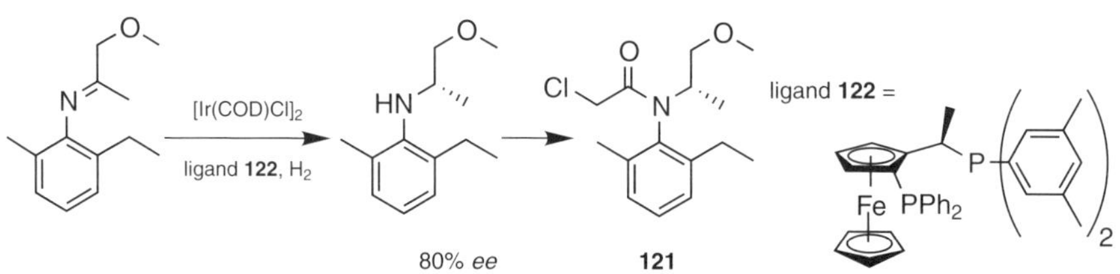

Scheme 10.28

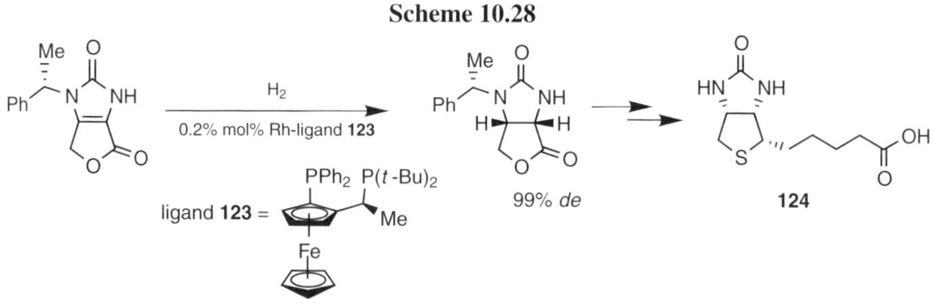

Scheme 10.29

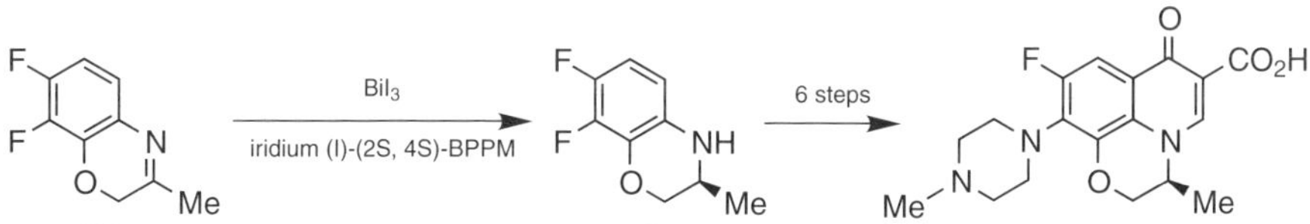

126 127 125

Scheme 10.30

Asymmetric Hydrogenation of Ketones Asymmetric hydrogenation of ketones is one of the most efficient methods for making chiral alcohols. This methodology has been used in the industrial production of synthetic intermediates for some important antibiotics: carbapenem (**128**) and sanfetrinem (**129**). Ru-BINAP catalysts are highly effective in the asymmetric hydrogenation of functionalized ketones. The Ru-BINAP complexes effect highly enantioselective hydrogenation of various β-keto esters to yield chiral β-hydroxy esters. By combining this asymmetric reaction with dynamic kinetic resolution, an important chiral intermediate for the synthesis of carbapenem antibiotics (Noyori and Ohkuma, 2001) was synthesized, as shown in Scheme 10.31. In the case of the preparation of sanfetrinem (**129**), the key step is the asymmetric hydrogenation of racemic 2-methoxycyclohexanone (**130**) with an Ru-(*S*) 3,5-xylyl-BINAP-(*S*, *S*)-DPEN-KOH catalyst leading to (1*R*, 2*S*)-2-methoxycyclohexanol (**131**) as the nearly sole stereoisomer (Matsumoto et al., 1999), as shown in Scheme 10.32.

Asymmetric Cyclopropanation Reaction The chiral cyclopropyl group is one of the most important structural motifs responsible for biological activity of molecules, such as curacin A (**132**) and plakoside A (**133**). Curacin A (**132**) is a novel antimitotic agent recently isolated from a Caribbean cyanobacterium, *Lyngbya majuscula* (Onoda et al., 1995). Therefore, asymmetric cyclopropanation has been used widely in synthetic organic chemistry. Plakoside A (**133**) is a prenylated galactosphingolipid isolated from the marine sponge *Plakortis simplex* and exhibits strongly immunosuppressive activity with rather low cytotoxicity (Seki et al., 2001). Asymmetric synthesis of the cyclopropane moiety of curacin A (**132**) is shown in Scheme 10.33. Double-asymmetric Simmons-Smith cyclopropanation of the diene

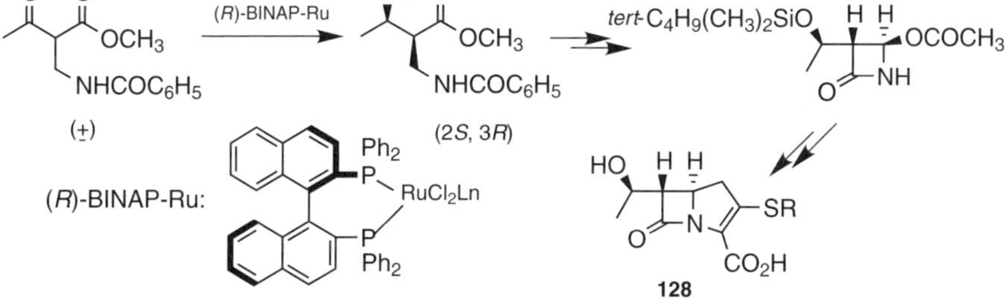

Scheme 10.31

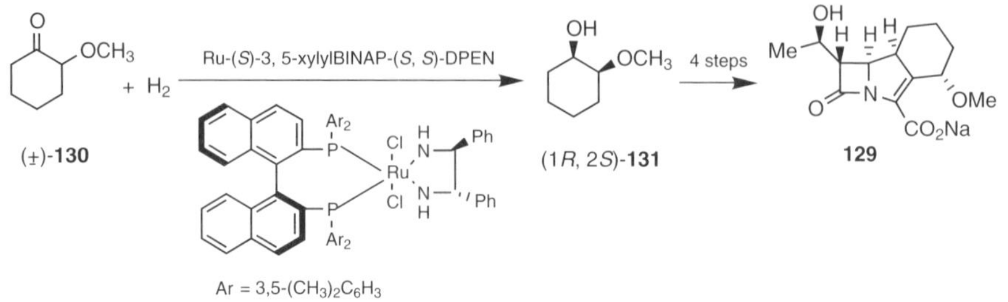

Scheme 10.32

(**134**) derived from diethyl L-tartrate in four steps proceeded with excellent diasterofacial selectivity (>99% *de*) to give the dicyclopropane (**135**), which was converted to the desired carboxylic acid (−)-**136** in three steps (Onoda et al., 1996).

Cilastatin (**137**), an excellent inhibitor of dehydropeptidase-I, which increases the *in vivo* stability of imipenem, has desired pharmacological properties (Graham et al., 1987). The (S)-(+)-2,2-dimethylcyclopropane carboxylic acid (+)-**138**, a key building block for the synthesis of cilastatin (**137**), was prepared from 2-methylpropene and chiral iron carbene complex via asymmetric cyclopropanation reaction and exhaustive ozonolysis with up to 92% *ee* (Wang et al. 1998) (Scheme 10.34).

Asymmetric Hydroformylation Reaction 1β-Methylcarbapenem (**139**) is an unnatural antibiotic that possesses an excellent antibacterial profile as well as enhanced chemical and metabolic stability. Monocyclic β-lactam (**140**), a valuable intermediate in the preparation of **139**, was synthesized by the asymmetric hydroformylation of the 4-vinyl β-lactam (**141**) catalyzed by rhodium aminophosphonite-phosphinite (**143**) and rhodium aminophosphine-phosphite (**144**) complexes. The stereoselectivity to the desired β-isomer was related to the presence of a substituent at the N atom of the β-lactam ring. The regioselectivity (branched/

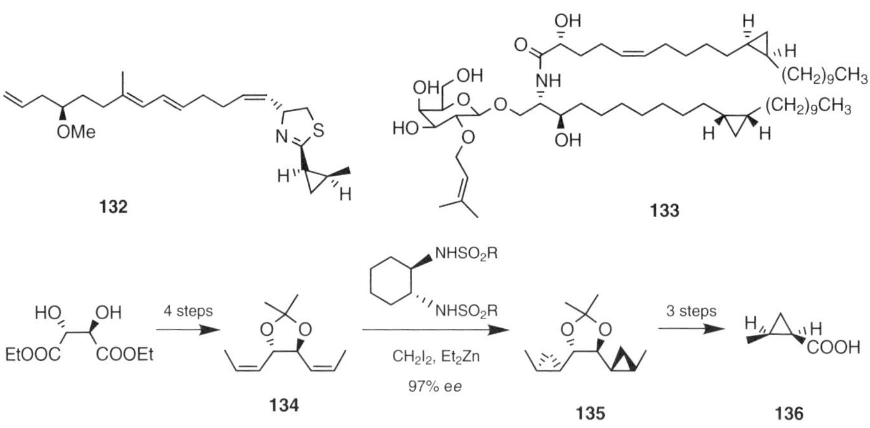

Scheme 10.33

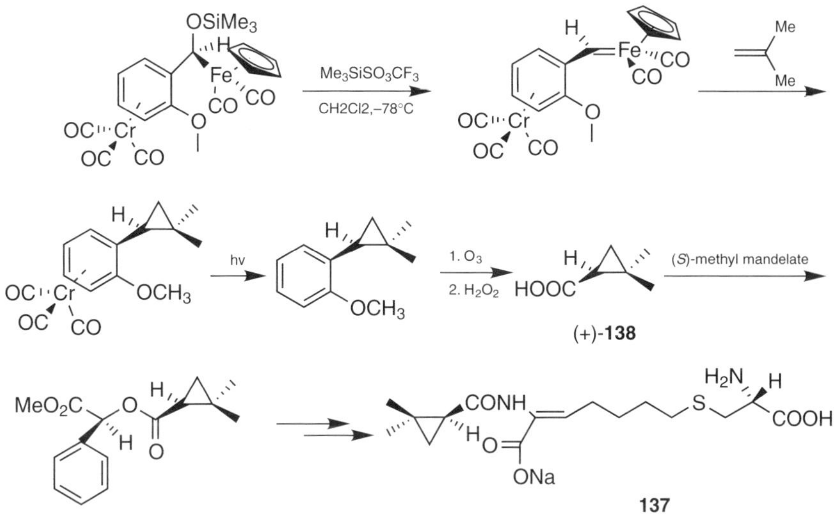

Scheme 10.34

linear) but not the stereoselectivity (β/α) was found to be dependent on the ratio of substrate to catalyst (Cesarotti and Rimoldi, 2004) (Scheme 10.35).

Asymmetric Dihydroxylation Reaction Since Sharpless's discovery of the asymmetric dihydroxylation reaction of alkenes mediated by osmium tetroxide-cinchona alkaloid complexes, continuous efforts have been made to utilize this reaction in preparation of chiral drugs or intermediates. For example, the

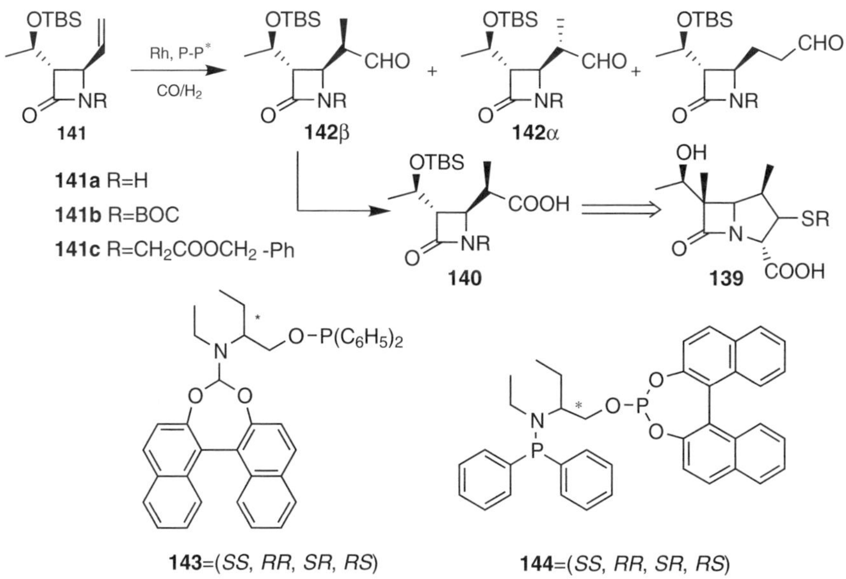

Scheme 10.35

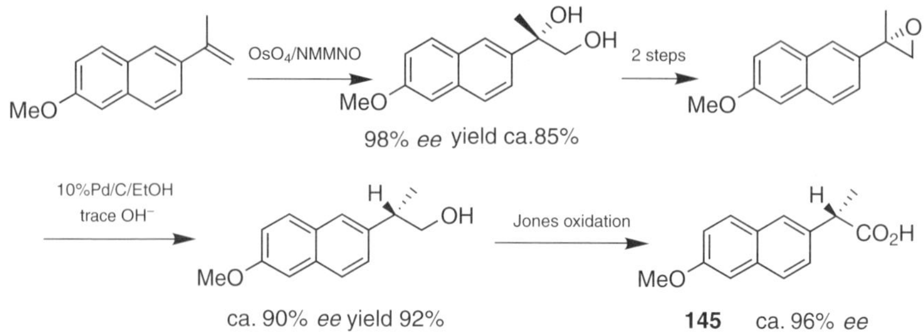

Scheme 10.36

non-steroidal anti-inflammatory drug (NSAID) naproxen (**145**) has been synthesized in high enantiomeric excess by Sharpless asymmetric dihydroxylation of the appropriate methyl styrenes (Griesbach et al., 1997) (Scheme 10.36).

Asymmetric Aminohydroxylation Reaction Loracarbef (**146**) is a carbacephalosporin antibiotic with extended chemical and serum stability, which is especially used in the treatment of pediatric ear infections. A formal synthesis of loracarbef was accomplished with high stereoselectivity. Stereoselective construction of the key intermediate *cis*-3,4-disubstituted azetidinone (**147**) was performed by employing intramolecular cyclization of *cis*-substituted azetidinone skeleton (**148**), which was obtained from Sharpless asymmetric aminohydroxylation of α,β-unsaturated ester **149** in good yield with high regioselectivity (>13:1) and enantioselectivity (89% *ee*) (Lee et al., 2001) (Scheme 10.37).

Asymmetric Sulfoxidation Reaction Sulfide (**150**) is a potent systematically available ACAT inhibitor, which is rapidly metabolized into the corresponding sulfoxide. Of the two-sulfoxide enantiomers, only one—**RP 73163**—exhibits useful biological activity. The asymmetric synthesis of sulfoxide RP 73163 is based on the enantioselective oxidation of a suitably designed prochiral methyl sulfide, followed by α alkylation of the resulting sulfoxide (Scheme 10.38).

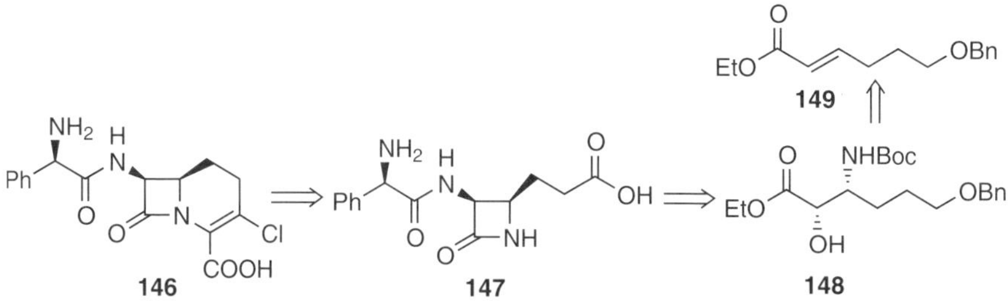

Scheme 10.37

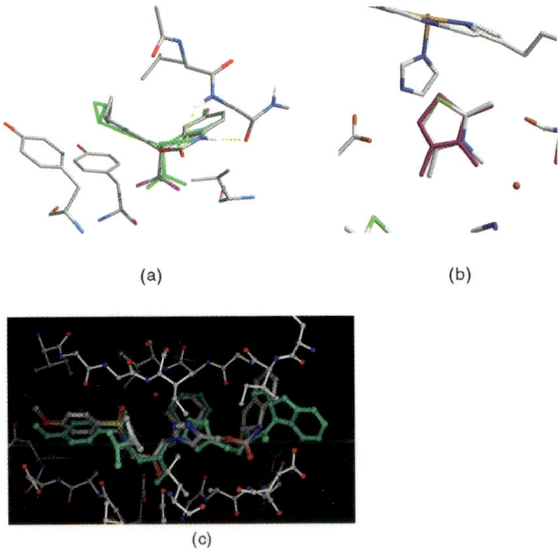

Figure 1.2 Predicted complexes versus X-ray crystallographic structures that were subsequently determined. See text for full caption.

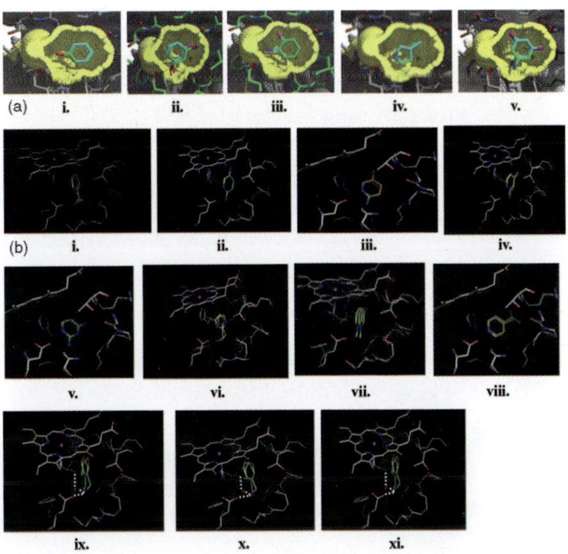

Figure 1.3 Docking predicted ligands from virtual screening against simple cavity sites. See text for full caption.

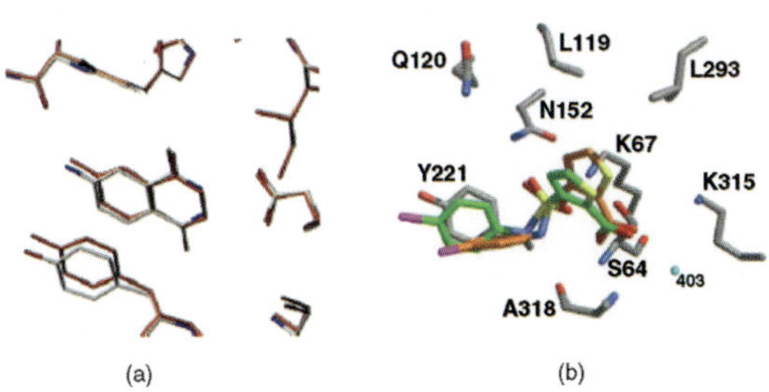

Figure 1.4 Predicted versus experimental structures from virtual screening. See text for full caption.

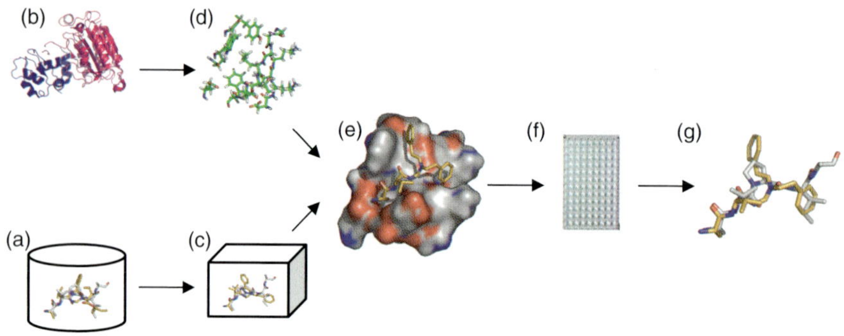

Figure 2.1 Typical receptor-based virtual screening scheme consisting of many steps in repeated cycles. See text for full caption.

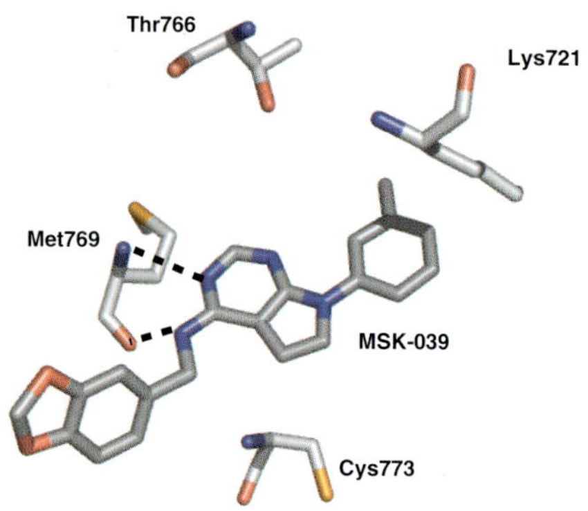

Figure 2.3 Structural model of MSK-039 docked at the EGFR kinase active site (only selected residues are shown). Dashed lines represent hydrogen bonds between ligand and active site residues.

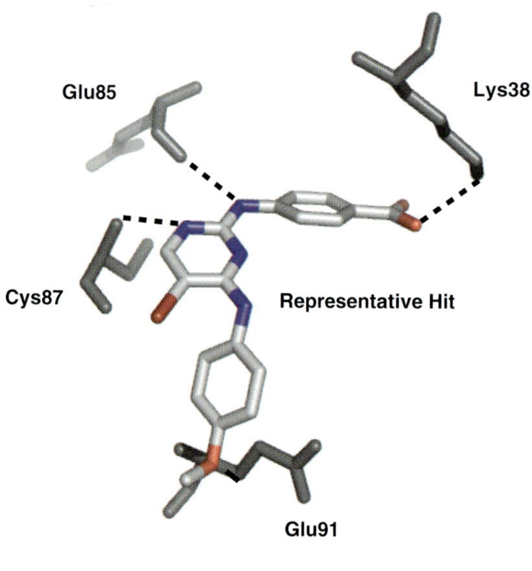

Figure 2.4 Structural model of a representative hit docked at the Chk-1 kinase active site (only selected residues are shown). Dashed lines represent hydrogen bonds between ligand and active site residues.

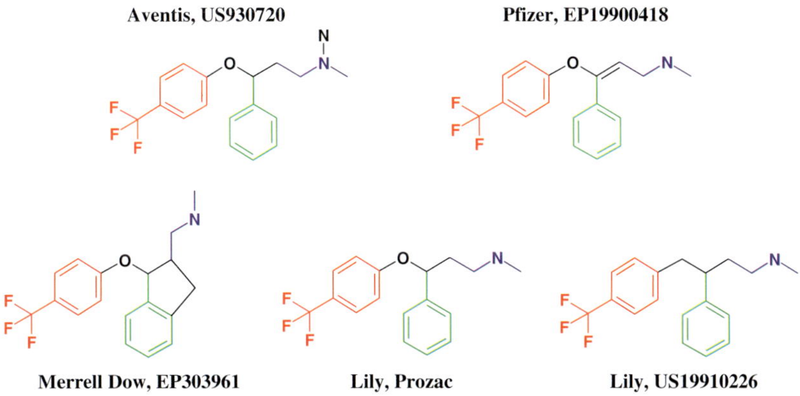

Figure 3.1 The chemical structures of a number of 5-HT reuptake inhibitors showing clear similarities in structure, illustrating the basis for the similarity and substructure methods described in the text. The common substructures are highlighted.

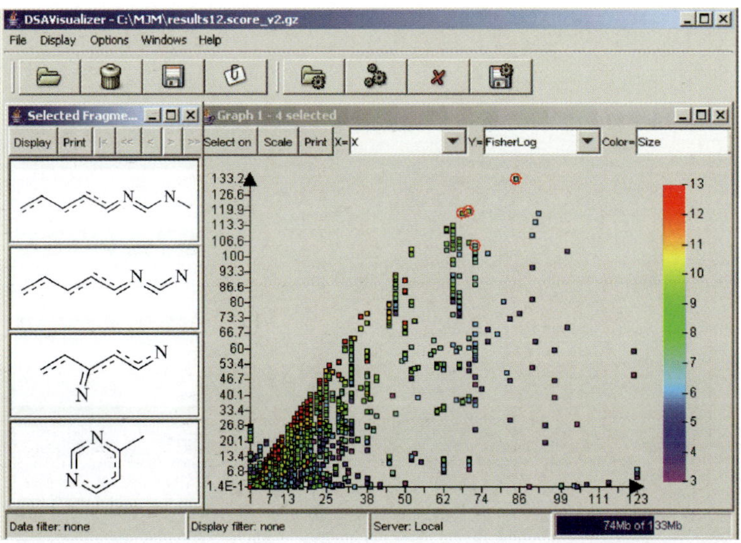

Figure 3.2 The DSA Visualizer for exploring substructure fragments in a set of compounds active for a particular target. See text for full caption.

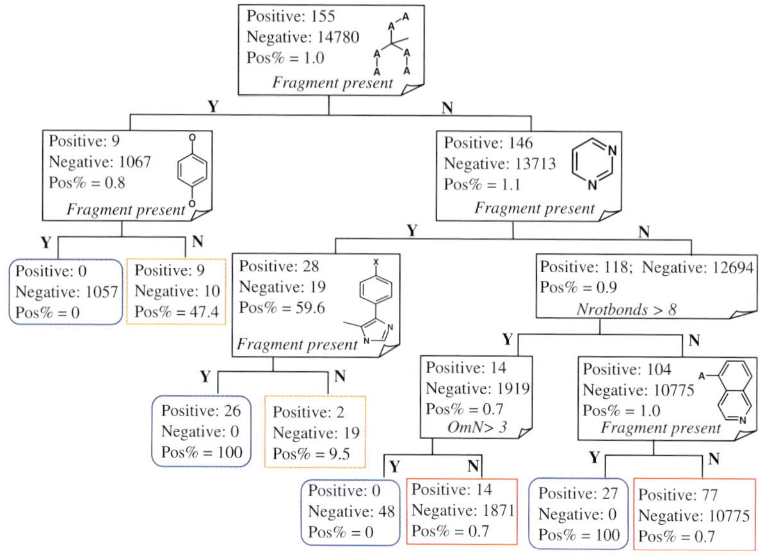

Figure 3.3 Example of a hypothetical recursive-partitioning tree. See text for full caption.

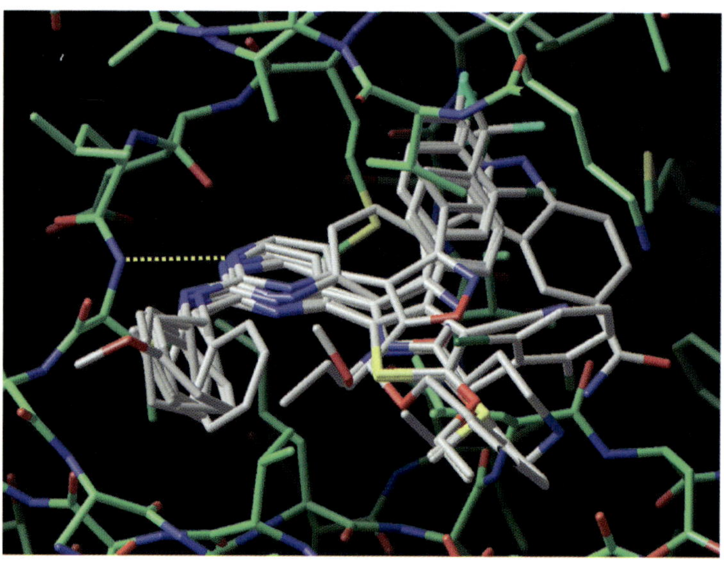

Figure 3.5 An example of molecular docking using the Glide program. See text for full caption.

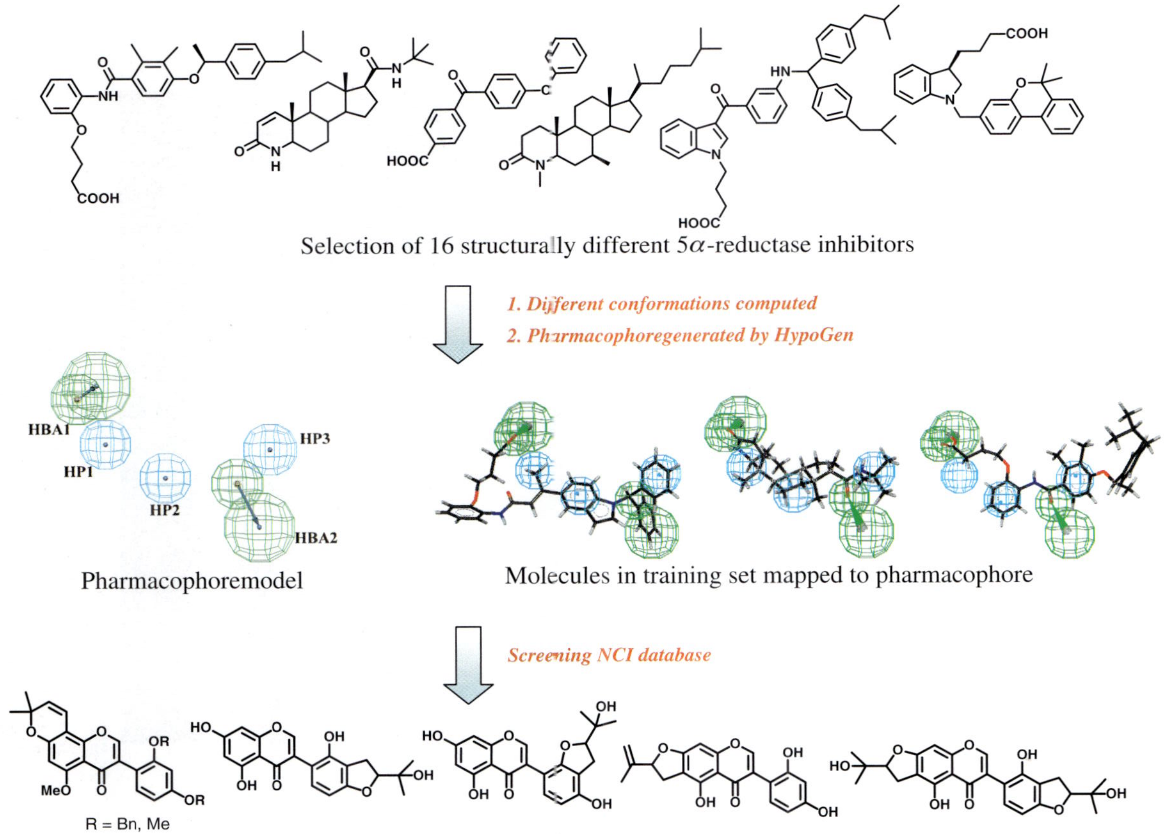

Figure 4.5 Isoflavone derivatives as 5α-reductase inhibitors from pharmacophore screening.

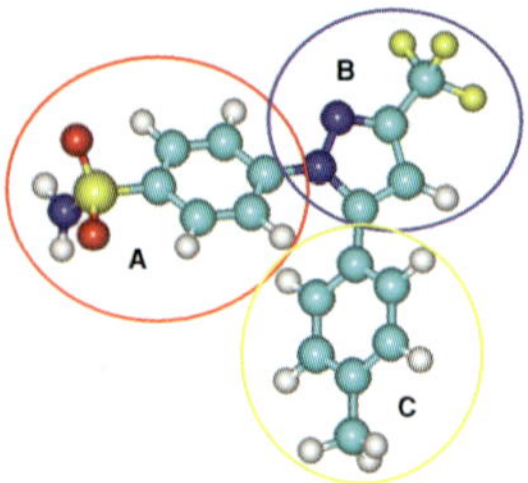

Figure 5.3 Structure of SC58635 and fragment division for COX-2 inhibitors.

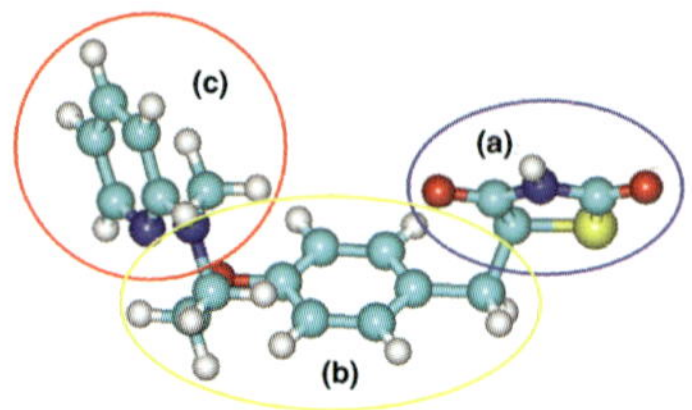

Figure 5.5 Common structural characteristic of PPARγ agonists and possibly structural fragment division.

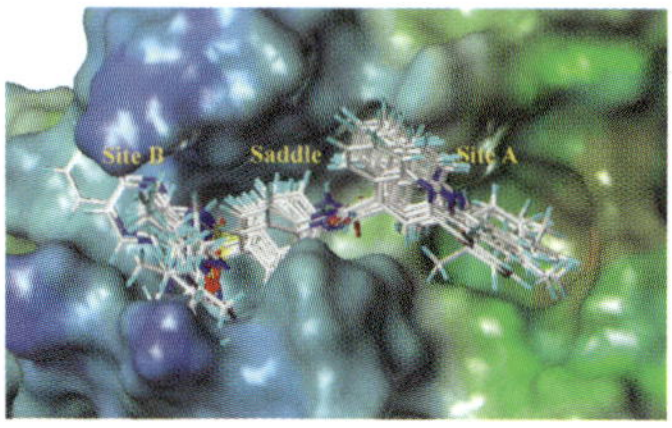

Figure 5.6 Three-dimensional structure of hCypA with designed inhibitors displayed. The CypA surface was colored by electrostatic potential.

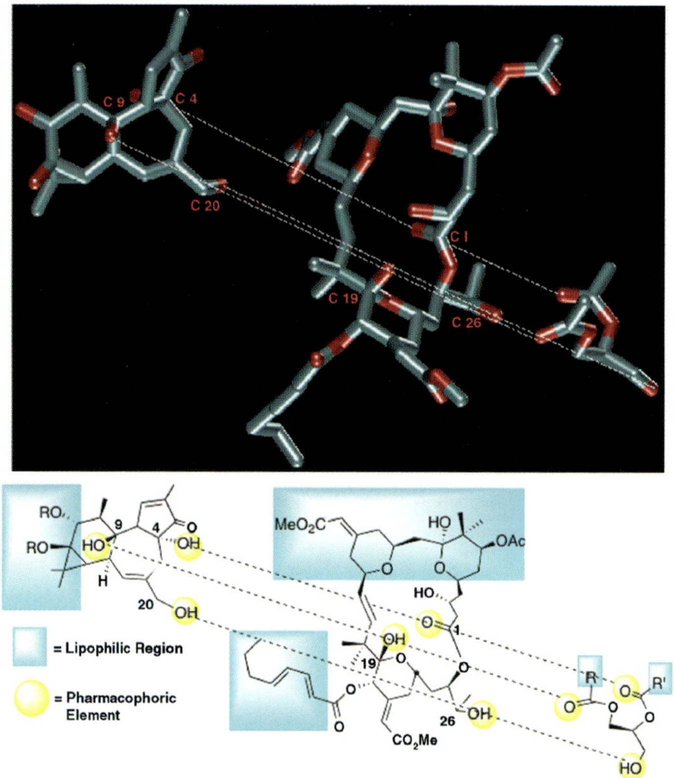

Figure 6.4 The preferred pharmacophoric correlation of 1,2-diacyl-*sn*-glycerol, the phorbol esters, and bryostatin 1.

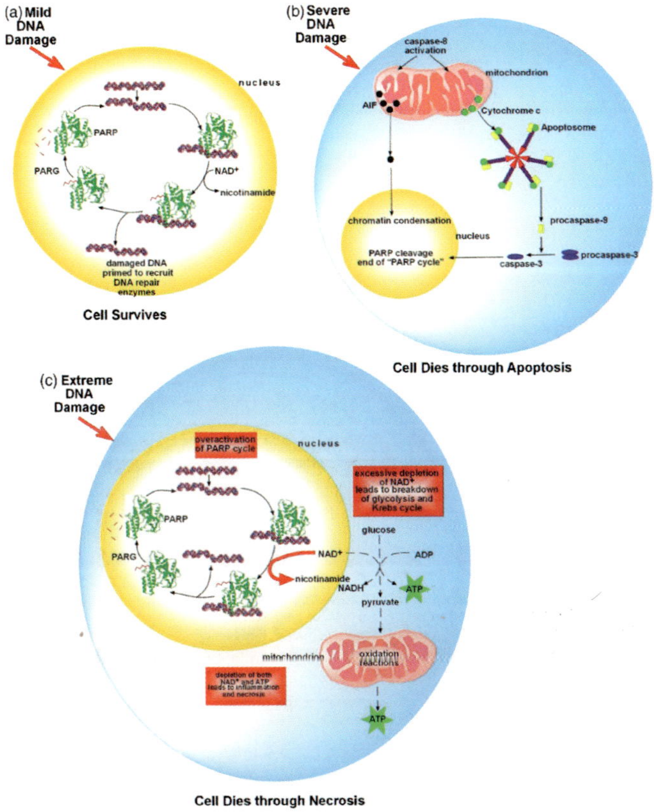

Figure 7.1 (a) In response to mild DNA damage, PARP-1 binds to DNA and catalyzes the formation of poly(ADP-ribose) polymers onto protein acceptors including itself. See text for full caption.

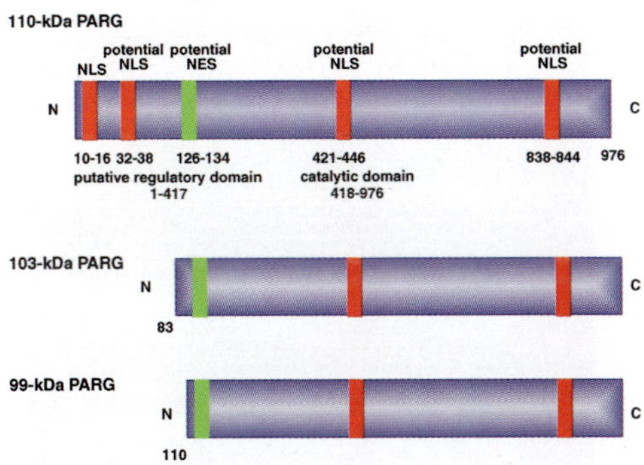

Figure 7.2 Human PARG110 has a two-domain structure, with both a regulatory domain (1–417) and a catalytic domain (418–976). See text for full caption.

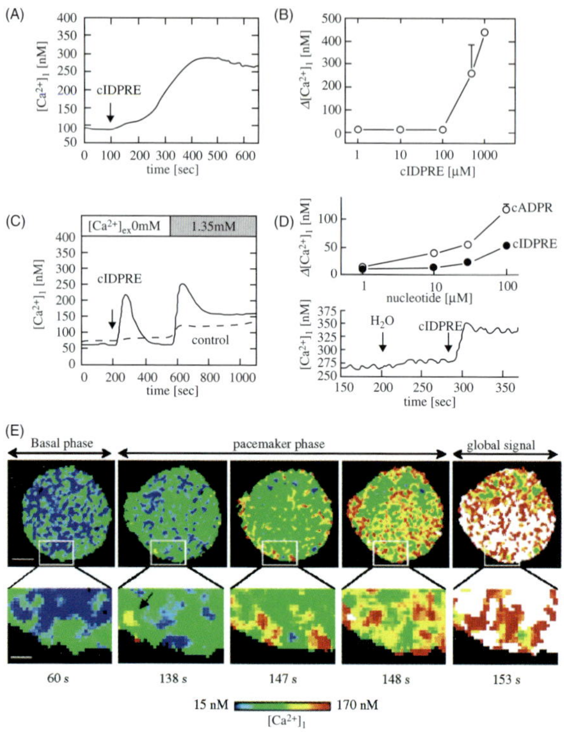

Figure 8.4 Effect of the cADPR mimic cIDPRE on Ca^{2+} signaling in T cells. See text for full caption.

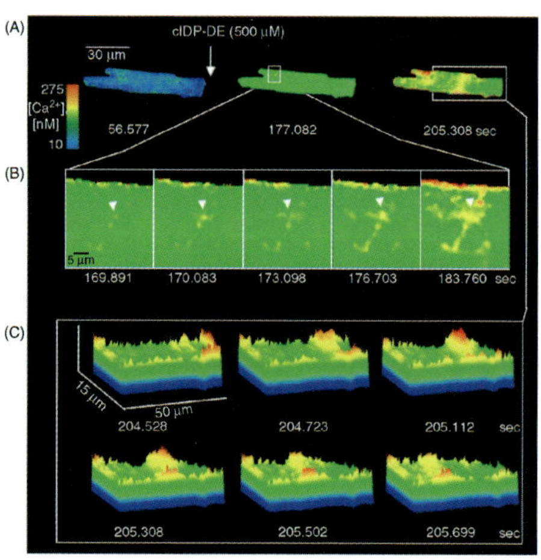

Figure 8.6 cIDP-DE-mediated local Ca^{2+}signaling and global Ca^{2+} waves in intact mouse cardiac myocytes. See text for full caption.

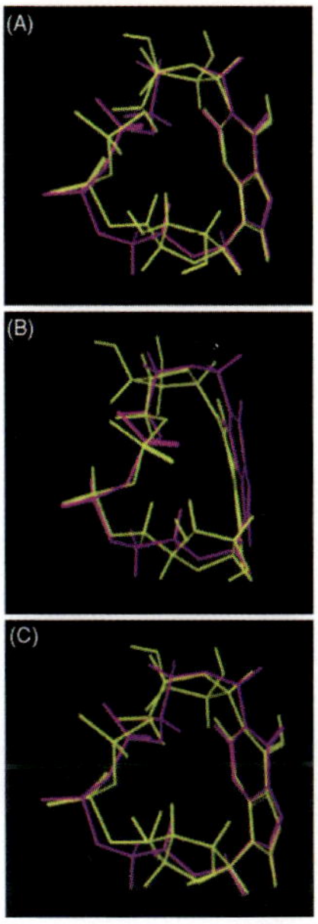

Figure 8.7 Superimposition of minimized conformations of cADPR and cIDP-DE. The conformations of cADPR (yellow) and cIDP-DE (pink) were minimized using the density functional theory (DFT) method. (**A**) Superimposition based on nucleobases; (**B**) superimposition based on diphosphates; (**C**) superimposition based on whole molecules.

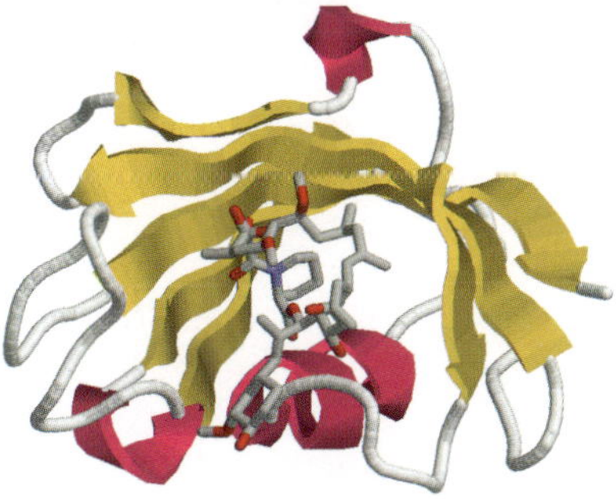

Figure 9.5 Complex structure of FKBP12-FK506; the protein is displayed in cartoon model, and FK506 is shown in stick model. See text for full caption.

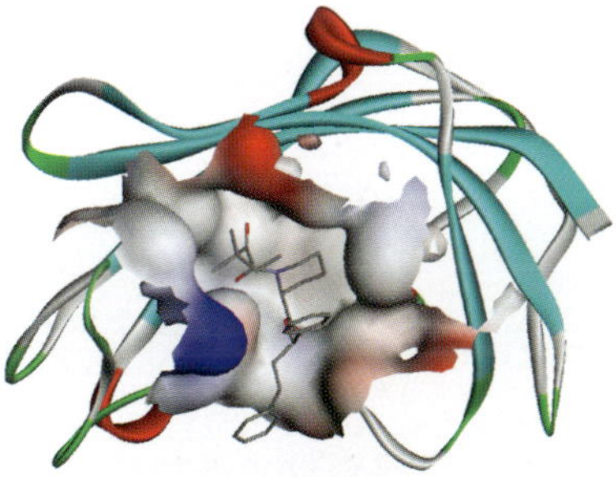

Figure 9.7 Complex structure of Holt's compound and FKBP12. See text for full caption.

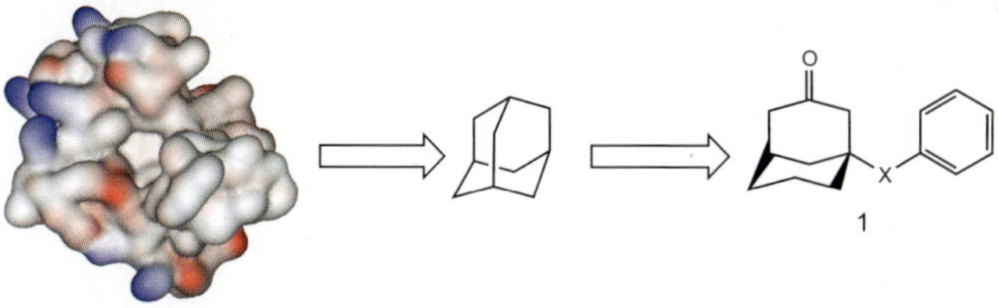

Figure 9.8 Ludi-based ligand design pipeline.

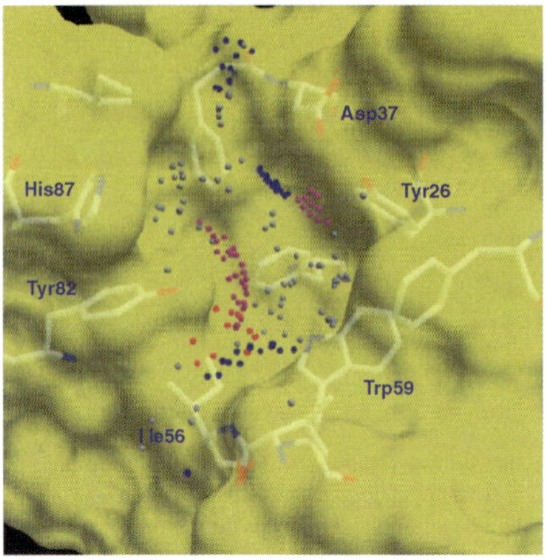

Figure 9.9 Site points in FKBP. See text for full caption.

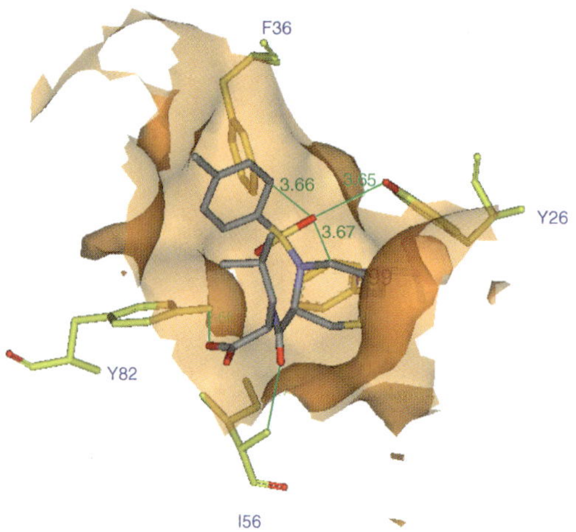

Figure 9.10 Complex of compound 308 and FKBP12. See text for full caption.

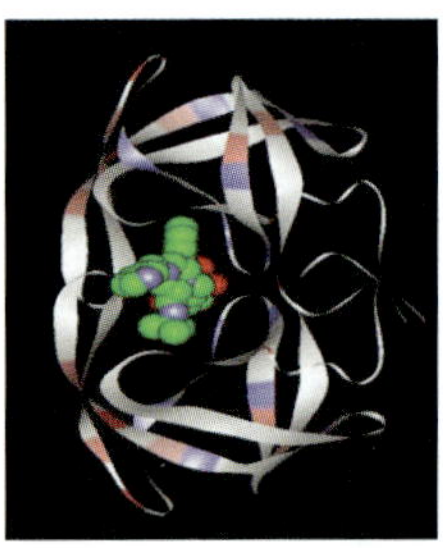

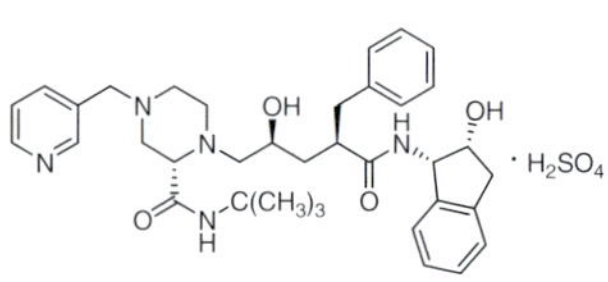

Crixivan 10

Figure 10.4 Crixivan bound to HIV protease (taken from www.biotech100.com).

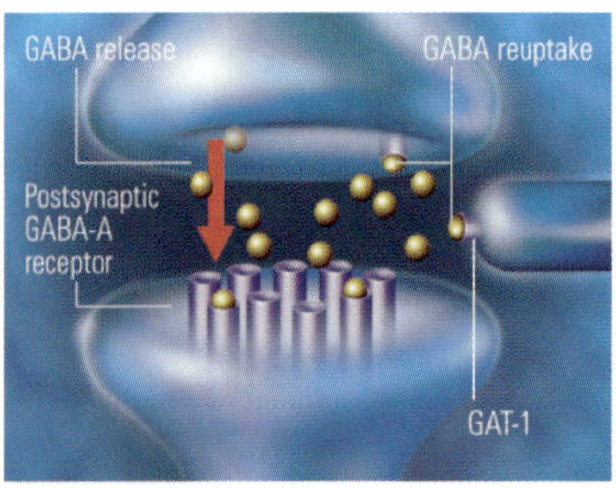

Figure 10.11 The process of normal release and reuptake of GABA.

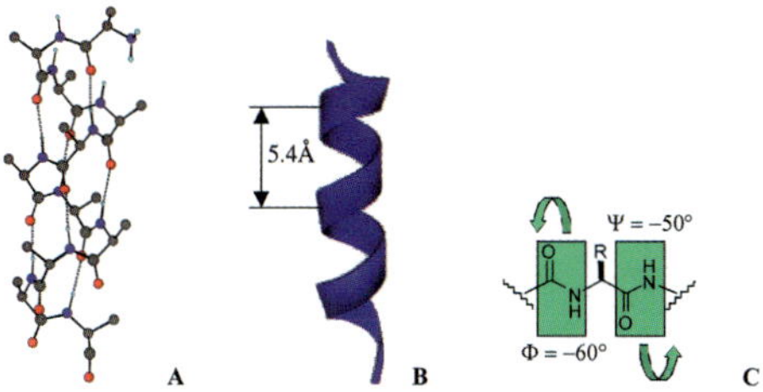

Figure 11.1 (A) An α-helix that is stabilized by internal hydrogen bonding. (B) Ribbon diagram showing rise per turn. (C) Amide backbone showing dihedral angles.

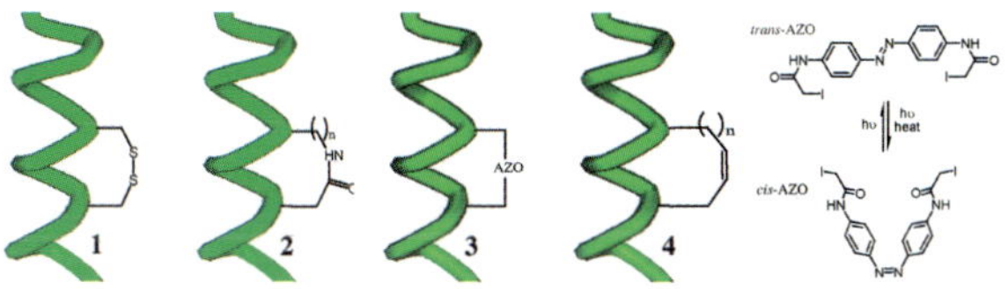

Figure 11.2 Some examples of helical stabilization by covalent linkages: disulfide (**1**), lactam (**2**), azobenzene (**3**), olefin (**4**).

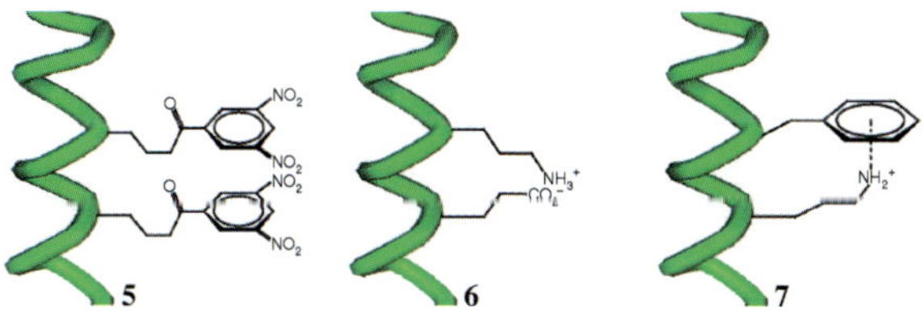

Figure 11.3 Some examples of helical stabilization by non-covalent interaction: π-π interaction (**5**), salt-bridge (**6**), cation-π interaction (**7**).

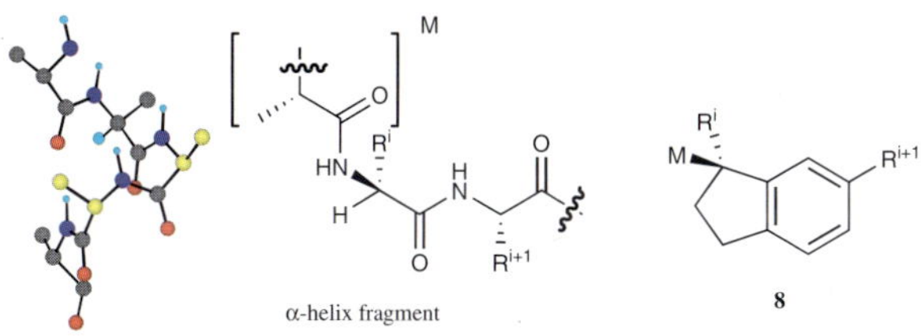

Figure 11.4 A 1,6-disubstituted indane molecule, **8**, that mimics of the i and $i + 1$ side chains (shown in yellow) of an α-helix.

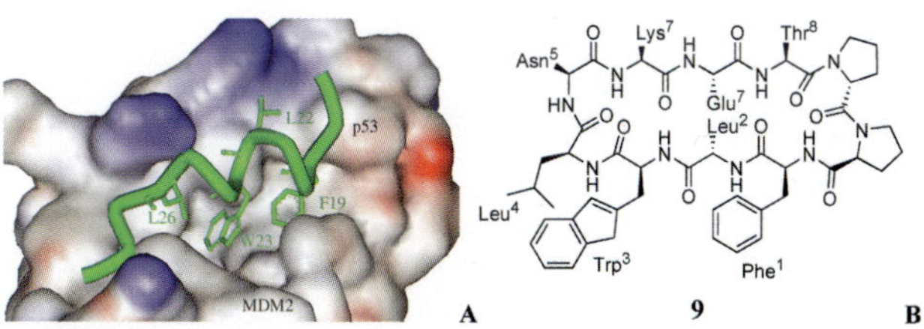

Figure 11.6 (A) X-ray crystal structure of the HDM2/p53 complex. (B) β-hairpin inhibitor that disrupts the HDM2-p53 interaction.

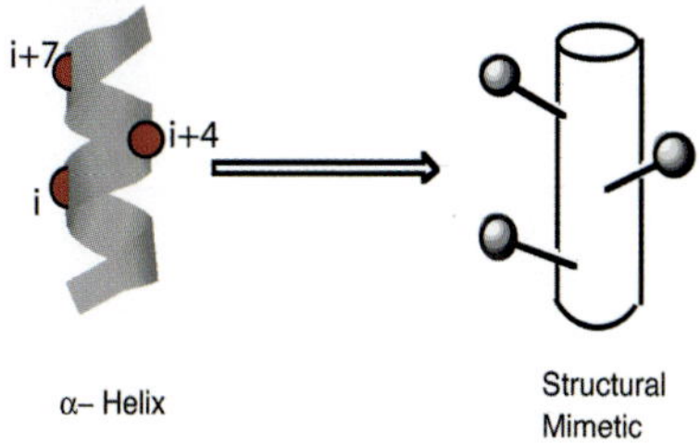

Figure 11.7 An α-helix backbone can be replaced by a synthetic scaffold that projects the functionalities in a similar fashion.

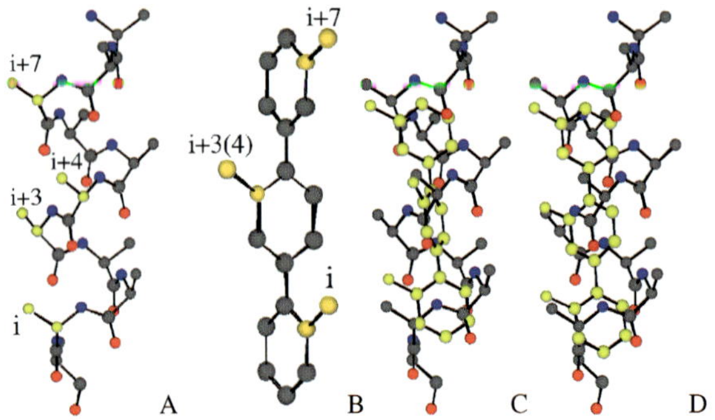

Figure 11.8 (A) A poly-alanine α-helix, with the i, $i + 3$, $i + 4$, and $i + 7$ side chains being highlighted in yellow. See text for full caption.

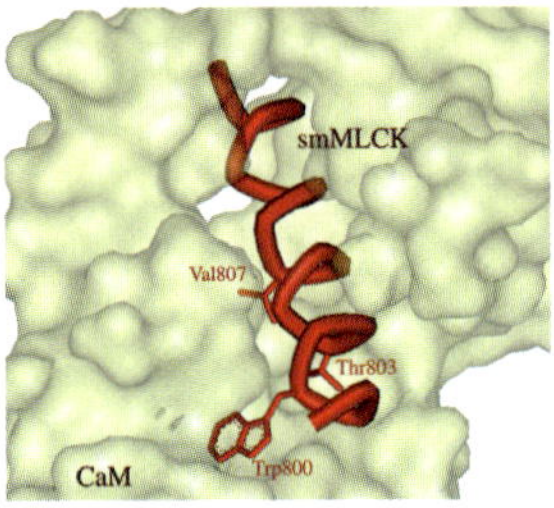

Figure 11.9 X-ray crystal structure of the CaM-binding region of smMLCK (residues 796-815, shown in red) in a complex with CaM. Side chains of the critical residues (Trp800, Thr803 and Val807) are shown in stick representation.

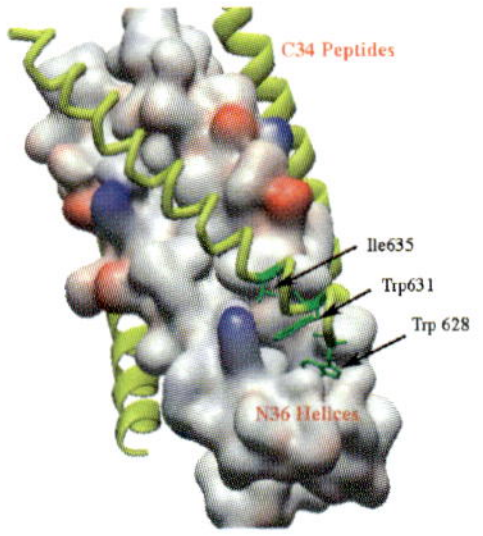

Figure 11.11 X-ray crystal structure of the gp41 fusion active core C34 (a.a. 628-661, shown in yellow helical tubes representation) bind to N36 (a.a. 546-581) by inserting the side chains of Trp628, Trp631 and Ile635 residues into a hydrophobic cleft on the protein surface.

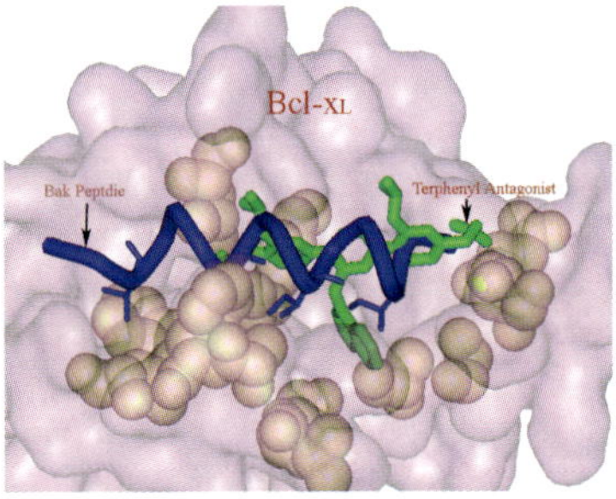

Figure 11.12 Results of the ^{15}N-HSQC experiments of **13** binding to Bcl-x_L. See text for full caption.

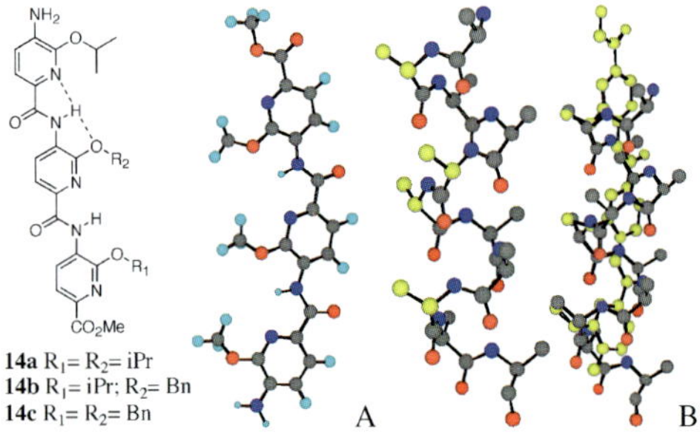

Figure 11.13 (A) X-ray crystal structure of an oligoamide foldamer **14**. (B) Overlay of a poly-alanine α-helix and **14**, showing good agreement of the side chains.

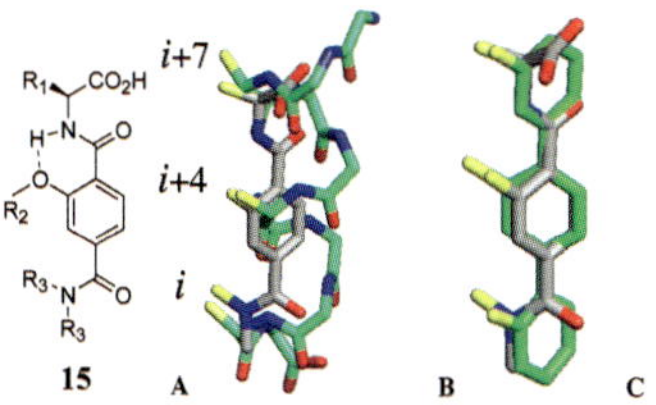

Figure 11.14 (A) Terephthalamide helical mimetics. (B) The superimposition of **15** on the i, $i + 4$, and $i + 7$ positions of a poly-alanine α-helix (RMS deviation = 1.03 Å). (C) The superimposition of terphenyl **13** and terephthalamide **15** (RMS deviation = 0.34 Å). The functional side chains are highlighted.

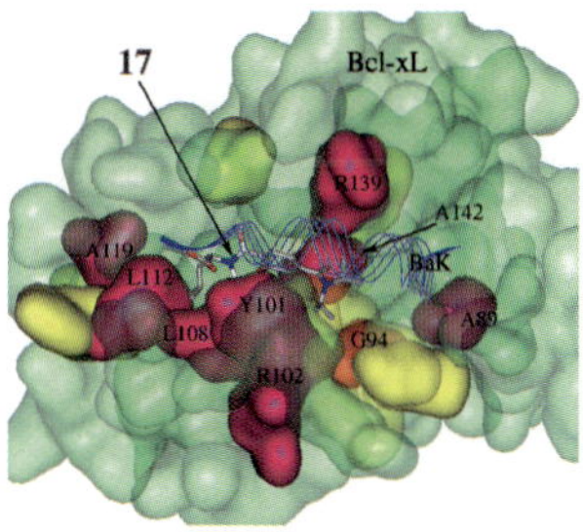

Figure 11.16 Superimposition of the top-ranked result of the molecular-docking studies of **17** and the Bak peptide. The residues that showed significant and moderate chemical shift changes upon the addition of **17** were shown in magenta and yellow, respectively.

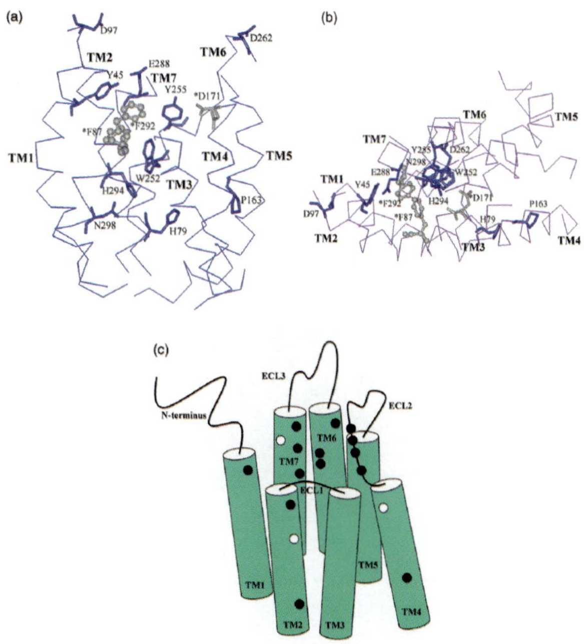

Figure 12.4 Distinct functional sites for HIV-1 gp120 and SDF-1α on a hypothetical structural model of CXCR4. See text for full caption.

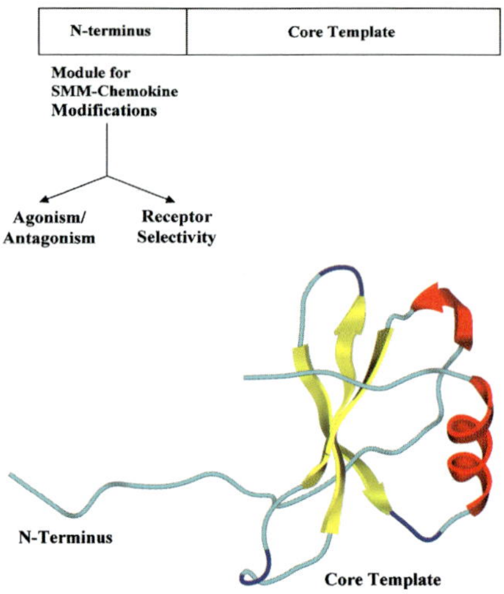

Figure 12.5 General design concept for SMM-chemokines. The N-terminal (1–10) sequence module of vMIP-II was modified by introducing unnatural D-amino acids or specific sequences grafted from other chemokines to chemically engineer receptor selectivity and/or signaling property.

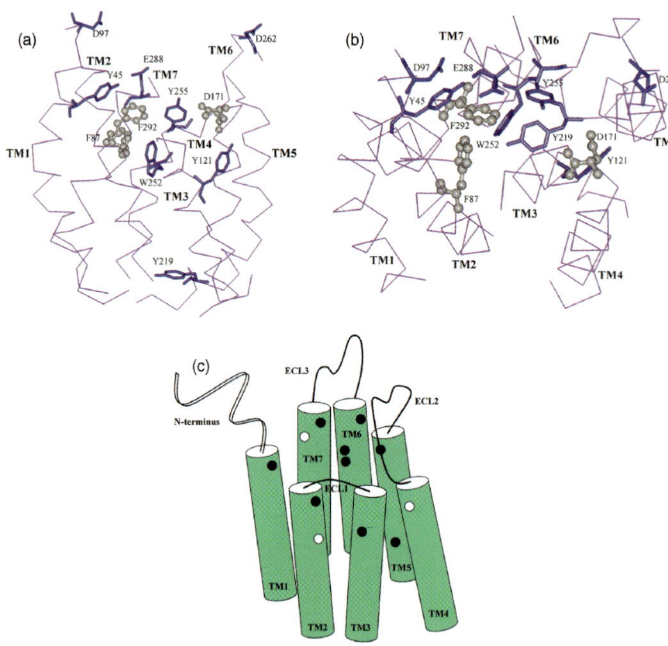

Figure 12.7 Distinct functional sites for SDF-1α and SMM-chemokines highlighted on a hypothetical structural model of CXCR4. See text For Full caption.

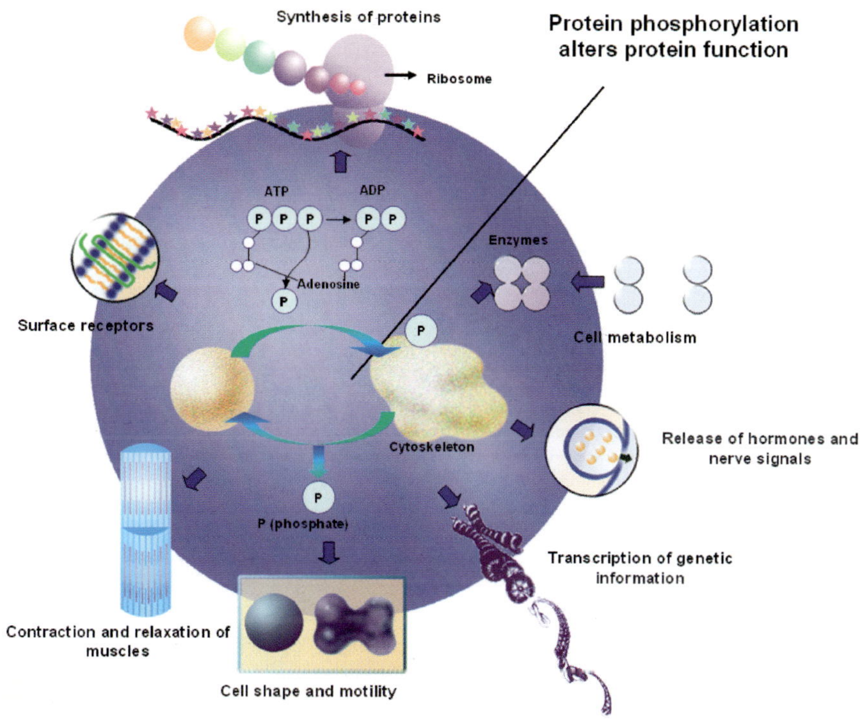

Figure 13.3 Protein phosphorylation regulating cellular processes.

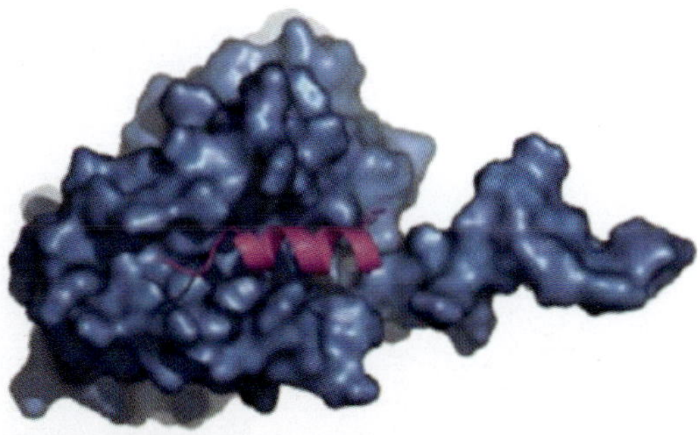

Figure 17.2 Structure of Bcl-2 and associated BH3 peptide. The 3D structure of Bcl-X_L is depicted (blue) with Bak BH3 peptide bound (red).

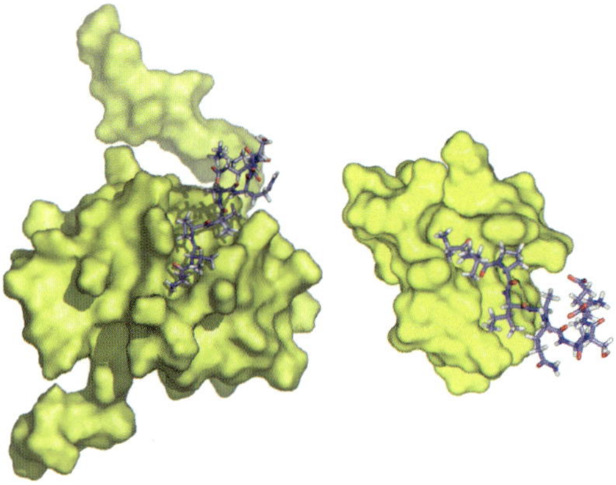

Figure 17.3 Structure of BIR domain and inhibitory SMAC peptide. BIR domain of XIAP is presented in space-filling model, while the SMAC tetra-peptide is presented in stick mode. The left structure gives the entire Bir3 domain, while the right structure is a subdomain of Bir3. The structure is derived from PDB #1G3F.

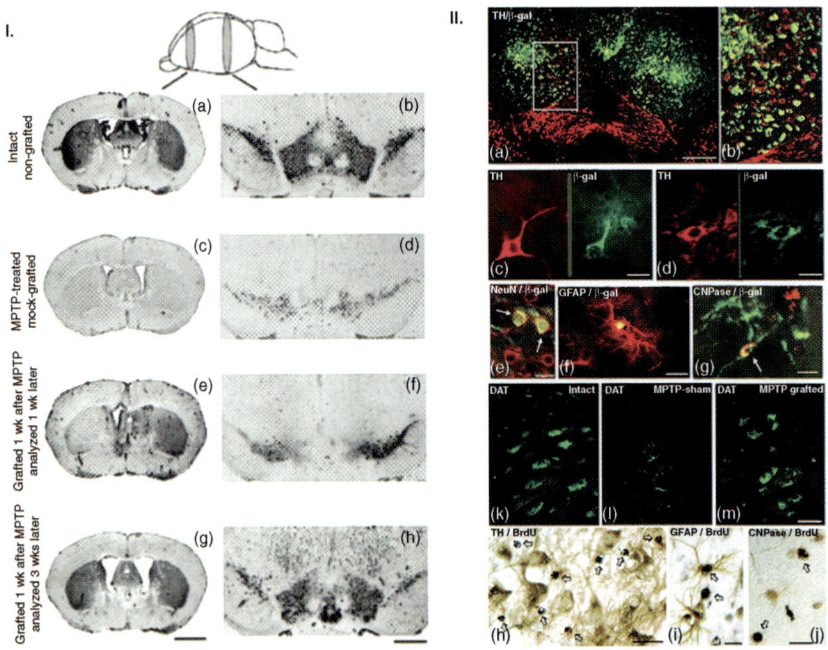

Figure 20.2 NSCs possess an inherent mechanism for rescuing dysfunctional neurons. Shown here is evidence from the effects of NSCs in the restoration of mesencephalic dopaminergic function. [Modified from Ourednik et al. (2002)]. See text for full caption.

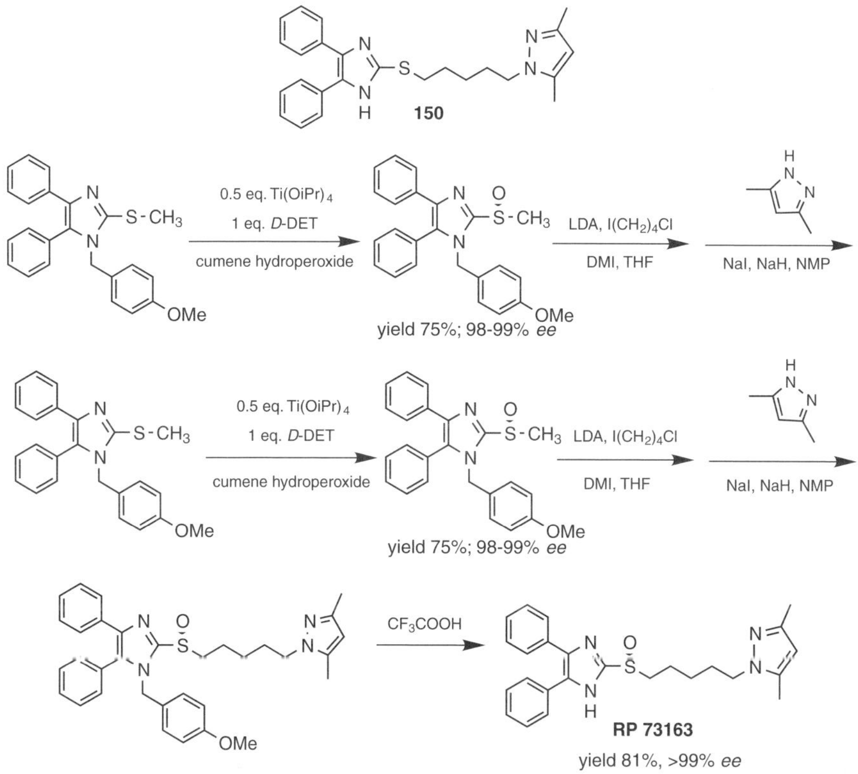

Scheme 10.38

Asymmetric Codimerization Reaction Substituted 8-oxabicyclo [3.2.1]octanes (**151**) are used as a key intermediate for the synthesis of tromboxane-A_2 (TXA_2) analogues. A novel and unexpected stereoselective synthesis of functionalized **151** has been finished by interaction α,β-unsaturated ketones **152** with γ,δ-unsaturated ketones

TABLE 10.7 Synthesis of Substituted 8-Oxabicyclo[3.2.1]octanes Catalyzed by Rh(I)-Sn(II) System[a]

R_1	R_2	Conversion of **153** (%)	Yield **151**[b] To converted **153** (%)	Yield **151**[b] To mol/g -at Rh
Me	Me	31	39	12
Ph	Me	54	26	11
t-Bu	Me	26	14	4
5-(2-methylfuryl)	Me	47	33	15
Me	Ph	94	13	12

[a] α,β-Unsaturated ketones (9.26 mmol), γ,δ-unsaturated ketones (9.26 mmol), $[RhCl(C_2H_4)_2]_2$ (0.046 mmol), $SnCl_2 \cdot 2H_2O$ (0.19 mmol), acetone 3 mL, 80°C, 10 h.
[b] The yield of isolated products.

Source: Data were taken from Pitchen et al. (1994).

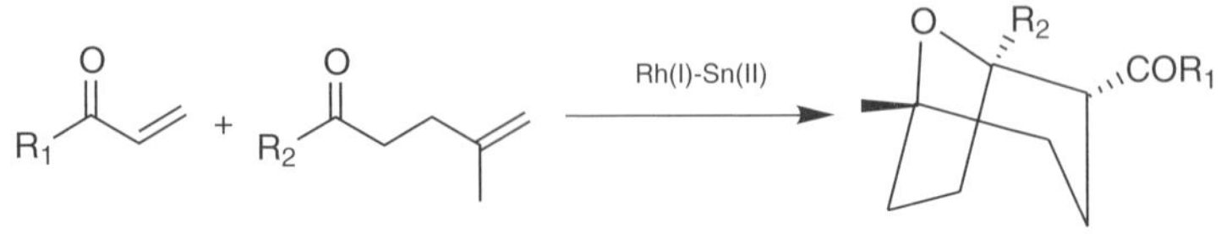

152 153 151

Scheme 10.39

153 catalyzed by the $[RhCl(C_2H_4)_2]_2$-$SnCl_2$ system (Kovalev et al., 1992). The reaction proceeds stereoselectively and gives only one *endo*-isomer with the carbonyl group in the equatorial position, as shown in Scheme 10.39. The results of the reaction are listed in Table 10.7.

10.4 BIOTARGETS-DIRECTED DESIGN AND DISCOVERY OF BIOACTIVE COMPOUNDS

It has long been recognized that knowledge of the 3 D structures of proteins has the potential to accelerate drug discovery. Many new protein targets have been identified from genome analyses and studied by X-ray analysis or NMR spectroscopy (Congreve et al., 2005). Described here is an example of using GABA transporters as targets for a drug discovery project.

10.4.1 GABA Transporter

γ-Aminobutyric acid (GABA), the major inhibitory neurotransmitter in the brain, has been the target of extensive pharmacological research. GABA reuptake from the synaptic cleft is an important mechanism in the regulation of GABA activity, which is effected by a selective transporter named a GABA transporter (GAT) (Radian et al., 1986). GAT, together with the betaine, taurine, and creatine transporters, belongs to a subfamily of Na^+- (and Cl^-) ion-dependent transporters. The human GAT proteins consist of 599 amino acids and a putative structure (Nelson et al., 1990; Kanner, 1994; Gao et al., 1999), as shown in Fig. 10.10.

In the central nervous system, GABA transport between neurons is effected by the release of neurotransmitters from the presynaptic neuron into the synaptic cleft to interact with GABA receptors in the membrane of the postsynaptic neuron. The GABA transport is ended by reuptake of the transmitter from the synaptic cleft by transporters located in the presynaptic membrane and is then partly inactivated enzymatically and partly stored in presynaptic vesicles for further use. The uptake in vesicles is caused by transporters located in the vesicular membrane. Glial cells surrounding the synapse also inactivate the transmitter by uptake from the synaptic cleft by transporters followed by enzymatic inactivation. The reuptake is effected by selective GABA transporters. Inhibitors of uptake transporters enhance the concentration of the transmitter in the synaptic cleft, thereby intensifying the action of the transmitter on its postsynaptic receptors, a useful property in case of transmitter malfunction (Soudijn, 2000). The process of normal release and reuptake of GABA is illuminated in Fig. 10.11.

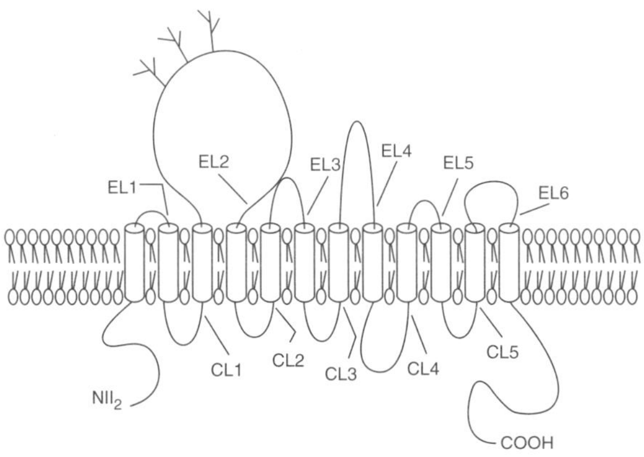

Figure 10.10 A putative structure of GAT.

In order to elucidate the functional roles of GABA transport processes, it is necessary to obtain detailed knowledge about the anatomical distribution of GABA transporters as well as their pharmacological properties. The cloning of four GABA transporters responsible for the cellular reuptake of GABA (mouse nomenclature GAT1-4) has allowed such studies to be performed over the past 15 years (Schousboe et al., 2004). Inhibition of the reuptake of GABA by potent and selective inhibitors of the GABA transporter enhances GABA activity. A therapeutic benefit is obtained by inhibiting these proteins in disorders in which GABA reserves are depleted—for instance, epilepsy or psychiatric disorders. So far, only GAT1 has convincingly been targeted in a selective manner, and little is known about the significance of the other subtypes (Dalby, 2003).

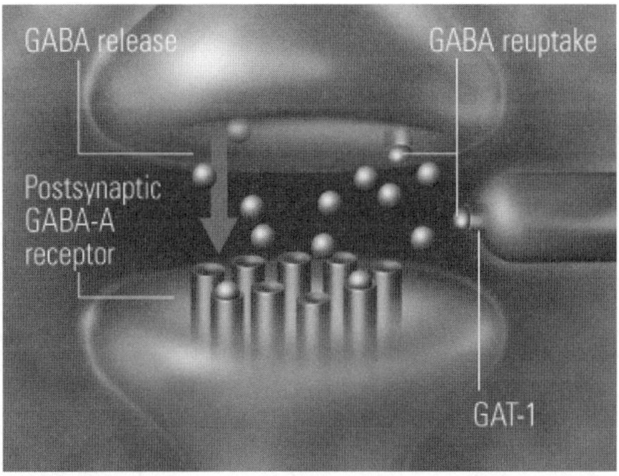

Figure 10.11 The process of normal release and reuptake of GABA (taken from www.gabitril.com). See color plates.

10.4.2 Inhibitors of the GABA Transporter

Some neuromedicinal chemistry groups have previously made significant contributions in the development of compounds that selectively interfere with the systems transporting this crucial neurotransmitter without affecting signaling mechanisms directly. GABA and several of its conformationally restricted analogues effectively inhibit the uptake of [^{3}H]-GABA into rat brain synaptosomes. The rank-order of potency is homo-β-proline **154** = (*R*)-nipecotic acid **155** > GABA **156** > *cis*-4-OH-nipecotic acid **157** > guvacine **158** > (*S*)-nipecotic acid **159** (Ali et al., 1985).

Although (*R*)-nipecotic acid and guvacine are effective inhibitors of the GABA transporter *in vitro*, the compounds are not centrally active after systemic administration. This is probably due to the hydrophilic character of these cyclic amino acids. Therefore, in order to develop more lipophilic GAT inhibitors, the nitrogen atom of nipecotic acid and guvacine was substituted with diarylalkenyl groups. Some optimal active compounds had been obtained, such as SKF89976-A (**160**), CI-966 (**161**), SKF100330- A (**162**), tiagabine (**163**), NO-711 (**164**), and other analogues (Andersen et al., 1999, 2001).

From the data of GABA transporter inhibitory activity, it can be seen that (*R*)-nipecotic acid derivatives have more potent effects than (*S*)-nipecotic acid derivatives. For example, the GABA transporter inhibitory activity of (*R*)-nipecotic acid (**155**) (IC_{50} = 1.69 μM) is more potent than that of (*S*)-nipecotic acid (**159**) (IC_{50} = 9.88 μM). (*R*)-SKF89976-A (**160**) (IC_{50} = 0.11 μM) has a more potent inhibitory effect than (*S*)-SKF89976-A (**165**) (IC_{50} = 1.91 μM).

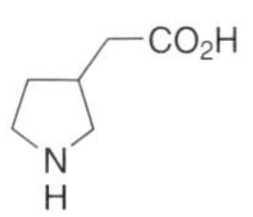

154

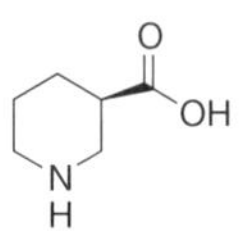

155

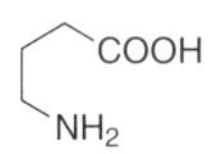

156

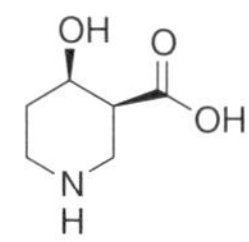

157

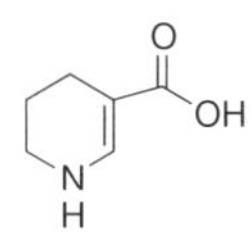

158

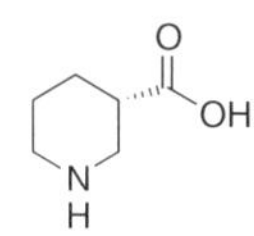

159

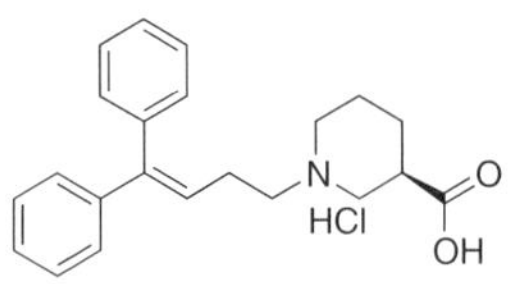

160

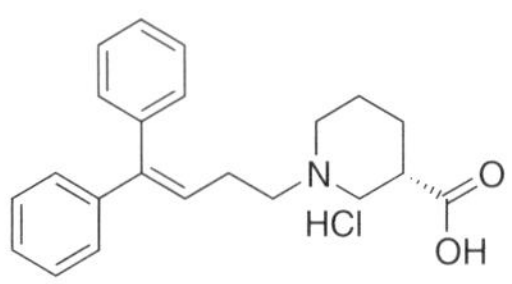

165

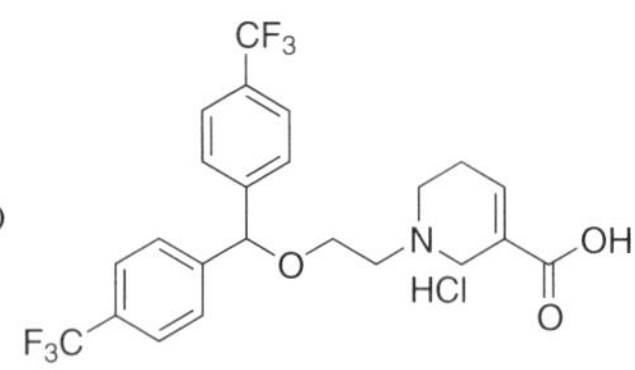

161

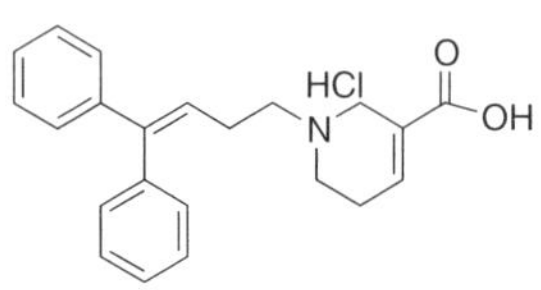

162

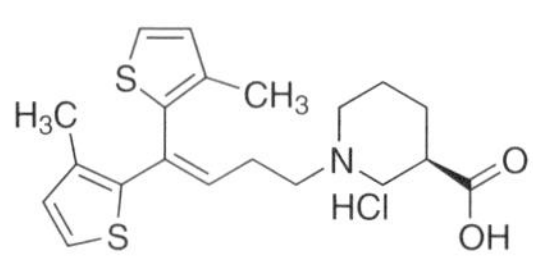

163

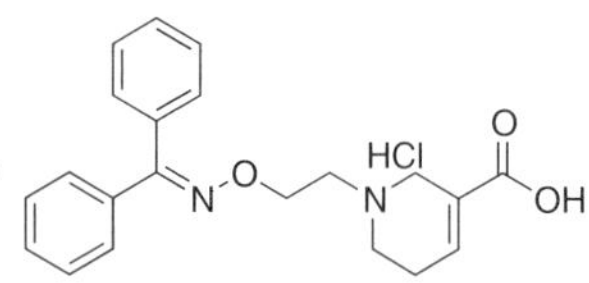

164

The (*R*)-nipecotic acid and guvacine derivatives have been the starting key for drug development programs in different companies and have led to a compound applied clinically as an antiepileptic drug as well as pointing to other potential therapeutic areas—that is, sleep disorders and pain. This drug is tiagabine hydrochloride (**163**) (TGB, Gabitril®), a selective γ-aminobutyric acid (GABA) reuptake inhibitor (SGRI), which increases synaptic GABA availability via inhibition of the

GAT-1 on presynaptic neurons and glial cells (Borden et al., 1994). TGB has been shown to be effective as add-on therapy in adults and children 12 years and older with refractory partial seizures with or without secondary generalization. To date, exposure to TGB amounts to 147,686 patient-years (Bialer et al., 2004).

10.4.3 Design of New Inhibitors of the GABA Transporter

Knud Erik Andersen and co-workders (Knutsen et al., 1999) have designed some novel GABA uptake inhibitors of remarkable potency, using a putative new model of ligand interaction at GAT-1, as shown in Fig. 10.12. This model involves the postulated interaction of an electronegative region in the small-molecule GABA uptake inhibitor with a positive domain in the protein structure of the site controlling GABA uptake. The increasing electronegative character of the linker is coupled to the increase in potency of the compounds as inhibitors of [^{3}H]-GABA uptake *in vitro* (Liu et al., 1993; Borden et al., 1994; Bennett and Kanner, 1997).

On the basis of the putative new model of ligand interaction at the GABA transporter type 1 and the efforts of our continued research in this area, we have designed and synthesized a series of new (*R*)-nipecotic acid derivatives containing thio/sulfinyl groups. The general structure of these compounds is shown in Fig. 10.13. Extensive animal studies with these compounds have revealed that some of them shown good inhibitory activity of [^{3}H]-GABA uptake *in vitro* of culture cells. Hopefully, through further research of these compounds, new compounds will be discovered that can help elucidate the mechanisms of ligand interaction at the GABA transporter type 1 as well as provide drug candidates.

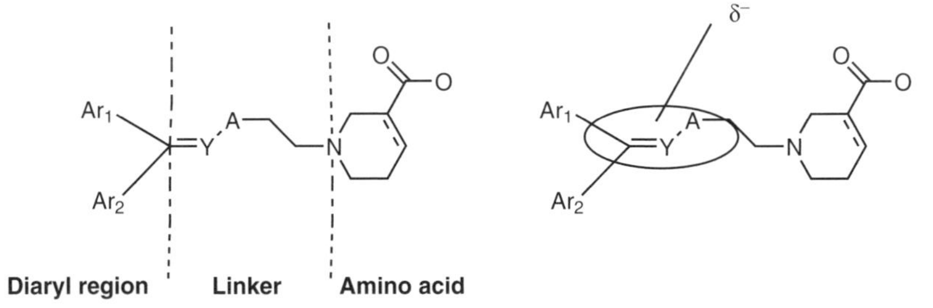

Figure 10.12 Putative new model of ligand interaction at GAT-1.

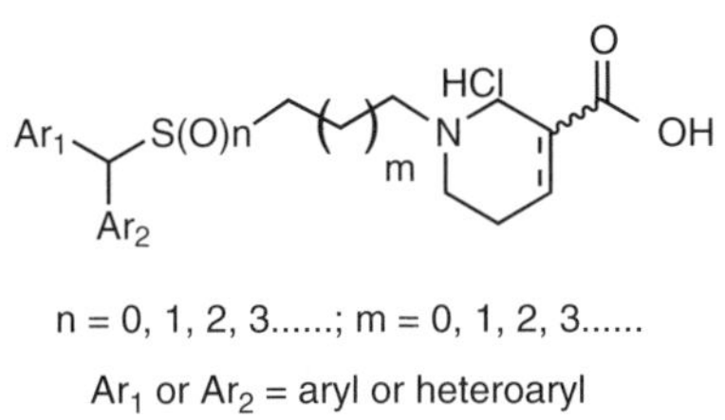

Figure 10.13 General structure of (*R*)-nipecotic acid derivatives containing thio/sulfinyl groups.

REFERENCES

Agranat, I., Caner, H., and Caldwell, J. (2002). Putting chirality to work: The strategy of chiral switches. *Nat Rev Drug Discov* **1**, 753–768.

Akutagawa, S. (1997). Enantioselective isomerization of allylamine to enamine: Practical asymmetric synthesis of (−)-methanol by Rh-BINAP catalysts. *Top Catal* **4**, 271–274.

Ali, F. E., Bondinell, W. E., Dandridge, P. A., Frazee, J. S., Garvey, E., Girard, G. R., Kaiser, C., Ku, T. W., Lafferty, J. J., Moonsammy, G. I., Oh, H. J., Rush, J. A., Setler, P. E., Stringer, O. D., Venslavski, J. W., Volpe, B. W., Yunger, L. M., and Zirkle C. L. (1985). Orally active and potent inhibitors of γ-aminobutyric acid uptake. *J Med Chem* **28** 653–660.

Andersen, K. E., Sorensen, J. L., Huusfeldt, P. O., Knutsen, L. J. S., Lau, J., Lundt, B. F., Petersen, H., Suzdak, P. D., and Swedberg, M. D. B. (1999). Synthesis of novel GABA uptake inhibitors. 4.[1] Bioisosteric transformation and successive optimization of known GABA uptake inhibitors leading to a series of potent anticonvulsant drug candidates. *J Med Chem* **42**, 4281–4291.

Andersen, K. E., Sorensen, J. L., Lau, J., Lundt, B. F., Petersen, H., Huusfeldt, P. O., Suzdak, P. D., and Swedberg, M. D. B. (2001). Synthesis of novel γ-aminobutyric acid (GABA) uptake inhibitors. 5.[1] Preparation and structure–activity studies of tricyclic analogues of known GABA uptake inhibitors. *J Med Chem* **44**, 2152–2163.

Bartsch, W., Sponer, G., Strein, K., Muller-Beckmann, B., Kling, L., Bohm, E., Martin, U., and Borbe, H. O. (1990). Pharmacological characteristics of the stereoisomers of carvedilol. *Eur J Clin Pharmacol* **38**, S104–S107.

Baumann, P., Zullino, D. F., and Eap, C. B. (2002). Enantiomers' potential in psycho pharmacology—a critical analysis with special emphasis on the antidepressant escitalopram. *Eur Neuropsychopharmacol* **12**, 433–444.

Bayley, C. R., and Vaidya, N. A. (1992). Resolution of racemates by diastereometric salt formation. In A. N. Collins, G. N. Sheldrake, and J. Crosby (Eds.), *Chirality in Industry*. New York: John Wiley & Sons, p. 71.

Beilles, S., Cardinael, P., Ndzie, E., Petit, S., and Coquerel, G. (2001). Preferential crystallisation and comparative crystal growth study between pure enantiomer and racemic mixture of a chiral molecule: 5-ethyl-5-methylhydantoin. *Chem Eng Sci* **56**, 2281–2294.

Bennett, E. R., and Kanner, B. I. (1997). The membrane topology of GAT-1, a (Na^+Cl^-)-coupled γ-aminobutyric acid transporter from rat brain. *J Biol Chem* **272**, 1203–1210.

Bialer, M., Johannessen, S. I., Kupferberg, H. J., Levy, R. H., Perucca, E., and Tomson T. (2004). Progress report on new antiepileptic drugs: A summary of the Seventh Eilat Conference (EILAT VII). *Epilepsy Res* **61**, 1–48.

Blaser, H. U., and Spindler, F. (1997). Enantioselective catalysis for agrochemicals. The case histories of (*S*)-metolachlor, (*R*)-metalaxyl and clozylacon. *Top Catal* **4**, 275–282.

Bonner, W. A. (2000). Parity violation and the evolution of biomolecular homochirality. *Chirality* **12**, 114–126.

Borden, L. A., Dhar, T. G. M., Smith, K. E., Weinshank, R. L., Branchek, T. A., and Gluchowski, C. (1994). Tiagabine, SK&F 89976A, CI-966, and NNC-711 are selective for the cloned GABA transporter GAT-1. *Eur J Pharmacol* **269**, 219–224.

Bornscheuer, U. T. (2002). Microbial carboxyl esterases: Classification, properties and application in biocatalysis. *FEMS Microbiol Rev* **26**, 73–81.

Brittain, H. G. (1990). Crystallographic consequences of molecular dissymmetry. *Pharmacol Res* **7**, 683–690.

Buckland, B. C., Drew, S. W., Connors, N. C., Chartrain, M. M., Lee, C., Salmon, P. M., Gbewonyo, K., Zhou, W. C., Gailliot, P., Singhvi, R., Jr., Olewinski, R. C., Sun, W. J., Reddy, J., Zhang, J. Y., Jackey, B. A., Taylor, C., Goklen, K. E., Junker, B., and Greasham, R. L. (1999). Microbial conversion of indene to indandiol: A key intermediate in the synthesis of crixivan. *Metab Eng* **1**, 63–74.

Caldwell, J. (1996). Importance of stereospecific bioanalytical monitoring in drug development. *J Chromatogr A* **719**, 3–13.

Caldwell, J. (1999). Through the looking glass in chiral drug development. *Mod Drug Discov* **2**, 51–60.

Caner, H., Groner, E., and Levy, L. (2004). Trends in the development of chiral drugs. *Drug Discov Today* **9**, 105–110.

Carreno, M. C. (1995). Applications of sulfoxides to asymmetric synthesis of biologically active compounds. *Chem Rev* **95**, 1717–1760.

Cesarotti, E., and Rimoldi, I. (2004). Stereoselective synthesis of 1-methylcarbapenem precursors: Studies on the diastereoselective hydroformylation of 4-vinyl β-lactam with aminophosphonite-phosphinite and aminophosphine-phosphite rhodium(I) complexes. *Tetrahedron Asymmetry* **15**, 3841–3845.

Challener, C. A. (2001a). Overview of chirality. In C. A. Challener (Ed.), *Chiral Drugs*. New York: John Wiley & Sons, p. 5.

Challener, C. A. (2001b). Sources of chiral compound. In C. A. Challener (Ed.), *Chiral Drugs*. New York: John Wiley & Sons, p. 41.

Chen, P. R., Gu, J. X., Wei, Z. L., Han, S. Q., Li, Z. Y., and Lin, G. Q. (2003). Asymmetric syntheses aided by biocatalysts. *Chin J Chem* **21**, 983–993.

Congreve, M., Murray, C. W., and Blundell, T. L. (2005). Keynote review: Structural biology and drug discovery. *Drug Discov Today* **10**, 895–907.

Corey, E. J., and Link, J. O. (1991). A catalytic enantioselective synthesis of denopamine, a useful drug for congestive heart failure. *J Org Chem* **56**, 442–444.

Corey, E. J., and Reichard G. A. (1989). Enantioselective and practical syntheses of *R*- and *S*-fluoxetines. *Tetrahedron Lett* **30**, 5207–5210.

Crosby, J. (1997). Introduction. In N.A. Collins, G. N. Sheldrake, and J. E. Crosby (Eds.), *Chirality in Industry II*. New York: John Wiley & Sons, pp. 3–4.

Dalby, N. O. (2003). Inhibition of γ-aminobutyric acid uptake: Anatomy, physiology and effects against epileptic seizures. *Eur J Pharmacol* **479**, 127–137.

Deng L., and Jacobsen, E. N. (1992). A practical highly enantioselective synthesis of the taxol side chain via asymmetric catalysis. *J Org Chem* **57**, 4320–4323.

Eap, C. B., Bender, S., Gastpar, M., Fischer, W., Haarmann, C., Powell, K., Jonzier-Perey, M., Cochard, N., and Baumann, P. (2000). Steady state plasma levels of the enantiomers of trimipramine and of its metabolites in CYP2D6-, CYP2C19- and CYP3A4/5-phenotyped patients. *Ther Drug Monit* **22**, 209–214.

Effenberger, F., and Jäger, J. (1997). Synthesis of the adrenergic bronchodilators (*R*)-terbutaline and (*R*)-salbutamol from (*R*)-cyanohydrins. *J Org Chem* **62**, 3867–3873.

Eichelbaum, M., and Gross, A. S. (1996). Stereochemical aspects of drug action and disposition. *Adv Drug Res* **28**, 1–64.

Eliel, E. L., and Wilen, S. H. (1994). Properties of racemates and of their enantiomer components. In E. L. Eliel and S. H. Wilen (Eds.), *Stereochemistry of Organic Compounds*. New York: John Wiley & Sons, pp. 162–208.

Fadel, A., and Garcia-Argote, S. (1996). Asymmetric construction of benzylic quaternary carbons from chiral malonates: formal synthesis of both (−)- and (+)- aminoglutethimides AG and analogues. *Tetrahedron Asymmetry* **7**, 1159–1166.

Faramarzi, M. A., Yazdi, M. T., Amini, G. Z., and Shafiee, A. (2003). Microbial hydroxylation of progesterone with *Acremonium strictum*. *FEMS Microbiol Lett* **222**, 183–186.

FDA's Policy Statement for the Development of New Stereoisomeric Drugs May 1, 1992; Corrections made on January 3, 1997.

Foster, J. R., Somogyi, A. A., Dyer, K. R., White, J. M., and Bochner F. (2000). Steady-state pharmacokinetics of (*R*)- and (*S*)-methadone in methadone maintenance patients. *Br J Clin Pharmacol* **50**, 427–440.

Gao, X. J., Mclean, H., Caveney, S., Donly, C. (1999). Molecular cloning and functional characterization of a GABA transporter from the CNS of the cabbage looper, Trichoplusia ni. *Insect Biochem Molec* **29**, 609–623.

Gao, Y., and Sharpless, K. B. (1988). Asymmetric synthesis of both enantiomers of tomoxetine and fluoxetine. Selective reduction of 2,3-epoxycinnamyl alcohol with Red-Al. *J Org Chem* **53**, 4081–4084.

García, I., Pascual, A., Ballesta, S., and Perea, E. J. (2000). Uptake and intracellular activity of ofloxacin isomers in human phagocytic and non-phagocytic cells. *Int J Antimicrob Agents* **15**, 201–205.

Gillard, M., Van der Perren, C., Moguilevsky, N., Massingham, R., and Chatelain P. (2002). Binding characteristics of cetirizine and levocetirizine to human H_1 histamine receptors: Contribution of Lys^{191} and Thr^{194}. *Mol Pharmacol* **61**, 391–399.

Grabowski, E. J. J. (2005). Enantiopure drug synthesis: from methyldopa to imipenem to efavirenz. *Chirality* **17**, S249–S259.

Graham, D. W., Ashton, W. T., Barash, L., Brown, J. E., Brown, R. D., Canning, L. F., Chen, A., Springer, J. P., and Rogers, E. F. (1987). Inhibition of the mammalian β-lactamase renal dipeptidase (dehydropeptidase-I) by Z-2-(acylamino)-3-substituted-propenoic acids. *J Med Chem* **30**, 1074–1090.

Griesbach, R. C., Hamon D. P. G., and Kennedy, R. J. (1997). Asymmetric dihydroxylation in an approach to the enantioselective synthesis of 2-arylpropanoic acid non-steroidal anti-inflammatory drugs. *Tetrahedron Asymmetry* **8**, 507–510.

Gu, J. X., Li, Z. Y., and Lin, G. Q. (1992). A microbially based approach for the asymmetric synthesis of both enantiomers of *R*- and *S*-fluoxetine. *Chin J Chem* **10**, 355–358.

Gu, J. X., Li, Z. Y., and Lin, G. Q. (1993). Reductive biotransformation of carbonyl compounds—application of fungus, *Geotrichum* sp. G38 in organic synthesis. *Tetrahedron* **49**, 5805–5816.

Gu, J. X., Li, Z. Y., and Lin G. Q. (1995). Microbial-chemical routes to *R*. and *S*-denopamine a useful drug for congestive heart failure. *Chin J Chem* **13**, 150–155.

Han, S. Q., Lin, G. Q., and Li, Z. Y. (1998). Synthesis of (*R*)-cyanohydrins by crude (*R*)-oxynitrilase catalyzed reactions in micro-aqueous medium. *Tetrahedron Asymmetry* **9**, 1835–1838.

Hanna, G. M., and Lau-Cam, C. A. (1995). Determination of the optical purity of timolol maleate by proton nuclear magnetic resonance spectroscopy with a chiral Pr(III) shift reagent. *J Pharm Biomed Anal* **13**, 1313–1319.

Hillier, M. C., and Reider P. J. (2002). Stereoselective synthesis from a process research perspective. *Drug Discov Today* **7**, 303–314.

Höll, V., Kouba, M., Dietel, M., and Vogt, G. (1992). Stereoisomers of calcium antagonists which differ markedly in their potencies as calcium blockers are equally effective in modulating drug transport by P-glycoprotein. *Biochem Pharmacol* **43**, 2601–2608.

Holland, H. L. (1982). The mechanism of the microbial hydroxylation of steroids. *Chem Soc Rev* **11**, 371–395.

Im, D. S., Cheong, C. S., Lee, S. H., Jung, Y. K., and Jeong, I. H. (2003). Chemo-enzymatic synthesis of (*R*)-(+)-aminoglutethimide by kinetic resolution of (±)-4-cyano-4-phenyl-1-hexanol. *J Mol Catal B Enzymol* **26**, 185–191.

Imwinkelried, R. (1997). Catalytic asymmetric hydrogenation in the manufacture of *d*-biotin and dextromethorphan. *Chimia* **51**, 300.

Inamasu, M., Totsuka, T., Ikeo, T., Nagao, T., and Takeyama S. (1987). β_1-adrenergic selectivity of the new cardiotonic agent denopamine in its stimulating effects on adenylate cyclase. *Biochem Pharmacol* **36**, 1947–1954.

Jacques, J., Collet, A., and Wilen, S. H. (1981). Resolution by direct crystallization. In J. Jacques, A. Collet, and S. H. Wilen (Eds.), *Enantiomers, Racemates, and Resolutions*. New York: John Wiley & Sons, pp. 217–250.

Jekkel, A., Ilkoy, E., Horvath, G., Pallagi, I., Suto, J., and Ambrus, G. (1998). Microbial hydroxylation of 13 β-ethyl-4-gonene-3,17-dione. *J Mol Catal B Enzymol* **5**, 385–387.

Jia, X., Li, X. S., Xu, L. J., Li, Y. M., Shi, Q., Au-Yeung, T. T. L., Yip, C. W., Yao, X. S., and Chan, A. S. C. (2004). Titanium-catalyzed tandem sulfoxidation-kinetic resolution process: A convenient method for higher enantioselectivities and yields of chiral sulfoxide. *Adv Synth Catal* **346**, 723–726.

Kanner, B. (1994). Sodium-coupled neurotransmitter transport: structure, function and regulation. *J Exp Biol* **196**, 237–249.

Kirinura, K. (2003). Selective α-glucosylation of eugenol by α-glucosyl transfer enzyme of *Xanthomonas campestris* WU-9701. *J Biosci Bioeng* **96**, 199–202.

Kinbara, K. (2005). Design of resolving agents based on crystal engineering. *Synlett* **5**, 732–743.

Kitoha, S., Kubotab, A., Kunimotoa, K. K., Kuwaec, A., and Hanaic K. (2005). Conglomerate formation and crystal structure of 4-phenyl-1,3-thiazolidin-2-one. *J Mol Struct* **737**, 277–282.

Knowles, W. S. (1986). Application of organometallic catalysis to the commercial production of L-DOPA. *J Chem Ed* **63**, 222–225.

Knutsen, L. J. S., Andersen, K. E., Lau, J., Lundt, B. F., Henry, R. F., Morton, H. E., Nærum, L., Petersen, H., Stephensen, H., Suzdak, P. D., Swedberg, M. D. B., Thomsen, C., and Sørensen P. O. (1999). Synthesis of novel GABA uptake inhibitors. 3. Diaryloxime and diarylvinyl ether derivatives of nipecotic acid and guvacine as anticonvulsant agents. *J Med Chem* **42**, 3447–3462.

Komatsu, N., Nishibayashi, Y., Sugita, T., and Uemura S. (1992). Catalytic asymmetric oxidation of sulfides to sulfoxides using *R*-(+)-binaphthol. *Tetrahedron Lett* **33**, 5391–5394.

Komatsu, N., Hashizume, M., Sugita, T., and Uemura S. (1993). Catalytic asymmetric oxidation of sulfides to sulfoxides with *tert*-butyl hydroperoxide using binaphthol as a chiral auxiliary. *J Org Chem* **58**, 4529–4533.

Koul, S., Parshad, R., Taneja, S. C., and Qazi, G. N. (2003). Enzymatic resolution of naproxen. *Tetrahedron Asymmetry* **14**, 2459–2465.

Kovalev, I. P., Ipatkin, V. V., Strelenko, Y. A., Ignatenko, A. V., and Nikishin, G. I. (1992). Catalytic codimerization of α,β- with γ,δ-unsaturated ketones: novel stereoselective method of the synthesis of functionalized 8-oxabicyclo[3.2.1]octanes. *Tetrahedron Lett* **33**, 1791–1794.

Kurosu, M., and Kishi, Y. (1998). A novel example for optical resolution of racemic ketones originating from Batrachotoxin synthesis. *J Org Chem* **63**, 6100–6101.

Kurosu, M., Marcin, L. R., Grinsteiner, T. J., and Kishi Y. (1998). Total synthesis of (±)-batrachotoxinin A. *J Am Chem Soc* **120**, 6627–6628.

Lamzin, V. S., Dauter, Z., and Wilson, K. S. (1995). How nature deals with stereoisomers. *Curr Opin Struct Biol* **5**, 830–836.

Lane, R. M., and Baker, G. B. (1999). Chirality and drugs used in psychiatry: Nice to know or need to know? *Cell Mol Neurobiol* **19**, 355–372.

Lee, J. C., Kim, G. T., Shim, Y. K., and Kang, S. H. (2001). An asymmetric aminohydroxylation approach to the stereoselective synthesis of *cis*-substituted azetidinone of loracarbef. *Tetrahedron Lett* **42**, 4519–4521.

Leksawasdi, N., Rosche, B., and Rogers, P. L. (2005). Mathematical model for kinetics of enzymatic conversion of benzaldehyde and pyruvate to (*R*)-phenylacetylcarbinol. *Biochem Eng J* **23**, 211–220.

Lin, G. Q., Han, S. Q., and Li, Z. Y. (1999). Enzymatic synthesis of (*R*)-cyanohydrins by three (*R*)-oxynitrilase sources in micro-aqueous organic medium. *Tetrahedron* **55**, 3531–3540.

Lin, G. Q., Li, Y. M., and Chan, A. S. C. (2001). Introduction. In G. Q. Lin, Y. M. Li, and A. S. C. Chan (Eds.), *Principles and Applications of Asymmetric Synthesis*. New York: John Wiley & Sons, pp. 1–4.

Liu, Q. R., López-Corcurera, B., Mandiyan, S., Nelson, H., and Nelson, N. (1993). Molecular characterization of four pharmacologically distinct α-aminobutyric acid transporters in mouse brain. *J Biol Chem* **268**, 2106–2112.

Liu, Z. S., and Fang, Z. L. (1997). Combination of flow injection with capillary electrophoresis. Part 2. Chiral separation of intermediate enantiomers in chloramphenicol synthesis. *Anal Chim Acta* **353**, 199–205.

Lopez-Vidriero, M. T., Jacobs, M. S., and Clarke, W. (1985). The effect of isoprenaline on the ciliary of an *in vitro* preparation of rat trachea. *Eur J Pharmacol* **112**, 429–432.

Maier, N. M., Franco, P., and Lindner, W. (2001). Separation of enantiomers: Needs, challenges, perspectives. *J Chromatogr A* **906**, 3–33.

Man, H. W., Corral, L. G., Stirling, D. I., and Muller, G. W. (2003). α-Fluoro-substituted thalidomide analogues. *Bioorg Med Chem Lett* **13**, 3415–3417.

Martin, G. J., Lavoine-Hanneguelle, S., Mabon, F., and Martin, M. (2004). The fellowship of natural abundance ^{2}H-isotopomers of monoterpenes. *Phytochemistry* **65**, 2815–2831.

Matsumoto, T., Murayama, T., Mitsuhashi, S., and Miura, T. (1999). Diastereoselective synthesis of a key intermediate for the preparation of tricyclic β-lactam antibiotics. *Tetrahedron Lett* **40**, 5043–5046.

Matsuura, T., and Koshima, H. (2005). Introduction to chiral crystallization of achiral organic compounds. *J Photochem Photobiol C Photochem Rev* **6**, 7–24.

McCoy, M. (1999). Biocatalysis grows for drug synthesis: fine chemical makers are increasingly using enzymatic methods to make chiral intermediates. *Chem Eng News* **77**, 11–64.

McCoy, M. (2001). Making drugs with little bugs. *Chem Eng News* **79**, 37–43.

Mehvar, R., and Brocks, D. R. (2001). Stereospecific pharmacokinetics and pharmacodynamics of β-adrenergic blockers in humans. *J Pharm Pharm. Sci* **4**, 185–200.

Miller, M., Schneider, J., Sathyanarayana, B. K., Toth, M. V., Marshall, G. R., Clawson, L., Selk, L., Kent, S. B. H., and Wlodawer A. (1989). Structure of complex of synthetic HIV-1 protease with a substrate-based inhibitor at 2.3 Å resolution. *Science* **246**, 1149–1152.

Milton, R. C., Milton, S. C. F., and Kent, S. S. H. (1992). Total chemical synthesis of a D-enzyme: the enantiomers of HIV-1 protease show demonstration of reciprocal chiral substrate specificity. *Science* **256**, 1445–1448.

Mislow, K. (1999). Molecular chirality. *Top Stereochem* **22**, 1–82.

Müller, T. A., and Kohler, H. P. (2004). Chirality of pollutants—Effects on metabolism and fate. *Appl Microbiol Biotechnol* **64**, 300–316.

Nelson, H., Mandiyan, S., and Nelson, N. (1990). Cloning of the human brain GABA transporter. *FEBS Lett* **269**, 181–184.

Noyori, R., and Ohkuma, T. (2001). Asymmetric catalysis by architectural and functional molecular engineering: Practical chemo- and stereoselective hydrogenation of ketones. *Angew Chem Int Ed* **40**, 40–73.

Onoda, T., Shirai, R., Koiso, Y., and Iwasaki, S. (1995). Synthetic study of curacin A: Synthesis of partial structure of curacin A and elucidation of the absolute stereochemistry and the methylcyclopropylthiazoline moiety. *Tetrahedron Lett* **36**, 5765–5768.

Onoda, T., Shirai, R., Koiso, Y., and Iwasaki, S. (1996). Synthetic study on curacin A: A novel antimitotic agent from the cyanobacterium *Lyngbya majuscula. Tetrahedron* **52**, 14543–14562.

Pallavicini, M., Valoti, E., Villa, L., and Piccolo, O. (2000). Resolution of β-unsaturated amines with isopropylidene glycerol hydrogen phthalate. *Tetrahedron Asymmetry* **11**, 4017–4025.

Patel, R. N. (2001a). Biocatalytic synthesis of intermediates for the synthesis of chiral drug substances. *Curr Opin Biotechnol* **12**, 587–604.

Patel, R. N. (2001b). Enzymatic synthesis of chiral intermediates for Omapatrilat, an antihypertensive drug. *Biomol Eng* **17**, 167–182.

Patel, R. N. (2002). Microbial/enzymatic synthesis of chiral intermediates for pharmaceuticals. *Enzyme Microb Technol* **31**, 804–826.

Patel, R. N., Banerjee, A., Chu, L., Brzozowski, D., Nanduri, V., and Szarka, L. J. (1998). Microbial synthesis of chiral intermediates for β_3-receptor agonists. *J Am Oil Chem Soc* **75**, 1473–1482.

Patel, R. N., Banerjee, A., and Nanduri, V. (2000). Enzymatic acetylation of 10-deacetylbaccatin III to baccatin III by C-10 deacetylase from *Nocardioides luteus* SC 13913. *Enzyme Microb Technol* **27**, 371–375.

Piperaki, S., Penn, S. G., and Goodall, D. M. (1995). Systematic approach to treatment of enantiomeric separations in capillary electrophoresis and liquid chromatography II. A study of the enantiomeric separation of fluoxetine and norfluoxetine. *J Chromatogr A* **700**, 59–67.

Pires de Abreu, L. R., Lanchote, V. L., Bertucci, C., Cesarino, E. J., and Bonato, P. S. (1999). Simultaneous determination of propafenone and 5-hydroxypropafenone enantiomers in plasma by chromatography on an amylose derived chiral stationary phase. *J Pharmacol Biomed Anal* **20**, 209–216.

Pitchen, P., France, C. J., McFarlane, I. M., Newton, C. G., and Thompson, D. M. (1994). Large scale asymmetric synthesis of a biologically active sulfoxide. *Tetrahedron Lett* **35**, 485–488.

Radian, R., Bendahan, A., and Kanner, B. I. (1986). Purification and identification of the functional sodium- and chloride-coupled γ-aminobutyric acid transport glycoprotein from rat brain. *J Biol Chem* **261**, 437–441.

Rasool, S., Johri, S., Riyaz-ul-Hassan, S., Maqbool, Q., Verma, V., Koul, S., Taneja, S. C., and Qazi, G. N. (2005). Molecular cloning of enantioselective ester hydrolase from *Bacillus pumilus* DBRL-191. *FEMS Microbiol Lett* **249**, 113–120.

Rentsch, K. M. (2002). The importance of stereoselective determination of drugs in the clinical laboratory. *J Biochem Biophys Methods* **54**, 1–9.

Rogers, R. S. (1999). Companies turn to biocatalysis: drive to meet customer demands for single-isomer intermediates takes on urgency. *Chem Eng News* **77**, 87–91.

Rojas-Corrales, M. O., Giber-t-Rahola, J., and Mica, J. A. (1998). Tramadol induces antidepressant-type effects in mice. *Life Sci* **63**, 175–180.

Rouhi, A. M. (2002). Chiral roundup. *Chem Eng News* **80**, 43–50.

Rouhi, A. M. (2003). Chiral at work: Drug developers can learn much from recent successful and failed chiral switches. *Chem Eng News* **81**, 56–61.

Rouhi, A. M. (2004). Chiral chemisty: Traditional methods thrive despite numerous hurdles, including tough luck, slow commercialization of catalytic processes. *Chem Eng News* **82**, 47–62.

Ruffolo, R. R., Jr., and Messick, K. (1985). Systemic hemodynamic effects of dopamine, (±)-dobutamine and the (+)- and (−)-enantiomers of dobutamine in anesthetized normotensive rats. *Eur J Pharmacol* **109**, 173–181.

Sacerdote, P., Bianchi, M., Gaspani, L., and Panerai, A. E. (1999). Effects of tramadol and its enantiomers on Concanavalin-A induced-proliferation and NK activity of mouse splenocytes: Involvement of serotonin. *Int J Immunopharmacol* **21**, 727–734.

Sakai, K., Sakurai, R., Yuzawa, A., and Hirayama, N. (2003). Practical continuous resolution of α-amino-ε-caprolactam by diastereomeric salt formation using a single resolving agent with a solvent switch method. *Tetrahedron Asymmetry* **14**, 3713–3718.

Sakuta, H., and Okamoto, K. (1994). Effects of Ca^{2+} channel antagonists and their isomers on glibenclamide-sensitive K^+ currents in follicle-enclosed *Xenopus* oocytes. *Eur J Pharmacol* **255**, 1–7.

Salunkhe, M. M., and Nair, R. V. (2001). Novel route for the resolution of both enantiomers of dropropizine by using oxime esters and supported lipases of *Pseudomonas cepacia. Enzyme Microb Technol* **28**, 333–338.

Sato, T., Takeuchi, H., Takahashi, K., Kurosu, J., Yoshida, K., Tsugane, T., Shimura, S., Kino, K., Collins, N. A., Schousboe, A., Sarup, A., Larsson, O. M., and White, H. S. (2004). GABA transporters as drug targets for modulation of GABAergic activity. *Biochem Pharmacol* **68**, 1557–1563.

Satoh, K., Minako, I., and Kanai, K. (1998). Synthesis of a key intermediate of levofloxacin via enantioselective hydrogenation catalyzed by iridium (I) complexes. *Tetrahedron Asymmetry* **9**, 2657–2662.

Schousboe, A., Sarup, A., Larsson, O. M., and White, H. S. (2004). GABA transporters as drug targets for modulation of GABAergic activity. *Biochem Pharmacol* **68**, 1557–1563.

Seki, M., Kayo, A., and Mori, K. (2001). Synthesis of (2*S*,3*R*,11*S*,12*R*,2ß*R*,11ß*S*,12ß*R*)-plakoside A, a prenylated and immunosuppressive marine galactosphingolipid with cyclopropane-containing alkyl chains. *Tetrahedron Lett* **42**, 2357–2360.

Sheldon, R. A. (1993). Racemate resolution via crystallization. In R. A. Sheldon (Ed.), *Chirotechnology: Industrial Synthesis of Optically Active Compounds*. New York: Marcel Dekker, pp. 191–192.

Soudijn, W., and van Wijngaarden, I. (2000). The GABA transporter and its inhibitors. *Curr Med Chem* **7**, 1063–1079.

Soudijn, W., van Wijngaarden, I., and Ijzerman, A. P. (2004). Stereoselectivity of drug-receptor interactions. *IDrugs* **6**, 43–56.

Stinson, S. C. (2000). Chiral drugs. *Chem Eng News* **78**, 55–78.

Tenero, D., Boike, S., Boyle, D., Ilson, B., Fesniak, H. F., Brozena, S., and Jorkasky, D. (2000). Steady-state pharmacokinetics of carvedilol and its enantiomer in patients with congestive heart failure. *J Clin Pharmacol* **40**, 844–853.

Tian, P., Peng, J. S., Li, Z. Y., and Lin, G. Q. (2003). Lipase-catalyzed resolutions of both enantiomers of ornidazole and secnidazole. *Chin J Chem* **21**, 853–857.

Tillementa, J. P., Testab, B., and Brée, F. (2003). Compared pharmacological characteristics in humans of racemic cetirizine and levocetirizine, two histamine H_1-receptor antagonists. *Biochem Pharmacol* **66**, 1123–1126.

Triggle, D. J. (1997). Stereoselectivity of drug action. *Drug Discov Today* **2**, 138–147.

Tripathi, C. M., Agarwal, S. C., and Basu, S. K. (1997). Production of L-phenylacetylcarbinol by fermentation. *J Ferment Bioeng* **84**, 487–492.

Vansal, S. S., and Feller, D. R. (1999). Direct effects of ephedrine isomers on human β-adrenergic receptor subtypes. *Biochem Pharmacol* **58**, 807–810.

Varesio, E., and Veuthey, J. L. (1995). Chiral separation of amphetamines by high-performance capillary electrophoresis. *J Chromatogr A* **717**, 219–228.

Waldeck, B. (2003). Three-dimensional pharmacology, a subject ranging from ignorance to overstatements. *Pharmacol Toxicol* **93**, 203–210.

Wang, Q. W., Yang, F., Du, H., Mahmun Hossain, M., Bennett, D., and Grubisha, D. S. (1998). The synthesis of *S*-(+)-2,2-dimethylcyclopropane carboxylic acid: A precursor for cilastatin. *Tetrohedron Asymmetry* **9**, 3971–3977.

Wellman, P. J., Miller, D. K., Livermore, C. L., Green, T. A., McMahon, L. R., and Nation, J. R. (1998). Effects of (−)-ephedrine on locomotion, feeding, and nucleus accumbens dopamine in rats. *Psychopharmacology* **135**, 133–140.

Ye, Q., Wang, X. S., Zhao, H., and Xiong, R. G. (2005). Spontaneous resolution of a novel chiral coordination polymer through supramolecular interactions and solvent symmetry breaking. *Tetrahedron Asymmetry* **16**, 1595–1602.

Young, R., and Glennon, R. (1998). Discriminative stimulus properties of (−)-ephedrine. *Pharmacol Biochem Behav* **60**, 771–775.

Zaitsev, V. G., and Mikhal'chuk, A. L. (2001). Enantioconvergent synthesis of (−)-(*S*)- and (+)-(*R*)-2-acetyl-3,6-dihydroxycyclohex-2-enone. Starting from rac-6-hydroxy-3-methoxy cyclohex-2-enone. *Chirality* **13**, 488–492.

Zaks, A. (2001). Industrial biocatalysis. *Curr Opin Chem Biol* **5**, 130–136.

Zeng, S., Zhong, J., Pan, L., and Li, Y. (1999). High-performance liquid chromatography separation and quantitation of ofloxacin enantiomers in rat microsomes. *J Chromatogr B Biomed Sci Appl* **728**, 151–155.

Zhang, Y. R., Wu, D. R., Wang-Iverson, D. B., and Tymiak, A. A. (2005). Enantioselective chromatography in drug discovery. *Drug Discov Today* **10**, 571–577.

Zimmerman, D. M., Smits, S. E., Hynes, M. D., Cantrell, B. E., Leander, J. D., Mendelsohn, L. G., and Nickander, R. (1985). Picenadol. *Drug Alcohol Depend* **14**, 381–401.

Zurer, P. (2005). Crixivan. *Chem Eng News* **83**, 3.

CHAPTER 11

ALPHA-HELIX MIMETICS IN DRUG DISCOVERY

*Hang Yin, Gui-In Lee, and Andrew D. Hamilton**

11.1 INTRODUCTION

The α-helix, first characterized by Linus Pauling in 1951 (Pauling et al., 1951), has been extensively studied due to its prevalence in structural biology (Brandon, 1991). α-Helices are the most common secondary conformation in natural proteins (~40% of amino acids adopt helical conformations) (Ruan et al., 1990). A typical α-helix completes one rotation with 3.6 amino acid residues, each with backbone dihedral angles of $\Psi = -50^\circ$ and $\Phi = -60^\circ$ (Fig. 11.1) (Pauling and Corey, 1954). The helix has a rise of 1.5 Å/residue or 5.4 Å/turn (Regan, 1993); as a result, the side chain of a certain residue at position i projects from the same face as those from the $i + 3$, $i + 4$, and $i + 7$ residues in the sequence. There is a large entropy cost in helix formation, which is thermodynamically compensated by the formation of intramolecular hydrogen bonds. The backbone of the α-helix is primarily stabilized by hydrogen bonds between the carbonyl oxygen at the i position and the carboxamide hydrogen at the $i + 4$ position (Pauling and Corey, 1954). Because all amides are oriented in the same direction and also the hydrogen bonding sites on the first and last turns are unfulfilled, a macro-dipole is produced (Galoppini and Fox, 1996; Scholtz et al., 1991). The positive end of the dipole is centered at the N-terminus, and the negative end is centered at the C-terminus. The total dipole is augmented if the peptide exists in conditions where both the termini are ionized. α-Helices composed of L-amino acids

*Corresponding author: andrew.hamilton@yale.edu

Drug Discovery Research: New Frontiers in the Post-Genomic Era, Edited by Ziwei Huang

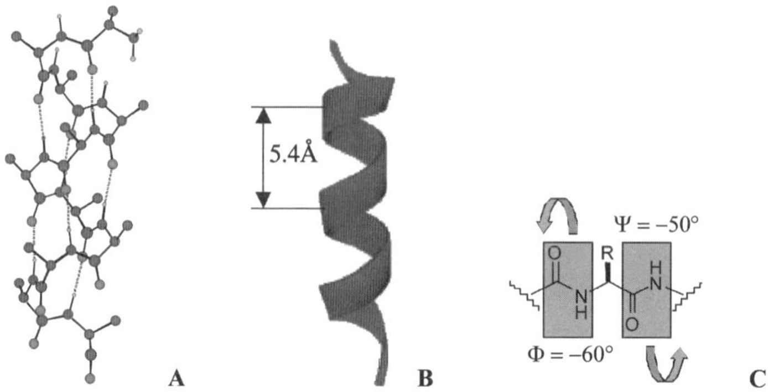

Figure 11.1 (A) An α-helix that is stabilized by internal hydrogen bonding. (B) Ribbon diagram showing rise per turn. (C) Amide backbone showing dihedral angles. See color plates.

almost always adopt a right-handed twist. The alternative left-handed conformation is destabilized due to steric crowding of the side chains and the backbone carbonyl groups. The helix propensities of the amino acids have been studied extensively in an attempt to predict secondary structure from primary sequences (Lifson, 1961; Zimm and Bragg, 1959). Certain amino acids, such as alanine, are more prevalent in α-helices, while other residues, such as proline, are rarely found. The principal factors that dictate helix propensities are conformational restrictions of the side chains and steric considerations. α-Helices are the bioactive regions of numerous proteins, playing critical roles in mediating many protein–protein, protein–DNA, and protein–RNA interactions (Fairlie et al., 1998). Therefore, they are attractive targets for the design of molecular mimetics, which could be very useful in the development of potential drugs and *de novo* proteins. In the past decade, major advances have been made in this field, evolving from strategies based on induced helix stabilization to the recent advent of helix proteomimetics, molecules that mimic the surface functionalities presented by α-helical secondary structures (Peczuh and Hamilton, 2000).

11.2 CONSTRAINED PEPTIDES THAT ADOPT α-HELICAL CONFORMATIONS

Peptides that adopt helical conformations would serve as good inhibitors of macromolecular interactions (Fairlie et al., 1998). However, peptide sequences derived from α-helical domains in proteins frequently lack structural integrity due to the loss of the stabilizing environment provided by the protein. The modest thermodynamic gain of internal hydrogen bonding in an isolated helix is simply not enough to compensate for the large entropic penalty of conformational restriction for short peptides. Therefore, the goal of much research has been to develop methods that address this phenomenon by designing molecular systems that either initiate or stabilize α-helicity in short peptides (Andrews and Tabor, 1999).

11.2.1 Stabilization of α-Helical Structures Using Covalent Interactions

The formation of covalent linkages between adjacent residues in peptides has been shown to impart stabilization to the helical form of the peptide (Fig. 11.2). Disulfide bonds, lactam linkages, hydrazones, and carbon–carbon bonds have all been used to stabilize helices (Fig. 11.2). Noteworthy examples include a redox-triggered disulfide bond (**1**) used by Jaskson et al. (1991) and the lactam tethers (**2**) developed by Yu and Taylor (1999) and by Osapay and Taylor (1992) to promote helicity. Flint et al. (2002) have shown that an azobenzene (AZO) (**3**) linkage through two cysteine thiols can be used to both stabilize and destabilize the helical form of a peptide under photocontrol of the AZO *trans/cis*-isomerization. Blackwell and Grubbs (1998) have demonstrated that olefin metathesis (**4**) of two *O*-allyl serines at the i and $i+4$ positions can facilitate the transition from an α-helix to a 3_{10}-helix in nonpolar solvents. In a more recent example, Schafmeister et al. (2000) have elaborated upon this work to show that enhanced and compromised helix stability can be conferred upon the C-terminal peptide of RNase A via incorporation and subsequent metathesis of α,α–disubstituted amino acids.

11.2.2 Stabilization of α-Helical Structures Using Noncovalent Interactions

Stabilization of α-helical structures has also been achieved using noncovalent interactions between appropriately spaced residues in a peptide chain. Metal–ligand interactions, designed host–guest interactions, salt bridges, cation–π interaction, and π–π stacking have resulted in helix stabilization (Fig. 11.3). Albert and Hamilton (1995) have studied hydrophobic interactions in the stabilization of helical structure in short peptides that incorporated two ε-(3,5-dinitrobenzoyl)Lys residues (**5**) at various positions. Solvent mixture titrations in TFE/water demonstrated that helical stability is highest for the peptide having a pair of modified residues at the i and $i+4$ positions. Using an intermolecular hydrogen-bonding strategy, the same group has shown in several studies that guanidinium-based receptors can recognize aspartate residues at multiple i and $i+3$ or $i+4$ positions and induce α-helix formation (Peczuh and Hamilton, 2000; Albert et al., 1995; Peczuh et al., 1997). Scholtz et al. (1993) showed that the (i, $i+3$) and (i, $i+4$) Glu–Lys interactions (**6**) are

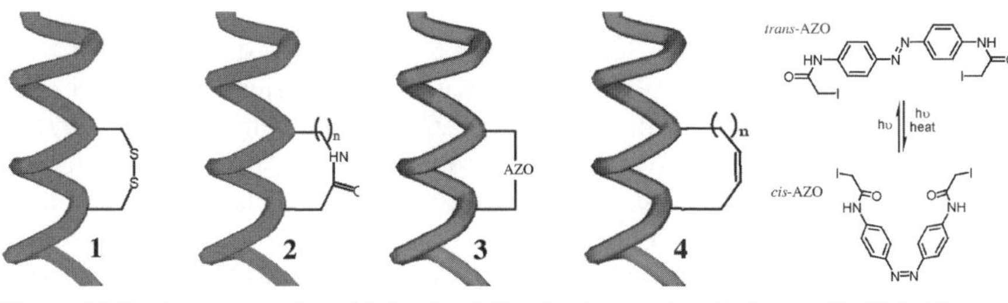

Figure 11.2 Some examples of helical stabilization by covalent linkages: disulfide (**1**), lactam (**2**), azobenzene (**3**), olefin (**4**). See color plates.

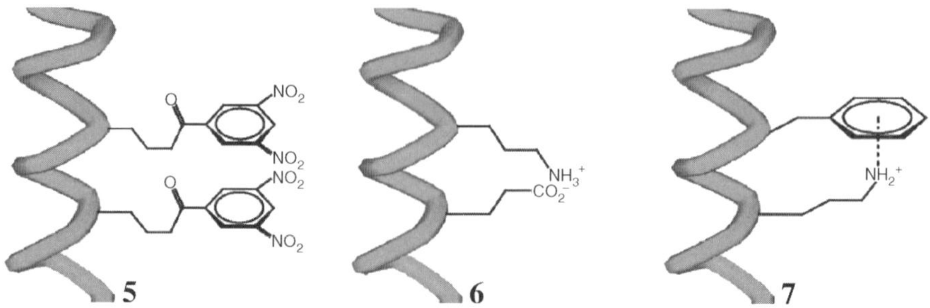

Figure 11.3 Some examples of helical stabilization by non-covalent interaction: π-π interaction (**5**), salt-bridge (**6**), cation-π interaction (**7**). See color plates.

helix-stabilizing and are similar to each other in strength, regardless of the orientation of the side chains. The same group also reported that an Arg–Asp salt bridge, spaced either i, $i + 3$, or i, $i + 4$ apart, stabilizes the α-helical conformation in a similar fashion (Huyghuesdespointes et al., 1993). Olson et al. (2001) demonstrated that the cation–π interaction of a Trp–Arg (i, $i + 4$) pair stabilizes the α-helical conformation with a quantified stabilization energy of −0.4 kcal/mol. Tsou et al. (2002) showed that the Phe–Lys cation–π interaction (**6**) contributed similar energy (−0.4 kcal/mol) to the stability of an α-helical peptide.

11.3 LOW-MOLECULAR-WEIGHT α-HELICAL PROTEOMIMETICS

Although many of these helical stabilization methods have been very successful, the use of such peptide systems for exerting biological effects is often compromised due to poor cellular uptake and proteolytic degradation. The structural characteristics of α-helices make them intriguing targets for low-molecular-weight proteomimetics. Such approaches offer the advantages of improved biostability, lower molecular weight (<750 Da), and better bioavailability (Lipinski et al., 1997). The challenge is to design synthetically viable scaffolds that project functionality with similar distance and angular constraints to the analogous side chains projecting from an α-helical surface.

The development of small molecule modulators of protein–protein and protein–peptide interactions has been widely regarded as a formidable goal. Large interfacial areas, typically more than 1600 Å^2 of buried area (around 170 atoms), are involved in forming a protein–protein interface, posing a serious challenge for any small molecule to be competitive (Stites, 1997). The binding regions of protein partners are often discontinuous and thus hard to be mimicked by simple synthetic peptides. Conventional methods for identifying inhibitors of protein–protein interactions require much input in the preparation and screening of a diverse chemical library to discover lead compounds. An alternative approach for identifying inhibitors is to design synthetic scaffolds that reproduce structural features of the protein secondary structure.

11.3.1 Indane-Based Mimetics that Target the Tachykinin Receptors

At this stage, only a few synthetic mimetics of α-helices have been reported. The pioneering work of Horwell et al. (1996) showed that 1,6-disubstituted indanes present functionalities in a similar spatial arrangement to the i and $i + 1$ residues of an α-helix (Horwell et al., 1996a,b). Trisubstituted indanes **8** (Fig. 11.4) were used to mimic the presentation of the i, $i + 1$ residues of an α-helix. Molecular modeling experiments involving the (S)-enantiomer of an Ala–Ala mimetic were conducted, overlaying the two C_α and the two C_β carbon atoms of the lowest energy conformation of the indane system with the corresponding positions in a polyalanine α-helix. A root mean square (RMS) deviation of 0.15 Å was reported, suggesting that the template was orientated within the same space occupied by the helix backbone. Derivatives with large hydrophobic peptide side chains (Phe–Phe and Trp–Phe) were found to bind to the tachykinin receptors, NK1, NK2, and NK3, with micromolar affinity. This work represents a peptidomimetic approach to the design of molecules based on templating the i and $i + 1$ side chains of an α-helix, elucidating nicely the potential of using nonpeptide scaffolds to present critical recognition functionality in a suitable spatial orientation for binding to a target protein. However, these mimics do not cover a surface large enough to sufficiently represent an α-helical mimetic. Although initial results suggest that this strategy is partly successful by producing compounds with biological activity similar to their dipeptide target, further application of this concept to the mimicry of other biologically active helices has not been reported. Also, extension of this strategy to mimic larger α-helical regions remains to be explored.

11.3.2 Pentasaccharide-Based Mimetic of GCN4

The work of Xuereb et al. (2000) represents the first example of a synthetic scaffold to mimic an extended surface area of an α-helix, in which a pentasaccharide scaffold (**9**, Fig. 11.5) was used to mimic the basic region of the transcription factor GCN4. GCN4 contains a bZIP protein that binds DNA in the major groove with one surface of an α-helical region (Ellenberger et al., 1992). A number of positively charged side chains emanating from an α-helical region of GCN4 are responsible for binding contacts to

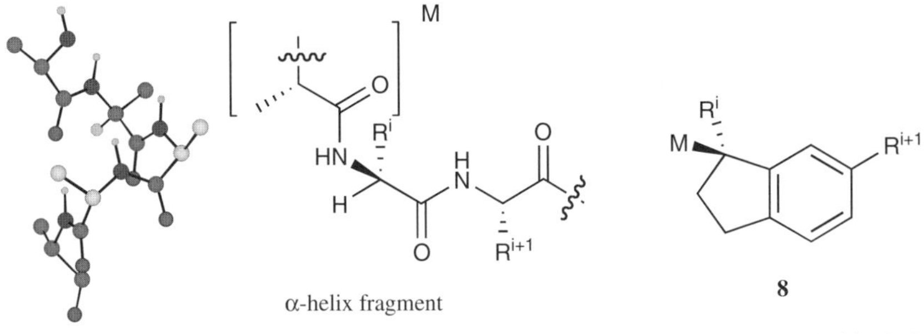

8

Figure 11.4 A 1,6-disubstituted indane molecule, **8**, that mimics of the i and $i + 1$ side chains (shown in yellow) of an α-helix. See color plates.

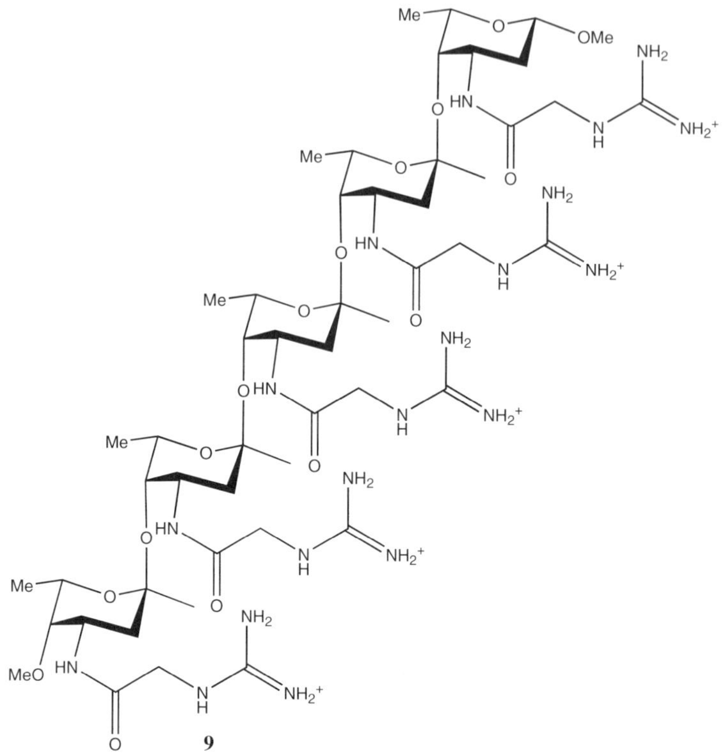

9

Figure 11.5 Structure of pentasaccharide **9**.

both the phosphate backbone and the nucleotide bases. To mimic the basic domain, a conformationally rigid oligosaccharide scaffold was designed in which positively charged guanidinium groups project from the surface in a similar array to that displayed by the α-helix target. The monomer chosen for the design is 2-deoxyfucose because it is more hydrophobic than fully oxygenated sugars, and this was expected to help facilitate groove binding (Kahne, 1995; Ding and Ellestad, 1991). Conformational analyses and molecular mechanics calculations showed that a pentasaccharide of this monomer with guanidinium groups attached at the C_3 positions was a suitable α-helix mimetic. The preferred conformation of this oligosaccharide in D_2O was calculated through a combination of 2D-NMR experiments and was found to match the predicted glycosidic linkage conformation. Using an ethidium bromide competition assay, pentasaccharide **9** was found to bind calf thymus DNA with a dissociation constant (K_d) of 10^{-6} M, which is similar to an unstructured pentalysine control peptide. However, the nonpolyelectrolyte contribution to binding was calculated to be 1 kcal/mol higher for the pentasaccharide, suggesting that there is a contribution to the binding that is not associated with electrostatics. Nuclear Overhauser effects (NOE) suggested that **9** binds to DNA in the minor groove with one surface of the oligosaccharide contacting the floor of the DNA. Preliminary experiments to determine selectivity suggested that **9** discriminates duplex DNA from other types of nucleic acid. Even at high concentrations, pentasaccharide **9** was unable to displace ethidium bromide from double-stranded regions of yeast tRNA. However, the limits of this approach are quite apparent. The mimetic binds relatively weakly in comparison to the natural protein, although it mimics a large percentage of the side chains responsible for the binding energy of GCN4 to DNA. More importantly, pentasaccharide **9** fails to bind in the major groove of DNA, a point that weakens the argument of the helix mimicry design, especially since naturally occurring proteins rarely use the α-helix motif in minor groove recognition.

11.3.3 Hairpin-Based Mimetics of p53 that Disrupt p53–HDM2 Interaction

Fasan et al. (2004) have recently reported structural α-helical mimetics as antagonists of p53–HDM2 interaction based on a β-hairpin scaffold. p53 is a transcription factor mutated in approximately 50% of human tumors (Hollstein et al., 1994). Overexpression of wild-type p53 induces a large number of downstream genes that lead to cell cycle arrest or apoptosis. In unstressed normal cells, p53 is present at a very low level due to rapid degradation through the ubiquitin-dependent proteasome pathway. Human double minute 2 (HDM2) is an important regulator of p53 and acts by binding p53 to promote its ubiquitination (Haupt et al., 1997; Honda et al., 1997; Kubbutat et al., 1997). The overexpression of HDM2 abrogates the ability of p53 to induce cell cycle arrest and apoptosis (Chen et al., 1993). Therefore, deregulation of the overexpression or hyperactivation of HDM2 plays a key role in tolerance of wild type p53, making it an attractive target for the development of antitumor agents.

Recently, several approaches have been pursued to disrupt the p53–HDM2 interaction. García-Echeverría et al. (2000) have reported that a 13-residue peptide derived from the p53 protein (a.a. 16–28) gave a potent inhibition ($IC_{50} = 8.7\ \mu M$) of HDM2. Blaydes et al. (1997) used microinjection of the 3G5 antibody to inhibit p53 binding to HDM2 and activated p53 function *in vivo*. Chen et al. (1998) showed that antisense oligonucleotides can inhibit MDM2, a murine analogue protein of HDM2, resulting in activation of p53 and cell growth arrest or apoptosis. These studies provided a proof-of-concept for activating p53 by targeting HDM2; yet the design of synthetic molecules that target the HDM2 exterior surface has been barely explored.

Robinson's helical mimetic design is based on the X-ray crystal structure of the HDM2/p53 complex (Fig. 11.6a), which reveals that the p53 peptide adopts a largely amphipathic α-helical conformation, with side chains of Phe19, Trp23, and Leu26 inserting into a hydrophobic cleft on the HDM2 surface. β-Hairpins (Fig. 11.6b) naturally occur in many protein–protein interfaces (Fasan et al., 2004). The distance between the C_α atoms of two residues i and $i + 2$ along one strand of a β-hairpin is similar to the α-helical side chains at the i and $i + 4$ residues (Fasan et al., 2004). Incorporated with the functional epitope, a β-hairpin can act as the scaffold to present the Phe19 and Trp23 in a similar fashion as the p53 peptide. To test this design, a series of β-hairpins with different peptide sequences grafted onto one of the hairpin strands was prepared by solid-phase synthesis. β-Hairpin **9**, which has a sequence of Phe1, Trp3, and Leu4, showed the ability to inhibit p53 binding to HDM2 with an IC_{50} of $125 \pm 8\ \mu M$ using a surface plasmon resonance (SPR) assay. The control compounds,

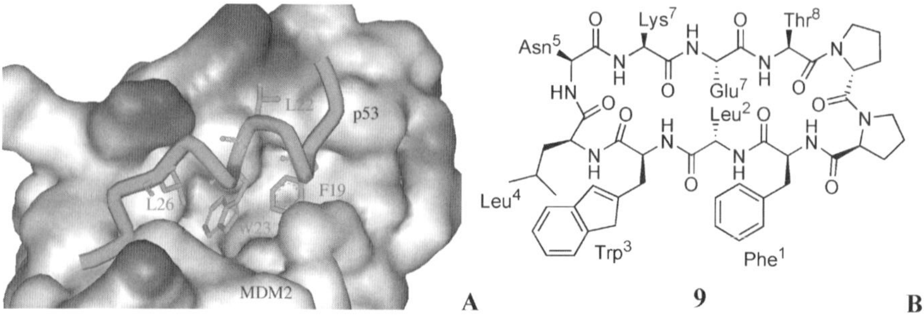

9

Figure 11.6 (A) X-ray crystal structure of the HDM2/p53 complex. (B) β-hairpin inhibitor that disrupts the HDM2-p53 interaction. See color plates.

with Ala replacing either the Phe or Trp residue, were not able to block the HDM2–p53 complexation (IC_{50}>1000 μM), indicating the importance of the side-chain complementarity. The most potent β-hairpin inhibitor, with a sequence of Phe1, (6Cl)-Trp (6-chlorotryptophan), and Leu4, showed an increased affinity over the initial lead **9** by a factor of almost 900-fold ($IC_{50} = 0.14 \pm 0.06$ μM). As a comparison, the linear p53-derived peptide (a.a. 11–29) showed an IC_{50} of 1.1 μM in the same assay. The binding modes of these β-hairpin mimetics were studied by ^{15}N-HSQC NMR experiments. The residues that showed chemical shift changes upon addition of the hairpin mimetics are consistent with the p53-binding site revealed by X-ray crystallography, confirming that β-hairpins indeed act as structural mimetics of the p53 peptide.

11.3.4 Terphenyl-Based Mimetics that Target Calmodulin, gp41, and Bcl-x_L Proteins

Hamilton and co-workers have reported a strategy of α-helix mimicry using oligophenylene scaffolds (Orner et al., 2001; Ernst et al., 2001; Kutzki et al., 2002). The goal was to reduce the complexity of the helical backbone while keeping the critical functionality in a similar spatial arrangement (Fig. 11.7). The first generation of the

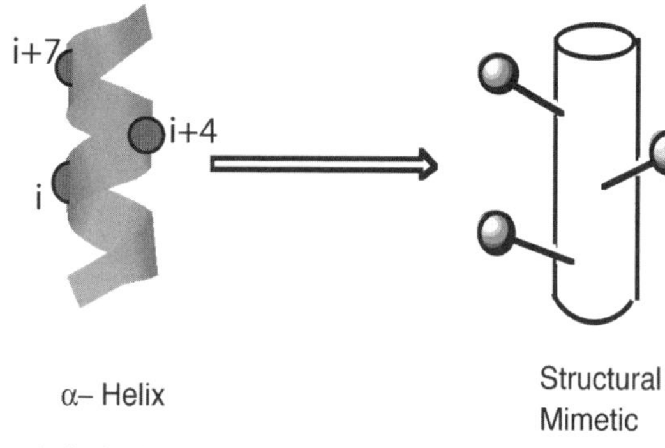

Figure 11.7 An α-helix backbone can be replaced by a synthetic scaffold that projects the functionalities in a similar fashion. See color plates.

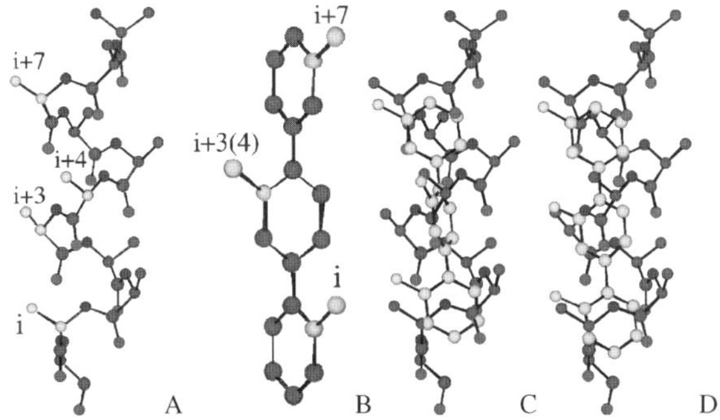

Figure 11.8 (A) A poly-alanine α-helix, with the i, $i+3$, $i+4$, and $i+7$ side chains being highlighted in yellow. (B) An x-ray crystal structure of 3,2′,2″-trimethyl-4-nitro-4″-hydroxy-terphenyl. The analogous carbon atoms between structures are highlighted in yellow. Overlay of an energy-minimized conformation of 3,2′,2″-trimethylterphenyl (yellow) and a poly-alanine α-helix at (C) the i, $i+4$, and $i+7$ or (D) the i, $i+3$, and $i+7$ positions. See color plates.

design was based on a 3,2′,2″-trisubstituted terphenyl scaffold (Orner et al., 2001; Ernst et al., 2001; Kutzki et al., 2002). This molecule was chosen for its low molecular weight, the modularity of its synthesis, and most importantly for its conformational characteristics. The ortho-functionalization serves to project the critical side chains that mimic the i, $i+3$ or $i+4$, and $i+7$ residues on one face of the helix (Fig. 11.8).

Due to the low rotational energy barrier about the phenyl-phenyl bonds in this system (Carreira and Towns, 1977), energetically minimized conformations are accessible that present the substituents in a way that mimics α-helices (Fig. 11.8). An X-ray crystal structure of 3,2′,2″-trimethyl-4-nitro-4″-hydroxy-terphenyl revealed that the molecule takes up the expected staggered conformation with dihedral angles of 59.1° and 120.7° (Fig. 11.8b) (Orner et al., 2001). Due to the low barrier of rotation, interconversion between different conformations in solution should be facile, thus easily populating the desired conformation for binding. The low rotational energy barrier (<15 kcal/mol), as exemplified by similarly substituted biphenyls, permits some desired conformational flexibility for adaptation to the target surfaces (Carreira and Towns, 1977). However, this flexibility will likely come with an entropic cost due to the recognition of the terphenyl on induced-fit-binding.

Computer modeling experiments comparing a polyalanine α-helix to 3,2′,2″-trimethylterphenyl indicated that the phenyl–phenyl torsion angles of 55° gave a conformation with close correspondence (RMS deviation = 0.9 Å) to the positions of the three methyl groups and the i, $i+4$, $i+7$ alanine methyl side chains (Fig. 11.8c). Overlay of the isomeric low-energy conformation (in which the middle ring has phenyl–phenyl torsion angles of −55°) with the i, $i+3$, and $i+7$ alanine methyl side chains gave an RMS deviation of 1.06 Å (Fig.11.8d).

11.3.4.1 Inhibitors of the Calmodulin/Smooth Muscle Myosin Light Chain Kinase Association

The complex of calmodulin (CaM) and smooth muscle myosin light chain kinase (smMLCK) was chosen as an initial system to study because of its extensive structural characterization (Meador et al., 1992, 1993). Calmodulin (calcium modulated protein) is a ubiquitous and essential Ca^{2+}-binding protein found in all eukaryotes (Vogel, 1994). It plays a pivotal role in the regulation of gene expression, protein syntheses, ion channel functions, cell motility, and chemotaxis by sensing the Ca^{2+} level in the cytoplasm. CaM is a 17-kDa acidic protein containing two globular EF-hand motifs, each of which binds two Ca^{2+} ions. Upon binding of Ca^{2+}, CaM undergoes a significant conformational change, resulting in exposure of two hydrophobic patches to recruit CaM-binding proteins (CaMBP). Many CaM-binding proteins recognize an α-helix of roughly 20 amino acids. A representative example is the binding of smMLCK (RS20) to CaM. The crystal structure of this complex is shown in Fig. 11.9 (Meador et al., 1992). Mutational studies of RS20 have established that an i, $i+3$, and $i+7$ arrangement of residues (Trp800, Thr803, and Val807) is critical for binding.

Compound **11a** was designed using this terphenyl mimetic strategy, and for synthetic simplicity the indole segment of Trp800 residue was changed (Fig. 11.10) to a phenyl group and the hydroxyl group of Thr803 was removed. Compound **11a** was determined to bind CaM using an enzymatic assay (Johnson et al., 1987). The enzyme 3′, 5′-cyclic nucleotide phosphodiesterase (PDE) is a CaM-activated enzyme, which can be inhibited by RS20 via competition binding to CaM. Compound **11a** effectively inhibits the ability of PDE to hydrolyze substrate with a comparable affinity ($IC_{50} = 0.8\ \mu M$) to RS20 ($IC_{50} = 0.08\ \mu M$). The control molecule **10**, possessing three methyl side chains, showed no inhibitory effect under the same conditions ($IC_{50} > 20\ \mu M$).

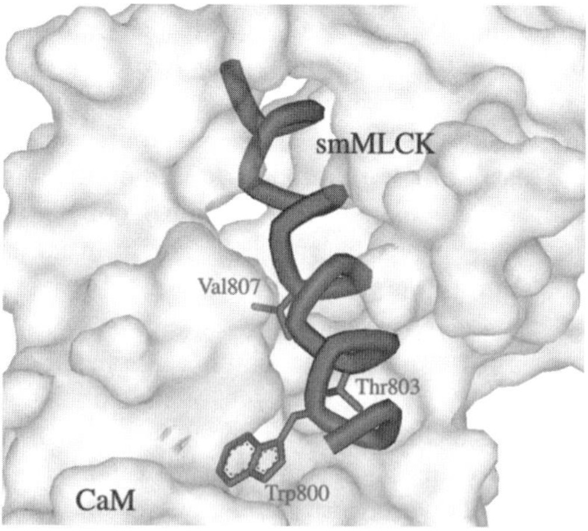

Figure 11.9 X-ray crystal structure of the CaM-binding region of smMLCK (residues 796-815, shown in red) in a complex with CaM. Side chains of the critical residues (Trp800, Thr803 and Val807) are shown in stick representation. See color plates.

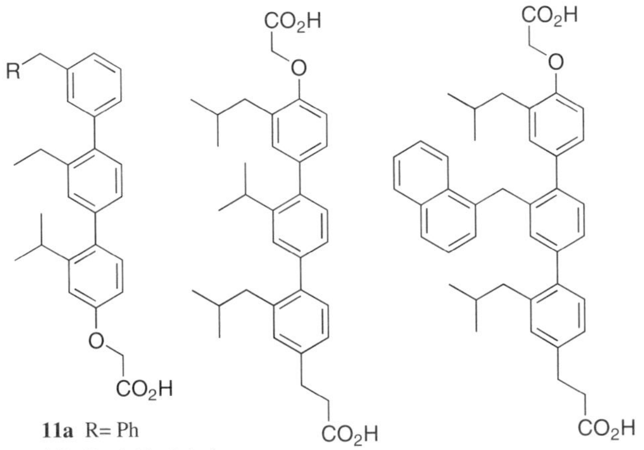

11a R= Ph
11b R= 1-Naphthyl
11c R= 2-Naphthyl

12

13

Figure 11.10 Terphenyl α-helical mimetics as antagonists of protein-protein interactions.

A competition experiment using fluorescence spectroscopy was used to elucidate the binding site of these inhibitors. An α-helical peptide derived from the plasma membrane calcium pump, C20W, has been shown to bind CaM at the same site as smMLCK RS20 (Elshorst et al., 1999). Addition of dansylated C20W to CaM resulted in an increase in fluorescence, which was indicative of complex formation. Addition of **11a** to the complex showed the reverse effect, suggesting that **11a** and C20W are competing for the same binding site. The results from these assays suggested that an improved representation of the indole moiety (Trp800) in the design could perhaps optimize the fit into the CaM binding pocket and enhance potency. Molecules **11b** and **11c** were prepared in which a naphthylmethlene side chain replaces the benzyl substituent in **11a**. These molecules showed more potency to inhibit PDE with IC_{50} values of 9 and 20 nM, respectively.

11.3.4.2 Inhibitors of the Assembly of the Hexameric Helical Core of the gp41 Protein

To show the generality of this terphenyl helical mimetic design, Hamilton et al. (2001) extended the use of the terphenyl compounds to the disruption of the assembly of the hexameric helical core of the gp41 protein. The envelope glycoproteins of HIV-1, gp120, and gp41 are generated from the polyprotein precursor gp160 by proteolytic cleavage. While gp120 is responsible for the target cell recognition and viral tropism through the CD4 and CCR5/CXCR4 cytokine coreceptors, the membrane spanning gp41 promotes the fusion of the viral and cell membrane. The helical domain of gp41 has an α_6-helical bundle composed of trimers of C34 and N36 peptides (Lu et al., 1995). During the target cell binding, gp41 undergoes a conformational change and exposes its hydrophobic N36, allowing the fusion peptide to insert into the host cell membrane. The surface N36 has a particularly deep cavity into which three hydrophobic residues from C34: Trp628, Trp631, and Ile635 (Chan et al., 1997).

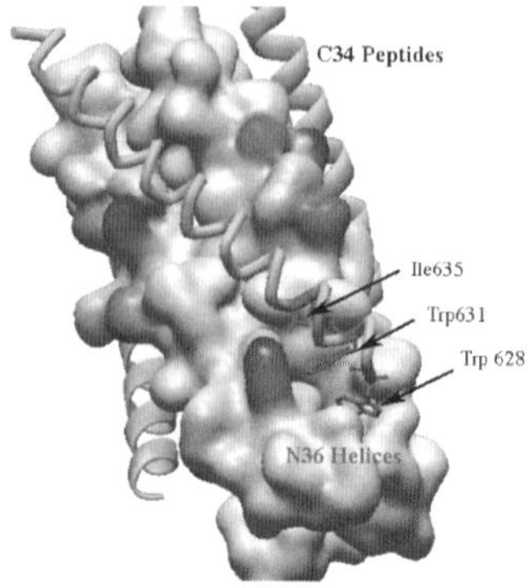

Figure 11.11 X-ray crystal structure of the gp41 fusion active core C34 (a.a. 628-661, shown in yellow helical tubes representation) bind to N36 (a.a. 546-581) by inserting the side chains of Trp628, Trp631 and Ile635 residues into a hydrophobic cleft on the protein surface. See color plates.

Synthetic molecules mimicking the interaction surface of C34 can be effective inhibitors to abolish the C34 and N36 interaction. Consequently, the C34 peptide of gp41 can be a useful target for inhibiting HIV-1 entry. A series of terphenyl derivatives was designed to mimic the hydrophobic functionality at the i, $i+4$, and $i+7$ positions of this α-helical region of the C34 peptide, to disrupt binding to the trimeric N-terminal core, and to prevent HIV-1 from entering into host cells. Terphenyl **12** was found to disrupt the interaction of two peptides (N36 and C34) to form a stable six-helix bundle that is analogous to the gp41 core (Fig. 11.11). Circular dichroism (CD) spectroscopy demonstrated that the C34 peptide forms a random coil on its own in solution while the N36 peptide forms concentration-dependent aggregates. Titration of **12** into a PBS-buffered solution of this model system resulted in a decrease of the CD signal at 222 nm and 208 nm corresponding to a reduction in helicity of the hexameric bundle that saturates at three equivalents of terphenyl. The melting point (T_m) value of the N36 core in the presence of excess **12** is significantly lower than that of the gp41 core and resembles that of the N36 region alone. These results suggested that the terphenyl **12** competes for the C34 binding site. The electrostatic and hydrophobic features of **12** were found to be critical for this behavior because both the terphenyls lacking alkyl substituents and those with positively charged substituents had little effect on the CD signal of the gp41 core. An ELISA experiment using an antibody that binds to the N36/C34 complex but not the individual peptides showed that **12** disrupts the formation of the hexameric bundle with an IC_{50} of $13.18 \pm 2.54\ \mu g \cdot mL^{-1}$, while a dye transfer fusion assay indicated that **12** inhibits HIV-1 mediated cell–cell fusion with an IC_{50} of $15.70 \pm 1.30\ \mu g \cdot mL^{-1}$.

11.3.4.3 Inhibitors of the Bcl-x_L/Bak Interaction

B-cell lymphoma-2 (Bcl-2) proteins are critical members in programmed cell death, also known as apoptosis (Adams and Cory, 1998; Reed, 1997). The Bcl-2 proteins include two groups: the proapoptotic and the prosurvival subfamilies. The proapoptotic subfamily proteins, such as Bak, Bad, and Bax, possess the death-promoting BH3 domain and act as a checkpoint downstream of p53 (Lane, 1992), and upstream of mitochondrial rupture and caspase cysteine proteases, which transduce the apoptotic signal (Adams and Corey, 1998, 2001, 2002). These molecules (Bak, Bax, Bad, and Bid) are able to induce apoptosis through heterodimerization with the prosurvival Bcl-2 family members (Chao and Korsmeyer, 1998). Previous studies showed that oncogenic mutations induced apoptotic defects through a Bcl-2-dependent pathway (Graeber et al., 1996; Fearon and Vogelstein, 1990). Overexpression of the prosurvival proteins, such as Bcl-2 and Bcl-x_L, can inhibit the potency of many currently available anticancer drugs by blocking the apoptotic pathway (Strasser et al., 1997). Several low-molecular-weight inhibitors of Bcl-2 (Bcl-x_L) have been identified by screening diverse chemical libraries (Enyedy et al., 2001; Wang et al., 2000; Lugovskoy et al., 2002; Tzung et al., 2001; Degterev et al., 2001). The rational design of agents that directly mimic the death-promoting BH3 domain of the proapoptotic subfamily of Bcl-2 proteins is an efficient alternative because it allows structure-based optimization of the initial hits (Adams and Cory, 2001).

A current strategy for inhibiting Bcl-x_L function is to block the Bak-recognition site on Bcl-x_L and disrupt the protein–protein contact. A Bcl-x_L/Bak complex structure determined by NMR spectroscopy showed that a helical region of Bak (a.a. 72–87, K_d – 340 nM) binds to a hydrophobic cleft on the surface of Bcl-x_L, yet in solution this isolated 16-mer peptide exists as a random coil (Sattler et al., 1997). Furthermore, the crucial residues for binding were shown by alanine scanning to be Val74, Leu78, Ile81, and Ile85, which project at the i, $i+4$, $i+7$, and $i+11$ positions along one face of the helix (Fig. 11.12). A series of terphenyl derivatives was prepared

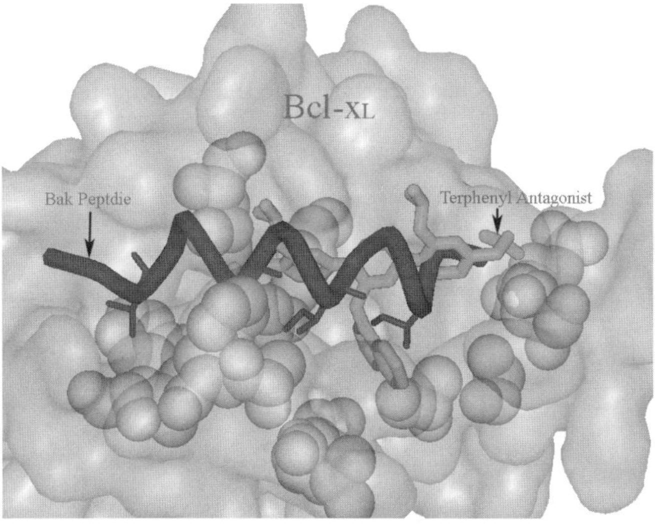

Figure 11.12 Results of the ^{15}N-HSQC experiments of **13** binding to Bcl-x_L. The residues that showed significant chemical shift changes in the presence of **13** are shown in yellow. The highest ranked binding mode of inhibitor **13** predicted from a computational docking simulation has been superimposed on the helical Bak BH3 domain for comparison. See color plates.

with different sequences and substitution patterns on the 3,2′,2″-positions. A fluorescence polarization assay was used to monitor the interaction between the inhibitor and the target protein. Terphenyl **13**, with two carboxyl groups and a substituent sequence of isobutyl, 1-naphthylenemethylene, and isobutyl groups, was identified as a potent inhibitor ($K_d = 114\,\text{nM}$). The binding specificity was confirmed by scrambling the sequence of the substitutions, which led to a 30-fold drop in K_i. The importance of the two carboxyl groups was confirmed by partial removal or conversion to positively charged groups. ^{15}N-HSQC NMR experiments with **13** indicated that these terphenyl derivatives target the hydrophobic cleft known to interact with the Bak peptide (shown in blue, Fig. 11.12).

11.3.5 Oligoamide Foldamers as Mimetics of the Bak Peptide

The hydrophobicity of these terphenyls compounds and their complex syntheses prompted us to search for simpler scaffolds that could similarly mimic the side chain presentation on an α-helix (Lipinski et al., 1991). The goal of these design modifications was to maintain the similarity between the arrangement of the i, $i+4$, and $i+7$ side chains of an α-helix and the substituents on 3,2′,2″-positions on terphenyl **13** (Kutzki et al., 2002), while minimizing the structural complexity and reducing the high hydrophobicity of the inhibitors. This strategy of simplifying a proven *proteomimetic* was accomplished by using oligoamide foldamer, **14**, as the scaffold (Ernst et al., 2003). The anticipated conformation of the molecule **14** is depicted in Fig. 11.13a and should result from a favorable intramolecular hydrogen bond between the amide-NH and the pyridine nitrogen in addition to the disfavored interaction that would occur between the amide carbonyl and the nitrogen of the pyridine. Further conformational biasing is induced by a hydrogen bond between the amide-NH and the oxygen of the alkoxy substituent. The intramolecularly hydrogen-bonded conformation was confirmed

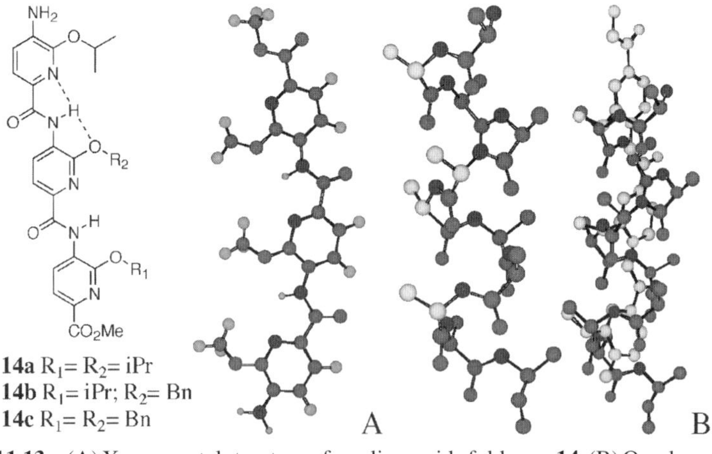

14a $R_1 = R_2 =$ iPr
14b $R_1 =$ iPr; $R_2 =$ Bn
14c $R_1 = R_2 =$ Bn

Figure 11.13 (A) X-ray crystal structure of an oligoamide foldamer **14**. (B) Overlay of a polyalanine α-helix and **14**, showing good agreement of the side chains. See color plates.

by X-ray crystallography, which further revealed that the alkoxy side chains were tilted at 45° to maximize interaction of the lone pair on the oxygen with the amide-NH. The foldamer was tested as a proteomimetic of Bak to disrupt its interaction with Bcl-x_L. Three inhibitors, **14a**, **14b**, and **14c**, were identified using fluorescence polarization assays with K_i values of 2.3, 9.8, and 1.6 μM, respectively. The foldamers with tris-benzyloxyl and tris-methoxyl groups showed weaker activities, indicating a requirement for size matching and sequence specificity between the protein and the mimetic.

11.3.6 Terephthalamide-Based Mimetics of the Bak Peptide

An alternative design for mimicking the binding surface of α-helices based on a novel terephthalamide scaffold was reported recently (Yin and Hamilton, 2004). The flanking phenyl rings in **13** were replaced by two functionalized carboxamide groups, which retain the planar geometry that is presented by the phenyl rings, due to the restricted rotation of the amide bonds. These carboxamide groups also increase the polarity of the inhibitors. Computational calculations predicted that terephthalamide **15** ($R_1 = R_2 = R_3 = Me$) has a log P (partition coefficient for n-octanol/water) of 2.73, compared to 10.84 for terphenyl **13**, suggesting that the terephthalamide derivatives may have favorable solubility in aqueous solution. Another conformational constraint in the molecule was imposed by an intramolecular hydrogen bond between the amide–NH and the alkoxy oxygen atom, to influence the position of the amino acid side chain R_1 (Fig. 11.14a) (Ernst et al., 2003). The increased rigidity of the terephthalamide

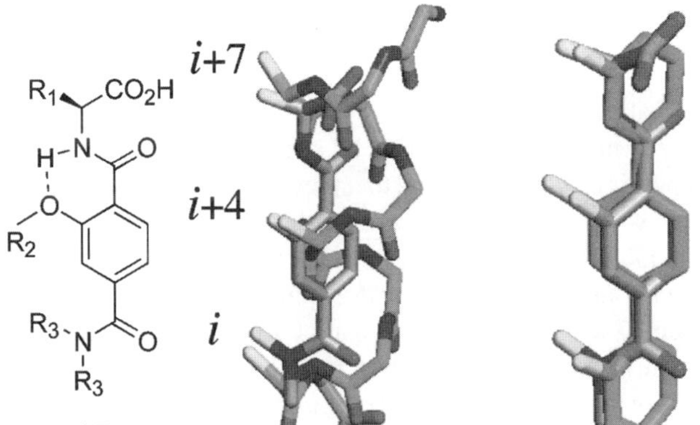

15 A B C

Figure 11.14 (A) Terephthalamide helical mimetics. (B) The superimposition of **15** on the i, $i+4$, and $i+7$ positions of a poly-alanine α-helix (RMS deviation = 1.03 Å). (C) The superimposition of terphenyl **13** and terephthalamide **15** (RMS deviation = 0.34 Å). The functional side chains are highlighted. See color plates.

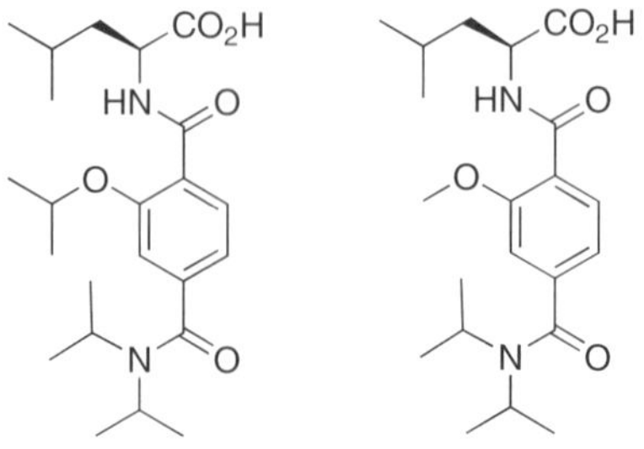

16 **17**

Figure 11.15 Terephthalamide inhibitors of Bcl-x_L.

avoids the entropy loss of the terphenyls upon binding to their protein targets. The modularity of inhibitor synthesis is also strengthened. Terphenyl **13** requires difficult C–C bond formations to attach the side chains to the scaffold and to assemble functionalized phenyl building blocks (Orner et al., 2001; Kutzski et al., 2002). Terephthalamide derivatives, on the other hand, were synthesized using *O*-alkylation and amide bond formation steps, accelerating the process of preparing these compounds. Computational energy minimization using MM2 force field within the Macromodel 7.0 program showed good stereochemical similarity among energy-minimized **15**, the i, $i+4$, and $i+7$ side chains of an α-helix (RMS deviation = 1.03 Å), and a comparable terphenyl scaffold (RMS deviation = 0.34 Å).

Terephthalamide **16** and **17** (Fig. 11.15) were identified as potent inhibitors of Bcl-x_L with K_i values of 0.78 ± 0.07 μM and 1.785 ± 0.388 μM, respectively, using the previously reported fluorescence polarization assay (Kutzki et al., 2002). Intact cell assays using human embryonic kidney 293 (HEK293) cells showed that terephthalamide derivatives inhibit the Bcl-x_L/BH3 complexation in whole cells. Treatment of HEK293 cells with terephthalamide **17** (100 μM) resulted in inhibition of the association of Bcl-x_L with Bax to an average extent of $81.1 \pm 13.5\%$, compared to the $68.7 \pm 20.3\%$ inhibition induced by the Bax BH3 peptide. This result demonstrates that terephthalamide **17** is able to perturb Bcl-x_L/Bax interaction in cells, suggesting that it has good uptake properties.

The displacement of the Bak peptide by a terephthalamide in the fluorescence polarization competition assay suggested that they both bind to the same hydrophobic pocket on the Bcl-x_L surface. Comparison of the Bak peptide and the top scored computer-docked **17** also suggested that the terephthalamide scaffold mimics the peptide helical backbone (Fig. 11.16). The docking result showed that the side chains of terephthalamide **17** have an analogous spatial arrangement to the three key side chains of Val74, Leu78, and Ile81 of the Bak peptide. The ^{15}N-HSQC chemical shift perturbation experiments have confirmed this binding mode, because the addition of **17** led to peak shifts in a number of residues found near this pocket. Furthermore, the residues affected by terephthalamide **17** and terphenyl **13** are similar (Gly94 of the BH3; Arg102 and Leu108 of the BH2; Arg139, Ile140, and Ala142 of the BH1 domains of Bcl-x_L), indicating the terephthalamide is a successful alternative scaffold to the terphenyl as α-helix mimetics.

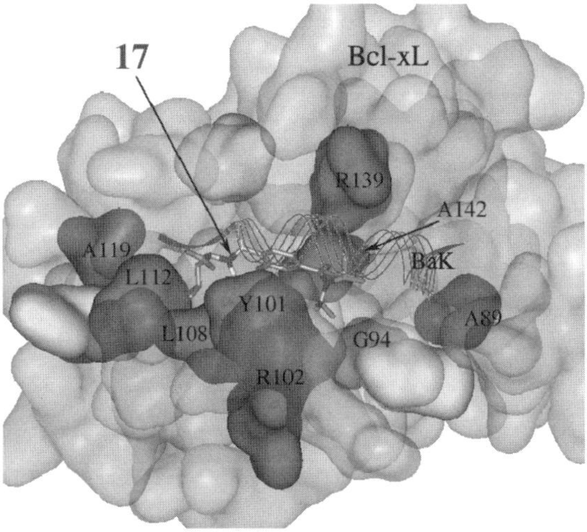

Figure 11.16 Superimposition of the top-ranked result of the molecular-docking studies of **17** and the Bak peptide. The residues that showed significant and moderate chemical shift changes upon the addition of **17** are shown in magenta and yellow, respectively. See color plates.

11.4 CONCLUSION AND PERSPECTIVES

A foremost goal in chemical biology and medicinal chemistry is to reduce the structural complexity of biomolecules by designing low-molecular-weight synthetic agents that reproduce their essential features (Mutter and Tuchscherer, 2000). Using synthetic agents to mimic α-helices, in particular, has immense interest in drug discovery because of the central role that the α-helical regions play in many biological processes. The advantages of using synthetic agents to target protein–protein interfaces include resistance to proteolytic degradation, lack of need to prepare large chemical libraries, and higher hit rates in preliminary screening. Although still at its early stage, the approach of rationally designing α-helix mimetics has made major strides in the last decade, substantially contributing to our understanding of fundamental principles underlying biological processes and serving as a first step in accessing molecules of therapeutic relevance. Selected examples of recent studies (most of which are based on shape and electrostatic complementarity) in this interdisciplinary research field have been reviewed. In view of the expected exponential growth of knowledge about biomolecular recognition along with improved technology in computer-aided simulations in the coming years (Jorgensen, 2004), the design of therapeutically relevant molecules with higher affinity and selectivity are within reach.

ACKNOWLEDGMENTS

We thank the National Institutes of Health (GM35208 and GM 69850) for financial support of this work.

REFERENCES

Adams, J. M., and Cory, S. (1998). *Science* **281**(5381), 1322–1326.

Adams, J. M., and Cory, S. (2001). *Trends Biochem Sci* **26**(1), 61–66.

Adams, J. M., and Cory, S. (2002). *Curr Opin Cell Biol* **14**(6), 715–720.

Albert, J. S., and Hamilton, A. D. (1995). *Biochemistry* **34**(3), 984–990.

Albert, J. S., Goodman, M. S., and Hamilton, A. D. (1995). *J Am Chem Soc* **117**(3), 1143–1144.

Andrews, M. J. I., and Tabor, A. B. (1999). *Tetrahedron* **55**(40), 11711–11743.

Blackwell, H. E., and Grubbs, R. H. (1998). *Angew Chem-Int Edit* **37**(23), 3281–3284.

Blaydes, J. P., Gire, V., Rowson, J. M., and WynfordThomas, D. (1997). *Oncogene* **14**(15), 1859–1868.

Brandon, C. T. (1991). *Introduction to Protein Structure*. New York: Garland.

Carreira, L. A., and Towns, T. G. (1977). *J Mol Struct* **41**(1), 1–9.

Chan, D. C., Fass, D., Berger, J. M., and Kim, P. S. (1997). *Cell* **89**(2), 263–273.

Chao, D. T., and Korsmeyer, S. J. (1998). *Annu Rev Immunol* **16**, 395–419.

Chen, J. D., Marechal, V., and Levine, A. J. (1993). *Mol Cell Biol* **13**(7), 4107–4114.

Chen, L. H., Agrawal, S., Zhou, W. Q., Zhang, R. W., and Chen, J. D. (1998). *Proc Natl Acad Sci USA* **95**(1), 195–200.

Degterev, A., Lugovskoy, A., Cardone, M., Mulley, B., Wagner, G., Mitchison, T., and Yuan, J. Y. (2001). *Nat Cell Biol* **3**(2), 173–182.

Ding, W. D., and Ellestad, G. A. (1991). *J Am Chem Soc* **113**(17), 6617–6620.

Ellenberger, T. E., Brandl, C. J., Struhl, K., and Harrison, S. C. (1992). *Cell* **71**(7), 1223–1237.

Elshorst, B., Hennig, M., Forsterling, H., Diener, A., Maurer, M., Schulte, P., Schwalbe, H., Griesinger, C., Krebs, J., Schmid, H., Vorherr, T., and Carafoli, E. (1999). *Biochemistry* **38**(38), 12320–12332.

Enyedy, I. J., Ling, Y., Nacro, K., Tomita, Y., Wu, X. H., Cao, Y. Y., Guo, R. B., Li, B. H., Zhu, X. F., Huang, Y., Long, Y. Q., Roller, P. P., Yang, D. J., and Wang, S. M. (2001). *J Med Chem* **44**(25), 4313–4324.

Ernst, J. T., Kutzki, O., Debnath, A. K., Jiang, S., Lu, H., and Hamilton, A. D. (2001). *Angew Chem-Int Edit* **41**(2), 278–282.

Ernst, J. T., Becerril, J., Park, H. S., Yin, H., and Hamilton, A. D. (2003). *Angew Chem-Int Edit* **42**(5), 535–550.

Fairlie, D. P., West, M. L., and Wong, A. K. (1998). *Curr Med Chem* **5**(1), 29–62.

Fasan, R., Dias, R. L. A., Moehle, K., Zerbe, O., Vrijbloed, J. W., Obrecht, D., and Robinson, J. A. (2004). *Angew Chem-Int Edit* **43**(16), 2109–2112.

Fearon, E. R., and Vogelstein, B. (1990). *Cell* **61**(5), 759–767.

Flint, D. G., Kumita, J. R., Smart, O. S., and Woolley, G. A. (2002). *Chem Biol* **9**(3), 391–397.

Galoppini, E., and Fox, M. A. (1996). *J Am Chem Soc* **118**(9), 2299–2300.

Garcia-Echeverria, C., Chene, P., Blommers, M. J. J., and Furet, P. (2000). *J Med Chem* **43**(17), 3205–3208.

Graeber, T. G., Osmanian, C., Jacks, T., Housman, D. E., Koch, C. J., Lowe, S. W., and Giaccia, A. J. *Nature*, 1996. **379**(6560), 88–91.

Haupt, Y., Maya, R., Kazaz, A., and Oren, M. (1997). *Nature* **387**(6630), 296–299.

Hollstein, M., Rice, K., Greenblatt, M. S., Soussi, T., Fuchs, R., Sorlie, T., Hovig, E., Smithsorensen, B., Montesano, R., and Harris, C. C. (1994). *Nucleic Acids Res* **22**(17), 3551–3555.

Honda, R., Tanaka, H., and Yasuda, H. (1997). *FEBS Lett* **420**(1), 25–27.

Horwell, D. C., Howson, W., Ratcliffe, G. S., and Willems, H. M. G. (1996a). *Bioorg Med Chem* **4**(1), 33–42.

Horwell, D., Pritchard, M., Raphy, J., and Ratcliffe, G. (1996b). *Immunopharmacology* **33**(1–3), 68–72.

Huyghuesdespointes, B. M. P., Scholtz, J. M., and Baldwin, R. L. (1993). *Protein Sci* **2**(1), 80–85.

Jackson, D. Y., King, D. S., Chmielewski, J., Singh, S., and Schultz, P. G. (1991). *J Am Chem Soc* **113**(24), 9391–9392.

Johnson, J. D., Walters, J. D., and Mills, J. S. (1987). *Anal Biochem* **162**(1), 291–295.

Jorgensen, W. L., (2004). *Science* **303**(5665), 1813–1818.

Kahne, D. (1995). *Chem Biol* **2**(1), 7–12.

Kubbutat, M. H. G., Jones, S. N., and Vousden, K. H. (1997). *Nature* **387**(6630), 299–303.

Kutzki, O., Park, H. S., Ernst, J. T., Orner, B. P., Yin, H., and Hamilton, A. D. (2002). *J Am Chem Soc* **124**(40), 11838–11839.

Lane, D. P., (1992). *Nature* **358**(6381), 15–16.

Lifson, S. (1961). *J Chem Phys* **34**(6), 1963–1974.
Lipinski, C. A., Lombardo, F., Dominy, B. W., and Feeney, P. J. (1997). *Adv Drug Deliver Rev* **23**(1–3), 3–25.
Lu, M., Blacklow, S. C., and Kim, P. S. (1995). *Nat Struct Biol* **2**(12), 1075–1082.
Lugovskoy, A. A., Degterev, A. I., Fahmy, A. F., Zhou, P., Gross, J. D., Yuan, J. Y., and Wagner, G. (2002). *J Am Chem Soc* **124**(7): 1234–1240.
Meador, W. E., Means, A. R., and Quiocho, F. A. (1992). *Science* **257**(5074), 1251–1255.
Meador, W. E., Means, A. R., and Quiocho, F. A. (1993). *Science* **262**(5140), 1718–1721.
Mutter, M., and Tuchscherer, G. (2000). *Chimia* **54**(10), 552–557.
Olson, C. A., Shi, Z. S., and Kallenbach, N. R. (2001). *J Am Chem Soc* **123**(26), 6451–6452.
Orner, B. P., Ernst, J. T., and Hamilton, A. D. (2001). *J Am Chem Soc* **123**(22), 5382–5383.
Osapay, G., and Taylor, J. W. (1992). *J Am Chem Soc* **114**(18), 6966–6973.
Pauling, L., and Corey, R. B. (1954). *Fortschr Chem Org Naturst* **11**, 180–239.
Pauling, L., Corey, R. B., and Branson, H. R. (1951). *Proc Natl Acad Sci USA* **37**(4), 205–211.
Peczuh, M. W., Hamilton, A. D., SanchezQuesada, J., deMendoza, J., Haack, T., and Giralt, E. (1997). *J Am Chem Soc* **119**(39), 9327–9328.
Peczuh, M. W., and Hamilton, A. D. (2000). *Chem Rev* **100**(7), 2479–2493.
Reed, J. C. (1997). *Nature* **387**(6635), 773–776.
Regan, L. (1993). *Proc Natl Acad Sci USA* **90**(23), 10907–10908.
Ruan, F. Q., Chen, Y. Q., and Hopkins, P. B. (1990). *J Am Chem Soc* **112**(25), 9403–9404.
Sattler, M., Liang, H., Nettesheim, D., Meadows, R. P., Harlan, J. E., Eberstadt, M., Yoon, H. S., Shuker, S. B., Chang, B. S., Minn, A. J., Thompson, C. B., and Fesik, S. W. (1997). *Science* **275**(5302), 983–986.
Schafmeister, C. E., Po, J., and Verdine, G. L. (2000). *J Am Chem Soc* **122**(24), 5891–5892.
Scholtz, J. M., York, E. J., Stewart, J. M., and Baldwin, R. L. (1991). *J Am Chem Soc* **113**(13), 5102–5104.
Scholtz, J. M., Qian, H., Robbins, V. H., and Baldwin, R. L. (1993). *Biochemistry* **32**(37), 9668–9676.
Stites, W. E. (1997). *Chem Rev* **97**(5), 1233–1250.
Strasser, A., Huang, D. C. S., and Vaux, D. L. (1997). *Biochim Biophys Acta-Rev Cancer* **1333**(2), F151–F178.
Tsou, L. K., Tatko, C. D., and Waters, M. L. (2002). *J Am Chem Soc* **124**(50), 14917–14921.
Tzung, S. P., Kim, K. M., Basanez, G., Giedt, C. D., Simon, J., Zimmerberg, J., Zhang, K. Y. J., and Hockenbery, D. M. (2001). *Nat Cell Biol* **3**(2), 183–191.
Vogel, H. J. (1994). *Biochem Cell Biol* **72**(9–10), 357–376.
Wang, J. L., Liu, D. X., Zhang, Z. J., Shan, S. M., Han, X. B., Srinivasula, S. M., Croce, C. M., Alnemri, E. S., and Huang, Z. W. (2000). *Proc Natl Acad Sci USA* **97**(13), 7124–7129.
Xuereb, H., Maletic, M., Gildersleeve, J., Pelczer, I., and Kahne, D. (2000). *J Am Chem Soc* **122**(9), 1883–1890.
Yin, H., and Hamilton, A. D. (2004). *Bioorg Med Chem Lett* **14**(6), 1375–1379.
Yu, C. X., and Taylor, J. W. (1999). *Bioorg Med Chem* **7**(1), 161–175.
Zimm, B. H., and Bragg, J. K. (1959). *J Chem Phys* **31**(2), 526–535.

CHAPTER 12

BASIC AND TRANSLATIONAL RESEARCH OF CHEMOKINE LIGANDS AND RECEPTORS AND DEVELOPMENT OF NOVEL THERAPEUTICS

Won-Tak Choi, Yohichi Kumaki, I.M. Krishna Kumar, Jing An, Douglas D. Richman, Joseph G. Sodroski, and Ziwei Huang[*]

12.1 INTRODUCTION

Chemokine (chemoattractant cytokine) receptors are a group of membrane proteins that belong to the superfamily of G-protein-coupled receptors (GPCRs). They possess seven transmembrane (TM) helices and transmit signals from extracellular ligands to

[*]Corresponding author: ziweihuang@burnham.org

Drug Discovery Research: New Frontiers in the Post-Genomic Era, Edited by Ziwei Huang

TABLE 12.1 A list of representative chemokine receptors and their ligands

Subfamily	Chemokine Receptors	Ligands
CXC	CXCR1 (IC-8R-α)	IL-8
	CXCR2 (IL-8R-β)	IL-8, GROα, β, γ, NAP-2, ENA78, GCP-2
	CXCR3	IP10, Mig
	CXCR4	SDF-1
CC	CCR1	RANTES, MIP-1α, MCP-2, MCP-3
	CCR2a/b	MCP-1, MCP-2, MCP-3, MCP-4
	CCR3	Eotaxin, RANTES, MCP-3, MCP-4
	CCR4	RANTES, MIP-1α, MCP-1
	CCR5	RANTES, MIP-1α, MIP-1β

intracellular biological pathways via heterotrimeric G-proteins (Kobilka, 1992; Murphy, 1994; Strader et al., 1994). As the natural ligands of chemokine receptors, chemokines are small soluble proteins of about 70 amino acid residues with a molecular weight of 8–10 kDa. They play prominent roles in leukocyte activation and trafficking to sites of inflammation by interacting with chemokine receptors. Based on the positions of two conserved cysteine residues in their amino (N)-termini, chemokines can be divided into four subfamilies: CC, CXC, CX3C, and C (Berger et al., 1999; Murphy, 1994). The two main subfamilies of chemokines are CXC and CC (Table 12.1). They are important for the selective activation and recruitment of a large variety of cell types in inflammation. CXC chemokines are primarily involved in the activation of neutrophils, whereas CC chemokines do not affect neutrophils and generally stimulate other leukocytes such as monocytes, lymphocytes, and basophils. In addition to important roles in many physiological processes, chemokines are implicated in a wide range of human acute and chronic inflammatory diseases (i.e., acute respiratory distress syndrome, allergic asthma, psoriasis, and arthritis), neurological disorders, cancer, and most notably acquired immune deficiency syndrome (AIDS) (Baggiolini et al., 1997; Berger et al., 1999; Murphy, 1994; Proudfoot, 2002) (Table 12.2).

12.2 CHEMOKINE RECEPTORS IN HIV-1 ENTRY

In recent years, chemokine receptors have been found to be involved in the pathogenesis of human immunodeficiency virus type 1 (HIV-1) infection. HIV-1 enters cells through a fusion process in which the HIV-1 envelope glycoprotein gp120 binds to CD4, the main receptor for HIV-1 on the target cell surface. However, it has long been known that CD4 alone is not sufficient for HIV-1 fusion and entry, and that additional receptors may be needed (Clapham et al., 1991; Maddon et al., 1986). In 1996, chemokine receptors CXCR4 and CCR5 were discovered to be the long-sought coreceptors for syncytium-inducing and non-syncytium-inducing HIV-1 strains, respectively (Alkhatib et al., 1996; Deng et al., 1996; Dragic et al., 1996; Feng et al., 1996). While all HIV-1 strains appear to require CXCR4, CCR5, or both

TABLE 12.2 Chemokine Receptors and Diseases

Chemokine Receptors	Associated Diseases
CXCR1	Sepsis, atherosclerosis
CXCR2	Sepsis, atherosclerosis
CXCR3	Multiple sclerosis, rheumatoid arthritis, transplant
CXCR4	AIDS, cancer
CCR1	Multiple sclerosis, rheumatoid arthritis, transplant, asthma, nephritis
CCR2	Multiple sclerosis, rheumatoid arthritis, transplant, asthma, atherosclerosis
CCR3	Asthma
CCR4	Asthma
CCR5	AIDS, multiple sclerosis, rheumatoid arthritis, transplant, asthma, nephritis, inflammatory bowel disease

(Simmons et al., 1996; Zhang et al., 1996), some strains can also use other chemokine receptors, such as CCR3 and CCR2b, as coreceptors for fusion and infection (Choe et al., 1996; Doranz et al., 1996). The viral fusion process may involve the initial binding of HIV-1 gp120 to its high-affinity receptor CD4, which results in conformational changes in gp120 and CD4 (Clements et al., 1991; Gershoni et al., 1993; Sattentau et al., 1993). The gp120–CD4 complex then interacts with a chemokine coreceptor such as CXCR4 or CCR5 to form a heterotrimeric complex of gp120–CD4–coreceptor (Lapham et al., 1996; Trkola et al., 1996; Wu et al., 1996) (Fig. 12.1).

During the asymptomatic stage of disease, macrophage (M)-tropic strains of HIV-1 involved in sexual transmission primarily use CCR5 as an entry coreceptor

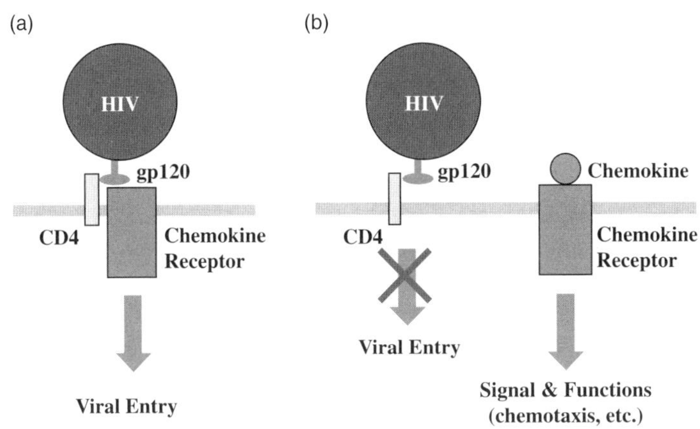

Figure 12.1 (**a**) The interactions of HIV-1 gp120, CD4, and a chemokine coreceptor, such as CXCR4 or CCR5, leading to the cell entry of the virus. (**b**) The binding of a chemokine ligand (such as SDF-1α) to its receptor (such as CXCR4) blocks the association of HIV-1 gp120/CD4 with CXCR4, thus preventing the viral entry. On the other hand, the chemokine receptor–ligand interaction can activate normal signaling and biological functions (such as chemotaxis).

(Alkhatib et al., 1996; Deng et al., 1996; Dragic et al., 1996). The crucial importance of CCR5 in HIV-1 transmission and feasibility of CCR5-specific agents to inhibit viral infection were further demonstrated by the observation that individuals with CCR5 mutations appear to be both healthy and highly resistant to HIV-1 infection (Samson et al., 1996). However, in 40–50% of HIV-1-infected individuals, T-cell (T)-tropic strains that predominantly use CXCR4 eventually replace M-tropic strains, leading to rapid disease progression (Cheng-Mayer et al., 1988; Schellekens et al., 1992; Tersmette et al., 1989). CXCR4 can also be used by human immunodeficiency virus type 2 (HIV-2) strains adapted to replication in CD4-negative cell lines (Endres et al., 1996). Dual-tropic strains are those HIV-1 isolates that are capable of using both CXCR4 and CCR5. In addition, an HIV-1 envelope can bind CXCR4 independently, and this interaction is enhanced by the presence of CD4 (Bandres et al., 1998).

Two theories have been proposed for the mechanism by which natural chemokines of CXCR4 or CCR5 can prevent chemokine receptor-dependent HIV-1 entry. First, natural chemokines of CXCR4 or CCR5 can inhibit HIV-1 infection (Bleul et al., 1996a; Oberlin et al., 1996) by blocking the binding of HIV-1 gp120 to CXCR4 or CCR5. The mere occupancy of HIV-1 coreceptors by chemokines even in the absence of G-protein-mediated signaling is known to be sufficient for inhibition of HIV-1 infection (Arenzana-Seisdedos et al., 1996; Oravecz et al., 1996). Alternatively, natural chemokines can inhibit HIV-1 entry by inducing receptor downregulation from the cell surface, thereby removing the essential coreceptors (Amara et al., 1997; Förster et al., 1998). In fact, these studies suggested that coreceptor internalization contributes to efficient chemokine inhibition of HIV-1 entry.

The discovery of distinct chemokine coreceptors for different HIV-1 strains has important implications for understanding HIV-1 pathogenesis and developing new therapeutic strategies. Among many new directions in AIDS research opened by this discovery, an important area of investigation is the biochemical and biophysical characterization of the interactions of chemokine receptors with HIV-1 and natural as well as synthetic chemokine ligands. The understanding of the precise structure–function relationship of chemokine receptors and ligands will be essential for providing the structural and biochemical basis to address basic questions of (a) mechanisms of HIV-1 viral entry, host cell tropism, and resistance to infection; (b) mechanisms of receptor recognition and signal transduction by natural chemokines and synthetic mimics and their role in the normal biological functions; (c) any common and/or distinct regions/sites on the receptors for HIV-1 and chemokine recognition that are important for devising selective intervening strategies; and (d) any general insights into the structure–function and mechanism of other membrane proteins that belong to the G-protein-coupled receptor (GPCR) superfamily.

To define precise regions involved in the molecular interactions of chemokine receptors with their natural ligands and HIV-1 gp120, an approach combining structural biology, molecular biology, and synthetic chemistry is being used. As described in the following section, molecular modeling techniques are applied to propose plausible structural models for chemokine receptors and their complexes with ligands. Such models reveal potential sites of interaction between a receptor and a ligand. To experimentally test the putative functional roles of such sites, peptide analogues targeting the functional epitopes on chemokine ligands are synthesized and tested in biological assays. In addition, site-directed mutagenesis of the chemokine receptor is carried out to test the predicted sites for ligand binding and signaling. These studies provide structural and chemical insights into receptor–ligand interactions and form the basis for the design of small molecular agents that can prevent HIV-1 entry.

12.3 STRUCTURE–FUNCTION RELATIONSHIP OF CHEMOKINE RECEPTORS

The information about the detailed three-dimensional structure of a chemokine receptor and its complex with a ligand is critical for understanding the role and mechanisms of the function of chemokine receptors and ligands in Human immunodeficiency virus type 1 (HIV-1) pathogenesis. However, no crystal structure is available for chemokine receptors or any other G-protein-coupled receptor (GPCR) proteins (Kobilka, 1992; Strader et al., 1994). Since the structure of the seven transmembrane (TM) helices in a similar protein, bacteriorhodopsin (bR), has been determined by electron microscopy (Henderson et al., 1990), many members of the GPCR superfamily have been modeled by computational techniques based on the structure of the TM segments of bR (Kontoyianni and Lybrand, 1993; Strader et al., 1994). Our laboratory extended the computational methodologies used for modeling other GPCR proteins to the family of chemokine receptors. A general approach for modeling the structure and interaction of chemokine receptors has been developed from our previously published study of interleukin-8 receptor type β, a member of the chemokine receptor family (Luo et al., 1997).

As shown in Fig. 12.2, the structure for the TM helices of a chemokine receptor is constructed based on that of bR. To further define the three-dimensional structure of the extracellular segments of the chemokine receptor and understand how they interact with a ligand, high-temperature molecular dynamics simulations are performed. The three-dimensional structure of the ligand, as determined by NMR experiment, is used as the geometric constraint in dynamics simulations. A nanosecond simulation is conducted for the receptor–ligand complex, and 500 structures are extracted from the trajectory of the simulation. These structures are clustered by the similarity in their backbone structures. Plausible structural models for the receptor–ligand complex are selected based on the examination of available experimental data and binding energy of the complex. Using this approach, we proposed a hypothetical structure of CCR5 in complex with a chemokine ligand such as macrophage inflammatory protein (MIP)-1β (Fig. 12.3a), which was useful for guiding subsequent site-directed mutagenesis studies of CCR5 to identify some key residues on CCR5 important for ligand binding and signal transduction (Zhou et al., 2000a). In a related study, the structure of CXCR4 in complex with stromal cell-derived factor (SDF)-1α was also proposed by using a similar method (Fig. 12.3b).

In conjunction with molecular modeling, chimeras and site-specific mutant studies were performed to study the role of CXCR4 and CCR5 in HIV-1 entry. These studies demonstrated that the amino (N)-terminus, the second extracellular loop (ECL2), and the third extracellular loop (ECL3) of CXCR4 (Brelot et al., 1997,

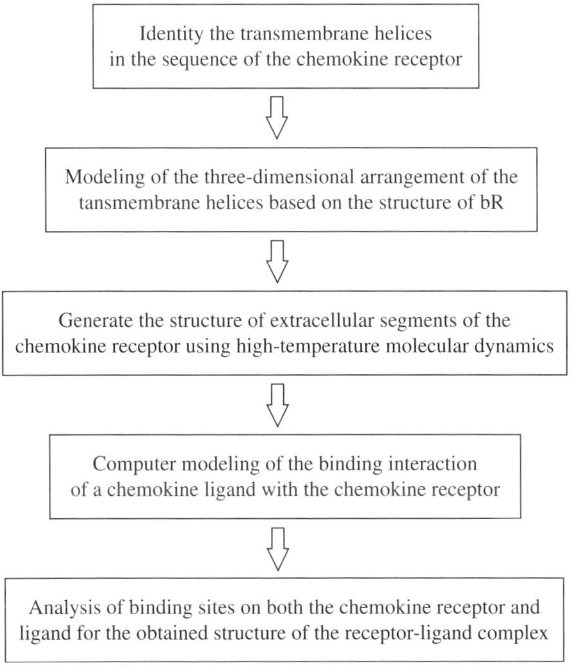

Figure 12.2 The general procedure for the modeling studies of a chemokine receptor–ligand complex.

2000; Chabot and Broder, 2000; Chabot et al., 1999; Doranz et al., 1999; Kajumo et al., 2000; Lu et al., 1997; Picard et al., 1997a,b; Wang et al., 1998; Zhou et al., 2001) or CCR5 (Alkhatib et al., 1997; Atchison et al., 1996; Bieniasz et al., 1997; Gosling et al., 1997; Rucker et al., 1996; Zhou et al., 2000a) are required for HIV-1 coreceptor activity. For instance, the previous structure–function studies of CXCR4 showed that there is a significant overlap between HIV-1 and chemokine functional sites on the extracellular regions of CXCR4 (Brelot et al., 1997, 2000; Chabot et al., 1999; Doranz et al., 1999; Kajumo et al., 2000; Zhou et al., 2001). Although there have been some discrepancies among these studies, the extracellular regions of the coreceptor have been the focus, with each extracellular region being implicated in coreceptor function. These studies also indicated that multiple extracellular domains of CXCR4 are required for chemokine interactions and receptor signaling (Alkhatib et al., 1997; Atchison et al., 1996; Bieniasz et al., 1997; Brelot et al., 1997; Chabot and Broder, 2000; Doranz et al., 1997; Lu et al., 1997; Rucker et al., 1996). These chimeric and mutational studies were used to validate the accuracy of a two-site model that was initially developed for the C5a chemoattractant and its receptor (Siciliano et al., 1994). According to the model, the motif composed of amino acids 12–17 of SDF-1α, RFFESH loop, first docks on the N-terminal domain of CXCR4, and this contact allows the subsequent interaction of the flexible N-terminus of SDF-1α with the receptor groove formed by TM domains and/or extracellular loops, thus triggering the receptor function (Crump et al., 1997; Gupta et al., 2001; Siciliano et al., 1994). The N-terminus of SDF-1α, being relatively flexible and unstructured in solution

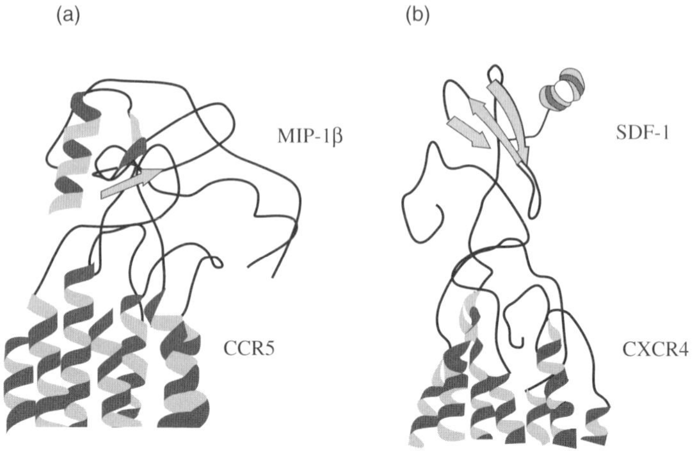

Figure 12.3 The proposed three-dimensional structures for (**a**) the complex between CCR5 and MIP-1β and (**b**) the complex between CXCR4 and SDF-1α.

(Crump et al., 1997), has been proven to be essential for CXCR4 recognition and signal transduction (Crump et al., 1997; Dealwis et al., 1998; Heveker et al., 1998). Taken together, these findings have provided structural and chemical insights into receptor–ligand interactions and formed the basis for the design of small-molecule agents that prevent HIV-1 entry.

Despite the previous studies, a number of areas remained largely unexplored, for which further investigations using structural modeling and site-directed mutagenesis of the receptors are critical, such as (a) structural and functional roles of other residues, especially those near the important residues already identified in above studies; (b) structural and functional roles of residues and sites located near or within the TM helices that have not been studied, yet may play an important role in signaling and controlling the function of extracellular receptor surfaces; (c) the mechanism of action of natural chemokines needs to be further elucidated in terms of their binding and signaling, whereas many of previous studies focused on HIV-1 interaction; and (d) novel synthetic agonistic or antagonistic ligands developed by us and others, whose specific binding sites on CXCR4 or CCR5 are unclear, require thorough characterization in order to understand their mechanism of action and develop them into more potent and specific therapeutics.

For instance, it has yet to be determined whether residues and sites located near or within the TM helices of CXCR4 might play any structural and functional roles for ligand interactions and, if they do, whether such roles are differentiating for different ligands (i.e., HIV-1 gp120 and SDF-1α). Also, based on our previous report that three residues in the ECL2—Arg^{183}, Tyr^{184}, and Tyr^{190}—are not involved in SDF-1α binding but play an important role in HIV-1 coreceptor activity (Zhou et al., 2001), we hypothesized that CXCR4–gp120 interaction might be potentially disrupted without impairing its natural chemokine receptor activity. To address these questions, we constructed a panel of mutations at residues near or within the TM helices and several mutants of the ECL2 residues to investigate the role of TM and ECL2 in HIV-1 entry versus natural ligand binding and signaling.

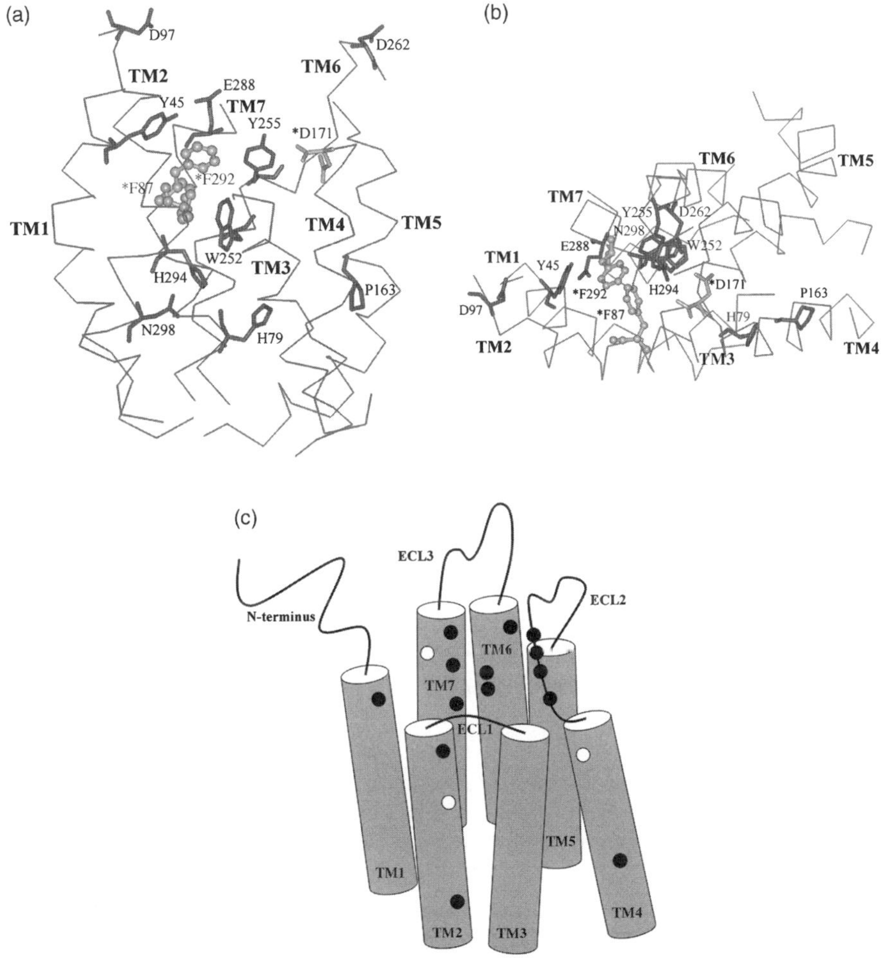

Figure 12.4 Distinct functional sites for HIV-1 gp120 and SDF-1α on a hypothetical structural model of CXCR4. The TM residues, Phe^{87} and Phe^{292}, required for SDF-1α binding only are highlighted in lighter color and represented as ball-and-stick with an asterisk. Asp^{171}, the only residue found to be involved in both HIV-1 gp120 and SDF-1α interactions, is highlighted in the lighter color with an asterisk. The TM residues that are involved in HIV-1 coreceptor activity only are highlighted in the darker color. Only the TM domains of CXCR4, with side (**A**) and top (**B**) views, are shown for simplicity. (**C**) Schematic illustration of the locations of residues found to be important for HIV-1 or SDF-1α on CXCR4 TM and ECL2 domains. The residues required for SDF-1α binding are shown as white spots, while those that are involved in HIV-1 coreceptor activity are shown as black spots. Asp^{171} is highlighted as a white spot. See color plates.

According to this study (Tian et al., 2005), many residues throughout CXCR4 TM and ECL2 domains are specifically involved in interaction with HIV-1 gp120, because most of these sites did not play a role in either SDF-1α binding or signaling (Fig. 12.4). As many of these distinct sites for either HIV-1 or SDF-1α function are located on the upper part of the TM barrel close to the extracellular side or in the

ECL2, their role is likely to be involved with direct interactions with different ligands. On the other hand, some residues predicted to be located deep inside the TM barrel and thus unlikely to contact gp120 directly were also able to reduce the coreceptor activity of CXCR4 without impairing the binding and signaling activities of SDF-1α. It is possible that these mutations presumably change the conformations of CXCR4 that are important for HIV-1 but not SDF-1α. Further work is needed to understand the mechanism of how the function of HIV-1 or SDF-1α is mediated differently by distinct sets of residues and possibly by conformational states. Nevertheless, the findings from the present study provide a basis for the development of new inhibitory agents that modulate the functional sites or conformations of CXCR4 for the purpose of reducing or avoiding the limitations and side effects caused by nonselective inhibitors of this important coreceptor.

The important functional roles of TM residues in CXCR4 are consistent with other lines of evidence previously reported in other GPCRs. For instance, studies of human neurokinin peptide receptor (NK1R) (Huang et al., 1995) and melanin-concentrating hormone (MCH) receptor (Macdonald et al., 2000) suggested that some residues in the TM regions contribute to ligand binding and activation. A recent study of CCR5 chimera also suggested that the N-termini of MIP-1α and RANTES (regulated on activation, normal T-cell expressed and secreted) mediate receptor activation by interacting with the TM helix bundle (Blanpain et al., 2003). Furthermore, according to a two-site model of the CXCR4-SDF-1α interaction, the motif consisting of amino acids 12–17 of SDF-1α first docks on the N-terminal domain of CXCR4, which allows its flexible N-terminus to interact with the receptor groove formed by TM domains and/or extracellular loops of CXCR4, thus triggering the receptor function (Crump et al., 1997; Gupta et al., 2001). When these results and models published by others are combined with the findings from the present study, one can hypothesize that the N-terminus of SDF-1α or certain flexible functional determinants of gp120 may readily interact with some residues in the TM domains of CXCR4. The interaction of the TM domains with HIV-1 gp120 should involve an extensive set of TM residues that are different from those residues required for the interaction with SDF-1α.

12.4 STRUCTURE–FUNCTION RELATIONSHIP OF CHEMOKINE LIGANDS

The only known natural ligand of CXCR4 is the stromal cell-derived factor (SDF)-1α, a CXC chemokine that plays critical roles in the migration, proliferation, and differentiation of leukocytes (Bleul et al., 1996b; Oberlin et al., 1996). CXCR4 can also be recognized by an antagonistic ligand, the viral macrophage inflammatory protein (vMIP)-II encoded by the Kaposi's sarcoma-associated herpes virus (Moore et al., 1996). vMIP-II displays a broader spectrum of receptor activities than any mammalian chemokine, because it binds with high affinity to a number of both CXC and CC chemokine receptors, including CXCR4, CCR5, and CCR2, and inhibits cell entry of HIV-1 mediated by these receptors (Boshoff et al., 1997; Kledal et al., 1997). Studies with knockout mice of CXCR4 and SDF-1α have demonstrated that these molecules play important roles in immunomodulation, organogenesis, hematopoiesis, and derailed cerebellar neuron migration (Ma et al., 1998; Nagasawa et al., 1996; Zou et al., 1998). The CC chemokine ligands of CCR5 include macrophage inflammatory protein (MIP)-1β, MIP-1α, and RANTES (regulated on activation, normal T-cell expressed and secreted). Among these CCR5 ligands, RANTES and MIP-1α can bind to other CC chemokine receptors, while MIP-1β is known to be most specific for CCR5 (Wells et al., 1996). The structures of several chemokines that bind CXCR4 or CCR5 have been determined by NMR or X-ray technique, including those of SDF-1α (Crump et al., 1997; Dealwis et al., 1998), vMIP-II (Fernandez et al., 2000; Liwang et al., 1999), MIP-1β (Lodi et al., 1994), and RANTES (Chung et al., 1995). Based on these structures, the structure–function relationship of these chemokines has been investigated mainly by a chemical approach of synthetic chemokine mutants and peptides.

The studies of the structures of chemokine receptors and their interactions with chemokine ligands and gp120 are critical for the development of new therapeutic agents that block the interaction between the HIV-1 envelope and chemokine receptors. Several CCR5 receptor antagonists have been reported to block HIV-1 infection (Arenzana-Seisdedos et al., 1996; Nardese et al., 2001; Simmons et al., 1997). For instance, aminoxypentane (AOP)-RANTES, a chemically modified analogue of RANTES, is a CCR5 antagonist that is effective in inhibiting CCR5-dependent HIV-1 strains (Simmons et al., 1997). However, since RANTES binds to a number of chemokine receptors in addition to CCR5, treatment with RANTES analogues that block the natural chemokines from functioning in these receptors could lead to potential problems. The development of more specific inhibitors of CCR5 clearly depends on further studies of chemokine ligand binding to CCR5 and other receptors. Regardless, coreceptor-based therapeutic strategies (Amara et al., 1997; Donzella et al., 1998; Doranz et al., 1997; Fenard et al., 2001; Heveker et al., 1998; Huang, 2000; Loetscher et al., 1998; Luo et al., 1999a,b, 2000; Murakami et al., 1997; Sachpatzidis et al., 2003; Schols et al., 1997; Zhou et al., 2000b, 2002) have the appeal of targeting relatively invariant host determinants, in contrast with anti-HIV-1 agents directed against components of the rapidly mutating virus population. The feasibility of this approach to inhibit viral infection is further demonstrated by the observation that individuals with CCR5 mutations appear to be both healthy and highly resistant to HIV-1 infection (Berger et al., 1999; Samson et al., 1996).

As for CXCR4, several groups reported that peptides and organic compounds (Donzella et al., 1998; Doranz et al., 1997; Fenard et al., 2001; Sachpatzidis et al., 2003; Zhou et al., 2000b, 2002), which were discovered much earlier through random screening and are unrelated to natural chemokines, have antagonistic activity in CXCR4. For instance, they include ALX40-4C (Doranz et al., 1997), T22 (Murakami et al., 1997), and AMD3100 (Schols et al., 1997). However, due to their high positive charge and unclear mechanism of action, the CXCR4 specificity and side effects of these molecules are still uncertain. Furthermore, there is cause for concern regarding undesired side effects of blocking the normal CXCR4-SDF-1α function, since knockout mice lacking either CXCR4 (Tachibana et al., 1998; Zou et al., 1998) or SDF-1α (Nagasawa et al., 1996) die during embryogenesis, with evidence of hematopoietic, cardiac, vascular, and cerebellar defects. As a result, it would be desirable to develop compounds that can target specific regions of CXCR4 that are selective for HIV-1 coreceptor function only, but not the normal function of SDF-1α. In fact, AMD3100 and ALX40-4C have entered clinical trials, but further development of AMD3100 was discontinued due to cardiac toxicity (Doranz et al., 2001). Even if adverse effects are not observed during clinical trials, any CXCR4 antagonist has a potential to do just that if it would be used chronically as highly active antiretroviral therapy.

An alternative route to designing specific CXCR4 inhibitors is to use natural chemokine ligands as design templates (Heveker et al., 1998; Huang, 2000; LaBarge and Blau, 2002; Loetscher et al., 1998; Luo et al., 1999a,b, 2000; Sachpatzidis et al., 2003; Zhou et al., 2000b, 2002) by mimicking specific regions of a chemokine ligand in order to study the structure–function relationship of the native molecule and to develop novel agonists or antagonists of chemokine receptors. This approach was first attempted on SDF-1α, the only known natural ligand of CXCR4 (Heveker et al., 1998; Loetscher et al., 1998; Luo et al., 1999a,b). The structure of SDF-1α consists of three major domains: the disordered amino (N)-terminus with mostly extended conformation, a central positively charged core region of three antiparallel β-strands, and the overlying carboxyl (C)-terminal α-helix (Crump et al., 1997; Dealwis et al., 1998). Peptides derived from the N-terminus of the SDF-1α sequence were proven to be essential for CXCR4 recognition, signal transduction, and antiviral activity, but they were less potent than native SDF-1α (Heveker et al., 1998; Loetscher et al., 1998). The exchange of K1 and P2 with other amino acids resulted in the generation of potent CXCR4 antagonists (Heveker et al., 1998). The functional role of the C-terminus helix of SDF-1α, which by itself had no activity in receptor binding and signaling, was also studied by joining the N- (residues 1–4) and C-terminus (residues 56–67) with a four-glycine linker (Luo et al., 1999a). The attachment of the C-terminus dramatically increased the effect of the N-terminal fragment by inducing chemotaxis and intracellular Ca^{2+} influx, but its activity was completely abolished by the addition of heparin, suggesting that the C-terminal fragment of the peptide binds glycosaminoglycans (GAGs) (Luo et al., 1999a). These findings demonstrated a way to enhance biological activity of CXCR4 antagonists, possibly by attaching a GAG-binding motif (Luo et al., 1999a). Furthermore, the experimental support for the role of the highly positive β-sheet region of SDF-1α in interacting with negatively charged residues in the extracellular domains of CXCR4 is also available (Luo et al., 1999b). The attachment of positively charged residues to the N-terminal peptide sequence of SDF-1α was found to enhance the ability of the peptides in CXCR4 binding and inhibiting CXCR4-mediated T-cell (T)-tropic HIV-1 entry (Luo et al., 1999b). Taken together with the observation that peptides and organic compounds, such as AMD3100 (Schols et al., 1997), T22 (Murakami et al., 1997), and ALX40-4C (Doranz et al., 1997), have high positive charges and affinity for CXCR4, the study indicates that the electrostatic interaction may play a role in CXCR4 recognition (Luo et al., 1999b).

CXCR4 can also be recognized by vMIP-II, which displays a broader spectrum of receptor activities than any mammalian chemokine, because it binds with high affinity to a number of both CC and CXC chemokine receptors (Boshoff et al., 1997;

Kledal et al., 1997). The overall topology of vMIP-II is very similar to that of SDF-1α; it consists of a disordered N-terminus, followed by a helical turn, which leads into the first strand of a three-stranded antiparallel β-sheet. Following the sheet is a C-terminal helix (Liwang et al., 1999; Sozzani et al., 1998). Recently, synthetic peptides derived from the N-terminus of vMIP-II were studied by our laboratory to elucidate the mechanism of vMIP-II interaction with its coreceptors. It was discovered that the N-terminus of vMIP-II is the major determinant for CXCR4 recognition (Luo et al., 2000). Only V1 peptide (1–21 residues) from the N-terminus of vMIP-II showed CXCR4 binding, whereas all other peptides derived from other regions of vMIP-II did not have any activity in CXCR4. On the other hand, none of these peptides showed binding activity for CCR5. Further studies showed that V1 selectively prevents CXCR4 signal transduction and coreceptor function in mediating the entry of T- and dual-tropic HIV-1 isolates (Luo et al., 2000).

Realizing that modified analogues based on the V1 peptide may lead to a new class of compounds with higher potency and stability, an all-D-amino acid analogue of V1 peptide, designated as DV1 peptide, was also synthesized (Zhou et al., 2002). Despite dramatically different conformations of side-chain groups as a result of being a mirror image of each other, DV1 displayed higher binding affinity toward CXCR4 than V1 (Zhou et al., 2002). Additional peptides composed of mixed D- and L-amino acids showed CXCR4 binding, further demonstrating the remarkable stereochemical flexibility of the CXCR4–peptide interface (Zhou et al., 2002). In addition, acting as CXCR4 antagonists, DV1 showed significant activity in inhibiting the replication of CXCR4-dependent HIV-1 strains (Zhou et al., 2002). Unnatural D-peptides can be highly desirable and advantageous over natural L-peptides for therapeutic development, because they are highly stable and resistant to proteolytic degradation (Zhou et al., 2002). As to CCR5 binding determinants, from studies of vMIP-II we hypothesized that CCR5 binding has a strict structural requirement and involves multiple sites other than the N-terminus of vMIP-II. This is consistent with the more recent report from another laboratory that peptides from the N-loop and first β-strand of RANTES showed CCR5 binding, and that this binding depends on a dimeric structure (Nardese et al., 2001).

More recently, Sachpatzidis et al., (2003), understanding the potential adverse effects from the use of antagonistic inhibitors of CXCR4 in HIV-1 therapy, identified RSVM and ASLW as novel allosteric agonists that are insensitive to the CXCR4 antagonist AMD3100 by screening a semi-randomized 17-mer library in a yeast strain expressing a functional CXCR4 receptor (Sachpatzidis et al., 2003). In chemotaxis assays, RSVM behaves as a partial agonist, while ASLW behaves as a superagonist, displaying a chemotactic index that is greater than the maximum observed with SDF-1α (Sachpatzidis et al., 2003). Allosteric agonists may be therapeutically useful in combination with small-molecule antagonists for anti-HIV therapy since they could maintain essential receptor function. The data also illustrated that other binding sites may exist for nonphysiological agonists (Sachpatzidis et al., 2003). Despite their early stage of development, allosteric agonists and other potent CXCR4 peptide antagonist or agonists discussed previously could serve as new leads for the development of new therapeutic agents for HIV-1 infection and other immune system diseases.

12.5 IMPORTANCE OF CHEMOKINES AND CHEMOKINE RECEPTORS IN OTHER PATHOLOGICAL PROCESSES

12.5.1 Chemokine Receptors and HIV-Associated Dementia

HIV-associated dementia (HAD) is a cognitive and motor dysfunction observed after infection with HIV-1 (Kaul et al., 2001). Although about one-half of children and one-fourth of adults infected with HIV-1 eventually develop the dementia (Lipton and Gendelman, 1995; McArthur et al., 2003), there is no specific treatment for HAD. In recent years, chemokine receptors have been found to play important roles in the pathogenesis of HIV-1 infection, including HAD. In a plausible model, HIV-1 enters target cells through a direct fusion process in which HIV-1 gp120 binds to CD4, the main receptor on the target cell surface, and to one of the two chemokine receptors, usually CXCR4 or CCR5 (Berger et al., 1999; Poignard et al., 2001). Natural chemokines that bind CXCR4 and CCR5 can inhibit HIV-1 infection (Bleul et al., 1996a; Oberlin et al., 1996), probably by blocking common binding sites on the chemokine receptors that are also required for gp120 interaction and/or by inducing receptor internalization (Amara et al., 1997; Förster et al., 1998). Although CXCR4 is expressed in virtually all tissues, including neurons, microglia, and astrocytes in the brain (McManus et al., 2000; Sanders et al., 1998), CCR5 is the primary receptor by which the cells in the nervous system get infected (He et al., 1997; Rottman et al., 1997).

There are two major mechanisms by which gp120-induced neuronal apoptosis occurs. The predominant pathway is an indirect process that results in the release of toxic factors from infected cells such as macrophages and microglia. HIV-1 can stimulate or infect microglia and macrophages by interacting with the chemokine receptors in conjunction with CD4. Macrophages and microglia can also be activated by factors, such as shed gp120, released from infected cells (Gartner, 2000; Kaul et al., 2001). The other mechanism is a direct interaction of gp120 with neurons via the chemokine receptors (Meucci et al., 1998). In fact, picomolar concentrations of CXCR4 (or X4)-preferring or CXCR4/CCR5 (or dual)-tropic gp120 are known to induce neurotoxicity through CXCR4 despite the fact that neurons do not express CD4 on the cell surface (Hesselgesser et al., 1998; Kaul and Lipton, 1999; Ohagen et al., 1999). However, the previous studies demonstrating that gp120-induced neuronal damage in rodent cultures was prevented by anti-gp120 antibodies but not by anti-CD4 antibodies implied that CD4 is not necessary for neurotoxicity (Dreyer et al., 1990; Kaiser et al., 1990). Whether or not neuronal apoptosis occurs via an indirect or a direct pathway, the chemokine receptors are involved in the development of HAD, which suggests that therapeutic interventions at the level of chemokine receptors may reverse or attenuate the symptoms associated with HAD.

12.5.2 Chemokine Receptors and Multiple Sclerosis

The etiology of multiple sclerosis (MS) remains a mystery and is passionately debated by clinical investigators. It is believed that MS pathology results from the transmigration and accumulation of leukocytes in the central nervous system (CNS) that consequently attack glia cells. Chemokines are involved in the pathogenesis of immune-mediated inflammation of the CNS, both in controlling leukocyte migration across the brain endothelium and in the activation and movement of cells within the brain parenchyma. The biological activities of chemokines are mediated by interactions with their corresponding chemokine receptors (Zlotnik, 2000). The β-chemokine receptors CCR3 and CCR5 have been detected in the normal control human CNS tissue associated with microglia (Xia et al., 1998) and are expressed by cultured microglia (Bonecchi et al., 1998). Investigations have shown that glial cells within the CNS have the capacity to express both chemokines and chemokine receptors during a variety of inflammatory and degenerative conditions (Simpson et al., 1998; Xia et al., 1998). Increased expressions of CCR3 and CCR5 have been detected in the CNS and their ligands in MS (Simpson et al., 1998). CCR2 has also been identified on mononuclear phagocytes. These findings imply that β-chemokines, such as RANTES, monocyte chemoattractant protein (MCP)-1α, -2, -3, and -4, and macrophage inflammatory protein (MIP)-1α and MIP-1β, could all potentially activate and recruit both resident glia and infiltration haematogenous cells to sites of CNS inflammation. Cytokine-stimulated glial cells have also been shown to express chemokines and chemokine receptors *in vitro* (Jiang et al., 1998).

MS is an autoimmune disease of the CNS characterized by relapse and remittance of clinical signs and the presence of inflammation and demyelination in the white matter. Accumulating evidence has suggested that these findings represent the first stage of the disease followed by a chronic irreversible phase due to degeneration of the myelin sheath and the underlying axon. As the suppression of the relapse during the course of MS with immunotherapy is extremely important to prevent the development of the subsequent irreversible stage of the disease, it is essential to clarify the mechanism of disease relapse using an animal model, that is, Encephalomyelitis (EAE). EAE is a T-cell-mediated inflammatory demyelineating disease of the CNS that functions as a mouse model for human MS. Upon immunization of susceptible mice with myelin proteins (McRae et al., 1992; Trotter et al., 1987), EAE neuroinflammation follows with the subsequent invasion of leukocytes into the CNS through the blood–brain barrier. Leukocytes in the EAE-affected brain derived primarily from infiltrating the blood monocytes and lymphocytes actively mediate tissue destruction in EAE by destroying the myelin sheath, thus leading to neurological disease similar to MS. MCP-1, a natural chemokine, has been identified not only in MS, but also in EAE (Miyagishi et al., 1997). CCR2, a receptor for MCP-1, is a seven-transmembrane spanning protein that is functionally linked to downstream signaling pathways through heterotrimeric G-proteins (Charo, 1999). In MS, CCR2 is expressed on microglia and macrophages in chronic active lesions and on perivascular mononuclear cells in both white matter lesions and unaffected cortex (Simpson et al., 2000). Furthermore, CCR2 expression on circulating CD4 T-cells is significantly elevated during MS relapse (Misu et al., 2001; Sorensen, 2001). In an animal model of MS, CCR2 knockout mice showed reduced mononuclear infiltration and decreased susceptibility to EAE (Fife et al., 2000; Izikson et al., 2000). Studies have been conducted in MS lesions, infiltrating cells, and cerebrospinal fluid (CSF) and the interrelationship among the levels of chemokines in CSF, expression of chemokine receptors on lymphocytes, MS activity, and the clinical character of MS have been examined (Balashov et al., 1999; Mahad and Woodroofe, 2002; Sorensen et al., 2002). The elevated expression of CCR2 in the CNS suggests that this β-chemokine receptor and its ligand MCP-1 play a major role in the pathogenesis of MS (Simpson, 2000).

Of particular interest are (a) the CCR2 receptor expressed on peripheral blood monocytes and activated T-cells (Gu L, 1999) and (b) MCP-1 secreted by a variety of cell types in response to proinflammatory stimuli. Mahad and Woodroofe (2002) indicated that expression and cellular localization of CCL2/MCP-1 and CCR2 in MS have been described in the brain, cerebralspinal fluid (CSF), and blood. Evidence from descriptive, transgenic, knockout, and neutralizing studies of EAE points toward a nonredundant role of CCL2 and CCR2 in the recruitment of inflammatory infiltrate into the CNS. Hence, CCL2 and CCR2 may be targets for specific and effective treatment of MS (Mahad and Woodroofe, 2002). Some evidence has suggested that MCP-1 and CCR2 are both essential for the induction and progression of EAE and has demonstrated that MCP-1 mRNA is significantly upregulated in EAE during the acute phase of the disease (Juedes et al., 2000; Kennedy et al., 1998). Vaccination of mice with naked MCP-1 DNA, resulting in the generation of anti-MCP-1 antibodies, inhibited the induction of EAE (Youssef et al., 1998). Moreover, MCP-1 knockout mice exhibited a reduction in clinical and histopathological EAE after immunization with significantly reduced accumulation of macrophages in the CNS (Huang et al., 2001). Furthermore, two independent studies of CCR2 knockout mice by Fife et al. (2000) and Izikson et al. (2000) have found a complete resistance to EAE induction and failure of monocyte/macrophage accumulation in the CNS, suggesting that CCR2 may play a specific nonredundant role in monocyte recruitment, and that the inhibition of the CCR2:MCP-1 axis may ameliorate EAE neuropathology.

12.5.3 Chemokine Receptors and Stem Cell Migration to the Sites of CNS Injury

Migration toward pathology is the first critical step in stem cell engagement during regeneration. Neural stem cells (NSCs), both mouse and human, have a capacity for precise migration to even widespread and distant areas of pathology in a number of experimental models of CNS disease, including at ages where extensive migration has conventionally been deemed to be limited (Imitola et al., Imitola et al., 2003, 2004; Park et al., 2002a,b; Snyder et al., 1997). Because stem cell engagement with a degenerating environment (LaBarge and Blau, 2002; Snyder et al., 1997) is the first critical step in regeneration, realizing the therapeutic promise of the NSC depends in part on understanding the mechanisms underlying its mobilization during injury. In this regard, Imitola et al. previously showed that human NSCs (hNSCs) migrate *in vivo* toward an infarcted area (a representative CNS injury), where local astrocytes and endothelium upregulate the inflammatory chemoattractant stromal cell-derived factor (SDF)-1α. NSCs express CXCR4. Exposure of hNSCs to SDF-1α and the consequent induction of CXCR4-mediated signaling triggered a series of intracellular processes associated with fundamental aspects of survival, proliferation, and, importantly, migration. These steps include rapid and sustained phosphorylation of p38^{MAPK} kinase [implicated in regulating cytokine-induced cell migration (Cara et al., 2001; Dechert et al., 2001; Sharma et al., 2003)], phosphorylation of the ribosomal S6 kinase (p90^{RSK}) [implicated in phosphorylation of cytoskeletal molecules involved in neurite outgrowth (Suzuki et al., 2001; Wong et al., 1996)], phosphorylation of c-Jun [a kinase involved in migratory responses (Li et al., 2003)], rapid activation of extracellular response kinase (involved in proliferative responses), and rapid phosphorylation of paxilin [a scaffold molecule critical for migration (Huang et al., 2003)]. These data implicate SDF-1α/CXCR4, representative of the inflammatory milieu characterizing many pathologies, as a pathway that activates NSC molecular programs during injury and suggest that inflammation may be viewed not simply as playing an adverse role, but also as providing stimuli that recruit cells with a regenerative homeostasis-promoting capacity. The detailed molecular mechanisms of how SDF-1α/CXCR4 and potentially other chemokine/receptor pairs mediate neural stem cell migration and repair in CNS pathologies remain to be elucidated.

12.5.4 Chemokine Receptors and Rheumatoid Arthritis

Rheumatoid arthritis (RA) is an autoimmune disease characterized by chronic inflammation and massive infiltration of leukocytes into the inflamed joints. In RA, a large number of CD4$^+$ memory T-cells infiltrate the inflamed synovium (Kohem, 1996; Morimoto et al., 1988; Morita et al., 1998), where they get activated as they express cytokines and activation makers (Cohen et al., 1995; Dolhain et al., 1996; Grom et al., 1996; MacDonald, 1997; Nanki et al., 2000a,b; Smeets et al., 1998). Inflammatory cytokines [including IL-1, IL-6, IFN-γ, and tumor necrosis factor (TNF)-α] and activation markers are known to play central roles in the pathogenesis of RA (Nanki et al., 2000a). Although the development of biological drugs targeting these cytokines has produced promising results in the clinical therapy of RA patients, the design of other drugs that are independent of cytokine's function is needed for the improvement of RA chemotherapy. Previously, Nanki et al. reported that the memory T-cells express a chemokine receptor CXCR4 at a high level and that the concentration of SDF-1α, an endogenous ligand of CXCR4, is extremely high in the synovium of RA patients. It is also found that SDF-1α stimulates the migration of memory T-cells and inhibits T-cell apoptosis, suggesting that the SDF-1α-CXCR4 interaction plays an important role in the accumulation of T-cells in the RA synovium (Nanki et al., 2000b). Since SDF-1α (Nagasawa and Kishimoto, 1994; Tashiro et al., 1993) is independent of the other inflammatory cytokines in terms of its expression and function, the development of inhibitors that blocks the SDF-1α-CXCR4 interaction might produce promising leads for RA therapy.

Several recent publications support the involvement of SDF-1α-CXCR4 interaction in memory T-cell migration in the inflamed RA synovium. For instance, several CXCR4 antagonists, including T22 (an 18-mer peptide) (Murakami et al., 1997), T140 (a 14-mer) (Tamamura et al., 1998), and FC131 (a cyclic pentapeptide) (Fujii et al., 2003), which have inhibitory activities against the entry of T-cell line-tropic (X4-) HIV-1 into the target cells (Murakami et al., 1997, 1999; Xu et al., 1999), are known to have anti-RA activities. De Clercq's group also reported that another

CXCR4 antagonist, AMD3100, has both anti-HIV-1 and anti-RA activities (Hendrix et al., 2000; Schols et al., 1997). AMD3100 is a bicyclam derivative first described for its potent activity against HIV-1 infection (De Clercq et al., 1992) and is currently under investigation for clinical applicability in AIDS patients (De Clercq et al., 1994). However, the clinical applications of AMD3100 are limited by the fact that this compound is rapidly cleared from circulation, and that the treatment schedule in mice requires the use of osmotic minipumps (Datema et al., 1996).

12.5.5 Chemokine Receptors and Allergic Asthma

Allergic asthma is often characterized by airway inflammation of different cell types, airway hyper-responsiveness (AHR), mucus production, and a variable airflow (Busse, 2001; O'Byrne and Parameswaran, 2001). The initiating phase is characterized by IgE and mast cells. In the propagation phase, Th_2-polarized T-lymphocytes and eosinophils are guided to a chronic inflammatory state by infiltrating the airways. The effector phase is characterized by the production of spasmogenic substances, AHR, and mucus hypersecretion. Anti-cytokine therapy has been explored as a potential therapeutic approach in the treatment of allergic asthma, as the cytokines IL-4, IL-5, and IL-13 are of fundamental importance in its pathophysiology. The initial study using monoclonal antibodies (MAbs) against IL-5 (Leckie et al., 2000) in human asthmatics demonstrated a decrease in the number of eosinophils, though no major change in the AHR was observed. More recent studies showed the importance of CCR3 for the recruitment of eosinophils into the lung tissues and AHR in a murine model of allergic asthma. Ma et al. (2002) showed that epicutaneously allergen-sensitized CCR3 knockout mice were protected from AHR after allergen challenge. This protection was accompanied by the absence of CCR3-positive cells infiltrating the lung, especially eosinophils and intraepithelial mast cells (Humbles et al., 2002; Ma et al., 2002). Eosinophils of the CCR3 knockout mice accumulated along the blood vessels of the lung and migrated through the endothelium (Humbles et al., 2002). However, the cells were not able to transfer the lamina elastica of the venules and failed to migrate into the lung tissues (Humbles et al., 2002). The inflammatory infiltrate in allergic diseases consists of eosinophils, Th_2 cells, basophils, and mast cells. All these cells are known to express CCR3 on their surfaces constitutively or upon activation by cytokines (Ponath et al., 1996; Sallusto et al., 1997). Among these cells, eosinophils express the highest amount of CCR3, which is the dominant chemokine receptor on their surfaces (Elsner et al., 2000). Several ligands have been described for CCR3, and they include eotaxin, eotaxin-2, eotaxin-3, RANTES, MCPs, and HCC-1. In addition to the induction of actin polymerization, chemotaxis, and intracellular calcium flux in eosinophils via CCR3, they also release toxic ROS upon activation of CCR3. After binding of eotaxin or related chemokines to CCR3, the receptor is internalized. Receptor recycling seems to be important for the activation of leukocytes by chemokines. As such, blocking CCR3, which is a principal receptor for eosinophil migration in allergic asthma, could be a promising therapeutic tool in the management of asthma. In fact, this is substantiated by the development of MAbs, antagonists derived from modified chemokines (peptide chemokine receptor antagonists), or small molecules targeting CCR3 (Schwarz, 2002).

12.6 SMM-CHEMOKINES: A NEW CLASS OF UNNATURAL SYNTHETIC MOLECULES AS CHEMICAL PROBES OF CHEMOKINE RECEPTOR BIOLOGY AND LEADS FOR THERAPEUTIC DEVELOPMENT

12.6.1 Concept and Development of SMM-Chemokines

Chemokine ligands and receptors are hotly investigated areas in biomedical research. The dissection of the biological roles of specific ligands and receptors has been challenging because of the lack of selectivity in chemokine ligand–receptor interactions, given that 50 chemokine ligands and 19 functional receptors have been described to date (Proudfoot, 2002). This also greatly limits direct applications of natural chemokines in the treatment of various diseases such as (HIV-1) infection. To overcome the limitation of natural chemokines, various approaches have been taken to develop synthetic chemokine analogues that have higher efficacy and improved properties, such as the synthetic modifications of various chemokines conducted in a number of elegant works by Clark-Lewis and colleagues (Crump et al., 1997; Jones et al., 1997; Loetscher et al., 1998) and the recent success in the application of medicinal chemistry to the generation of RANTES analogues reported by Offord and colleagues (Hartley et al., 2004). The lack of selectivity among chemokine ligands is best exemplified by viral morphage inflammatory protein (vMIP)-II, which recognizes a variety of CC and CXC chemokine receptors, including CXCR4, CCR5, and CCR2 (Moore et al., 1996). Although the potential benefits of chemokine receptor inhibitors for AIDS and other diseases have been demonstrated, the lack of selectivity in the chemokine system has made it problematic to use natural, nonspecific chemokines in clinical applications due to their potential side effects. Consequently, the development of new inhibitors engineered with higher selectivity for targeted receptors and reduced toxicity is clearly desirable, because they can be used both in clinical applications and as specific probes of receptor biology to study the role of a particular ligand or receptor.

There has been intensive work in the development of new chemokine analogues by other groups in the field. For instance, Clark-Lewis and his colleagues have studied extensively the structure–function relationship of interleukin 8 and other chemokine receptors using chemically synthesized chemokine analogues (Crump et al., 1997; Jones et al., 1997; Loetscher et al., 1998). Also, on the basis of these previous works by others, we have been working toward the development of a systematic strategy based on chemokine structures to synthesize a new family of unnatural chemokines that, unlike natural chemokines, have higher receptor binding selectivity. We recently reported our progress in developing such a strategy by employing SMM (synthetically and modularly modified)-chemokines as a potential general method for the *de novo* design of novel ligands selective for any chemokine receptor of interest (Fig. 12.5). For this approach, the word "synthetically" refers to the use of total chemical synthesis to incorporate an almost unlimited range of unnatural amino acids and chemical modifications at any specific site(s). The word "modularly" refers to the changes of short important sequence modules, rather than the entire sequences, to achieve efficiency and be cost-effective in the generation of specificity and diversity.

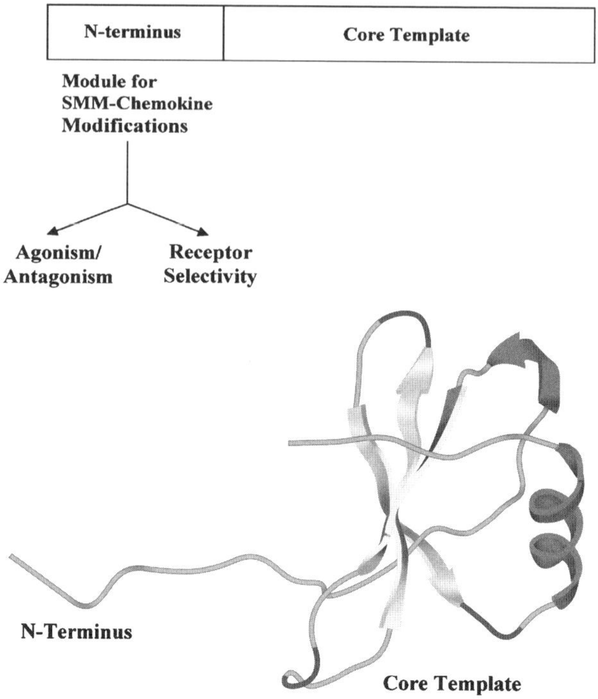

Figure 12.5 General design concept for SMM-chemokines. The N-terminal (1–10) sequence module of vMIP-II was modified by introducing unnatural D-amino acids or specific sequences grafted from other chemokines to chemically engineer receptor selectivity and/or signaling property. See color plates.

The word "modified" refers to the ability through various chemical modifications to improve biological properties (e.g., low toxicity, high receptor selectivity, and affinity) that are more desirable in clinical applications and in basic research of receptor biology. A proof of the concept was shown by applying this strategy to transform vMIP-II, a very nonselective chemokine, into new analogues with significantly enhanced selectivity and potency for CXCR4 or CCR5, two principal coreceptors for HIV-1 entry, through modifying only a small Amino (N)-terminal module of 10 residues (Kumar et al., 2006). Two representative SMM-chemokines, RCP168 and RCP188, selective for CXCR4 and CCR5, respectively, showed similar or significantly enhanced binding affinities for their corresponding target receptors but drastically decreased or even completely abolished cross-binding activities for other receptors (Table 12.3).

In addition to high receptor selectivity, another important biological property of these *de novo*-designed ligands is signaling activity. RCP168, a vMIP-II analogue with its N-terminal (1–10) residues replaced with D-amino acids, did not trigger either Ca^{2+} signaling or receptor internalization, which is distinct from the natural ligand of CXCR4, stromal cell-derived factor (SDF)-1α (Figs. 12.6a and 12.6b). More interestingly, RCP168 did not interfere with the Ca^{2+} signaling induced by SDF-1α at its effective CXCR4 binding concentration and only showed its effect at over 20 times

TABLE 12.3 Binding Affinity and Selectivity of SMM-Chemokines for CXCR4 and CCR5

Analogue	CXCR4 Binding (nM)	CCR5 Binding (nM)
RCP111	22	4
RCP112	>2700	146
RCP168	5	43
RCP169	141	>2700
RCP188	>2700	6
RCP189	>2700	106

higher concentrations. RCP168 also potently inhibits HIV-1 entry, in contrast to its much weaker activity in interfering with SDF-1α signaling. Thus, these disparate inhibitory activity profiles of RCP168 in differentiating HIV-1 coreceptor function versus the normal function of CXCR4 may prove to be advantageous in clinical applications, because RCP168 may not induce unwanted Ca^{2+} signaling or interfere with SDF-1α signaling important for the normal physiological functions at the concentrations used for inhibiting HIV-1 infection. The mechanistic basis for the disparate activities of RCP168 was investigated and shown by the mutational mapping analysis of binding sites of RCP168 and other D-amino acid-containing SMM-chemokines on CXCR4, revealing that RCP168 binding sites on CXCR4 overlap significantly with HIV-1 but differ from SDF-1α (Choi et al., 2005).

The CCR5-specific SMM-chemokine, RCP188, with its grafting of the N-terminal module of a CCR5 agonist macrophage inflammatory protein (MIP)-1β, was able to mimic the signaling activity of MIP-1β in activating Ca^{2+} release in CCR5-expressing 293 cells (Fig. 12.6c). This demonstrates that CCR5-selective

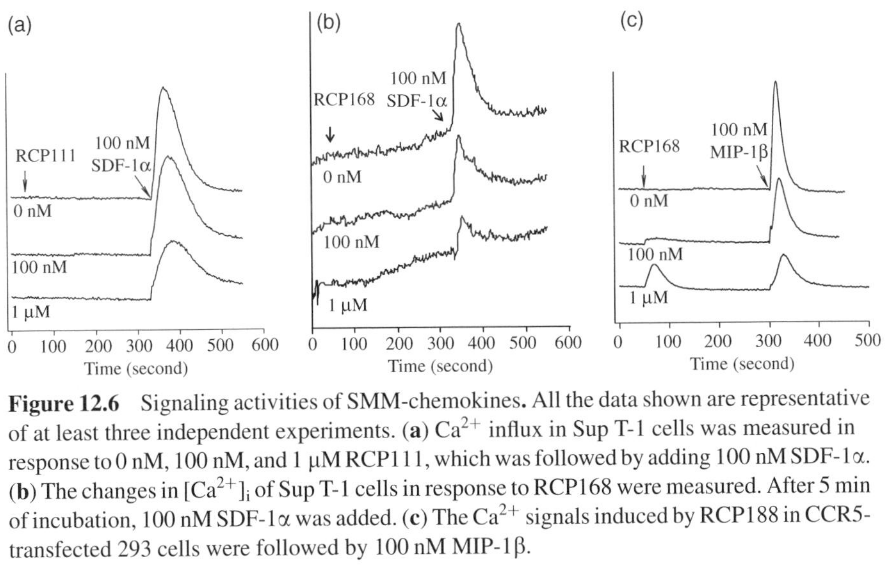

Figure 12.6 Signaling activities of SMM-chemokines. All the data shown are representative of at least three independent experiments. (**a**) Ca^{2+} influx in Sup T-1 cells was measured in response to 0 nM, 100 nM, and 1 μM RCP111, which was followed by adding 100 nM SDF-1α. (**b**) The changes in $[Ca^{2+}]_i$ of Sup T-1 cells in response to RCP168 were measured. After 5 min of incubation, 100 nM SDF-1α was added. (**c**) The Ca^{2+} signals induced by RCP188 in CCR5-transfected 293 cells were followed by 100 nM MIP-1β.

binding and signaling activities are encoded in the imported 10-residue sequence from MIP-1β. As such, we are able to apply the concept of SMM-chemokines to generate specific agonists and antagonists for either CCR5 or CXCR4, thus demonstrating that both receptor selectivity and signaling property can be designed at will into our synthetic ligands. While antagonists are often used for therapeutic applications, agonists can also be of great interest as molecular probes to study a particular receptor-mediated signaling pathway. The results from the first generation of SMM-chemokines reported here seem remarkable and promising considering that only the N-terminal 10-residue module was modified, yet a significant effect on binding or signaling selectivity was achieved. This illustrates the simplicity and robustness of this method and at the same time raises the possibility of its continued refinement and sophistication.

Furthermore, the high-resolution crystal structure of RCP168 (unpublished data) provided the structure-based mechanism for the selective interaction between the ligand and chemokine receptors, which guided the design and synthesis of even more selective inhibitory analogs. The high-resolution crystal structure of RCP168, when compared with that of vMIP-II previously reported by others (Fernandez et al., 2000; Liwang et al., 1999), revealed that the enhanced selectivity of RCP168 was associated with the structural changes at the N-terminus due to the D-amino acid modification and, surprisingly, the 30's loop due to a conformational change propagation from the distal N-terminus through a disulfide bridge. The observed structural changes in the N-terminal and the 30's loop regions of RCP168 in association with the enhanced selectivity of RCP168 suggest that these regions are important for the selectivity of chemokine–receptor interactions. Based on this structural insight, new analogues containing modifications at the 30's loop were designed by replacing the 30's loop of RCP168 with the corresponding region of SDF-1α, the only natural ligand of CXCR4, with the aim of generating higher CXCR4 selectivity. Indeed, this analogue, RCP303, exhibited more CXCR4 selective binding profile than the parent molecule, because the 30's loop substitution led to a substantial reduction in binding to both CCR5 and CCR2, further confirming the role of the 30's loop in affecting the receptor binding selectivity. In light of this finding, one may rationalize a structural basis for the conformational change cascade in chemokine–receptor interactions, which might include (1) the initial binding of the N-terminus of a chemokine to the receptor, (2) the resulting conformational changes in the N-terminus (including the N-loop) and subsequently the 30's loop as facilitated by the disulfide bridge, and finally (3) the triggered recognition between the 30's loop and the receptor, leading to the multipoint (at least including the N-terminus and the 30's loop) contact between the chemokine and its receptor.

In addition to suggesting a potential mechanism for ligand-receptor selectivity based on the conformation changes of the ligand at the N-terminus and the 30's loop, this study unveiled an interesting property of CXCR4—that is, the flexibility of the CXCR4 surface in recognizing ligands of different chiralities or, more precisely, different conformations. We previously found that all-D-amino acid peptides derived from the N-terminus of vMIP-II selectively bind to CXCR4 (Zhou et al., 2002). The results of RCP168 provided further support to the notion that CXCR4 is capable of interacting with diverse conformations of a ligand. The feature that CXCR4 can recognize both D- and L-amino acid-containing ligands is unprecedented among the

G-protein-coupled receptors (GPCR) superfamily, because we are not aware of other similar cases reported in the literature. The subsequent studies to map the receptor binding sites using site-directed mutants of CXCR4 have shown distinctive regions involved in D- versus L-ligand binding (Choi et al., 2005). Interestingly, D-amino acid-containing ligands such as RCP168 recognize sites on CXCR4 shared by HIV-1 gp120, but not by SDF-1α. In light of these results, the unusual flexibility of the CXCR4–ligand binding surface raises an intriguing question about whether HIV-1 gp120 might exploit this feature of CXCR4–ligand interaction for its entry into the target cell. It is well known that gp120 can frequently mutate itself, especially in the V3 loop region that is thought to recognize the coreceptor. The sequence mutations presumably bring about changes in the conformation of gp120, which could serve as a strategy for the virus to evade the recognition of HIV-1-neutralizing antibodies. However, it is puzzling that despite these changes, gp120 retains the ability to recognize the host cell receptors such as CXCR4. Our findings may shed some new light on a possible mechanism of how the varied structures of gp120 could be accommodated by the flexible CXCR4 ligand binding surface.

RCP168 is a prototype molecule for a novel family of SMM-chemokines that are highly selective and potent CXCR4 receptor inhibitors as compared with the natural chemokines. An important goal of the SMM-chemokine approach is to engineer *de novo* receptor selectivity into nonselective, natural chemokines. This was clearly demonstrated by the generation of RCP168. In contrast to the broad activity of vMIP-II, RCP168 has higher selectivity and affinity in CXCR4 binding. As a result, RCP168 is more potent in inhibiting HIV-1 infection than SDF-1α. The anti-HIV activity of RCP168 is also comparable to that of T-20. These results strongly suggest that RCP168 may serve as a prototype molecule for the development of highly selective and effective anti-HIV agents to be used in combination with other currently available drugs such as T-20 and/or drugs targeted to HIV-1 protease or reverse transcriptase. In this regard, it is encouraging to note that the structural basis for RCP168 activity as revealed by its high-resolution crystal structure has led to the design of a new generation of analogues such as RCP303 that showed even higher CXCR4 selectivity than RCP168, demonstrating that the combined structural and chemical biology approach can be used to develop chemokine analogues with desired high potency and selectivity.

12.6.2 Use of SMM-Chemokines to Probe Mechanism of Receptor Signaling

In another study (Dong et al., 2005), we applied the concept of SMM-chemokines containing D-amino acids to study the chemical basis of CXCR4–ligand interactions. Specifically, we asked the question of whether the binding and signaling functions of a chemokine ligand via its receptor are mediated by the same or distinct domains on the receptor. For this purpose, we used SDF-1α as a model to study the chemical mechanism of chemokine ligand–receptor binding and signaling. By replacing the N-terminal (1–8) residues of SDF-1α with all D-forms of (1–10) sequence module of vMIP-II, we found different stereochemical requirements in binding and signaling of the ligand, because the D-amino acid replacement only affected its signaling activity but not CXCR4 binding. In addition, the new compound was shown to be an effective HIV-1 inhibitor, indicating that its receptor binding, but not signaling, contributes to its anti-HIV activity. This study provided new insight into the stereochemical basis of CXCR4–ligand binding and signaling and suggests that further studies of the chemical basis of other chemokine ligand–receptor interactions may be of interest for understanding the mechanism of action of this important family of small protein ligands.

12.6.3 Application of SMM-Chemokines to Probe Mechanism of CXCR4–Ligand Interactions

In a recent work accepted for publication (Tian et al., 2005), we found that many residues throughout CXCR4 transmembrane (TM) and extracellular loop 2 (ECL2) domains are specifically involved in interaction with HIV-1 gp120, because most of these sites did not play a role in either SDF-1α binding or signaling. These results provided the direct experimental evidence for the distinct functional sites on CXCR4 for HIV-1 and the normal ligand SDF-1α. To further understand the CXCR4–ligand interaction and to develop new CXCR4 inhibitors to block HIV-1 entry, we used CXCR4-targeting SMM-chemokines, derived from the native sequence of SDF-1α or vMIP-II, as receptor probes and conducted ligand binding site-mapping experiments on a panel of site-directed mutants of CXCR4 (Choi et al., 2005). This study provided the first experimental evidence demonstrating that SMM-chemokines interact with many residues on CXCR4 TM and extracellular domains that are important for HIV-1 entry, but not SDF-1α binding or signaling (Fig. 12.7). The preferential overlapping in the CXCR4 binding residues of SMM-chemokines with HIV-1 over SDF-1α illustrates a mechanism for the potent HIV-1 inhibition by these SMM-chemokines and suggests that these chemically engineered molecules have interesting and unique receptor binding mechanisms distinct from those of the natural chemokines and may be used to selectively disrupt the coreceptor activity of CXCR4. The discovery of distinct functional sites or conformational states influenced by these receptor sites mediating different functions of the natural ligand versus the viral or synthetic ligands has important implications for drug discovery, since the sites shared by SMM-chemokines and HIV-1 but not by SDF-1α can be targeted for the development of selective HIV-1 inhibitors devoid of interference with the normal SDF-1α function.

12.6.4 Application of SMM-Chemokines to Probe Signaling Pathways Involved in HIV-Associated Dementia

To understand the mechanism of CXCR4 or CCR5 signaling in neuronal apoptosis associated with HIV-associated dementia (HAD), we have also applied SMM-chemokine analogues derived from natural SDF-1α or vMIP-II as chemical probes of the mechanism(s) whereby these SMM-chemokines prevent or promote neuronal apoptosis. Because of the profound activities of chemokine receptors in HAD, developing selective, potent inhibitors of chemokine receptors and understanding the physiological or pathological processes of HAD are crucial in devising novel strategies for clinical interventions.

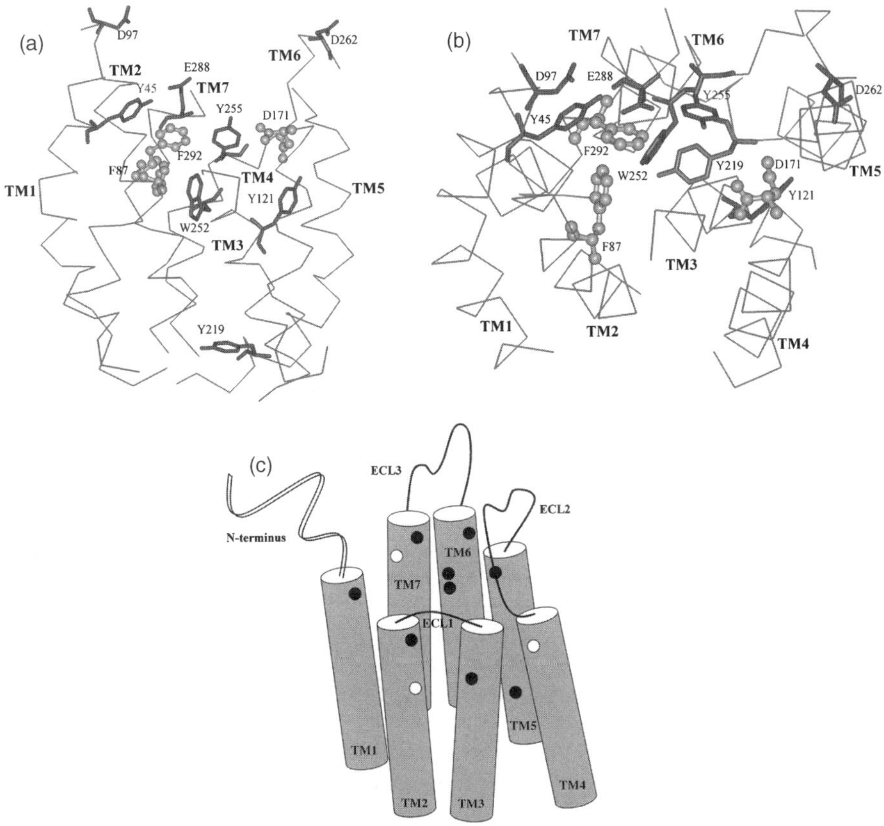

Figure 12.7 Distinct functional sites for SDF-1α and SMM-chemokines highlighted on a hypothetical structural model of CXCR4. The residues involved in both SDF-1α and SMM-chemokine binding are highlighted in the lighter color and represented as ball-and-stick, whereas those selectively involved in SMM-chemokine binding are highlighted in the darker color. Only the TM domains, with side (**a**) and top (**b**) views, are shown for simplicity. (**c**) Schematic illustration of the locations of residues important for ligand binding on CXCR4 TM and extracellular domains. The residues involved in the binding activities of both SDF-1α and SMM-chemokines are highlighted in white spots, whereas those selectively involved in SMM-chemokine binding (most of which overlap with HIV-1 binding) are highlighted in black spots. Such overlapping sites between HIV-1 and SMM-chemokines may serve as a potential target recognized by new selective anti-HIV inhibitors. See color plates.

We previously demonstrated how this SMM-chemokine approach can be applied to convert the nonselective vMIP-II into highly selective ligands of CXCR4 or CCR5 in terms of their binding, signaling, and/or antiviral activities (Kumar et al., 2006). We also used a similar strategy to modify biological and pharmacological properties of SDF-1α, which is of particular importance because SDF-1α, the only natural ligand of CXCR4, is neurotoxic with or without gp120 (Kaul and Lipton, 1999). In this unpublished work (Fig. 12.8), using rodent cerebrocortical cultures (RCCs) that not only contain the type and proportion of cells

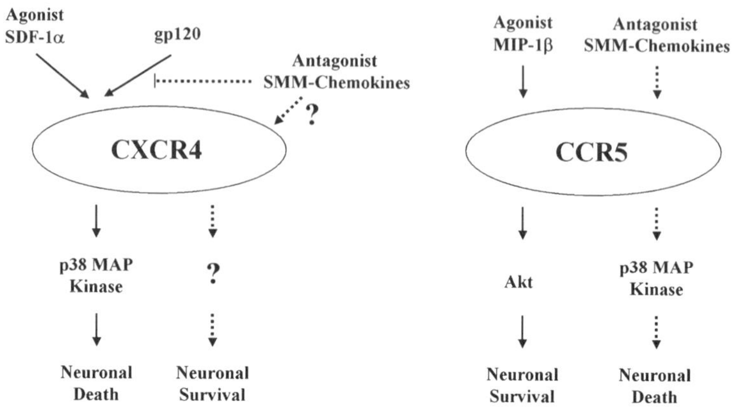

Figure 12.8 A model for the molecular mechanisms of action of neuroprotective CXCR4-selective SMM-chemokines versus neurotoxic CCR5-specific SMM-chemokines. Antagonistic CXCR4-selective SMM-chemokines prevent $gp120_{IIIB}$-induced neuronal apoptosis most likely via steric hindrance without activating Akt, ERK, or JNK, whereas natural CXCR4 agonist SDF-1α (and HIV-1 gp120) activates the pathway involving p38 MARK. Antagonistic CCR5-selective SMM-chemokines lead to neuronal apoptosis via the p38 MAPK pathway, whereas natural CCR5 agonist MIP-1β activates a different pathway involving Akt.

normally found in brain, i.e., neurons, astrocytes, and macrophages/microglia, but also express CXCR4, CCR5, and other chemokine receptor homologues (Heesen et al., 1998; Hesselgesser et al., 1997; Lavi et al., 1997; Rottman et al., 1997) that, like the human chemokine receptors, are capable of mediating HIV-1 infection via gp120 binding (Parolin et al., 1998; Plesoff et al., 1997), we demonstrated that neurotoxic SDF-1α (Kaul and Lipton, 1999) and vMIP-II can be modified to protect neurons from the neuronal apoptosis induced by CXCR4-preferring $gp120_{IIIB}$ without activating the Ser/Thr kinase Akt, extracellular regulated kinase (Erk), or c-JUN-N-terminal kinase (JNK). We also found that the inhibition of CCR5 by antagonistic SMM-chemokines leads to neurotoxicity, which is in sharp contrast to the neuroprotection induced by natural CCR5 ligands (Kaul et al., 2001; Kaul and Lipton, 1999), by activating p38 mitogen-activated protein kinase (MAPK). Furthermore, the simultaneous treatment of RCCs with a neuroprotective CXCR4 ligand and a neurotoxic CCR5-selective SMM-chemokine/CCR5-preferring $gp120_{BAL}$ led to neurotoxicity, suggesting the predominant role of steric hindrance in the neuroprotective mechanisms of antagonistic CXCR4-specific SMM-chemokines. These results illustrate a new strategy of using novel chemically engineered inhibitors of chemokine receptors to study the molecular mechanisms and intervention methods of the neuronal cell death caused by HIV-1 infection. One important lesson from HAD is that the development of new therapeutics for HAD is likely to have significant impact on several other important neurodegenerative diseases, because pathways to neuronal protection or damage may, at least in part, be common to other CNS disorders including stroke, spinal cord injury, and Alzheimer's disease.

12.7 CONCLUSIONS AND FUTURE PROSPECTS

Specific SMM (synthetically and modularly modified)-chemokines for a designated chemokine receptor can be potentially used in many areas of basic research and therapeutic development. For instance, one can use these SMM-chemokines as highly specific molecular probes to study the biology of chemokine receptors at the cellular level—that is, to characterize specific biological roles of a receptor in a physiological or pathological process out of the complicated and potentially overlapping receptor network. Also, the mechanism of receptor–ligand interactions at the molecular or atomic level can be investigated by making various artificial changes in ligand probes and testing how the receptor responds. This was demonstrated in another study in which the SMM-chemokines as reported here were used as chemical biology probes to discover new functional sites on CXCR4 important for the selective interactions with (HIV-1) but not with the normal physiological ligand of CXCR4, SDF-1α, providing a new basis for the development of more selective antiviral therapies than the conventional drugs that interrupt both pathological and physiological pathways (Choi et al., 2005). As chemokine receptors belong to the superfamily of G-protein-coupled receptors (GPCRs), which represent the largest class of drug targets and yet are still poorly understood in terms of their structure-function relationship and ligand-receptor interactions, the development of chemical biology probes such as SMM-chemokines may have general implications for the study of these important membrane receptors.

In addition to being valuable research tools, receptor-selective SMM-chemokines are promising leads for developing highly effective pharmaceutics. For clinical applications, SMM-chemokines can be highly desirable and advantageous for their greater selectivity, higher potency, and reduced toxicity compared with their natural counterparts (Kumar et al., 2006). In fact, biological studies of RCP168 in the blockade of HIV-1 entry demonstrated its potency and selectivity better than SDF-1α or viral macrophage inflammatory protein (vMIP)-II. Most importantly, not only was RCP168 found, in the study of drug-nonresistant HIV-1 strains, to be comparable in efficacy to some of the most well-known and commonly used HIV-1 drugs such as AZT, which targets HIV-1 reverse transcriptase as shown in this study, or T-20, which is the only marketed drug targeting HIV-1 gp41 protein and its mediated viral entry, but also RCP168 directed at an HIV-1 coreceptor expressed on the host cells was capable of overcoming the drug-resistance problem in drug-resistant viral isolates, including AZT- and 3TC-resistant HIV-1 isolates. This highlights the advantage of targeting a nonviral protein over virus-encoded proteins in terms of tackling the drug-resistance issue. Since drug resistance is a major problem in the treatment of not only AIDS but also other viral infections, the results shown in this study may have broader implications for the development of therapeutic strategies for diseases caused by other viruses.

ACKNOWLEDGEMENTS

We thank all former and current members of the Huang laboratory who have contributed to these projects and collaborators at Raylight Corporation/Chemokine Pharmaceutical Inc. and other institutions. This work has been supported by grants from the National Institutes of Health.

REFERENCES

Alkhatib, G., Ahuja, S. S., Light, D., Mummidi, S., Berger, E. A., and Ahuja, S. K. (1997). CC chemokine receptor 5-mediated signaling and HIV-1 co-receptor activity share common structure determinants. *J Biol Chem* **272**, 19771–19776.

Alkhatib, G., Combadiere, C., Broder, C. C., Feng, Y., Kennedy, P. E., and Murphy, P. M. (1996). CC CKR5: A RANTES, MIP-1alpha, MIP-1beta receptor as a fusion cofactor for macrophage-tropic HIV-1. *Science* **272**, 1955–1958.

Amara, A., Gall, S. L., Schwartz, O., Salamero, J., Montes, M., Loetscher, P., Baggiolini, M., Virelizier, J.-L., and Arenzana-Seisdedos, F. (1997). HIV coreceptor downregulation as antiviral principle: SDF-1α-dependent internalization of the chemokine receptor CXCR4 contributes to inhibition of HIV replication. *J Exp Med* **186**, 139–146.

Arenzana-Seisdedos, F., Virelizier, J. L., Rousset, D., Clark-Lewis, I., Loetscher, P., Moser, B., and Baggiolini, M. (1996). HIV blocked by chemokine antagonist. *Nature* **383**, 400.

Atchison, R. E., Gosling, J., Monteclaro, F. S., Franci, C., Digilio, L., Charo, I. F., and Goldsmith, M. A. (1996). Multiple extracellular elements of CCR5 and HIV-1 entry: Dissociation from response to chemokines. *Science* **274**, 1924–1926.

Baggiolini, M., Dewald, B., and Moser, B. (1997). Human chemokines: An update. *Annu Rev Immunol* **15**, 675–705.

Balashov, K. E. R., Weiner, H. L., and Hancock, W. W. (1999). CCR5(+) and CXCR3(+) T cells are increased in multiple sclerosis and their ligands MIP-1alpha and IP-10 are expressed in demyelinating brain lesions. *Proc Natl Acad Sci USA* **96**, 6873–6878.

Bandres, J. C., Wang, Q. F., O'Leary, J., Baleaux, F., Amara, A., Hoxie, J. A., Zolla-Pazner, S., and Gorny, M. K. (1998). Human immunodeficiency virus (HIV) envelope binds to CXCR4 independently of CD4, and binding can be enhanced by interaction with soluble CD4 or by HIV envelope deglycosylation. *J Virol* **72**, 2500–2504.

Berger, E. A., Murphy, P. M., and Farber, J. M. (1999). Chemokine receptors as HIV-1 coreceptors: Roles in viral entry, tropism, and disease. *Annu Rev Immunol* **17**, 657–700.

Bieniasz, P. D., Fridell, R. A., Aramori, I., Ferguson, S. S., Caron, M. G., and Cullen, B. R. (1997). HIV-1-induced cell fusion is mediated by multiple regions within both the viral envelope and the CCR-5 co-receptor. *EMBO J* **16**, 2599–2609.

Blanpain, C., Doranz, B. J., Bondue, A., Govaerts, C., De Leener, A., Vassart, G., Doms, R. W., Proudfoot, A., and Parmentier, M. (2003). The core domain of chemokines binds CCR5 extracellular domains while their amino terminus interacts with the transmembrane helix bundle. *J Biol Chem* **278**, 5179–5187.

Bleul, C. C., Farzan, M., Choe, H., Parolin, C., Clark-Lewis, I., and Sodroski, J. (1996a). The lymphocyte chemoattractant SDF-1 is a ligand for LESTR/fusin and blocks HIV-1 entry. *Nature* **382**, 829–833.

Bleul, C. C., Fuhlbrigge, R. C., Casasnovas, J. M., Aiuti, A., and Springer, T. A. (1996b). A highly efficacious lymphocyte chemoattractant, stromal cell-derived factor 1 (SDF-1). *J Exp Med* **184**(3), 1101–1109.

Bonecchi, R. B., Bordignon, P. P., D'Ambrosio, D., Lang, R., Borsatti, A., Sozzani, S., Allavena, P., Gray, P. A., Mantovani, A., and Sinigaglia, F. (1998). Differential expression of chemokine receptors and chemotactic responsiveness of type 1 T helper cells (Th1s) and Th2s. *J Exp Med* **187**, 129–134.

Boshoff, C., Endo, Y., Collins, P. D., Takeuchi, Y., Reeves, J. D., Schweickart, V. L., Siani, M. A., Sasaki, T., Williams, T. J., Gray, P. W., Moore, P. S., Chang, Y., and Weiss, R. A. (1997). Angiogenic and HIV-inhibitory functions of KSHV-encoded chemokines. *Science* **278**, 290–294.

Brelot, A., Heveker, N., Pleskoff, O., Sol, N., and Alizon, M. (1997). Role of the first and third extracellular domains of CXCR-4 in human immunodeficiency virus coreceptor activity. *J Virol* **71**, 4744–4751.

Brelot, A., Heveker, N., Montes, M., and Alizon, M. (2000). Identification of residues of CXCR4 critical for human immunodeficiency virus coreceptor and chemokine receptor activities. *J Biol Chem* **275**, 23736–23744.

Busse, W. W. L., (2001). Advances in immunology: Asthma. *N Engl J Med* **344**, 350–362.

Cara, D. C., Kaur, J., Forster, M., McCafferty, D. M., and Kubes, P. (2001). Role of p38 mitogen-activated protein kinase in chemokine-induced emigration and chemotaxis *in vivo*. *J Immunol* **167**, 6552–6558.

Chabot, D. J., and Broder, C. C. (2000). Substitutions in a homologous region of extracellular loop 2 of CXCR4 and CCR5 alter coreceptor activities for HIV-1 membrane fusion and virus entry. *J Biol Chem* **275**, 23774–23782.

Chabot, D. J., Zhang, P. F., Quinnan, G. V., and Broder, C. C. (1999). Mutagenesis of CXCR4 identifies important domains for human immunodeficiency virus type 1 X4 isolate envelope-mediated membrane fusion and virus entry and reveals cryptic coreceptor activity for R5 isolates. *J Virol* **73**, 6598–6609.

Charo, I. F. (1999). CCR2: From cloning to the creation of knockout mice. *Chem Immunol* **72**, 30–41.

Cheng-Mayer, C., Seto, D., Tateno, M., and Levy, J. A. (1988). Biologic features of HIV-1 that correlate with virulence in the host. *Science* **240**, 80–82.

Choe, H., Farzan, M., Sun, Y., Sullivan, N., Rollins, B., Ponath, P. D., Wu, L., Mackay, C. R., LaRosa, G., Newman, W., Gerard, N., Gerard, C., and Sodroski, J. S. (1996). The beta-chemokine receptors CCR3 and CCR5 facilitate infection by primary HIV-1 isolates. *Cell* **85**, 1135–1148.

Choi, W. T., Tian, S., Dong, C. Z., Kumar, S., Liu, D., Madani, N., An, J., Sodroski, J. G., and Huang, Z. (2005). Unique ligand binding sites on CXCR4 probed by a chemical biology approach: Implications for the design of selective human immunodeficiency virus type 1 inhibitors. *J Virol* **79**, 15398–15404.

Chung, C. W., Cooke, R. M., Proudfoot, A. E., and Wells, T. N. (1995). The three-dimensional solution structure of RANTES. *Biochemistry* **34**, 9307–9314.

Clapham, P. R., Blanc, D., and Weiss, R. A. (1991). Specific cell surface requirements for the infection of CD4-positive cells by human immunodeficiency virus types 1 and 2 and by Simian immunodeficiency virus. *Virology* **181**, 703–715.

Clements, G. J., Proce-Jones, M. J., Stephens, P. E., Sutton, C., Schulz, T. F., Clapham, P. R., McKeating, J. A., McClure, M. O., Thomson, S., Marsh, M., Kay, J., Weiss, R. A., and Moore, H. P. (1991). The V3 loops of the HIV-1 and HIV-2 surface glycoproteins contain proteolytic cleavage sites: A possible function in viral fusion? *AIDS Res Hum Retroviruses* **7**, 3–16.

Cohen, S. B. K., Chu, P. D., Thomssen, C. Q., Webb, H., Maini, L. M., Londei, R. N., and Feldmann, M. (1995). High level of interleukin-10 production by the activated T cell population within the rheumatoid synovial membrane. *Arthritis Rheum* **38**, 946–952.

Crump, M. P., Gong, J. H., Loetscher, P., Rajarathnam, K., Amara, A., Arenzana-Seisdedos, F., Virelizier, J. L., Baggiolini, M., Sykes, B. D., and Clark-Lewis, I. (1997). Solution structure and basis for functional activity of stromal cell-derived factor-1; dissociation of CXCR4 activation from binding and inhibition of HIV-1. *EMBO J* **16**, 6996–7007.

Datema, R. L. R., Hincenberghs, M.-B., Moreno, S., Warren, V., Linquist, B., Rosenwirth, J., Seifert, J., Seifert, J., and McCune, M. (1996). Antiviral efficacy *in vivo* of the anti-human immunodeficiency virus bicyclam SDZ SID 791 (JM 3100), an inhibitor of infectious cell entry. *Antimicrob Agents Chemother* **40**, 750.

De Clercq, E. Y., Pauwels, R., Baba, M., Schols, D., Nakashima, H., Balzarini, J., Debyser, Z., Murrer, B. A., Schwartz, D., et al. (1992). Potent and selective inhibition of human immunodeficiency virus (HIV)-1 and HIV-2 replication by a class of bicyclams interacting with a viral uncoating event. *Proc Natl Acad Sci USA* **89**, 5286–5290.

De Clercq, E. Y., Pauwels, R., Balzarini, J., Witvrouw, M., De Vreese, K., Debyser, Z., Rosenwirth, B. Peichl, P., Datema, R., et al. (1994). Highly potent and selective inhibition of human immunodeficiency virus by the bicyclam derivative JM3100. Highly potent and selective inhibition of human immunodeficiency virus by the bicyclam derivative JM3100. *Antimicrob Agents Chemother* **38**, 668–674.

Dealwis, C., Fernandez, E. J., Thompson, D. A., Simon, R. J., Siani, M. A., and Lolis, E. (1998). Crystal structure of chemically synthesized [N33A] stromal cell-derived factor 1alpha, a potent ligand for the HIV-1 "fusin" coreceptor. *Proc Natl Acad Sci USA* **95**, 6941–6946.

Dechert, M. A., Holder, J. M., and Gerthoffer, W. T. (2001). p21-activated kinase 1 participates in tracheal smooth muscle cell migration by signaling to p38 Mapk. *Am J Physiol Cell Physiol* **281**, C123–132.

Deng, H., Liu, R., Ellmeier, W., Choe, S., Unutmaz, D., Burkhart, M., Marzio, P. D., Marmon, S., Sutton, R. E., Hill, C. M., Davis, C. B., Peiper, S. C., Schall, T. J., Littman, D. R., and Landau, N. R. (1996). Identification of a major co-receptor for primary isolates of HIV-1. *Nature* **381**, 661–666.

Dolhain, R. J. t. H., Hoefakker, S., Tak, P. P., de Ley, M., Claassen, E., Breedveld, F. C., and Miltenburg, A. M. (1996). Increased expression of interferon (IFN)-gamma together with IFN-gamma receptor in the rheumatoid synovial membrane compared with synovium of patients with osteoarthritis. *Br J Rheumatol* **35**, 24–32.

Dong, C. Z., Kumar, S., Choi, W. T., Madani, N., Tian, S., An, J., Sodroski, J. G., and Huang, Z. (2005). Different stereochemical requirements for CXCR4 binding and signaling functions as revealed by an anti-HIV, D-amino acid-containing SMM-chemokine ligand. *J Med Chem* **48**, 7923–7924.

Donzella, G. A., Schols, D., Lin, S. W., Este, J. A., Nagashima, K. A., Maddon, P. J., Allaway, G. P., Sakmar, T. P., Henson, G., De Clercq, E., and Moore, J. P. (1998). AMD3100, a small molecule inhibitor of HIV-1 entry via the CXCR4 co-receptor. *Nat Med* **4**, 72–77.

Doranz, B., Orsini, M., Turner, J., Hoffman, T., Berson, J., Hoxie, J., Peiper, S., Brass, L., and Doms, R. (1999). Identification of CXCR4 domains that support coreceptor and chemokine receptor functions. *J Virol* **73**, 2752–2761.

Doranz, B. J., Rucker, J., Yi, Y., Smyth, R. J., Samson, M., Peiper, S. C., Parmentier, M., Collman, R. G., and Doms, R. W. (1996). A dual-tropic primary HIV-1 isolate that uses fusin and the beta-chemokine receptors CKR-5, CKR-3, and CKR-2b as fusion cofactors. *Cell* **85**, 1149–1158.

Doranz, B. J., Grovit-Ferbas, K., Sharron, M. P., Mao, S. H., Goetz, M. B., Daar, E. S., Doms, R. W., and O'Brien, W. A. (1997). A small-molecule inhibitor directed against the chemokine receptor CXCR4 prevents its use as an HIV-1 coreceptor. *J Exp Med* **186**, 1395–1400.

Doranz, B. J., Filion, L. G., Diaz-Mitoma, F., Sitar, D. S., Sahai, J., Baribaud, F., Orsinis, M. J., Benovic, J. L., Cameron, W., and Doms, R. W. (2001). AIDS Res Hum Retroviruses **17**, 475–486.

Dragic, T., Litwin, V., Allaway, G. P., Martin, S. R., Huang, Y., Nagashima, K. A., Cayanan, C., Maddon, P. J., Koup, R. A., Moore, J. P., and Paxton, W. A. (1996). HIV-1 entry into CD4+ cells is mediated by the chemokine receptor CC-CKR-5. *Nature* **381**, 667–673.

Dreyer, E. B., Kaiser, P. K., Offermann, J. T., and Lipton, S. A. (1990). HIV-1 coat protein neurotoxicity prevented by calcium channel antagonists. *Science* **248**, 364–367.

Elsner, J. M., Bruhl, H., Dulkys, Y., Kimmig, D., Simmons, G., Clapham, P. R., Schlondorff, D., Kapp, A., Wells, T. N. and Proudfoot, A. E. (2000). Differential activation of CC chemokine receptors by AOP-RANTES. *J Biol Chem* **275**, 7784–7794.

Endres, M. J., Clapham, P. R., Marsh, M., Ahuja, M., Turner, J. D., McKnight, A., Thomas, J. F., Stoebenau-Hag- garty, B., Choe, S., Vance, P. J., Wells, T. N., Power, C. A., Sutterwala, S. S., Doms, R. W., Landau, N. R., and Hoxie, J. A. (1996). CD4-independent infection by HIV-2 is mediated by fusin/CXCR4. *Cell* **87**, 745–756.

Fenard, D., Lambeau, G., Maurin, T., Lefebvre, J. C., and Doglio, A. (2001). A peptide derived from bee venom-secreted phospholipase A2 inhibits replication of T-cell tropic HIV-1 strains via interaction with the CXCR4 chemokine receptor. *Mol Pharmacol* **60**, 341–347.

Feng, Y., Broder, C. C., Kennedy, P. E., and Berger, E. A. (1996). HIV-1 entry cofactor: Functional cDNA cloning of a seven-transmembrane, G protein-coupled receptor. *Science* **272**, 872–877.

Fernandez, E. J., Wilken, J., Thompson, D. A., Peiper, S. C., and Lolis, E. (2000). Comparison of the structure of vMIP-II with eotaxin-1, RANTES, and MCP-3 suggests a unique mechanism for CCR3 activation. *Biochemistry* **39**, 12837–12844.

Fife, B. T. H., Kuziel, W. A., and Karpus, W. J. (2000). CC chemokine receptor 2 is critical for induction of experimental autoimmune encephalomyelitis. *J Exp Med* **192**, 899–905.

Förster, R., Kremmer, E., Schubel, A., Breitfeld, D., Kleinschmidt, A., Nerl, C., Bernhardt, G., and Lipp, M. (1998). Intracellular and surface expression of the HIV-1 coreceptor CXCR4/fusin on various leukocyte subsets: Rapid internalization and recycling upon activation. *J Immunol* **160**, 1522–1531.

Fujii, N. O., Hiramatsu, K., Araki, T., Ueda, S., Tamamura, H., Otaka, A., Kusano, S., Terakubo, S., Nakashima, H., Broach, J. A., Trent, J. O., Wang, Z. X., and Peiper, S. C. (2003). Molecular-size reduction of a potent CXCR4-chemokine antagonist using orthogonal combination of conformation- and sequence-based libraries. *Angew Chem Int Ed Engl* **42**, 3251–3253.

Gartner, S. (2000). HIV infection and dementia. *Science* **287**, 602–604.

Gershoni, J. M., Denisova, G., Raviv, D., Smorodinsky, N. I., and Buyaner, D. (1993). HIV binding to its receptor creates specific epitopes for the CD4/gp120 complex. *FASEB J* **7**, 1185–1187.

Gosling, J., Monteclaro, F. S., Atchison, R. E., Arai, H., Tsou, C.-l., Goldsmith, M. A., and Charo, I. F. (1997). Molecular uncoupling of C-C chemokine receptor 5-induced chemotaxis and signal transduction from HIV-1 coreceptor activity. *Proc Natl Acad Sci USA* **94**, 5061–5066.

Grom, A. A. M., Luyrink, L., Emery, H., Passo, M. H., Glass, D. N., Bowlin, T., and Edwards, C. 3rd (1996). Patterns of expression of tumor necrosis factor alpha, tumor necrosis factor beta, and their receptors in synovia of patients with juvenile rheumatoid arthritis and juvenile spondylarthropathy. *Arthritis Rheum* **39**, 1703–1710.

Gu L, T. S., and Rollins, B. J. (1999). Monocyte chemoattractant protein-1. *Chem Immunol* **72**, 7–29.

Gupta, S. K., Pillarisetti, K., Thomas, R. A., and Aiyar, N. (2001). Pharmacological evidence for complex and multiple site interaction of CXCR4 with SDF-1alpha: Implications for development of selective CXCR4 antagonists. *Immunol Lett* **78**, 29–34.

Hartley, O., Gaertner, H., Wilken, J., Thompson, D., Fish, R., Ramos, A., Pastore, C., Dufour, B., Cerini, F., Melotti, A., Heveker, N., Picard, L., Alizon, M., Mosier, D., Kent, S., and Offord, R. (2004). Medicinal chemistry applied to a synthetic protein: Development of highly potent HIV entry inhibitors. *Proc Natl Acad Sci USA* **101**, 16460–16465.

He, J., Chen, Y., Farzan, M., Choe, H., Ohagen, A., Gartner, S., Busciglio, J., Yang, X., Hofmann, W., Newman, W., Mackay, C. R., Sodroski, J., and Gabuzda, D. (1997). CCR3 and CCR5 are coreceptors for HIV-1 infection of microglia. *Nature* **385**, 645–649.

Heesen, M., Berman, M. A., Charest, A., Housman, D., Gerard, C., and Dorf, M. E. (1998). Cloning and chromosomal mapping of an orphan chemokine receptor: Mouse RDC1. *Immunogenetics* **47**, 364–370.

Henderson, R., Baldwin, J. M., Ceska, T. A., Zemlin, F., Beckmann, E., and Downing, K. H. (1990). Model for the Structure of Bacteriorhodopsin based on high-resolution electron cryo-microscopy. *J Mol Biol* **213**, 899–929.

Hendrix, C. W. F., MacFarland, R. T., Giandomenico, C., Fuchs, E. J., Redpath, E., Bridger, G., and Henson, G. W. (2000). Pharmacokinetics and safety of AMD-3100, a novel antagonist of the CXCR-4 chemokine receptor, in human volunteers. *Antimicrob Agents Chemother* **44**, 1667–1673.

Hesselgesser, J., Halks-Miller, M., DelVecchio, V., Peiper, S. C., Hoxie, J., Kolson, D. L., Taub, D., and Horuk, R. (1997). CD4-independent association between HIV-1 gp120 and CXCR4: Functional chemokine receptors are expressed in human neurons. *Curr Biol* **7**, 112–121.

Hesselgesser, J., Taub, D., Baskar, P., Greenberg, M., Hoxie, J., Kolson, D. L., and Horuk, R. (1998). Neuronal apoptosis induced by HIV-1 gp120 and the chemokine SDF-1 alpha is mediated by the chemokine receptor CXCR4. *Curr Biol* **8**, 595–598.

Heveker, N., Montes, M., Germeroth, L., Amara, A., Trautmann, A., Alizon, M., and Schneider-Mergener, J. (1998). Dissociation of the signalling and antiviral properties of SDF-1-derived small peptides. *Curr Biol* **8**, 369–376.

Huang, C., Rajfur, Z., Borchers, C., Schaller, M. D., and Jacobson, K. (2003). JNK phosphorylates paxillin and regulates cell migration. *Nature* **424**, 219–223.

Huang, D. R. W., Kivisakk, P., Rollins, B. J. and Ransohoff, R. M. (2001). Absence of monocyte chemoattractant protein 1 in mice leads to decreased local macrophage recruitment and antigen-specific T helper cell type 1 immune response in experimental autoimmune encephalomyelitis. *J Exp Med* **193**, 713–726.

Huang, R. R., Vicario, P. P., Strader, C. D., and Fong, T. M. (1995). Identification of residues involved in ligand binding to the neurokinin-2 receptor. *Biochemistry* **34**, 10048–10055.

Huang, Z. (2000). Structural chemistry and therapeutic intervention of protein–protein interaction in immune response, HIV entry and apoptosis. *Pharmacol Therapeut* **86**, 201–215.

Humbles, A. A. L., Friend, D. S., Okinaga, S., Lora, J., Al-Garawi, A., Martin, T. R., Gerard, N. P., and Gerard, C. (2002). The murine CCR3 receptor regulates both the role of eosinophils and mast cells in allergen-induced airway inflammation and hyperresponsiveness. *Proc Natl Acad Sci USA* **99**, 1479–1484.

Imitola, J., Snyder, E. Y., and Khoury, S. J. (2003). Genetic programs and responses of neural stem/progenitor cells during demyelination: Potential insights into repair mechanisms in multiple sclerosis. *Physiol Genomics* **14**, 171–197.

Imitola, J., Park, K. I., Teng, Y. D., Nisim, S., Lachyankar, M., Ourednik, J., Mueller, F. J., Yiou, R., Atala, A., Sidman, R. L., Tuszynski, M., Khoury, S. J., and Snyder, E. Y. (2004). Stem cells: Cross-talk and developmental programs. *Philos Trans R Soc Lond B Biol Sci* **359**, 823–837.

Izikson, L. K., R. S. Charo, I. F. Weiner, H. L. and Luster, A. D. (2000). Resistance to experimental autoimmune encephalomyelitis in mice lacking the CC chemokine receptor (CCR)2. *J Exp Med* **192**, 1075–1080.

Jiang, Y. S., Adhikari, S., Xia, Y., Feng, L., Sonntag, M. K., deFiebre, C. M., Pennell, N. A., Streit, W. J. and Harrison, J. K. (1998). Chemokine receptor expression in cultured glia and rat experimental allergic encephalomyelitis. *J Neuroimmunol* **86**, 1–12.

Jones, S. A., Dewald, B., Clark-Lewis, I., and Baggiolini, M. (1997). Chemokine antagonists that discriminate between interleukin-8 receptors. Selective blockers of CXCR2. *J Biol Chem* **272**, 16166–16169.

Juedes, A. E. H., Bergman, C. M., Neild, A. L. and Ruddle, N. H. (2000). Kinetics and cellular origin of cytokines in the central nervous system: insight into mechanisms of myelin oligodendrocyte glycoprotein-induced experimental autoimmune encephalomyelitis. *J Immunol* **164**, 419–426.

Kaiser, P. K., Offermann, J. T., and Lipton, S. A. (1990). Neuronal injury due to HIV-1 envelope protein is blocked by anti-gp120 antibodies but not by anti-CD4 antibodies. *Neurology* **40**, 1757–1761.

Kajumo, F., Thompson, D. A., Guo, Y., and Dragic, T. (2000). Entry of R5X4 and X4 human immunodeficiency virus type 1 strains is mediated by negatively charged and tyrosine residues in the amino-terminal domain and the second extracellular loop of CXCR4. *Virology* **271**, 240–247.

Kaul, M., Garden, G. A., and Lipton, S. A. (2001). Pathways to neuronal injury and apoptosis in HIV-associated dementia. *Nature* **410**, 988–994.

Kaul, M., and Lipton, S. A. (1999). Chemokines and activated macrophages in HIV gp120-induced neuronal apoptosis. *Proc Natl Acad Sci USA* **96**, 8212–8216.

Kennedy, K. J. S., Kunkel, S. L., Lukacs, N. W., and Karpus, W. J. (1998). Acute and relapsing experimental autoimmune encephalomyelitis are regulated by differential expression of the CC chemokines macrophage inflammatory protein-1alpha and monocyte chemotactic protein-1. *J Neuroimmunol* **92**, 98–108.

Kledal, T. N., Rosenkilde, M. M., Coulin, F., Simmons, G., Johnsen, A. H., Alouani, S., Power, C. A., Luttichau, H. R., Gerstoft, J., Clapham, P. R., Clark-Lewis, I., Wells, T. N. C., and Schwartz, T. W. (1997). A broad-spectrum chemokine antagonist encoded by Kaposi's sarcoma-associated herpesvirus. *Science* **277**, 1656–1659.

Kobilka, B. (1992). Adrenergic receptors as models for G protein-coupled receptors. *Annu Rev Neurosci* **15**, 87–114.

Kohem, C. L. B., Wisbey, H., Tortorella, C., Lipsky, P. E., and Oppenheimer-Marks, N. (1996). Enrichment of differentiated CD45RBdim,CD27-memory T cells in the peripheral blood, synovial fluid, and synovial tissue of patients with rheumatoid arthritis. *Arthritis Rheum* **39**, 844–854.

Kontoyianni, M., and Lybrand, T. P. (1993). Computer modeling studies of G protein coupled receptors. *Med Chem Res* **3**, 407–418.

Kumar, S., Choi, W. T., Dong, C. Z., Madani, N., Tian, S., Liu, D., Wang, Y., Pesavento, J., Wang, J., Fan, X., Yuan, J., Fritzsche, W. R., An, J., Sodroski, J. G., Richman, D. D., and Huang, Z. (2006). SMM-Chemokines: A class of unnatural synthetic molecules as chemical probes of chemical receptor biology and leads for therapeutic development. *Chem Biol* **13**, 69–79.

LaBarge, M. A., and Blau, H. M. (2002). Biological progression from adult bone marrow to mononucleate muscle stem cell to multinucleate muscle fiber in response to injury. *Cell* **111**, 589–601.

Lapham, C. K., Ouyang, J., Chandrasekhar, B., Nguyen, N. Y., Dimitrov, D. S., and Golding, H. (1996). Evidence for cell-surface association between fusin and the CD4-gp120. *Science* **274**, 602–605.

Lavi, E., Strizki, J. M., Ulrich, A. M., Zhang, W., Fu, L., Wang, Q., O'Connor, M., Hoxie, J. A., and Gonzalez-Scarano, F. (1997). CXCR-4 (Fusin), a co-receptor for the type 1 human immunodeficiency virus (HIV-1), is expressed in the human brain in a variety of cell types, including microglia and neurons. *Am J Pathol* **151**, 1035–1042.

Leckie, M. J. t. B., Khan, J., Diamant, Z. O'Connor, B. J., Walls, C. M., Mathur, A. K., Cowley, H. C., Chung, K. F. Djukanovic, R., Hansel, T. T., Holgate, S. T., Sterk, P. J., and Barnes, P. J. (2000). Effects of an interleukin-5 blocking monoclonal antibody on eosinophils, airway hyper-responsiveness, and the late asthmatic response. *Lancet* **356**, 2144–2148.

Li, G., Gustafson-Brown, C., Hanks, S. K., Nason, K., Arbeit, J. M., Pogliano, K., Wisdom, R. M., and Johnson, R. S. (2003). c-Jun is essential for organization of the epidermal leading edge. *Dev Cell* **4**, 865–877.

Lipton, S. A., and Gendelman, H. E. (1995). Seminars in medicine of the Beth Israel Hospital, Boston. Dementia associated with the acquired immunodeficiency syndrome. *N Engl J Med.* **332**, 934–940.

Liwang, A. C., Wang, Z. X., Sun, Y., Peiper, S. C., and Liwang, P. J. (1999). The solution structure of the anti-HIV chemokine vMIP-II. *Protein Sci* **8**, 2270–2280.

Lodi, P. J., Garrett, D. S., Kuszewski, J., Tsang, M. L., Weatherbee, J. A., Leonard, W. J., Gronenborn, A. M., and Clore, G. M. (1994). High-resolution solution structure of the beta chemokine hMIP-1 beta by multidimensional NMR. *Science* **263**, 1762–1767.

Loetscher, P., Gong, J. H., Dewald, B., Baggiolini, M., and Clark-Lewis, I. (1998). N-terminal peptides of stromal cell-derived factor-1 with CXC chemokine receptor 4 agonist and antagonist activities. *J Biol Chem* **273**, 22279–22283.

Lu, Z., Berson, J. F., Chen, Y., Turner, J. D., Zhang, T., Sharron, M., Jenks, M. H., Wang, Z., Kim, J., Rucker, J., Hoxie, J. A., Peiper, S. C., and Doms, R. W. (1997). Evolution of HIV-1 coreceptor usage through interactions with distinct CCR5 and CXCR4 domains. *Proc Natl Acad Sci USA* **94**, 6426–6431.

Luo, J., Luo, Z., Zhou, N., Hall, J. W., and Huang, Z. (1999a). Attachment of C-terminus of SDF-1 enhances the biological activity of its N-terminal peptide. *Biochem Biophys Res Commun* **264**, 42–47.

Luo, Z., Butcher, D., and Huang, Z. (1997). Molecular modeling of interleukin-8 receptor β and analysis of the receptor–ligand interaction. *Protein Eng* **10**, 1039–1045.

Luo, Z., Zhou, N., Luo, J., Hall, J. W., and Huang, Z. (1999b). The role of positively charged residues in CXCR4 recognition probed with synthetic peptides. *Biochem Biophy Res Commun* **263**, 691–695.

Luo, Z., Fan, X., Zhou, N., Hiraoka, M., Luo, J., Kaji, H., and Huang, Z. (2000). Structure–function study and anti-HIV activity of synthetic peptide analogues derived from viral chemokine vMIP-II. *Biochemistry* **39**, 13545–13550.

Ma, Q., Jones, D., Borghesani, P. R., Segal, R. A., Nagasawa, T., Kishimoto, T., Bronsoni, R. T., and Springer, T. A. (1998). Impaired B-lymphopoiesis, myelopoiesis, and derailed cerebellar neuron migration in CXCR4- and SDF-1-deficient mice. *Proc Natl Acad Sci USA* **95**, 9448–9453.

Ma, W. B., Humbles, P. J., Laouini, A. A., Yalcindag, D., Alenius, A., Friend, H., Oettgen, D. S., Gerard, H. C., Geha, C., and R. S. (2002). CCR3 is essential for skin eosinophilia and airway hyperresponsiveness in a murine model of allergic skin inflammation. *J Clin Invest* **109**, 621–628.

Macdonald, D., Murgolo, N., Zhang, R., Durkin, J. P., Yao, X., Strader, C. D., and Graziano, M. P. (2000). Molecular characterization of the melanin-concentrating hormone/receptor complex: Identification of critical residues involved in binding and activation. *Mol Pharmacol* **58**, 217–225.

MacDonald, K. P. N., Lipsky, P. E., and Thomas, R. (1997). Functional CD40 ligand is expressed by T cells in rheumatoid arthritis. *J Clin Invest* **100**, 2404–2414.

Maddon, P. J., Dalgleish, A. G., McDougal, J. S., Clapham, P. R., Weiss, R. A., and Axel, R. (1986). The T4 gene encodes the AIDS virus receptor and is expressed in the immune system and the brain. *Cell* **47**, 333–348.

Mahad, D. J. H., and Woodroofe, M. N. (2002). Expression of chemokines in cerebrospinal fluid and serum of patients with chronic inflammatory demyelinating polyneuropathy. *J Neurol Neurosurg Psychiatry* **73**, 320–303.

McArthur, J. C., Haughey, N., Gartner, S., Conant, K., Pardo, C., Nath, A., and Sacktor, N. (2003). Human immunodeficiency virus-associated dementia: An evolving disease. *J Neurovirol* **9**, 205–221.

McManus, C. M., Weidenheim, K., Woodman, S. E., Nunez, J., Hesselgesser, J., Nath, A., and Berman, J. W. (2000). Chemokine and chemokine–receptor expression in human glial elements: induction by the HIV protein, Tat, and chemokine autoregulation. *Am J Pathol* **156**, 1441–1453.

McRae, B. L. K., Tan, L. J. Dal Canto, M. C., Picha, K. S., and Miller, S. D. (1992). Induction of active and adoptive relapsing experimental autoimmune encephalomyelitis (EAE) using an encephalitogenic epitope of proteolipid protein. *J Neuroimmunol* **38**, 229–240.

Meucci, O., Fatatis, A., Simen, A. A., Bushell, T. J., Gray, P. W., and Miller, R. J. (1998). Chemokines regulate hippocampal neuronal signaling and gp120 neurotoxicity. *Proc Natl Acad Sci USA* **95**, 14500–14505.

Misu, T. O., Fujihara, K., Matsushima, K., Yoshie, O., Okita, N., Takase, S., and Itoyama, Y. (2001). Chemokine receptor expression on T cells in blood and cerebrospinal fluid at relapse and remission of multiple sclerosis: Imbalance of Th1/Th2-associated chemokine signaling. *J Neuroimmunol* **114**, 207–212.

Miyagishi, R. K., Takayama, C., Inoue, Y., and Tashiro, K. (1997). Identification of cell types producing RANTES, MIP-1 alpha and MIP-1 beta in rat experimental autoimmune encephalomyelitis by *in situ* hybridization. *J Neuroimmunol* **77**, 17–26.

Moore, P. S., Boshoff, C., Weiss, R. A., and Chang, Y. (1996). Molecular mimicry of human cytokine and cytokine response pathway genes by KSHV. *Science* **274**, 1739–1744.

Morimoto, C. R., Fox, D. A., Anderson, P., DiMaggio, M., Levine, H., and Schlossman, S. F. (1988). Abnormalities in CD4+ T-lymphocyte subsets in inflammatory rheumatic diseases. *Am J Med* **84**, 817–825.

Morita, Y. Y., Kawashima, M., Harada, S., Tsuji, K., Shibuya, K., Maruyama, K., and Makino, H. (1998). Flow cytometric single-cell analysis of cytokine production by CD4+ T cells in synovial tissue and peripheral blood from patients with rheumatoid arthritis. *Arthritis Rheum* **41**, 1669–1676.

Murakami, T., Nakajima, T., Koyanagi, Y., Tachibana, K., Fujii, N., Tamamura, H., Yoshida, N., Waki, M., Matsumoto, A., Yoshie, O., Kishimoto, T., Yamamoto, N., and Nagasawa, T. (1997). A small molecule CXCR4 inhibitor that blocks T cell line-tropic HIV-1 infection. *J Exp Med* **186**, 1389–1393.

Murakami, T. Z., Koyanagi, Y., Tanaka, Y., Kim, J., Suzuki, Y., Minoguchi, S., Tamamura, H., Waki, M., Matsumoto, A., Fujii, N., Shida, H., Hoxie, J. A., Peiper, S. C., and Yamamoto, N. (1999). Inhibitory mechanism of the CXCR4 antagonist T22 against human immunodeficiency virus type 1 infection. *J Virol* **73**, 7489–7496.

Murphy, P. M. (1994). The molecular biology of leukocyte chemoattractant receptors. *Annu Rev Immunol* **12**, 593–633.

Nagasawa, T., Hirota, S., Tachibana, K., Takakura, N., Nishikawa, S., Kitamura, Y., Yoshida, N., Kikutani, H., and Kishimoto, T. (1996). Defects of B-cell lymphopoiesis and bone-marrow myelopoiesis in mice lacking the CXC chemokine PBSF/SDF-1. *Nature* **382**, 635–638.

Nagasawa, T. K., and Kishimoto, T. (1994). Molecular cloning and structure of a pre-B-cell growth-stimulating factor. *Proc Natl Acad Sci USA* **91**, 2305–2309.

Nanki, T. a. L., P. E. (2000a). Cytokine, activation marker, and chemokine receptor expression by individual CD4(+) memory T cells in rheumatoid arthritis synovium. *Arthritis Res* **2**, 415–423.

Nanki, T. H., El-Gabalawy, H. S., Suson, S., Shi, K., Girschick, H. J., Yavuz, S., and Lipsky, P. E. (2000b). Stromal cell-derived factor-1-CXC chemokine receptor 4 interactions play a central role in CD4+ T cell accumulation in rheumatoid arthritis synovium. *J Immunol* **165**, 6590–6598.

Nardese, V., Longhi, R., Polo, S., Sironi, F., Arcelloni, C., Paroni, R., DeSantis, C., Sarmientos, P., Rizzi, M., Bolognesi, M., Pavone, V., and Lusso, P. (2001). Structural determinants of CCR5 recognition and HIV-1 blockade in RANTES. *Nat Struct Biol* **8**, 611–615.

Oberlin, E., Amara, A., Bachelerie, F., Bessia, C., Virelizier, J. L., Arenzana-Seisdedos, F., Schwartz, O., Heard, J. M., Clark-Lewis, I., Legler, D. F., Loetscher, M., Baggiolini, M., and Moser, B. (1996). The CXC chemokine SDF-1 is the ligand for LESTR/fusin and prevents infection by T-cell-line-adapted HIV-1. *Nature* **382**, 833–835.

O'Byrne, P. I., and Parameswaran, K. (2001). The trials and tribulations of IL-5, eosinophils, and allergic asthma. *J Allergy Clin Immunol* **108**, 503–508.

Ohagen, A., Ghosh, S., He, J., Huang, K., Chen, Y., Yuan, M., Osathanondh, R., Gartner, S., Shi, B., Shaw, G., and Gabuzda, D. (1999). Apoptosis induced by infection of primary brain cultures with diverse human immunodeficiency virus type 1 isolates: evidence for a role of the envelope. *J Virol* **73**, 897–906.

Oravecz, T., Pall, M., and Norcross, M. A. (1996). Beta-chemokine inhibition of monocytotropic HIV-1 infection. Interference with a postbinding fusion step. *J Immunol* **157**, 1329–1332.

Park, K. I., Ourednik, J., Ourednik, V., Taylor, R. M., Aboody, K. S., Auguste, K. I., Lachyankar, M. B., Redmond, D. E., and Snyder, E. Y. (2002a). Global gene and cell replacement strategies via stem cells. *Gene Ther* **9**, 613–624.

Park, K. I., Teng, Y. D., and Snyder, E. Y. (2002b). The injured brain interacts reciprocally with neural stem cells supported by scaffolds to reconstitute lost tissue. *Nat Biotechnol* **20**, 1111–1117.

Parolin, C., Borsetti, A., Choe, H., Farzan, M., Kolchinsky, P., Heesen, M., Ma, Q., Gerard, C., Palu, G., Dorf, M. E., Springer, T., and Sodroski, J. (1998). Use of murine CXCR-4 as a second receptor by some T-cell-tropic human immunodeficiency viruses. *J Virol* **72**, 1652–1656.

Picard, L., Simmons, G., Power, C. A., Meyer, A., Weiss, R. A., and Clapham, P. R. (1997a). Multiple extracellular domains of CCR-5 contribute to human immunodefiency virus type 1 entry and fusion. *J Virol* **71**, 5003–5011.

Picard, L., Wilkinson, D. A., McKnight, A., Gray, P. W., Hoxie, J. A., Clapham, P. R., and Weiss, R. A. (1997b). Role of the amino-terminal extracellular domain of CXCR-4 in human immunodeficiency virus type 1 entry. *Virology* **231**, 105–111.

Plesoff, O., Sol, N., Labrosse, B., and Alizon, M. (1997). Human immunodeficiency virus strains differ in their ability to infect CD4+ cells expressing the rat homolog of CXCR-4 (fusin). *J Virol* **71**, 3259–3262.

Poignard, P., Saphire, E. O., Parren, P. W., and Burton, D. R. (2001). gp120: Biologic aspects of structural features. *Annu Rev Immunol* **19**, 253–274.

Ponath, P. D. Q., Post, T. W., Wang, J., Wu, L., Gerard, N. P., Newman, W., Gerard, C., and Mackay, C. R. (1996). Molecular cloning and characterization of a human eotaxin receptor expressed selectively on eosinophils. *J Exp Med* **183**, 2437–2448.

Proudfoot, A. (2002). Chemokine receptors: Multifaceted therapeutic targets. *Nat Rev/Immunol* **2**, 106–115.

Rottman, J. B., Ganley, K. P., Williams, K., Wu, L., Mackay, C. R., and Ringler, D. J. (1997). Cellular localization of the chemokine receptor CCR5. Correlation to cellular targets of HIV-1 infection. *Am J Pathol* **151**, 1341–1351.

Rucker, J., Samson, M., Doranz, B. J., Libert, F., Berson, J. F., Yi, Y., Smyth, R. J., Collman, R. G., Broder, C. C., Vassart, G., Doms, R. W., and Parmentier, M. (1996). Regions in beta-chemokine receptors CCR5 and CCR2b that determine HIV-1 cofactor specificity. *Cell* **87**, 437–446.

Sachpatzidis, A., Benton, B. K., Manfredi, J. P., Wang, H., Hamilton, A., Dohlman, H. G., and Lolis, E. (2003). Identification of allosteric peptide agonists of CXCR4. *J Biol Chem* **278**, 896–907.

Sallusto, F. M., and Lanzavecchia, A. (1997). Selective expression of the eotaxin receptor CCR3 by human T helper 2 cells. *Science* **277**, 2005–2007.

Samson, M., Libert, F., Doranz, B. J., Rucker, J., Liesnard, C., Farber, C. M., Saragosti, S., Lapoumeroulie, C., Cognaux, J., Forceille, C., Muyldermans, G., Verhofstede, C., Burtonboy, G., Georges, M., Imai, T., Rana, S., Yi, Y., Smyth, R. J., Collman, R. G., Doms, R. W., Vassart, G., and Parmentier, M. (1996). Resistance to HIV-1 infection in caucasian individuals bearing mutant alleles of the CCR-5 chemokine receptor gene. *Nature* **382**, 722–725.

Sanders, V. J., Pittman, C. A., White, M. G., Wang, G., Wiley, C. A., and Achim, C. L. (1998). Chemokines and receptors in HIV encephalitis. *AIDS* **12**, 1021–1026.

Sattentau, Q. J., Moore, J. P., Vignaux, F., Traincard, F., and Poignard, P. (1993). Conformational changes induced in the envelope glycoproteins of the human and simian immunodeficiency viruses by soluble receptor binding. *J Virol* **67**, 7383–7393.

Schellekens, P. T., Tersmette, M., Roos, M. T., Keet, R. P., de Wolf, F., Coutinho, R. A., and Miedema, F. (1992). Biphasic rate of CD4+ cell count decline during progression to AIDS correlates with HIV-1 phenotype. *AIDS* **6**, 665–669.

Schols, D., Struyf, S., van Damme, J., Este, J. A., Henson, G., and De Clercq, E. (1997). Inhibition of T-tropic HIV strains by selective antagonization of the chemokine receptor CXCR4. *J Exp Med* **186**, 1383–1388.

Schwarz, M. K. W., (2002). New therapeutics that modulate chemokine networks. *Nat Rev Drug Discov* **1**, 347–358.

Sharma, G. D., He, J., and Bazan, H. E. (2003). p38 and ERK1/2 coordinate cellular migration and proliferation in epithelial wound healing: Evidence of cross-talk activation between MAP kinase cascades. *J Biol Chem* **278**, 21989–21997.

Siciliano, S. J., Rollins, T. E., DeMartino, J., Konteatis, Z., Malkowitz, L., Riper, G. V., Bondy, S., Rosen, H., and Springer, M. S. (1994). Two-site binding of C5a by its receptor: An alternative binding paradigm for G protein-coupled receptors. *Proc Natl Acad Sci USA* **91**, 1214–1218.

Simmons, G., Wilkinson, D., Reeves, J. D., Dittmar, M. T., Beddows, S., and Weber, J. (1996). Primary, syncytium-inducing human immunodeficiency virus type 1 isolates are dual-tropic and most can use either Lestr or CCR5 as coreceptors for virus entry. *J Virol* **70**, 8355–8360.

Simmons, G., Clapham, P. R., Picard, L., Offord, R. E., Rosenkilde, M. M., Schwartz, T., Buser, R., Wells, T. N. C., and E., P. A. (1997). Potent inhibition of HIV-1 infectivity in macrophages and lymphocytes by a novel CCR5 antagonist. *Science* **276**, 276–279.

Simpson, J. E. N., Cuzner, M. L. and Woodroofe, M. N. (1998). Expression of monocyte chemoattractant protein-1 and other beta-chemokines by resident glia and inflammatory cells in multiple sclerosis lesions. *J Neuroimmunol* **84**, 238–249.

Simpson, J. R., Newcombe, J., Cuzner, M. L., Male, D., and Woodroofe, M. N. (2000). Expression of the beta-chemokine receptors CCR2, CCR3 and CCR5 in multiple sclerosis central nervous system tissue. *J Neuroimmunol* **108**, 192–200.

Smeets, T. J. D., Miltenburg, A. M., de Kuiper, R., Breedveld, F. C., and Tak, P. P. (1998). Poor expression of T cell-derived cytokines and activation and proliferation markers in early rheumatoid synovial tissue. *Clin Immunol Immunopathol* **88**, 84–90.

Snyder, E. Y., Yoon, C., Flax, J. D., and Macklis, J. D. (1997). Multipotent neural precursors can differentiate toward replacement of neurons undergoing targeted apoptotic degeneration in adult mouse neocortex. *Proc Natl Acad Sci USA*. **94**, 11663–11668.

Sorensen, T. L. S., F. (2001). Distinct chemokine receptor and cytokine expression profile in secondary progressive MS. *Neurology* **57**, 1371–1376.

Sorensen, T. L. T., Kivisakk, P., Klaege, K. L., Majmudar, A., Ravid, R., Lassmann, H., Olsen, D. B., Strieter, R. M., Ransohoff, R. M., and Sellebjerg, F. (2002). Multiple sclerosis: A study of CXCL10 and CXCR3 co-localization in the inflamed central nervous system. *J Neuroimmunol* **127**, 59–68.

Sozzani, S., Luini, W., Bianchi, G., Allavena, P., Wells, T. N. C., Napolitano, M., Bernardini, G., Vecchi, A., D'Ambrosio, D., Sinigagli, F., Santoni, A., Maggi, E., Romagnani, S., and Mantovani, A. (1998). The viral chemokine macrophage inflammatory protein-II is a selective Th2 chemoattractant. *Blood* **92**, 4036–4039.

Strader, C. D., Fong, T. M., Tota, M. R., and Underwood, D. (1994). Structure and function of G protein-coupled receptors. *Annu Rev Biochem* **63**, 101–132.

Suzuki, Y., Rahman, M., and Mitsuya, H. (2001). Diverse transcriptional response of CD4(+) T cells to stromal cell-derived factor (SDF)-1: Cell survival promotion and priming effects of SDF-1 on CD4(+) T cells. *J Immunol* **167**, 3064–3073.

Tachibana, K., Hirota, S., Iizasa, H., Yoshida, H., Kawabata, K., Kataoka, Y., Kitamura, Y., Matsushima, K., Yoshida, N., Nishikawa, S., Kishimoto, T., and Nagasawa, T. (1998). The chemokine receptor CXCR4 is essential for vascularization of the gastrointestinal tract. *Nature* **393**, 591–594.

Tamamura, H., Xu, Y., Hattori, T., Zhang, X., Arakaki, R., Kanbara, K., Omagari, A., Otaka, A., Ibuka, T., Yamamoto, N., Nakashima, H., and Fujii, N. (1998). A low-molecular-weight inhibitor against the chemokine receptor CXCR4: A strong anti-HIV peptide T140. *Biochem Biophys Res Commun* **253**, 877–882.

Tashiro, K. T., Heilker, R., Shirozu, M., and Nakano, T., and Honjo, T. (1993). Signal sequence trap: A cloning strategy for secreted proteins and type I membrane proteins. *Science* **261**, 600–603.

Tersmette, M., Lange, J. M., de Goede, R. E., de Wolf, F., Eeftink-Schattenkerk, J. K., Schellekens, P. T., Coutinho, R. A., Huisman, J. G., Goudsmit, J., and Miedema, F. (1989). Association between biological properties of human immunodeficiency virus variants and risk for AIDS and AIDS mortality. *Lancet*. **1**, 983–985.

Tian, S., Choi, W. T., Liu, D., Pesavento, J., Wang, Y., An, J., Sodroski, J. G., and Huang, Z. (2005). Distinct functional sites for human immunodeficiency virus type 1 and stromal cell-derived factor 1a on CXCR4 transmembrane helical domains. *J Virol* **79**, 12667–12673.

Trkola, A., Dragic, T., Arthos, J., Binley, J. M., Olson, W. C., Allaway, G. P., Cheng-Mayer, C., Robinson, J., Maddon, P. J., and Moore, J. P. (1996). CD4-dependent, antibody-sensitive interactions between HIV-1 and its co-receptor CCR-5. *Nature* **384**, 184–187.

Trotter, J. L., Clark, H. B., Collins, K. G., Wegeschiede, C. L., and Scarpellini, J. D. (1987). Myelin proteolipid protein induces demyelinating disease in mice. *J Neurol Sci* **79**, 173–188.

Wang, Z. X., Berson, J. F., Zhang, T. Y., Cen, Y. H., Sun, Y., Sharron, M., Lu, Z. H., and Peiper, S. C. (1998). CXCR4 sequences involved in coreceptor determination of human immunodeficiency virus type-1 tropism. Unmasking of activity with M-tropic Env glycoproteins. *J Biol Chem* **273**, 15007–15015.

Wells, T. N. C., Power, C. A., Lusti-Narasimhan, M., Hoogewert, A. J., Cooke, R. M., Chung, C.-W., Peitsch, M. C., and Proudfoot, A. E. I. (1996). Selectivity and antagonism of chemokine receptors. *J Leuk Biol* **59**, 53–60.

Wong, E. V., Schaefer, A. W., Landreth, G., and Lemmon, V. (1996). Involvement of p90rsk in neurite outgrowth mediated by the cell adhesion molecule L1. *J Biol Chem* **271**, 18217–18223.

Wu, L., Gerard, N. P., Wyatt, R., Choe, H., Parolin, C., Ruffing, N., Borsetti, A., Cardoso, A. A., Desjardin, E., Newman, W., Gerard, C., and Sodroski, J. (1996). CD4-induced interaction of primary HIV-1 gp120 glycoproteins with the chemokine receptor CCR-5. *Nature* **6605**, 179–183.

Xia, M. Q. Q., Wu, L. J., Mackay, C. R., and Hyman, B. T. (1998). Immunohistochemical study of the beta-chemokine receptors CCR3 and CCR5 and their ligands in normal and Alzheimer's disease brains. *Am J Pathol* **153**, 31–37.

Xu, Y. T., Arakaki, R., Nakashima, H., Zhang, X., Fujii, N., Uchiyama, T., and Hattori, T. (1999). Marked increase in anti-HIV activity, as well as inhibitory activity against HIV entry mediated by CXCR4, linked to enhancement of the binding ability of tachyplesin analogs to CXCR4. *AIDS Res Hum Retroviruses* **15**, 419–427.

Youssef, S. W., Maor, G., Lanir, N., Gour-Lavie, A., Grabie, N., and Karin, N. (1998). Long-lasting protective immunity to experimental autoimmune encephalomyelitis following vaccination with naked DNA encoding C-C chemokines. *J Immunol* **161**, 3870–3879.

Zhang, L., Huang, Y., He, T., Cao, Y., and Ho, D. D. (1996). HIV-1 subtype and second-receptor use. *Nature* **383**, 768.

Zhou, N., Luo, Z., Hall, J. W., Luo, J., Han, X., and Huang, Z. (2000a). Molecular modeling and site-directed muagenesis of CCR5 reveal residues critical for chemokine binding and signal transduction. *Eur J Immunol* **30**, 164–173.

Zhou, N., Luo, Z., Luo, J., Hall, J. W., and Huang, Z. (2000b). A novel peptide antagonist of CXCR4 derived from the N-terminus of the viral chemokine vMIP-II. *Biochemistry* **39**, 3782–3787.

Zhou, N., Luo, Z., Luo, J., Liu, D., Hall, J. W., Pomerantz, R. J., and Huang, Z. (2001). Structural and functional characterization of human CXCR4 as a chemokine receptor and HIV-1 co-receptor by mutagenesis and molecular modeling studies. *J Biol Chem* **276**, 42826–42833.

Zhou, N., Luo, Z., Luo, J., Fan, X., Cayabyab, M., Hiraoka, M., Liu, D., Han, X., Pesavento, J., Dong, C. Z., Wang, Y., An, J., Kaji, H., Sodroski, J. G., and Huang, Z. (2002). Exploring the stereochemistry of CXCR4-peptide recognition and inhibiting HIV-1 entry with D-peptides derived from chemokines. *J Biol Chem* **277**, 17476–17485.

Zlotnik, A. Y. (2000). Chemokines: A new classification system and their role in immunity. *Immunity* **12**, 121.

Zou, Y., Kottmann, A., Kuroda, M., Taniuchi, I., and Littman, D. (1998). Function of the chemokine receptor CXCR4 in haematopoiesis and in cerebellar development. *Nature* **393**, 595–599.

CHAPTER 13

RESEARCH PROGRESS IN PROTEIN POST-TRANSLATIONAL MODIFICATION

Jia Hu, Yanting Guo, and Yanmei Li[*]

13.1 INTRODUCTION

A. Ciechanover and A. Hershko, Israeli scientists, and I. Rose, an American scientist, received the 2004 Nobel prize in chemistry because they discovered the mechanism of ubiquitin-regulated proteolysis. Their work indicates the research direction of proteolysis, helps people understand more about their immune systems, and is of great significance in DNA repair and regulation and in treatment of human diseases.

Actually, protein ubiquitylation is a type of protein post-translational modification. Protein post-translational modification refers to the chemical modification process on specific amino acid residues that occurs after mRNA has been translated into protein. Protein post-translational modification plays an important role in organisms. The accomplishment of the Human Genome Project is one of the greatest scientific and technological achievements of our time. Upon close inspection of the first complete draft of the human genome, it is surprising to find that only about 30,000–50,000 genes have been found (Venter et al., 2001), which is only about 3–5

[*]Corresponding author: liym@mail.tsinghua.edu.cn

Drug Discovery Research: New Frontiers in the Post-Genomic Era, Edited by Ziwei Huang

times those in the eelworm or the drosophila. It is far from enough to regulate such a complex life process depending only on such a small number of genes. Therefore, the protein post-translational modification process is extremely important. It makes the protein obtain more complicated structures, perfect functions, more accurate regulations, and more specific operations. Functions of many intercellular proteins are regulated by dynamic post-translational modification, and many physiological functions of cells, such as cell response to environment (Guo et al., 2004), are carried out by dynamic post-translational modification. Human complexity is not simply a result of the direct protein products of genes. It is the protein post-translational modification that allows one gene to correspond to not only one protein, which therefore, to some extent, grants our human life its diversity.

13.2 PROTEIN POST-TRANSLATIONAL MODIFICATIONS

There are more than 20 types of protein post-translational modifications in the eukaryote cell. The most common modifications include ubiquitylation, phosphorylation, glycosylation, lipodation, methylation, and acetylation. Recently, as the development of human genomics and proteomics studies, research on protein post-translational modifications has made great progress.

13.2.1 Ubiquitylation

Regulation of cell functions via proteolysis of its proteome members was neglected for a long time. The discovery of ubiquitin and thereafter the proteasome completely revolutionized the thinking of the scientific community. Ubiquitin is a highly conserved 76-amino-acid polypeptide and is expressed in all eukaryotes. Protein tagged with ubiquitin can be specifically recognized and then degraded by the proteasome, which is a common pathway for degradation of short-life or aberrant proteins. Unlike the protein degradation processing in the enteron, this proteolytic device also requires much of the cell's energy to function. The ubiquitin combining with protein requires energy, and so does the proteolysis process itself (Hilt and Wolf, 2004). It is necessary to realize that the ubiquitin proteasome system is a regulatory system, vital to all eukaryotic cells.

13.2.1.1 Ubiquitin Proteasome System

The ubiquitin proteasome system is the regulatory mechanism in eukaryotes (Fang and Weissman, 2004). In the 1970s and 1980s, the enigma of ubiquitin-mediated proteolysis was uncovered (Fig. 13.1). Ubiquitylation occurs as a result of the sequential action of three classes of enzymes: ubiquitin activating enzyme (E1), ubiquitin conjugating enzyme (E2), and ubiquitin protein ligase (E3) (Hershko and Ciechanover, 1998; Pickart, 2001; Weissman, 2001). Substrate specificity is largely determined by the E3. The work of E2s and E3s can be reversed by the action of DUBs (Fig. 13.2). Two major classes of DUBs have been described so far: UCHs and UBPs, which are both cysteine hydrolases (Wilkinson, 2000). In general, UCHs hydrolyze

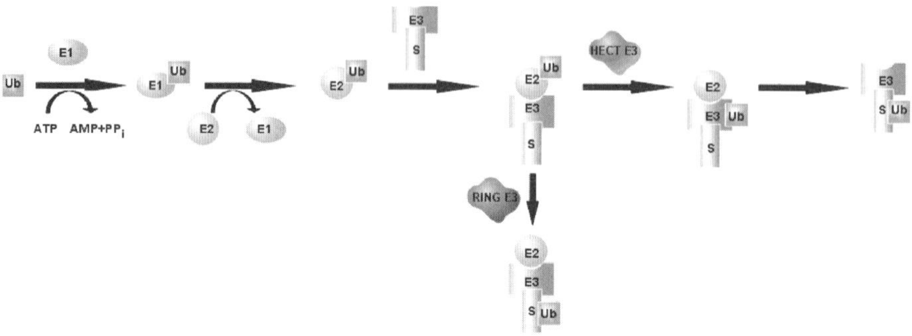

Figure 13.1 Ubiquitin-mediated proteolysis.

primarily C-terminal ester and amide bonds of ubiquitin and are capable of cleaving ubiquitin precursors to generate active ubiquitin. UBPs cleave and disassemble polyubiquitin chains.

13.2.1.2 Role of Ubiquitylation in Organisms

Ubiquitylation plays an essential role in ultimately all basic cellular functions—for instance, cellular differentiation, organelle biogenesis, apoptosis, DNA repair, new protein synthesis, regulated cell proliferation, protein transportation, antigen processing, and stress responses.

Bence et al. (2001) reported that protein aggregation directly impaired the function of the ubiquitin–proteasome system. Transient expression of two unrelated aggregation-prone proteins causes nearly complete inhibition of the ubiquitin-proteasome system. Because of the central role of ubiquitin-dependent proteolysis in regulating fundamental cellular events, a potential mechanism linking protein aggregation leads to cellular dysfunction and death.

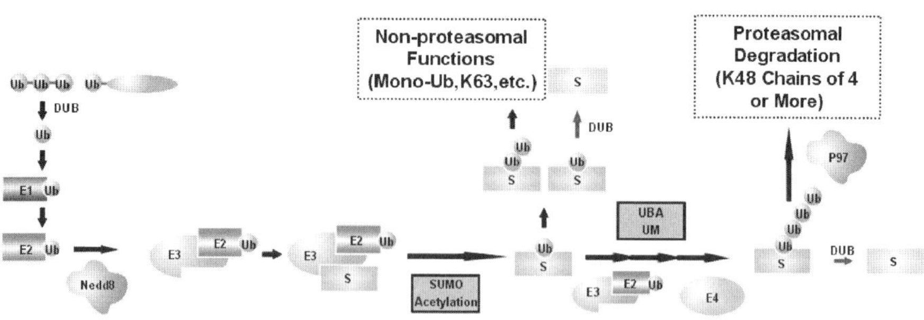

Figure 13.2 DUBs-mediated deubiquitylation.

Intraneuronal inclusions containing ubiquitylated filamentous protein aggregates are a common feature of many major human neurodegenerative disorders, including Alzheimer's and Parkinson's diseases (Spillantini et al., 1997; Engelender et al., 1999). Engelender et al. (1999) reported the interaction between synphilin-1 and ubiquitin protein ligase E3 SIAH-1 and SIAH-2 in the Lewy body. The synphilin-1/SIAH complex cannot be hydrolyzed, which leads to a high level of ubiquitylated cellular inclusion. As in the case of synphilin-1, in intact cells SIAH-2 interacts with α-synuclien and makes it monoubiquitylated. Monoubiquitylated proteins cannot be specifically degraded by a proteasome, which leads to the formation of cellular inclusion. Layfield et al. (2003) gave a new explanation regarding the presence of ubiquitylated protein in inclusion through close examination of ubiquitin-positive inclusions at the protein level. The presence of ubiquitin within inclusions may indicate a secondary, presumably protective cellular response, rather than a primary dysfunction of ubiquitin-mediated proteolysis. Ubiquitylation is probably a following process. Within this model, other factors are likely to initiate in inclusion biogenesis.

Ubiquitylation is also an important modification on histones and mainly occurs on histones H2A and H2B. Recent studies have uncovered the enzymes involved in histone H2B ubiquitylation (Robzyk et al., 2000; Hwang et al., 2004; Wood et al., 2003) and a "cross-talk" between H2B ubiquitylation and histone methylation (Sun and Allis, 2002). Wang et al. (2004a) reported the purification and functional characterization of an E3 ubiquitin ligase hPRC1Lf. It is found that hPRC1L is specific for histone H2A and monoubiquitylated nucleosomal histone H2A at Lys^{119}. In addition, Lys^{123} on the *S. cerevisiae* H2B C-terminal domain is the substrate of E3 Rad6, and this modification is of great importance in miosis and mitosis. The TBP-associated complex, $TAF_{II}250$, mediates monoubiquitylation of histone H1, which may contribute to transcription (Pham and Sauer, 2000).

Activator ubiquitylation is essential for transcriptional activation. In 2001, Salghtti et al. (2001) proposed that ubiquitylation regulated TAD function by serving as a dual signal for activation and activator restriction and that VP16 TAD signals ubiquitylation through the E3 Met30.

In the studies on p53, it is found that HAUSP is identified as a DUB for p53 in vitro and in vivo and stabilizes p53 even in the presence of excess Mdm2 (Li et al., 2004), while COP1 serves as an E3 ubiquitin ligase for p53 *in vitro* and *in vivo*. p53 was targeted for degradation by the ubiquitin proteasome system, which inhibits p53-dependent transcription and apoptosis (Dornan et al., 2004).

13.2.2 Phosphorylation

Phosphorylation is the process in which the phosphoryl group is transferred from ATP onto specific sites of proteins by protein kinase. Actually, most cellular processes are regulated by reversible phosphorylation, and at least 30% of proteins have such modifications (Ficarro et al., 2002; Krupa et al., 2004). Phosphorylation always occurs on Ser, Thr, and Tyr residues of proteins. The shape and function of cells will be changed under phosphorylation regulation.

Reversible phosphorylation is related to almost all physiological and pathological processes, including cellular signal conduction, tumor occurrence, metabolism,

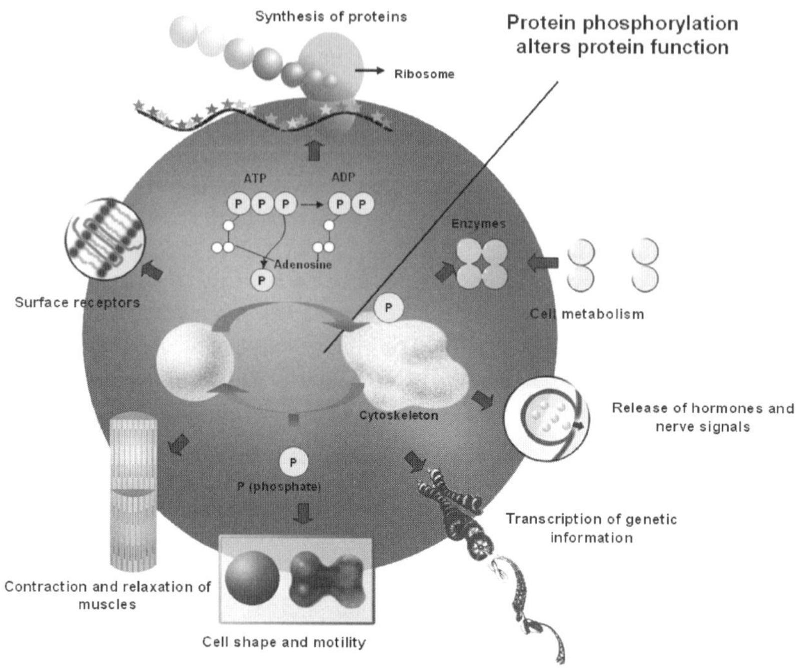

Figure 13.3 Protein phosphorylation regulating cellular processes. See color plates.

nervous activity, muscle contraction and proliferation, and development and differentiation of cells (Fig. 13.3). In 1992, E. H. Fischer and E. G. Krebs received the Nobel prize in physiology or medicine because of their discoveries concerning protein reversible phosphorylation as a biological regulatory mechanism.

In the process of cellular signal conduction, incretion and cytokine acting as cellular signals combine with membrane or cytoplasmic receptors and are activated by kinase. Incretion and cytokine are phosphorylated as the phosphorylation of kinase and result in intracellular signaling.

In the study of cancer, it is found that tubulin phosphorylation may lead to the occurrence of cancer. The cell possesses an LCP apoptotic step in the form of assembly of nitrotyrosinated α-tubulin onto microtubules. This leads to microtubule dysfunction and ultimately apoptosis. Phosphorylation of TTL is postulated to cause escape from LCP apoptosis. Cells that escape this LCP apoptotic step develop into cancer (Idriss, 2004).

In the study of DNA metabolism, it is found that cellular DNA damage causes the N-terminus of the 32-kDa subunit of human RPA to become hyperphosphorylated, which helps regulate DNA metabolism and promote DNA repair. Current data indicate that hyperphosphorylation causes a change in RPA conformation that down-regulates activity in DNA replication but does not affect DNA repair processes (Binz et al., 2004).

Cabrejos et al. (2004) investigated the role of phosphorylation by vertebrate protein kinase CK2 on the activity of the general transcription factors TFIIA, TFIIE,

TFIIF, and RNAPII. Data showed that the largest subunits of TFIIA, TFIIE, and TFIIF were phosphorylated by the CK2 holoenzyme. Also, RNA polymerase II was phosphorylated by CK2 in the 214,000- and 20,500-Da subunits. Phosphorylation of TFIIA, TFIIF, and RNAPII increases the formation of complexes on the TATA box of the Ad-MLP promoter. Phosphorylation of TFIIF increases the formation of transcripts, whereas phosphorylation of RNA polymerase II dramatically inhibits transcript formation.

In the studies of phosphorylation on histones, Maile et al. (2004) discovered that histone H2B was phosphorylated at evolutionarily conserved Ser33 (H2B-S33) by the CTK of the *Drosophila* general transcription factor TFIID subunit TAF1. Phosphorylation of H2B-S33 at the promoter of the cell-cycle regulatory gene *string* and the segmentation gene *giant* coincides with transcriptional activation.

In the studies of the HCV NS5A, Evans et al. (2004) reported that the interaction between HCV NS5A and hVAP-A was required for efficient RNA replication. Further analyses revealed that adaptive mutations suppressed NS5A hyperphosphorylation and promoted hVAP-A binding. NS5A hyperphosphorylation negatively regulates viral RNA replication.

Jones et al. (2004) described a new model for regulation of SHP-1 involving phosphorylation of its C-terminal Ser^{591} by associated protein kinase C_A. Phosphorylation of SHP-1 negatively regulated the activity of SHP-1 as demonstrated by a decrease in the in vitro ability of SHP-1 to dephosphorylate Vav1 on tyrosine, which leads to an increase in the tyrosine phosphorylation status of its substrates.

Li and co-workers (Luo et al., submitted; Luo et al., 2004) demonstrated that phosphorylation may influence local configuration of protein through hydrogen-bond interaction. In studies of the C-terminal domain of c-Fos protein, it is found that phosphorylation on S_{362} causes alteration in local configuration and further brings instability to turn structure (Li et al., 2004). From their studies, it is concluded that the MALDI-TOF MS is a reliable method to analyze the phosphorylation site on peptides or proteins (Huang et al., 2004)

13.2.3 Glycosylation

Protein glycosylation refers to the process in which oligosaccharides combine with specific amino acid residues on protein in the form of indican.

Protein glycosylation can be divided into four main categories, mainly depending on the linkage between the amino acid and the sugar: *O*-linked glycosylation, *N*-linked glycosylation, *C*-mannosylation, and GPI anchor attachments (Blom et al., 2004).

O-linked glycosylation mainly occurs on Ser and Thr nearing Pro and mostly on the β configuration of protein. *O*-glycans are built up in a stepwise fashion with monosaccharides added one at a time to a growing branch. At present there is no acceptor motif defined for *O*-linked glycosylation. *O*-linked glycosylation may happen at two cellular locations in the cell. One is in the Golgi; the other is in the nucleus and cytoplasm of cells (Hart, 1997). Those taking place in the Golgi are initiated by the addition of various reducing terminal linkages such as *N*-acetylgalactosamine, *N*-acetylglucosamine, mannose, and fucose. *O*-glycosylation of secreted and membrane-binding proteins is a post-translational event, taking place after *N*-glycosylation and folding of the protein (Asker et al., 1995), in the *cis*-Golgi compartment (Roth et al., 1994). *O*-glycosylation occurring in the nucleus and cytoplasm of cells is characterized by the attachment of a monosaccharide, *N*-acetylglucosamine, to a Ser or Thr residue (Blom et al., 2004). The best-known form of *O*-glycosylation in mammals is the addition of GalNAc, catalyzed by GalNAc-transferases, to Ser or Thr residues of secreted and cell surface proteins and further addition of Gal, GalNAc, or GlcNAc moieties.

O-GlcNAc residues are found in two conformations attached to the polypeptide backbone: *O*-α-GlcNAc and *O*-β-GlcNAc (Blom et al., 2004). This *O*-GlcNAcylation has three key features: *O*-GlcNAcylation occurs at the sites similar to those modified by protein kinases; *O*-GlcNAcylation is reciprocal to phosphorylation and may have the reverse function of phosphorylation on many proteins; and *O*-GlcNAcylation is highly dynamic, with rapid cycling in response to cellular signals (Hart, 1997).

Oligosaccharides attached to Asn residues of secreted or membrane-binding proteins are described as *N*-linked. This process is catalyzed by glycotransferase on endoplasmic reticulum. The sequence motif Asn-Xaa-Ser/Thr (Xaa is any amino acid except Pro) has been defined as a prerequisite for *N*-glycosylation (Gavel and von Heijne, 1990). Although rare, the sequence motif Asn-Xaa-Cys has also been shown to act as an acceptor site (Blom et al., 2004).

C-mannosylation is the attachment of an α-mannopyranosyl residue to the indole C2 of Trp via a C-C link and occurs on the first Trp in the motif W-X-X-W (or in some cases, W-X-X-C and W-X-X-F). The GPI anchor refers to glycophosphatidylinositol groups attached to a protein chain near the *C*-terminal, which anchors the protein to the cell membrane (Blom et al., 2004; Hart, 1997).

13.2.3.1 Analysis of Glycosylation Sites

The identification of glycosylation sites in proteins is possible through a combination of proteolytic digestion, separation, MS, and MS/MS. LC combined with MS/MS has been a reliable method for detecting glycopeptides in digestion mixtures and for assigning glycosylation sites. Krokhin et al. (2004) identified and confirmed the glycosylation site by MALDI and had set out to develop the software for automatically identificating glycopeptide precursors.

13.2.3.2 Synthesis of Glycoprotein

The synthetic challenges associated with generating homogeneous populations of selectively modified proteins have hindered detailed studies of the effects of these modifications on protein structure and function.

Zhang et al. (2004) demonstrated an approach to the cotranslational synthesis of selective glycoproteins in which the modified amino acid is genetically encoded. This approach should be generally applicable to the protein with other post-translational modifications. Also, it provides a potent strategy for producing homogeneous glycoforms of therapeutic glycoprotein.

In a recent study, the bacterial *N*-linked glycosylation pathway is discovered in the human gastrointestinal pathogen *Campylobacter jejuni*. In addition, an *O*-linked glycosylation pathway has been identified and characterized in *C. jejuni* and the related species *Campylobacter coli*. These provide model systems for the elucidation and exploitation of glycoprotein biosynthesis. Both pathways show similarities to the respective *N*- and *O*-linked glycosylation processes in eukaryotes, and these modifications could play similar biological roles. Homologues of the genes in both pathways are found in other organisms. The complex glycans linked to the glycoprotein share common biosynthetic precursors (Szymanski et al., 2003).

13.2.3.3 Role of Glycosylation in Organisms

Protein glycosylation influences the function of protein (Cassey, 1995) and is of great importance in many cellular processes such as immunoprotection, virus replication, cell growth, intracellular adhesion, and inflammation. Known *O*-glycosylated proteins (Comer and Hart, 2000) include transcription factors, nuclear-pore and heat-shock proteins, RNA polymerase II, and the transcription product of carcinogen and enzymes (Fig. 13.4). Aberrant glycosylation always results in the occurrence of diseases.

Higher glycosylation on transferrin has been demonstrated in patients with Alzheimer's disease, rheumatoid arthritis, and other diseases associated with a free radical etiology. Transferrin is a glycosylated metal-carrying serum protein. One of the biological functions of glycosylation is to stabilize the transferrin and indirectly influence iron homeostasis (van Rensburg et al., 2004). Also, there are many studies concerning the relationship between glycosylation and rapid-developed muscular dystrophy. Muntoni et al. (2004) demonstrated a novel mechanism responsible for muscular dystrophy. Aberrant glycosylation of α-dystroglycan appears to be a common finding in all these cases.

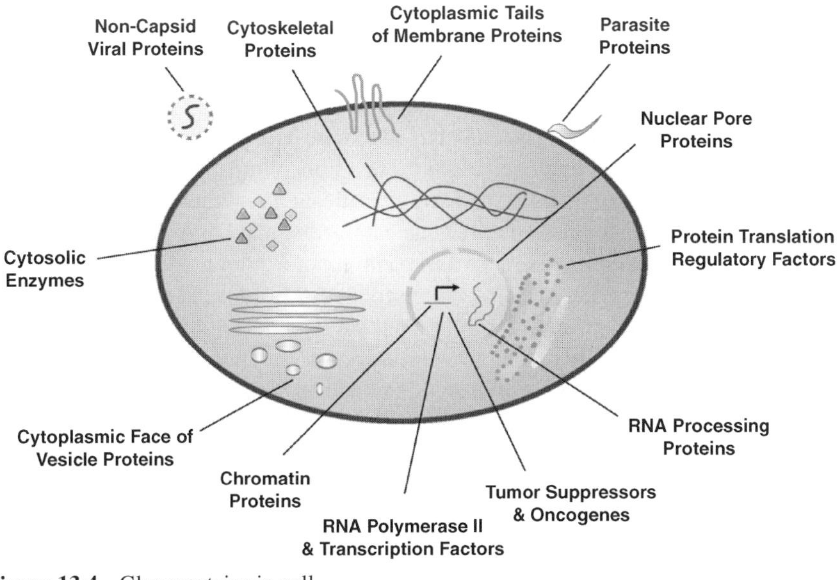

Figure 13.4 Glycoproteins in cell.

Chen et al. (2004) synthesized the *O*-GlcNAcylated polypeptides and studied the effects of *O*-GlcNAcylation on the configuration and biological function of the polypeptides. They also studied the different regulatory mechanisms between *O*-GlcNAcylation and *O*-phosphorylation.

Since protein glycosylation is vital to fundamental cellular processes, targeting glycosylation for drug design is likely to be a necessity. Therapeutic strategies are to regulate relative enzymes. Imino sugars are monosaccharide mimics that have a nitrogen atom in place of the ring oxygen. Members of the imino-sugar family can inhibit several glycosylation enzymes, including ER α-glucosidases I and II and the ceramide-specific glucosyltransferase. Targeting ER α-glucosidases at a low level could be a potential strategy for treating virus infections without compromising the host cell. Imino sugar *N*-nonyl-DNI has antivirus activity to an animal model of hepatitis B virus, a model of neonatal calf diarrhea coronavirus, and an *in vitro* model of HCV (Dwek et al., 2002).

13.2.4 Lipodation

Protein lipodation refers to the process in which long-chain aliphatic hydrocarbons combine with protein to form conjugates on the O or S atom. Usually in the process, Cys is acetylated by a palmitoylethyl group or alkylated by a farnesyl group on the S atom. These two kinds of aliphatic hydrocarbon chains always modify the same protein, and the protein is anchored onto the cell membrane by the compatibility between the aliphatic hydrocarbon chain and biological phospholipids membrane (Cassey, 1995).

Lipid proteins are a type of membrane-binding protein. The specific aliphatic-hydrocarbon modification helps them anchor on the cell membrane and further helps the proteins function. Recent studies in biophysics find that only when lipid proteins have anchored onto the membrane could they obtain the activity to participate in the biological process (Guo et al., 2004).

13.2.4.1 Synthesis of Lipoprotein

In the last 10 years, genetic engineering has been introduced to organic synthesis, and a new approach has been developed to synthesize lipoconjugates. Waldmann and co-workers (Bader et al., 2000) have made significant breakthroughs in this field. Enzymes and noble metals are involved as catalyzers, and solid-phase synthesis is used in synthesizing lipopeptides. In the reaction, an artificial connecting group is introduced, and MIC is used as a protecting group. The polypeptide is then combined with C-terminal interrupted protein obtained by genetic engineering through the reactions of the MIC and hydrosulphide groups on Cys. These proteins were tested *in vitro* and have biological activities. SPR showed that lipoproteins had strong interactions with membranes. Using fluorescent microscopy to examine the distribution of these lipoproteins in the cell, it was also found that they exhibited exactly the same properties as that *in vivo*. Waldmann and co-workers (Völkert et al., 2003) described a modular strategy for the assembly of farnesylated N-Ras heptapeptides carrying a photoactivatable BP group within the lipid residue. This strategy is based on the fragment condensation of an *N*-terminal hexapeptide synthesized on the solid support with a Cys methyl ester modified with different farnesyl analogues. Using this strategy, 24 peptides were synthesized, incorporating farnesyl analogues with four different chain lengths. Two of these photoactivatable conjugates were ligated to oncogenic human N-RasG12VΔ181. A cellular transformation assay revealed that the semisynthetic proteins retained their biological activity despite the photolabel.

13.2.4.2 Role of Lipodation in Organisms

Protein lipodation is vital to signal conduction in life. Lipid protein acts as the switch of signal conduction. In signal conduction in the human body, a series of processes is needed from cytokine to regulation of gene expression. Signals from cytokine receptors are transduced to SH2-containing domain adaptors and then to the guanine nucleotide exchange factor and the Ras protein. The Ras protein combines with GTPd, which acts as the switch of the entire signal conduction process. The circulation of Ras protein in life is shown in Fig. 13.5. Recently, Ras protein was proposed as an effective drug target.

Aberrantly modified lipid protein will influence signal conduction. Ras protein mutants are found in 30% of all human tumor tissues. Among these tumors, 80% are malignant (Peters et al., 2002). The cause of this aberrant modification is spot mutation on the Ras protein. It is now unclear whether chemical signals or gene mutations result in Ras mutation. Nevertheless, progress has been made in the research of targeting protein lipodation for drug design. Farnesyltransferase inhibitor has pronounced inhibitory activity (Thutewohl et al., 2003), while there is no cytotoxic effect on normal cells. Also, palmitoylethyltransferase inhibitor demonstrates antitumor activity and also has an antitumor effect on breast and prostate cancer.

13.2.5 Methylation

Protein methylation is a complicated post-translational modification process like phosphorylation and glycosylation and plays an important role in the regulatory process. Methylation on the Lys or Arg side-chain amido is catalyzed by methyltransferase. In addition, another type of methylation occurs on the Asp or Glu side-chain carboxyl to form a methyl ester. We will focus on the former methylation type here. Methylation will strengthen the steric hindrance and influence the formation of a hydrogen bond by substitution of a hydrogen atom on nitrogen. Therefore, methylation has the ability to regulate intermolecular interaction or interaction between small molecule and target proteins.

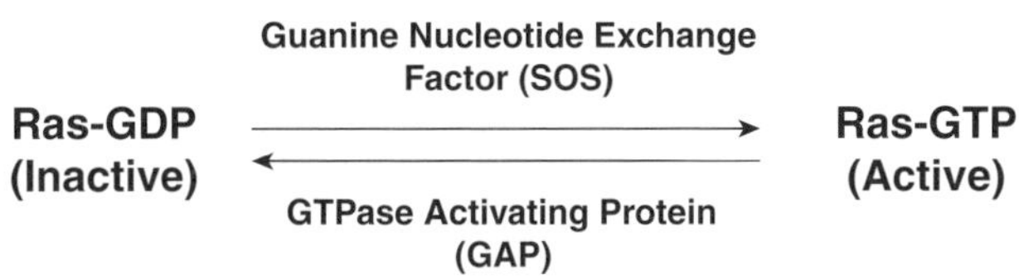

Figure 13.5 Circulation of Ras protein in life.

13.2.5.1 Classification of Arginine Methylation

In 1967, Paik and Kim discovered methylation on arginine and its role in cellular processes including signal conduction, transcription activation, and protein sorting. Arginine methylation may occur on many proteins. In eukaryotic cells, arginine methylation can be divided into three categories: N^G- MMA, N^GN^G-aDMA, and $N^GN'^G$-sDMA. Arginines in different proteins are differently methylated, and many proteins may even have more than one type of methylation. Methylations on hnRNPs and other RNA-binding proteins always occur on the RGG motif. Also, all arginine methylations on the RGG motif are monomethylations or asymmetric dimethylations, but not symmetric dimethylations. On the RXR and RG motifs, only asymmetric dimethylation may occur. However, arginine on myelin basic protein can be not only monomethylated but also symmetrically dimethylated. Also, arginine methylations on SmD1 and SmD3 are symmetric dimethylations. Unlike methylation of hnRNPs, that of myelin basic protein, SmD1 protein, and SmD3 protein occurs on the GRG motif, which indicates that the difference in the former or latter position of arginine may influence the methylation form.

13.2.5.2 Methylation on Histone

Histone is vital to transcription. It anticipates cellular processes by modifications such as phosphorylation, acetylation, or methylation on its tail. Histone arginine and lysine methylation are responsible for transcription regulation and heterochromosome formation (Trievel, 2004). A high level of histone acetylation is related to enhanced transcription activity, while histone methylation works in a much more complicated manner because methylation can both enhance and inhibit transcription.

Histone Lysine Methylation Histone lysine methylations occur on H3-K4, H3-K9, H3-K27, H3-K3, H3-K79, and H4-K20 N-terminals and sometimes on the H1 N-terminal. Methylations on H3-K9, H3-K27, and H4-K20 are related to chromosome inactivation and methylation on H4-K9 may lead to inhibition in the chromatin level on a large scale. Methylations on H3-K4, H3-K36, and H3-K79 are related to chromosome activation, and monomethylation on H3-K4 may act against gene restraint caused by methylation on H4-K9 (Sims et al., 2003).

In 2001, T. Hisashi et al. [cited in Tamaru and Selker (2001)] first proved the relationship between histone H3-K9 methylation and DNA methylation. In a gene scan of an *N. crassa* mutant with a DNA methylation defect, *dim-5* containing the SET-domain gene was identified. Dim-5 is H3 HMTaes. H3-K9 methylation by Dim-5 may help combine DNA methyltransferase by direct or indirect affinity of the mediator to methylated H3-K9. Jackson et al. (2002) gave new evidence of this relationship by function analysis of *Arabidopsis kryptonite* mutant.

In Shi et al. (2004), it was found that a nuclear homologue of amine oxidases, LSD1 (KIAA0601), functions as a histone demethylase and transcriptional corepressor. LSD1 specifically demethylates histone H3 lysine 4. Lysine demethylation occurs via an oxidation reaction that generates formaldehyde. Importantly, RNAi inhibition of LSD1 causes an increase in H3 Lys^4 methylation and concomitant derepression of target genes, suggesting that LSD1 represses transcription via histone demethylation. The results reveal dynamic regulation of histone methylation by both histone methylases and demethylases.

Histone Arginine Methylation Histone arginine methylation occurs on H3-R2, H3-R4, H3-R17, and H3-R26, and all these modifications can enhance transcription.

Methyltransferases have already been discovered, but enzymes that demethylate histones have not been identified for years. Cuthbert et al. (2004) proposed the deimination process, which converted histone arginine to citrulline and antagonized arginine methylation. Later, Wang et al. (2004) showed that human PAD4 regulated histone arginine methylation by converting methyl-arginine to citrulline and releasing methylamine. It has been found that PADI4 (refers to PAD4 in Wang's paper) specifically deiminates arginine residues R2, R8, R17, and R26 in the H3 tail. Deimination by PAD4 prevents arginine methylation by CARM1. Dimethylation of arginine prevents deimination by PAD4, although monomethylation still allows deimination to take place. *In vivo* targeting experiments on ecdogenous promoter demonstrate that PAD4 can repress hormone receptor-mediated gene induction. PAD4 is recruited to the *pS2* promoter following hormone induction when the gene is transcriptionally down-regulated. These results define a novel mechanism for antagonizing the transcriptional induction mediated by arginine methylation (Cuthbert et al., 2004; Wang et al., 2004b).

Trojer et al. (2004) studied filamentous fungus *Aspergillus nidulans* and identified three distinct PRMTs. The existence of *in vivo* arginine methylation on histones as demonstrated by site-specific antibodies and the high level and specificity of PRMTs for individual core histones in *A. nidulans* suggest an important role of these enzymes for chromatin modulating activities. All these enzymes exhibit intrinsic HMTaes activity when expressed as GST fusion proteins. Two of these proteins, termed RmtA and RmtC, reveal significant sequence homology to the well-characterized human proteins PRMT1 and PRMT5, respectively. The second methyltransferase, named RmtB, is only distantly related to human/rat PRMT3. Native as well as recombinant RmtA is specific for histone H4 with arginine 3 as the methylation site. Furthermore, methylation of histone H4 by recombinant RmtA affects the acetylation by p300/CBP and exhibits a unique substrate specificity in catalyzing the methylation of histones H4, H3, and H2A.

13.2.5.3 Role of Methylation in Organisms

Histone methylation plays a central role not only in epigenetic chromatin modification in eukaryote cells, but also in cell differentiation, development, gene expression, genome stability, and cancer research. Other types of protein methylations and methyltransferases are also vital to organisms. Aberrant protein methylation or mutation in methyltransferase always leads to disease. For instance, monomethylated arginine and asymmetric dimethylated arginine are NOS inhibitors. These kinds of modifications have already been discovered in many patients with heart disease. Arginine methylation is not reversible, which will influence NOS activity. Several methlylated proteins are found in autoimmune diseases. For example, the antibody against myelin basic protein was detected in patients with multiple sclerosis diseases. Chuikov et al. (2004) found a novel mechanism to regulate p53 function through lysine methylation by Set9 methyltransferase. Set9 specifically methylates p53 at one residue within the C-terminus regulatory region. Methylated p53 is restricted to the nucleus, and the modification positively affects its stability. Set9 regulates the expression of p53 target genes in a manner dependent on the p53-methylation site. The crystal structure of a ternary complex of Set9 with a p53 peptide and the cofactor product provides the molecular basis for recognition of p53 by this lysine methyltransferase.

13.2.6 Acetylation

Protein acetylation is another important type of protein post-translational modification in cells. Acetylation occurs on many proteins, such as histone.

13.2.6.1 Dynamic Histone Acetylation

Recently, more research has been carried out on histone acetylation. Histone acetylation, discovered more than 40 years ago, is a reversible modification of lysine within the N-terminal domain of core histones. Histone acetylation is catalyzed by HATs, and deacetylation is catalyzed by HDs or HDACs. The N-terminal tails of the core histones are rich in lysines and are therefore positively charged under physiological conditions. It is assumed that this allows an intimate interaction either with the negatively charged backbone of the DNA and/or with adjacent nucleosomes, leading to a "tight" nucleosome formation and higher-order chromatin folding. Acetylation on histone may weaken its interactions with the DNA, resulting in an "open" chromatin conformation (Hansen et al., 1998; Walia et al., 1998). Such a conformation facilitates the access for transcriptional regulators and the combination to transcription factor and promotes gene transcription. It is believed that localized relaxation of chromatin caused by histone N-terminal domain acetylation is a necessary but not sufficient condition for processes that repackage DNA. Until now, several types of HATs and HDs have been discovered, and CBP/p300, which may interact with several transcription regulators, is probably the most important. The function of the rapid, continuous, or repetitive acetylation and deacetylation reactions with half-lives of just a few minutes remains unknown. It is still unclear whether turnover itself is linked to chromatin transcription beyond its contribution to rapid changes toward hyper- or hypoacetylation of nucleosomes. However, recent experiments suggest that turnover may be linked directly to steps in gene transcription (Waterborg, 2002).

Hyperacetylation of normally silenced regions or deacetylation of normally actively transcribed regions can all lead to various disorders, including developmental and proliferative diseases such as leukemia, epithelial cancers, fragile X syndrome, and Rubinstein–Taybi syndrome. Timmermann et al. (2001) reviewed diseases induced by aberrant acetylation or deacetylation, so we will not focus on this here. *In vivo* V(D)J recombination is developmentally regulated by enhancer-dependent changes in the accessibility of chromosomal recombination signal sequences to the recombinase, but the molecular nature of these changes is unknown. McMurry and Krangel (2000) reported that histone H3 acetylation was measured along versions of a transgenic V(D)J recombination reporter and the endogenous T-cell receptor A/Δ locus. Enhancer activity was shown to impart long-range and developmentally regulated changes in H3 acetylation, and H3 acetylation status was tightly linked to V(D)J recombination. H3 hyperacetylation is proposed as a molecular mechanism coupling enhancer activity to accessibility for V(D)J recombination.

Polyglutamine diseases are hereditary neurodegenerative disorders caused by the expansion of a CAG repeat in the disease gene. Recent studies show that the altered balance between protein acetylation and deacetylation may be a key process contributing to expanded polyglutamine-induced pathogenesis. The restoration of this balance is possibly made by the genetic or pharmacological reduction of the opposing enzyme group, for example, the HDACs. Recent progress in HDAC research has made the development of inhibitors of specific HDAC family proteins possible, and these compounds could be effective candidates for treatment of these devastating diseases (Bodai et al., 2003).

13.2.6.2 Acetylation on Other Proteins

In the studies of tubulin acetylation, Hubbert et al. (2002) discovered that the enzyme HDAC6 reversed the post-translational acetylation of tubulin. This finding provided evidence that reducing tubulin acetylation enhanced cell motility. It was also suggested that decreasing tubulin acetylation reduced microtubule stability. However, in Palazzo et al. (2003), it was found that alteration in cell motility observed by Hubbert et al. in cells overexpressing HDAC6 resulted not from changes in the formation of stable microtubules, but from alterations in the degree of tubulin acetylation.

Nguyen et al. (2004) provided evidence that pRb underwent acetylation upon cellular differentiation, including skeletal myogenesis, and demonstrated that acetylation regulated the differentiation-specific function(s) of the pRb. The pRb is known to induce growth arrest and cellular differentiation. Studies have shown that the co-activator p300 was found to acetylate pRb. The biological significance of pRb acetylation, however, remains unclear. The p300-associated factor (P/CAF) can mediate pRb acetylation as pRb interacts directly with the acetyltransferase domain of P/CAF *in vitro* and can associate with P/CAF in differentiated cells. Experiments showed that acetylation did not affect pRb-dependent growth arrest or the repression of E2F transcriptional activity. Instead, acetylation was required for pRb-mediated terminal cell cycle exit and the induction of late myogenic gene expression.

13.3 INTERACTION OF DIFFERENT TYPES OF PROTEIN POST-TRANSLATIONAL MODIFICATIONS

Different post-translational modification processes do not occur isolatedly *in vivo*. In cellular activities, cooperation between different modifications is greatly needed. In signal conduction, for example, outside-membrane receptors and corresponding effectors, commonly glycoproteins, bind proper ligands. The stimuli from surroundings are conducted to the membrane through glycoprotein, and then to the membrane-binding lipoprotein; finally, they activate downstream proteins or kinases. Meanwhile, lipid protein always acts as the switch of phosphorylation on a series of proteins in almost the entire signal conduction. These phosphorylations are regulated by a unique kinase and are the main body of signal conduction.

A single protein can possess more than one modification. Different types of modifications influence each others' functions and cooperate with each other. Phosphorylation is parallel with glycosylation in many cases, especially in dynamic characteristics and in its ubiquitous existence in cells. These two modifications are found in transcription factors, carcinogenic products, and enzymes. From the data obtained, the addition and elimination of *N*-acetylglucosamine (GlcNAc) may act as a regulatory mechanism. Besides the fact that phosphorylation and glycosylation have a similar process, the two types of modifications are reciprocal, which is the same as the concept of "Yin–Yang" in ancient China (Qian et al., 2003). Phosphorylation and glycosylation on RNA polymerase II, for example, play different roles in gene expression. The RNA polymerase C-terminal domain is hyperglycosylated. As polymerase enters into the nucleolus and interacts with the transcription factor, the protein is immediately and totally deglycosylated and at the same time phosphorylated, which indicates that RNA polymerase can possess a high level of GlcNAc only when nonphosphorylated. Phosphorylation and glycosylation occur in the inner and outer parts of the cell membrane, respectively. Thus, it is concluded that glycosylation shares the same sites with phosphorylation and influences the phosphorylation degree. For example, a microtubule-associated protein, tau, is multiglycosylated. There are more than 12 glycosylation sites, including about 4 GlcNAcyl-modified sites, in one molecule. In the brains of patients with Alzheimer's disease, tau proteins form a kind of hyperphosphorylated double-helix tangle. The phosphorylated Ser^{262} on tau protein, which should be glycosylated in its normal condition, blocks the interaction with microtubule. Therefore, the aberrant phosphorylation is induced by glycosylation defection (Qian et al., 2003).

Histone can be methylated and acetylated at the same time (Fig. 13.6) (Rice and Allis, 2001).

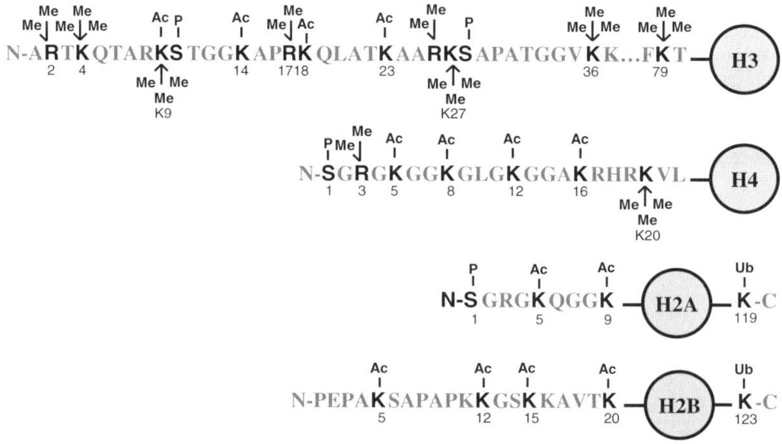

Figure 13.6 Post-translational modification on histones.

The major targets for post-translational acetylation and methylation are conserved lysine residues located in the N-terminal tails of histones H3 and H4. Histone acetylation occurs throughout the cell cycle, whereas histone methylation peaks in G2 phase and during heterochromatin assembly (Rice and Allis, 2001). Observations have suggested that certain combinations of acetyl and methyl modifications of lysines in histone tails may have antagonistic or cooperative biological effects. For example, hyperacetylated H4 from transcriptionally active chromatin preparations is a preferential target of histone H3 methylation, suggesting that these modifications may act synergistically to promote transcription in a way that remains unclear.

13.4 SUMMARY

Because protein post-translational modification is not directly decided by genes, studies on protein post-translational modification are of great significance for future research on proteomics. Thus, the post-translational modification proteomics is established and now becomes an intense focus of research in the world. Recent advances in these modifications are helping elucidate the role of these modifications in life and hold great promise for future pharmaceuticals. Targeting the mutantly modified protein in aberrant cells will help determine how post-translational modification regulates its function. Understanding what influences the post-translational modification will help uncover cellular processes and functions of protein networks at the molecular level and will enable more precision in directly targeting molecules for drug design. Post-translational modification mimics are set to dominate the next wave of protein therapeutics and become powerful medicinal tools in the twenty-first century (Benjamin, 2004).

ACKNOWLEDGMENTS

This work was supported by the National Natural Science Foundation of China (Grant Nos. 20272032 and NSFCBIC 20320130046) and the Teaching and Research Award Program for Outstanding Young Teachers in Higher Education Institutions of MOE, China.

REFERENCES

Asker, N., Baeckstrom, D., Axelsson, M. Carlstedt, I., and Hansson, G. C. (1995). The human MUC2 mucin apoprotein appears to dimerize before *O*-glycosylation and shares epitopes with the "insoluble" mucin of rat small intestine. *Biochem J* **308**, 873–880.

Bader, B., Kuhn, K., Owen, D. J., Waldmann, H., Wittinghofer, A., and Kuhlmann, J. (2000). Bioorganic synthesis of lipid-modified proteins for the study of signal transduction. *Nature* **403**, 223–226.

Bence, N. F., Sampat, R. M., and Kopito, R. R. (2001). Impairment of the ubiquitin–proteasome system by protein aggregation. *Science* **292**, 1552–1555.

Benjamin, G. D. (2004). Mimicking posttranslational modifications of proteins, *Science* **303**, 480–482.

Binz, S. K., Sheehan, A. M., and Wold, M. S. (2004). Replication protein A phosphorylation and the cellular response to DNA damage. *DNA Repair* **3**, 1015–1024.

Blom, N., Sicheritz-Pontén, T., Gupta, R. Gammeltoft, S., and Brunak, S. (2004). Prediction of post-translational glycosylation and phosphorylation of proteins from the amino acid sequence. *Proteomics* **4**, 1633–1649.

Cabrejos, M. E., Allende, C. C., and Maldonado, E. (2004). Effects of phosphorylation by protein kinase CK2 on the human basal components of the RNA polymerase II transcription machinery. *J Cell Biochem* **93**, 2–10.

Cassey, P. J. (1995). Protein lipodation in cell signaling. *Science* **268**, 221–225.

Chen, Y. X., Li, Y. M., Schlummer, S., Waldmann H., and Zhao Y. F. (2004). Synthesis and function of *O*-GlcNAcylated polypeptides. *Collection 3rd National Chembiology Congress (Chinese), Changsha* **2004**, 3–15.

Chuikov, S., Kurash, J. K., Wilson, J. R. Xiao, B., Justin, N., Ivanov, G. S., McKinney, K., Tempst, P., and Prives, C. (2004). Regulation of p53 activity through lysine methylation. *Nature* **432**, 353–360.

Comer, F. I., and Hart, G. W. (2000). O-Glycosylation of nuclear and cytosolic proteins. *J Biol Chem* **275**, 29179–29182.

Cuthbert, G. L., Daujat, S., Snowden, A. W., Erdjument-Bromage, H., Hagiwara, T., Yamada, M., Schneider, R., Gregory, P. D., Tempst, P., Bannister, A. J., and Kouzarides, T. (2004). Histone deimination antagonizes arginine methylation. *Cell* **118**, 545–553.

Dornan, D., Wertz, I., Shimizu, H., Arnott, D., Frantz, G. D., Dowd, P., O' Rourke, K., Koeppen, H., and Dixit, V. M. (2004). The ubiquitin ligase COP1 is a critical negative regulator of p53. *Nature* **429**, 86–92.

Dwek, R. A., Butters, T. D., Platt, F. M., and Zitzmann, N. (2002). Targeting glycosylation as a therapeutic approach. *Nat Rev Drug Discov* **1**, 65–75.

Engelender, S., Kaminsky, Z., Guo, X., Sharp, A. H., Amaravi, R. K., Kleiderlein, J. J., Margolis, R. L., Troncoso, J. C., Lanahan, A. A., Worley, P. F., Dawson, V. L., Dawson, T. M., and Ross, C. A. (1999). Synphilin-1 associates with α-synuclein and promotes the formation of cytosolic inclusions. *Nat Genet* **22**, 110–114.

Evans, M. J., Rice, C. M., and Goff, S. P. (2004). Phosphorylation of hepatitis C virus nonstructural protein 5A modulates its protein interactions and viral RNA replication. *Proc Natl Acad Sci USA* **101**, 13038–13043.

Fang, S., and Weissman, A. M. (2004). A field guide to ubiquitylation. *Cell Mol Life Sci* **61**, 1546–1561.

Ficarro, S. B., McCleland, M. L., Stukenberg, P. T., Burke, D. J., Ross, M. M., Shabanowitz, J., Hunt, D. F., and White, F. M. (2002). Phosphoproteome analysis by mass spectrometry and its application to *Saccharomyces cerevisiae*. *Nat Biotechnol* **20**, 301–305.

Gavel, Y., and von Heijne, G.. (1990). Statistical studies of *N*-glycosylated proteins have indicated that the frequency of nonglycosylated Asn-Xaa-(Thr/Ser) sequons increases toward the C terminus. *Protein Eng* **3**, 433–442.

Guo, Y. T., Li, Y. M., and Zhao, Y. F. (2004). New progress in lipid protein synthesis. *Org Chem (Chinese)* **24**, 722–727.

Hansen, J. C., Tse, C., and Wolffe, A. P. (1998). Structure and function of the core histone N-termini: more than meets the eye. *Biochemistry* **37**, 17637–17641.

Hart, G. W. (1997). Dynamic O-linked glycosylation of nuclear and cytoskeletal proteins. *Annu Rev Biochem* **66**, 315–335.

Hershko, A., and Ciechanover, A. (1998). The ubiquitin system. *Annu Rev Biochem* **67**, 425–479.

Hilt, W., and Wolf, D. H. (2004). The ubiquitin–proteasome system: Past, present and future. *Cell Mol Life Sci* **61**, 1545.

Huang, Z. P., Li, Y. M., Luo, S. Z., and Zhao, Y. F. (2004). Identification of phosphorylation site of phosphopeptide by MALDI-PSD mass spectrometry. *Mol Cell Proteomics* **3**, 131.

Hubbert, C., Amaris, G., Shao, R., Kawaguchi, Y., Ito, A., Nixon, A., Yoshida, M., Wang, X. F., and Yao, T. P. (2002). HDAC6 is a microtubule-associated deacetylase. *Nature* **417**, 455–458.

Hwang, W. W., Venkatasubrahmanyam S., Ianculescu A. G., Tong, A., Boone, C., and Madhani, H. D. (2003). A conserved RING finger protein required for histone H2B monoubiquitination and cell size control. *Mol Cell* **11**, 261–266.

Idriss, H. T. (2004). Three steps to cancer: How phosphorylation of tubulin, tubulin tyrosine ligase and P-glycoprotein may generate and sustain cancer. *Cancer Chemother Pharmacol* **54**, 101–104.

Jackson, J. P., Lindroth, A. M., Cao, X. F., and Jacobsen? S. E. (2002). Control of CpNpG methylation by the KRYPTONITE histone H3 methyltransferase. *Nature* **416**, 556–560.

Jones, M. L., Craik, J. D., Gibbins, J. M., and Poole, A.W. (2004). Regulation of SHP-1 tyrosine phosphatase in human platelets by serine phosphorylation at its C terminus. *J Biol Chem* **279**, 40475–40483.

Krokhin, O., Ens W., Standing, K. G., Wilkins, J., and Perreault, H. (2004). Site-specific *N*-glycosylation analysis: Matrix-assisted laser desorption/ionization quadrupole–quadrupole time-of-flight tandem mass spectral signatures for recognition and identification of glycopeptides. *Rapid Commun Mass Spectrom* **18**, 2020–2030.

Krupa, A., Preethi, G., and Srinivasan, N. (2004). Structural modes of stabilization of permissive phosphorylation sites in protein kinases: Distinct strategies in Ser/Thr and Tyr kinases. *J Mol Biol* **339**, 1025–1039.

Layfield, R., Cavey, J. R., and Lowe, J. (2003). Role of ubiquitin-mediated proteolysis in the pathogenesis of neurodegenerative disorders. *Ageing Res Rev* **2**, 343–356.

Li, M., Chen, D., Shiloh, A., Luo, J. Y., Nikolaev, A. Y., Qin, J., and Gu, W. (2002). Deubiquitination of p53 by HAUSP is an important pathway for p53 stabilization. *Nature* **416**, 648–653.

Li, Y. M., Luo, S. Z., Huang, Z. P., and Zhao Y. F. (2004). Identification of post-translational modification site by MALDI-TOF. *Collection in Proteomics 2nd Annual China Congress (Chinese), Dalian*, 66–67.

Luo S. Z., Li Y. M., Chen Z. Z., Abe H., Cui L. P., Nakanishi H., Qin X. R., and Zhao Y. F. (2004). Synthesis and matrix assisted laser desorption/ionization time of flight (MALDI-TOF) mass spectrometry study of phosphopeptide. *Lett Peptide Sci* **10**, 57–62.

Luo, S. Z., Li, Y. M., Su, X. Y., Huang Z. P., and Zhao Y. F. (Submitted). Hydrogen-bonding interactions induced by phosphorylation stereochemically influences the local structure of peptides.

Maile, T., Kwoczynski, S., Katzenberger, R. J., Wassarman, D. A., and Sauer, F. (2004). TAF1 activates transcription by phosphorylation of Serine 33 in histone H2B. *Science* **304**, 1010–1014.

McMurry, M. T., and Krangel, M. S. (2000). A role for histone acetylation in the developmental regulation of V(D)J recombination. *Science* **287**, 495–498.

Muntoni, F., Brockington, M., Torelli, S., and Brown, S. C. (2004). Defective glycosylation in congenital muscular dystrophies, *Curr Opin Neurol* **17**, 205–209.

Nguyen, D. X., Baglia, L. A., Huang, S. M., Baker, C. M., and McCance, D. J. (2004). Acetylation regulates the differentiation-specific functions of the retinoblastoma protein. *EMBO J* **23**, 1609–1618.

Palazzo, A., Ackerman, B., and Gundersen, G. G. (2003). Tubulin acetylation and cell motility. *Nature* **421**, 230.

Peters, C., Wagner, M., Völkert, M., and Waldmann, H. (2002). Bridging the gap between cell biology and organic chemistry: chemical synthesis and biological application of lipidated peptides and proteins. *Naturwissenschaften* **89**, 381–390.

Pham, A. D., and Sauer, F. (2000). Ubiquitin-activating/conjugating activity of TAFII250, a mediator of activation of gene expression in drosophila. *Science* **289**, 2357–2360.

Pickart, C. M. (2001). Mechanisms underlying ubiquitination. *Annu Rev Biochem* **70**, 503–533.

Qian, W., Liu, F., Zhu, L., Gong, C. X., and Jin, S. Y. (2003). Influence of protein *O*-GlcNAcylation to protein phosphorylation on Tau protein. *Prog Biochem Biophys (Chinese)* **30**, 623–628.

Rice, J. C., and Allis, C. D. (2001). Histone methylation versus histone acetylation: New insights into epigenetic regulation. *Curr Opin Cell Biol* **13**, 263–273.

Robzyk, K., Recht, J. and Osley, M. A. (2000). Rad6-dependent ubiquitination of histone H2B in yeast. *Science* **287**, 501–504.

Roth, J., Wang, Y., Eckhardt, A. E., Maccallum, D. K., Song, Z. W., and Goldstein, I. J. (1994). Differential expression of cell surface sialoglycoconjugates on wild-type and cultured ehrlich tumor cells as revealed by quantitative lectin-gold ultrastructural cytochemistry. *Proc Natl Acad Sci USA* **91**, 8935–8939.

Salghetti, S. E., Caudy, A. A., Chenoweth, J. G., and Tansey, W. P. (2001). Regulation of transcriptional activation domain function by ubiquitin. *Science* **293**, 1651–1653.

Shi, Y., Lan, F., Matson, C., Mulligan, P., Whetstine, J. R., Cole, P. A., Casero, R. A., and Shi, Y. (2004). Histone demethylation mediated by the nuclear amine oxidase homolog LSD1. *Cell* **119**, 941–953.

Sims, R. J., III, Nishioka, K., and Reinberg, D. (2003). Histone lysine methylation: A signature for chromatin function. *Trends Genet* **19**, 629–639.

Spillantini, M. G., Schmidt, M. L., Lee, V. M., Trojanowski, J. Q., Jakes, R., and Goedert, M. (1997). Alpha-synuclein in Lewy bodies. *Nature* **388**, 839–840.

Sun, Z. W., and Allis, C. D. (2002). Ubiquitination of histone H2B regulates H3 methylation and gene silencing in yeast. *Nature* **418**, 104–108.

Szymanski, C. M., Logan, S. M., Linton, D., and Wren, B. W. (2003). Campylobacter—A tale of two protein glycosylation systems. *Trends Microbiol* **11**, 233–238.

Tamaru, H., and Selker, E. U. (2001). A histone H3 methyltransferase controls DNA methylation in *Neurospora crassa. Nature* **414**, 277–283.

Thutewohl, M., Kissau, L., Popkirova, B., Karaguni, I. M., Nowak, T., Bate, M., Kuhlmann, J., Muller, O., and Waldmann, H. (2003). Identification of mono- and bisubstrate inhibitors of protein farnesyltransferase and inducers of apoptosis from a pepticinnamin E library. *Bioorg Med Chem* **11**, 2617–2626.

Timmermann S., Lehrmann H., Polesskaya A., and Harel-Bellan A. (2001). Histone acetylation and disease. *Cell Mol Life Sci* **58**, 728–736.

Trievel, R. C. (2004). Structure and function of histone methyltransferases. *Crit Rev Eukaryot Gene Expr* **14**, 147–169.

Trojer, P., Dangl, M., Bauer, I., Graessle, S., Loidl, P., and Brosch, G. (2004). Histone methyltransferases in Aspergillus nidulans: Evidence for a novel enzyme with a unique substrate specificity. *Biochemistry* **43**, 10834–10843.

van Rensburg, S. J., Berman, P., Potocnik, F., MacGregor, P., Hon, D., and de Villiers, N. (2004). 5- and 6-glycosylation of transferrin in patients with Alzheimer's disease. *Metab Brain Dis* **19**, 89–96.

Venter, J. C. (2001). The sequence of the human genome. *Science* **291**, 1304–1351.

Völkert, M., Uwai, K., Tebbe, A., Popkirova, B., Wagner, M., Kuhlmann, J., and Waldmann, H. (2003). Synthesis and biological activity of photoactivatable N-Ras peptides and proteins. *J Am Chem Soc* **125**, 12749–12758.

Walia, H., Chen, H. Y., Sun J. M., Holth, L. T., and Davie, J. R. (1998). Histone acetylation is required to maintain the unfolded nucleosome structure associated with transcribing DNA. *J Biol Chem* **17**, 2865–2876.

Wang, H. B., Wang, L. J., Erdjument-Bromage, H., Vidal, M., Tempst, P., Jones, R. S., and Zhang, Y. (2004a). Role of histone H2A ubiquitination in polycomb silencing. *Nature* **431**, 873–878.

Wang, Y., Wysocka, J., Sayegh, J., Lee, Y. H., Perlin, J. R., Leonelli, L., Sonbuchner, L. S., McDonald, C. H., Cook, R. G., Dou, Y., Roeder, R. G., Clarke, S., Stallcup, M. R., Allis, C. D., and Coonrod, S. A. (2004b). Human PAD4 regulates histone arginine methylation levels via demethylimination. *Science* **306**, 279–283.

Waterborg, J. H. (2002). Dynamics of histone acetylation *in vivo*. A function for acetylation turnover? *Biochem Cell Biol* **80**, 363–378.

Weissman, A. M. (2001). Themes and variations on ubiquitylation. *Nat Rev Mol Cell Biol* **2**, 169–178.

Wilkinson, K. D. (2000). Ubiquitination and deubiquitination: targeting of proteins for degradation by the proteasome. *Semin Cell Dev Biol* **11**, 141–148.

Wood, A., Krogan, N. J., Dover, J., Schneider, J., Heidt, J., Boateng, M. A., Dean, K., Golshani, A., Zhang, Y., Greenblatt, J. F., Johnston, M., and Shilatifard, A. (2003). An E3 ubiquitin ligase required for recruitment and substrate selection of Rad6 at a promoter. *Mol Cell* **11**, 267–274.

Zhang, Z. W., Gildersleeve, J., Yang Y. Y., Xu, R, Loo, J. A., Uryu, S., Wong, C., and Schultz, P. G. (2004). A new strategy for the synthesis of glycoproteins. *Science* **303**, 371–373.

CHAPTER 14

CONTROLLED RELEASE OF ANTIBIOTICS ENCAPSULATED IN NANOPOROUS SOL–GEL MATERIALS WITH TUNABLE PORE PARAMETERS

*Houping Yin, Laura Zheng, and Yen Wei**

14.1 INTRODUCTION

While new drug discovery remains a high priority in pharmaceutical industries, research on innovative drug-delivery methodologies is also drawing comparable attention. Successful delivery technology can ensure a full use of the drugs with higher efficiency, longer duration, and fewer side effects. Economic consideration is also a driving force in the research of high-performance drug-delivery systems. Now drug delivery is one of the fastest-growing segments of the pharmaceuticals market (Barbe et al., 2004). New drug-delivery methods aim at the development of capabilities of delivering precise quantities of a drug at the right time and as close as possible to the treatment site. Biomaterials suitable for local delivery are currently one of the most important topics in medical research (Meseguer-Olmo et al., 2002). A local release system of drugs has some advantages, such as decreasing the risks of systemic

*Corresponding author: weiyen@drexel.edu

Drug Discovery Research: New Frontiers in the Post-Genomic Era, Edited by Ziwei Huang

toxicity and side effects associated with oral and parenteral therapies, and higher drug concentrations in the desired area could be achieved to improve the efficacy of the treatment (Falaize et al., 1999).

Interest in controlled release has grown greatly and rapidly. Controlled release is concerned with the control in time and space of the biological effects of drugs. The ideal controlled-release system is one that can release drugs in a controlled manner and that has a long enough time span to provide optimum effectiveness of the drugs, thus to minimize adverse reactions, or to maximize efficacy (Falaize et al., 1999). These objectives may be achieved by controlling diffusion, reaction rates, or other physicochemical parameters by using some rate-controlling materials. There are a number of mechanisms that provide timed release of drugs, such as microencapsulation, transdermal patches, and implants (Freiberg and Zhu, 2004). Sustained drug delivery provides the same medicinal effect with higher efficiency, longer duration, and fewer side effects than traditional delivery methods.

A variety of materials have been studied as drug carriers. The desired carriers should be able to reduce the toxicity of drugs, increase their absorption, and/or improve their release profile. In recent years, numerous polymeric carriers and controlled-release systems have been developed (Acemoglu, 2004). Biodegradable and bioerodible polymers with nontoxic degradation products have been well studied as drug-delivery systems (Grolleman et al., 1986; Laurencin et al., 1987; Leong et al., 1985). One drawback with these polymer delivery systems is that possible damage could be done to the loaded drug caused by the exothermic polymerization reaction. Sol–gel-derived glasses have also been studied extensively in the past decade as controlled-release materials (Ahola et al., 2000; Bottcher et al., 1998; Falaize et al., 1999; Kortesuo et al., 1999, 2001; Nicoll et al., 1997; Radin et al., 2002; Santos et al., 1999). The sol–gel materials could be used to host various synthetic or natural therapeutic agents for controlled-delivery applications. It was shown that these materials are biocompatible *in vivo*. The materials cause no adverse tissue reactions and degrade into $Si(OH)_4$, which is eliminated from the body through the kidneys (Kortesuo et al., 2000). The drug molecules are simply entrapped within the silica matrix with no chemical bonding and remain in a biologically active form. The release is achieved through the dissolution and diffusion of the drug molecules from the sol–gel matrix. Upon release, the drug can immediately exert its effect upon the body. It has been reported that the bactericidal effect of antibiotics is retained upon their release from silica materials (Aughenbaugh et al., 2001; Radin et al., 2001). Their biological effect is the same as that when used alone and could be predicted based upon their release rate. So far these materials have been found to be very promising in bone therapies (Arcos et al., 2003; Hall et al., 2003; Meseguer-Olmo et al., 2002). Because of poor blood circulation in bone tissue, systemic antibiotic administration often cannot provide efficient concentration (Arcos et al., 2003). It was found that sol–gel synthesis parameters, such as sol pH, water/alkoxide molar ratio, or drug concentration, have an effect on the drug release rate (Aughenbaugh et al., 2001; Kortesuo et al., 2001). The silica matrix delivery approach also provides a sustained delivery. The release of the drug to the physiological environment is extended over a longer period than if the drug is used in its native form. However, most of these sol–gel materials are microporous with typical pore sizes smaller than 1.5 nm and very broad pore size distribution.

In 1998, we reported a novel, low-cost, environmentally friendly, biocompatible, nonsurfactant templated method for the preparation of mesoporous or, more generally speaking, nanoporous materials via sol–gel reactions (Wei et al., 1998; for a recent review, see Wei and Qiu, 2004). By using nonsurfactant organic molecules as template, mesoporous materials can be prepared under mild conditions. In contrast to most surfactant templated synthesis of mesoporous materials, the template removal is achieved by solvent extraction at room temperature (Wei et al., 1999). The materials usually have high surface areas (e.g., 1000 m^2/g) and large pore volumes (e.g., 0.5–1.0 cm^3/g) as well as controllable pore sizes in the range of 2–12 nm with a narrow size distribution. The pore diameters are often tunable by simply varying the template concentration. One of the most unique advantages of our nonsurfactant approach to mesoporous materials is its biocompatibility. The entire preparation can be performed at room temperature and near neutral pH. The nonsurfactant templates could be sugar compounds, such as glucose, maltose, fructose, soluble starch, or cyclodextrins, which are highly biocompatible.

In general, the synthesis starts with the hydrolysis of inorganic precursors—for example, tetramethyl orthosilicate (TMOS) or tetraethyl orthosilicate (TEOS) for silica. The template is then added into the sol. Upon gelation and drying, the template-silica nanocomposites are obtained usually as transparent and monolithic solids. The template molecules can be completely removed by solvent extraction. The removal of the templates results in a mesoporous silica matrix with interconnected channels of regular diameters. These channels usually do not have any discernible packing order or orientation. Thus the mesoporosity is isotropic and accessible in all directions. The pore parameters, such as pore size, pore volume, and surface area, are adjustable to a certain extent by varying the concentration of template in the preparation. Generally, the values of pore parameters increase with the template concentration. The mesoporous materials prepared with nonsurfactant templates have been successfully used for enzyme and protein immobilization (Wei et al., 2000, 2001, 2002a,b). These studies showed that these macromolecules could retain their biological activities within the mesoporous silica matrices. Through proper design of the channel diameter, these incorporated macromolecules would not leach out of the matrices. These bioactive sol–gel materials could be used as biocatalysts and biosensors (Dave et al., 1994; Ellerby et al., 1992). More importantly, the characteristics of the tailorable physical (e.g., pore size and volume) and chemical (e.g., hydrophilicity and other functionalities) structures of the nanopores could provide a unique system for the stabilization of proteins against denaturation and foreign agents such as protease. The nanoporous matrices may also function like rigid matrix artificial chaperones to provide designable assistance in protein folding processes (Wei and Qiu, 2004). Recently, the single protein molecule that may function as a catalytic enzyme is thought to undergo constant unfolding-refolding conformational changes. The catalytic activity measured is the average of all conformations within the measurement period (Xie, 2001). The controlled biocompatible nanoporous materials containing a single catalytic protein may provide an excellent and a unique system for studying and establishing the effect of such a conformational change on the bioactivity of the protein molecules because the conformational change depends on available space around the protein molecules (Wei and Qiu, 2004).

In this work, sol–gel silica materials with designed pore parameters were synthesized to host various antibiotics for controlled-delivery applications. Various amounts of fructose were used as template to tailor the pore structure. The *in vitro* release properties were investigated in relation to the silica pore structures. The sol–gel synthesis parameters other than template concentration, such as pH of the sol, water/alkoxide molar ratio, and drug concentration, were more or less fixed in this study to avoid further complication in the pore structure variations. With certain pore sizes, an initial large concentration release followed by a long-term steady-state release was observed. This is desirable for some treatment modalities. We found that the amount of antibiotics released and the release rate were dependent on the silica nanostructure that was controlled by the amount of fructose template employed in the sol–gel reactions.

14.2 EXPERIMENTAL

14.2.1 Materials

Antibiotics hygromycin B, mycophenolic acid, and vancomycin (vancomycin-HCl) were purchased from Sigma. Tetramethoxysilane (TMOS, Aldrich), hydrochloric acid (HCl, Fisher), and D-fructose (98%, Aldrich) were used as received without further purification. Doubly distilled, deionized water was used to make all aqueous solutions.

14.2.2 Synthesis of Mesoporous Silica Materials Loaded with Antibiotics

Immobilization of antibiotics in silica gels was carried out by mixing the antibiotics solution with prehydrolyzed TMOS sol. For a typical example, at room temperature, to a homogeneous solution of 1.52 g of TMOS, 15 μL of 40-mM HCl, and 0.35 g of distilled and deionized H_2O in a 10-mL glass vial, an appropriate amount of 50 wt% fructose aqueous solution was added under magnetic stirring. Amounts of 50 wt% fructose solution were designed to yield 0–60% by weight of template in the final dry gel products. The sol mixture was prepared in a relatively large quantity for the convenience of preparing a series of samples with different concentrations. Ten milligrams of antibiotics in 1.0 mL distilled and dionized water were added into each vial in a period of 10 min under stirring. Upon completion of the dissolution, the stirring was continued for another 3–5 min, and then the stirring bar was removed. The vial was then sealed with a piece of paraffin film. Upon gelation of the system within a few minutes to a few hours at room temperature, 2–4 holes were pinned in the paraffin film with a syringe needle to allow for the evaporation of solvent and reaction byproducts (i.e., methanol and water). After 7 days under fume hoods, the sample-containing vials were placed in a dynamic vacuum oven and dried to reach a constant weight at room temperature in about 1 week. The silica gel samples containing both the antibiotic and template molecules were obtained in the form of transparent, dry, hard, and glassy disklike monoliths.

14.2.3 Instrumentation and Characterization

Characterization of the microstructure parameters (e.g., pore size and pore size distribution, specific surface area, and pore volume) of the silica matrices was carried out with N_2 adsorption–desorption isotherms at $-196°C$ on a Micromeritics ASAP 2010 system (Norcross, GA) on the samples after completion of the releasing study. Prior to measurements, the samples were degassed at 100°C and 1 Pa for 6–7 h. Infrared spectra of KBr powder-pressed pellets were recorded on a Perkin–Elmer Model 1600 FTIR spectrophotometer. Thermal gravimetric analysis (TGA) was carried out with a TA Q50 Instrument under oxygen atmosphere at a programmed heating rate of 10°C/min.

14.2.4 Drug Releasing Study

A 0.01 mM phosphate buffered saline (PBS) solution with a pH of 7.4 was used for the releasing study following the similar protocols in the literature (US FDA, 1998). The antibiotics-containing silica disks were immersed separately in 3 mL of PBS in 5-mL glass vials and placed in a 37°C water bath. The disks usually cracked within a few minutes after immersion. The immersion solutions were exchanged after a designated time period. At the end of the time period, the solution was completely removed from each vial, and 3 mL of fresh PBS was immediately added. Immersion was continued until release stopped. To confirm the releasing pattern, the procedure was repeated with two more independent preparations.

14.2.5 Antibiotics Release Assay

The antibiotics concentration of each elution sample was analyzed using a UV/VIS spectrometer (Perkin Elmer, Lambda 2) at different wavelengths, namely 260, 250, and 280 nm, respectively, for hygromycin B, mycophenolic acid, and vancomycin. Antibiotics-PBS standard solutions were used for calibration under identical conditions.

14.3 RESULTS AND DISCUSSION

14.3.1 Materials Characterization

The as-synthesized sol–gel silica materials loaded with template and antibiotic molecules were transparent, glassy disks with typical dimensions of 8 mm in diameter, 3–8 mm in thickness, and 0.7–1.7 g in weight depending on the fructose content. The good optical transparency indicates that the template and antibiotics molecules were homogeneously distributed within the silica without macroscopic phase separations. The synthesis of the silica–template–antibiotics composite has been achieved under mild conditions such as at room temperature and low acidity, which should not cause denaturation and damage of bioactive molecules. When the silica biogels were immersed into PBS solution, they usually cracked into several smaller pieces, which is typical of bulk sol–gel materials.

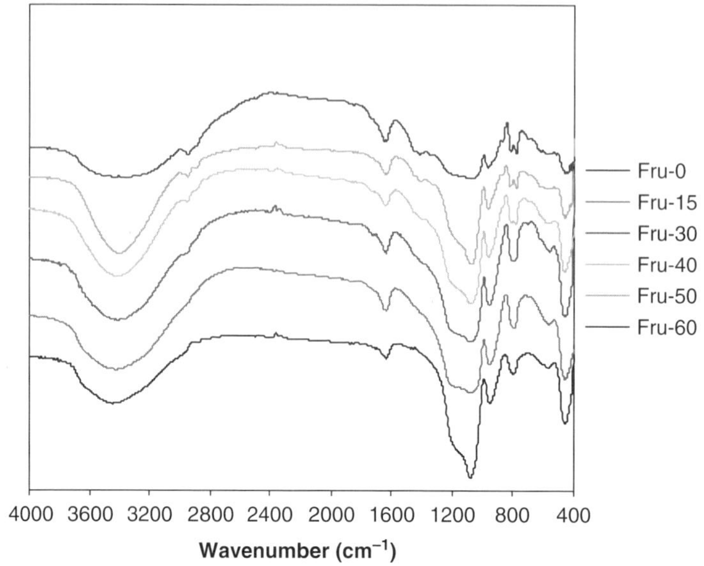

Figure 14.1 IR spectra of silica with various amounts of fructose template before the drug release study.

Figure 14.1 shows some representative FT-IR spectra of the silica materials with increasing template content. FT-IR spectral series of the materials loaded with 10 mg of each antibiotics were similar to those in Fig. 14.1. The bottom spectrum in Fig. 14.1 is from the sample without fructose template but loaded with 10-mg antibiotics, which is the same as the IR spectrum of pure silica. Incorporation of a very small amount (i.e., 10 mg) of antibiotics did not show any appreciable effect on the spectral features. In contrast, as the template concentration is increased, the spectra exhibit some noticeable changes in the 2900–2800 cm^{-1}, 1500–1400 cm^{-1}, and 820–780 cm^{-1} regions. With the increasing content of template, these changes become more significant. These regions correspond to aliphatic C–H bending and stretching vibrations of the fructose molecules. Hence, these vibration bands are indicative of the presence and content of fructose template in the sample series.

Figure 14.2 shows the FT-IR spectra of the same silica materials as in Fig. 14.1, but after the drug release study. The spectral series do not show any discernible absorption in the 2900–2800 cm^{-1}, 1500–1400 cm^{-1}, and 820–780 cm^{-1} regions as observed earlier in Fig. 14.1. All the spectra show the typical absorption bands as those in pure silica, which indicates that all fructose template has been completely removed during the drug release studies. The complete removal of fructose template from the silica matrices is further evidenced by the fact that calcination of all the silica materials after drug release study to 750°C shows no appreciable weight loss in the TGA measurements. These observations indicate that during the drug release studies, the templates were completely extracted out of the template–silicate nanocomposites and the silica matrices were nanoporous.

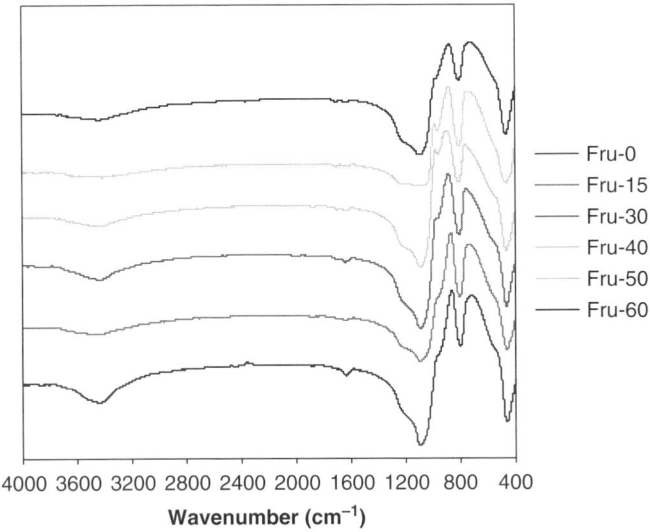

Figure 14.2 IR spectra of silica with various amounts of fructose as template after the drug release study.

Nitrogen adsorption–desorption isotherms of the nanoporous silicas after the removal of fructose template were recorded over the relative pressure (P/P_0) from 0 to 1 at −196°C. The typical isotherms of silica samples prepared at various amounts of the fructose template from 0 to 60 wt% are shown in Fig. 14.3. Figure 14.3a shows the isotherms of the silica materials loaded with antibiotics mycophenolic acid. Isotherms of the silica loaded with the antibiotic hygromycin B are similar. Figure 14.3b shows the isotherms of the silica materials loaded with vancomycin. At a fructose content ranging from 15 to 60 wt% in the sample preparation, the isotherms have similar features. According to the IUPAC classification (Sing et al., 1985), these isotherms are typical type IV isotherms. They have a characteristic hysteresis loop at around 0.5–0.9 P/P_0, which is associated with capillary condensation in mesoporous structure. With increasing concentration of fructose, the hysteresis loops become greater in magnitude and shift to higher P/P_0.

However, without using fructose template, the isotherms of silica loaded with only antibiotics seem to differ when a different antibiotic compound was entrapped. This is demonstrated by the difference between the two bottom isotherms of Fig. 14.3a and Fig. 14.3b. The bottom isotherm in Fig. 14.3a has no hysteresis loop, which usually indicates that the material is microporous. The bottom isotherm in Fig. 14.3b resembles a type IV isotherm, indicating that the material might have mesoporosity to a certain extent. The difference could be attributed to the molecular size of the antibiotics. Vancomycin's molecular structure is much larger than the other two antibiotics. Hygromycin B ($C_{20}H_{37}N_3O_{13}$) has a formula weight of 527.5 and mycophenolic acid ($C_{17}H_{20}O_6$) has a formula weight of 320.3, while vancomycin ($C_{66}H_{75}Cl_2N_9O_{24}$) has a formula weight of 1448.5. Mesopores might form with

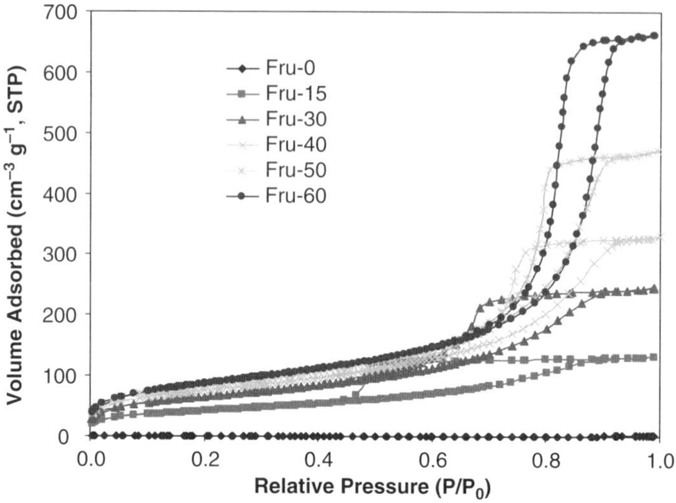

(a)

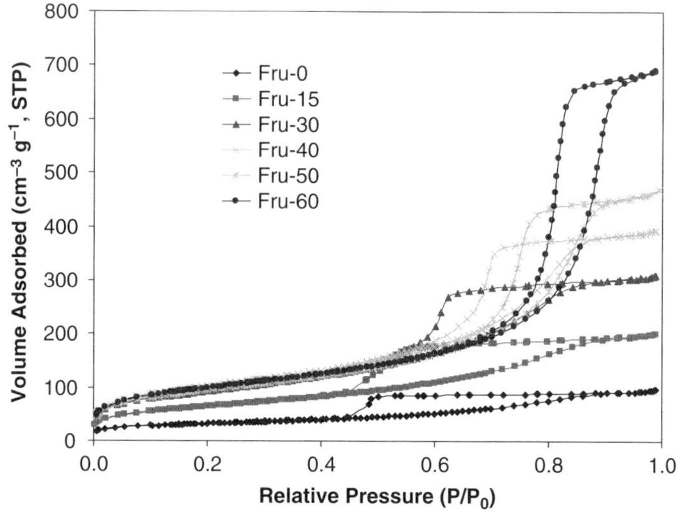

(b)

Figure 14.3 Nitrogen adsorption–desorption isotherms at −196°C for porous silica samples prepared with various amounts of fructose and after the drug release study (i.e., the fructose was completely removed from the samples): (**a**) With antibiotics hygromycin B or mycophenolic acid; (**b**) with vancomycin.

vancomycin, probably because of its large size, but with the other two antibiotics, only micropores formed in the silica matrix.

The composition of the fructose-templated silica materials loaded with vancomycin and the pore parameters of the mesoporous silica after the release study are summarized in Table 14.1. The data of materials loaded with two other antibiotics were quite similar except the pore diameters for sample Fru-0, in which

TABLE 14.1 Composition and Pore Parameters of the Sol–Gel Silica Samples Prepared in the Presence of 0–60 wt% Fructose and 10 mg of Vancomycin After Release Study

Sample Code[a]	Fructose (wt%) Calcd.	Fructose (wt%) Found[b]	BET Surface Area (m^2g^{-1})	Pore Volume (cm^3g^{-1})	D_{BJH}[c] (Å)	D_{BET}[d] (Å)
Fru-60	60	56	345	1.07	116	124
Fru-50	50	47	336	0.72	82	86
Fru-40	40	39	360	0.61	66	68
Fru-30	30	31	329	0.48	57	58
Fru-15	15	21	228	0.31	57	55
Fru-0	0	0	111	0.15	58	54

[a]The numerical figure in the sample code denotes the template fructose concentration (wt%) in the materials before the release study.
[b]Values from the weight loss of the samples, which represent total content of volatile compounds at 750°C based on TGA measurement.
[c]Average pore diameter calculated from $4V/A$, where V is the total pore volume and A is the BET surface area.
[d]Average pore diameter calculated from BJH desorption branch of the isotherms.

no fructose template was added. For the Fru-0 sample of vancomycin, the pore diameter is about 5 nm. For the Fru-0 sample of the other two antibiotics, the pore diameter is about 3–4 nm. The compositions measured based on the total weight loss at 750°C from TGA are comparable with the designed feeding concentrations. The discrepancies could be attributed to incomplete sol–gel reactions. The silica materials exhibit a large pore volume up to 1.1 cm^3/g, and the materials have tunable pore diameters of 5–12 nm. The effects of template concentrations on pore parameters resemble those we observed for other nonsurfactant compounds (Wei et al., 1998, 1999). As the fructose concentration increases, the pore volume and pore diameter tend to increase. The Brunauer–Emmett–Teller (BET) surface area increases with the fructose content increase up to 40 wt%; after that, it tends to decrease. As reported earlier, this is a result of competition of the effect of pore diameter increase over the effect of pore volume increase (Pang et al., 2000, 2001).

As shown in Fig. 14.4, the Barrett–Joyner–Halenda (BJH) pore size distribution curves from the desorption branches of the samples encapsulated with hygromycin B or mycophenolic acid (shown in Fig. 14.4a) and vancomycin (shown in Fig. 14.4b) exhibit relatively narrow distributions. In the case of the vancomycin sample (Fig. 14.4b), the pore diameter is in the range of 3–11 nm. A larger pore size is observed for a sample with higher fructose content. The Fru-0 curve exemplified in Fig. 14.4b is from the sample loaded with vancomycin. For the other two antibiotics, there is no identifiable distribution peak for samples in the absence of fructose template (i.e., Fru-0) as shown in Fig. 14.4a.

Isotherms analysis indicates that the removal of the template after the release study results in a mesoporous silica matrix with interconnected channels of regular diameters. This feature was also confirmed by transmission electron microscopic and X-ray diffraction studies. These interconnected channels allow the surrounding fluid

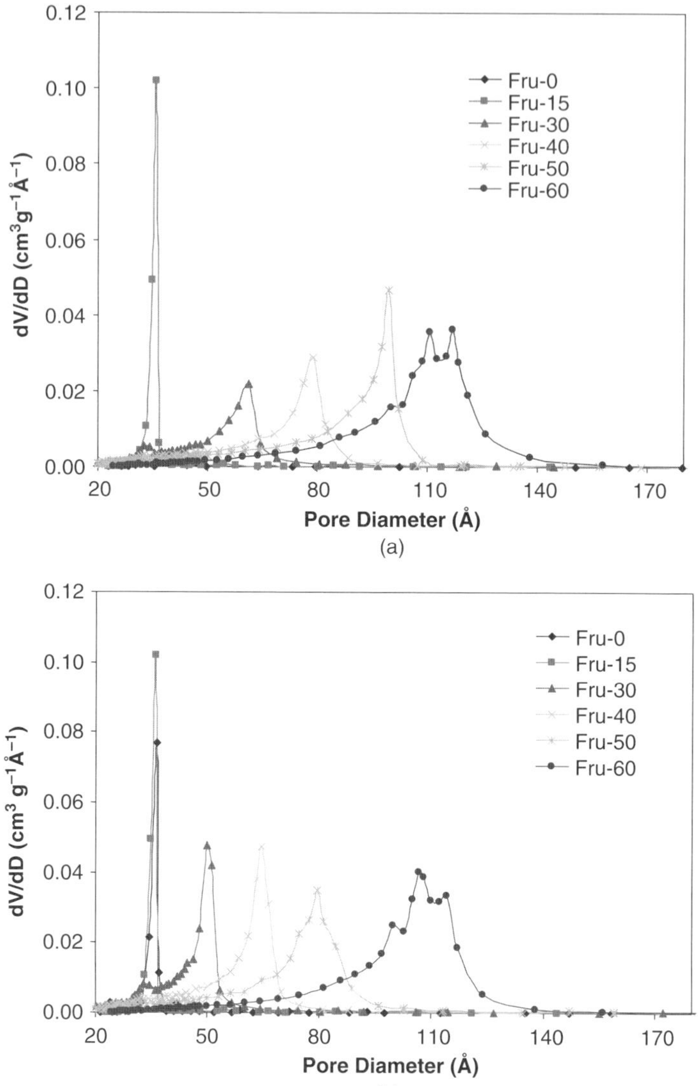

Figure 14.4 BJH pore size distributions obtained from the desorption branches of nitrogen sorption isotherms at −196°C for porous silica samples prepared with various amounts of fructose after the drug release study: (**a**) With antibiotics hygromycin B or mycophenolic acid; (**b**) with vancomycin.

to penetrate and to dissolve and release the template molecules and the incorporated antibiotics molecules. Thus, the antibiotics could elute from the silica matrix through these channels. Because the diffusion rate of drug molecules should be related to the channel diameter and structures, the drug release could, therefore, be controlled by tailoring the pore parameters.

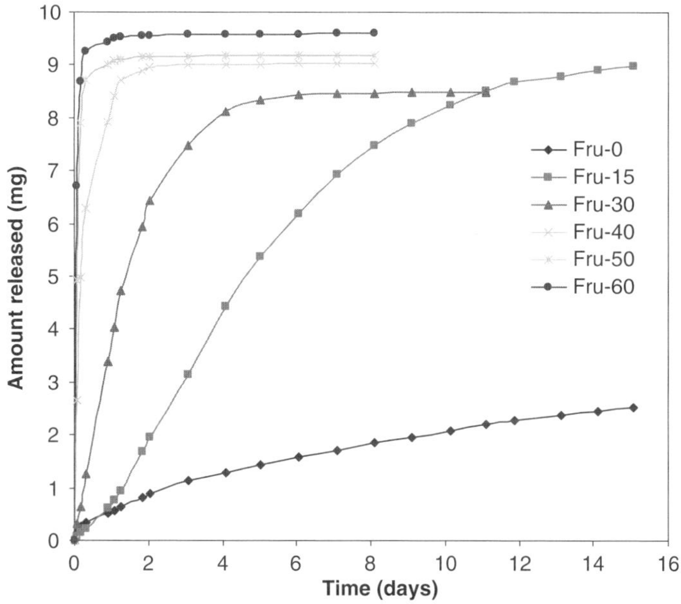

Figure 14.5 Release profile of vancomycin in silica materials with various template content.

14.3.2 *In Vitro* Release

In vitro cumulative release of vancomycin is plotted in Fig. 14.5. The release study was repeated with two more independent sample preparations. The same release profile was obtained each time. The solutions were exchanged at the designated time points. All the samples showed a faster initial release followed by a slower, steady release. The shape of the release profile was found to be dependent on the template content. The results indicate that the release is faster when the template content is higher. The data show that the release was gradual and sustained when the amount of template was no more than 30%. From Table 14.1, one can see that when the template content is no more than 30%, the interconnected channels formed by the template removal have a similar pore diameter (~5 nm) to the Fru-0 sample. By increasing the template content, the channels are better connected (increased pore volumes). A larger pore volume allows more of the immersion solution to enter the silica matrix in a given period of time. Thus, vancomycin could be released faster than from the Fru-0 sample, although these materials have the same pore diameters. For the nontemplated Fru-0 sample, probably because of the limited interconnections of channels as implied from small pore volumes, the immersion solution could not readily penetrate through the silica matrix to the interior; therefore, the release of antibiotics from the silica matrix was difficult. Samples Fru-40 to Fru-60 have much larger pore diameters and volumes. The antibiotics can easily leach out through the interconnected channels. This is likely the reason for the observed fast release. The templated silica materials showed a very good recovery of loaded antibiotics. The amount of antibiotics released was more than 85% for all templated samples, while only about 25% of vancomycin was released after 15 days for the nontemplated Fru-0 sample. The good recovery of loaded antibiotics indicates that there was no significant loss of the antibiotics during the synthesis and that the antibiotics were not chemically bonded to the materials.

The release kinetics were also dependent on the amount of template. When template was introduced into the silica matrix, the samples all showed a faster release compared to the Fru-0 sample. When the cumulative release versus time curve shows some linearity, that part of the process is considered as near-zero-order. The last part of the release is characterized by near-zero-order kinetics for all the samples as shown in Fig. 14.5. The cumulative release was also plotted versus the square root of time (in hours). A linear relationship shows in part of curves for samples Fru-0 to Fru-30 as shown in Fig. 14.6. This indicates that a first-order release occurred during part of the process. For sample Fru-0, the curve was linear up to 2 days of immersion with r^2 value 0.99. The fast first-order release was followed by a slow near-zero-order release from day 2 onward. For sample Fru-15, a three-stage release process was observed. That is, first an initial faster release stage in day 1, followed by a relatively fast stage with first-order kinetics until day 10 with r^2 value 0.99 and a subsequent slower release of near-zero-order. For sample Fru-30, a faster initial stage was observed for the first half day, followed by a first-order release up to day 5 with r^2 value 0.97 and a subsequent slow near-zero-order release. The transition from the first- to the zero-order process (5 and 10 days for samples Fru-30 and Fru-15, respectively) occurred when about 80% of vancomycin had been released. But for samples Fru-40 to Fru-60, a much faster initial release occurred in the first day, and the kinetics could not be determined. The release kinetics were dependent on the template content. By varying the template content, the release kinetics of silica materials could be tailored with respect to specific therapeutic goals.

In contrast, for the other two antibiotics—that is, hygromycin B and mycophenolic acid—the release profiles did not show a controlled and sustained release. Figure 14.7 shows, for example, the release profile of hygromycin B. About 90% of the two antibiotics were released within 3 days. The release kinetics were not template content-dependent. The release rate and amount do not have a relation to the silica structure parameters, such as pore diameters. Such a significant difference from vancomycin could be attributed to the size effect. For the two antibiotics hygromycin B and mycophenolic acid, their molecular sizes are smaller than the diameters of the interconnected channels with the smallest amount of template (15 wt%) employed in the study. They can easily leach out of the silica matrix. Therefore, no sustained release was observed.

14.4 CONCLUSIONS

Nanoporous silicas containing antibiotics have been successfully prepared via the nonsurfactant-templated sol-gel reactions of tetraalkoxysilane in the presence of fructose as the template and the antibiotics, including hygromycin B, mycophenolic acid, and vancomycin, followed by removal of the template by extraction during the drug release. The pore parameters, such as pore diameter, pore volume, and specific area, can be tailored by varying the template concentration. In the case of vancomycin,

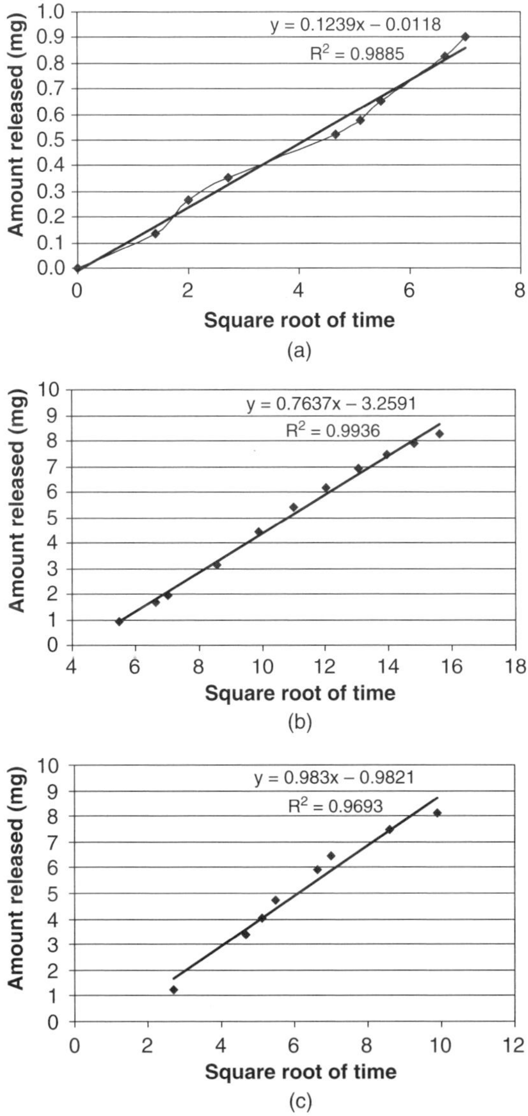

Figure 14.6 Part of cumulative release (in mg) versus the square root of time (in hours) shows linearity in samples Fru-0 (**a**), Fru-15 (**b**), and Fru-30 (**c**).

the biomaterials show a long-term controlled release. The amount of antibiotics released and the release rate are dependent on the silica pore parameters. From the samples prepared with the template, most of the antibiotics (>85%) were released in a short time period. In contrast, for the samples prepared with less or no template, only about 25% of the antibiotics were released by day 15. Since the template provides isotropically distributed interconnected channels inside the silica matrix with tunable pore diameters, the release rate and amount are greatly improved. Variation of the

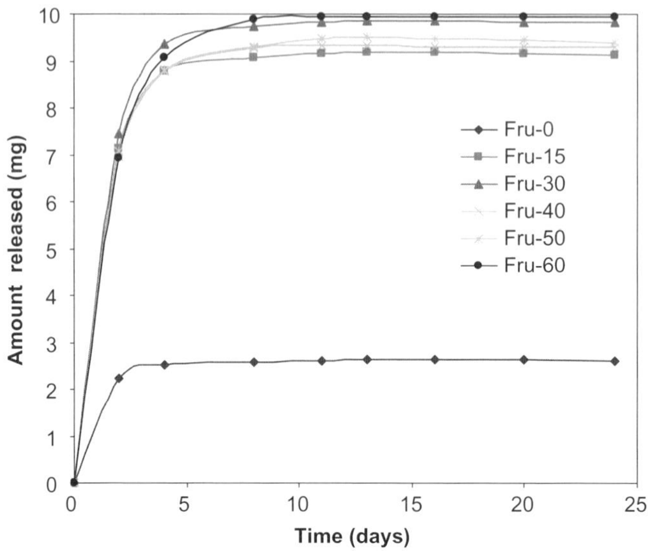

Figure 14.7 Release profile of hygromycin B in silica materials with various template content.

template content is a very simple, effective technique for the control of the rate and amount of drug release. With a high antibiotic recovery and with the release kinetics that can be tailored with respect to the requirements for a particular therapeutic treatment, porous silica materials with tunable pore diameters would make a good controlled drug release system.

ACKNOWLEDGMENTS

We are grateful to the National Institutes of Health (Grant No. R01-DE09848), the Nanotechnology Institute of Southeastern Pennsylvania, and the U.S. Army Research Office for their support of this work. We thank A. A. and G. B., respectively, for insightful suggestions and intellectual encouragement.

REFERENCES

Acemoglu, M. (2004). Chemistry of polymer biodegradation and implications on parenteral drug delivery. *Int J Pharm* **277**, 133–139.

Ahola, M., Kortesuo, P., Kangasniemi, I., Kiesvaara, J., and Yli-Urpo, A. (2000). Silica xerogel carrier material for controlled release of toremifene citrate. *Int J Pharm* **195**(1–2), 219–227.

Arcos, D., Pena, J., and Vallet-Regi, M. (2003). Influence of a SiO_2-CaO-P_2O_5 sol–gel glass on the bioactivity and controlled release of ceramic/polymer/antibiotic mixed materials. *Chem Mater* **15**(21), 4132–4138.

Aughenbaugh, W., Radin, S., and Ducheyne, P. (2001). Silica sol–gel for the controlled release of antibiotics. II. The effect of synthesis parameters on the *in vitro* release kinetics of vancomycin. *J Biomed Mater Res* **57**(2), 321–326.

Barbe, C., Bartlett, J., Kong, L., Finnie, K., Lin, H. Q., Larkin, M., Calleja, S., Bush, A., and Calleja, G. (2004). Silica particles: A novel drug-delivery system. *Adv Mater* **16**, 1959–1966.

Bottcher, H., Slowik, P., and Suss, W. (1998). Sol–gel carrier systems for controlled drug delivery. *J Sol–Gel Sci Tech* **13**(1/2/3), 277–281.

Dave, B. C., Dunn, B., Valentine, J. S., and Zink, J. I. (1994). Sol–gel encapsulation methods for biosensors. *Anal Chem* **66**(22), 1120A–1127A.

Ellerby, L. M., Nishida, C. R., Nishida, F., Yamanaka, S. A., Dunn, B., Valentine, J. S., and Zink, J. I. (1992). Encapsulation of proteins in transparent porous silicate glasses prepared by the sol–gel method. *Science* **255**(5048), 1113–1115.

Falaize, S., Radin, S., and Ducheyne, P. (1999). *In vitro* behavior of silica-based xerogels intended as controlled release carriers. *J Am Ceram Soc* **82**(4), 969–976.

Freiberg, S., and Zhu, X. X. (2004). Polymer microspheres for controlled drug release. *Int J Pharm* **282**, 1–18.

Grolleman, C. W. J., De Visser, A. C., Wolke, J. G. C., Van der Goot, H., and Timmerman, H. (1986). Studies on a bioerodible drug carrier system based on polyphosphazene. Part I. Synthesis. *J Contr Rel* **3**(2–3), 143–154.

Hall, S. R., Walsh, D., Green, D., Oreffo, R., and Mann, S. (2003). A novel route to highly porous bioactive silica gels. *J Mater Chem* **13**(2), 186–190.

Kortesuo, P., Ahola, M., Karlsson, S., Kangasniemi, I., Kiesvaara, J., and Yli-Urpo, A. (1999). Sol–gel-processed sintered silica xerogel as a carrier in controlled drug delivery. *J Biomed Mater Res* **44**(2), 162–167.

Kortesuo, P., Ahola, M., Karlsson, S., Kangasniemi, I., Yli-Urpo, A., and Kiesvaara, J. (2000). Silica xerogel as an implantable carrier for controlled drug delivery-evaluation of drug distribution and tissue effects after implantation. *Biomaterials* **21**(2), 193–198.

Kortesuo, P., Ahola, M., Kangas, M., Yli-Urpo, A., Kiesvaara, J., and Marvola, M. (2001). *In vitro* release of dexmedetomidine from silica xerogel monoliths: Effect of sol–gel synthesis parameters. *Int J Pharm* **221**(1–2), 107–114.

Laurencin, C. T., Koh, H. J., Neenan, T. X., Allcock, H. R., and Langer, R. (1987). Controlled release using a new bioerodible polyphosphazene matrix system. *J Biomed Mater Res* **21**(10), 1231–1246.

Leong, K. W., Brott, B. C., and Langer, R. (1985). Bioerodible polyanhydrides as drug-carrier matrixes. I: Characterization, degradation, and release characteristics. *J Biomed Mater Res* **19**(8), 941–955.

Meseguer-Olmo, L., Ros-Nicolas, M. J., Clavel-Sainz, M., Vicente-Ortega, V., Alcaraz-Banos, M., Lax-Perez, A., Arcos, D., Ragel, C. V., and Vallet-Regi, M. (2002). Biocompatibility and *in vivo* gentamicin release from bioactive sol-gel glass implants. *J Biomed Mater Res* **61**(3), 458–465.

Nicoll, S. B., Radin, S., Santos, E. M., Tuan, R. S., and Ducheyne, P. (1997). In vitro release kinetics of biologically active transforming growth factor- β1 from a novel porous glass carrier. *Biomaterials* **18**(12), 853–859.

Pang, J.-B., Qiu, K.-Y., Wei, Y., Lei, X.-J., and Liu, Z.-F. (2000). A facile preparation of transparent and monolithic mesoporous silica materials. *Chem Commun* (6), 477–478.

Pang, J.-B., Qiu, K.-Y., and Wei, Y. (2001). A new nonsurfactant pathway to mesoporous silica materials based on tartaric acid in conjunction with metal chloride. *Chem Mater* **13**(7), 2361–2365.

Radin, S., Ducheyne, P., Kamplain, T., and Tan, B. H. (2001). Silica sol–gel for the controlled release of antibiotics. I. Synthesis, characterization, and *in vitro* release. *J Biomed Mater Res* **57**(2), 313–320.

Radin, S., Falaize, S., Lee, M. H., and Ducheyne, P. (2002). *In vitro* bioactivity and degradation behavior of silica xerogels intended as controlled release materials. *Biomaterials* **23**, 3113–3122.

Santos, E. M., Radin, S., and Ducheyne, P. (1999). Sol–gel derived carrier for the controlled release of proteins. *Biomaterials* **20**, 1695–1700.

Sing, K. S. W., Everett, D. H., Haul, R. A. W., Moscou, L., Pierotti, R. A., Rouquerol, J., and Siemieniewska, T. (1985). Reporting physisorption data for gas/solid systems with special reference to the determination of surface area and porosity (Recommendations 1984). *Pure Appl Chem* **57**(4), 603–619.

US Food and Drug Administration (1998). *Bacteriological Analytical Manual*, 8th edition, Revision A.

Wei, Y., Jin, D. L., Ding, T. Z., Shih, W. H., Liu, X. H., Cheng, S. Z. D., and Fu, Q. (1998). A non-surfactant templating route to mesoporous silica materials. *Adv Mater* **10**(4), 313–316.

Wei, Y., Xu, J., Dong, H., Dong, J. H., Qiu, K., and Jansen-Varnum, S. A. (1999). Preparation and physisorption characterization of D-glucose-templated mesoporous silica sol–gel materials. *Chem Mater* **11**(8), 2023–2029.

Wei, Y., Xu, J., Feng, Q., Dong, H., and Lin, M. (2000). Encapsulation of enzymes in mesoporous host materials via the nonsurfactant-templated sol–gel process. *Mater Lett* **44**(1), 6–11.

Wei, Y., Xu, J., Feng, Q., Lin, M., Dong, H., Zhang, W.-J., and Wang, C. (2001). A novel method for enzyme immobilization: Direct encapsulation of acid phosphatase in nanoporous silica host materials. *J Nanosci Nanotech* **1**(1), 83–93.

Wei, Y., Dong, H., Xu, J., and Feng, Q. (2002a). Simultaneous immobilization of horseradish peroxidase and glucose oxidase in mesoporous sol–gel host materials. *ChemPhysChem* **3**(9), 802–808.

Wei, Y., Sun, Z., Zheng, J., Dong, H., Yuan, J. M., and Ping, G. (2002b). Rigid matrix artificial chaperone (RMAC)-mediated refolding of heme protein. *Poly Mater Sci Eng* **87**, 252–253.

Wei, Y., and Qiu, K.-Y. (2004). A novel nonsurfactant route to nanoporous materials and its biological applications. *Series on Chemical Engineering*, Vol. 4 (Nanoporous Materials), pp. 873–892.

Xie, S. (2001). Single-molecule approach to enzymology. *Single Mol* **2**, 229–236.

PART III

BIOLOGICAL APPROACHES AND TRANSLATIONAL RESEARCH IN DRUG DISCOVERY

CHAPTER 15

RETINOIC ACID AND ARSENIC TRIOXIDE TREATMENT IN ACUTE PROMYELOCYTIC LEUKEMIA: A MODEL OF ONCOPROTEIN TARGETED THERAPY

*Jian-Hua Tong, Sai-Juan Chen, and Zhu Chen**

15.1 INTRODUCTION

Acute promyelocytic leukemia (APL) is a distinct subtype of acute myeloid leukemia (AML), and it accounts for 10–15% of AML cases in adults. According to the FAB (French–American–British) nomenclature of acute leukemia, APL is classified as M3 subtype, characterized by the block of differentiation of clonal hemopoietic precursors at the promyelocytic stage of the myelopoiesis and the presence of a specific cytogenetic marker. In addition to the common clinical characteristics of acute leukemia (such

*Corresponding author: zchen@stn.sh.cn

Drug Discovery Research: New Frontiers in the Post-Genomic Era, Edited by Ziwei Huang

as anemia or fever), most APL patients present a severe bleeding syndrome, often accompanied by a hypofibrinogenopenia and a disseminated intravascular coagulation (DIC), resulting in relatively high early mortality (Avvisati et al., 2001).

The therapeutic approach of APL has been greatly improved over the last three decades. In the late 1970s and early 1980s, conventional cancer chemotherapy, consisting of an anthracycline and cytosine arabinoside, was mainly used in the treatment of APL. Although APL patients are usually sensitive to chemotherapy, and a relatively high complete remission (CR) rate (60–70%) is observed, use of high-dose chemotherapy drugs can produce severe side effects—such as bone marrow suppression, immunological dysfunction, gastrointestinal reaction, hemorrhage exacerbation, and even a secondary tumor occurrence—by killing normal cells. In the mid-1980s, the application of all-*trans* retinoic acid (ATRA), first performed at the Shanghai Institute of Hematology (SIH), for the treatment of APL achieved a significant breakthrough in cancer differentiation induction therapy (Huang et al., 1988). Instead of killing the malignant cells as chemotherapy does, ATRA can induce APL cells to undergo terminal differentiation, with which a high CR rate of 85–90% is achieved by rapidly ameliorating the severe coagulopathy of APL patients and reducing the cytotoxic effects and multiple drug resistance caused by chemotherapy. However, although very effective in remission induction of APL patients, the treatment with ATRA alone is not sufficient to maintain a long disease-free remission, and relapse is inevitable if remission is not consolidated with chemotherapy. Moreover, in the majority of cases, relapsed patients often lose their sensitivity to ATRA (Degos and Wang, 2001). Since the 1990s, arsenic trioxide (As_2O_3), an ancient traditional Chinese drug, has been used as an effective drug in the treatment of APL, not only in newly diagnosed cases, but also in relapsed and/or refractory patients after ATRA or chemotherapy (Shen et al., 1997; Sun et al., 1992). The principal side effects of As_2O_3 are dermatological and gastrointestinal symptoms, which can be tolerated and/or can disappear rapidly with symptomatic treatment. Recently, another arsenic derivative, As_4S_4, has also proven to be highly effective and safe for both remission induction and maintenance in all stages of APL (Lu et al., 2002). Besides, it has been reported that a dual treatment with ATRA and As_2O_3 for APL can further improve the CR rate as well as the overall survival of patients (Shen et al., 2004). This ATRA/As_2O_3 dual treatment may open a new era for the real cure of APL, a previously life-threatening subtype of leukemia.

With the revolutionary achievement in the treatment of APL, significant progress has also been made in understanding this malignancy during the last decades.

15.2 CYTOGENETIC CHARACTERISTICS OF APL CELLS AND THEIR SENSITIVITY TO ATRA

APL is distinct from other subtypes of leukemia because of its two biological features: (1) its exclusive cytogenetic marker and (2) its responsiveness to ATRA.

In the 1970s, it was discovered that APL is associated with a disease-specific karyotype. In the great majority of APL patients, abnormal promyelocytes harbored a reciprocal chromosomal translocation t(15;17)(q22;q21) (Rowley et al., 1977). Two genes involved were subsequently cloned by virtue of the development of molecular biological techniques. The retinoic acid receptor α (RARα) gene on chromosome 17 and the promyelocytic leukocyte (PML) gene on chromosome 15 are fused to generate a PML-RARα chimeric protein (Melnick and Licht, 1999). It is noteworthy that APL patients with t(15;17) exhibit an exquisite sensitivity to ATRA and that complete remission can be induced (Huang et al., 1988). Molecular studies on a large series of APL cases revealed a great heterogeneity of PML-RARα transcripts due to variable breakpoint cluster regions within PML genes as well as alternative splicing (Pandolfi et al., 1992; Tong et al., 1992). Three main PML-RARα isoforms—type L, type S, and type V—were found to exist in APL patients. Clinical data showed that the patients who contained type S PML-RARα had a high likelihood of early death or relapse after being treated with ATRA (Pandolfi et al., 1992).

In addition to the PML-RARα rearrangement, four other APL-associated translocations have been characterized to date at the molecular level. They are t(11;17)(q23;21), t(11;17)(q13;q21), t(5;17)(q35;q21), and del(17)(q11q21), which fuse RARα to promyelocytic leukemia zinc finger (PLZF), nuclear mitotic apparatus (NuMA), nucleophosmin (NPM), and STAT5b genes, respectively, and lead to expression of PLZF-RARα, NuMA-RARα, NPM-RARα, and STAT5b-RARα fusion proteins (Arnould et al., 1999; Chen et al., 1993; Redner et al., 1996; Wells et al., 1997). The leukemic blasts from the patients having variant translocations present similar morphology, but they differ on their sensitivity to ATRA. It was reported that the patients with NuMA-RARα or NPM-RARα could achieve remission after ATRA/chemotherapy, whereas those with PLZF-RARα or STAT5b-RARα had a generally poor response to ATRA (Zelent et al., 2001). Nevertheless, it should be noticed that the breakpoints in the RARα locus of all these translocations are localized in its second intron, and common features of all these fusion proteins are invariable associations with the B through F regions of RARα that contain its DNA and ligand-binding domains. These observations indicated that the translocation-induced impairment of RARα signaling was probably responsible for both leukemogenesis and sensitivity to ATRA.

15.3 RETINOIC ACID SIGNALING PATHWAY IN APL CELL MATURATION

15.3.1 Retinoic Acid and Its Receptors

Retinoic acids (RAs) are a group of derivatives from vitamin A that regulate a wide range of complex physiological events in embryonic development and homeostasis maintenance in vertebrates. Moreover, RAs are capable of triggering differentiation not only in normal cells, but also in cancer cells.

The multiple physiological functions of RAs are generally mediated by their cognate receptors. So far at least two sorts of receptors are found to be able to bind RA: cellular RA binding proteins (CRABPs) and nuclear RA receptors.

The CRABPs are small-size proteins that belong to the β-clamp protein family binding small hydrophobic ligands. It has long been known that CRABPs are implicated in the biosynthesis, storage, metabolism, and transport of RA. By modulating the concentration of free RA available to the nuclear receptors, CRABPs can indirectly influence the gene expression required for "RA response." However, several recent studies demonstrated that CRABP II could also be associated with nuclear RA receptors in mammalian cells and that this protein complex could bind the DR5 responsive element of RA-target genes. In the presence of RA, CRABP II could enhance the transactivation of nuclear RA receptors, strongly suggesting a potential role of CRABP II as a novel ligand-dependent transcriptional regulator involved in the RA signaling pathway in eukaryotic cells (Bastie et al., 2001; Delva et al., 1999).

The nuclear receptors for RA include two classes: retinoic acid receptors (RARs) and retinoid X receptors (RXRs). Both belong to the steroid/thyroid hormone receptor superfamily, which act as nuclear ligand-inducible transcription factors. In general, RARs function by heterodimerization with RXRs; the RAR/RXR complex then binds to specific retinoic acid response elements (RAREs) present in the promoters of their target genes. In the absence of RA, the heterodimer binds to the corepressor (CoR) complex containing SMRT, NcoR, and mSin3, which in turn interacts with histone deacetylase, leading to the silencing of target genes. In the presence of RA, the conformation of RAR/RXR dimer changes, resulting in dissociation of the corepressors and recruitment of the coactivators (CoAs), such as CBP/P300, ACTR, and P/CAF, which contain histone acetylase activity. It is believed that the histone acetylation can decondense the chromatin and stimulate the gene transcription (Chambon, 2005; Rosenfeld and Glass, 2001).

15.3.2 Dominant-Negative Effect of PML-RARα in APL Cells

In the case of APL, the fusion protein PML-RARα due to t(15;17) retains the DNA and ligand-binding domains. A series of data showed that PML-RARα might impair the RA signaling pathway by binding as homodimer or heterodimer with RXR to RARα target genes in competition with RARα or by binding as homodimers to a novel set of target genes. The identification of some PML-RARα-specific responsive elements further suggested a major gain of function of this fusion protein as a new transcription factor that probably controls the expression of certain genes responsible for APL cell proliferation (Kamashev et al., 2004). Besides, high levels of PML-RARα in APL cells could also sequester RXR and/or some RARα cofactors, acting as a dominant-negative inhibitor against normal RXR function in APL. RXR is not only an important component in the RA-responsive pathway, but also a promiscuous heterodimerization partner for a great number of nuclear receptors such as the vitamin D_3 receptor (VDR) and thyroid hormone receptor (TR) (Melnick and Licht, 1999; Rosenfeld and Glass, 2001).

Regarding transcriptional properties, by comparison with wild-type RARα, PML-RARα has an altered ability to bind the corepressors more tightly than wild-type RARα, requiring pharmacological doses of ATRA (10^{-6} mol/L) for the release, whereas RARα/RXR releases corepressors only at 10^{-9} mol/L ATRA (Lin et al., 2001). This can be one of the reasons of deregulation of some key myeloid genes' expressions at physiological doses of ATRA, and can account for the need for high doses of ATRA to induce differentiation in APL. These results are further supported by the fact that RA combined with sodium butyrate or trichostatin A (TSA), inhibitors of histone deacetylase, can accelerate APL cell differentiation (Kosugi et al., 1999).

On the other hand, PML-RARα can yet exert its dominant effects on wild-type PML. PML is a phosphoprotein tightly associated with nuclear matrix. More and more studies indicate that PML has multiple biological activities in cell growth suppression, apoptosis, and immune response pathways of many cytokines (Melnick and Licht, 1999). The normal function of PML is closely related with its cellular localization. The macro-speckled nuclear distribution, referred to as nuclear bodies (NBs) or PML oncogenic domains (PODs), is one of the most important features of PML. In normal cells, the NB structure can vary in number, size, and protein composition depending on cell type, cell cycle, and external hormonal stimulation. It is indeed a complex of multiproteins. Until now, more than 20 proteins with diverse natures have been found to exist in PML-associated NB, including Sp100, SUMO-1, pRB, CBP, NDP55, ISG20, PLZF, Daxx, and p53. This makes the NB hold a multiplicity in its function (Zhong et al., 2000a). It has been shown that PML can form not only homodimers through its RING finger/B-box motif and coiled-coil motif, but also multimeric complexes with some partners present in the NB. For example, Daxx is identified as a molecule important for the cytoplasmic transduction of the Fas-dependent proapoptotic stimulus; it can physically interact with PML in NB. But such ability of Daxx to induce apoptosis is completely abrogated in $PML^{-/-}$ cells, suggesting PML and Daxx cooperate in a novel NB-dependent pathway for apoptosis (Wang et al., 1998b; Zhong et al., 2000b). Also, the interaction between PML and CBP may reflect the regulation effect of PML on gene transcription to some extent (Melnick and Licht, 1999). SUMO-1 (small ubiquitin modifier) is an ubiquitin-like protein that can covalently modify NB proteins such as PML and Sp100 (Boddy et al., 1996; Seeler and Dejean, 2001). Some recent studies confirmed that sumoylation of PML was a prerequisite for the proper formation of the NB, and PML might exert some of its biological functions through the regulation of NB formation (Muller et al., 1998; Zhong et al., 2000a).

In APL cells, PML-RARα and PML display an aberrant microspeckled nuclear pattern, probably due to the ability of PML-RARα to heterodimerize with PML. Besides PML, other NB components, such as Sp100, CBP, Daxx, and SUMO-1, are also found to delocalize to nuclear micro-speckles from their usual sites (Melnick and Licht, 1999; Zhong et al., 2000a). Such disruption of the NB may explain why the fusion protein fails to inhibit cell growth and induce apoptosis. ATRA administration of APL cells can gradually relocalize PML and other NB members into their wild-type NB configuration. This relocalization is highly correlated with induction of morphological differentiation in APL cells (Duprez et al., 1996). Mechanistic studies show that RA-caused PML-RARα degradation is responsible for the reorganization of PML-associated NB (Yoshida et al., 1996). The molecular details of this degradation involve the ubiquitin/proteasome system. Briefly, the AF-2 domain of RA-activated PML-RARα could bind the SUG-1 component of the proteasome 19S complex, leading to its catabolism (Zhu et al., 1999a, 2002a). Proteasome inhibitors can block RA-induced degradation, and PML-RARα remains microspeckled, further confirming the above conclusion. It is also found that the degradation of PML-RARα not only restores the normal nuclear distribution of PML-associated NB, but also releases its heterodimerization partner RXR, thus stabilizing RAR/RXR and other nuclear receptor heterodimers required for RXR and recovering the regulation control of gene transcription in favor of APL cell maturation. However, the idea that the degradation of PML-RARα and the restoration of NB distribution are indispensable to APL cell differentiation is controversial. Because some reports demonstrate that the RXR-specific ligands can, through the RXR/RXR homodimer, synergize with cAMP to induce maturation of NB4 cells without PML-RARα degradation and NB reorganization (Benoit et al., 1999), implying the degradation of fusion protein may be simply a concomitant phenomenon during APL cell differention instead of a requirement.

Nevertheless, there is no doubt that the PML-RARα exhibits its dominant-negative activity on RARα and PML. The results that PML-RARα disturbs cell differentiation and apoptosis in hematopoiesis are further reinforced by the observation of in vitro transfection experiments and in vivo transgenic mice studies (Melnick and Licht, 1999; Wang et al., 1998a). Paradoxically, the exquisite sensitivity of APL cells to ATRA likely results from their ability to target PML-RARα. This observation has been supported by the fact that a mutated PML-RARα confers ATRA resistance in APL cells (Duprez et al., 2000) as well as that the leukemias arising in PML-RARαm4 mutant transgenic mice do not differentiate in response to RA treatment (Kogan et al., 2000).

15.3.3 Cross-Talk Between RA and Other Signaling Pathways

It has long been accepted that cell maturation is a process with great complexity. Although numerous lines of evidences show that the pharmacological doses of ATRA can exert their differentiating effects on APL cells by binding to PML-RARα, dissociating the nuclear corepressors complex from PML-RARα to relieve the transcriptional repression, triggering the degradation of PML-RARα oncoprotein and restoring PML-associated NB structure, several reports on the integration of signal cross talk in the context of APL cell differentiation should hold our attention (Benoit et al., 2001).

15.3.3.1 RA and cAMP/PKA

The second messenger, cyclic adenosine monophosphophate (cAMP), plays an important role in the response to hormonal signals for cell proliferation, differentiation, and apoptosis, including that in hematopoietic development. The interplay between RA and cAMP signaling for granulocyte maturation is observed in their synergistic effects on APL cells. Some cAMP agonists can potentiate maturation of NB4 cells, an APL cell line expressing PML-RARα (Lanotte et al., 1991), and render the cells responsive even to physiological levels of retinoids, whereas disrupting the cAMP signal pathway with cAMP-PKA (protein kinase A) antagonist will result in a reversible resistance to pharmacological levels of retinoids (Quenech'Du et al., 1998; Ruchaud et al.,

1994), indicating that the cAMP-PKA pathway may be an integrated part of the RA signal network. This possibility is further reinforced by the report of cross-talk between RXR-specific agonist "rexinoid" and PKA signaling (Benoit et al., 1999).

Our recent studies showed a rapid increase in intracellular cAMP levels and PKA activities during ATRA-induced NB4 cell maturation, which was mediated through adenylate cyclase (Zhao et al., 2004). More strikingly, we found that the phosphylation of RARα by cAMP-PKA signaling could enhance its transcriptional response to RA, suggesting the RA receptor as a unique node of cross talk between cAMP and RA pathways. This result was then strengthened by the demonstration of de The's group that cAMP could also reverse the normal silencing of the transactivating function of RXR when bound to RARα or PML-RARα (Kamashev et al., 2004). Thus, a cytoplasm-to-nucleus pathway can be constructed from the activation of adenylate cyclase to the phosphorylation of RA receptors by PKA, followed by reprogramming the cellular transcriptional machinery. However, the precise mechanism underlying the activation of the adenylate cyclase by ATRA, which is generally considered to be a nuclear hormone, remains unclear. The fact that the cAMP response occurs within a few minutes, which is not abrogated by actinomycin D, is strongly against the conventional concept that this cAMP response is due to the activation of nuclear RA receptors followed by gene transcription/translation regulation and then signaling from nucleus to cytoplasm. Is there any receptor/binding protein for RA existing on the cell surface? As a matter of fact, the activation of pleiotropic cytosolic signaling transduction dependent on rapid nongenomic action of steroid hormones has been indeed reported in the experiments concerning glucocorticoid (Chen and Qiu, 1999) and bile acid (Kawamata et al., 2003; Maruyama et al., 2002). Identification of a G-protein coupled receptor (GPCR) responsive to bile acid strongly suggests that there may be other new GPCRs for nuclear hormones (Im, 2004). The exploration of cytoplasmic-membrane-type receptor/binding protein for ATRA, which is distinct from the known nuclear RA receptors, would be helpful in understanding signal transduction of RA. Meanwhile, it would enrich and complete the theory regarding the mechanism of action of RA and provide interesting possibilities for drug discovery.

15.3.3.2 RA and Interferons

Another important signaling cross-talk in APL cell differentiation is characterized by synergism between RA and interferons (IFNs). IFNs are a family of cytokines mainly with tumor growth suppressive and anti-viral activities. By interacting with their cognate receptors, IFNs can activate Jak/STAT pathways through phosphorylation. The activated STAT proteins then translocate to the nucleus where they stimulate the transcription of a set of genes, leading to different signal cascades and various biological effects (Brivanlou and Darnell, 2002). The combination of ATRA and IFNs represents a useful therapeutic approach in the treatment of leukemias and other malignancies (Gallagher et al., 1987; Nason-Burchenal et al., 1996).

It has been found that IFNs and RA can synergistically inhibit growth and induce differentiation in myeloid leukemia cell lines, such as NB4, HL-60, and U937 (Chelbi-Alix and Pelicano, 1999; Gaboli et al., 1998; Garattini et al., 1998). How do

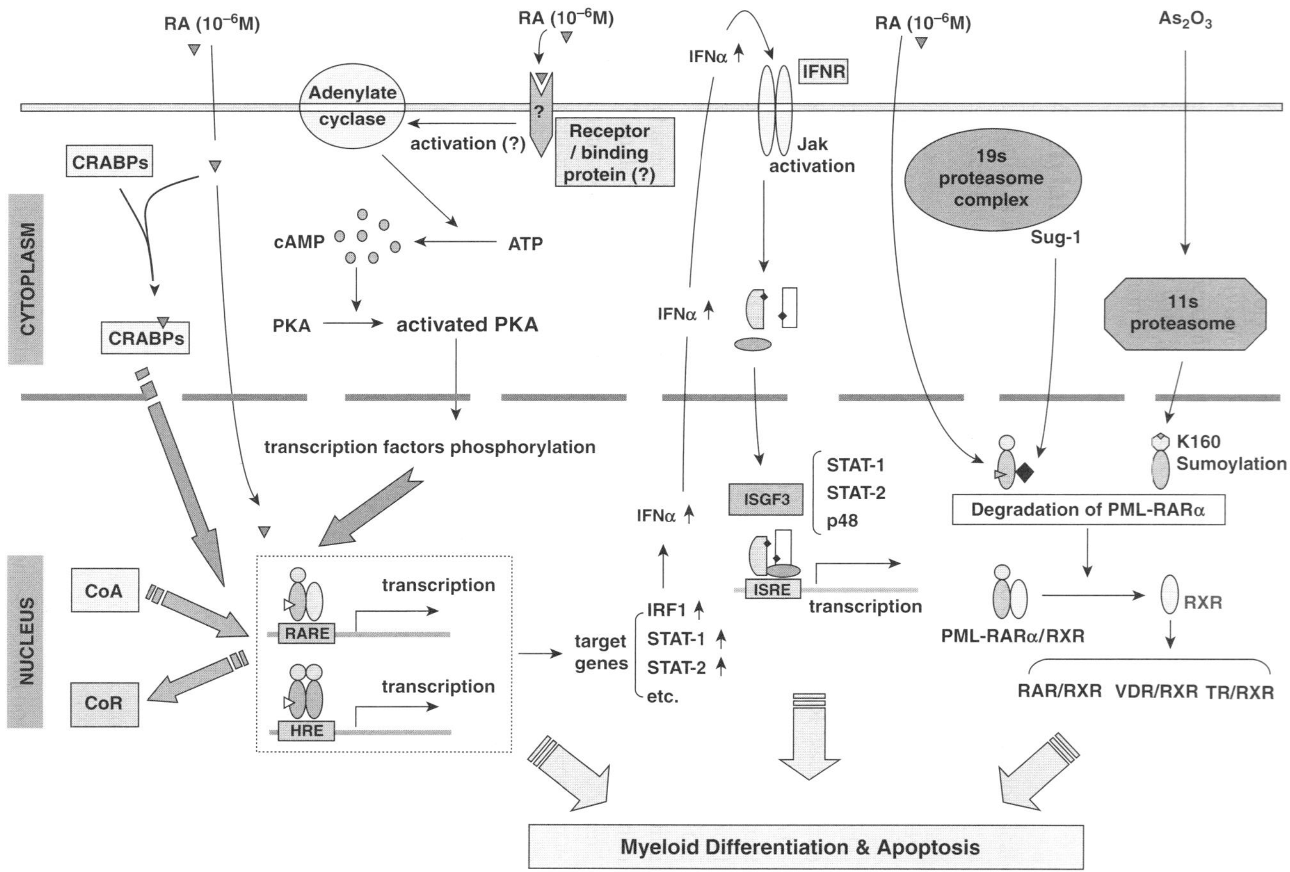

Figure 15.1 An interplay between cell membrane and nuclear signaling in RA/As_2O_3 therapy for APL.

these two agents cooperate at the molecular level? With the recent efforts made in experimental studies, the mechanisms involved in cross-talk between RA and IFNs are being gradually unveiled: (1) IFNs are found to be capable of modulating the expresssion of RA receptors in APL cells and thus ultimately control the expression of ATRA-regulated genes, including the levels and the state of activation of members of the Jak-STAT pathway. This process leads to cross-regulation of genes normally under the control of RA and IFNs respectively. (2) Some NB-associated proteins such as PML, Sp100, and ISG20 are targets of IFNs because the promoters of these genes contain ISRE and/or GAS sequences that bind to STAT proteins (Grotzinger et al., 1996; Regad and Chelbi-Alix, 2001; Stadler et al., 1995). IFNs can cooperate with ATRA to induce PML expression through induction and activation of STATs. Since PML is a growth suppressor, it may confer in part the antiproliferative activity of IFNs/RA combination. (3) It is reported that ATRA can induce IFNα synthesis and secretion and increase the expression of STAT1 in NB4 cells, but not in ATRA-resistant cells NB4-LR1 and NB4-LR2 (Pelicano et al., 1997, 1999). Thus, numerous genes that have been isolated as "retinoic acid-induced genes" are indeed IFN-targeted genes, such as the RIG-G gene identified in our previous work from ATRA-induced NB4 cells (Yu et al., 1997). A database search shows that RIG-G, also called IFI60 (de Veer et al., 1998), belongs to an IFN-induced gene family. In fact, while RIG-G mRNA is dramatically upregulated in NB4 cells after 72 h of ATRA treatment, it is obviously induced after 4 h of exposure to IFNα (Xiao et al., 2006). Our current studies demonstrated that overexpression of RIG-G could significantly inhibit cell proliferation by accumulating cells in G_1/S transition, probably through upregulation of p27 and p21 levels. We therefore hypothesize that induction of RIG-G might represent one of the key molecular nodes in which the synergistic effect of RA and IFN signaling pathways can be converged in the course of APL cell differentiation.

To sum up, no single signal is likely to work optimally. Responses apparently triggered by a single signal indeed result from cross-talk of diverse signaling pathways, just as in the case of ATRA-induced NB4 cell maturation, in which the potential role of cAMP/PKA or IFNs is masked to some degree (see Fig. 15.1). Besides the above-mentioned signaling cross-talks, there are many other arguments sustaining the notion that multiple signaling pathways are required for hematopoietic differentiation. For exemple, combination of RA with certain interleukins, G-CSF, vitamin D_3, and TPA also generate exploitable synergistic responses (Benoit et al., 2001). The recent finding that Cobalt chloride-mimicking hypoxia favors retinoic acid-induced differentiation suggests a potential link between HIF-1α protein and leukemic cell differentiation (Huang et al., 2003; Jiang et al., 2005). Use of TSA to target chromatin remodeling, or interfering with DNA methyltransferase by azacytidine to restore chromatin structure (Fazi et al., 2005), can potentiate the effect of RA, developing a novel transcriptional/differentiation therapy of leukemia.

15.3.4 Limitations of Retinoic Acid in APL Therapy

Despite the great advantage of ATRA in both high rate of CR and low cytotoxic reaction, certain side effects of ATRA have emerged in APL therapy, mainly those including retinoic acid syndrome (RAS) during treatment and retinoic acid resistance following treatment.

15.3.4.1 Retinoic Acid Syndrome

Retinoic acid syndrome (RAS), also called hyperleukocytosis, is observed in about one-third of APL patients during treatment with ATRA alone. The major clinical features of RAS combine fever, respiratory distress, weight gain, lower-extremity edema, pleural or pericardial effusions, and sometimes renal failure, frequently preceded by an increase in white blood cell counts in peripheral blood. The pathophysiology of RAS is still not clearly understood and seems to be related to the induction of G-CSF, IL-1β, IL-6, IL-8, and TNFα secretion by APL cells during ATRA treatment. In addition, ATRA-induced modification of the adhesive properties of APL cells might also play a role in RAS. At the present time, prevention of RAS involves using chemotherapy in combination with ATRA and adding high-dose corticosteroids (Degos and Wang, 2001; Fenaux et al., 2001).

15.3.4.2 Retinoic Acid Resistance

Acquired retinoic acid resistance occurs in most, if not all, APL patients who are treated with ATRA alone and achieve CR for only a short time (Degos and Wang, 2001). This is because the differentiation induction therapy is unable to eliminate leukemic cells completely. Furthermore, these residual cells often exhibit unresponsiveness to ATRA, explaining why relapsed patients are usually resistant or refractory to ATRA. In order to reveal the molecular mechanisms of APL cells' resistance to ATRA, numerous studies were made over the last decade. Several defects at different levels have been described to account for the resistance of APL cells to ATRA:

1. A significant decrease of ATRA levels in plasma due to ATRA-induced upregulation of cytochrome P450 and CRABP II implicated in the metabolism of ATRA (Kizaki et al., 1997), thereby reducing ATRA concentration in the nucleus as well as reducing the sensitivity of APL cells to ATRA (Chen et al., 1996a). However, this deduction is likely doubtful because of the data that the principal ATRA metabolites are as active as ATRA at inducing NB4 cell maturation (Idres et al., 2001, 2002; Marill et al., 2003).

2. No PML-RARα expression or expression of a dominant-negative PML-RARα mutant without ligand-binding activity. In NB4-LR2 cells (Duprez et al., 2000), a truncated PML-RARα due to a nonsense mutation in the E domain is found to block RARα and wild-type PML-RARα transcriptional activities, rendering the cells resistant to ATRA. However, this resistance can be reversed by cooperation of an RXR/RXR-specific pathway and cAMP signaling (Benoit et al., 1999). A similar phenomenon is also observed in other ATRA-resistant cells, such as HL-60R (Robertson et al., 1992) and NB4-306 (Dermime et al., 1993).

3. Increased affinity of fusion protein to nuclear corepressor complex. In a few APL patients with type V PML-RARα isoform, some intron 2 sequences of RARα gene are found to insert between the truncated PML and RARα mRNA fusion partners, providing potential sites for binding CoR so that these patients exhibit less sensitivity to a pharmacological dose of ATRA (Gu et al., 2002). Deletion of these "inserted" sequences can restore the sensitivity to ATRA, suggesting that the aberrant binding of nuclear CoR mediates RA resistance (Gu et al., 2002). This observation is in accordance with the case of APL patients bearing t(11;17), in which the presence of a second CoR binding site on PLZF moiety resists the effect of ATRA (Guidez et al., 1998; He et al., 1998; Hong et al., 1997).

4. Failure of cooperation between ATRA and cAMP signaling pathways. NB4-LR1 cells, because of uncoupling RA and cAMP action, display no maturation in the presence of RA alone. cAMP is required for triggering RA-primed NB4-LR1 cells to undergo terminal maturation (Ruchaud et al., 1994). The fact that the distinct changes of intracellular cAMP concentration in NB4 and NB4-LR1 cells upon ATRA treatment are attributed to the role of adenylate cyclase suggests that the defaults in the activation of adenylate cyclase may contribute to the resistance of NB4-LR1 cells to ATRA-induced maturation (Zhao et al., 2004).

5. Deregulation of telomerase activity in APL cells probably involved in an eventual novel mechanism of retinoic acid resistance (Pendino et al., 2002).

Since the retinoic acid resistance is inevitable and the mechanism involved is so complex, it is necessary to explore new approaches for APL therapy. Currently, intravenous administration of As_2O_3 has become the first-line approach of therapy for APL patients with relapses or refractory disease.

15.4 ARSENIC TRIOXIDE IN THE TREATMENT OF APL

15.4.1 Clinical Practice of Arsenic in APL Therapy

Arsenic has long been known to act as a carcinogen, involved in human skin, lung, liver, and kidney tumors. Paradoxically, it has also been demonstrated to contain antitumor activity in some cases. In traditional Chinese medicine, arsenic compounds have been used for more than 500 years in the treatment of asthma and certain skin diseases. Use of arsenic for the treatment of chronic myeloid leukemia can be traced to the middle of the nineteenth century. But chemotherapeutic agents later replaced this treatment because of the toxicity of arsenic (Zhang et al., 2001; Zhou et al., 2005). In the 1970s, the discovery of the therapeutic effect of As_2O_3 in APL revived this ancient drug. A Chinese group reported in 1992 for the first time that a solution containing As_2O_3 and a trace mount of mercury chloride could induce CR in APL patients; the CR rate, the 5-year survival rate, and the 10-year survival rate were 65.5%, 50.0%, and 18.8%, respectively (Sun et al., 1992). In 1996, SIH was able to conduct a clinical trial using a pure As_2O_3 solution at a dose of 0.16 mg/kg/day to treat a group of relapsed APL patients previously treated with ATRA and chemotherapy. A complete remission was obtained after 28–54 days among 14 out of 15 cases (Chen et al., 1996b; Shen et al., 1997). This suggested that As_2O_3 displayed no obvious cross-resistance with ATRA. Many groups from China and western countries then confirmed these results (Dombret et al., 2002; Niu et al., 1999; Soignet et al., 1998). Morphologically, after 3–4 weeks of As_2O_3 treatment, the leukemic promyelocytes decrease, while a lot of myelocyte-like cells and degenerative cells appear in the bone marrow and peripheral blood of patients (Chen et al., 1997). Furthermore, recent studies on the clinical efficacy of low-dose As_2O_3 demonstrated that low-dose As_2O_3 seemed to have the same effect as the conventional dosage (Shen et al., 2001). The major side effects of arsenic, including abnormal increase in hepatic enzymes, hyperleucocytosis, skin reaction, and gastrointestinal disturbances, are moderate and tolerable and are less frequent in low-dose-treated patients. No significant bone marrow suppression is so far observed. These results suggest that arsenic is an effective and reletively safe drug in APL treatment, especially for relapsed and refractory patients (Shen et al., 2001).

15.4.2 Molecular and Cellular Mechanisms of Arsenic in APL Therapy

The clinical success of As_2O_3 in APL treatment has been stimulating the research on its mechanisms of action. Many significant advances in this regard have been made during the past few years.

The ability of As_2O_3 to induce APL cells to undergo apoptosis was first disclosed in some *in vitro* studies. It was found that 1 μM of As_2O_3 could trigger apoptosis of NB4 cells with downregulation of Bcl-2 gene expression (Chen et al., 1996b). Mechanistic studies showed that this As_2O_3-induced apoptotic effect was associated with the collapse of mitochondrial transmembrane potentials due to thiol-group oxidation and with the activation of caspase 3 (Cai et al., 2000). It was further found that As_2O_3 could also modulate the microspeckled pattern of PML nuclear distribution in NB4 cells by degradation of PML-RARα (Zhu et al., 1997). Yet, unlike retinoic acid, which basically targets the RARα moiety of the fusion protein, As_2O_3 targets its PML moiety. As_2O_3-induced PML sumoylation on a specific lysine residual within PML moiety is required for the recruitment of the 11S proteasome regulatory complex, which is responsible for PML-RARα degradation (Lallemand-Breitenbach et al., 2001). It is now considered that As_2O_3 exerts its competence of apoptosis induction in APL cells through rapidly degrading PML-RARα. In fact, in contrast to ATRA, As_2O_3 can degrade both PML-RARα fusion protein and wild-type PML (Zhu et al., 2001).

Aside from apoptosis-inducing activity, dose effect of As_2O_3 indicates that at low concentration (0.1–0.25 μM) it can also induce a certain degree of differentiation of NB4 cells (Chen et al., 1997). However, this differentiation is far from terminal because most cells cannot mature beyond the myelocyte–metamyelocyte stage and the NBT reduction test remains negative. Moreover, this partial differentiation can be observed only after long (>10 days) exposure of the cells to the drug. Since the low-dose As_2O_3 exerts no obvious effect on modulation of the interaction between CoR and RAR/RXR or PML-RARα, it is unlikely that the partial differentiation induction of low-dose As_2O_3 in NB4 cells is mediated directly through the RARα pathway. The mechanisms underlying this As_2O_3-induced incomplete or partial differentiation remain to be explored.

On the other hand, pharmacokinetics analysis of APL patients receiving As_2O_3, whether in standard doses or low doses, reveals that the plasma As_2O_3 level is rapidly eliminated after a peak level superior to 1 μM caused by intravenous drip. During the treatment, the plasma As_2O_3 levels fluctuated mostly between 0.1 and 0.5 μM, and the continuous administration of As_2O_3 did not alter its pharmacokinetic behaviors (Shen et al., 1997, 2001). According to this finding and the morphological observation on bone marrow smears of As_2O_3-treated APL patients, it can be thus speculated that APL cells are *in vivo* primarily exposed to a drug concentration in favor of differentiation rather than apoptosis. Moreover, the apoptotic effect of As_2O_3 can be observed in a large range of cell types, covering not only leukemic cells but also some solid tumor cells, whereas the differentiating activity is relatively limited to APL (Chen et al., 2002; Zhu et al., 1999b).

The discrepancy between the differentiation states observed *in vivo* and *in vitro* suggests some differentiating factors present in bone marrow microenvironments of APL patients cooperating with As_2O_3 to induce APL cell maturation *in vivo*, including cytokines and other signal molecules. A significant increase in IL-1β and G-CSF expression is observed in arsenic-treated APL cells (Jiang et al., 2003). Our recent data showed existence of a strong synergy between cAMP and the low-dose of As_2O_3 in fully inducing differentiation of NB4, NB4-LR1, and fresh APL cells (Zhu et al., 2002b). We found that cAMP could facilitate the As_2O_3-induced degradation of PML-RARα and significantly induce G_1 arrest and cell-cycle exit of NB4 and NB4-LR1 cells by modulating several major players in G_1/S transition regulation. The cooperation might result from the concomitant PML-RARα degradation induced by As_2O_3 and the cell growth inhibition by cAMP. Our findings concerning this novel signaling cross-talk of As_2O_3 and cAMP was soon confirmed by the demonstration that *in vivo* activation of cAMP signaling was beneficial in mouse models of APL as well as in an RA/As_2O_3-resistant APL patient (Guillemin et al., 2002). These results may deepen our understanding of As_2O_3-induced differentiation *in vivo* and increase the potency of other differentiation inducers *in vivo*.

15.4.3 Dual Treatment with ATRA and Arsenic Trioxide in APL Therapy

Though the reports on the combined effects of ATRA and As_2O_3 in APL cell lines were conflicting (Gianni et al., 1998; Jing et al., 2001; Shao et al., 1998), our latest clinical data showed that the ATRA and As_2O_3 combination for remission/maintenance therapy of APL manifested much better results in newly diagnosed APL than either of the two drugs used alone in terms of the high quality of remission and the status of the disease-free survival (Shen et al., 2004). This observation was reminiscent of the results previously obtained in mice transplanted with PML-RARα transgene cells—that the association of ATRA with As_2O_3 could lead to a much faster decrease in the leukemic population and cure the induced leukemia through enhanced differentiation and apoptosis (Lallemand-Breitenbach et al., 1999). In order to understand the complexity and dynamics of *in vivo* synergy between RA and As_2O_3 in APL, we identified a number of pathways with temporospatial relationships by conducting an investigation incorporating advanced technologies of genomics, proteomics, and computational biology (Zheng et al., 2005). These pathways are regulated in concert and harmony, forming a network. It seems that RA exerts its effects on APL cells mainly through nuclear-receptor-mediated transcriptional regulation, whereas As_2O_3 exercises its impact through targeting multiple pathways/cascades at the levels of proteome, transcriptome, and probably metabolome. The finding of a strong activation of the ubiquitin/proteasome system strongly suggests that the synergistic effects of ATRA and As_2O_3 on degradation of PML-RARα oncoprotein and on differentiation/apoptosis contribute to the mechanism of superior efficacy of RA/As_2O_3 dual treatment in APL. Although the protocol of ATRA/As_2O_3 combination in APL treatment still needs to validate and the long-term effect of this treatment on APL patients remains to be illustrated, such an inspiring discovery will furnish insights for new therapeutic strategies in the treatment of APL. With the ATRA/As_2O_3 dual treatment followed by obligatory consolidating therapy for maintenance, it is worth expecting that APL will become a curable disease in the near future.

15.5 CONCLUSIONS

Altogether, in the case of APL, PML-RARα fusion protein due to a specific translocation t(15;17) has been considered as a dominant-negative receptor that blocks myeloid differentiation most likely through the impairment of retinoid reponse. In order to restore retinoid-signaling pathways, two strategies can be taken into account: (1) a targeted degradation or inactivation of PML-RARα and (2) an activation of some alternative signaling pathways bypassing PML-RARα. During the last two decades, studies on the mechanisms of differentiation and apoptosis induction tharapy of APL have significantly impacted understanding the pathogenesis of APL and identifying new approaches in the treatment of cancer. It is now recognized that PML-RARα fusion protein is a molecule basis of both APL pathogenesis and therapy. ATRA and As_2O_3, though the mechanisms and the sites of action of the two drugs are widely divergent, both target the same fusion protein PML-RARα. Accordingly, it should be considered that the APL treatment, whether with ATRA, As_2O_3, or both, is a PML-RARα oncoprotein targeted therapy. Establishment and realization of molecule target-based cancer therapy in another leukemia model beyond APL is our desire and research orientation.

ACKNOWLEDGMENTS

The authors were supported in part by the Chinese National Key Basic Research Project 973 (2002CB512805, 2004CB518600), the Chinese National High Tech Program 863 (2002BA711A04, 2004AA227050), the National Natural Science Foundation of China (90209007, 30271466, 30570778, 30670882), the Shanghai Municipal Commission for

Science and Technology, and the Shanghai Municipal Commission for Education. Jian-Hua Tong was also supported by the "Shu Guang" Program of Shanghai Municipal Commission for Education. The authors appreciate Professor Zhen-Yi Wang and Professor Zhi-Xiang Shen from SIH, and Dr. Michel Lanotte from Hospital St. Louis, Paris, France, for their continuous support. The authors are also grateful to Dr. Ji Zhang for help in reviewing the manuscript.

REFERENCES

Arnould, C., Philippe, C., Bourdon, V., Gr goire, M. J., Berger, R., and Jonveaux, P. (1999). The signal transducer and activator of transcription STAT5b gene is a new partner of retinoic acid receptor alpha in acute promyelocytic-like leukaemia. *Hum Mol Genet* **8**, 1741–1749.

Avvisati, G., Lo Coco, F., and Mandelli, F. (2001). Acute promyelocytic leukemia: clinical and morphologic features and prognostic factors. *Semin Hematol* **38**, 4–12.

Bastie, J. N., Despouy, G., Balitrand, N., Rochette-Egly, C., Chomienne, C., and Delva, L. (2001). The novel co-activator CRABPII binds to RARalpha and RXRalpha via two nuclear receptor interacting domains and does not require the AF-2 'core'. *FEBS Lett* **507**, 67–73.

Benoit, G., Altucci, L., Flexor, M., Ruchaud, S., Lillehaug, J., Raffelsberger, W., Gronemeyer, H., and Lanotte, M. (1999). RAR-independent RXR signaling induces t(15;17) leukemia cell maturation. *EMBO J* **18**, 7011–7018.

Benoit, G., Roussel, M., Pendino, F., Segal-Bendirdjian, E., and Lanotte, M. (2001). Orchestration of multiple arrays of signal cross talk and combinatorial interactions for maturation and cell death: Another vision of t(15;17) preleukemic blast and APL-cell maturation. *Oncogene* **20**, 7161–7177.

Boddy, M. N., Howe, K., Etkin, L. D., Solomon, E., and Freemont, P. S. (1996). PIC 1, a novel ubiquitin-like protein which interacts with the PML component of a multiprotein complex that is disrupted in acute promyelocytic leukaemia. *Oncogene* **13**, 971–982.

Brivanlou, A. H., and Darnell, J. E., Jr. (2002). Signal transduction and the control of gene expression. *Science* **295**, 813–818.

Cai, X., Shen, Y. L., Zhu, Q., Jia, P. M., Yu, Y., Zhou, L., Huang, Y., Zhang, J. W., Xiong, S. M., Chen, S. J., Wang, Z. Y., Chen, Z., and Chen, G. Q. (2000). Arsenic trioxide-induced apoptosis and differentiation are associated respectively with mitochondrial transmembrane potential collapse and retinoic acid signaling pathways in acute promyelocytic leukemia. *Leukemia* **14**, 262–270.

Chambon, P. (2005). The nuclear receptor superfamily: A personal retrospect on the first two decades. *Mol Endocrinol* **19**, 1418–1428.

Chelbi-Alix, M. K., and Pelicano, L. (1999). Retinoic acid and interferon signaling cross talk in normal and RA-resistant APL cells. *Leukemia* **13**, 1167–1174.

Chen, G. Q., Shen, Z. X., Wu, F., Han, J. Y., Miao, J. M., Zhong, H. J., Li, X. S., Zhao, J. Q., Zhu, J., Fang, Z. W., Chen, S. J., Chen, Z., and Wang, Z. Y. (1996a). Pharmacokinetics and efficacy of low-dose all-*trans* retinoic acid in the treatment of acute promyelocytic leukemia. *Leukemia* **10**, 825–828.

Chen, G. Q., Zhu, J., Shi, X. G., Ni, J. H., Zhong, H. J., Si, G. Y., Jin, X. L., Tang, W., Li, X. S., Xong, S. M., Shen, Z. X., Sun, G. L., Ma, J., Zhang, P., Zhang, T. D., Gazin, C., Naoe, T., Chen, S.J., Wang, Z. Y., and Chen, Z. (1996b). *In vitro* studies on cellular and molecular mechanisms of arsenic trioxide (As_2O_3) in the treatment of acute promyelocytic leukemia: As_2O_3 induces NB4 cell apoptosis with downregulation of Bcl-2 expression and modulation of PML-RAR alpha/PML proteins. *Blood* **88**, 1052–1061.

Chen, G. Q., Shi, X. G., Tang, W., Xiong, S. M., Zhu, J., Cai, X., Han, Z. G., Ni, J. H., Shi, G. Y., Jia, P. M., Liu, M. M., He, K. L., Niu, C., Ma, J., Zhang, P., Zhang, T. D., Paul, P., Naoe, T., Kitamura, K., Miller, W., Waxman, S., Wang, Z. Y., de The, H., Chen, S. J., and Chen, Z. (1997). Use of arsenic trioxide (As_2O_3) in the treatment of acute promyelocytic leukemia (APL): I. As_2O_3 exerts dose-dependent dual effects on APL cells. *Blood* **89**, 3345–3353.

Chen, Y. Z., and Qiu, J. (1999). Pleiotropic signaling pathways in rapid, nongenomic action of glucocorticoid. *Mol Cell Biol Res Commun* **2**, 145–149.

Chen, Z., Brand, N. J., Chen, A., Chen, S. J., Tong, J. H., Wang, Z. Y., Waxman, S., and Zelent, A. (1993). Fusion between a novel Kruppel-like zinc finger gene and the retinoic acid receptor-alpha locus due to a variant t(11;17) translocation associated with acute promyelocytic leukaemia. *EMBO J* **12**, 1161–1167.

Chen, Z., Chen, G. Q., Shen, Z. X., Sun, G. L., Tong, J. H., Wang, Z. Y., and Chen, S. J. (2002). Expanding the use of arsenic trioxide: Leukemias and beyond. *Semin Hematol* **39**, 22–26.

de Veer, M. J., Sim, H., Whisstock, J. C., Devenish, R. J., and Ralph, S. J. (1998). IFI60/ISG60/IFIT4, a new member of the human IFI54/IFIT2 family of interferon-stimulated genes. *Genomics* **54**, 267–277.

Degos, L., and Wang, Z. Y. (2001). All *trans* retinoic acid in acute promyelocytic leukemia. *Oncogene* **20**, 7140–7145.

Delva, L., Bastie, J. N., Rochette-Egly, C., Kraiba, R., Balitrand, N., Despouy, G., Chambon, P., and Chomienne, C. (1999). Physical and functional interactions between cellular retinoic acid binding protein II and the retinoic acid-dependent nuclear complex. *Mol Cell Biol* **19**, 7158–7167.

Dermime, S., Grignani, F., Clerici, M., Nervi, C., Sozzi, G., Talamo, G. P., Marchesi, E., Formelli, F., Parmiani, G., Pelicci, P. G., and Gambacorti-Passerini, C. (1993). Occurrence of resistance to retinoic acid in the acute promyelocytic leukemia cell line NB4 is associated with altered expression of the pml/RAR alpha protein. *Blood* **82**, 1573–1577.

Dombret, H., Fenaux, P., Soignet, S. L., and Tallman, M. S. (2002). Established practice in the treatment of patients with acute promyleocytic leukemia and the introduction of arsenic trioxide as a novel therapy. *Semin Hematol* **39**, 8–13.

Duprez, E., Benoit, G., Flexor, M., Lillehaug, J. R., and Lanotte, M. (2000). A mutated PML/RARA found in the retinoid maturation resistant NB4 subclone, NB4-R2, blocks RARA and wild-type PML/RARA transcriptional activities. *Leukemia* **14**, 255–261.

Duprez, E., Lillehaug, J. R., Naoe, T., and Lanotte, M. (1996). cAmp signalling is decisive for recovery of nuclear bodies (Pods) during maturation of Ra-resistant t(15;17) promyelocytic leukemia Nb4 cells expressing Pml-Rar alpha. *Oncogene* **12**, 2451–2459.

Fazi, F., Travaglini, L., Carotti, D., Palitti, F., Diverio, D., Alcalay, M., McNamara, S., Miller, W. H., Jr., Lo Coco, F., Pelicci, P. G., and Nervi, C. (2005). Retinoic acid targets DNA-methyltransferases and histone deacetylases during APL blast differentiation *in vitro* and *in vivo*. *Oncogene* **24**, 1820–1830.

Fenaux, P., Chomienne, C., and Degos, L. (2001). All-*trans* retinoic acid and chemotherapy in the treatment of acute promyelocytic leukemia. *Semin Hematol* **38**, 13–25.

Gaboli, M., Gandini, D., Delva, L., Wang, Z. G., and Pandolfi, P. P. (1998). Acute promyelocytic leukemia as a model for cross-talk between interferon and retinoic acid pathways: from molecular biology to clinical applications. *Leuk Lymphoma* **30**, 11–22.

Gallagher, R. E., Lurie, K. J., Leavitt, R. D., and Wiernik, P. H. (1987). Effects of interferon and retinoic acid on the growth and differentiation of clonogenic leukemic cells from acute myelogenous leukemia patients treated with recombinant leukocyte-alpha A interferon. *Leuk Res* **11**, 609–619.

Garattini, E., Mologni, L., Ponzanelli, I., and Terao, M. (1998). Cross-talk between retinoic acid and interferons: Molecular mechanisms of interaction in acute promyelocytic leukemia cells. *Leuk Lymphoma* **30**, 467–475.

Gianni, M., Koken, M. H., Chelbi-Alix, M. K., Benoit, G., Lanotte, M., Chen, Z., and de The, H. (1998). Combined arsenic and retinoic acid treatment enhances differentiation and apoptosis in arsenic-resistant NB4 cells. *Blood* **91**, 4300–4310.

Grotzinger, T., Jensen, K., and Will, H. (1996). The interferon (IFN)-stimulated gene Sp100 promoter contains an IFN-gamma activation site and an imperfect IFN-stimulated response element which mediate type I IFN inducibility. *J Biol Chem* **271**, 25253–25260.

Gu, B. W., Xiong, H., Zhou, Y., Chen, B., Wang, L., Dong, S., Yu, Z. Y., Lu, L. F., Zhong, M., Yin, H. F., Zhu, G. F., Huang, W., Ren, S. X., Gallagher, R. E., Waxman, S., Chen, G. Q., Wang, Z. G., Chen, Z., Fu, G., and Chen, S. J. (2002). Variant-type PML-RAR(alpha) fusion transcript in acute promyelocytic leukemia: Use of a cryptic coding sequence from intron 2 of the RAR(alpha) gene and identification of a new clinical subtype resistant to retinoic acid therapy. *Proc Natl Acad Sci USA* **99**, 7640–7645.

Guidez, F., Ivins, S., Zhu, J., Soderstrom, M., Waxman, S., and Zelent, A. (1998). Reduced retinoic acid-sensitivities of nuclear receptor corepressor binding to PML- and PLZF-RARalpha underlie molecular pathogenesis and treatment of acute promyelocytic leukemia. *Blood* **91**, 2634–2642.

Guillemin, M. C., Raffoux, E., Vitoux, D., Kogan, S., Soilihi, H., Lallemand-Breitenbach, V., Zhu, J., Janin, A., Daniel, M. T., Gourmel, B., Degos, L., Dombret, H., Lanotte, M., and de The, H. (2002). *In vivo* activation of cAMP signaling induces growth arrest and differentiation in acute promyelocytic leukemia. *J Exp Med* **196**, 1373–1380.

He, L. Z., Guidez, F., Tribioli, C., Peruzzi, D., Ruthardt, M., Zelent, A., and Pandolfi, P. P. (1998). Distinct interactions of PML-RARalpha and PLZF-RARalpha with co-repressors determine differential responses to RA in APL. *Nat Genet* **18**, 126–135.

Hong, S. H., David, G., Wong, C. W., Dejean, A., and Privalsky, M. L. (1997). SMRT corepressor interacts with PLZF and with the PML-retinoic acid receptor alpha (RARalpha) and PLZF-RAR-alpha oncoproteins associated with acute promyelocytic leukemia. *Proc Natl Acad Sci USA* **94**, 9028–9033.

Huang, M. E., Ye, Y. C., Chen, S. R., Chai, J. R., Lu, J. X., Zhoa, L., Gu, L. J., and Wang, Z. Y. (1988). Use of all-*trans* retinoic acid in the treatment of acute promyelocytic leukemia. *Blood* **72**, 567–572.

Huang, Y., Du, K. M., Xue, Z. H., Yan, H., Li, D., Liu, W., Chen, Z., Zhao, Q., Tong, J. H., Zhu, Y. S., and Chen, G. Q. (2003). Cobalt chloride and low oxygen tension trigger differentiation of acute myeloid leukemic cells: possible mediation of hypoxia-inducible factor-1alpha. *Leukemia* **17**, 2065–2073.

Idres, N., Benoit, G., Flexor, M. A., Lanotte, M., and Chabot, G. G. (2001). Granulocytic differentiation of human NB4 promyelocytic leukemia cells induced by all-*trans* retinoic acid metabolites. *Cancer Res* **61**, 700–705.

Idres, N., Marill, J., Flexor, M. A., and Chabot, G. G. (2002). Activation of retinoic acid receptor-dependent transcription by all-*trans*-retinoic acid metabolites and isomers. *J Biol Chem* **277**, 31491–31498.

Im, D. S. (2004). Discovery of new G protein-coupled receptors for lipid mediators. *J Lipid Res* **45**, 410–418.

Jiang, G., Bi, K., Tang, T., Zhang, Y., Ren, H., Jiang, F., Ren, Q., Zhen, G., Liu, C., Peng, J., Guo, G., Liu, X., and Tian, Z. (2003). Effect of arsenic trioxide on cytokine expression by acute promyelocytic leukemia cells. *Chin Med J (Engl)* **116**, 1639–1643.

Jiang, Y., Xue, Z. H., Shen, W. Z., Du, K. M., Yan, H., Yu, Y., Peng, Z. G., Song, M. G., Tong, J. H., Chen, Z., Huang, Y., Lubbert, M., and Chen, G. Q. (2005). Desferrioxamine induces leukemic cell differentiation potentially by hypoxia-inducible factor-1 alpha that augments transcriptional activity of CCAAT/enhancer-binding protein-alpha. *Leukemia* **19**, 1239–1247.

Jing, Y., Wang, L., Xia, L., Chen, G. Q., Chen, Z., Miller, W. H., and Waxman, S. (2001). Combined effect of all-trans retinoic acid and arsenic trioxide in acute promyelocytic leukemia cells *in vitro* and *in vivo*. *Blood* **97**, 264–269.

Kamashev, D., Vitoux, D., and de The, H. (2004). PML-RARA-RXR oligomers mediate retinoid and rexinoid/cAMP cross-talk in acute promyelocytic leukemia cell differentiation. *J Exp Med* **199**, 1163–1174.

Kawamata, Y., Fujii, R., Hosoya, M., Harada, M., Yoshida, H., Miwa, M., Fukusumi, S., Habata, Y., Itoh, T., Shintani, Y., Hinuma, S., Fujisawa, Y., and Fujino, M. (2003). A G protein-coupled receptor responsive to bile acids. *J Biol Chem* **278**, 9435–9440.

Kizaki, M., Ueno, H., Matsushita, H., Takayama, N., Muto, A., Awaya, N., and Ikeda, Y. (1997). Retinoid resistance in leukemic cells. *Leuk Lymphoma* **25**, 427–434.

Kogan, S. C., Hong, S. H., Shultz, D. B., Privalsky, M. L., and Bishop, J. M. (2000). Leukemia initiated by PMLRARalpha: the PML domain plays a critical role while retinoic acid-mediated transactivation is dispensable. *Blood* **95**, 1541–1550.

Kosugi, H., Towatari, M., Hatano, S., Kitamura, K., Kiyoi, H., Kinoshita, T., Tanimoto, M., Murate, T., Kawashima, K., Saito, H., and Naoe, T. (1999). Histone deacetylase inhibitors are the potent inducer/enhancer of differentiation in acute myeloid leukemia: A new approach to anti-leukemia therapy. *Leukemia* **13**, 1316–1324.

Lallemand-Breitenbach, V., Guillemin, M. C., Janin, A., Daniel, M. T., Degos, L., Kogan, S. C., Bishop, J. M., and de The, H. (1999). Retinoic acid and arsenic synergize to eradicate leukemic cells in a mouse model of acute promyelocytic leukemia. *J Exp Med* **189**, 1043–1052.

Lallemand-Breitenbach, V., Zhu, J., Puvion, F., Koken, M., Honore, N., Doubeikovsky, A., Duprez, E., Pandolfi, P. P., Puvion, E., Freemont, P., and de The, H. (2001). Role of promyelocytic leukemia (PML) sumolation in nuclear body formation, 11S proteasome recruitment, and As_2O_3-induced PML or PML/retinoic acid receptor alpha degradation. *J Exp Med* **193**, 1361–1371.

Lanotte, M., Martin-Thouvenin, V., Najman, S., Balerini, P., Valensi, F., and Berger, R. (1991). NB4, a maturation inducible cell line with t(15;17) marker isolated from a human acute promyelocytic leukemia (M3). *Blood* **77**, 1080–1086.

Lin, R. J., Sternsdorf, T., Tini, M., and Evans, R. M. (2001). Transcriptional regulation in acute promyelocytic leukemia. *Oncogene* **20**, 7204–7215.

Lu, D. P., Qiu, J. Y., Jiang, B., Wang, Q., Liu, K. Y., Liu, Y. R., and Chen, S. S. (2002). Tetra-arsenic tetrasulfide for the treatment of acute promyelocytic leukemia: A pilot report. *Blood* **99**, 3136–3143.

Marill, J., Idres, N., Capron, C. C., Nguyen, E., and Chabot, G. G. (2003). Retinoic acid metabolism and mechanism of action: A review. *Curr Drug Metab* **4**, 1–10.

Maruyama, T., Miyamoto, Y., Nakamura, T., Tamai, Y., Okada, H., Sugiyama, E., Nakamura, T., Itadani, H., and Tanaka, K. (2002). Identification of membrane-type receptor for bile acids (M-BAR). *Biochem Biophys Res Commun* **298**, 714–719.

Melnick, A., and Licht, J. D. (1999). Deconstructing a disease: RARalpha, its fusion partners, and their roles in the pathogenesis of acute promyelocytic leukemia. *Blood* **93**, 3167–3215.

Muller, S., Matunis, M. J., and Dejean, A. (1998). Conjugation with the ubiquitin-related modifier SUMO-1 regulates the partitioning of PML within the nucleus. *Embo J* **17**, 61–70.

Nason-Burchenal, K., Gandini, D., Botto, M., Allopenna, J., Seale, J. R., Cross, N. C., Goldman, J. M., Dmitrovsky, E., and Pandolfi, P. P. (1996). Interferon augments PML and PML/RAR alpha expression in normal myeloid and acute promyelocytic cells and cooperates with all-*trans* retinoic acid to induce maturation of a retinoid-resistant promyelocytic cell line. *Blood* **88**, 3926–3936.

Niu, C., Yan, H., Yu, T., Sun, H. P., Liu, J. X., Li, X. S., Wu, W., Zhang, F. Q., Chen, Y., Zhou, L., Li, J. M., Zeng, X. Y., Yang, R. R., Yuan, M. M., Ren, M. Y., Gu, F. Y., Cao, Q., Gu, B. W., Su, X. Y., Chen, G. Q., Xiong, S. M., Zhang, T. D., Waxman, S., Wang, Z. Y., Chen, Z., Hu, J., Shen, Z. X., and Chen, S. J. (1999). Studies on treatment of acute promyelocytic leukemia with arsenic trioxide: Remission induction, follow-up, and molecular monitoring in 11 newly diagnosed and 47 relapsed acute promyelocytic leukemia patients. *Blood* **94**, 3315–3324.

Pandolfi, P. P., Alcalay, M., Fagioli, M., Zangrilli, D., Mencarelli, A., Diverio, D., Biondi, A., Lo Coco, F., Rambaldi, A., Grignani, F., Rochette-Egly, C., Gaube, M. P., Chambon, P., and Peliccci, P. G. (1992). Genomic variability and alternative splicing generate multiple PML/RAR alpha transcripts that encode aberrant PML proteins and PML/RAR alpha isoforms in acute promyelocytic leukaemia. *EMBO J* **11**, 1397–1407.

Pelicano, L., Li, F., Schindler, C., and Chelbi-Alix, M. K. (1997). Retinoic acid enhances the expression of interferon-induced proteins: evidence for multiple mechanisms of action. *Oncogene* **15**, 2349–2359.

Pelicano, L., Brumpt, C., Pitha, P. M., and Chelbi-Alix, M. K. (1999). Retinoic acid resistance in NB4 APL cells is associated with lack of interferon alpha synthesis Stat1 and p48 induction. *Oncogene* **18**, 3944–3953.

Pendino, F., Sahraoui, T., Lanotte, M., and Segal-Bendirdjian, E. (2002). A novel mechanism of retinoic acid resistance in acute promyelocytic leukemia cells through a defective pathway in telomerase regulation. *Leukemia* **16**, 826–832.

Quenech'Du, N., Ruchaud, S., Khelef, N., Guiso, N., and Lanotte, M. (1998). A sustained increase in the endogenous level of cAMP reduces the retinoid concentration required for APL cell maturation to near physiological levels. *Leukemia* **12**, 1829–1833.

Redner, R. L., Rush, E. A., Faas, S., Rudert, W. A., and Corey, S. J. (1996). The t(5;17) variant of acute promyelocytic leukemia expresses a nucleophosmin-retinoic acid receptor fusion. *Blood* **87**, 882–886.

Regad, T., and Chelbi-Alix, M. K. (2001). Role and fate of PML nuclear bodies in response to interferon and viral infections. *Oncogene* **20**, 7274–7286.

Robertson, K. A., Emami, B., and Collins, S. J. (1992). Retinoic acid-resistant HL-60R cells harbor a point mutation in the retinoic acid receptor ligand-binding domain that confers dominant negative activity. *Blood* **80**, 1885–1889.

Rosenfeld, M. G., and Glass, C. K. (2001). Coregulator codes of transcriptional regulation by nuclear receptors. *J Biol Chem* **276**, 36865–36868.

Rowley, J. D., Golomb, H. M., and Dougherty, C. (1977). 15/17 translocation, a consistent chromosomal change in acute promyelocytic leukaemia. *Lancet* **1**, 549–550.

Ruchaud, S., Duprez, E., Gendron, M. C., Houge, G., Genieser, H. G., Jastorff, B., Doskeland, S. O., and Lanotte, M. (1994). Two distinctly regulated events, priming and triggering, during retinoid-induced maturation and resistance of NB4 promyelocytic leukemia cell line. *Proc Natl Acad Sci USA* **91**, 8428–8432.

Seeler, J. S., and Dejean, A. (2001). SUMO: Of branched proteins and nuclear bodies. *Oncogene* **20**, 7243–7249.

Shao, W., Fanelli, M., Ferrara, F. F., Riccioni, R., Rosenauer, A., Davison, K., Lamph, W. W., Waxman, S., Pelicci, P. G., Lo Coco, F., Avvisati, G., Testa, U., Peschle, C., Gambacorti-Passerini, C., Nervi, C., and Miller, W. H., Jr. (1998). Arsenic trioxide as an inducer of apoptosis and loss of PML/RAR alpha protein in acute promyelocytic leukemia cells. *J Natl Cancer Inst* **90**, 124–133.

Shen, Y., Shen, Z. X., Yan, H., Chen, J., Zeng, X. Y., Li, J. M., Li, X. S., Wu, W., Xiong, S. M., Zhao, W. L., Tang, W., Wu, F., Liu, Y. F., Niu, C., Wang, Z. Y., Chen, S. J., and Chen, Z. (2001). Studies on the clinical efficacy and pharmacokinetics of low-dose arsenic trioxide in the treatment of relapsed acute promyelocytic leukemia: A comparison with conventional dosage. *Leukemia* **15**, 735–741.

Shen, Z. X., Chen, G. Q., Ni, J. H., Li, X. S., Xiong, S. M., Qiu, Q. Y., Zhu, J., Tang, W., Sun, G. L., Yang, K. Q., Chen, Y., Zhou, L., Fang, Z. W., Wang, Y. T., Ma, J., Zhang, P., Zhang, T. D., Chen, S. J., Chen, Z., and Wang, Z. Y. (1997). Use of arsenic trioxide (As_2O_3) in the treatment of acute promyelocytic leukemia (APL): II. Clinical efficacy and pharmacokinetics in relapsed patients. *Blood* **89**, 3354–3360.

Shen, Z. X., Shi, Z. Z., Fang, J., Gu, B. W., Li, J. M., Zhu, Y. M., Shi, J. Y., Zheng, P. Z., Yan, H., Liu, Y. F., Chen, Y., Shen, Y., Wu, W., Tang, W., Waxman, S., De The, H., Wang, Z. Y., Chen, S. J., and Chen, Z. (2004). All-*trans* retinoic acid/As_2O_3 combination yields a high quality remission and survival in newly diagnosed acute promyelocytic leukemia. *Proc Natl Acad Sci USA* **101**, 5328–5335.

Soignet, S. L., Maslak, P., Wang, Z. G., Jhanwar, S., Calleja, E., Dardashti, L. J., Corso, D., DeBlasio, A., Gabrilove, J., Scheinberg, D. A., Pandolfi, P. P., and Warrell, R. P., Jr. (1998). Complete remission after treatment of acute promyelocytic leukemia with arsenic trioxide. *N Engl J Med* **339**, 1341–1348.

Stadler, M., Chelbi-Alix, M. K., Koken, M. H., Venturini, L., Lee, C., Saib, A., Quignon, F., Pelicano, L., Guillemin, M. C., Schindler, C., and de The, H. (1995). Transcriptional induction of the PML growth suppressor gene by interferons is mediated through an ISRE and a GAS element. *Oncogene* **11**, 2565–2573.

Sun, H. D., Ma, L., Hu, X. C., and Zhang, T. D. (1992). Ai-Lin 1 treated 32 cases of acute promyelocytic leukemia. *Chin. J. Integrat. Chin. West. Med* **12**, 170–171.

Tong, J. H., Dong, S., Geng, J. P., Huang, W., Wang, Z. Y., Sun, G. L., Chen, S. J., Chen, Z., Larsen, C. J., and Berger, R. (1992). Molecular rearrangements of the MYL gene in acute promyelocytic leukemia (APL, M3) define a breakpoint cluster region as well as some molecular variants. *Oncogene* **7**, 311–316.

Wang, Z. G., Delva, L., Gaboli, M., Rivi, R., Giorgio, M., Cordon-Cardo, C., Grosveld, F., and Pandolfi, P. P. (1998a). Role of PML in cell growth and the retinoic acid pathway. *Science* **279**, 1547–1551.

Wang, Z. G., Ruggero, D., Ronchetti, S., Zhong, S., Gaboli, M., Rivi, R., and Pandolfi, P. P. (1998b). PML is essential for multiple apoptotic pathways. *Nat Genet* **20**, 266–272.

Wells, R. A., Catzavelos, C., and Kamel-Reid, S. (1997). Fusion of retinoic acid receptor alpha to NuMA, the nuclear mitotic apparatus protein, by a variant translocation in acute promyelocytic leukaemia. *Nat Genet* **17**, 109–113.

Xiao, S., Li, D., Zhu, H. Q., Song, M. G., Pan, X. R., Jia, P. M., Peng, L. L., Dou, A. X., Chen, G. Q., Chen, S. J., Chen, Z., and Tong, J. H. (2006). Rig-G as a key mediator of the antiproliferative activity of interferon-related pathways through enhancing p21 and p27 proteins. *PNAS* **103**, 16448–16453.

Yoshida, H., Kitamura, K., Tanaka, K., Omura, S., Miyazaki, T., Hachiya, T., Ohno, R., and Naoe, T. (1996). Accelerated degradation of PML-retinoic acid receptor alpha (PML-RARA) oncoprotein by all-*trans*-retinoic acid in acute promyelocytic leukemia: Possible role of the proteasome pathway. *Cancer Res* **56**, 2945–2948.

Yu, M., Tong, J. H., Mao, M., Kan, L. X., Liu, M. M., Sun, Y. W., Fu, G., Jing, Y. K., Yu, L., Lepaslier, D., Lanotte, M., Wang, Z. Y., Chen, Z., Waxman, S., Wang, Y. X., Tan, J. Z., and Chen, S. J. (1997). Cloning of a gene (RIG-G) associated with retinoic acid-induced differentiation of acute promyelocytic leukemia cells and representing a new member of a family of interferon-stimulated genes. *Proc Natl Acad Sci USA* **94**, 7406–7411.

Zelent, A., Guidez, F., Melnick, A., Waxman, S., and Licht, J. D. (2001). Translocations of the RARalpha gene in acute promyelocytic leukemia. *Oncogene* **20**, 7186–7203.

Zhang, T. D., Chen, G. Q., Wang, Z. G., Wang, Z. Y., Chen, S. J., and Chen, Z. (2001). Arsenic trioxide, a therapeutic agent for APL. *Oncogene* **20**, 7146–7153.

Zhao, Q., Tao, J., Zhu, Q., Jia, P. M., Dou, A. X., Li, X., Cheng, F., Waxman, S., Chen, G. Q., Chen, S. J., Lanotte, M., Chen, Z., and Tong, J. H. (2004). Rapid induction of cAMP/PKA pathway during retinoic acid-induced acute promyelocytic leukemia cell differentiation. *Leukemia* **18**, 285–292.

Zheng, P. Z., Wang, K. K., Zhang, Q. Y., Huang, Q. H., Du, Y. Z., Zhang, Q. H., Xiao, D. K., Shen, S. H., Imbeaud, S., Eveno, E., Zhao, C. J., Chen, Y. L., Fan, H. Y., Waxman, S., Auffray, C., Jin, G., Chen, S. J., Chen, Z., and Zhang, J. (2005). Systems analysis of transcriptome and proteome in retinoic acid/arsenic trioxide-induced cell differentiation/apoptosis of promyelocytic leukemia. *Proc Natl Acad Sci USA* **102**, 7653–7658.

Zhong, S., Muller, S., Ronchetti, S., Freemont, P. S., Dejean, A., and Pandolfi, P. P. (2000a). Role of SUMO-1-modified PML in nuclear body formation. *Blood* **95**, 2748–2752.

Zhong, S., Salomoni, P., Ronchetti, S., Guo, A., Ruggero, D., and Pandolfi, P. P. (2000b). Promyelocytic leukemia protein (PML) and Daxx participate in a novel nuclear pathway for apoptosis. *J Exp Med* **191**, 631–640.

Zhou, G. B., Zhao, W. L., Wang, Z. Y., Chen, S. J., and Chen, Z. (2005). Retinoic acid and arsenic for treating acute promyelocytic leukemia. *PLoS Med* **2**,.0035–0038.

Zhu, J., Koken, M. H., Quignon, F., Chelbi-Alix, M. K., Degos, L., Wang, Z. Y., Chen, Z., and de The, H. (1997). Arsenic-induced PML targeting onto nuclear bodies: Implications for the treatment of acute promyelocytic leukemia. *Proc Natl Acad Sci USA* **94**, 3978–3983.

Zhu, J., Gianni, M., Kopf, E., Honore, N., Chelbi-Alix, M., Koken, M., Quignon, F., Rochette-Egly, C., and de The, H. (1999a). Retinoic acid induces proteasome-dependent degradation of retinoic acid receptor alpha (RARalpha) and oncogenic RARalpha fusion proteins. *Proc Natl Acad Sci USA* **96**, 14807–14812.

Zhu, J., Lallemand-Breitenbach, V., and de The, H. (2001). Pathways of retinoic acid- or arsenic trioxide-induced PML/RARalpha catabolism, role of oncogene degradation in disease remission. *Oncogene* **20**, 7257–7265.

Zhu, J., Chen, Z., Lallemand-Breitenbach, V., and de The, H. (2002a). How acute promyelocytic leukaemia revived arsenic. *Nat Rev Cancer* **2**, 705–713.

Zhu, Q., Zhang, J. W., Zhu, H. Q., Shen, Y. L., Flexor, M., Jia, P. M., Yu, Y., Cai, X., Waxman, S., Lanotte, M., Chen, S. J., Chen, Z., and Tong, J. H. (2002b). Synergic effects of arsenic trioxide and cAMP during acute promyelocytic leukemia cell maturation subtends a novel signaling cross-talk. *Blood* **99**, 1014–1022.

Zhu, X. H., Shen, Y. L., Jing, Y. K., Cai, X., Jia, P. M., Huang, Y., Tang, W., Shi, G. Y., Sun, Y. P., Dai, J., Wang, Z. Y., Chen, S. J., Zhang, T. D., Waxman, S., Chen, Z., and Chen, G. Q. (1999b). Apoptosis and growth inhibition in malignant lymphocytes after treatment with arsenic trioxide at clinically achievable concentrations. *J Natl Cancer Inst* **91**, 772–778.

CHAPTER 16

2-CHLORODEOXYADENOSINE (CLADRIBINE): RATIONAL DEVELOPMENT OF A NOVEL CHEMOTHERAPEUTIC AGENT

Howard B. Cottam and Dennis A. Carson[*]

16.1 INTRODUCTION

The discovery and development of effective chemotherapeutic agents to treat patients with cancer have been major endeavors of academic, government, and industrial laboratories for more than 50 years. Without question, the greatest progress to date in chemotherapy has been made against the leukemias and lymphomas, although significant advances are being made in the solid tumor areas as well. The ultimate goal of treatment approaches in cancer chemotherapy is the elimination of cancer cells while sparing normal cells. A better understanding of how fundamental cancer biology can be exploited to effectively eliminate cancer cells has emerged with the concept that cancer is more complex than simply a disorder of cell growth regulation, but involves the regulation of cell death and critical factors in the tumor–host microenvironment. The majority of drugs developed for cancer chemotherapy have been DNA damaging agents or those that inhibit cell division and consequently work best in rapidly dividing

[*]Corresponding author: dcarson@ucsd.edu

Drug Discovery Research: New Frontiers in the Post-Genomic Era, Edited by Ziwei Huang

cancers. However, many tumors have large fractions of cells that are not dividing for extended periods and are thus difficult to treat selectively with these cell-cycle-specific agents. In these tumor types, the most effective agents will be those that can destroy the cancer cells whether they are actively dividing or resting. The successful development of such a drug, 2-chlorodeoxyadenosine (2CDA) for lymphoproliferative diseases, is a specific example of this concept and is the subject of this review.

The discovery of 2CDA as a useful chemotherapeutic agent is unique in that it was not the result of the random screening of large numbers of candidate compounds or high-throughput screening of compound collections such as combinatorial libraries or a member of a large virtual screening database of compounds. Rather, it was the product of a rational attempt to exploit the biochemical basis for lymphocyte depletion in a rare immunodeficiency disease in children caused by the functional lack of the enzyme adenosine deaminase (ADA). The development of 2CDA resulted from the insights into the mechanism by which ADA deficiency leads to immunodeficiency (Carson et al., 1977). The key observation was that children deficient in ADA due to an inborn error of metabolism suffered from severe combined immunodeficiency disease (Giblett et al., 1972; Ochs et al., 1973) and that this was caused by an accumulation of 2′-deoxyadenosine triphosphate (dATP) to levels that proved to be lethal to lymphocytes (Cohen et al., 1978). This lymphospecific toxicity in ADA deficiency results from the high deoxycytidine kinase (dCK) to 5′-nucleotidase ratio found in lymphocytes that favors the production and accumulation of the phosphorylated forms of deoxyadenosine (Carson et al., 1983; Kawasaki et al., 1993). Thus, the possibility became attractive that one might exploit this newfound sensitivity of lymphocytes to dATP exposure for applications in lymphoproliferative and autoimmune diseases. This insight served as a foundation for two different therapeutic strategies: (1) direct pharmacological inhibition of ADA and (2) use of an analogue of deoxyadenosine that might mimic the lethal effects observed in ADA deficiency. An example of the first strategy is the development of ADA inhibitors such as deoxycoformycin, which has met with limited success (Escoda et al., 1990; Keating 1993). The second strategy requires a consideration of the structure–activity aspects of possible candidate compounds in that they should be (1) a close analogue of deoxyadenosine and serve as a substrate for activation, (2) stable to degradation by functional ADA, and (3) stable to phosphorolysis by cellular purine nucleoside phosphorylase. 2CDA was found to be an excellent candidate in every respect. Thus, the simple replacement of the hydrogen by a chlorine atom at the 2-position of the adenine ring in 2CDA renders this nucleoside resistant to degradation by ADA while maintaining its substrate activity for phosphorylation by deoxycytidine kinase. It was this ADA resistance that initially qualified 2CDA to be included with other nucleoside analogues, known to be deaminase-resistant, for evaluation *in vitro* against a number of malignant cell lines and *in vivo* against L1210 in mice in hopes of finding analogues more potent and selective than the endogenous dATP. 2CDA was thus identified as the most potent among the initial group of 25 deoxyadenosine analogues tested (Carson et al., 1980).

16.2 CHEMISTRY AND METHODS OF PREPARATION

2CDA is 2′-deoxyadenosine that bears a chlorine atom at the 2-position (Fig. 16.1). 2CDA has been prepared by a number of methods involving (1) transformation of preformed nucleosides and (2) glycosylation procedures, both chemical and enzymatic. A brief description of these methods is presented here.

16.2.1 Classical Synthetic Methods

2CDA was first reported in 1963 as an intermediate in the synthesis of 2′-deoxyadenosine via a "cyclonucleoside" of adenosine (Ikehara and Tada 1963, 1965), an example of transformation of a preformed nucleoside. Early glycosylation procedures used in deoxyribonucleoside synthesis include the coupling under fusion conditions of a suitably protected deoxyribose with a versatile aglycone, as well as solution phase procedures (Ts'o, 1974; Walker et al., 1979). The first reported synthesis of 2CDA utilizing a glycosylation strategy was accomplished via fusion coupling of 2,6-dichloropurine and 1,3,5-tri-*O*-acetyl-2-deoxy-D-*erythro*-pentofuranose followed by amination at the 6-position with concomitant deprotection of the sugar ester groups as shown in Fig. 16.2 (Christensen et al., 1972). This glycosylation procedure is not stereospecific and unavoidably results in the formation of anomers of both α- and β-configuration, which must be separated to provide the desired β-anomeric product. Thus, the yield of the N-9 β-anomer is low, usually less than 20% overall for both steps.

16.2.2 Modern Stereospecific Methods

In view of the difficulties encountered by the formation of anomeric mixtures in the classical glycosylation procedures, we developed a stereospecific method in which 1-chloro-2-deoxy-3,5-di-*O*-toluoyl-D-*erythro*-pentofuranose (Hoffer, 1960) is reacted with the sodium salt of the chloroheterocycle, in this case 2,6-dichloropurine, resulting in the formation of the desired *N*-9 β-isomeric product in 59% yield (Kazimierczuk et al., 1984) (Fig. 16.3). No α-anomeric product was observed. This

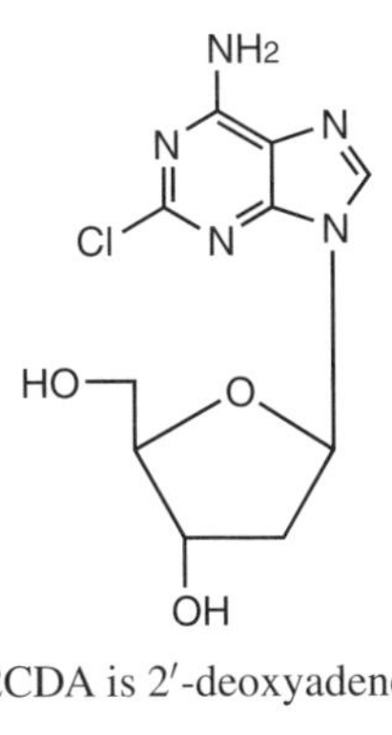

Figure 16.1 Structure of 2CDA. 2CDA is 2′-deoxyadenosine substituted with a chlorine at the 2-position.

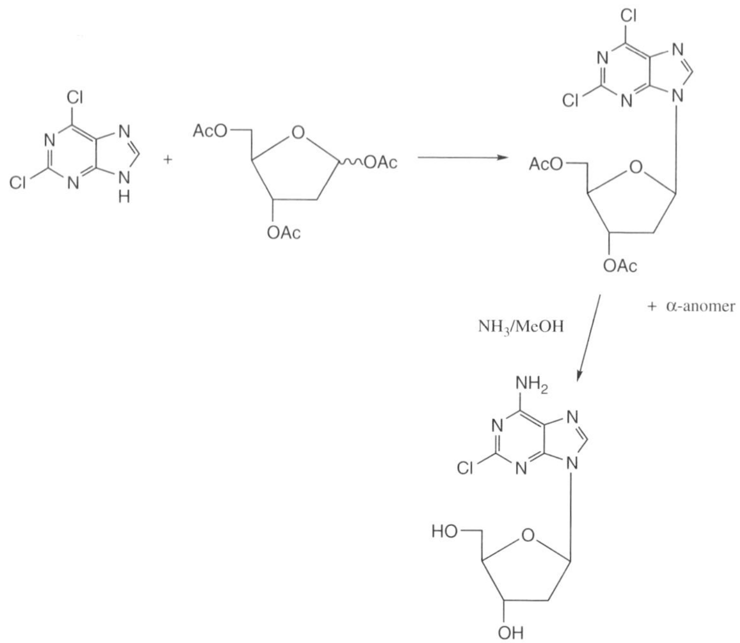

Figure 16.2 Classical preparation of 2CDA starting with 2,6-dichloropurine and fusing with esterified 2-deoxy-D-ribose, followed by amination-deprotection with methanolic ammonia.

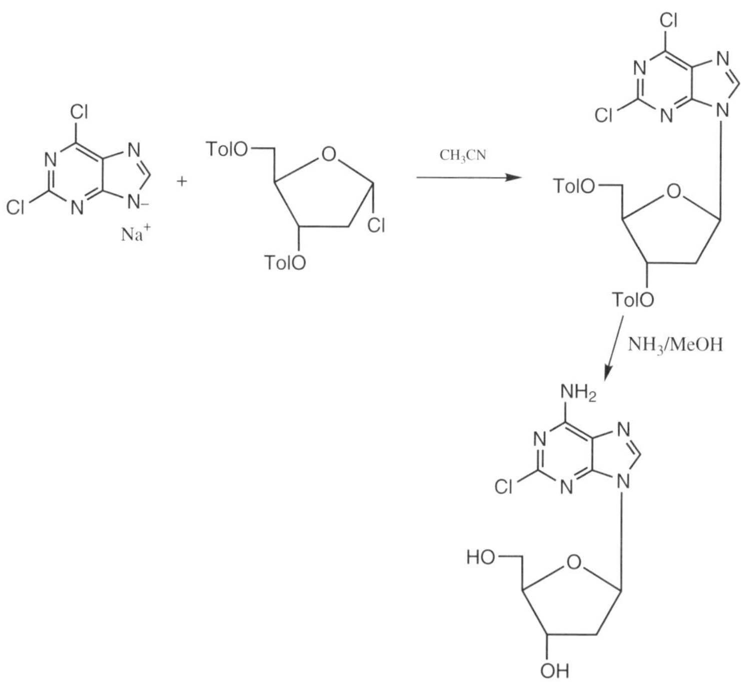

Figure 16.3 Modern stereospecific preparation of 2CDA utilizing the sodium salt of 2,6-dichloropurine and a 1-α-chloro-esterified 2-deoxy-D-ribose in an S_N2 reaction.

stereospecificity was achieved due to the fact that the starting chloro sugar has the α-configuration in the solid state, and the exclusive formation of the protected β-nucleosides is viewed as the result of a direct Walden inversion (S_N2) at the C_1 carbon by the anionic heterocyclic nitrogen. Thus, the incoming heterocycle approaches from the β-face of the sugar as the chloro group is leaving from the α-face, resulting in the nucleoside product having the desired β-configuration. This strategy is still used today for the manufacture of clinical grade 2CDA.

Variations of this sodium salt procedure have recently been used to prepare 2CDA under phase transfer conditions (Kazimierczuk and Kaminski, 2000) and from 2-choroadenine directly (Gerszberg and Alonso, 2000). Other recent methods include the transformation of 2′-deoxyguanosine (Janeba et al., 2003) and heterocyclic modifications to avoid the formation of unwanted N-7 isomeric product (Gupta and Munk, 2004).

16.2.3 Enzymatic Methods

The first human clinical trials were conducted using 2CDA that had been prepared by enzymatic methods. Thus, Carson et al. (1980) prepared 2CDA by transglycosylation using a bacterial (*Lactobacillus helviticus*) enzyme with 2-chloroadenine as a substrate and thymidine as the deoxyribose donor. Later, other enzymatic transglycosylations were reported, for example, with 2-chloroadenine as substrate and using whole cells of *E. coli* BMT-1D/1A as the biocatalyst and 2′-deoxyguanosine as the glycosyl donor (Mikhailopulo et al., 1993), as well as using 2′-deoxyadenosine as a co-substrate with 2-chloroadenosine in an *E. coli* system (Blank et al., 1992). In addition, a purine nucleoside phosphorylase enzyme system that utilizes 1-phosphorylated sugars and nucleobase derivatives has been reported (Komatsu et al., 2001).

16.3 PHARMACOLOGY

Although 2CDA was originally described in 1963, no report of its biological properties appeared until 1970 (Simon et al., 1970), wherein Robins and co-workers showed it to be a weak inhibitor of calf intestine ADA. As mentioned, the presence of a chlorine atom at the 2-position prevents 2CDA from being deaminated by ADA. In addition, 2CDA is not a substrate for mammalian purine nucleoside phosphorylase (Bzowska and Kazimierczuk, 1995), the enzyme that catalyzes the cleavage of the sugar moiety from the nucleobase. This resistance to enzyme-catalyzed degradation thus allows for the subsequent formation of cellular intermediates essential for biological activity.

16.3.1 Mechanism of Drug Activation

Following transport into the cell, 2CDA must be phosphorylated eventually to the triphosphate form to be biologically active. Phosphorylation of 2CDA to its monophosphate is mediated by deoxycytidine kinase (Carson et al., 1980) and mitochondrial deoxyguanosine kinase (Arner, 1996) while dephosphorylation is catalyzed by cytosolic 5′-nucleotidase. Cells with a high dCK to 5′-nucleotidase ratio are particularly sensitive to deoxyadenosine analogue toxicity, including 2CDA and fludarabine, while cells with lower ratios are resistant to these agents (Dumontet et al., 1999). Transport of 2CDA into the cell by one of several specific nucleoside transporter systems, although not strictly an activation mechanism, is essential for subsequent activation to take place (Beck et al., 2000). Nucleoside permeation across mammalian cell membranes is complex with at least four distinct transporters known. Two of these (es and ei) are equilibrative (facilitated diffusion) carriers that have been studied in considerable detail. The other two (cif and cit) are concentrative, Na^+-dependent carriers. Studies appear to indicate that 2CDA is transported into the cell by a concentrative cif carrier while the equilibrative es carrier is responsible for efflux of the drug (Crawford et al., 1990).

16.3.2 Mechanism of Action

The cytotoxicity of 2CDA mainly depends on the accumulation of 2CDA-triphosphate after phosphorylation of 2CDA-monophosphate by nucleoside monophosphate kinase and nucleoside diphosphate kinase. The triphosphate form of 2CDA has a number of activities. In dividing cells, it inhibits ribonucleotide reductase (Griffig et al., 1989) and DNA polymerase α. It can be incorporated into DNA resulting in impairment of DNA synthesis (Chunduru et al., 1993). In resting cells, DNA single strand breaks occur and accumulate since cells are unable to properly repair their DNA due to the inhibition of DNA polymerization and ligation by excess 2CDA-triphosphate (Seto et al., 1985). The enzyme poly (ADP-ribose) polymerase is activated, which then results in depletion of NAD and ATP pools and disruption of cellular metabolism. At the same time, Bcl-2 (the oncogene known to enhance cell survival) is down-regulated (Petersen et al., 1996). The scene is thus set for cell death by apoptosis. The molecular details of the 2CDA-mediated apoptotic process in chronic lymphocytic leukemia (CLL) cells have recently been reported from our laboratory (Leoni et al., 1998; Genini et al., 2000). A scheme depicting the activation of the apoptotic pathway by deoxyadenosines such as 2CDA is shown in Fig. 16.4.

The deoxynucleotide analogue, in this case 2CDA, enters the cell and is converted progressively to its active 2CDATP. The 2CDATP causes DNA strand break formation, activating poly (ADP ribose) polymerase and p53, with resultant depletion of NAD and adenine nucleotides and a concomitant increase in oxidative stress. In untreated cells, the concentrations of ATP and cytochrome *c* in the cytoplasm are insufficient to trigger the caspase cascade. However, the binding of APAF-1 to caspase-9 in the presence of 2CDATP and small amounts of released cytochrome *c* leads to the cleavage of caspase-3, converting it to an active autocatalytic protease in a process analogous to the blood-clotting and complement cascades. The active caspase-3 then stimulates, in turn, the caspase-activated DNase (CAD) endonuclease, which irreversibly degrades DNA. Moreover, we found that ADP appears to act as a physiological intracellular inhibitor of the cytochrome *c* and APAF-1-mediated caspase pathway in both normal and CLL cells (Genini et al., 2000).

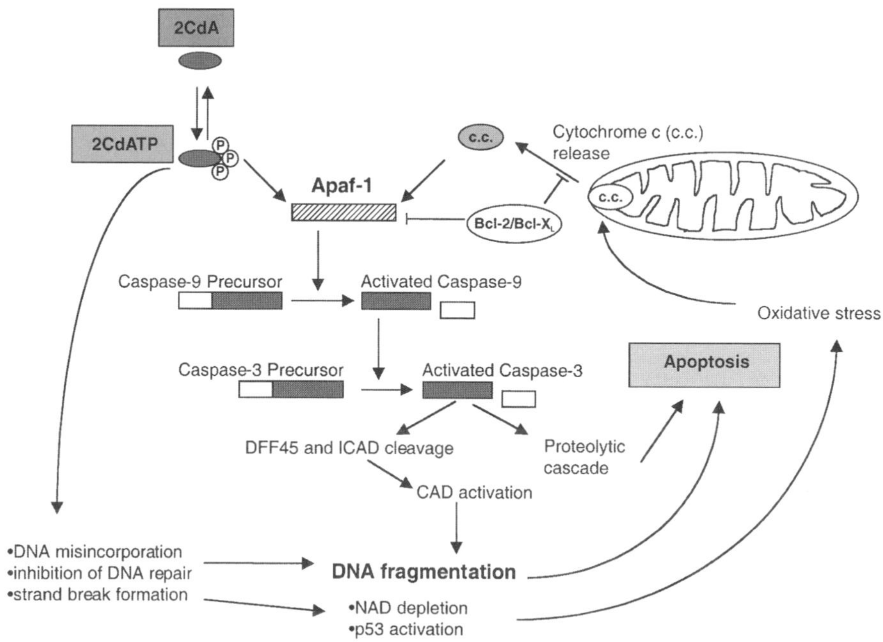

Figure 16.4 Mechanism of action of 2CDA: Activation of the apoptotic pathway. 2CDA enters the cell and is converted to its active 2CDATP. The 2CDATP causes DNA strand break formation, activating poly(ADP ribose) polymerase and p53, with resultant depletion of NAD and adenine nucleotides and a concomitant increase in oxidative stress. In untreated cells, the concentrations of ATP and cytochrome *c* in the cytoplasm are insufficient to trigger the caspase cascade. However, the binding of APAF-1 to caspase-9 in the presence of 2CDATP and small amounts of released cytochrome *c* leads to the cleavage of caspase-3, converting it to an active autocatalytic protease. The active caspase-3 then stimulates, in turn, the caspase-activated DNase (CAD) endonuclease that irreversibly degrades DNA.

Thus, 2CDA is distinguished from other chemotherapeutic agents that affect purine metabolism in that it is cytotoxic to both actively dividing and quiescent lymphocytes and monocytes, inhibiting both DNA synthesis and repair.

16.3.3 Preclinical Evaluation

Preclinical studies showed that both B- and T-lymphoblastoid cell lines were sensitive to 2CDA and that prolonged exposure of resting peripheral blood lymphocytes to this agent in vitro resulted in greater lymphotoxicity than did brief incubations (Carson et al., 1984). This observation led to the selection of a continuous intravenous infusion schedule for the initial clinical trials of 2CDA. Reviews of the preclinical and clinical pharmacokinetic studies of 2CDA using various schedules and routes of administration have appeared (Saven and Piro, 1994; Liliemark, 1997). The pharmacokinetics of 2CDA best fit a two-compartment, first-order-elimination model. Recently, detailed *in vivo* metabolism studies of 2CDA in mice, monkeys, and humans have been reported (Wu et al., 2004).

16.4 CLINICAL ACTIVITY

Relative to traditional antimetabolites, 2CDA is unique in that it is equally active against dividing and resting lymphocytes, which is essential for the observed activity of 2CDA in indolent lymphoproliferative disorders. However, 2CDA has also shown significant activity in certain solid tumors and autoimmune conditions.

16.4.1 Hematologic Malignancies

Of the lymphoproliferative conditions that have been treated with 2CDA, hairy cell leukemia (HCL) has shown the most dramatic response. It is administered over a 7-day period by intravenous infusion at a dose of 0.09 mg/kg. Overall response rates in clinical studies have ranged from 80% to 100%, with a large majority of these being complete remissions. 2CDA is approved by the U.S. FDA for the treatment of active HCL and is indicated for symptomatic patients, those with significant neutropenia, leukocytosis, anemia, thrombocytopenia, splenomegaly, constitutional symptoms due to HCL, recurrent serious infections, and painful lymphadenopathy. A review of the guidelines for clinical use of 2CDA in HCL has recently appeared (Goodman et al., 2003). 2CDA also has activity in CLL (Juliusson et al., 1992; Saven, 1995), non-Hodgkin lymphoma (Kay et al., 1992; Hickish et al., 1993), Waldenstrom macroglobulinemia (Dimopoulos et al., 1993), cutaneous T-cell lymphoma (Saven et al., 1992), acute myelogenous leukemia (AML) (Carson et al., 1984; Santana et al., 1992), and chronic myelogenous leukemia (CML) (Beutler, 1993; Saven et al., 1994).

A particularly difficult-to-treat lymphoproliferative disorder is CLL, a disease caused by the proliferation and accumulation of B lymphocytes. It is the most common form of leukemia in North America and Europe. The alkylating agent chlorambucil, with or without prednisone, is the standard initial treatment (Foon et al., 1990). 2CDA has also been used as initial therapy for CLL patients. Patients who received purine nucleoside analogues, such as 2CDA or fludarabine, as their initial therapy and achieved long-lasting response can be successfully retreated with the same agent. Purine nucleoside analogues administered in combination with other chemotherapeutic agents and/or monoclonal antibodies may produce higher response rates, including complete response, compared with purine nucleoside analogs alone or other treatment regimens (Robak, 2005a,b). A study of the activity of 2CDA was conducted in 90 CLL patients, mostly Binet stage C disease who had failed alkylator therapy, and resulted in 4% complete responses and 40% partial responses (Saven, 1996). Here, the ratio of dCK to cytoplasmic 5′-nucleotidase was predictive of 2CDA responsiveness (Kawasaki et al., 1993). In combination studies in fludarabine-resistant CLL patients, pretreatment with bryostatin 1 has been shown to potentiate the efficacy of 2CDA by increasing the ratio of dCK to 5′-nucleotidase activity (Ahmad et al., 2000; Beck et al., 2004). However, a study in 20 previously untreated CLL patients who were administered 2CDA as a 7-day continuous infusion for four courses achieved 25% complete responses and 60% partial responses for an overall response rate of 85% (Saven et al., 1995). In another study, oral 2CDA was evaluated in 126 previously untreated CLL patients and compared with 40 previously treated patients (Karlsson et al., 2002). The overall responses were 77% and 39% in these groups, respectively. Interestingly, in these studies the achievement of complete remission was not a prerequisite for long-term survival. Combination preclinical and clinical studies using 2CDA with other cytotoxic agents and/or monoclonal antibodies show synergistic actions (Robak et al., 2005a,b). These and other studies establish the significant activity of 2CDA, used alone or in combination with other agents, in patients with previously treated and untreated CLL (Robak, 2005a,b).

Patients with low-grade non-Hodgkin lymphoma have been responsive to 2CDA as a single agent (Kay et al., 1992; Hickish et al., 1993) and in combination with mitoxantrone (Saven et al., 1996; Rummel et al., 2002) or cyclophosphamide (Van Den Neste et al., 2000; Robak et al., 2001). 2CDA is also active in advanced, untreated low-grade lymphomas, but myelosuppression is a cumulative adverse effect and limits the number of courses of treatment that can be given (Canfield et al., 1997).

In a study of 15 patients with cutaneous T-cell lymphoma treated with 2CDA after having failed standard topical and/or systemic therapy, the overall response rate was 47%. Three of 15 patients (20%) achieved complete responses, and four of 15 patients (27%) achieved partial responses (Saven et al., 1992). The median duration of response was 5 months.

Several studies conducted with 2CDA as a single agent in patients with Waldenstrom macroglobulinemia, a B-cell neoplastic condition, showed good overall response rates from 55% to 100%, with the higher response rates noted in the previously untreated patient groups (Dimopoulos et al., 1993; O'Brien et al., 1996; Fridrik et al., 1997; Liu et al., 1998).

Modest activity has been seen with 2CDA in myeloid leukemias. In CML, one study showed an overall response rate of 47% in patients previously treated with combination therapies (Dann et al., 1998). In another study, 2CDA was also active in chronic-phase CML, inducing complete hematologic responses in these patients, but the absence of cytogenetic responses and severe immunosuppression limit its clinical use (Saven et al., 1994). In AML, 2CDA has single-drug activity and may enhance the formation of the active metabolite (ara-CTP) of cytosine arabinoside (ara-C) in combination (Juliusson et al., 2003). In fact, this combination along with idarubicin showed an overall complete response rate of 62% with 30% long-term survival (>2 years). In relapsed/refractory AML, however, 2CDA is ineffective (Gordon et al., 2000). The mechanism of activity in myeloid leukemias is unclear since only low levels of dCK are expressed in granulocytes (Carson et al., 1980, 1983).

In other studies, 2CDA was devoid of activity in patients with multiple myeloma (Dimopoulos et al., 1992) but has shown promise in the treatment of Langerhans histiocytosis (Dimopoulos et al., 1997) and B-cell prolymphocytic leukemia, inducing high overall and complete response rates (Saven et al., 1997).

16.4.2 Solid Tumors

Because 2CDA and other similar purine nucleoside analogs are active antimetabolites in both proliferating and resting lymphoid cells, it was thought that these agents might have significant activity against certain solid tumors as well. A dose-escalation study of 2CDA in patients with various solid tumors was conducted to determine the maximum-tolerated dose (MTD) and define its toxicity profile at higher doses (Saven et al., 1993). In these studies, two of seven patients (28.6%) with malignant astrocytomas obtained partial responses with a median duration of 8 months. Interestingly, there was no clinical correlation with dCK expression and response to 2-CdA. Another phase 1 clinical trial was conducted to determine the MTD in 18 patients with solid tumors (non-small-cell lung and colorectal) given 2CDA by 1-hour intermittent infusion, repeated daily for 5 days. No responses were observed (Kobayashi et al., 1994). There is some *in vitro* evidence that 2CDA could be useful against hepatocellular carcinomas (Graziadei et al., 1998), but no clinical trials for this solid tumor type have been reported. With limited exceptions, 2CDA as a single agent has not shown significant activity in solid tumors.

16.4.3 Multiple Sclerosis and Other Nonmalignant Diseases

It is not surprising, given the antilymphocyte activity of 2CDA, that the use of this agent to suppress an inappropriate immune response has been investigated. Multiple sclerosis (MS) is characterized by clinical signs and symptoms of CNS demyelination. The disease can be chronic and progressive and have relapsing and remitting stages. Although MS is thought to be a lymphocyte-dependent autoimmune disease, the specific antigens and triggering agents involved in the disease are unknown. The rationale for the use of 2CDA in MS is that autoantigen-specific T-lymphocytes are thought to be activated peripherally before migrating to the CNS, where they mediate damage to myelin (Hafler and Weiner, 1989). This suggests that 2CDA-induced reduction in the number of lymphocytes may help slow the progression of the disease. The use of 2CDA in MS has been studied, and its potential use in this disease has been reviewed (Langtry and Lamb, 1998). In a 2-year placebo-controlled, double-blind, crossover study to evaluate 2CDA in the treatment of chronic progressive multiple sclerosis, analysis of the results revealed a favorable influence on the neurological performance scores in both the Kurtze extended disability status and the Scripps neurological rating scale and also on MRI findings in patients treated with 2CDA. In the first year, the most striking finding was that while clinical deterioration continued in the placebo-treated patients, the condition of patients who received 2CDA stabilized or even improved slightly (Beutler et al., 1996). In another study, a double-blind, placebo-controlled, randomized trial of 2CDA in relapsing-remitting MS was conducted in which 2CDA given by subcutaneous injection had a favorable effect on the frequency and severity of relapses and MRI findings. MRI-enhancing lesions were completely suppressed in the 2CDA patients by the sixth month of treatment (Romine et al., 1999). Thus, 2CDA has shown promise in the treatment of both progressive and relapsing-remitting MS.

In addition to MS, other nonmalignant disorders have responded to 2CDA treatment (Beutler et al., 1997a,b). Among them, autoimmune hemolytic anemia seems to respond to 2CDA in some patients but may be induced in others, usually in those with concomitant CLL after several courses of treatment (Robak et al., 1997), but has also been reported in some patients with Waldenstrom macroglobulinemia (Tetreault and Saven, 2000). The finding that human monocytes were as sensitive to 2CDA toxicity as lymphoid cells suggested that these analogs could offer a novel therapeutic strategy for chronic inflammatory and autoimmune diseases characterized by inappropriate monocyte deployment or function (Carrera et al., 1990). Rheumatoid arthritis (RA) and inflammatory bowel disease are two inflammatory diseases for which 2CDA therapy has shown some efficacy, although 2CDA is clearly not an anti-inflammatory agent. Combined analysis of three studies involving 23 patients with longstanding RA who were given either continuous or 4-hour infusions, or were given oral 2CDA, indicated that more than half of the patients had >20% improvement in joint swelling during 2CDA treatment (Beutler et al., 1997a,b). The effect of the drug on pain and tenderness was less dramatic. Collectively, these data support the hypothesis that 2CDA may retard the pannus formation in patients with longstanding RA. Whether or not 2CDA can decrease joint erosions in RA patients remains to be established.

Primary sclerosing cholangitis (PSC) is a chronic hepatic autoimmune disorder of unknown etiology, thought to be mediated by biliary autoreactive cytotoxic lymphocytes. A study was conducted in four patients with stages I and II PSC who were given 2CDA subcutaneously in an open-label pilot trial of 6-month duration and 2-year followup (Duchini et al., 2000). Patients had a quantifiable decrease in the hepatic inflammatory infiltrate on liver biopsy. No significant changes were found in symptom scores, liver panel tests, or cholangiograms. However, the drug was well-tolerated, and two of four patients reported remission of their inflammatory bowel disease symptoms.

Psoriasis and psoriatic arthritis are conditions for which an immune pathogenesis is indicated by the association of the disease with particular HLA types, by the dense lymphocyte infiltrates seen in both the skin and affected joints, and by beneficial effects of treatment with cyclosporine A and anti-CD4 antibodies (Bos et al., 1989). A study of 2CDA given orally to six patients weekly for 12 weeks followed by monthly maintenance therapy showed that after 6 months, four out of six patients had improved joint disease, and five out of six had improved psoriasis (Eibschutz et al., 1995).

Finally, because the only toxicity of 2CDA noted in the early studies was bone-marrow suppression, it has been used in preparing patients with lymphomas and leukemias for bone-marrow transplantation because destruction of the marrow in such patients is a part of the therapeutic aim (Beutler et al., 1997a,b).

16.5 CONCLUSIONS AND FUTURE DIRECTIONS

The development of the purine nucleoside analog 2CDA as a novel anticancer and immunosuppressive agent was accomplished by Carson and colleagues based on a greater understanding of the biochemical basis of a rare human immunological disease caused by an inborn error of metabolism. The result was the introduction into the clinic of an agent that could function by virtue of its activity in both actively dividing and resting lymphoid and myeloid cells. Thus, 2CDA and similar analogs have found great utility in the indolent hematological malignancies. Purine analogues such as 2CDA as single agents in the indolent lymphoproliferative diseases are likely to be used for many years. The extended role will be in combinations in both hematological malignancies and solid tumors. In addition, the immunosuppressive effect of the purine analogues will continue to lead to increased use of these drugs in preparative regimens for bone-marrow transplantation and perhaps use in enhancing cytotoxicity in association with preparative regimens in autologous and allogeneic bone-marrow transplants (Keating, 1997).

Newer, next-generation purine nucleoside analogues, such as the arabino-fluoro derivative of 2CDA (clofarabine), may have improved pharmacokinetic properties and exhibit broader single-agent antitumor activity than 2CDA (Carson et al., 1992; Faderl et al., 2005). Thus, seemingly minor changes in the structure of nucleoside analogues can result in major differences in pharmacokinetics, bioavailability, and biological activity. Future studies of 2CDA and the next-generation purine nucleoside analogues will include combination strategies of rational biochemical modulation with other agents such as cytarabine (ara-C).

REFERENCES

Ahmad, I., Al-Katib, A. M., et al. (2000). Sequential treatment of a resistant chronic lymphocytic leukemia patient with bryostatin 1 followed by 2-chlorodeoxyadenosine: case report. *Clin Cancer Res: Official J Am Assoc Cancer Res* **6**(4), 1328–1332.

Arner, E. S. (1996). On the phosphorylation of 2-chlorodeoxyadenosine (CdA) and its correlation with clinical response in leukemia treatment. *Leuk Lymphoma* **21**(3–4), 225–231.

Beck, F. W., Al-Katib, A. M., et al. (2000). Bryostatin 1-induced modulation of nucleoside transporters and 2-chlorodeoxyadenosine influx in WSU-CLL cells. *Int J Mol Med* **5**(4), 341–347.

Beck, F. W. J., Eilender, D. S., et al. (2004). Factors affecting bryostatin 1-enhanced 2-CdA cytotoxicity in resistant B-cell chronic lymphocytic leukemia. *Int J Mol Med* **14**(1), 113–119.

Beutler, E. (1993). Use of 2-halo adenosine derivatives as therapeutic agents against chronic myelogenous leukemia. *PCT International Application Wo*, Scripps Research Institute, USA, 35 pp.

Beutler, E., Sipe, J. C., et al. (1996). The treatment of chronic progressive multiple sclerosis with cladribine. *Proc Nat Acad Sci USA* **93**(4), 1716–1720.

Beutler, E., Carson, D., et al. (1997a). 2-Chlorodeoxyadenosine in nonmalignant disorders. *Basic Clin Oncol* **12**(Nucleoside Analogs in Cancer Therapy), 359–382.

Beutler, E., McMillan, R., et al. (1997b). Development of 2-chlorodeoxyadenosine (2-CdA) as a chemotherapeutic and immunosuppressive drug. *Basic Clin Oncol* **12**(Nucleoside Analogs in Cancer Therapy), 159–173.

Blank, W. A., Elder, K. J., et al. (1992). Synthesis of 2-chloro-2′-deoxyadenosine by washed cells of *E. coli*. *Biotechnol Lett* **14**(8), 669–672.

Bos, J. D., Meinardi, M. M., et al. (1989). Use of cyclosporin in psoriasis. *Lancet* **2**(8678–8679), 1500–1502.

Bzowska, A., and Kazimierczuk, Z. (1995). 2-Chloro-2′-deoxyadenosine (cladribine) and its analogs are good substrates and potent selective inhibitors of Escherichia coli purine-nucleoside phosphorylase. *Eur J Biochem* **233**(3), 886–890.

Canfield, V. A., Vose, J., et al. (1997). 2-Chlorodeoxyadenosine as initial therapy for advanced low grade lymphomas. *Leuk Lymphoma* **24**(3–4), 335–339.

Carrera, C. J., Terai, C., et al. (1990). Potent toxicity of 2-chlorodeoxyadenosine toward human monocytes *in vitro* and *in vivo*. A novel approach to immunosuppressive therapy. *Journal Clin Invest* **86**(5), 1480–1408.

Carson, D. A., Kaye, J., et al. (1977). Lymphospecific toxicity in adenosine deaminase deficiency and purine nucleoside phosphorylase deficiency: Possible role of nucleoside kinase(s). *Proc Nat Acad Sci USA* **74**(12), 5677–5681.

Carson, D. A., Wasson, D. B., et al. (1980). Deoxycytidine kinase-mediated toxicity of deoxyadenosine analogs toward malignant human lymphoblasts *in vitro* and toward murine L1210 leukemia *in vivo*. *Proceedings of the National Academy of Sci USA* **77**(11), 6865–6869.

Carson, D. A., Wasson, D. B., et al. (1983). Specific toxicity of 2-chlorodeoxyadenosine toward resting and proliferating human lymphocytes. *Blood* **62**(4), 737–743.

Carson, D. A., Wasson, D. B., et al. (1984). Antileukemic and immunosuppressive activity of 2-chloro-2′-deoxyadenosine. *Proc Nat Acad Sci USA* **81**(7), 2232–2236.

Carson, D. A., Wasson, D. B., et al. (1992). Oral antilymphocyte activity and induction of apoptosis by 2-chloro-2′-arabino-fluoro-2′-deoxyadenosine. *Proc Nat Acad Sci USA* **89**(7), 2970–2974.

Christensen, L. F., Broom, A. D., et al. (1972). Synthesis and biological activity of selected 2,6-disubstituted(2-deoxy-a- and -b-D-erythro-pentofuranosyl)purines. *J Med Chem* **15**(7), 735–739.

Chunduru, S. K., Appleman, J. R., et al. (1993). Activity of human DNA polymerases alpha and beta with 2-chloro-2′-deoxyadenosine 5′-triphosphate as a substrate and quantitative effects of incorporation on chain extension. *Arch Biochem Biophys* **302**(1), 19–30.

Cohen, A., Hirschhorn, R., et al. (1978). Deoxyadenosine triphosphate as a potentially toxic metabolite in adenosine deaminase deficiency. *Proc Nat Acad Sci USA* **75**(1), 472–476.

Crawford, C. R., Ng, C. Y., et al. (1990). Isolation and characterization of an L1210 cell line retaining the sodium-dependent carrier cif as its sole nucleoside transport activity. *J Biol Chem* **265**(23), 13730–13734.

Dann, E. J., Anastasi, J., et al. (1998). High-dose cladribine therapy for chronic myelogenous leukemia in the accelerated or blast phase. *J Clin Oncol: Official J Am Soc Clin Oncol* **16**(4), 1498–1504.

Dimopoulos, M. A., Kantarjian, H., et al. (1993). Treatment of Waldenstrom macroglobulinemia with 2-chlorodeoxyadenosine. *Ann Internal Med* **118**(3), 195–198.

Dimopoulos, M. A., Kantarjian, H. M., et al. (1992). 2-Chlorodeoxyadenosine in the treatment of multiple myeloma. *Blood* **80**(6), 1626.

Dimopoulos, M. A., Theodorakis, M., et al. (1997). Treatment of Langerhans cell histiocytosis with 2 chlorodeoxyadenosine. *Leuk Lymphoma* **25**(1–2), 187–189.

Duchini, A., Younossi, Z. M., et al. (2000). An open-label pilot trial of cladribine (2-chlorodeoxyadenosine) in patients with primary sclerosing cholangitis. *J Clin Gastroenterol* **31**(4), 292–296.

Dumontet, C., Fabianowska-Majewska, K., et al. (1999). Common resistance mechanisms to deoxynucleoside analogs in variants of the human erythroleukemic line K562. *Br J Haematol* **106**(1), 78–85.

Eibschutz, B., Baird, S. M., et al. (1995). Oral 2-chlorodeoxyadenosine in psoriatic arthritis: A preliminary report. *Arthritis Rheum* **38**(11), 1604–1609.

Escoda, L., Lopez-Guillermo, A., et al. (1990). Treatment of various lymphoproliferative syndromes with deoxycoformycin: Results in 6 patients. *Sangre* **35**(6), 421–424.

Faderl, S., Gandhi, V., et al. (2005). The role of clofarabine in hematologic and solid malignancies-development of a next-generation nucleoside analog. *Cancer (New York, NY, United States)* **103**(10), 1985–1995.

Foon, K. A., Rai, K. R., et al. (1990). Chronic lymphocytic leukemia: New insights into biology and therapy. *Ann Intern Med* **113**(7), 525–539.

Fridrik, M. A., Jager, G., et al. (1997). First-line treatment of Waldenstrom's disease with cladribine. Arbeitsgemeinschaft Medikamentose Tumortherapie. *Ann Hematol* **74**(1), 7–10.

Genini, D., Adachi, S., et al. (2000). Deoxyadenosine analogs induce programmed cell death in chronic lymphocytic leukemia cells by damaging the DNA and by directly affecting the mitochondria. *Blood* **96**(10), 3537–3543.

Gerszberg, S., and Alonso, D. (2000). Method for the production of of antitumor 2-chloro-2′-deoxyadenosine (cladribine) and its 3,5-di-*o*-*p*-toluoyl derivative. *PCT International Application Wo*, Sterrenbeld Biotechnologie North America, Inc., USA, 22 pp.

Giblett, E. R., Anderson, J. E., et al. (1972). Adenosine-deaminase deficiency in two patients with severely impaired cellular immunity. *Lancet* **2**(7786), 1067–1069.

Goodman, G. R., Beutler, E., et al. (2003). Cladribine in the treatment of hairy-cell leukaemia. *Best Practice Res Clin Haematol* **16**(1), 101–116.

Gordon, M. S., Young, M. L., et al. (2000). Phase II trial of 2-chlorodeoxyadenosine in patients with relapsed/refractory acute myeloid leukemia: A study of the Eastern Cooperative Oncology Group (ECOG), E5995. *Leuk Res* **24**(10), 871–875.

Graziadei, I., Kelly, T., et al. (1998). Antitumor effect of the nucleoside analogs 2-chlorodeoxyadenosine and 2′,2′-difluorodeoxycytidine on human hepatoma HepG2 cells. *J Hepatol* **28**(3), 504–509.

Griffig, J., Koob, R., et al. (1989). Mechanisms of inhibition of DNA synthesis by 2-chlorodeoxyadenosine in human lymphoblastic cells. *Cancer Res* **49**(24 Pt 1), 6923–6928.

Gupta, P. K., and Munk, S. A. (2004). Process for the regioselective and stereoselective preparation of 9-b-anomeric nucleoside analogs as antitumor and antiviral prodrugs. *U.S. Patent Application Publication Us*, Ash Stevens, Inc., USA, 10 pp.

Hafler, D. A., and Weiner, H. L. (1989). MS: A CNS and systemic autoimmune disease. *Immunol Today* **10**(3), 104–107.

Hickish, T., Serafinowski, P., et al. (1993). 2′-Chlorodeoxyadenosine: Evaluation of a novel predominantly lymphocyte selective agent in lymphoid malignancies. *Br J Cancer* **67**(1), 139–143.

Hoffer, M. (1960). a-Thymidine. *Chem Berichte* **93**, 2777–2781.

Ikehara, M., and Tada, H. (1963). New type of cyclonucleoside derived from 2-chloro-8-mercapto-9-*b*-D-xylofuranosyladenine. *J Am Chem Society* **85**(15), 2344–2345.

Ikehara, M., and Tada, H. (1965). Nucleosides and nucleotides. XXIV. Purine cyclonucleosides. 1. 8,2′-Cyclonucleoside derived from 2-chloro-8-mercapto-9-*b*-D-xylofuranosyladenine. *J Am Chem Soc* **87**(3), 606–610.

Janeba, Z., Francom, P., et al. (2003). Efficient syntheses of 2-chloro-2′-deoxyadenosine (Cladribine) from 2′-deoxyguanosine. *J Org Chem* **68**(3), 989–992.

Juliusson, G., Elmhorn-Rosenborg, A., et al. (1992). Response to 2-chlorodeoxyadenosine in patients with B-cell chronic lymphocytic leukemia resistant to fludarabine. *N Engl J Med* **327**(15), 1056–1061.

Juliusson, G., Hoeglund, M., et al. (2003). Increased remissions from one course for intermediate-dose cytosine arabinoside and idarubicin in elderly acute myeloid leukaemia when combined with cladribine. a randomized population-based phase II study. *Br J Haematol* **123**(5), 810–818.

Karlsson, K., Stromberg, M., et al. (2002). Oral cladribine for B-cell chronic lymphocytic leukaemia: Report of a phase II trial with a 3-*d*, 3-weekly schedule in untreated and pretreated patients, and a long-term follow-up of 126 previously untreated patients. *Br J Haematol* **116**(3), 538–548.

Kawasaki, H., Carrera, C. J., et al. (1993). Relationship of deoxycytidine kinase and cytoplasmic 5′-nucleotidase to the chemotherapeutic efficacy of 2-chlorodeoxyadenosine. *Blood* **81**(3), 597–601.

Kay, A. C., Saven, A., et al. (1992). 2-Chlorodeoxyadenosine treatment of low-grade lymphomas. *J Clin Oncol: Official J Am Soc Clin Oncol* **10**(3), 371–377.

Kazimierczuk, Z., and Kaminski, J. (2000). Preparation of 2-chloro-2′-deoxyadenosine. *Pol.* Pl (Pol.), 8 pp.

Kazimierczuk, Z., Cottam, H. B., et al. (1984). Synthesis of 2′-deoxytubercidin, 2′-deoxyadenosine, and related 2′-deoxynucleosides via a novel direct stereospecific sodium salt glycosylation procedure. *J Am Chem Soc* **106**(21), 6379–6382.

Keating, M. J. (1993). Chronic lymphoproliferative disorders: Chronic lymphocytic leukemia and hairy-cell leukemia. *Curr Opinion Oncol* **5**(1), 35–41.

Keating, M. J. (1997). Future role of purine analogs in oncology. *Haematol Blood Transfusion* **38**(Acute Leukemias VI), 380–387.

Kobayashi, K., Vogelzang, N. J., et al. (1994). A phase I study of intermittent infusion cladribine in patients with solid tumors. *Cancer* **74**(1), 168–173.

Komatsu, H., Awano, H., et al. (2001). Process for selectively producing 1-phosphorylated sugar derivative anomer and process for producing nucleoside. *PCT International Application Wo*, Mitsui Chemicals, Inc., Japan, 82 pp.

Langtry, H. D., and Lamb, H. M. (1998). Cladribine: A review of its use in multiple sclerosis. *BioDrugs* **9**(5), 419–433.

Leoni, L. M., Chao, Q., et al. (1998). Induction of an apoptotic program in cell-free extracts by 2-chloro-2′-deoxyadenosine 5′-triphosphate and cytochrome c. *Proc Nat Acad Sci USA* **95**(16), 9567–9571.

Liliemark, J. (1997). The clinical pharmacokinetics of cladribine. *Clin Pharmacokin* **32**(2), 120–131.

Liu, E. S., Burian, C., et al. (1998). Bolus administration of cladribine in the treatment of Waldenstrom macroglobulinemia. *Br J Haematol* **103**(3), 690–695.

Mikhailopulo, I. A., Zinchenko, A. I., et al. (1993). Synthesis of 2-chloro-2′-deoxyadenosine by microbiological transglycosylation. *Nucleosides Nucleotides* **12**(3–4), 417–422.

O'Brien, S., Kantarjian, H., et al. (1996). Purine analogs in chronic lymphocytic leukemia and Waldenstrom's macroglobulinemia. *Annals Oncol: Official J Eur Soc Med Oncol/ESMO* **7**(Suppl 6): S27–S33.

Ochs, H. D., Yount, J. E., et al. (1973). Adenosine-deaminase deficiency and severe combined immunodeficiency syndrome. *Lancet* **1**(7816), 1393–1394.

Petersen, A. J., Brown, R. D., et al. (1996). Nucleoside transporters, bcl-2 and apoptosis in CLL cells exposed to nucleoside analogues *in vitro*. *Eur J Haematol* **56**(4), 213–220.

Robak, T. (2005a). The place of cladribine in the treatment of chronic lymphocytic leukemia: A 10-year experience in Poland. *Ann Hematol* **84**(2), 63–70.

Robak, T. (2005b). Therapy of chronic lymphocytic leukaemia with purine nucleoside analogues: Facts and controversies. *Drugs Aging* **22**(12), 983–1012.

Robak, T., Blasinska-Morawiec, M., et al. (1997). Autoimmune hemolytic anemia in patients with chronic lymphocytic leukemia treated with 2-chlorodeoxyadenosine (cladribine). *Eur J Haematol* **58**(2), 109–113.

Robak, T., Gora-Tybor, J., et al. (2001). Cladribine in combination with mitoxantrone and cyclophosphamide(CMC) in the treatment of heavily pre-treated patients with advanced indolent lymphoid malignancies. *Eur J Haematol* **66**(3), 188–194.

Robak, T., Korycka, A., et al. (2005). Purine nucleoside analogues for the treatment of hematological malignancies: Pharmacology and clinical applications. *Curr Cancer Drug Targets* **5**(6), 421–444.

Romine, J. S., Sipe, J. C., et al. (1999). A double-blind, placebo-controlled, randomized trial of cladribine in relapsing-remitting multiple sclerosis. *Proc Assoc Am Physicians* **111**(1), 35–44.

Rummel, M. J., Chow, K. U., et al. (2002). Reduced-dose cladribine (2-CdA) plus mitoxantrone is effective in the treatment of mantle-cell and low-grade non-Hodgkin's lymphoma. *European J Cancer (Oxford, England: 1990)* **38**(13), 1739–1746.

Santana, V. M., Mirro, J., Jr., et al. (1992). 2-Chlorodeoxyadenosine produces a high rate of complete hematologic remission in relapsed acute myeloid leukemia. *J Clin Oncol: Official J Am Soc Clin Oncol* **10**(3), 364–370.

Saven, A. (1996). The Scripps Clinic experience with cladribine (2-CdA) in the treatment of chronic lymphocytic leukemia. *Seminars Hematol* **33**(1 Suppl 1), 28–33.

Saven, A., Lemon, R. H., et al. (1995). 2 Chlorodeoxyadenosine activity in patients with untreated chronic lymphocytic leukemia. *J Clin Oncol: Official J the Am Soc Clin Oncol* **13**(3), 570–574.

Saven, A., Carrera, C. J., et al. (1992). 2-Chlorodeoxyadenosine: An active agent in the treatment of cutaneous T-cell lymphoma. *Blood* **80**(3), 587–592.

Saven, A., Kawasaki, H., et al. (1993). 2-Chlorodeoxyadenosine dose escalation in nonhematologic malignancies. *J Clin Oncol: Official J the Am Soc Clin Oncol* **11**(4), 671–678.

Saven, A., and Piro, L. D. (1994). 2-Chlorodeoxyadenosine: a newer purine analog active in the treatment of indolent lymphoid malignancies. *Ann Intern Med* **120**(9), 784–791.

Saven, A., Piro, L. D., et al. (1994). Complete hematologic remissions in chronic-phase, Philadelphia-chromosome-positive, chronic myelogenous leukemia after 2-chlorodeoxyadenosine. *Cancer* **73**(12), 2953–2963.

Saven, A., Lee, T., et al. (1996). Cladribine and mitoxantrone dose escalation in indolent non-Hodgkin's lymphoma. *J Clin Oncol: Official J Am Soc Clin Oncol* **14**(7), 2139–2144.

Saven, A., Lee, T., et al. (1997). Major activity of cladribine in patients with *de novo* B-cell pro-lymphocytic leukemia. *J Clin Oncol* **15**(1), 37–43.

Seto, S., Carrera, C. J., et al. (1985). Mechanism of deoxyadenosine and 2-chlorodeoxyadenosine toxicity to nondividing human lymphocytes. *J Clin Invest* **75**(2), 377–383.

Simon, L. N., Bauer, R. J., et al. (1970). Calf intestine adenosine deaminase. Substrate specificity. *Biochemistry* **9**(3), 573–577.

Tetreault, S. A., and Saven, A. (2000). Delayed onset of autoimmune hemolytic anemia complicating cladribine therapy for Waldenstrom macroglobulinemia. *Leuk Lymphoma* **37**(1–2), 125–130.

Ts'o, P. O. P. (Ed.) (1974). *Basic Principles in Nucleic Acid Chemistry*, Vol. 1.

Van Den Neste, E., Louviaux, I., et al. (2000). Phase I/II study of 2-chloro-2′-deoxyadenosine with cyclophosphamide in patients with pretreated B cell chronic lymphocytic leukemia and indolent non-Hodgkin's lymphoma. *Leukemia: official journal of the Leukemia Society of America, Leukemia Research Fund, U.K* **14**(6), 1136–1142.

Walker, R. T., De Clercq, E., et al. (1979). *NATO Advanced Study Institutes Series, Series A: Life Sciences*, Vol. 26: *Nucleoside Analogs: Chemistry, Biology, and Medical Applications*.

Wu, W. N., McKown, L. A., et al. (2004). Metabolism of the antineoplastic and immunosuppressive drug 2-CdA (Leustatin) in animals and humans. *Xenobiotica* **34**(6), 591–606.

CHAPTER 17

APOPTOSIS-BASED DRUG DISCOVERY FOR CANCER

John C. Reed, Dayong Zhai, Marc Hyer, and Kate Welsh*

17.1 INTRODUCTION

Apoptosis is caused by the activation of intracellular proteases, known as caspases. These cysteine proteases cleave their cellular targets at aspartic acid residues residing in the context of tetrapeptide motifs within polypeptide substrates (Boatright and Salvesen, 2003). The human genome encodes 11 or 12 caspases, depending on certain hereditary polymorphisms (Reed et al., 2004; Saleh et al., 2004). Numerous cellular targets of caspases have been identified, which when cleaved produce in aggregate the characteristic morphology we call "apoptosis". Several pathways for triggering caspase activation exist, although two have been elucidated in great detail and have been the center of much attention in recent years. These two pathways for apoptosis are commonly referred to as the intrinsic and the extrinsic pathways (Fig. 17.1) (Salvesen, 2002).

The intrinsic pathway centers on mitochondria as initiators of cell death. Multiple signals converge on mitochondria, including DNA damage, hypoxia, and oxidative stress, causing these organelles to release cytochrome *c* (cyt *c*) and other apoptogenic proteins into cytosol. In the cytosol, cyt *c* binds caspase-activating

*Corresponding author: jreed@burnham.org

Drug Discovery Research: New Frontiers in the Post-Genomic Era, Edited by Ziwei Huang

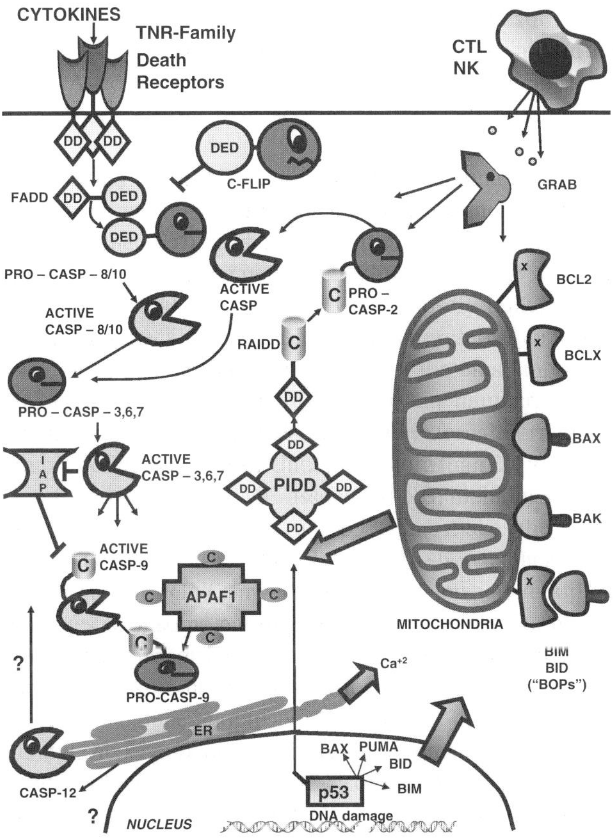

Figure 17.1 Caspase-activation pathways. See text for details.

protein Apaf1, triggering its oligomerization into a hepatmeric complex that binds pro-caspase-9, forming a multiprotein structure known as the "apoptosome" (Salvesen and Renatus, 2002). Physical binding of Apaf1 to pro-caspase-9 is mediated by their caspase recruitment domains (CARDs) through homotypic CARD-CARD binding. Activation of apoptosome-associated cell death protease caspase-9 then initiates a proteolytic cascade, where activated caspase-9 cleaves and activates downstream effector proteases, such as pro-caspase-3.

In contrast, the extrinsic apoptotic pathway relies on TNF (tumor necrosis factor)-family death receptors for triggering apoptosis. A subgroup of the TNF-family receptors contain a cytosolic death domain that enables their intracellular interaction with downstream adapter proteins, which link these receptors to specific caspases. Upon ligand binding, TNF-family receptors containing cytosolic death domains (e.g., Fas; TNFR; TRAIL-R1; TRAIL-R2) cluster in membranes, recruiting caspase-binding adapter proteins, including the bipartite adapter FADD that contains both a death domain (DD) and a death effector domain (DED) (Wallach et al., 1999). The DED of FADD binds DED-containing pro-caspases (e.g. caspase-8 and -10), forming a "death inducing signaling complex" (DICS) and resulting in caspase activation by an "induced proximity" mechanism (Boatright and Salvesen, 2003).

Aside from the intrinsic and extrinsic pathways, multiple additional routes to caspase activation are possible, although for some the pathophysiological relevance is less well established (Fig. 17.1). For example, a pathway activated by cytolytic T-cells (CTLs) and natural killer (NK) cells involves injection of the contents of cytotoxic granules into target cells, at least in part through perforin-mediated mechanisms. Among the contents of the cytotoxic granules is granzyme B, a serine protease with specificity for aspartyl residues that cleaves and activates multiple members of the caspase family [reviewed in Lieberman (2003)]. A caspase-activation pathway linked to stress in the Golgi and endoplasmic reticulum (ER) has also been proposed, although details are unclear at present—particularly for human cells that seem to differ from their rodent counterparts with respect to production of ER-associated caspases (Saleh et al., 2004; Breckenridge et al., 2003). A p53-inducible protein, PIDD, recently was demonstrated to activate caspase-2 via bipartite adapter protein RAIDD (Tinel and Tschopp, 2004). The RAIDD protein has a DD that binds a corresponding DD in PIDD, and a CARD that binds the CARD found within the N-terminal prodomain of pro-caspase-2. In addition, a variety of proteins are capable of activating pro-caspase-1 and possibly related members of the caspase family (e.g., caspase-4 and -5 in humans), but their relevance to apoptosis is unclear (Reed et al., 2004; Tinel and Schopp, 2004). This branch of the caspase family is involved principally in cytokine maturation rather than apoptosis, cleaving and activating pro-interleukin-1β (pro-IL-1β) among other cytokines (Thornberry and Lazebnik, 1998). However, under certain pathological circumstances, particularly in the setting of host–pathogen interactions and ischemia–reperfusion injury, excessive caspase-1 activity may trigger apoptosis (Reed, 1999a).

It is important to note that while the pathways described above can all lead to caspase activation and apoptosis, some of them also trigger parallel caspase-independent cell-death mechanisms, which nevertheless kill cells. For example, mitochondria not only release caspase-activating proteins such as cyt *c*, but also release endonuclease G (Endo G) and a chromatin-modifying protein apoptosis inducing factor (AIF) that promote genome digestion and cell death independent of caspases (Penninger and Kroemer, 2003). Similarly, ER stress typically promotes a caspase-independent cell death mechanism, which may involve disturbances in intracellular Ca^{2+} homeostasis (Demaurex and Distelhorst, 2003). Also, granzyme B as well as other proteins found within the cytotoxic granules of CTLs and NK cells can invoke caspase-independent cell-death mechanisms [reviewed in (Lieberman (2003)]. Thus, some types of cell-death stimuli can induce parallel paths to apoptotic (caspase-dependent) and non-apoptotic (caspase-independent) cell demise.

17.2 APOPTOSIS BLOCKERS

Given the critical importance of making the correct choices about cell life–death decisions in complex multicellular organisms, it is not surprising that the pathways governing caspase activation are under exquisite control by networks of proteins that directly or indirectly communicate with these proteases. A delicate balance between pro-apoptotic and anti-apoptotic regulators of apoptosis pathways is at play on a continual basis, ensuring the survival of long-lived cells and the proper turnover of short-lived cells in a variety of tissues, including the bone marrow, thymus, and peripheral lymphoid tissues. However, imbalances in this delicate dance of pro- and anti-apoptotic proteins occur in disease scenarios, including cancer where the scales tip in favor of anti-apoptotic proteins and endow cells with a selective survival advantage that promotes neoplasia and malignancy.

The anti-apoptotic proteins responsible for creating roadblocks to apoptosis have been mapped to specific pathways, providing insights into the defective cell-death mechanisms that contribute to malignancy. Among these apoptosis blockers are members of the Bcl-2 family, a large group of proteins ($n = 27$ in humans) that control mitochondria-dependent steps in cell-death pathways, including dictating whether cyt *c* is or is not released from these organelles (Fig. 17.1). These proteins control the intrinsic pathway (Kroemer and Reed, 2000), although effects on the ER pathway for cell death have also been documented (Demaurex and Distelhorst, 2003). Another anti-apoptotic protein known to create apoptosis roadblocks in cancer cells is FLIP, a DED-containing protein that competes with pro-caspase-8 and -10 for interactions with TNF/Fas-family death receptor complexes (Tschopp et al., 1998). FLIP thus can create blockades in the extrinsic pathway. Downstream at the convergence of the apoptosis pathways, IAP-family proteins thwart cell death by directly binding to and suppressing the activity of certain caspases. The specific caspases and mechanisms of inhibition vary, depending on the IAP ($n = 8$ in humans), but both effector (e.g., caspase-3 and -7) and initiator (e.g., caspase-9) proteases are among the known targets of IAP-mediated suppression (Deveraux and Reed, 1999; Salvesen and Duckett, 2002).

The explanations for overexpression of anti-apoptotic proteins in cancer and leukemia cells vary, but documented mechanisms include chromosomal translocations, gene amplification, gene hypomethylation, transcriptional upregulation, and perhaps altered protein stability [reviewed in (Kitada et al., (2002)]. Conversely, pro-apoptotic genes that oppose these cytoprotective proteins can become inactivated in malignant cells through gene deletion, somatic mutation, gene hypermethylation, transcriptional downregulation, and probably other mechanisms.

The apoptosis-blocking proteins that thwart signaling through specific apoptosis pathways provide targets for possible drug discovery as do agents that stimulate pro-apoptotic proteins within these cell-death pathways. Theoretically, all drug targets can be addressed in multiple ways, focusing on agents that modulate the activity of the target at the levels of the DNA (gene expression), mRNA, or protein. A summary of some of the more promising strategies for modulating the activity of apoptosis genes and proteins for cancer therapy is provided below, according to the apoptosis pathways they affect.

17.3 EXTRINSIC PATHWAY ACTIVATORS

The extrinsic pathway is activated *in vivo* by TNF-family ligands that engage death receptors such as TNF, Fas, or TRAIL, resulting in the activation of caspases and inducing apoptosis. Interest has emerged in developing therapeutics that kill cancer cells via the extrinsic pathway, particularly since chemorefractory cells tend to have defects in their intrinsic pathway, given the predominant reliance of cytotoxic drugs and X-irradiation on the mitochondrial pathway for cell death (Green and Reed, 1998). In this regard, the TNF cytokine family consists of 18 members in humans, with 29 counter receptors [reviewed in Reed et al. (2004) and Locksley et al. (2001)]. Some of these TNF-family receptors transduce signals predominantly for cell survival, particularly those that bind intracellular TRAF-family adapter proteins that link to downstream protein kinase pathways. Blocking these receptors represents a potentially attractive strategy for a variety of lymphoid malignancies but will not be discussed here. Other members of the TNF family directly trigger apoptosis, particularly those that contain the death domain (DD) structure within their cytosolic tails ($n = 8$ in humans). Of these receptors and their corresponding ligands, four receptors (three ligands) have been studied in detail with respect to their potential exploitation as cancer therapeutics—namely, TNF (TNFR1), Fas (FasL), and TRAIL (DR4; DR5). Strategies for exploiting apoptosis-inducing TNF-family ligands and receptors for cancer therapy mostly focus on (a) production of recombinant ligands (biologicals), expressing only the extracellular domain of these type II integral membrane proteins, or (b) on use of agonistic monoclonal antibodies that bind the receptors and trigger apoptosis. In addition, however, emerging knowledge about intracellular modulators of the apoptosis pathways utilized by TNF-family death receptors has suggested opportunities for engaging the extrinsic pathway at other levels, improving sensitivity of malignant cells to biologicals that activate the extrinsic pathway.

17.3.1 TNF-Family Ligands and Receptor-Targeting Antibodies

Over three decades ago, TNF was reported to selectively kill tumor but not normal cells, raising hopes that this or similar cytokines might be exploited for improved treatment of cancer [reviewed in Bazzoni and Beutler (1996)]. Unfortunately, the proinflammatory actions of this cytokine precluded its systemic administration and squelched efforts to apply it clinically. Then, we knew little about how TNF transduces signals into cells through its specific receptors, thus making it difficult to propose alternative strategies. Today, in contrast, we have an in-depth knowledge of the pathways that TNF and related cytokines activate, and from this knowledge have come several potential targets for drug discovery. We also understand far better how TNF-family cytokines regulate pathways that control apoptosis, and we can begin to envision ways of exploiting that knowledge for sensitizing tumors to these cytokines that our immune cells use to do battle with cancer.

Binding of TNF to one of its principal cellular receptors expressed widely on cells, TNFR1, triggers two parallel signaling pathways. These pathways bifurcate at the adapter protein TRADD. One of these pathways results in activation of caspase-family cell-death proteases, triggering apoptosis. The other parallel pathway triggers activation of NFκB-family transcription factors. NFκB influences the expression of many target genes involved in host defenses and immune regulation, among which are several genes that suppress apoptosis. As a result, this NFκB pathway nullifies the caspase pathway, negating apoptosis [reviewed in Karin and Lin (2002)] in addition to accounting for the untoward inflammatory actions of this cytokine.

The TNF-family cytokines Fas-ligand (FasL) and TRAIL (Apo2 ligand) trigger activation of the caspase pathway without concomitant induction of NFκB. These death ligands are expressed on CTLs, NK cells, and other types of immune-relevant cells (activated monocytes/macrophages and dendritic cells) and are used as weapons for eradication of virus-infected and transformed cells [reviewed in Locksley et al. (2001), Ashkenazi and Dixit (1998), and Nagata (1996)]. Various studies using mice with genetic alterations in genes encoding these death ligands or their receptors, as well as use of neutralizing antibodies and Fc-fusion proteins, have provided evidence that FasL and TRAIL play important roles in tumor suppression by cellular immune mechanisms. For example, FasL is important for CTL-mediated killing of some tumor targets, and TRAIL is critical for NK-mediated tumor suppression [see, for example, Rosen et al. (2000) and Johnsen et al. (1999)].

The absence of pro-inflammatory effects of FasL and TRAIL has raised hopes that they might be successfully applied for cancer therapy, where TNF failed due to toxicity. While agonistic antibodies that trigger the receptor, Fas (CD95), are unfortunately highly toxic to liver (Ogasawara et al., 1993), TRAIL and agonistic antibodies that bind TRAIL receptors appear to be well-tolerated *in vivo*. Indeed, Phase I clinical trials have recently been completed using an agonistic monoclonal antibody ETR1 (Human Genome Science, Inc.) directed against TRAIL-receptor-1 (TRAIL-R1) [also known as death receptor-4 (DR4)], revealing little toxicity (Le, 2004). In mouse xenograft models using selected tumor human cell lines, TRAIL and agonistic antibodies directed against TRAIL receptors have been demonstrated to possess potent antitumor activity (Ashkenazi et al., 1999), raising hopes of using these biological agents as a novel approach to cancer treatment and thereby mimicking some of the effector mechanisms normally employed by the immune system in its defense against transformed cells.

Preclinical animal models also suggest that TRAIL-based therapies may be synergistic with conventional cytotoxic anticancer drugs. The molecular basis for this apparent synergy can differ among tumor cell lines tested, including: (a) chemotherapy-induced increases in TRAIL receptors, particularly DR5, which is a direct transcriptional target of p53 (Wu et al., 1997); (b) p53-mediated increases in Bid, a pro-apoptotic Bd-2 family protein that is cleaved and activated by the apical protease in the TRAIL-induced pathway (e.g., caspase-8), thus representing a point of cross-talk between the extrinsic and intrinsic pathways (Sax et al., 2002); (c) p53-independent increases in Bak (LeBlanc et al., 2002); and probably others [reviewed in Roth and Reed (2003)]. Thus, knowledge of the ways that conventional cytotoxic drugs modify pathways relevant to TRAIL receptor signaling may provide a rational basis for selecting combination therapies and identifying subgroups of patients for inclusion in clinical trials.

17.3.2 NFκB Inhibitors

Abnormal elevations in NFκB activity are found in many types of tumors, including many lymphoid malignancies and several types of solid tumors [reviewed in

Karin et al. (2002), Rayet and Gelinas (1999)]. In fact, the first member of the NFκB family identified, *c-REL*, is the cellular counterpart of a viral transforming gene, *v-rel*, discovered in the avian leukosis virus. The importance of NFκB-family transcription factors for suppression of TNF-induced apoptosis is well established (Baud and Karin, 2001). Consequently, interfering with the sequence of signal-transduction events that generates active NFκB has emerged as an attractive approach to sensitizing cancer cells to TNF and related cytokines. Several anti-apoptotic genes are among the direct transcriptional targets of Rel-family proteins, including the genes encoding c-FLIP, cIAP2, Bcl-X_L, and Bfl-1. Thus, suppression of NFκB activity is an attractive strategy for restoring apoptosis sensitivity to several apoptosis pathways simultaneously.

Efforts to elucidate the signaling events responsible for activation of NFκB by TNF have revealed several candidate targets that are potentially amenable to small-molecule drug discovery approaches, particularly protein kinases responsible for eradicating endogenous inhibitors of NFκB. For example, although alternative pathways exist (Fan and Maniatis, 1991), an evolutionarily conserved pathway for NFκB activation has been revealed involving various upstream kinases funneling signals into a pivotal kinase complex that consists of two homologous serine-kinases, IKKα and IKKβ, and a scaffold protein, IKKγ (also known as NEMO) (DiDonato et al., 1997; Zandi et al., 1997, 1998; Rothwarf et al., 1998). This IKK complex is responsible for phosphorylation of endogenous inhibitors of NFκB, the IκB proteins, with phosphorylation targeting these NFκB inhibitors for ubiquitination and subsequent proteasome-dependent degradation. Degradation of IκB releases NFκB, allowing its translocation into the nucleus [reviewed in Karin and Ben-Neriah (2000)]. Thus, the protein kinases involved in this pathway have emerged as promising drug targets, as well as possibly the F-box protein that recognizes phosphorylated IκB and recruits it to E3 ubiquitin ligase complexes, and the proteasome itself (Karin and Ben-Neriah, 2000; Maniatis, 1999).

In this regard, small-molecule inhibitors of IKKs have been described, which display pro-apoptotic activity against some types of cultured malignant cells. These IKK inhibitors include BMS-345541 (Burke et al., 2003); Bay 11-7082 (Dai et al., 2004); SC-514 (Kishore et al., 2003); Sulfasalazine (Weber et al., 2000); Silibinin, a dietary flavonoid that seems to directly inhibit IKKs (Dhanalakshmi et al., 2003); pyrolidine dithiocarbamate (Sumitono et al., 1999); natural product Wedelolactone and semi-synthetic derivatives of B-carboline natural products (Li et al., 2003; Castro et al., 2003); and others. The specificity of some of these inhibitors for IKK is questionable, but toxicology studies will determine whether they might be clinically useful.

Proteasome inhibitors include bortezomib, recently approved for second-line treatment of myeloma. This compound is in clinical testing for several other types of cancer and leukemia (Richardson et al., 2003; Wright et al., 2003) (Millennium Pharmaceuticals, Inc.). Bortezomib (PS-341) (Velcade™) inhibits the chymotrypsin-like protease activity of the proteasome (Wright et al., 2003; Paramore and Frantz, 2003). Pharmacological inhibition of proteasome activity not only suppresses IκB degradation, but also has effects on the stability of a large repertoire of cellular proteins, and thus the molecular basis for the antitumor activity of proteasome inhibitors remains unclear and may only partly be related to NFκB activity.

17.3.3 FLIP

Many tumors exhibit intrinsic resistance to TRAIL, anti-Fas antibodies, or both of these death stimuli, despite expressing the necessary cell-surface receptors (Snell et al., 1997). Multiple antagonists of the extrinsic pathway have been identified, including several DED-containing proteins that compete for binding to the adapter proteins or pro-caspases that participate in TNF-family death receptor signaling, including c-FLIP, BAR, and possibly Bap31 [reviewed in Krammer (2000) and Griffith et al. (1998)]. Among these, FLIP has received the most attention for its role in producing Fas- and TRAIL-resistant states in tumor cells (Tschopp et al., 1998; Irmter et al., 1997).

The c-FLIP protein is highly similar in overall sequence to pro-caspase-8 and -10, containing tandem copies of the DED, followed by a pseudo-caspase domain that lacks enzymatic activity. The *c-FLIP* gene resides in a tandem gene cluster on human chromosome 2, which contains the genes encoding pro-caspase-8 and -10, suggestive of gene-duplication events. Two isoforms of c-FLIP are produced from a single gene, including the long form described above and a shorter isoform that consists of only the DED domains. The shorter version of FLIP thus resembles analogous proteins encoded in the genomes of some mammalian viruses (Tschopp et al., 1998b). FLIP-S is exclusively anti-apoptotic, while FLIP-L can be either pro- or anti-apoptotic, depending on its levels of expression relative to pro-caspase-8 and -10 (Chang et al., 2002). FLIP proteins form complexes with pro-caspase-8 and -10, preventing their effective activation, as well as competing for binding to adapter proteins required for caspase recruitment to death receptor complexes [reviewed in Tschopp et al., (1998a, 1999)]. Overexpression of FLIP occurs commonly in cancers, implying *in vivo* selection, probably as a result of confrontations of neoplastic cells with the immune system.

Although no drugs have been described to date that directly bind FLIP proteins, experimental agents have been found that reduce FLIP protein expression. Among these are synthetic triterpenoids 2-cyano-3,12-dioxoolean-1,9-bien-28-oic acid (CDDO), CDDO-methyl ester (CDDO-me), and CDDO-imidazole (CDDO-Im), which induce ubiquitination and proteasome-dependent destruction of FLIP proteins in cultured cancer cells, sensitizing solid tumor lines to TRAIL apoptosis induction by TNF, Fas, and TRAIL (Kim et al., 2002). CDDO-Im was shown to synergize with recombinant TRAIL in a tumor xenograft model study using mice, revealing promising antitumor activity without toxicity to normal tissues (Hyer et al., 2005). Moreover, these and other CDDO analogs exhibit single agent activity in terms of inducing apoptosis *in vitro* of some types of cancer and leukemia cells (Pedersen et al., 2002b; Ikeda et al., 2004). The mechanism of CDDO and related compounds is still under investigation, but several studies have found activation of caspase-8 as a proximal event, indicative of stimulation of the extrinsic pathway (Pedersen et al., 2002b; Ito et al., 2000), although other mechanisms have also been proposed (Konopleva et al., 2002; Lapillone et al., 2003; Hail et al., 2004). Given its ability to induce activation of components of the extrinsic pathway, CDDO and its analogues may provide an option for bypassing the blocks to apoptosis that typically arise within the mitochondrial apoptosis pathway in chemorefractory tumors. CDDO and CDDO-me are undergoing final preclinical development steps in preparation for human clinical trials, in collaboration with NCI DTP.

Additional possible routes to suppression of FLIP expression include NFκB pathway inhibitors (see above) and HDAC inhibitors (Aron et al., 2002). Also, two chemicals (4-(4-chloro-2-methylphenoxy)-*N*-hydroxybutanamide and 1-(3,5-di-*tert*-butyl-4-hydroxyphenyl)-2-(3-ethyl-2-imino-2,3-dihydro-benzoimidazol-1-yl)ethanone) were recently identified by cell-based high-throughput screening that reduce FLIP expression and sensitize tumor cells to TRAIL and FAS at concentrations of 5–50 μM (Schimmer et al., 2005). The cellular target(s) of these compounds are unknown, but siRNA experiments support a FLIP-dependent mechanism of action.

17.3.4 Additional Strategies for Overcoming Extrinsic Pathway Resistance

Other routes to restoring integrity of the extrinsic pathway have been revealed that may apply to leukemia and cancer. Protein kinase C (PKC) modulators such as byrostatin induce TNF production and apoptosis of myeloid leukemia cell lines via a caspase-8-dependent mechanism (Cartee et al., 2003). Also, all *trans*-retinoid acid (ATRA) induces TRAIL production and triggers autocrine induction of apoptosis in acute promyelomonocytic leukemia (APML) cells that harbor the RARα-PML fusion protein resulting from t(15;17) chromosomal translocations (Altucci et al., 2001).

17.4 INTRINSIC PATHWAY ACTIVATORS

Bcl-2-family proteins are the most prominent of the intrinsic pathway targets for cancer drug discovery. Though the human genome encodes 27 Bcl-2-family proteins, only six of these are anti-apoptotic and thus represent logical targets for therapy. The six anti-apoptotic members of the family are Bcl-2, Bcl-X_L, Mcl-1, Bcl-W, Bfl-1, and Bcl-B. Overexpression of several of these anti-apoptotic Bcl-2-family proteins has been documented in various malignancies, although data are most extensive for the founding member of the family Bcl-2. Attempts to overcome the cytoprotective effects of Bcl-2 or related anti-apoptotic proteins in cancer and leukemia include three strategies: (1) shutting off gene transcription, (2) inducing mRNA degradation with antisense oligonucleotides, and (3) directly attacking the proteins with small-molecule drugs. A fourth strategy attempts to invoke endogenous antagonists of anti-apoptotic Bcl-2-family proteins to nullify their cytoprotective functions.

17.4.1 Drugs Regulating *BCL-2* Gene Expression

Some synthetic retinoids are known to reduce levels of *BCL-2* or *BCL-XL* mRNA in leukemia cells, suggesting a potential explanation for the pro-apoptotic effect of these agents that are already approved and in clinical use (Reed, 1999b). Estrogen receptor antagonists also reduce *BCL-2* expression in breast cancer cell lines (Pratt, 1998), providing at least a partial explanation for the pro-apoptotic activity of these compounds in ER-positive tumors of mammary origin. Steroid hormone, 1α,25$(OH)_2$

vitamin D also suppresses expression of anti-apoptotic genes *BCL-2*, *BCL-X*, and *MCL-1* in prostate cancer cell lines that contain the vitamin D receptor (VDR) (Guzey et al., 2002) and may underlie in part the antitumor activity observed recently in clinical trials of high dose 1α,25(OH)$_2$ vitamin D (calcitriol) for treatment of hormone refractory prostate cancer (Johnson et al., 2002). Compounds that inhibit histone deactylases (HDACs), chromatin-modifying enzymes that interact with many transcriptions factors, also reduce the expression of *BCL-2*, *BCL-X*, or *MCL-1* at a transcriptional level (Khan et al., 2004; Mori et al., 2004; Rosato et al., 2003). Clinical trials of HDAC inhibitors are progressing, with hints of activity documented for lymphoma and some solid tumors (Kelly et al., 2003). PPARγ-modulating drugs also can reduce expression of *BCL-2* or other anti-apoptotic *BCL-2* family genes (83).

17.4.2 Drugs Attacking Bcl-2 mRNA

Antisense oligodeoxynucleotides (ODNs) targeting the *BCL-2* mRNA are undergoing clinical evaluation, with Phase III trials underway or recently completed for relapsed or refractory CLL, AML, and myeloma and also for hormone refractory prostate cancer (Buchele, 2003). The molecules in clinical trials are composed of nuclease-resistant phosphorothioates, which hybridize via Watson–Crick base-pairing to the first 18 nucleotides within the open reading frame of Bcl-2 mRNAs and induce RNaseII-mediated degradation (Reed et al., 1990). This DNA-based drug (G3139) (oblimersen sodium) (Genasense™) is being developed in combination with cytotoxic chemotherapy, in an attempt to exploit the chemosensitizing properties of Bcl-2-based therapies. The first reported Phase I trial of *BCL-2* antisense therapy showed promising activity against chemorefractory non-Hodgkins lymphoma (NHL) (Webb et al., 1997). Uncontrolled Phase II data also suggest that this *BCL-2* antisense molecule (Oblimersen sodium) (Genasense™) has promising bioactivity against B-cell malignancies and adult AML, while producing little toxicity at doses <4 mg/kg. A recently completed trial of oblimersen sodium at 7 mg/kg for advanced melanoma (in combination with alkylating agent dacarbazine) demonstrated improved response rates and time to progression (TTP) but failed to meet its primary endpoint of prolonging patient survival. Further follow-up of the melanoma patient cohort, however, suggests a survival benefit for patients with lower pretreatment tumor burden, as defined by serum levels of lactate dehydrogenase (LDH). A recently completed Phase III trial of oblimersen sodium at 3 mg/kg in patients with refractory CLL met its primary endpoint objective of demonstrating improved responses to cytotoxic drug therapy (cyclophosphamide plus fludarabine). These clinical studies further validate *BCL-2* as a promising target for cancer therapy.

Demonstration of a correlation between clinical responses and antisense-mediated knockdown of Bcl-2 mRNA or protein levels in patients undergoing treatment with oblimersen sodium has been documented in some but not all studies, raising questions about the mechanism of this novel targeted therapy (Rudin et al., 2004; Marcucci et al., 2003). However, it has been argued that cells with successful antisense-mediated reductions in Bcl-2 mRNA are difficult to recover from patients because of their rapid *in vivo* clearance induced by apoptosis. The Bcl-2 protein also exhibits a very slow spontaneous degradation rate ($\sim$12–36 hr), suggesting that prolonged suppression of mRNA accumulation is necessary to effect a decline in Bcl-2 protein [reviewed in Reed (1996a)]. A pro-inflammatory effect of oblimersan sodium, seen particularly at higher doses, suggests an additional non-antisense bioactivity, which may or may not be relevant to tumor responses. In this regard, CpG motifs in synthetic DNA molecules can trigger inflammatory responses by engaging toll-like receptor-9 (TLR9), tapping into a mechanism of innate immunity intended to sense the presence of unmethylated bacterial DNA (Kimbrell and Beutler, 2001). Interestingly, mismatched ODNs containing CpG motifs were shown to reduce *BCL-2* expression and promote apoptosis of prostate cancer cell lines through an antisense-independent mechanism (Lai et al., 2003). Nevertheless, tumor xenograft studies of oblimersen sodium in immunocompromised mice provide strong support for an antisense-based mechanism, whereby sequence-specific reductions in *BCL-2* expression either directly trigger tumor apoptosis or sensitize malignant cells to traditional cytotoxic anticancer drugs (Cotter et al., 1994; Klasa et al., 2000; Guinness et al., 2000; Jansen et al., 1998; Miyake et al., 1999).

The genetic lesions complementing *BCL-2* expression in tumors probably dictate whether inhibiting *BCL-2* is sufficient to induce apoptosis (single agent activity) versus merely sensitizing cells to chemotherapy. For example, B-cell malignancies with simultaneous activation of *C-MYC* and *BCL-2* have been described, arising from transformed follicular lymphomas where a t(14;18) activating *BCL-2* is combined with a t(8;14) activating *C-MYC* (Gauwerky et al., 1988). Activated *C-MYC* drives both cell proliferation and cell death, the latter therefore requiring complementation by *BCL-2* (Evan and Vousden, 2001). Thus, depriving cells with activated *C-MYC* of their Bcl-2 protein is tantamount to a death sentence, suggesting a possible strategy for patient selection based on simple biomarker profiling that demonstrates overexpression of both *BCL-2* and *C-MYC*.

Aside from oblimersen sodium, no nucleic acid-based inhibitors of Bcl-2 or its anti-apoptotic relatives are presently in clinical testing, but multiple preclinical programs have been reported. For example, antisense DNA molecules that target Bcl-X_L mRNA and induce its RNaseH-mediated destruction (ISIS16009) have been generated and shown to reduce resistance of tumor and leukemia cell lines to cytotoxic anticancer drugs in culture and in xenograft models in mice (Guensberg et al., 2002; Fennell et al., 2001). Mcl-1-targeting antisense ODNs (Derenne et al., 2002) also reportedly overcome the survival-promoting activity of follicular dendritic cells on cultured CLL B-cells, which induce Mcl-1 expression when co-cultured with these leukemic B-cells (Pedersen et al., 2002a). Mcl-1 antisense also induces apoptosis of multiple myeloma cells more efficiently than antisense targeting either Bcl-2 or Bcl-X_L (Derenne, 2002), and displays activity in preclinical animal models of myeloma (Thallinger et al., 2003). A clever strategy using antisense chemistry that fails to support RNaseH-based mRNA destruction has also been described, wherein a 2'methoxyethyl-based antisense ODN hybridized to a splice acceptor site in the Bcl-X_L pre-mRNA forced splicing to a downstream exon to produce a shorter Bcl-X protein with pro-apoptotic rather than anti-apoptotic activity (Taylor et al., 1999). A dual inhibitor of both Bcl-2 and Bcl-X_L has also been reported that takes advance of nucleotide sequence conservation, targeting both of these genes simultaneously (Zangemeister-Wittke et al., 2000).

17.4.3 Drugs Attacking the Bcl-2 Protein

Small-molecule inhibitors that directly bind Bcl-2 or related anti-apoptotic proteins have also entered clinical trials for cancer. The most advanced of these is a natural product, gossypol, whose ability to bind and inhibit Bcl-2 was unknown at the time initial clinical testing began (Stein et al., 1992; Kitada et al., 2003) (Table 17.1). Gossypol is a compound found in cottonseeds originally used as an herbal medicine in China. Gossypol promotes apoptosis of tumor cells *in vitro* and shows antitumor activity in tumor xenograft studies in mice [reviewed in Pellecchia (2004)]. Gossypol binds a hydrophobic pocket found on the surface of anti-apoptotic Bcl-2-family proteins. This binding pocket represents a regulatory site where endogenous antagonists dock onto Bcl-2 and related anti-apoptotic proteins, negating their cytoprotective activity (Fesik, 2000). The endogenous antagonists bind via a conserved 16 amino-acid motif called a Bcl-2 homology-3 (BH3) domain (Fig. 17.2). Proof-of-concept experiments using BH3 peptides have suggested that compounds docking at this regulatory site on Bcl-2 and its related anti-apoptotic proteins could provide a route to effective suppression of these proteins and thereby promote apoptosis of malignant cells (Holinger et al., 1999).

Gossypol interacts with the BH3-binding pockets of five of the six anti-apoptotic Bcl-2-family proteins, displacing BH3 peptides with an IC_{50} of ~0.5–2 μM from Bcl-2, Bcl-X_L, Bcl-B, Bcl-W, and Mcl-1 *in vitro* (Zhai et al., 2005), thus making it a broad-spectrum inhibitor. An enantiomer of gossypol is touted to have superior activity compared to the naturally occurring racemer and is in clinical development by ASCENTA, Inc., in collaboration with the Developmental Therapeutic Branch (DTB) of the National Cancer Institute (NCI-USA) (Qiu et al., 2002). However, gossypol is a highly reactive compound, containing two aldehydes that may explain some of the toxicities originally seen in Phase I trials of this natural product, in addition to producing unfavorable pharmacological properties. For this reason, attempts to produce semi-synthetic analogues that retain activity against Bcl-2 have begun. The most advanced of these is the compound called apogossypol, in which the two aldehydes were eliminated (Becattini et al., 2004). This compound is also under evaluation by the DTB of NCI in the hope of achieving stable formulations that can be advanced into human clinical trials.

GX-15-070 is a synthetic broad-spectrum antagonist of anti-apoptotic Bcl-2-family proteins currently being evaluated in Phase I trials by GEMINEX, Inc. This inhibitor is a cycloprodigiosin derivative touted as a broad-spectrum inhibitor of anti-apoptotic Bcl-2-family proteins (unpublished information presented by sponsoring company, GeminX, Inc.), which is said to have roughly 1 μM affinity for Bcl-2 and several anti-apoptotic Bcl-2 family proteins. Little information about GX-15-070, however, is available in peer-reviewed literature.

Several other small-molecule antagonists of Bcl-2, Bcl-X_L, and Mcl-1 have been reported, all of which are still in preclinical evaluation at this time. Among these are: (a) HA14-1, a chromene derivative identified by a computational modeling approach (Wang et al., 2000); (b) BH3I-1 and -2, a thiazolidin derivative and benzene sulfonyl derivative, respectfully, identified by screening using a BH3 peptide displacement assay (Degterev et al., 2001); (c) antimycin analogues, identified by

TABLE 17.1 Structures of Bcl-2 Antagonists

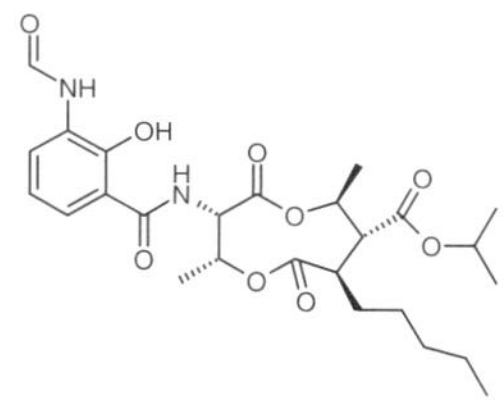

Antimycin A3

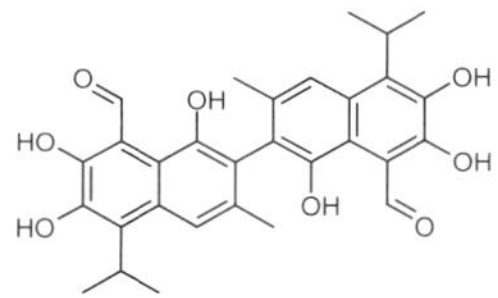

Gossypol

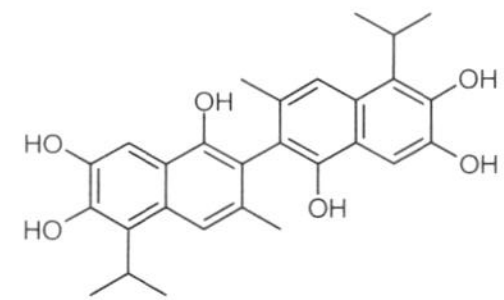

Apogossypol

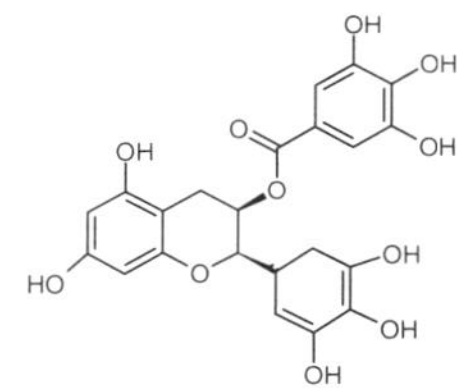

EGCG

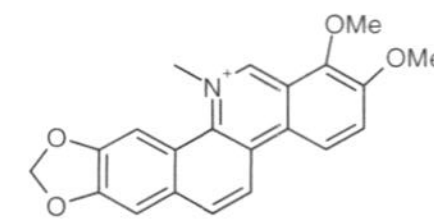

Chelerythrine

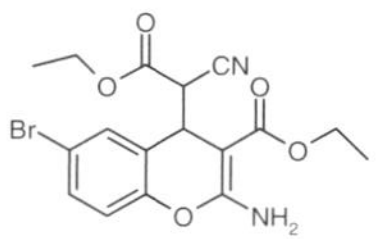

HA14-1

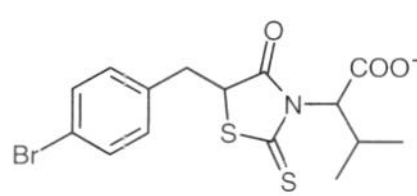

BH3I-1

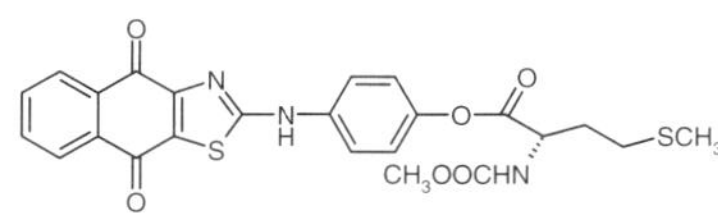

YC137

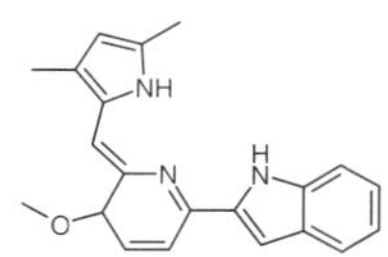

GX-15

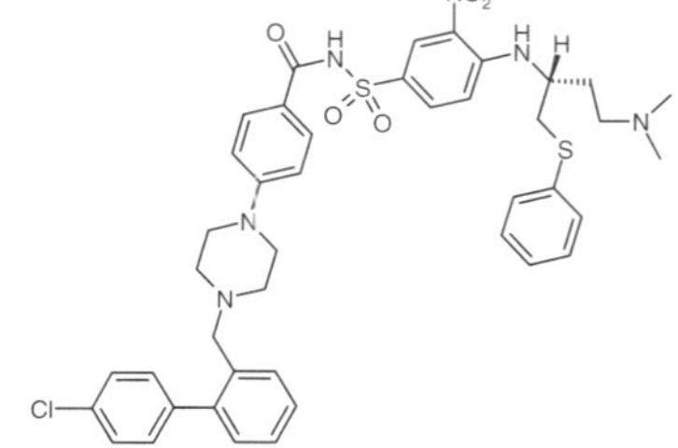

ABT-737

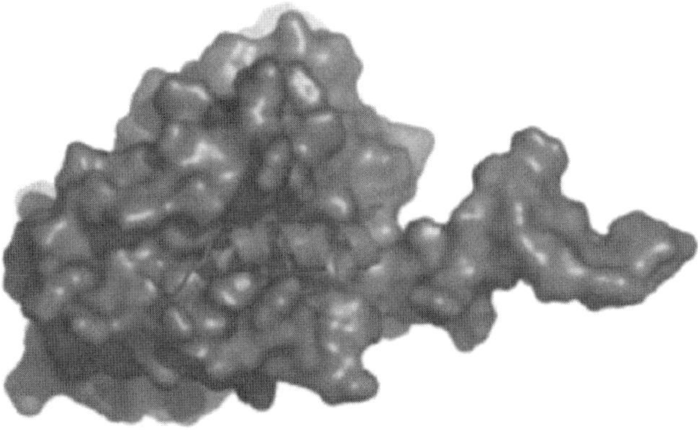

Figure 17.2 Structure of Bcl-2 and associated BH3 peptide. The 3D structure of Bcl-X_L is depicted (blue) with Bak BH3 peptide bound (red). See color plates.

computational modeling (Tzung et al., 2001); (d) certain theaflavins, a class of natural products abundant in black tea, some epigallechatechins, natural products abundant in green tea (Leone et al., 2003), and the natural product chelerythrine (Chan et al., 2003); (e) ABT737, a synthetic small-molecule inhibitor produced by NMR-guided, structure-based drug design (Abbott Laboratories, Inc.) (Olfersdorf et al., 2005) (Table 17.1). Side-by-side comparisons of these small-molecule inhibitors of anti-apoptotic Bcl-2-family proteins have recently been reported, suggesting a rank-order potency with respect to affinity for the BH3 pocket of Bcl-2 or Bcl-X_L of ABT-737 > EGCG > theaflavins > gossypol/apogossypol > chelerythrine > HA14-1 and antimycin (Zhai 2006).

Several issues will be encountered as these small-molecule antagonists of Bcl-2 move forward with their preclinical and clinical development, including: (a) compound stability and formulation; (b) pharmacokinetics (PK) and metabolism; (c) toxicity; and (d) off-target mechanisms. For example, some of the natural products that target Bcl-2 are inherently unstable molecules prone to rapid inactivation by oxidation *in vitro*, in addition to rapid metabolism to inactive products *in vivo*. With respect to toxicity, since some of these small molecules attack multiple anti-apoptotic members of the Bcl-2 family, toxicity to normal tissues is more likely than with antisense-based drugs that selectively target only one family member. The potential advantage of broad-spectrum inhibitors of anti-apoptotic Bcl-2-family proteins, however, is that they may be more effective due to redundancy within the Bcl-2 family and the tendency of tumors to express more than one anti-apoptotic member of the family. Finally, most of the natural products that bind the BH3 pocket of Bcl-2 have additional cellular targets, and these off-target activities may or may not be important to the antitumor activity of the compounds. Provided these off-target activities do not add untoward toxicity, this may be an advantage rather than a disadvantage for anticancer drug development. Nevertheless, multitarget activity potentially complicates structure-activity relation (SAR) studies in which one might envision producing semi-synthetic derivatives that retain activity against Bcl-2 since chemical modifications that enhance Bcl-2 binding might simultaneously reduce interaction with other cancer-relevant targets.

17.4.4 Activating Endogenous Antagonists of Bcl-2

In addition to Bcl-2-targeting strategies that attack at the level of gene expression, mRNA, or proteins, a fourth possible route to dealing with the problem of pathological overexpression of Bcl-2 in malignant cells has recently been revealed. This strategy seeks to induce expression or activation of endogenous proteins that bind and negate the cytoprotective action of Bcl-2. For example, an orphan member of the retinoid/steroid family of nuclear receptors, TR3 (Nur77) has been shown to translocate from the nucleus to the cytosol in response to certain cell-death stimuli in some types of cells (Li et al., 2000). In the cytosol, TR3 binds a regulatory domain within the Bcl-2 protein, thus accounting for accumulation of TR3 on the surface of mitochondria. TR3 induces a profound conformational change of Bcl-2, causing exposure of this BH3 domain and converting Bcl-2 from a protector to a killer (Lin et al., 2004). The BH3 domain of Bcl-2 binds both pro-apoptotic Bcl-2-family members Bax and Bak as well as anti-apoptotic Bcl-2-family proteins, probably activating the pro-apoptotic while inactivating the anti-apoptotic.

Recently, a class of compounds representing analogs of the retinoid APHN has been identified that induces TR3 expression and translocation into the cytosol. The prototype compound, 3Cl-AHPC, lacks activity against retinoid receptors although having evolved from attempts to synthesize highly selective retinoid ligands (Dawson et al., 2004). 3Cl-AHPC has demonstrated activity against cultured leukemia cells and in preclinical animal models of AML (Zhang et al., 2002; Dawson et al., 2001) and is active even against retinoid refractory leukemia cell lines. 3Cl-AHPC appears to induce phosphorylation of TR3, possibly via a Jun N-terminal kinase (JNK) pathway, correlating with its exodus from the nucleus (Kolluri et al., 2003). This compound may therefore provide a mechanism to invoke the TR3 pathway, converting Bcl-2 from a protector to a killer and thus making Bcl-2 a liability rather than an advantage for malignant cells.

Opposing this TR3-mediated pathway for apoptosis is Akt/PKB, which appears to nullify pro-apoptotic actions of TR3 by phosphorylation at serine 530 (Pekarsky et al., 2001; Masuyama et al., 2001). Thus, TR3 joins a long list of apoptosis-relevant substrates of Akt, which includes Bcl-2 antagonist BID, human caspase-9, pro-apoptotic kinase Ask1, p53 antagonist Mdm2, and others [reviewed in Testa and Bellacosa (2001)]. Small wonder then that Akt has emerged as a prominent target for cancer drug discovery [reviewed in Mitsiades et al., (2004)]. In this regard, relatively selective but fairly weak inhibitors of Akt such as 1 *L*-hydroxymethyl-chiro-inositol-2-((*R*)-2-methyl-3-*O*-octadecyl/carbonate) have been described, which has preclinical activity against multiple myeloma cells (Mitsiades et al., 2002). Interestingly, 3Cl-AHPC may also reduce Akt activity in tumor cells (Farhana et al., 2004), further promoting the TR3 pathway for cell death. It remains to be determined what effect the Tcl-1 oncoprotein has on nuclear export of TR3, but it may be of interest that this protein overexpressed in CLLs promotes nuclear targeting of Akt and regulates the transcriptional activity of TR3 (Pekarsky et al., 2001).

The revelation that the Bcl-2 protein can have two opposing phenotypes, depending upon its interactions with other proteins such as TR3, suggests a possible explanation for the paradoxical association of higher Bcl-2 with favorable clinical outcome in some types of cancer (Reed, 1996b). Interestingly, TR3 is among a small set of genes identified by DNA microarray analysis demonstrating a strong association with longer survival among patients with diffuse large B-cell lymphoma (Shipp et al., 2002). Examination of Bcl-2 and TR3 expression in clinical specimens may help select patients most likely to benefit from therapeutic strategies designed to stimulate the TR3 pathway for Bcl-2 conversion. Clearly, one would not want to combine TR3-based therapies with drugs affecting *BCL-2* expression, given that antisense-mediated suppression of *BCL-2* expression protects tumor cell lines in culture from TR3-induced apoptosis (Lin et al., 2004).

Binding of TR3 to Bcl-2 requires the presence of an ~50 amino-acid flexible loop region located between the first two α-helices in the Bcl-2 protein (Lin et al., 2004). Prior studies have shown that this domain in Bcl-2 and Bcl-X_L negatively regulates the anti-apoptotic functions of these proteins (Chang et al., 1997). It is currently unknown which of the six anti-apoptotic Bcl-2 family proteins can bind TR3, but all of them have predicted "loop" domains. Interesting, at least one anti-apoptotic Bcl-2-family member, mouse A1 (whose human counterpart is Bfl1) reportedly resides in the nucleus (Somogyi et al., 2001), raising suspicions that it may target nuclear proteins such as TR3. The loop of Bcl-2 is a documented site of phosphorylation, correlating with inactivation of the anti-apoptotic function of Bcl-2 in some scenarios (Haldar et al., 1998). It remains to be explored whether phosphorylation of Bcl-2 affects TR3 binding, but microtubule-modulating drugs such as the taxanes are the most potent inducers of Bcl-2 loop phosphorylation identified thus far. Of note, somatic mutations have been documented in rearranged *BCL-2* genes in lymphomas, corresponding to amino-acid substitutions in the loop of the Bcl-2 protein (Tanaka et al., 1992). This observation implies that Bcl-2 may acquire mutations that abolish TR3 sensitivity, a hypothesis that awaits experimental confirmation.

To some extent, the concept of invoking endogenous antagonists of Bcl-2 is not novel. Anti-apoptotic Bcl-2-family proteins are normally held in check by endogenous proteins that contain a BH3 domain, especially the so-called BH3-only proteins (BOPs). Several of these BOPs are linked to pathways that help explain how the currently available cytotoxic anticancer drugs trigger cell death, including: (a) the genes encoding Puma, Noxa, and Bid, whose expression is directly induced by p53; (b) isoforms of the Bim proteins, which are normally sequestered on microtubules; and (c) Bax, which is held in an inactive conformation in the cytosol by the regulatory subunit (Ku70) of the DNA-dependent kinase, becoming released during responses to DNA damage (Huang and Strasser, 2000; Sawada et al., 2003). Unfortunately, defects in these specific pathways involving BOPs or pathological overexpression of Bcl-2 that overwhelms the activity of these endogenous antagonists all too frequently create roadblocks to apoptosis that must be overcome by alternative means. The observation that commonly used cytotoxic anticancer drugs are capable of modulating endogenous pathways impinging on Bcl-2 and related anti-apoptotic proteins underscores why Bcl-2-targeted therapies are often synergistic with conventional cytotoxic agents in terms of inducing apoptosis of malignant cells. Knowledge of the molecular details of the mechanisms by which traditional cytotoxic anticancer drugs affect Bcl-2 family proteins thus may help in selecting optimal combination therapies for clinical trials of Bcl-2 antagonists, where the Bcl-2 antagonists are used to sensitize cancer and leukemia cells to apoptosis induction by conventional cytotoxic drugs or radiation therapy. Underscoring the importance of this knowledge base are recent studies suggesting that alkylating agents are capable of inducing cell death via a necrotic rather than apoptotic mechanism, which is independent of Bcl-2/Bax-family proteins (Zong et al., 2004). It is therefore unfortunate that many of the early clinical trials of Bcl-2 antagonists (e.g., oblimersen sodium [Genasense™]) have involved combination with alkylating agents since this newfound knowledge suggests that this would not be a synergistic combination.

As we continue to learn more about apoptosis mechanisms involving Bcl-2-family proteins, novel strategies for therapeutic intervention emerge. Another recent example is the discovery that a cytoplasmic pool of p53 associates with mitochondria, directly inducing activation of the Bax protein through a transcription-independent mechanism (Chipuk et al., 2004). Importantly, even mutant p53 is capable of activating this cell-death pathway, raising hopes of finding pharmacological interventions that would entice mutant p53 to attack mitochondria and trigger apoptosis of cancer cells in which this important tumor suppressor gene product has suffered somatic mutations that inactivate its nuclear functions.

17.5 CONVERGENCE PATHWAY

The intrinisic and extrinsic pathways for apoptosis converge on downstream effector caspases. Certain effector caspases are targets of suppression by an endogenous family of anti-apoptotic proteins called IAPs (inhibitor of apoptosis proteins). The IAPs represent an evolutionarily conserved family of proteins, originally identified in insect viruses (Miller, 1999). IAPs contain one or more copies of a domain called the baculoviral IAP repeat (BIR). BIR domains are sometimes accompanied by other domains, including RING domains and ubiquitin-conjugating enzyme domains (E2s), which endow these proteins with additional properties such as an ability to target themselves and associated proteins for ubiquitation and proteasome-dependent degradation. The human genome encodes eight IAP-family members: XIAP, cIAP1, cIAP2, Naip, Survivin, ML-IAP (Livin; K-IAP), ILP2 (Ts-IAP), and Apollon (Bruce) (Reed et al., 2004).

Pathological overexpression of IAPs has been documented in cancer and leukemia [reviewed in Deveraux and Reed (1999) and Prefontaine et al., (1998)]. For example, the IAP-family member Survivin is overexpressed in most cancers (Ambrosini et al., 1997) and has become a topic of considerable attention for its dual role as a regulator of cell division (chromosome segregation and cytokinesis) and apoptosis [reviewed in Altieri et al., (1999), Reed and Reed (1999), and Reed and Bischoff (2000)]. Similarly, the IAP-family member ML-IAP is rarely expressed in normal tissues but is found at elevated levels in melanomas and some renal cancers (Vucic et al., 2000; Lin et al., 2000; Kasof and Gomes, 2001). Chromosomal translocations activating the *cIAP2* gene are found in some mucosal-associated lymphomas (Motegi et al., 2000). The *XIAP* gene is overexpressed in a substantial proportion of AMLs (Tamm et al., 2000), CLLs (Schimmer et al., 2003), hepatocellular carcinomas (Shiraki et al., 2003), ovarian cancers (Li et al., 2005), and prostate cancers (Krajewska et al., 2003).

The functional importance of overexpressed IAPs for apoptosis suppression in cancers has been supported by antisense experiments in which knocking down expression of *SURVIVIN, XIAP,* or other IAPs family genes has been shown to induce apoptosis of tumor cell lines in culture or to sensitize tumor cell lines to apoptosis induced by anticancer drugs [Chen, 2000; Gordon, 2002; Li, 2001; Sasaki, 2000; Holcik, 2000]. In contrast, gene knockout studies in mice imply that most normal cells are potentially less dependent on IAPs compared to tumor cells, inasmuch as targeted disruption of the genes for *xiap*, *ciap1*, and *ciap2* produces little phenotype (Harlin et al., 2001).

Routes to small-molecule drug discovery have been revealed through knowledge of the structural and molecular details of how IAPs inhibit caspases, enlightened by insights gained from studies of endogenous inhibitors of the IAPs. For example, specific segments within IAP-family proteins have been identified that bind specific caspases, providing structural information useful for envisioning generations of compounds that disrupt these protein interactions, freeing caspases to induce apoptosis (Deveraux and Reed, 1999; Salvesen and Duckett, 2002; Fesik and Shi, 2001). XIAP, for example, which contains three tandem BIRs, utilizes different domains to suppress both downstream effector caspases that operate at points of convergence of apoptosis pathways (e.g., caspase-3 and -7), and caspase-9, the apical protease in the mitochondrial pathway for apoptosis. The BIR2 domain and a flanking upstream linking segment of XIAP bind active caspase-3 and -7, while the BIR3 domain is responsible for binding active caspase-9. In contrast, ILP2, which contains a single BIR domain, is a potent suppressor only of caspase-9 but not caspase-3 and -7.

17.5.1 Drugs Targeting IAP-Family Proteins

Using an enzyme derepression assay in which XIAP-mediated suppression of caspase-3 is overcome by chemical compounds, at least two classes of XIAP antagonists have been described that target the vicinity of the BIR2 domain and that free caspase-3, including diphenyl and triphenylureas and benzenesulfonamide derivatives (Schimmer et al., 2004; Wu et al., 2003) (Table 17.2). The phenylureas exhibit broad-spectrum apoptosis-inducing activity against cultured human tumor cell lines and primary leukemias and suppress growth of tumor xenografts in mice without apparent toxicity to normal tissues (Schimmer et al., 2004). Importantly, these compounds induce apoptosis through a Bcl-2/Bcl-X_L-independent mechanism, which provides encouragement that they may prove useful for chemorefractory cancers and leukemias.

Attempts to mimic the actions of endogenous antagonists of IAPs define another strategy for achieving specific inhibitors. Endogenous proteins such as SMAC and OMI (HtrA2) bind and suppress IAPs, releasing caspases to kill cells (Du et al., 2000; Verhagen et al., 2000; Suzuki et al., 2000). A 4'mer peptide corresponding to the N-terminus of the mature SMAC protein is reported to be sufficient to bind IAPs and block their association with caspases (Fesik and Shi, 2001;

Chai et al., 2000), which is analogous to similar mechanisms previously defined in *Drosophila* where IAPs are kept in check by endogenous antagonists Reaper, Hid, Grim, and Sickle [reviewed in Goyal (2001)]. By fusing membrane-penetrating peptides onto SMAC, OMI, Hid, or Reaper peptides, it is possible to induce apoptosis of human cancer cell lines in culture as well as to suppress tumor growth in xenograft models in mice (Deng et al., 2002; Guo et al., 2002; Carson et al., 2002; Chauhan et al., 2001). These data thus provide proof-of-concept evidence that small molecules that mimic the effects of these IAP-binding peptides could potentially be exploited as drugs for cancer treatment.

The three-dimensional structure of the BIR3 domain of XIAP complexed with SMAC reveals that N-terminal 4-amino acids of the mature SMAC protein bind in the same crevice normally occupied by the N-terminus of the small subunit of processed caspase-9, competing for binding (Fesik and Shi, 2001) (Fig. 17.3). Consequently, small-molecule compounds that mimic the SMAC 4'mer peptide dislodge active caspase-9 from BIR3, promoting apoptosis. Analogous SMAC-peptide binding crevices are anticipated on the BIRs of many other IAP-family proteins, based on molecular modeling, suggesting SMAC peptide-mimicking compounds would be rather broad-spectrum in their activity. Several groups have initiated preclinical studies of SMAC-mimics. Among the reported antagonists that target the SMAC binding site are embelin, a natural product (Nikolovska-Coleska et al., 2004), as well as several synthetic peptidyl and nonpeptidyl SMAC mimics (Li et al., 2004; Sun et al., 2005), with the most potent binding with low nanomolar affinity (Oost et al., 2004) .

In addition to SMAC and OMI, additional endogenous antagonists of IAPs have been described, including ARTS, XAF1, and NRAGE (Gottfried et al., 2004; Liston

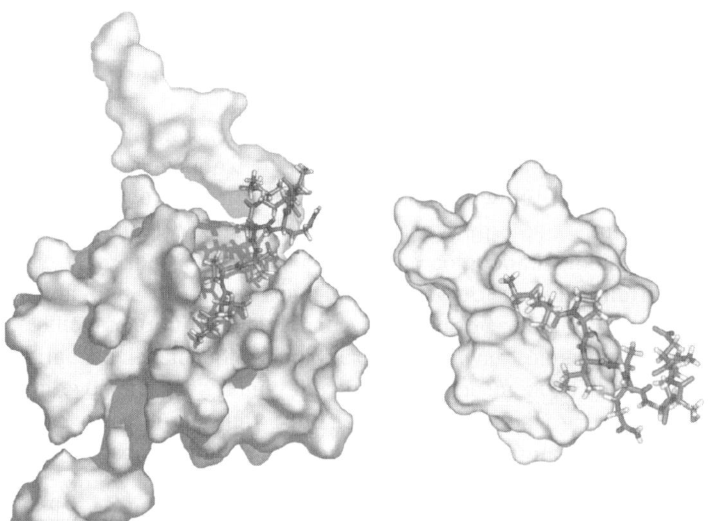

Figure 17.3 Structure of BIR domain and inhibitory SMAC peptide. BIR domain of XIAP is presented in space-filling model, while the SMAC tetra-peptide is presented in stick mode. The left structure gives the entire Bir3 domain, while the right structure is a subdomain of Bir3. The structure is derived from PDB #1G3F. See color plates.

TABLE 17.2 Structures of IAP Antagonists

SMAC		Non-SMAC
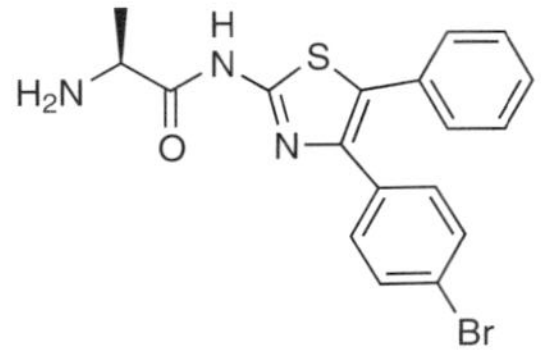 **ABT-Thiazol derivative 20 g** 2-Amino-*N*-[4-(4-bromo-phenyl)-5-phenyl-thiazol-2-yl]-propionamide	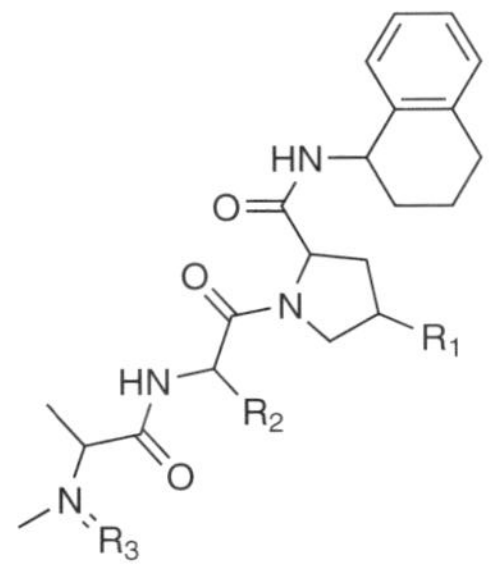 **Capped Tripeptides** R_1 = H or phenyl; R_2 = *t*-butyl or Cyclohexyl; R_3 = H or methyl	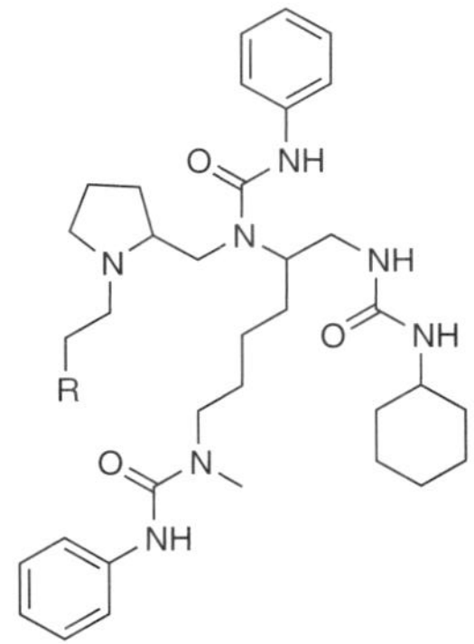 **1396-11**R = Adamantyl (*N*-({(2*S*)-1-[2-(1-adamantyl)ethyl] pyrrolidin-2-yl}methyl)-*N*-{(1*R*)-1-{[(anilinocarbonyl)amino]methyl}-5-[(anilinocarbonyl)(methyl)amino]pentyl}-*N'*-phenylurea); **1396-12** R = 2-Cyclohexyl-ethyl (*N*-[(5*R*)-6-[(anilinocarbonyl)amino]-5-((anilinocarbonyl){[(2S)-1-(4-cyclohexylbutyl)pyrrolidin-2-yl]methyl}amino)hexyl]-*N*-methyl-*N'*-phenylurea)
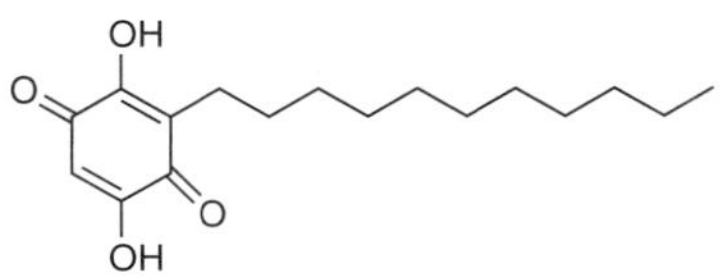 **Embelin** 2,5-Dihydroxy-3-undecyl-[1,4] benzoquinone	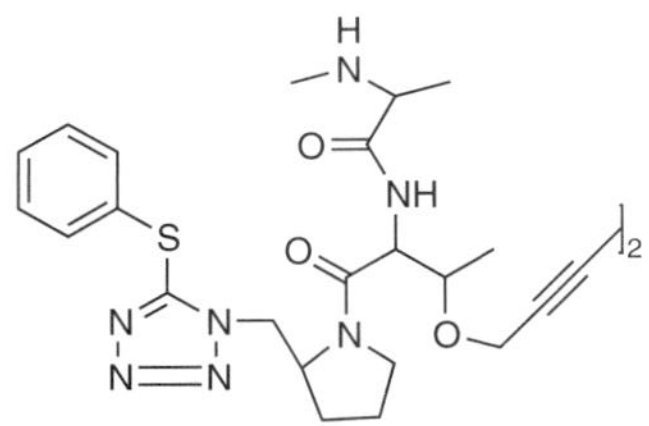 **Compound 3** Dimer of *N*-{2-But-2-ynyloxy-1-[2-(5-phenylsulfanyl-tetrazol-1-ylmethyl)-pyrrolidine-1-carbonyl]-propyl}-2-methylamino-propionamide.	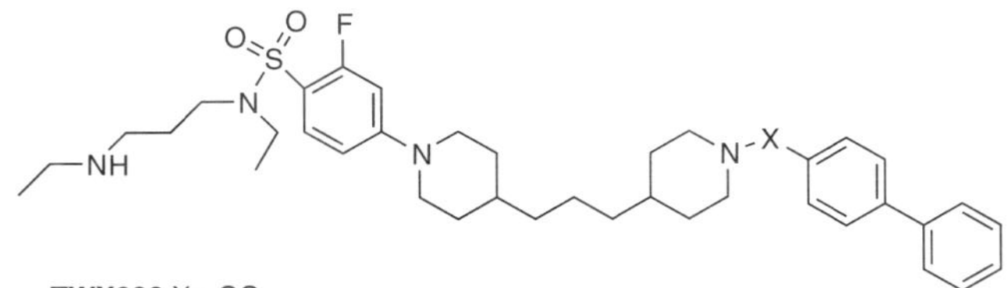 **TWX006** X=-CO- (4-(4-{3-[1-(Biphenyl-4-carbonyl)-piperidin-4-yl]-propyl}-piperidin-1-yl)-*N*-ethyl-*N*-(3-ethylamino-propyl)-2-fluorobenzenesulfonamide **TWX024** X= -CONH- (4-[3-(1-{4-[Ethyl-(3-ethylamino-propyl)-sulfamoyl]-3-fluoro-phenyl}-piperidin-4-yl)-propyl]-piperidine-1-carboxylic acid biphenyl-4-ylamide)

et al., 2001; Jordan et al., 2001). The structural details of how these proteins interact with their IAP targets is thus far undefined but, when known, may provide additional strategies for generation of small-molecule antagonists that word through non-SMAC mechanisms. Given that the aforementioned phenylurea-base antagonists of XIAP have been reported to target a non-SMAC site on XIAP (Schimmer et al., 2004), it will be interesting to determine whether those compounds in fact mimic ARTS, NRAGE, or XAF1 in terms of their docking site on XIAP.

Finally, a variety of indirect strategies for negating the activity of IAPs are beginning to emerge. For example, studies have shown that phosphorylation of Survivin on threonine 34 is required for execution of cell division and suppression of apoptosis (O'Connor et al., 2000). The predominant kinases responsible for phosphorylation of Survivin appear to be Cdc2 or related cyclin-dependent kinases. Compounds inhibiting Cdc2-family kinases include flavopiridol, currently in clinical trials.

17.5.2 Drugs Inhibiting IAP-Family Gene Expression

In addition to targeting IAP-family proteins directly with small-molecule drugs, efforts to reduce expression using antisense-based drugs are also well underway. Antisense ODNs targeting Survivin mRNA (Chen et al., 2000) have begun clinical development using second-generation antisense chemistry (ISIS Pharmaceuticals/Lilly Inc). An antisense-based approach to Survivin is particularly attractive because the mechanism by which this protein suppresses caspases differs from most other IAPs (Marusawa et al., 2003), and a clear path forward for small-molecule drug discovery remains uncertain based on available protein structure data (Fesik, 2000). In addition, a phosphorothioate-based antisense ODN targeting XIAP (AEG35156/GEM640) has been designated for human clinical trials (Agera Pharmaceuticals, Inc.). The classical dilemma of specificity versus redundancy is highly germane to the question of whether small-molecule compounds versus nucleic-acid-based drugs are preferable. Most strategies for small-molecule drugs are likely to target multiple members of the IAP family and thus could afford advantages in terms of addressing more drug targets but also might suffer disadvantages with respect to possible toxicities. In this regard, the observation that some cancers overexpress four or more IAP members simultaneously argues in favor of a small-molecule strategy involving broad-spectrum inhibitors (Krajewska et al., 2003). Yet, experiments using antisense or siRNA to knock-down expression of individual IAPs have suggested that highly selective inhibitors could also be successful. Future work will empirically define the best paths forward for cancer drug discovery efforts that target single or multiple members of the IAP family.

17.6 CONCLUSIONS

Emerging knowledge about the molecular mechanism of apoptosis dysregulation in cancer and leukemia has revealed a plethora of potential drug discovery targets. Detailed structural analysis of apoptosis proteins and studies of their biochemical mechanisms have suggested strategies for lead generation, resulting in numerous novel chemical entities (NCEs) with mechanism-based activity. Although providing encouraging proof-of-principle data that help validate several apoptosis targets, much work lies ahead in terms of optimizing the spectrum of activity of compounds that interact with multiple members of apoptosis protein families, improving their stability and pharmacological properties, establishing optimal formulations for stability and delivery, and elucidating attendant rate-limiting toxicities. Many of the most logical targets for promoting apoptosis of cancer and leukemia cells are technically challenging, often involving either disrupting protein interactions or altering gene expression, as opposed to traditional pharmaceuticals that typically attack the active sites of enzymes. Modern techniques of structure-based drug optimization render this task feasible, but still challenging. Such targets require long-term commitments, often outstripping the usual drug discovery and development cycle time incorporated into the practices of pharmaceutical companies. For those with the stamina, the rewards are likely to be great, creating a new era in cancer therapy where the intrinsic or acquired resistance of malignant cells to apoptosis can be pharmacologically reversed, reinstating natural pathways for cell suicide.

There is good reason to suspect that malignant cells will be preferentially susceptible to restoration of apoptosis sensitivity, compared to normal cells. Cancer cells exhibit a wide variety of abnormal behaviors and molecular processes that normally trigger an apoptosis response, including cell-cycle checkpoint dysregulation, oncogene activation, chromosome segregation defects, cell detachment from substratum, and outgrowth of blood supply (hypoxia). These defects render tumor cells more dependent on apoptosis-suppressing genes and proteins, and thus withdrawing this crutch from malignant cells may promote self-destruction of transformed cells while sparing normal cells. The full validity of this hypothesis awaits experimental verification in human clinical trials, but the present insights from animal studies and the current forays into the clinic provide encouragement. Altogether, therefore, promises are high that apoptosis-based strategies for cancer drug discovery will yield effective weapons for use in the war on cancer, and thus these efforts should be supported and encouraged.

ACKNOWLEDGMENT

We thank M. Hanaii for manuscript preparation.

REFERENCES

Altieri, D. C., Marchisio, P. C., and Marchisio, P. C. (1999). Survivin apoptosis: An interloper between cell death and cell proliferation in cancer. *Lab Invest* **79**, 1327–1333.

Altucci, L., Rossin, A., Raffelsberger, W., Reitmair, A., Chomienne, C., and Gronemeyer, H. (2001). Retinoic acid-induced apoptosis in leukemia cells is mediated by paracrine action of tumor-selective death ligand TRAIL. *Nat Med* **7**, 680–686.

Ambrosini, G., Adida, C., and Altieri, D. C. (1997). A novel anti-apoptosis gene, Survivin, expressed in cancer and lymphoma. *Nat Med* **3**, 917–921.

Aron, J. L., Parthun, M. R., Marcucci, G., Kitada, S., Mone, A. P., Davis, M., Shen, T., Kitada, S., Murphy, T., Kanakry, C., Lucas, D. M., Reed, J. C., Grever, M. R., and Byrd, J. C. (2002). Depsipeptide (FR901228) induces histone acetylation and inhibition of histone deacetylase in chronic lymphocytic Leukemia cells concurrent with activation of caspase-8 mediated apoptosis and down regulation of c-FLIP protein. *Blood* **102**, 652–658.

Ashkenazi, A., and Dixit, V. M. (1998). Death receptors: Signaling and modulation. *Science* **281**, 1305–1308.

Ashkenazi, A., Pai, R. C., Fong, S., Leung, S., Lawrence, D. A., Marsters, S. A., Blackie, C., Chang, L., McMurtrey, A. E., Hebert, A., DeForge, L., Koumenis, I. L., Lewis, D., Harris, L., Bussiere, J., Koeppen, H., Shahrokh, Z., and Schwall, R. H. (1999). Safety and antitumor activity of recombinant soluble Apo2 ligand. *J Clin Invest* **104**, 155–162.

Baldwin, A. S. (2001). Control of oncogenesis and cancer therapy resistance by the transcription factor NF-kappaB. *J Clin Invest* **107**, 241–246.

Baud, V., and Karin, M. (2001) Signal transduction by tumor necrosis factor and its relatives. *Trends Cell Biol* **11**, 372–377.

Bazzoni, F., and Beutler, B. (1996). The tumor necrosis factor ligand and receptor families. *N Engl J Med* **334**, 1717–1725.

Becattini, B., Kitada, S., Leone, M., Monosov, E., Chandler, S., Dayong, Z., Kipps, T. J., Reed, J. C., and Pellecchia, M. (2004). Rational design and real time in-cell detection of the pro-apoptotic activity of a novel compound targeting Bcl-Xl. *Chem Biol* **11**, 389–395.

Boatright, K. M., and Salvesen, G. S. (2003). Mechanisms of caspase activation. *Curr Opin Cell Biol* **15**, 725–731.

Breckenridge, D. G., Germain, M., Mathai, J. P., Nguyen, M., and Shore, G. C. (2003). Regulation of apoptosis by endoplasmic reticulum pathways. *Oncogene* **22**, 8608–8618.

Buchele, T. (2003). Proapoptotic therapy with oblimersen (bcl-2 antisense oligonucleotide)—Review of preclinical and clinical results. *Onkologie* **26** (Suppl 7), 60–69.

Burke, J. R., Pattoli, M. A., Gregor, K. R., Brassil, P. J., MacMaster, J. F., McIntyre, K. W., Yang, X., Iotzova, V. S., Clarke, W., Strnad, J., Qui, Y., and Zusi, F. C. (2003). BMS-345541 is a highly selective inhibitor of IkB kinase that binds at an allosteric site of the enzyme and blocks NF-κB-dependent transcription in mice. *J Biol Chem* **278**, 1450.

Carson, J. P., Behnam, M., Sutton, J. N., Du, C., Wang, X., Hunt, D. F., Weber, M. J., and Kulik, G. (2002). Smac is required for cytochrome c-induced apoptosis in prostate cancer LNCaP cells. *Cancer Res* **62**, 18–23.

Cartee, L., Maggio, S. C., Smith, R., Sankala, H. M., Dent, P., and Grant, S. (2003). Protein kinase C-dependent activation of the tumor necrosis factor receptor-mediated extrinsic cell death pathway underlies enhanced apoptosis in human myeloid leukemia cells exposed to bryostatin 1 and flavopiridol. *Mol Cancer Ther* **2**, 83–93.

Castro, A. C., Dang, L. C., Soucy, F., Grenier, L., Mazdiyasni, H., Hottelet, M., Parent, L., Pien, C., Palombella, V., and Adams, J. (2003). Novel IKK inhibitors: Beta-carbolines. *Bioorg Med Chem Lett* **13**, 2419–2422.

Chai, J., Du, C., Wu, J. W., Kyin, S., Wang, X., and Shi, Y. (2000). Structural and biochemical basis of apoptotic activation by Smac/DIABLO. *Nature* **406**, 855–862.

Chan, S. L., Lee, M. C., Tan, K. O., Yang, L. K., Lee, A. S., Flotow, H., Fu, N. Y., Butler, M. S., Soejarto, D. D., Buss, A. D., and Yu, V. C. (2003). Identification of chelerythrine as an inhibitor of BclXL function. *J Biol Chem* **6**, 20453–20456.

Chang, B. S., Minn, A. J., Muchmore, S. W., Fesik, S. W., and Thompson, C. B. (1997). Identification of a novel regulatory domain in Bcl-x_L and Bcl-2. *EMBO J* **16**, 968–977.

Chang, D. W., Xing, Z., Pan, Y., Algeciras-Schimnich, A., Barnhart, B. C., Yaish-Ohad, S., Peter, M. E., and Yang, X. (2002). c-$FLIP_L$ is a dual function regulator for caspase-8 activation and CD95-mediated apoptosis. *EMBO J* **21**, 3704–3714.

Chauhan, D., Hideshima, T., Rosen, S., Reed, J. C., Kharbanda, S., and Anderson, K. C. (2001). Apaf-1/Cytochrome c-independent and SMAC-dependent induction of apoptosis in multiple myeloma (MM) cells. *J Biol Chem* **276**, 24453–24456.

Chen, J., Wu, W., Tahir, S. K., Kroeger, P. E., Rosenberg, S. H., Bennett, F., Krajewski, S., Krajewska, M., Welsh, K., Reed, J. C., and Ng, S. C. (2000). Down-regulation of survivin by antisense oligonucleotides increases apoptosis, inhibits cytokinesis and anchorage-independent growth. *Neoplasia* **2**, 235–241.

Chipuk, J. E., Kuwana, T., Bouchier-Hayes, L., Droin, N. M., Newmeyer, D. D., Schuler, M., and Green, D. R. (2004). Direct activation of Bax by p53 mediates mitochondrial membrane permeabilization and apoptosis. *Science* **303**, 1010–1014.

Cotter, F. E., Johnson, P., Hall, P., Pocock, C., Al Mahdi, N., Cowell, J. K., and Morgan, G. (1994). Antisense oligonucleotides suppress B-cell lymphoma growth in a SCID-hu mouse model. *Oncogene* **9**, 3049–3055.

Dai, Y., Pei, X. Y., Rahmani, M., Conrad, D. H., Dent, P., and Grant, S. (2004). Interruption of the NF-kappaB pathway by Bay 11–7082 promotes UCN-01-mediated mitochondrial dysfunction and apoptosis in human multiple myeloma cells. *Blood* **103**, 2761–2770.

Dawson, M. I., Hobbs, P. D., Peterson, V. J., Leid, M., Lange, C. W., Feng, K. C., Chen, G., Gu, J., Li, H., Kolluri, S. K., Zhang, X., Zhang, Y., and Fontana, J. A. (2001). Apoptosis induction in cancer cells by a novel analogue of 6-[3-(1-adamantyl)-4-hydroxyphenyl]-2-naphthalenecarboxylic acid lacking retinoid receptor transcriptional activation activity. *Cancer Res* **61**, 4723–4730.

Dawson, M. I., Harris, D., Liu, G., Hobbs, P., Lange, C., Jong, L., Bruey-Sedano, N., James, S., Zhang, X.-K., Peterson, V., Leid, M., Farhana, L., Rishi, A., and Fontana, J. (2004). Antagonist analogue of 6-[3'-(1-Adamantyl)-4'-hydroxyphenyl]-2-naphthalenecarboxylic acid (AHPN) family of apoptosis inducers that effectively blocks AHPN-induced apoptosi but not cell-cycle arrest. *J Med Chem* 3518–3536.

Degterev, A., Lugovskoy, A., Cardone, M., Mulley, B., Wagner, G., Mitchison, T., and Yuan, J. (2001). Identification of small-molecule inhibitors of interaction between the BH3 domain and Bcl-xL. *Nat Cell Biol* **3**, 173–182.

Demaurex, N., and Distelhorst, C. (2003). Apoptosis—The calcium connection. *Science* **300**, 65–67.

Deng, Y., Lin, Y., and Wu, X. (2002) TRAIL-induced apoptosis requires Bax-dependent mitochondrial release of Smac/DIABLO. *Genes Dev* **16**, 33–45.

Derenne, S., Monia, B., Dean, N. M., Taylor, J. K., Rapp, M. J., Harousseau, J. L., Bataille, R., and Amiot, M. (2002). Antisense strategy shows that Mcl-1 rather than Bcl-2 or Bcl-xL is an essential survival protein of human myeloma cells. *Blood* **100**, 194–199.

Deveraux, Q. L., and Reed, J. C. (1999). IAP family proteins: Suppressors of apoptosis. *Genes Dev* **13**, 239–252.

Dhanalakshmi, S., Singh, R. P., Agarwal, C., and Agarwal, R. (2002). Silibinin inhibits constitutive and TNFalpha-induced activation of NF-kappaB and sensitizes human prostate carcinoma DU145 cells to TNFalpha-induced apoptosis. *Oncogene* **21**, 1759–1767.

DiDonato, J., Hayakawa, M., Rothwarf, D., Zandi, E., and Karin, M. (1997). A cytokine-responsive IkB kinase that activates the transcription factor NF-κB. *Nature* **388**, 548–554.

Du, C., Fang, M., Li, Y., Li, L., and Wang, X. (2000). SMAC, a mitochondrial protein that promotes cytochrome c-dependent caspase activation by eliminating IAP inhibition. *Cell* **102**, 33–42.

Evan, G. I., and Vousden, K. H. (2001). Proliferation, cell cycle and apoptosis in cancer. *Nature* **411**, 342–348.

Fan, C. M., and Maniatis, T. (1991). Generation of p50 subunit of NF-κB by processing of p105 through an ATP-dependent pathway. *Nature* **354**, 395–398.

Farhana, L., Dawson, M. I., Huang, Y., Zhang, Y., Rishi, A. K., Reddy, K. B., Freeman, R. S., and Fontana, J. A. (2004). Apoptosis signaling by the novel compound 3-Cl-AHPC involves increased EGFR proteolysis and accompanying decreased phosphatidylinositol 3-kinase and AKT kinase activities. *Oncogene* **23**, 1874–1884.

Fennell, D. A., Corbo, M. V., Dean, N. M., Monia, B. P., and Cotter, F. E. (2001). *In vivo* suppression of Bcl-XL expression facilitates chemotherapy-induced leukaemia cell death in a SCID/NOD-Hu model. *Br J Haematol* **112**, 706–713.

Fesik, S. W. (2000). Insights into programmed cell death through structural biology. *Cell* **103**, 273–282.

Fesik, S. W., and Shi, Y. (2001). Structural biology. Controlling the caspases. *Science* **294**, 1477–1478.

Gauwerky, C. E., Hoxie, J., Nowell, P. C., and Croce, C. M. (1988). Pre-B-cell leukemia with a t(14;18) translocation is preceded by follicular lymphoma. *Oncogene* **2**, 431–435.

Gottfried, Y., Rotm, A., Lotan, R., Steller, H., and Larisch, S. (2004). The mitochondrial ARTS protein promotes apoptosis through targeting XIAP. *EMBO J* **23**, 1627–1635.

Goyal, L. (2001). Cell death inhibition: Keeping caspases in check. *Cell* **104**, 805–808.

Green, D. R., and Reed, J. C. (1998). Mitochondria and apoptosis. *Science* **281**, 1309–1312.

Griffith, T., Chin, W., Jackson, G., Lynch, D., and Kubin, M. (1998). Intracellular regulation of TRAIL-induced apoptosis in human melanoma cells. *J Immunol* **161**, 2833–2840.

Guensberg, P., Wacheck, V., Lucas, T., Monia, B., Pehamberger, H., Eichler, H. G., and Jansen, B. (2002). Bcl-xL antisense oligonucleotides chemosensitize human glioblastoma cells. *Chemotherapy* **48**, 189–195.

Guinness, M. E., Kenney, J. L., Reiss, M., and Lacy, J. (2000). Bcl-2 antisense oligodeoxynucleotide therapy of Epstein–Barr virus-associated lymphoproliferative disease in severe combined immunodeficient mice. *Cancer Res* **60**, 5354–5358.

Guo, F., Nimmanapalli, R., Paranawithana, S., Wittman, S., Griffin, D., Bali, P., O'Bryan, E., Fumero, C., Wang, H. G., and Bhalla, K. (2002). Ectopic overexpression of second mitochondria-derived activator of caspases (Smac/DIABLO) or cotreatment with N-terminus of Smac/DIABLO peptide potentiates epothilone B derivative-(BMS 247550) and Apo- 2L/TRAIL-induced apoptosis. *Blood* **99**, 3419–3426.

Guzey, M., Kitada, S., and Reed, J. C. (2002). Apoptosis induction by $1a_2 25(OH)_2$ in vitamin-D_3 prostate cancer cell lines. *Mol Cell Ther* **1**, 667–677.

Hail, N., Jr., Konopleva, M., Sporn, M., Lotan, R., and Andreeff, M. (2004). Evidence supporting a role for calcium in apoptosis induction by the synthetic triterpenoid 2-cyano-3,12-dioxooleana-1,9-dien-28-oic acid (CDDO). *J Biol Chem* **279**, 11179–11187.

Haldar, S., Basu, A., and Croce, C. M. (1998). Serine-70 is one of the critical sites for drug-induced Bcl2 phosphorylation in cancer cells. *Cancer Res* **58**, 1609–1615.

Harlin, H., Reffey, S. B., Duckett, C. S., Lindsten, T., and Thompson, C. B. (2001). Characterization of XIAP-deficient mice. *Mol Cell Biol* **21**, 3604–3608.

Holinger, E., Chittenden, T., and Lutz, R. (1999). Bak BH3 Peptides antagonize Bcl-x_L Function and induce apoptosis through cytochrome c-independent activation of caspases. *J Biol Chem* **274**, 13298–13304.

Huang, D. C., and Strasser, A. (2000). BH3-only proteins—Essential initiators of apoptotic cell death. *Cell* **103**, 839–842.

Hyer, M. L., Croxton, R., Krajewski, S., Krajewski, S., Kress, C. L., Lu, M., Suh, N., Sporn, M., Cryns, V. L., Zapata, J. M., and Reed, J. C. (2005). Synthetic triterpenoids cooperate with TRAIL to induce apoptosis of breast cancer cells. *Cancer Res* **65**, 4799–4808.

Ikeda, T., Nakata, Y., Kimura, F., Sato, K., Anderson, K., Motoyoshi, K., Sporn, M., and Kufe, D. (2004). Induction of redox imbalance and apoptosis in multiple myeloma cells by the novel triterpenoid 2-cyano-3,12-dioxoolean-1,9-dien-28-oic acid. *Mol Cancer Ther* **3**, 39–45.

Irmler, M., Thome, M., Hahne, M., Schneider, P., Hofmann, K., Steiner, V., Bodmer, J.-L., Schröter, M., Burns, K., Mattmann, C., Rimoldi, D., French, L. E., and Tschopp, J. (1997). Inhibition of death receptor signals by cellular FLIP. *Nature* **388**, 190–195.

Ito, Y., Pandey, P., Place, A., Sporn, M. B., Gribble, G. W., Honda, T., Kharbanda, S., and Kufe, D. (2000). The novel triterpenoid 2-cyano-3,12-dioxoolean-1,9-dien-28-oic acid induces apoptosis of human myeloid leukemia cells by a caspase-8-dependent mechanism. *Cell Growth Differ* **11**, 261–267.

Jansen, B., Wadl, H. S., Brown, B. D., Bryan, R. N., Elsas, A. V., Muller, M., Wolff, K., Eichler, H. G., and Pehamberger, H. (1998). Bcl-2 antisense therapy chemosensitizes human melanoma in SCID mice. *Nat Med* **4**, 232–234.

Johnsen, A. C., Haux, J., B., S., Nonstad, U., Egeberg, K., Sundan, A., Ashkenazi, A., and Espevik, T. (1999). Regulation of APO-2 ligand/TRAIL expression in NK cells—Involvement in NK cell-mediated cytotoxicity. *Cytokine* **11**, 664–672.

Johnson, C. S., Hershberger, P. A., and Trump, D. L. (2002). Vitamin D-related therapies in prostate cancer. *Cancer Metastasis Rev* **21**, 147–158.

Jordan, B. W., Dinev, D., LeMellay, V., Troppmair, J., Gotz, R., Wixler, L., Sendtner, M., Ludwig, S., and Rapp, U. R. (2001). Neurotrophin receptor-interacting mage homologue is an inducible inhibitor of apoptosis protein-interacting protein that augments cell death. *J Biol Chem* **276**, 39985–39989.

Karin, M., and Ben-Neriah, Y. (2000). Phosphorylation meets ubiquitination: The control of NF-κB activity. *Annu Rev Immunol* **18**, 621–663.

Karin, M., and Lin, A. (2002). NF-kappaB at the crossroads of life and death. *Nat Immunol* **3**, 221–227.

Karin, M., Cao, Y., Greten, F. R., and Li, Z.-W. (2002). NF-κB in cancer: From innocent bystander to major culprit. *Nat Rev* **2**, 301–310.

Kasof, G. M., and Gomes, B. C. (2001). Livin, a novel inhibitor-of-apoptosis (IAP) family member. *J Biol Chem* **276**, 3238–3246.

Kelly, W. K., Richon, V. M., O'Connor, O., Curley, T., MacGregor-Curtelli, B., Tong, W., Klang, M., Schwartz, L., Richardson, S., Rosa, E., Drobnjak, M., Cordon-Cordo, C., Chiao, J. H., Rifkind, R., Marks, P. A., and Scher, H. (2003). Phase I clinical trial of histone deacetylase inhibitor: suberoylanilide hydroxamic acid administered intravenously. *Clin Cancer Res* **9**, 3578–3588.

Khan, S. B., Maududi, T., Barton, K., Ayers, J., and Alkan, S. (2004). Analysis of histone deacetylase inhibitor, depsipeptide (FR901228), effect on multiple myeloma. *Br J Haematol* **125**, 156–161.

Kim, Y., Suh, N., Sporn, M., and Reed, J. C. (2002). An inducible pathway for degradation of FLIP protein sensitizes tumor cells to TRAIL-induced apoptosis. *J Biol Chem* **277**, 22320–22329.

Kimbrell, D. A., and Beutler, B. (2001). The evolution and genetics of innate immunity. *Nat Rev Genet* **2**, 256–267.

Kishore, N., Sommers, C., Mathialagan, S., Guzova, J., Yao, M., Hauser, S., Huynh, K., Bonar, S., Mielke, C., Albee, L., Weier, R., Graneto, M., Hanau, C., Perry, T., and Tripp, C. S. (2003). A selective IKK-2 inhibitor blocks NF-kappa B-dependent gene expression in interleukin-1 beta-stimulated synovial fibroblasts. *J Biol Chem* **278**, 32861–32871.

Kitada, S., Pedersen, I. M., Schimmer, A., and Reed, J. C. (2002). Dysregulation of apoptosis genes in hematopeietic malignancies. *Oncogene* **21**, 3459–3474.

Kitada, S., Leone, M., Sareth, S., Zhai, D., Reed, J. C., and Pellecchia, M. (2003). Discovery, characterization and structure-activity relationships studies of pro-apoptotic polyphenols targeting B-cell lymphocyte/leukemia-2 proteins. *J Med Chem* **46**, 4259–4264.

Klasa, R. J., Bally, M. B., Ng, R., Goldie, J. H., Gascoyne, R. D., and Wong, F. M. (2000). Eradication of human non-Hodgkin's lymphoma in SCID mice by BCL-2 antisense oligonucleotides combined with low-dose cyclophosphamide. *Clin Cancer Res* **6**, 2492–2500.

Kolluri, S. K., Bruey-Sedano, N., Cao, X., Lin, B., Lin, F., Han, Y. H., Dawson, M. I., and Zhang, X. K. (2003). Mitogenic effect of orphan receptor TR3 and its regulation by MEKK1 in lung cancer cells. *Mol Cell Biol* **23**, 8651–8667.

Konopleva, M., Tsao, T., Ruvolo, P., Stiouf, I., Estrov, Z., Leysath, C. E., Zhao, S., Harris, D., Chang, S., Jackson, C. E., Munsell, M., Suh, N., Gribble, G., Honda, T., May, W. S., Sporn, M. B., and Andreeff, M. (2002). Novel triterpenoid CDDO-Me is a potent inducer of apoptosis and differentiation in acute myelogenous leukemia. *Blood* **99**, 326–335.

Krajewska, M., Krajewski, S., Turner, B., Bubendorf, L., Kallioniemi, O.-P., Shabaik, A., Vitiello, A., Goddison, S., and Reed, J. C. (2003). Elevated expression of inhibitor of apoptosis proteins (IAPs) in prostate cancer. *Clin Can Res* **9**, 4914–4925.

Krammer, P. H. (2000). CD95's deadly mission in the immune system. *Nature* **407**, 789–795.

Kroemer, G., and Reed, J. C. (2000). Mitochondrial control of cell death. *Nature Med* **6**, 513–519.

Lai, J. C., Benimetskaya, L., Santella, R. M., Wang, Q., Miller, P. S., and Stein, C. A. (2003). G3139 (oblimersen) may inhibit prostate cancer cell growth in a partially bis-CpG-dependent non-antisense manner. *Mol Cancer Ther* **2**, 1031–1043.

Lapillonne, H., Konopleva, M., Tsao, T., Gold, D., McQueen, T., Sutherland, R. L., Madden, T., and Andreeff, M. (2003). Activation of peroxisome proliferator-activated receptor gamma by a novel synthetic triterpenoid 2-cyano-3,12-dioxooleana-1,9-dien-28-oic acid induces growth arrest and apoptosis in breast cancer cells. *Cancer Res* **63**, 5926–5939.

Leone, M., Zhai, D., Sareth, S., Kitada, S., Reed, J. C., and Pellecchia, M. (2003). Cancer prevention by tea polyphenols is linked to their direct inhibition of anti-apoptotic Bcl-2-family proteins. *Cancer Res* **63**, 8118–8121.

Le, L. (2004). Phase 1 study of a fully human monoclonal antibody to the tumor necrosis factor-related apoptosis-inducting ligand death receptor 4 (TRAIL-R1) in subjects with advanced solid malignanicies or non-Hodgkin's lymphoma. American Society of Clinical Oncology Annual Meeting.

LeBlanc, H., Lawrence, D., Varfolomeev, E., Totpal, K., Morlan, J., Schow, P., Fong, S., Schwall, R., Sinicropi, D., and Ashkenazi, A. (2002). Tumor-cell resistance to death receptor—Induced apoptosis through mutational inactivation of the proapoptotic Bcl-2 homolog Bax. *Nat Med* **8**, 274–281.

Li, C. C., Xie, Z. X., Zhang, Y. D., Chen, J. H., and Yang, Z. (2003). Total synthesis of wedelolactone. *J Org Chem* **68**, 8500–8504.

Li, H., Kolluri, S. K., Gu, J., Dawson, M. I., Cao, X., Hobbs, P. D., Lin, B., Chen, G.-Q., Lu, J.-S., Lin, F., Xie, Z., Fontana, J. A., Reed, J. C., and Zhang, X.-K. (2000). Cytochrome c release and apoptosis induced by mitochondrial targeting of nuclear orphan receptor TR3. *Science* **289**, 1159–1164.

Li, J., Feng, Q., Kim, J.-M., Schneiderman, D., Liston, P., Li, M., Vanderhyden, B., Faught, W., Fung Kee Fung, M., Senterman, M., Korneluk, R. G., and Tsang, B. K. (2005). Human ovarian cancer and cisplatin resistance: Possible role of inhibitor of apoptosis proteins. *Endocrinology* **142**, 370–380.

Li, L., Thomas, R. M., Suzuki, H., De Brabander, J. K., Wang, X., and Harran, P. G. (2004). A small molecule Smac mimic potentiates TRAIL- and TNFalpha-mediated cell death. *Science* **305**, 1471–1474.

Lieberman, J. (2003). The ABCs of granule-mediated cytotoxicity: New weapons in the arsenal. *Nat Rev Immunol* **3**, 361–370.

Lin, B., Kolluri, S. K., Lin, F., Liu, W., Han, Y.-H., Cao, X., Dawson, M. I., Reed, J. C., and Zhang, X.-K. (2004). Conversion of Bcl-2 from protector to killer by interaction with nuclear orphan receptor TR3/NGFI-B/Nur77. *Cell* **116**, 527–540.

Lin, J. H., Deng, G., Huang, Q., and Morser, J. (2000). KIAP, a novel member of the inhibitor of apoptosis protein family. *Biochem Biophys Res Commun* **279**, 820–831.

Liston, P., Fong, W. G., Kelly, N. L., Toji, S., Miyazaki, T., Conte, D., Tamai, K., Craig, C. G., McBurney, M. W., and Korneluk, R. G. (2001). Identification of XAF1 as an antagonist of XIAP anti-Caspase activity. *Nat Cell Biol* **3**, 128–133.

Locksley, R. M., Killeen, N., and Lenardo, M. J. (2001). The TNF and TNF receptor superfamilies: Integrating mamalian biology. *Cell* **104**, 487–501.

Maniatis, T. (1999). A ubiquitin ligase complex essential for the NF-κB, Wnt/Wingless, and hedgehog signaling pathways. *Genes Dev* **13**, 505–510.

Marcucci, G., Byrd, J. C., Dai, G., Klisovic, M. I., Kourlas, P. J., Young, D. C., Cataland, S. R., Fisher, D. B., Lucas, D., Chan, K. K., Porcu, P., Lin, Z. P., Farag, S. F., Frankel, S. R., Zwiebel, J. A., Kraut, E. H., Balcerzak, S. P., Bloomfield, C. D., Grever, M. R., and Caligiuri, M. A. (2003). Phase 1 and pharmacodynamic studies of G3139, a Bcl-2 antisense oligonucleotide, in combination with chemotherapy in refractory or relapsed acute leukemia. *Blood* **101**, 425–432.

Marusawa, H., Matsuzawa, S.-I., Welsh, K., Zou, H., Armstrong, R., Tamm, I., and Reed, J. C. (2003). HBXIP functions as a cofactor of survivin in apoptosis suppression. *EMBO J* **22**, 2729–2740.

Masuyama, N., Oishi, K., Mori, Y., Uneo, T., Takahama, Y., and Gotoh, Y. (2001). Akt inhibits the orphan nuclear receptor Nur77 and T-cell apoptosis. *J Biol Chem* **276**, 32799–32805.

Miller, L. (1999). An exegesis of IAPs: Salvation and surprises from BIR motifs. *Trends Cell Biol* **9**, 323–328.

Mitsiades, C. S., Mitsiades, N., Poulaki, V., Schlossman, R., Akiyama, M., Chauhan, D., Hideshima, T., Treon, S. P., Munshi, N. C., Richardson, P. G., and Anderson, K. C. (2002). Activation of NF-kappaB and upregulation of intracellular anti-apoptotic proteins via the IGF-1/Akt signaling in human multiple myeloma cells: Therapeutic implications. *Oncogene* **21**, 5673–5683.

Mitsiades, C. S., Mitsiades, N., and Koutsilieris, M. (2004). The Akt pathway: Molecular targets for anti-cancer drug development. *Curr Cancer Drug Targets* **4**, 235–256.

Miyake, H., Tolcher, A., and Gleave, M. E. (1999). Antisense Bcl-2 oligodeoxynucleotides inhibit progression to androgen-independence after castration in the Shionogi tumor model. *Cancer Res* **59**, 4030–4034.

Mori, N., Matsuda, T., Tadano, M., Kinjo, T., Yamada, Y., Tsukasaki, K., Ikeda, S., Yamasaki, Y., Tanaka, Y., Ohta, T., Iwamasa, T., Tomonaga, M., and Yamamoto, N. (2004). Apoptosis induced by the histone deacetylase inhibitor FR901228 in human T-cell leukemia virus type 1-infected T-cell lines and primary adult T-cell leukemia cells. *J Virol* **78**, 4582–4590.

Motegi, M., Yonezumi, M., Suzuki, H., Suzuki, R., Hosokawa, Y., Hosaka, S., Kodera, Y., Morishima, Y., Nakamura, S., and Seto, M. (2000). API2-MALT1 chimeric transcripts involved in mucosa-associated lymphoid tissue type lymphoma predict heterogeneous products. *Am J Pathol* **156**, 807–812.

Nagata, S. (1996). Fas-mediated apoptosis. *Adv Exp Med Biol* **406**, 119–124.

Nikolovska-Coleska, Z., Xu, L., Hu, Z., Tomita, Y., Li, P., Roller, P. P., Wang, R., Fang, X., Guo, R., Zhang, M., Lippman, M. E., Yang, D., and Wang, S. (2004). Discovery of embelin as a cell-permeable, small-molecular weight inhibitor of XIAP through structure-based computational screening of a traditional herbal medicine three-dimensional structure database. *J Med Chem* **47**, 2430–2440.

O'Connor, D. S., Grossman, D., Li, F., Plescia, J., Zhang, H., Villa, A., Tognin, S., Marchisio, P. C., and Altieri, D. C. (2000). Regulation of Apoptosis at cell division by p34^{cdc2} phosphorylation of survivin. *Proc Natl Acad Sci USA* **97**, 13103–13107.

Ogasawara, J., Watanabe-Fukunaga, R., Adachi, M., Matsuzawa, A., Kasugai, T., Kitamura, Y., Itoh, N., Suda, T., and Nagata, S. (1993). Lethal effect of the anti-Fas antibody in mice. *Nature* **364**, 806–809.

Oltersdorf, T., Elmore, S. W., Shoemaker, A. R., Armstrong, R. C., Augeri, D. J., Belli, B. A., Bruncko, M., Deckwerth, T. L., Dinges, J., Hajduk, P. J., Joseph, M. K., Kitada, S., Korsmeyer, S. J., Kunzer, A. R., Letai, A., Li, C., Mitten, M. J., Nettesheim, D. G., Ng, S. C., Nimmer, P. M., O'Connor, J. M., Oleksijew, A., Petros, A. M., Reed, J. C., Shen, W., Tahir, S. K., Thompson, C. B., Tomaselli, K. J., Wang, B., Wendt, M. D., Zhang, H., Fesik, S. W., and Rosenberg, S. H. (2005). An inhibitor of Bcl-2-family proteins induces regression of solid tumors. *Nature* **435**, 677–681.

Oost, T. K., Sun, C., Armstrong, R. C., Al-Assaad, A. S., Betz, S. F., Deckwerth, T. L., Ding, H., Elmore, S. W., Meadows, R. P., Olejniczak, E. T., Oleksijew, A., Oltersdorf, T., Rosenberg, S. H., Shoemaker, A. R., Tomaselli, K. J., Zou, H., and Fesik, S. W. (2004). Discovery of potent antagonists of the antiapoptotic protein XIAP for the treatment of cancer. *J Med Chem* **47**, 4417–4426.

Paramore, A., and Frantz, S. (2003). Bortezomib. *Nat Rev Drug Discov* **2**, 611–612.

Pedersen, I. M., Kitada, S., Leoni, L. M., Zapata, J. M., Karras, J., Tsukada, N., Kipps, T. J., Choi, Y. S., Bennet, F., and Reed, J. C. (2002a). Protection of CLL B-cells by a follicular dendritic cell line is dependent on induction of Mcl-1. *Blood* **100**, 1795–1801.

Pedersen, I. M., Kitada, S., Schimmer, A., Kim, Y., Zapata, J. M., Charboneau, L., Rassenti, L., Andreeff, M., Bennett, F., Sporn, M., Liotta, L., Kipps, T. J., and Reed, J. C. (2002b). The triterpenoid CDDO induces apoptosis in refractory CLL B-cells. *Blood* **100**, 2965–2972.

Pekarsky, Y., Hallas, C., Palamarchuk, A., Koval, A., Bullrich, F., Hirata, Y., Bichi, R., Letofsky, J., and Croce, C. M. (2001). Akt phosphorylates and regulates the orphan nuclear receptor Nur77. *Proc Natl Acad Sci USA* **98**, 3690–3694.

Pellecchia, M., and Reed, J. C. (2004). Inhibition of anti-apoptotic Bcl-2 family proteins by natural polyphenols: New avenues for cancer chemoprevention and chemotherapy. *Curr Pharm Design* **10**, 1387–1398.

Penninger, J. M., and Kroemer, G. (2003). Mitochondria, AIF and caspases—Rivaling for cell death execution. *Nat Cell Biol* **5**, 97–99.

Pratt, M. A. C., Krajewski, S., Menard, M., Krajewska, M., Macleod, H., and Reed, J. C. (1998). Estrogen withdrawal-induced human breast cancer tumour regression in nude mice is prevented by Bcl-2. *FEBS Lett* **440**, 403–408.

Prefontaine, G. G., Lemieux, M. E., Giffin, W., Schild-Poulter, C., Pope, L., LaCasse, E., Walker, P., and Hache, R. J. (1998). Recruitment of octamer transcription factors to DNA by glucocorticoid receptor. *Mol Cell Biol* **18**, 3416–3430.

Qiu, J., Levin, L. R., Buck, J., and Reidenberg, M. M. (2002). Different pathways of cell killing by gossypol enantiomenrs. *Exp Biol Med* **227**, 398–401.

Rayet, B., and Gelinas, C. (1999). Aberrant rel/nfkb genes and activity in human cancer. *Oncogene* **18**, 6938–6947.

Reed, J. C. (1996a). A day in the life of the Bcl-2 protein: Does the turnover rate of Bcl-2 serve as a biological clock for cellular lifespan regulation? *Leuk Res* **20**, 109–111.

Reed, J. C. (1996b). Balancing cell life and death: Bax, apoptosis, and breast cancer. *J Clin Invest* **97**, 2403–2404.

Reed, J. C. (1999a). Caspases and cytokines: Roles in inflammation and autoimmunity. *Adv Immunol* **73**, 265–299.

Reed, J. (1999b). Fenretinide: The death of a tumor cell. *J Natl Cancer Inst* **91**, 1099–1100.

Reed, J. C., and Bischoff, J. R. (2000). BIRinging chromosomes through cell division—and surviving the experience. *Cell* **102**, 545–548.

Reed, J., and Reed, S. (1999) Survivin' cell-separation anxiety. *Nat Cell Biol* **1**, E199–E200.

Reed, J. C., Stein, C., Subasinghe, C., Haldar, S., Croce, C. M., Yum, S., and Cohen, J. (1990). Antisense-mediated inhibition of BCL2 proto-oncogene expression and leukemic cell growth and survival: Comparisons of phosphodiester and phosphorothioate oligodeoxynucleotides. *Cancer Res* **50**, 6565–6570.

Reed, J. C., Doctor, K. S., and Godzik, A. (2004). The domains of apoptosis: A genomics perspective. *Sci STKE* **2004**, RE9.

Richardson, P. G., Barlogie, B., Berenson, J., Singhal, S., Jagannath, S., Irwin, D., Rajkumar, S. V., Srkalovic, G., Alsina, M., Alexanian, R., Siegel, D., Orlowski, R. Z., Kuter, D., Limentani, S. A., Lee, S., Hideshima, T., Esseltine, D. L., Kauffman, M., Adams, J., Schenkein, D. P., and Anderson, K. C. (2003). A phase 2 study of bortezomib in relapsed, refractory myeloma. *N Engl J Med* **348**, 2609–2617.

Rosato, R. R., Almenara, J. A., and Grant, S. (2003). The histone deacetylase inhibitor MS-275 promotes differentiation or apoptosis in human leukemia cells through a process regulated by generation of reactive oxygen species and induction of p21CIP1/WAF1 1. *Cancer Res* **63**, 3637–3645.

Rosen, D., Li, J.-H., Keidar, S., Markon, I., Orda, R., and Berke, G. (2000). Tumor immunity in perforin-deficient mice: A role for CD95 (Fas/APO-1). *J Immunol* **164**, 3229–3235.

Roth, W., and Reed, J. C. (2003). FLIP protein and TRAIL-induced apoptosis (review). *Vitam Horm* **67**, 189–206.

Rothwarf, D., Zandi, E., Natoli, G., and Karin, M. (1998). IKK-γ is an essential regulatory subunit of the IκB kinase complex. *NATURE* **395**, 297–300.

Rudin, C. M., Kozloff, M., Hoffman, P. C., Edelman, M. J., Karnauskas, R., Tomek, R., Szeto, L., and Vokes, E. E. (2004). Phase I study of G3139, a bcl-2 antisense oligonucleotide, combined with carboplatin and etoposide in patients with small-cell lung cancer. *J Clin Oncol* **22**, 1110–1117.

Sawada, M., Sun, W., Hayes, P., Leskov, K., Boothman, D. A., and Matsuyama, S. (2003). Ku70 suppresses the apoptotic translocation of Bax to mitochondria. *Nat Cell Biol* **5**, 320–329.

Saleh, M., Vaillancourt, J. P., Graham, R. K., Huyck, M., Srinivasula, S. M., Alnemri, E. S., Steinberg, M. H., Nolan, V., Baldwin, C. T., Hotchkiss, R. S., Buchman, T. G., Zehnbauer, B. A., Hayden, M. R., Farrer, L. A., Roy, S., and Nicholson, D. W. (2004). Differential modulation of endotoxin responsiveness by human caspase-12 polymorphisms. *Nature* **429**, 75–79.

Salvesen, G. S. (2002). Caspases: Opening the boxes and interpreting the arrows. *Cell Death Differ* **9**, 3–5.

Salvesen, G. S., and Duckett, C. S. (2002). IAP proteins: Blocking the road to death's door. *Nat Rev Mol Cell Biol* **3**, 401–410.

Salvesen, G. S., and Renatus, M. (2002). Apoptosome: The seven-spoked death machine. *Dev Cell* **2**, 256–257.

Sax, J. K., Fei, P., Murphy, M. E., Bernhard, E., Korsmeyer, S. J., and El-Deiry, W. S. (2002). BID regulation by p53 contributes to chemosensitivity. *Nat Cell Biol* **4**, 842–849.

Schimmer, A. D., Munk-Pedersen, I., Kitada, S., Demiralp, E., Minden, M. D., Pinto, R., Mah, K., Andreeff, M., Kipps, T., and Reed, J. C. (2003). Functional blocks in caspase activation pathways are common in leukemia and predict patient-response to induction chemotherapy. *Cancer Res* **63**, 1242–1248.

Schimmer, A. D., Welsh, K., Pinilla, C., Bonneau, M., Wang, Z., Pedersen, I. M., Scott, F. L., Glinsky, G. V., Scudiero, D. A., Sausville, E., Salvesen, G., Nefzi, A., Ostresh, J. M., Houghten, R. A., and Reed, J. C. (2004). Small-molecule antagonists of apoptosis-suppressor XIAP exhibit broad anti-tumor activity. *Cancer Cell* **5**, 25–35.

Schimmer, A. D., Thomas, M. P., Hurren, R., Gronda, M., Pellecchia, M., Pond, G. R., Konopleva, M., Gurfinkel, D., Mawji, I. A., Brown, E. J., and Reed, J. C. (2005). Identification of small-molecules that sensitize resistant tumor cells to TNF-family death receptors. *Cancer Res, submitted.*

Shipp, M. A., Ross, K. N., Tamayo, P., Weng, A. P., Kutok, J. L., Agular, R. C. T., Gaasenbeek, M., Angelo, M., Reich, M., Pinkus, G. S., Ray, T. S., Koval, M. A., Last, K. W., Norton, A., Lister, T. A., Mesirov, J., Neuberg, D. S., Lander, E. S., Aster, J. C., and Golub, T. R. (2002). Diffuse large b-cell lymphoma outcome prediction by gene-expression profiling and supervised machine learning. *Nat Med* **8**, 68–74.

Shiraki, K., Sugimoto, K., Yamanaka, Y., Yamaguchi, Y., Saitou, Y., Ito, K., Yamamoto, N., Yamanaka, T., Fujikawa, K., Murata, K., and Nakano, T. (2003). Overexpression of X-linked inhibitor of apoptosis in human hepatocellular carcinoma. *Int J Mol Med* **12**, 705–708.

Snell, V., Clodi, K., Zhao, S., Goodwin, R., Thomas, E. K., Morris, S. W., Kadin, M. E., Cabanillas, F., Andreeff, M., and Younes, A. (1997). Activity of TNF-related apoptosis-inducing ligand (TRAIL) in haematological malignancies. *Br J Haematol* **99**, 618–624.

Somogyi, R. D., Wu, Y., Orlofsky, A., and Prystowsky, M. B. (2001). Transient expression of the Bcl-2 family member, A1-a, results in nuclear localization and resistance to staurosporine-induced apoptosis. *Cell Death Differ*, 785–793.

Stein, C. A., Robert, A. E., and Joseph, A. (1992). A preliminary clinical study of gossypol in advanced human cancer. *Cancer Chemother Pharmacol* **30**, 480–482.

Sumitomo, M., Tachibana, M., Nakashima, J., Murai, M., Miyajima, A., Kimura, F., Hayakawa, M., and Nakamura, H. (1999). An essential role for nuclear factor kappa B in preventing TNF-alpha-induced cell death in prostate cancer cells. *J Urol* **161**, 674–679.

Sun, H., Nikolovska-Coleska, Z., Chen, J., Yang, C. Y., Tomita, Y., Pan, H., Yoshioka, Y., Krajewski, K., Roller, P. P., and Wang, S. (2005). Structure-based design, synthesis and biochemical testing of novel and potent Smac peptido-mimetics. *Bioorg Med Chem Lett* **15**, 793–797.

Suzuki, Y., Imai, Y., Nakayama, H., Takahashi, K., Takio, K., and Takahashi, R. (2001). A serine protease HtrA2/Omi, is released from the mitochondria and interacts with XIAP, induces cell death. *Mol Cell* **8**, 613–621.

Tamm, I., Kornblau, S. M., Segall, H., Krajewski, S., Welsh, K., Kitada, S., Scudiero, D. A., Tudor, G., Qui, Y. H., Monks, A., Andreeff, M., and Reed, J. C. (2000). Expression and prognostic significance of IAP-family genes in human cancers and myeloid leukemias. *Clin Cancer Res* **6**, 1796–1803.

Tanaka, S., Louie, D., Kant, J., and Reed, J. (1992). Frequent somatic mutations in translocated BcL2 genes of non-Hodgkin's lymphoma patients. *Blood* **79**, 229–237.

Taylor, J., Zhang, Q., Wyatt, R., and Dean, N. (1999). Induction of endogenous bcl-xS through the control of bcl-x pre-mRNA splicing by antisense oligonucleotides. *Nat Biotechnol* **17**, 1097–1100.

Testa, J. R., and Bellacosa, A. (2001). AKT plays a central role in tumorigenesis. *Proc Natl Acad Sci USA* **98**, 10983–10985.

Thallinger, C., Wolschek, M. F., Wacheck, V., Maierhofer, H., Gunsberg, P., Polterauer, P., Pehamberger, H., Monia, B. P., Selzer, E., Wolff, K., and Jansen, B. (2003). Mcl-1 antisense therapy chemosensitizes human melanoma in a SCID mouse xenotransplantation model. *J Invest Dermatol* **120**, 1081–1086.

Thornberry, N. A., and Lazebnik, Y. (1998). Caspases: Enemies within. *Science* **281**, 1312–1316.

Tinel, A., and Tschopp, J. (2004). The PIDDosome, a protein complex implicated in activation of caspase-2 in response to genotoxic stress. *Science* **304**, 843–846.

Tschopp, J., Irmler, M., and Thome, M. (1998a). Inhibition of Fas death signals by FLIPs. *Curr Opin Immunol* **10**, 552–558.

Tschopp, J., Thome, M., Hofmann, K., and Meinl, E. (1998b). The fight of viruses against apoptosis. *Curr Opin Genet Dev* **8**, 82–87.

Tschopp, J., Martinon, F., and Hofmann, K. (1999). Apoptosis: Silencing the death receptors. *Curr Biol* **9**, R381-R384.

Tzung, S., Kim, K. M., Basanez, G., Giedt, C. D., Simon, J., Zimmerberg, J., Zhang, K. Y. J., and Hockenbery, D. M. (2001). Antimycin: A mimics a cell-death-inducing Bcl-2 homology domain 3. *Nat Cell Biol* **3**, 183–192.

Verhagen, A. M., Ekert, P. G., Pakusch, M., Silke, J., Connolly, L. M., Reid, G. E., Moritz, R. L., Simpson, R. J., and Vaux, D. L. (2000). Identification of DIABLO, a mammalian protein that promotes apoptosis by binding to and antagonizing IAP proteins. *Cell* **102**, 43–53.

Vucic, D., Stennicke, H. R., Pisabarro, M. T., Salvesen, G. S., and Dixit, V. M. (2000). ML-IAP, a novel inhibitor of apoptosis that is perferentially expressed in human melanomas. *Curr Biol* **10**, 1359–1366.

Wallach, D., Varfolomeev, E. E., Malinin, N. L., Goltsev, Y. V., Kovalenko, A. V., and Boldin, M. P. (1999). Tumor necrosis factor receptor and Fas signaling mechanisms. *Annu Rev Immunol* **17**, 331–367.

Wang, J.-L., Liu, D., Zhang, Z.-J., Shan, S., Han, X., Srinivascula, A. M., Croce, C. M., Alnemri, E. S., and Huang, Z. (2000). Structure-based discovery of an organic compound that binds Bcl-2 protein and induces apoptosis of tumor cells. *Proc Natl Acad Sci USA* **97**, 7124–7129.

Webb, A., Cunningham, D., Cotter, F., Clarke, P. A., di Stefano, F., Ross, P., Corbo, M., and Dziewanowska, Z. (1997). BCL-2 antisense therapy in patients with non-Hodgkin lymphoma. *Lancet* **349**, 1137–1141.

Weber, C. K., Liptay, S., Wirth, T., Adler, G., and Schmid, R. M. (2000). Suppression of NF-kappaB activity by sulfasalazine is mediated by direct inhibition of IkappaB kinases alpha and beta. *Gastroenterology* **119**, 1209–1218.

Wright, J. J., Zerivitz, K., and Schoenfeldt, M. (2003). Clinical trials referral resource. Current clinical trials of bortezomib. *Oncology (Huntingt)* **17**, 677–680, 683–676, 691–672.

Wu, G. S., Burns, T. F., McDonald, E. R. R., Jiang, W., Meng, R., Krantz, I. D., Kao, G., Gan, D. D., Zhou, J. Y., Muschel, R., Hamilton, S. R., Spinner, N. B., Markowitz, S., Wu, G., and El-Deiry, W. S. (1997). KILLER/DR5 is a DNA damage-inducible p53-regulated death receptor gene. *Nat Genet* **17**, 141–143.

Wu, T. Y., Wagner, K. W., Bursulaya, B., Schultz, P. G., and Deveraux, Q. L. (2003). Development and characterization of nonpeptidic small molecule inhibitors of the XIAP/caspase-3 interaction. *Chem Biol* **10**, 759–767.

Yokoyama, Y., Okubo, T., Kano, I., Sato, S., and Kano, K. (2003). Induction of apoptosis by mono(2-ethylhexyl)phthalate (MEHP) in U937 cells. *Toxicol Lett* **144**, 371–381.

Zandi, E., Rothwarf, D. M., Delhase, M., Hayakawa, M., and Karin, M. (1997). The IKB kinase complex (IKK) contains two kinase subunits, IKKα and IKKβ, necessary for IKB phosphorylation and NF-KB activation. *Cell* **91**, 243–252.

Zandi, E., Chen, Y., and Karin, M. (1998). Direct phosphorylation of IκB by IKKα and IKKα: Discrimination between free and NF-κB-bound substrate. *Science* **281**, 1360–1363.

Zangemeister-Wittke, U., Leech, S. H., Olie, R. A., Simoes-Wust, A. P., Gautschi, O., Luedke, G. H., Natt, F., Haner, R., Martin, P., Hall, J., Nalin, C. M., and Stahel, R. A. (2000). A novel bispecific antisense oligonucleotide inhibiting both bcl-2 and bcl-xL expression efficiently induces apoptosis in tumor cells. *Clin Cancer Res* **6**, 2547–2555.

Zhai, D., Jin, C., and Reed, J. C. (2005). Comparison of current Bcl-2 compounds to all the six human Bcl-2 family proteins. *Submitted.*

Zhang, Y., Dawson, M. I., Mohammad, R., Rishi, A. K., Farhana, L., Feng, K.-C., Leid, M., Peterson, V., Zhang, X.-K., Edelstein, M., Eilander, D., Biggar, S., Wall, N., Reichert, U., and Fontana, J. A. (2002). Induction of apoptosis of human B-CLL and ALL cells by a novel retinoid and its non-retinoidal analog. *Blood* **100**, 2917–2925.

Zong, W. X., Ditsworth, D., Bauer, D. E., Wang, Z. Q., and Thompson, C. B. (2004). Alkylating DNA damage stimulates a regulated form of necrotic cell death. *Genes Dev* **18**, 1272–1282.

CHAPTER 18

MECHANISM-BASED DEVELOPMENT OF MEMANTINE AS A THERAPEUTIC AGENT IN TREATING ALZHEIMER'S DISEASE AND OTHER NEUROLOGIC DISORDERS: LOW-AFFINITY, UNCOMPETITIVE ANTAGONISM WITH FAST OFF-RATE

Huei-Sheng Vincent Chen and Stuart A. Lipton**

*Corresponding author: hsv_chen@burnham.org; slipton@burnham.org

Drug Discovery Research: New Frontiers in the Post-Genomic Era, Edited by Ziwei Huang

18.1 INTRODUCTION

Neurologic diseases are among the leading causes of death, disability, and economic expense in the world. For example, Alzheimer's disease (AD) ranks fourth as a cause of mortality in the United States. In fact, it has been estimated that as the population continues to age, treatment of patients with dementia will consume our entire gross national product by the latter decades of this century. Excitotoxic cell death, also termed excitotoxicity, is thought to contribute to neuronal cell injury and death in this and other neurodegenerative disorders. Excitotoxicity is defined as excessive exposure to the neurotransmitter glutamate or overstimulation of its membrane receptors, leading to neuronal injury or death. Glutamate, however, is the major excitatory neurotransmitter in the brain and mediates critical synaptic transmission for the normal functioning of the nervous system. Based on the pharmacology of agonist sites, there are three classes of glutamate-gated ion (or ionotropic) channels, known as α-amino-3-hydroxy-5-methyl-4-isoxazolepropionate (AMPA), kainate, and *N*-methyl-D-aspartate (NMDA) receptors. Among these, the ion channels coupled to classical NMDA receptors (NMDARs) are generally the most permeable to calcium (Ca^{2+}). Excitotoxic neuronal cell death is mediated in part by overactivation of NMDARs, which results in excessive Ca^{2+} influx through the receptor's associated ion channel. As a consequence of Ca^{2+} accumulation, excessive activation of the NMDARs leads to production of damaging free radicals and other enzymatic processes contributing to cell death (Fig.18.1) (Lipton and Rosenberg, 1994; Lipton and Nicotera, 1998). Many neurodegenerative diseases, including AD, Parkinson's disease, Huntington's disease, HIV-associated dementia, multiple sclerosis, amyotrophic lateral sclerosis (ALS), and glaucoma, are caused by different mechanisms but may share a final common pathway to neuronal injury due to the overstimulation of glutamate receptors, especially of the NMDA subtype (Lipton and Rosenberg, 1994). Acute disorders, such as stroke, CNS trauma, and epilepsy, also manifest a component of excitotoxicity. Hence, NMDAR antagonists could potentially be of therapeutic benefit in a number of acute and chronic neurological disorders manifesting excessive NMDA receptor activities (Kemp and McKernan, 2002).

NMDAR is made up of different subunits: NR1 (whose presence is mandatory), NR2A-D, and, in some cases, NR3A or B subunits. The receptor is probably composed of a tetramer of these subunits. The subunit composition determines the pharmacology and other parameters of the receptor-ion channel complex (Wollmuth and Sobolevsky, 2004). Alternative splicing of some subunits, such as NR1, further contributes to the diversity of pharmacological properties of the receptor (Zukin and Bennett, 1995). The subunits are differentially expressed both regionally in the brain and temporally during development. Physiological NMDAR activity is therefore essential for normal neuronal function (Cull-Candy et al., 2001). Potential neuroprotective agents that block virtually all NMDA receptor activity will therefore very likely have unacceptable clinical side effects. For this reason, many previous NMDAR antagonists have disappointingly failed advanced clinical trials for a number of neurodegenerative disorders. In contrast, studies in our laboratory have shown that the adamantane derivative memantine preferentially blocks excessive NMDA receptor activity without disrupting normal activity. Memantine does this

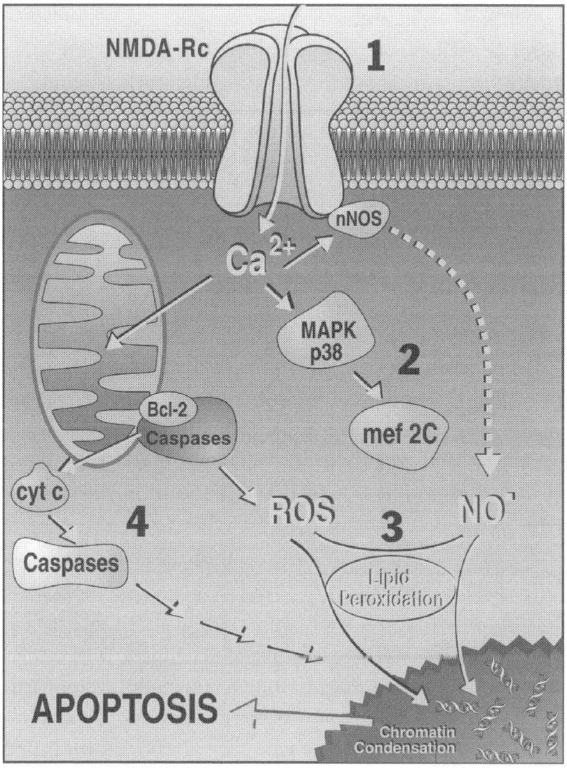

Figure 18.1 Schematic illustration of the apoptotic-like cell death pathways triggered by excessive NMDA receptor activity. The cascade of steps leading to neuronal cell death include: (**1**) NMDA receptor (NMDA-R) hyperactivation; (**2**) activation of the p38 mitogen activated kinase (MAPK)–MEF2C (transcription factor) pathway. MEF2 is subsequently cleaved by caspases to form an endogenous dominant-interfering form that contributes to neuronal cell death (Okamoto et al., 2002); (**3**) toxic effects of free radicals such as nitric oxide (NO) and reactive oxygen species (ROS); and (**4**) activation of apoptosis-inducing enzymes including caspases and apoptosis-inducing factor. nNOS: nitric oxide synthase; Cyt c: cytochrome c. (Adapted from the Lipton Web site at www.burnham.org)

through its action as a low-affinity, uncompetitive open-channel blocker with a relatively rapid off-rate from the channel. Here we review the molecular mechanism of memantine's clinically tolerated action and also the basis for the drug's development in treating several neurological disorders.

Of note, other calcium-permeable channels and routes of calcium entry, such as transient receptor potential (TRP) channels and calcium-permeable AMPA receptors, are also known to contribute to excitotoxicity. In addition, there are alternative approaches to providing protection against NMDAR-mediated excitotoxicity, such as methods uncoupling NMDARs from their downstream effectors (e.g., nitric oxide synthase; also see Kemp and McKernan, 2002). It is not our intention here to cover all possible strategies against calcium-mediated excitotoxicity. The purpose of the present review is to provide a brief and perhaps somewhat surprising primer on excitotoxicity as a promising target of neuroprotective strategies and to present a scientific and clinical overview of the excitotoxicity blocker memantine. Some preliminary information on second-generation memantine derivatives, termed NitroMemantines, is also provided.

18.2 EXCITOTOXICITY

18.2.1 Definition and Clinical Relevance

The ability of the nervous system to rapidly convey sensory information and complex motor commands from one part of the body to another, and to form thoughts and memories, is largely dependent on a single powerful excitatory neurotransmitter, glutamate. There are other excitatory neurotransmitters in the brain, but glutamate is the most common and widely distributed. Most neurons (and also glia) contain high concentrations of glutamate (~10 mM) (Lipton and Rosenberg, 1994); after sequestration inside synaptic vesicles, glutamate is released for very brief amounts of time (milliseconds) to communicate with other neurons via synaptic endings. Because glutamate is so powerful, however, its presence in excessive amounts or for prolonged periods of time can literally excite cells to death. This phenomenon was first documented when Lucas and Newhouse (1957) observed that subcutaneously injected glutamate selectively damaged the inner layer of the retina (representing primarily the retinal ganglion cells). John Olney, in seminal work, later coined the term "excitotoxicity" to describe this phenomenon (Olney, 1969; Olney and Ho, 1970).

A large variety of insults can lead to the excessive release of glutamate within the nervous system and, thus, excitotoxicity. When the nervous system suffers a severe mechanical insult, as in head or spinal cord injury, large amounts of glutamate are released from injured cells. These high levels of glutamate reach thousands of nearby cells that had survived the original trauma, causing them to depolarize, swell, lyse, and die by necrosis. The lysed cells release more glutamate, leading to a cascade of autodestructive events and progressive cell death that can continue for hours or even days after the original injury. A similar phenomenon occurs in stroke; the ischemic event deprives many neurons of the energy they need to maintain ionic homeostasis, causing them to depolarize and propagate the same type of autodestructive events seen in traumatic injury (Lipton, 1993a; Lipton and Rosenberg, 1994). This acute form of cell death occurs by a necrotic-like mechanism, although a slower component leading to an apoptotic-like death can also be present, as well as a continuum of events somewhere between the two (see below).

A slower, subtler form of excitotoxicity is implicated in a variety of chronic and slowly progressing neurodegenerative disorders as well as in the penumbra of stroke damage. In disorders such as AD, Huntington's disease, Parkinson's disease, multiple sclerosis, HIV-associated dementia, ALS, and glaucoma, it is hypothesized that chronic exposure to moderately elevated glutamate concentrations or glutamate receptor hyperactivity for longer periods of time than occur during normal neurotransmission trigger cellular processes in neurons that eventually lead to apoptotic-like cell death, a form of cell death related to the programmed cell death that occurs during normal development (Fig. 18.1) (Lipton and Nicotera, 1998; Bonfoco et al., 1995; Quigley et al., 1995; Vorwerk et al., 1996; Naskar et al., 1999). More subtle or incipient insults can lead to synaptic and dendritic damage in the early stages of disease. This process may be reversible and so is of considerable therapeutic interest.

Importantly, elevations in extracellular glutamate are not necessary to invoke an excitotoxic mechanism. Excitotoxicity can come into play even with normal levels of glutamate if NMDAR activity is increased— for example, when neurons are injured and thus become depolarized (more positively charged); this condition relieves the normal block of the ion channel by magnesium (Mg^{2+}) and thus abnormally increases NMDAR activity (Zeevalk and Nicklas, 1992). In addition, increased activity of the enzyme nitric oxide synthase (NOS) is associated with excitotoxic cell death. The neuronal isoform of the enzyme is physically tethered to the NMDA receptor and activated by Ca^{2+} influx via the receptor-associated ion channel, and increased levels of nitric oxide (NO) have been detected in animal models of stroke and several neurodegenerative diseases (Mullins et al., 1996).

18.2.2 Links Between Vascular Dementia and Excitotoxic Damage

As stated earlier, the glutamate content of most cells in the brain is approximately 10 mM. Because of the activity of glutamate transporters, most of this glutamate is intracellular. The extracellular glutamate concentration in the brain has been estimated to be approximately 0.6 μM. The sensitivity to excitotoxicity of cultured cortical neurons isolated away from astrocytes or of hippocampal neurons in intact tissue is approximately 2–5 μM glutamate (Lipton and Rosenberg, 1994). Therefore, the ambient concentrations of glutamate are close to those that can cause neuronal death, and it is important that extracellular glutamate concentration and compartmentalization be exquisitely controlled to prevent excitotoxicity. On the other hand, with 10 mM glutamate present in cells, the potential for disaster is obviously great.

Extracellular glutamate levels have been shown to rise in the face of hypoxic–ischemic insults. There will probably prove to be several mechanisms for the excess accumulation of glutamate, even in a single disorder such as ischemia (Lipton and Rosenberg, 1994). Energy failure might cause abnormal accumulations of glutamate either by impairment of uptake (into neurons and especially astrocytes) mediated via glutamate transporters or by reversal of the direction of transport. This series of events would be followed by injury to some neurons and abnormal potentiation of glutamate release from others. With glutamate release from injured neurons and excess physiological release from otherwise intact neighboring neurons, the process might then develop a self-propagating, vicious cycle that extends the area of neuronal damage.

Deprivation of oxygen and glucose—for example, during ischemia—cause a decrement in the production of high-energy phosphate compounds and "energy failure." However, short-term energy failure per se is not particularly toxic to neurons. What does make energy failure highly neurotoxic is the activation of glutamate receptor-dependent mechanisms. If suitable glutamate antagonists block these mechanisms, then neurons can survive a period of oxygen and metabolic substrate deprivation (reviewed by Lipton and Rosenberg, 1994).

18.2.3 Possible Links Between Excitotoxic Damage and Alzheimer's Disease

There are several potential links between excitotoxic damage and the primary insults of Alzheimer's disease, which, based on rare familial forms of the disease, are believed to involve toxicity from misfolded mutant proteins (reviewed by Rogawski and Wenk, 2003). These proteins include soluble oligomers of β-amyloid peptide (Aβ) and hyperphosphorylated tau proteins (Selkoe, 2001). For example, oxidative stress and increased intracellular Ca^{2+} generated by Aβ have been reported to enhance glutamate-mediated neurotoxicity *in vitro*. Additional experiments suggest that Aβ can increase NMDA responses and thus excitotoxicity (Koh et al., 1990; Mattson et al., 1992; Wu et al., 1995). Another potential link comes from recent evidence that glutamate transporters are downregulated in Alzheimer's disease and that Aβ can inhibit glutamate reuptake or even enhance its release (Topper et al., 1995; Harkany et al., 2000). Finally, excessive NMDAR activity has been reported to increase the hyperphosphorylation of tau, which contributes to neurofibrillary tangles and is involved in NMDA-mediated neurotoxicity (Couratier et al., 1996). The NMDAR antagonist memantine has been found to offer protection from these aforementioned neurotoxic processes (see below).

18.2.4 Pathophysiology of Excitotoxicity: Role of the NMDA Receptor

NMDA receptors are implicated in neuronal survival and maturation (Simon et al., 1984; Balazs et al., 1989), neuronal migration (Komuro and Rakic, 1993), induction of long-term potentiation (LTP, a cellular/electrophysiological correlate of learning and memory formation; Bliss and Collingridge, 1993), formation of sensory maps (Cline et al., 1987; Simon et al., 1992), and neurodegeneration (Meldrum and Garthwaite, 1990; Lipton, 1993b; Lipton and Rosenberg 1994, Cull-Candy et al., 2001). In contrast to AMPA- and kainate-type glutamate receptors, all functional NMDARs are heteromultimers (Wollmuth and Sobolevsky, 2004). Conventional NMDARs composed of NR1 and NR2A-D subunits require dual agonists, glutamate and glycine, for activation (Fig. 18.2A). The activity of the NMDAR-associated channel is modulated by a voltage-dependent block of Mg^{2+} (Nowak et al., 1984; Mayer et al., 1984), and the channel manifests high permeability to Ca^{2+} (Dingledine et al., 1999). NMDA is generally not thought to be an endogenous substance in the body; it is an experimental tool that is highly selective for this subtype of glutamate receptor, hence its name. Under normal conditions of synaptic transmission, the NMDAR-channel is gated by extracellular Mg^{2+} sitting in the channel and is only activated for brief periods of time. This brief opening of the NMDARs allows Ca^{2+} (and other cations) to move into the cell for the subsequent physiological functions. Under pathological conditions, however, overactivation of the receptor relieves the Mg^{2+} block and causes an excessive amount of Ca^{2+} influx into the nerve cell, which

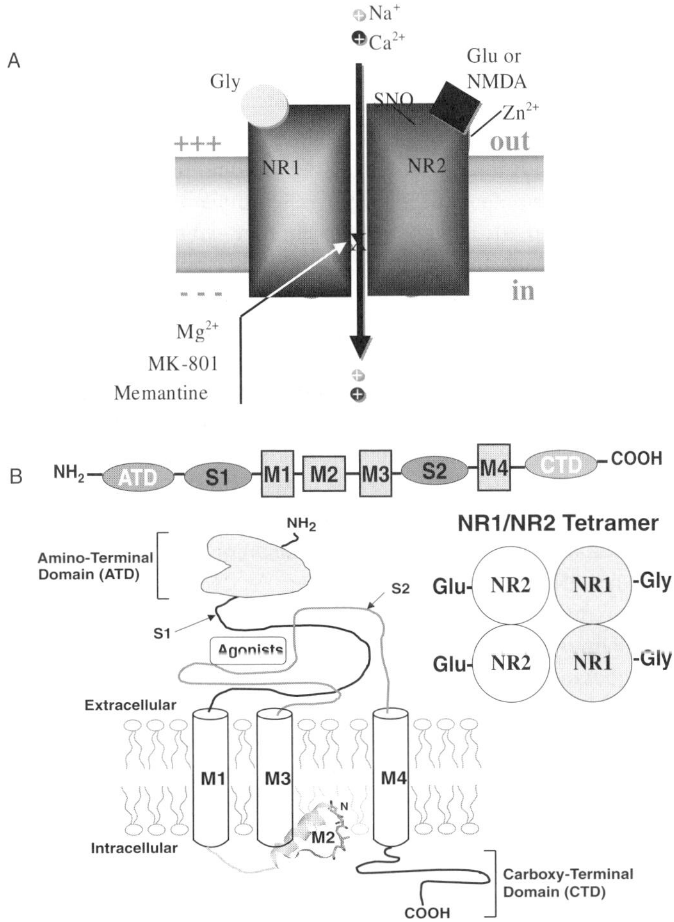

Figure 18.2 NMDA receptor model illustrating important binding and modulatory sites. (**A**). Glu or NMDA, glutamate or NMDA binding site. Gly, glycine binding site. Zn^{2+}, zinc binding site. NR1, NMDAR subunit 1. NR2, NMDAR subunit 2A. SNO, cysteine sulfhydryl group (—SH) reacting with nitric oxide species (NO). X, Mg^{2+}, MK-801, and memantine binding sites within the ion channel pore region. (**B**) Schematic representation of various domains of the NMDAR subunit. *Top*: Linear sequence. *Bottom left*: Proposed 3D folding. *Bottom right*: Proposed tetrameric structure of the classical NMDAR. ATD, amino terminal domain; S1 and S2, agonist binding domains; M1–M4, the four transmembrane domains; CTD, carboxyl terminal domain.

then triggers a variety of processes that can lead to necrosis, apoptosis, or dendritic/synaptic damage. These detrimental processes include (a) Ca^{2+} overload of mitochondria, resulting in oxygen free radical formation, activation of caspases, and release of apoptosis-inducing factor; (b) Ca^{2+}-dependent activation of neuronal NOS, leading to increased NO production and the formation of toxic peroxynitrite ($ONOO^-$) and *S*-nitrosylated glyceraldehyde-3-phosphate dehydrogenase (GAPDH); and (c) stimulation of mitogen-activated protein kinase p38 (MAPK p38), which activates transcription factors that can go into the nucleus to influence neuronal injury and apoptosis (Fig. 18.1; Dawson et al., 1991, 1993; Lipton et al., 1993; Bonfoco et al., 1995; Yun et al., 1998; Budd et al., 2000; Okamoto et al., 2002; Hara et al., 2005).

In NMDARs, the following membrane topology has been proposed (Fig. 18.2B, reviewed by Wollmuth and Sobolevsky, 2004): (1) The N-terminal domain contains the first ~380 amino acids that are related to the bacterial periplasmic binding protein sequence designated leucine/isoleucine/valine-binding protein (LIVBP), the Zn^{2+} binding site, the proton site, and other modulatory sites; (2) four transmembrane domains (M1-M4) are present, and the selectivity filter of the channel pore is formed by M2 (a P loop region); (3) the ligand-binding domains are formed by the pre-M1 (S1) and M3-M4 linker region (S2); (4) a cytoplasmic C-terminal domain interacts with intracellular proteins (Dingledine et al., 1999); and (5) the pre-M1 segment, the C-terminal portion of the M3 segment, and the N-terminal region of the M4 segment form the extracellular channel vestibule (Beck et al., 1999).

In general, NMDAR antagonists can be categorized pharmacologically into four major groups according to site of action on the receptor-channel complex (Wong and Kemp, 1991): Drugs acting at the (1) NMDA (agonist) recognition site, (2) glycine (co-agonist) site, (3) channel pore, and (4) modulatory sites, such as the redox modulatory site, the proton sensitive site, the high-affinity Zn^{2+} site, and the polyamine site. The degree of NMDAR activation and consequent influx of Ca^{2+} and Na^{+} into the cell can be altered by higher levels of agonists and by substances binding to one of the modulatory sites on the receptor. The two modulatory sites most relevant to this review are open-channel blocker sites within the ion channel pore and *S*-nitrosylation site(s) located toward the N-terminus (and hence extracellular region) of the receptor. Note that *S*-nitrosylation reactions represent transfer of NO to a thiol or sulfhydryl group (—SH) of a critical cysteine residue (Lipton et al., 2002). This reaction modulates protein function, in this case decreasing channel activity associated with stimulation of the NMDA receptor. Each of these sites can be considered potential targets for therapeutic intervention to block excitotoxicity, as explained below. Moreover, other modulatory sites also exist on the NMDA receptor and may in the future prove to be of therapeutic value. These include binding sites for Zn^{2+}, polyamines, the drug ifenprodil, and a pH (i.e., proton)-sensitive site (Kemp and McKernan, 2002). Additionally, three pairs of cysteine residues at extracellular domains contribute to the redox sites and can modulate NMDAR function by virtue of their redox sensitivity (Lipton et al., 2002). These redox-sensitive cysteine residues may constitute a unique "NO-reactive molecular oxygen sensor" in the brain, enhancing the degree of downregulation of NMDA receptor function by *S*-nitrosylation in the presence of low pO_2 levels and thus dictating the pathological effects of hypoxia that are mediated via the receptor (Takahashi et al., 2007).

18.3 RATIONAL DRUG DESIGN OF CLINICALLY TOLERATED NMDA RECEPTOR ANTAGONISTS

Excitotoxicity is a particularly attractive target for neuroprotective efforts because it is implicated in the pathophysiology of a wide variety of acute and chronic neurodegenerative disorders (Lipton and Rosenberg, 1994). The challenge facing those trying to devise strategies for combating excitotoxicity is that the same processes which, in excess, lead to excitotoxic cell death are, at lower levels, absolutely critical for normal neuronal function. To be clinically acceptable, an anti-excitotoxic therapy must block excessive activation of the NMDAR while leaving normal function relatively intact in order to avoid side effects. Drugs that simply compete with glutamate or glycine at the agonist binding sites block normal function and therefore do not meet this requirement, and thus they have failed in clinical trials to date because of side effects (drowsiness, hallucinations, and even coma) (Hickenbottom and Grotta, 1998; Lutsep and Clark, 1999; Palmer, 2000). In fact, competitive antagonists compete one for one with the agonist (glutamate or glycine) and therefore will block healthy areas of the brain (where lower, more physiological levels of these agonists exist) before they can affect pathological areas (where higher levels of agonists accumulate). Thus, such drugs would preferentially block normal activity and would most likely be displaced from the receptor by the high concentrations of glutamate for prolonged periods that can exist under excitotoxic conditions.

18.3.1 Advantages of Uncompetitive Antagonism and Open-Channel Block

The term "open-channel blocker of the NMDAR" means that the drug enters the receptor-associated ion channel only when it is open. Importantly, this type of drug will be most effective in the face of excessive (pathological) activity because statistically more channels are open and available to be blocked. This mechanism of inhibition, whose action is contingent upon prior activation of the receptor by the agonist, is defined as "uncompetitive" antagonism. Open-channel block is a most appealing strategy for therapeutic intervention during excessive NMDAR activation since the action of blockade requires prior activation of the receptors. This property, in theory, leads to a higher degree of channel blockade in the presence of excessive levels of glutamate and little blockade at relatively lower levels—for example, during physiological neurotransmission (see Section 18.4 for details; Rang, 1981; Chen, 1992; Chen et al., 1992; 1998; Lipton, 1993b; Lipton and Rosenberg, 1994; Chen and Lipton, 1997).

In fact, an uncompetitive open-channel blocker would prevent more severe excitotoxic processes better than mild disease. One would predict, for example, that moderate-to-severe dementia, involving excessive NMDAR activity leading to neuronal injury or death, would be treated more effectively than mild dementia or other excitotoxic disorders, involving only somewhat increased physiological firing. (Of course, these drugs will not reverse very severe disease because the neurons will already be lost.) Although preferential neuroprotection from moderate-to-severe excitotoxic processes seems counterintuitive, the drug's uncompetitive mechanism of action readily explains this uncanny phenomenon.

18.3.2 The Importance of Off-Rate from Channel Block

Importantly, we have shown that among open-channel blockers of the NMDAR, the details of the kinetics of blockade are important in order to avoid side effects

(Chen, 1992; Chen et al., 1992; Chen and Lipton, 1997). A relatively fast off-rate (and hence short dwell time in the channel) would prevent the drug from accumulating in open channels. This avoids progressive blockade of normal synaptic transmission. In contrast, a drug with a slow off-rate would build up in the ion channels that underlie synaptic events and consequently interfere with normal neurological function. The apparent affinity of a channel-blocking drug is related to its off-rate divided by its on-rate. At a given membrane potential, the on-rate is not only a property of diffusion and channel open probability, but also the drug's concentration. In contrast, the off-rate is an intrinsic property of the drug–receptor complex, unaffected by drug concentration. A relatively fast off-rate is a major contributor to a drug's low affinity for the channel pore. Thus, we propose that a clinically tolerated neuroprotective drug would consist of a low-affinity, open-channel blocker with a relatively fast off-rate. Hence, the drug would not substantially interfere with normal synaptic neurotransmission in an accumulative fashion. As a result, the drug would be both effective and well-tolerated.

As a useful analogy, the NMDA receptor can be thought of like a television set. The agonist sites are similar to the "on/off" switch of the television. Drugs that block here cut off all normal NMDAR function. What we need to find is the equivalent of the "volume" control (or in biophysical terms, the gain) of the receptor. Then, excessive Ca^{2+} influx through the NMDAR-associated ion channel would be prevented by simply turning down the "volume" of the Ca^{2+} flux toward normal values. A blocker that binds at a site within the channel, similar to the action of physiological levels of Mg^{2+}, could act as a sensor and provide an "automatic" volume control. Importantly, the automatic volume control needs to reach an optimal level. In the case of Mg^{2+} itself, the block is too ephemeral, a so-called "flickery block," and the cell continues to depolarize (become positively charged because of Ca^{2+} and Na^{+} entry) until Mg^{2+} is repelled, and the block is totally relieved. Hence, in most cases Mg^{2+} does not effectively block excessive Ca^{2+} influx to the degree needed to prevent neurotoxicity. If, on the other hand, a channel blocker binds with too high an affinity, it will accumulate in the channels, block normal activation, and thus prove clinically unacceptable. Following the television set analogy, turning the volume all the way down is as bad as turning off the "on/off" switch in terms of normal functioning of the television. This is the case with MK-801: It is a very good blocker of excitotoxicity, but because its dwell time in the ion channel is so long (reflecting its slow off-rate and high affinity), it progressively blocks critical normal functions. MK-801 can thus produce coma. Drugs with slightly shorter but still excessive dwell times (slow off-rates) make patients hallucinate (e.g., phencyclidine, also known as Angel Dust) or feel drowsy so that they classify as anesthetics (e.g., ketamine).

A clinically tolerated NMDAR antagonist would not make a patient feel drowsy, hallucinate, or become comatose, and in fact it should spare normal neurotransmission while blocking the ravages of excessive NMDA receptor activation. An uncompetitive, open-channel mechanism of blockade coupled with a longer dwell time in the channel (and consequently a slower off-rate) than Mg^{2+} but a substantially shorter dwell time (faster off-rate) than MK-801 would yield a drug that blocks NMDAR-operated channels only when they are excessively open, while relatively sparing normal neurotransmission.

18.3.3 Targeting Therapeutic Agents to the Pathologically Active Area

An NMDAR open-channel blocker binds increasingly well to neurons manifesting excessive channel activity; these are by definition potentially vulnerable neurons. Hence, we can attach other anti-apoptotic or pro-survival moieties to an open-channel blocker to form a new adduct that is targeted to vulnerable neurons, giving the second moiety specificity of action to the sick neuron. Using this concept, we propose that a series of second-generation drugs can be formulated that will have even greater neuroprotective properties than the original. These second-generation drugs can take advantage of the fact that the NMDAR has other modulatory sites (or "volume" controls) in addition to its ion channel that offer safe but effective clinical intervention.

One example of an additional modulatory site on the NMDAR that we can take therapeutic advantage of involves the action of nitric oxide (NO). We have shown that transfer of NO to thiol (—SH) groups on critical cysteine residues of the NMDAR (a reaction termed *S*-nitrosylation) decreases excessive receptor activity (Lei et al., 1992; Lipton et al., 1993; Choi et al., 2000; Takahashi et al., 2007). However, if administered systemically, NO can cause serious side effects, including severe hypotension (low blood pressure) by virtue of its ability to produce vasodilatation, and may even be toxic by nitrosylating other targets such as GAPDH or reacting with superoxide anion (O_2^-) to form peroxynitrite ($ONOO^-$). To avoid this problem, we can tether the NO group to an appropriate open-channel blocker in order to specifically target NO to the NMDAR nitrosylation sites. To date, such combinatorial drugs, linking the principles of open-channel block and *S*-nitrosylation of the NMDAR to provide two "volume controls," show great clinical potential. We propose that this newly recognized mode of action for drugs be designated "pathologically activated therapeutics" (PAT, meaning a gentle tap). By virtue of their relatively gentle binding, PAT drugs work best under pathological conditions, while exerting minimal effects on normal brain activity. We believe that these simple concepts embody the future of clinically tolerated neuroprotective drug design. Interestingly, we were the first to show that memantine fulfills these mechanistic criteria (Chen, 1992; Chen et al., 1992; Chen and Lipton, 1997).

18.4 MEMANTINE

18.4.1 Uncompetitive Open-Channel Block of NMDARs

Memantine (MEM, 1-amino-3, 5-dimethyl-adamantane, Fig. 18.3) was first synthesized by Eli Lilly and Company and patented in 1968, as documented in the Merck Index. It is a derivative of amantadine, an anti-influenza agent (Tominack and Hayden, 1987). Memantine has a three-ring (adamantane) structure with a bridgehead amine ($-NH_2$) that under physiological conditions carries a positive charge ($-NH_3^+$). Memantine has been used clinically with an excellent safety record for over 20 years in Europe to treat Parkinson's disease, spasticity, convulsions, vascular dementia (Ditzler, 1991), and Alzheimer's disease (Fleischhacker et al., 1986). The

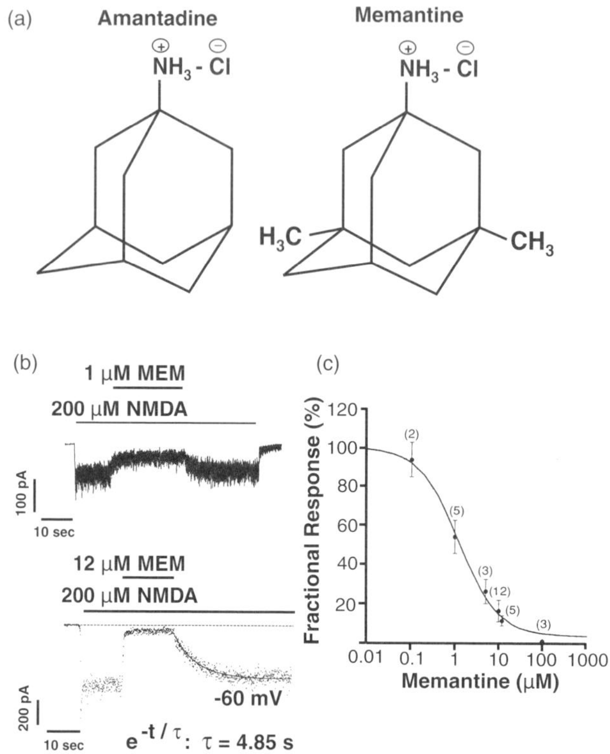

Figure 18.3 Structure, kinetic, and dose–response analysis of memantine (MEM). (**A**) *Left*: Chemical structure of amantadine. *Right*: Chemical structure of memantine, which has methyl group (—CH_3) side chains (unlike amantadine). (**B**) Blockade of 200 μM NMDA-activated currents by 1 and 12 μM MEM recorded from a solitary neuron at a holding potential of −60 mV. Kinetic analysis with single exponential fitting revealed that the on-time (time until peak blockade) of 12 μM memantine was approximately 1 sec, while the off-time constant (recovery time constant, τ) from the effect was ~5 sec. (**C**) Dose–response curve for MEM constructed using $I_{MEM}/I_{control}$ (%) versus MEM concentration.

reported efficacy of amantadine and memantine in Parkinson's disease, which was discovered by serendipity in a patient taking amantadine for influenza, led scientists to believe that these compounds were dopaminergic or possibly anticholinergic drugs. Work at a small German company named Merz first suggested that the drug might be an NMDA receptor inhibitor (Bormann, 1989; Kornhuber et al., 1989). We are the first group to characterize that memantine acts as an open-channel blocker of the NMDAR-coupled channel pore. We found that the drug blocks NMDA-evoked responses via uncompetitive antagonism with a 50% inhibition constant (IC_{50}) of ~1 μM at −60 mV. This blocking site senses <50% of the transmembrane potential from the outside surface and interacts with the Mg^{2+} blocking site in NMDA-gated channels (Chen et al., 1992; Chen and Lipton, 1997). Additionally, Wesemann et al. (1980) reported that the concentration of memantine in the parenchyma of a postmortem brain of a Parkinsonian patient who had received the standard dose of 20 mg memantine per day was approximately 1–12 μM. At 1–12 μM, memantine is

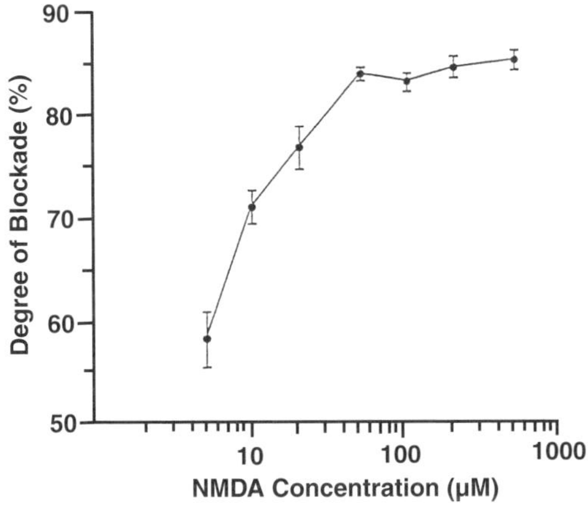

Figure 18.4 Paradoxically, a fixed dose of memantine (i.e., 6 μM) blocks the effect of increasing concentrations of NMDA to a greater degree than lower concentrations of NMDA. This finding is characteristic of an uncompetitive antagonist.

within the effective range of its NMDA-antagonistic action ($K_i = \sim 1$ μM) but is below the effective level of memantine at any other known receptor or ligand-gated channel (Chen et al. 1992, but see below). The antagonistic action of memantine on NMDARs is therefore thought to be the principal mechanism in the therapy of Parkinson's disease and possibly cerebral ischemia, dementia, and epilepsy (Lipton, 1993b; Lipton and Rosenberg, 1994; Rogawski, 2000).

Most importantly, we also showed why memantine could be clinically tolerated as an NMDA receptor antagonist; namely, it was an uncompetitive open-channel blocker with a dwell time/off-rate from the channel that limited pathological activity of the NMDA receptor while sparing normal synaptic activity (Chen et al., 1992; 1998; Chen and Lipton, 1997). We illustrate this most astonishing property of memantine in Fig. 18.4 and Fig. 18.5 (Chen et al., 1992). In this experiment, the concentration of memantine was held constant (at a clinically achievable low micromolar level) while the concentration of NMDA was increased over a wide range. It was found that the degree to which this fixed concentration of memantine blocked NMDA receptor activity actually increased as the NMDA concentration was increased to pathological levels. This is classical "uncompetitive" antagonist behavior. In fact, the component of the excitatory postsynaptic current due to physiological activation of NMDA receptors is inhibited by only 10 to 30% (Chen et al., 1998). During prolonged activation of the receptor, however, as occurs under excitotoxic conditions, memantine becomes a very effective blocker. In essence, memantine only acts under pathological conditions without much affecting normal function, thus relatively sparing synaptic transmission, preserving long-term potentiation, and maintaining physiological function on behavioral tests such as the Morris water maze (Chen et al., 1998). This notion is supported by the safety and efficacy profiles of memantine in two recent clinical trials for the treatment of Alzheimer's disease (Reisberg et al., 2003; Tariot et al., 2004). This strategy of selecting NMDAR antagonists of low-affinity/fast off-rate is in contrast to most drug discovery by big

Pharma, which uses high-affinity screens of the target. Neuroprotective agents that work by high-affinity binding to the NMDAR result in blockade of virtually all receptor activity; thus, these drugs manifest unacceptable clinical side effects. Instead, several years ago, we had proposed to protect the brain with drugs that do not bind very well under physiological conditions but are nevertheless selective under pathological conditions for a particular target, such as the NMDAR (Chen et al., 1992). Importantly, one should not confuse affinity with selectivity; as long as a drug acts selectively and specifically on the target of interest and the effective concentration can be achieved, a high affinity per se is not the key issue.

18.4.2 Memantine Interacts with the Intracellular Mg^{2+} Blocking Site in the NMDAR Channel Pore

Our earlier studies indicated that memantine exerts its effect on NMDA receptor activity by binding at or near the Mg^{2+} site within the ion channel (Chen, 1992; Chen et al., 1992; Chen and Lipton, 1997). Because of its interaction with external Mg^{2+} and based on mutational analysis of the NMDAR by others (Kashiwagi et al., 2002), the specific site of memantine action was assumed to be near the external Mg^{2+} blocking site at the selectivity filter region of the NMDAR-associated channel (Danysz and Parsons, 2003). This region is formed by asparagine (N) residues at the "N site" of NR1 and "N + 1 site" of NR2 subunits (reviewed by Dingledine et al., 1999). Compared to physiological block by external Mg^{2+}, a common explanation for the safety and effectiveness of memantine has been that memantine represented a "better magnesium," manifesting a somewhat slower unblocking rate, moderate voltage dependence, and slightly higher affinity (Danysz and Parsons, 2003). However, when applied from the intracellular versus extracellular surface, Mg^{2+} interacts differently on the N-site residues of NR1 and NR2 subunits (Wollmuth et al., 1998a,b). The N-site asparagine of the NR1 subunit represents the dominant blocking site for intracellular Mg^{2+}, whereas the N and N + 1 site asparagines of the NR2A subunit form the critical blocking site for extracellular Mg^{2+}. We recently performed a series of experiments using point mutations and the substituted cysteine accessibility method (SCAM) to show that the *N*-site asparagine of the NR1 subunit, located at the selectivity filter of the NMDAR-associated channel, is the specific and predominant blocking site for memantine (Chen and Lipton, 2005). The N and N + 1 sites of NR2A subunits provide the major electrostatic interaction with memantine upon binding to this deep, specific site (Fig. 18.6). The differential contribution to memantine block by the N- and N + 1 site asparagines in NR1 and NR2 subunits is reminiscent of their effects on intracellular Mg^{2+} blockade (Wollmuth et al., 1998b). The distinct patterns of interaction of memantine with the channel selectivity filter may confer upon memantine unique kinetic features leading to the drug's excellent clinical tolerability. In line with these results, memantine, in the absence of extracellular Mg^{2+}, displays minimal differences in blocking NMDARs containing various NR2 subunits (Bresink et al., 1996). We also found that none of the NR1 splice variants differentially affected memantine antagonism in the absence of external Mg^{2+} (Chen and Lipton, 2005).

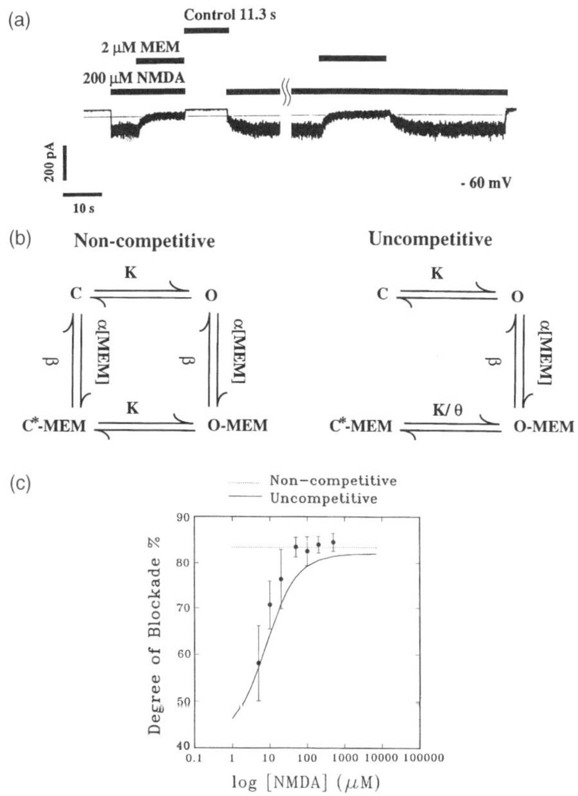

Figure 18.5 A trapping, uncompetitive scheme for memantine action. (**A**) Response to 200 μM NMDA was first blocked by 2 μM MEM at −60 mV, and then both agonists and antagonists were washed out by rapidly exposing the cell to a solution containing only the control solution for 11.3 sec to demonstrate MEM trapping (*left*). After washout with control solution, a second application of NMDA alone displayed a fast-rising phase of channel activation followed by slow relaxation, representing recovery from MEM blockade. This slow relaxation was similar to the regular recovery phase from MEM block that occurred without trapping (*right*). (**B**) Difference in predicted degrees of blockade between noncompetitive and uncompetitive antagonist action of memantine (MEM). *Top left*: Scheme for noncompetitive antagonism with C representing the closed channel; O, the open channel; C*-MEM, the blocked and closed channel; O-MEM, the open but blocked channel; α, the microscopic on-rate; β, the microscopic off-rate; [MEM], the concentration of the blocker; K, the equilibrium constant for opening from the closed state; and P_o, the open probability of the channel. The affinity of the blocker for the closed and open channel is the same. The open probability of the unblocked channel is the same as that of the blocked channel. *Top right*: Scheme for uncompetitive antagonism. K/θ ($\theta = 0.146$) is the equilibrium constant for opening from the C*-MEM state; P'_o is the open probability of the blocked channel, and the rest of the symbols have the same meaning as above. The blocker does not bind to the closed channel in this paradigm. (**C**) Computer-simulated degree of blockade for a noncompetitive antagonist (dotted line) and uncompetitive antagonist (solid curve) with the models and parameters indicated in **A** and **B**. The degree of blockade of NMDA-elicited current in Fig. 18.4 is replotted here. The inhibition equilibrium constant (K_i) for the blocker is 1.2 μM, and [MEM] is 6 μM. The empirical data points were very close to those predicted theoretically for pure uncompetitive antagonism (Chen and Lipton, 1997).

18.4.3 A Second Binding Site for Memantine in NMDA-Gated Channels

Several studies have also reported a second binding site for memantine in NMDA-gated channels (Antonov and Johnson, 1996; Blanpied et al., 1997; Bresink et al., 1996; Sobolevsky et al., 1998). This second site was reported to have a much lower affinity, minimal voltage dependence, and a noncompetitive mechanism of block. There were, however, several differences among these studies, including the estimated IC_{50}, the precise degree of voltage dependence, and the location of this blocking site in the NMDA-gated channel. Using point mutations and SCAM, we found that the second (superficial) memantine-blocking site, located at the extracellular vestibule of the channel, appears to be nonspecific and overlaps the site occupied by the nonspecific pore blocker hexamethonium. Residues in the post-M3 segment of the NR1 subunit are not directly involved in memantine binding (Chen and Lipton, 2005).

There is an important therapeutic implication of our recent description of the location of the second memantine binding site that concerns the uncompetitive mechanism of antagonism displayed by memantine. Uncompetitive, unlike competitive or noncompetitive, antagonists can block excessive activation of NMDARs while sparing normal neurotransmission; this is the property most likely responsible for the clinically tolerated mechanism of action of memantine, as explained above. Memantine blocks NMDARs via an uncompetitive mechanism at low micromolar concentrations, yet possesses a noncompetitive component (sometimes called "partial trapping" in the channel) at higher concentrations (Blanpied et al., 1997; Chen and Lipton, 1997). Lipophilic leak of memantine from its blocking site cannot explain this noncompetitive component (Mealing et al., 2001; Bolshakov et al., 2003). Instead, the noncompetitive behavior may be explained by the second site of memantine binding that we identified, which has very low affinity and is located at the channel vestibule. Occupancy by memantine of this shallow site may allow dissociation of the drug in either the open or closed conformation, resulting in a form of noncompetitive antagonism. This superficial site of memantine action, however, represents nonspecific binding and may also explain the noncompetitive component of many other very low-affinity, open-channel blockers.

Most importantly, we showed that a large difference in the affinity between these two sites of memantine binding is crucial for maintaining the selectivity of this type of "low-affinity" NMDAR open-channel blocker (Chen and Lipton, 2005). We believe that the distinct patterns of interaction and the relative degree of affinity of memantine for these two binding sites contribute to the drug's excellent pharmacological profile of clinical tolerability.

18.4.4 Voltage Dependence, Partial Trapping, and Lipophilic Leak of Memantine

Many possible factors have also been suggested for memantine's clinical tolerability, including moderate-to-low affinity, moderate voltage dependence, fast blocking and unblocking kinetics, and partial trapping in the NMDAR-associated channel

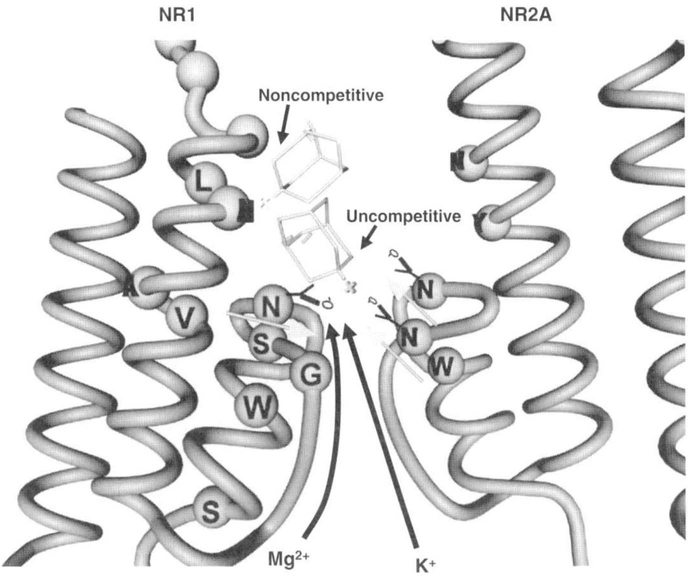

Figure 18.6 Atomic model showing two memantine binding sites in the channel permeation pathway of the NMDAR. Locations of memantine binding sites in the channel permeation pathway are shown at the level of the channel selectivity filter (the specific site) and at the L651 residue of the NR1 subunit (the nonspecific site). Internal permeant monovalent cation (K^+ or Cs^+) can compete with the externally applied MEM for binding, and MEM binding interacts with the intracellular Mg^{2+} binding site (Chen and Lipton, 1997, 2005).

(Rogawski and Wenk, 2003). All these macroscopic explanations are based on the assumption that memantine and other open-channel blockers bind at the same site as extracellular Mg^{2+} in the channel selectivity filter. We were recently the first to show that this assumption is incorrect. Instead, memantine interacts with the intracellular Mg^{2+} blocking site, which is located slightly deeper than the extracellular Mg^{2+} blocking site (Chen and Lipton, 2005). We also showed that the second, superficial site of memantine action is nonspecific and may explain the noncompetitive (or nontrapping) component of memantine at near millimolar concentrations. For so called "low-affinity" NMDA open-channel blockers, the apparent affinity, unblocking rate, and voltage dependence as well as the lipophilic leak (or closed-channel egress) (Blanpied et al., 1997) and degree of trapping of each open-channel blocker will represent "mixed" properties of both sites if the relative affinities are not too far apart. Therefore, despite prior reports, none of these properties at one site alone can explain the variable clinical tolerability of low-affinity NMDAR antagonists. In the case of memantine, however, the affinities of the two sites are sufficiently distinct so that the pharmacological properties of the specific site may account for its lack of side effects (Chen and Lipton, 2005). In our hands, memantine at therapeutic concentrations displays minimal closed-channel block or egress (minimal lipophilic leak) and therefore behaves like a perfect uncompetitive blocker [Fig. 18.6; see Chen and Lipton, (1997, 2005) for details].

18.4.5 Other Possible Effects of Memantine

At various concentrations, many different mechanisms of action of memantine have been reported: (1) antiviral action by inhibition of viral coat protein function (Tominack and Hayden, 1987), (2) inhibition of muscle-type nicotinic acetylcholine (ACh) receptors at the frog neuromuscular junction (Masuo et al., 1986), (3) antagonistic effects on NMDARs, (4) potentiation of strychnine-sensitive, glycine-activated currents (Lampe and Bigalke, 1991) (however, this result has not been substantiated by us or others), and (5) at concentrations above 100 μM, memantine may block voltage-sensitive sodium channels (Netzer and Bigalke, 1990) and has nonspecific effects on cell membranes (Osborne et al., 1982; Wesemann et al., 1983). In rat brain preparations, memantine does not interact directly with dopamine, opioid, GABA, or α1- and α2-adrenergic receptors and has no effect on the uptake of the neurotransmitters noradrenaline or serotonin [reviewed in Osborne et al. (1982)].

Memantine has been found to block serotonin (5-HT_3) receptor channels at concentrations approaching those that block NMDA receptor channels (Rammes et al., 2001; Reiser et al., 1988). Memantine's effect on 5-HT_3 receptors may possibly further enhance cognitive performance. Memantine has also been reported to inhibit α7 nicotinic acetylcholine (ACh) receptors in the *Xenopus* oocyte expression system and in rat cultured hippocampal neurons with an affinity similar to that reported for native NMDA receptors (Maskell et al., 2003; Aracava et al., 2005). These results, however, remain controversial, in part because desensitizing currents were evaluated rather than steady-state currents, and their significance thus needs to be confirmed (Banerjee et al., 2005).

Concerning its use in Alzheimer's disease, in the rat, memantine at therapeutic concentrations reduced the loss of cholinergic neurons in the nucleus basalis caused by NMDA-mediated toxicity or mitochondrial toxins [reviewed in Rogawski and Wenk (2003)]. Memantine has been reported to offer protection from neurotoxicity engendered by intrahippocampal injection of Aβ (Miguel-Hidalgo et al., 2002) and to enhance the processing of non-amyloidogenic β-amyloid precursor protein (Rogawski and Wenk, 2003). Memantine also improved performance on behavioral tests (T-maze and Morris water maze) in a transgenic mouse model of familial AD consisting of a mutant form of amyloid precursor protein and presenilin 1 (Tanila et al., 2003). Additionally, memantine was recently found to reduce tau hyperphosphorylation in culture (Iqbal et al., 2003). Chronic infusion of memantine also attenuated neuronal loss, improved short-term memory impairment, and reduced learning deficits and neurotoxicity caused by quinolinic acid-induced entorhinal cortex lesions in the rat [reviewed by Parsons et al. (1999)]. However, the exact mechanism(s) of protection by memantine from these animal or culture models of AD remains to be elucidated.

18.4.6 Neuroprotective Efficacy

The neuroprotective properties of memantine have been studied in a large number of *in vitro* and *in vivo* animal models by several laboratories [reviewed in Parsons et al. (1999)]. Among neurons protected in this manner both in culture and *in vivo* are cerebrocortical neurons, cerebellar neurons, and retinal neurons (Chen et al., 1992, 1998; Lipton, 1992; Osborne, 1999). Additionally, in a rat model of stroke, memantine, given as long as 2 hours after the ischemic event, reduces the amount of brain damage by approximately 50% (Chen et al., 1992, 1998).

A series of human clinical trials that investigated the efficacy of memantine for the treatment of AD, vascular dementia, HIV-associated dementia, diabetic neuropathic pain, depression, and glaucoma have recently been completed. Some of these studies have only recently been completed and remain unpublished at this time except in abstract form. One recent high-profile publication reported the result of a U.S. phase 3 (final) clinical study showing that memantine (20 mg/day) is efficacious for moderate-to-severe AD (Reisberg et al., 2003). Another study reported that, in combination with Aricept, memantine treatment offers some improvement of memory and function in moderate-to-severe Alzheimer's patients (Tariot et al., 2004). Both studies revealed excellent clinical tolerance and minimal side effects from therapeutic doses of memantine. These positive results of clinical studies for treating AD convinced the European Union and the U.S. FDA to approve memantine for the treatment of this form of dementia. Concerning other forms of dementia, one European multicenter, randomized, controlled trial reported that memantine was beneficial in severely demented patients, probably representing both AD and vascular dementia (Winblad and Poritis, 1999). Another recent publication of a randomized, placebo-controlled clinical trial also described significant benefit from memantine therapy (20 mg/day) in mild-to-moderate vascular dementia (Orgogozo et al., 2002). These clinical trials are summarized in Table 18.1. Most trials have reported minimal adverse effects from memantine. The only memantine-induced side effects encountered were rare dizziness and occasional restlessness/agitation at higher doses (40 mg/day), but these effects were mild and dose-related. Memantine is also under investigation as a potential treatment for other neurodegenerative disorders, including HIV-associated dementia, neuropathic pain, and glaucoma as well as depression and movement disorders.

Importantly, as we discovered and outlined above, the uncompetitive mode of memantine action would predict that, at a fixed dose, memantine would work better for a severe condition (e.g., excessive glutamate receptor activity to the point of causing cell death) than for more mild conditions manifested by slightly elevated synaptic transmission. Bearing this out, recent studies support that memantine may have a larger effect in moderate-to-severe dementia than in mild dementia. Another case in point is neuropathic pain, which is thought to be mediated at least in part by excessive NMDAR activity. Given the uncompetitive antagonism of memantine, more severe pain (e.g., the nocturnal pain of diabetic neuropathy) might be expected to benefit from memantine to a greater extent than more mild forms of neuropathic pain. In fact, a phase 2B clinical trial suggested that this is indeed the case, whereas preliminary reports indicate that more

TABLE 18.1 Clinical Trials with Memantine

• German/Merz phase 3 trial for vascular dementia and Alzheimer's disease	+
• Karolinska/Italian phase 3 trial for vascular dementia	+
• Two USA multicenter phase 3 trials for Alzheimer's disease	+
• United Kingdom phase 3 trial for vascular dementia	+
• French phase 3 trial for Vascular dementia	+
• USA phase 2 trials for neuropathic pain and HIV-associated dementia	+/−

mild pain conditions were not statistically benefited by memantine in phase 2/3 clinical trials. Along these same lines, one would predict that a higher concentration of memantine would be needed to combat pain than to prevent neuronal cell death because greater NMDA receptor activity is associated with cell death (a greater proportion of channels will be blocked by memantine in the face of increasing NMDAR activity). Again, clinical trials have suggested that this is indeed the case because 40 mg/day of memantine have been needed in successful pain studies but only 20 mg/day in severe dementia studies. However, further clinical trials will be necessary to prove the efficacy of memantine for severe neuropathic pain.

As promising as the results with memantine are, we are continuing to pursue ways to use additional modulatory sites (the "volume controls") on the NMDA receptor to block excitotoxicity even more effectively and safely than memantine alone. New approaches in this regard are explored below.

18.5 NITROMEMANTINES

NitroMemantines are second-generation memantine derivatives designed to have enhanced neuroprotective efficacy without sacrificing safety. As mentioned earlier, a nitrosylation site(s) is located on the N-terminus or extracellular domain of the NMDA receptor, and *S*-nitrosylation of this site (NO reaction with the sulfhydryl group of the cysteine residue) downregulates (but does not completely shut off) receptor activity (Fig. 18.2). The drug nitroglycerin, which generates NO-related species, can act at this site to limit excessive NMDA receptor activity. In fact, in rodent models, nitroglycerin can limit ischemic damage (Lipton and Wang, 1996), and there is some evidence that patients taking nitroglycerin for other medical reasons may be resistant to glaucomatous visual field loss as well (Zurakowski et al., 1998).

Consequently, we carefully characterized *S*-nitrosylation sites on the NMDA receptor in order to determine if we could design a nitroglycerin-like drug that could be more specifically targeted to the receptor. In brief, we found that five different cysteine residues on the NMDA receptor could interact with NO. One of these, located at cysteine residue #399 (C399) on the NR2A subunit of the NMDA receptor, mediates approximately 90% of the effect of NO under our experimental conditions (Choi et al., 2000). From crystal structure models and electrophysiological experiments, we further found that NO binding to the NMDAR at C399 apparently induces a conformational change in the receptor protein that makes glutamate and Zn^{2+} bind more tightly to the receptor. The enhanced binding of glutamate and Zn^{2+} in turn causes the receptor to desensitize and, consequently, causes the ion channel to close (Lipton et al., 2002). Electrophysiological studies have demonstrated this effect of NO on the NMDA channel (Lei et al., 1992; Lipton et al., 1993; Choi et al., 2000).

Unfortunately, nitroglycerin is not very attractive as a neuroprotective agent. The same cardiovascular vasodilator effect that makes it useful in the treatment of angina could cause dangerously large drops in blood pressure in patients with dementia, stroke, traumatic injury, or glaucoma. However, the open-channel block mechanism of memantine not only leads to a higher degree of channel blockade in the presence of excessive levels of glutamate, but also can be used as a homing signal for targeting drugs (e.g., the

NO group) to hyperactivated, open NMDA-gated channels. We have been developing combinatorial drugs (NitroMemantines) that theoretically should be able to use memantine to target NO to the nitrosylation sites of the NMDAR in order to avoid the systemic side effects of NO. Two sites of modulation are analogous to having two volume controls on your television set for fine-tuning the audio signal.

Preliminary studies have shown NitroMemantines to be highly neuroprotective in both *in vitro* and *in vivo* animal models. In fact, it appears to be more effective than memantine. Moreover, because of the targeting effect of memantine moiety, NitroMemantines appear to lack the blood pressure lowering effect typical of nitroglycerin. More research still needs to be performed on NitroMemantine drugs, but by combining two clinically tolerated drugs (memantine and nitroglycerin) we created a new, improved class of PAT drugs that are both clinically tolerated and neuroprotective.

18.6 CONCLUSIONS

Necrosis- and apoptosis-mediated excitotoxic cell death is implicated in the pathophysiology of many neurological diseases. This type of excitotoxicity is caused, at least in part, by excessive activation of NMDA-type glutamate receptors. Intense insults, such as that occurring in the ischemic core after a stroke, trigger massive stimulation of NMDA receptors, leading to neuronal cell swelling and lysis (necrosis). In contrast, more moderate NMDAR hyperactivity, such as that occurring in the ischemic penumbra of a stroke and in many slow-onset neurodegenerative diseases, results in a moderately excessive influx of calcium ions into nerve cells, which, in turn, triggers free radical formation and multiple pathways leading to the initiation of synaptic damage and apoptotic-like neuronal cell loss (Lipton and Nicotera, 1998). However, NMDA receptor activity is also required for normal neural function. Until recently, all drugs that showed promise as inhibitors of excitotoxicity also blocked normal neuronal function and consequently had severe and unacceptable side effects, so clinical trials for stroke, traumatic brain injury, and Huntington's disease all failed (Lees et al., 2000; Sacco et al., 2001). In the past decade, we have shown that memantine represents a class of drugs that have relatively low affinity and act as an uncompetitive, open-channel blocker. Due to its uncompetitive antagonism and relatively fast off-rate, memantine blocks excessive NMDAR activation but spares low (physiological) levels of NMDAR activity seen during normal neurotransmission. Importantly, memantine binds at the "intracellular" Mg^{2+} site in the channel pore and displays differential affinity for specific and nonspecific binding sites on the NMDAR. These molecular interactions confer upon memantine favorable kinetic properties that contribute to the drug's clinical tolerability as well as its neuroprotective profile (Chen and Lipton, 2005). The discovery that memantine, a low-affinity but still highly selective agent with a mechanism of uncompetitive antagonism, is neuroprotective yet clinically tolerated has triggered a paradigm shift in the history of drug development by the pharmaceutical industry. Clinical studies have borne out our hypothesis that low-affinity/fast off-rate memantine is a safe NMDA receptor antagonist in humans and beneficial in the treatment of neurological disorders mediated, at least in part, by excitotoxicity.

The NitroMemantines are second-generation NMDA receptor antagonists that may work even better than memantine. They use the memantine moiety as a homing signal for the targeted delivery of NO to a second modulatory site on the NMDAR. Work is progressing rapidly in this area of investigation.

Further clinical studies of the efficacy of memantine in the treatment of AD, vascular dementia, HIV-associated dementia, glaucoma, and severe neuropathic pain are currently underway, and there is every reason to expect the results to be positive, although this is, of course, not yet proven except in the case of Alzheimer's disease and possibly vascular dementia (Table 18.1). The efficacy of memantine in neurodegenerative diseases and its ability to protect neurons in animal models of both acute and chronic neurological disorders suggest that memantine and drugs acting in a similar manner could become very important new weapons in the fight against neuronal damage.

ACKNOWLEDGMENTS

We would like to thank our colleagues for their contributions to this work, which is updated here with an emphasis on Alzheimer's disease. We are especially grateful to Drs. Joachim Bormann, Yun-Beom Choi, and Jonathan S. Stamler for their discussions or collaborations. This work was supported in part by an AHA SDG and NIH grants P01 HD29587, R01 EY50477, and R01 EY09024. Dr. Lipton was a Senior Scholar in Aging of the Ellison Medical Foundation.

REFERENCES

Antonov, S. M., and Johnson, J. W. (1996). Voltage-dependent interaction of open-channel blocking molecules with gating of NMDA receptors in rat cortical neurons. *J Physiol (Lond)* **493**, 425–445.

Aracava, Y., Pereira, E. F., Maelicke, A., and Albuquerque, E. X. (2005). Memantine blocks alpha7 nicotinic acetylcholine receptors more potently than *N*-methyl-D-aspartate receptors in rat hippocampal neurons. *J Pharmacol Exp Ther* **312**, 1195–1205.

Balazs, R., Hack, N., Jorgensen, O. S., and Cotman, C. W. (1989). *N*-methyl- D-aspartate promotes the survival of cerebellar granule cells: pharmacological characterization. *Neurosci Lett* **101**, 241–246.

Banerjee, P., Samoriski, G., and Gupta, S. (2005). Comments on "Memantine blocks alpha7-nicotinic acetylcholine receptors more potently than *N*-methyl-D-aspartate receptors in rat hippocampal neurons." *J Pharmacol Exp Ther* **313**, 928–929.

Beck, C., Wollmuth, L. P., Seeburg, P. H., Sakmann, B., and Kuner, T. (1999). NMDAR channel segments forming the extracellular vestibule inferred from the accessibility of substituted cysteines. *Neuron* **22**, 559–570.

Blanpied, T. A., Boeckman, F. A., Aizenman, E., and Johnson, J. W. (1997). Trapping channel block of NMDA-activated responses by amantadine and memantine. *J Neurophysiol* **77**, 309–323.

Bliss, T. V., and Collingridge, G. L. (1993). A synaptic model of memory: long-term potentiation in the hippocampus. *Nature* **361**, 31–39.

Bolshakov, K. V., Gmiro, V. E., Tikhonov, D. B., and Magazanik, L. G. (2003). Determinants of trapping block of *N*-methyl-D-aspartate receptor channels. *J Neurochem* **87**, 56–65.

Bonfoco, E., Krainc, D., Ankarcrona, M., Nicotera, P., and Lipton, S. A. (1995). Apoptosis and necrosis: Two distinct events induced, respectively, by mild and intense insults with *N*-methyl-D-aspartate or nitric oxide/superoxide in cortical cell cultures. *Proc Natl Acad Sci USA* **92**, 7162–7166.

Bormann, J. (1989). Memantine is a potent blocker of *N*-methyl-D-aspartate (NMDA) receptor channels. *Eur J Pharmacol* **166**, 591–592.

Bresink, I., Benke, T. A., Collett, V. J., Seal, A. J., Parsons, C. G., Henley, J. M., and Collingridge, G. L. (1996). Effects of memantine on recombinant rat NMDA receptors expressed in HEK 293 cells. *British J Pharmacol* **119**, 195–204.

Budd, S. L., Tenneti, L., Lishnak, T., and Lipton, S. A. (2000). Mitochondrial and extramitochondrial apoptotic signaling pathways in cerebrocortical neurons. *Proc Natl Acad Sci USA* **97**, 6161–6166.

Chen, H.-S. V. (1992). Memantine: Open-channel block of *N*-methyl-D-aspartate (NMDA) elicited responses [Thesis] Cambridge: Harvard University. Available from University Microfilms Inc., Ann Arbor, MI.

Chen, H.-S. V., and Lipton, S. A. (1997). Mechanism of memantine block of NMDA-activated channels in rat retinal ganglion cells: Uncompetitive antagonism. *J Physiol (Lond)* **499**, 27–46.

Chen, H.-S. V., and Lipton, S. A. (2005). Pharmacological implications of two distinct mechanisms of interaction of memantine with *N*-methyl-D-aspartate-gated channels. *J Pharmacol Exp Ther* **314**, 961–971.

Chen, H.-S. V., Pellegrini, J. W., Aggarwal, S. K., Lei, S. Z., Warach, S., Jensen, F. E., and Lipton, S. A. (1992). Open-channel block of *N*-methyl-D-aspartate (NMDA) responses by memantine: Therapeutic advantage against NMDA receptor-mediated neurotoxicity. *J Neurosci* **12**, 4427–4436.

Chen, H.-S. V., Wang, Y. F., Rayudu, P. V., Edgecomb, P., Neill, J. C., Segal, M. M., Lipton, S. A., and Jensen, F. E. (1998). Neuroprotective concentrations of the *N*-methyl-D-aspartate open-channel blocker memantine are effective without cytoplasmic vacuolation following post-ischemic administration and do not block maze learning or long-term potentiation. *Neuroscience* **86**, 1121–1132.

Choi, Y. B., Tenneti, L., Le, D. A., Ortiz, J., Bai G., Chen, H.-S. V., and Lipton, S. A. (2000). Molecular basis of NMDA receptor-coupled ion channel modulation by *S*-nitrosylation. *Nat Neurosci* **3**, 15–21.

Cline, H. T., Debski, E. A., and Constantine-Paton, M. (1987). *N*-methyl-D-aspartate receptor antagonist desegregates eye-specific stripes. *Proc Natl Acad Sci USA* **84**, 4342–4345.

Couratier, P., Lesort, M., Sindou, P., Esclaire, F., Yardin, C., and Hugon, J. (1996). Modifications of neuronal phosphorylated tau immunoreactivity induced by NMDA toxicity. *Mol Chem Neuropathol* **27**, 259–273.

Cull-Candy, S., Brickley, S., and Farrant, M. (2001). NMDA receptor subunits: Diversity, development and disease. *Curr Op Neurobiology* **11**, 327–335.

Danysz, W., and Parsons, C. G. (2003). The NMDA receptor antagonist memantine as a symptomatological and neuroprotective treatment for Alzheimer's disease: Preclinical evidence. *Int J Geriatr Psychiatry* **18**, S23–S32.

Dawson, V. L., Dawson, T. M., London, E. D., Bredt, D. S., and Snyder, S. H. (1991). Nitric oxide mediates glutamate neurotoxicity in primary cortical cultures. *Proc Natl Acad Sci USA* **88**, 6368–6371.

Dawson, V. L., Dawson, T. M., Bartley, D. A., Uhl, G. R., and Snyder, S. H. (1993). Mechanisms of nitric oxide-mediated neurotoxicity in primary brain cultures. *J Neurosci* **13**, 2651–2661.

Dingledine, R., Borges, K., Bowie, D., and Traynelis, S. F. (1999). The glutamate receptor ion channels. *Pharmacol Rev* **51**, 7–61.

Ditzler, K. (1991). Efficacy and tolerability of memantine in patients with dementia syndrome. A double-blind, placebo controlled trial. *Arzneimittelforschung* **41**, 773–780.

Fleischhacker, W. W., Buchgeher, A., and Schubert, H. (1986). Memantine in the treatment of senile dementia of the Alzheimer type. *Prog Neuropsychopharmacol Biol Psychiatry* **10**, 87–93.

Hara, M. R., Agrawal, N., Kim, S. F., Cascio, M. B., Fujimuro, M., Ozeki, Y., Takahashi, M., Cheah, J. H., Tankou, S. K., Hester, L. D., Ferris, C. D., Hayward, S. D., Snyder, S. H., and Sawa, A. (2005). *S*-nitrosylated GAPDH initiates apoptotic cell death by nuclear translocation following Siah1 binding. *Nat Cell Biol* **7**, 665–674.

Harkany, T., Abraham, I., Timmerman, W., Laskay, G., Toth, B., Sasvari, M., Konya, C., Sebens, J. B., Korf, J., Nyakas, C., Zarandi, M., Soos, K., Penke, B., and Luiten, P. G. (2000). Beta-amyloid neurotoxicity is mediated by a glutamate-triggered excitotoxic cascade in rat nucleus basalis. *Eur J Neurosci* **12**, 2735–2745.

Hickenbottom, S. L., and Grotta J. (1998). Neuroprotective therapy. *Semin Neurol* **18**, 485–492.

Iqbal, K., Li, L., Sengupta, A., and Grundke-Iqbal, I. (2003). Memantine restores okadaic acid-induced changes in protein phosphatase-2A, CAMKII and tau hyperphosphorylation in rat. *J Neurochem* **85**, 42.

Kashiwagi, K., Masuko, T., Nguyen, C. D., Kuno, T., Tanaka, I., Igarashi, K., and Williams, K. (2002). Channel blockers acting at *N*-methyl-D-aspartate receptors: differential effects of mutations in the vestibule and ion channel pore. *Mol Pharmacol* **61**, 533–545.

Kemp, J. A., and McKernan, R. M. (2002). NMDA receptor pathways as drug targets. *Nat Neurosci* **5**, 1039–1042.

Koh, J.Y., Yang, L. L., and Cotman, C. W. (1990). Beta-amyloid protein increases the vulnerability of cultured cortical neurons to excitotoxic damage. *Brain Res* **533**, 315–320.

Komuro H., and Rakic, P. (1993). Modulation of neuronal migration by NMDA receptors. *Science* **260**, 95–97.

Kornhuber, J., Mack-Burkhardt, F., Riederer, P., Hebenstreit, G. F., Reynolds, G. P., Andrews, H. B., and Beckmann, H. (1989). [3H]MK-801 binding sites in postmortem brain regions of schizophrenic patients. *J Neural Transm* **77**, 231–236.

Koroshetz, W. J., and Moskowitz, M. A. (1996). Emerging treatments for stroke in humans. *Trends Pharmacol Sci* **17**, 227–233.

Lampe, H., and Bigalke, H. (1991). Modulation of glycine-activated membrane current by adamantane derivatives. *NeuroReport* **2**, 373–376.

Lees, K. R., Asplund, K., Carolei, A., Davis, S. M., Diener, H. C., Kaste, M., Orgogozo, J. M., and Whitehead, J. (2000). Glycine antagonist (gavestinel) in neuroprotection (GAIN International) in patients with acute stroke: a randomized controlled trial. GAIN International Investigators. *Lancet* **355**, 1949–1954.

Lei, S. Z., Pan, Z.-H., Aggarwal, S. K., Chen, H.-S. V., Hartman, J., Sucher, N. J., and Lipton, S. A. (1992). Effect of nitric oxide production on the redox modulatory site of the NMDA receptor–channel complex. *Neuron* **8**, 1087–1099.

Lipton, S. A. (1992). Memantine prevents HIV coat protein-induced neuronal injury *in vitro*. *Neurology* **42**, 1403–1405.

Lipton, S. A. (1993a). Molecular mechanisms of trauma-induced neuronal degeneration. *Curr Opin Neurol Neurosurg* **6**, 588–596.

Lipton, S. A. (1993b). Prospects for clinically-tolerated NMDA antagonists: Open-channel blockers and alternative redox states of nitric oxide. *Trends Neurosci* **16**, 527–532.

Lipton, S. A., and Chen, H.-S. (2004). Paradigm shift in neuroprotective drug development: Clinically tolerated NMDA receptor inhibition by memantine. *Cell Death Diff* **11**, 18–20.

Lipton, S. A., and Nicotera, P. (1998). Calcium, free radicals and excitotoxins in neuronal apoptosis. *Cell Calcium* **23**, 165–171.

Lipton, S. A., and Rosenberg, P. A. (1994). Mechanisms of disease: Excitatory amino acids as a final common pathway for neurologic disorders. *N Engl J Med* **330**, 613–622.

Lipton, S. A., and Wang, Y. F. (1996). NO-related species can protect from focal cerebral ischemia/reperfusion. In J. Krieglstein and H. Oberpichler-Schwenk (Eds.), *Pharmacology of Cerebral Ischemia.* Stuttgart: Wissenschaftliche Verlagsgesellschaft mbH, pp. 183–191.

Lipton, S. A., Choi, Y. B., Pan, Z. H., Lei, S. Z., Chen, H.-S. V., Sucher, N. J., Loscalzo, J., Singel, D. J., and Stamler, J. S. (1993). A redox-based mechanism for the neuroprotective and neurodestructive effects of nitric oxide and related nitroso-compounds. *Nature* **364**, 626–632.

Lipton, S. A., Choi, Y. B., Takahashi, H., Zhang, D., Li, W., Godzik, A., and Bankston, L. A. (2002). Cysteine regulation of protein function—as exemplified by NMDA–receptor modulation. *Trends Neurosci* **25**, 474–480.

Lucas, D. R., and Newhouse, J. P. (1957). The toxic effect of sodium L-glutamate on the inner layers of the retina. *Ama Arch Opthalmol* **58**, 193–201.

Lutsep, H. L., and Clark, W. M. (1999). Neuroprotection in acute ischemic stroke. Current status and future potential. *Drugs R D* **1**, 3–8.

Maskell, P. D., Speder, P., Newberry, N. R., and Bermudez, I. (2003). Inhibition of human alpha 7 nicotinic acetylcholine receptors by open channel blockers of *N*-methyl-D-aspartate receptors. *Br J Pharmacol* **140**, 1313–1319.

Masuo, K., Enomoto, K.-H., and Maeno, T. (1986). Effects of memantine on the frog neuromuscular junction. *Eur J Pharmacol* **130**, 187–195.

Mattson, M. P., Cheng, B., Davis, D., Bryant, K., Lieberburg, I., and Rydel, R. E. (1992). Beta-Amyloid peptides destabilize calcium homeostasis and render human cortical neurons vulnerable to excitotoxicity. *J Neurosci* **12**, 376–389.

Mayer, M. L., Westbrook, G. L., and Guthrie, P. B. (1984). Voltage-dependent block by Mg^{2+} of NMDA responses in spinal cord neurons. *Nature* **309**, 261–263.

Mealing, G. A., Lanthorn, T. H., Small, D. L., Murray, R. J., Mattes, K. C., Comas, T. M., and Morley, P. (2001). Structural modifications to an *N*-methyl-D-aspartate receptor antagonist result in large differences in trapping block. *J Pharmacol Exp Ther* **297**, 906–914.

Meldrum, B., and Garthwaite, J. (1990). Excitatory amino acid neurotoxicity and neurodegenerative disease. *Trends Pharmacol Sci* **11**, 379–387.

Miguel-Hidalgo, J. J., Alvarez, X. A., Cacabelos, R. and Quack, G. (2002). Neuroprotection by memantine against neurodegeneration induced by beta-amyloid(1–40). *Brain Res* **958**, 210–221.

Mullins, M. E., Sondheimer, N. J., Huang, Z., Singel, D. J, Huang, P. L, Fishman, M. C., Jensen, F. E., Lipton, S. A., Stamler, J. S., and Moskowitz, M. A. Spin-trapping NO in *n*NOS-deficient mice: Indications for stroke therapy. In J. S. Stamler, S. Gross, and S. Moncada (Eds.), *The Biology of Nitric Oxide*, Part 5. London: Portland Press, 1996, p. 9.

Naskar R., Vorwerk C. K., and Dreyer E. B. (1999). Saving the nerve from glaucoma: memantine to caspases. *Semin Ophthalmol* **14**, 152–158.

Netzer, R., and Bigalke, H. (1990). Memantine reduces repetitive action potential firing in spinal cord nerve cell cultures. *Eur J Pharmacol* **186**, 149–155.

Nowak, L., Bregestovski, P., Ascher, P., Herbert, A., and Prochiantz, A. (1984). Magnesium gates glutamate-activated channels in mouse central neurons. *Nature* **307**, 462–465.

Okamoto, S., Li, Z., Ju, C., Scholzke, M. N., Mathews, E., Cui, J., Salvesen, G. S., Bossy-Wetzel, E., and Lipton, S. A. (2002). Dominant-interfering forms of MEF2 generated by caspase cleavage contribute to NMDA-induced neuronal apoptosis. *Proc Natl Acad Sci USA* **99**, 3974–3979.

Olney, J. W. (1969). Glutamate-induced retinal degeneration in neonatal mice. Electron microscopy of the acutely evolving lesion. *J Neuropathol Exp Neurol* **28**, 455–474.

Olney, J. W., and Ho, O. L. (1970). Brain damage in infant mice following oral intake of glutamate, aspartate or cysteine. *Nature* **227**, 609–611.

Orgogozo, J. M., Rigaud, A. S., Stoffler, A., Mobius, H. J., and Forette, F. (2002). Efficacy and safety of memantine in patients with mild to moderate vascular dementia: A randomized, placebo-controlled trial (MMM 300). *Stroke* **33**, 1834–1839.

Osborne, N. N. (1999). Memantine reduces alterations to the mammalian retina, in situ, induced by ischemia. *Vis Neurosci* **16**, 45–52.

Osborne, N. N., Beale, R., Golomibiowska-Nikitin K., and Sontag K.-H. (1982). The effect of memantine on various neurobiological processes. *Arzneimittel Forschung* **32**, 1246–1255.

Palmer, G. C. (2001). Neuroprotection by NMDA receptor antagonists in a variety of neuropathologies. *Curr Drug Targets* **2**, 241–271.

Parsons, C. G., Danysz, W., and Quack, G. (1999). Memantine is a clinically well tolerated *N*-methyl-D-aspartate (NMDA) receptor antagonist—A review of preclinical data. *Neuropharmacology* **38**, 735–767.

Quigley, H. A., Nickells, R. W., Kerrigan, L. A., Pease, M. E., Thibault, D. J., and Zack, D. J. (1995). Retinal ganglion cell death in experimental glaucoma and after axotomy occurs by apoptosis. *Invest Ophthalmol Vis Sci* **36**, 774–786.

Rammes, G., Rupprecht, R., Ferrari, U., Zieglgansberger, W., and Parsons, C. G. (2001). The *N*-methyl-D-aspartate receptor channel blockers memantine, MRZ 2/579 and other amino-alkyl-cyclohexanes antagonize 5-HT(3) receptor currents in cultured HEK-293 and N1E-115 cell systems in a non-competitive manner. *Neurosci Lett* **306**, 81–84.

Rang, H. P. (1981). Drugs and ionic channels: Mechanisms and implications. *Postgrad Med J* **89**, 89–97.

Reisberg, B., Doody, R., Stoffler, A., Schmitt, F., Ferris S., and Mobius, H. J. (2003). Memantine in moderate-to-severe Alzheimer's disease. *N Engl J Med* **348**, 1333–1341.

Reiser, G., Binmoller, F. J., and Koch, R. (1988). Memantine (1-amino-3,5-dimethyladamantane) blocks the serotonin-induced depolarization response in a neuronal cell line. *Brain Res* **443**, 338–344.

Rogawski, M. A. (2000). Low affinity channel blocking (uncompetitive) NMDA receptor antagonists as therapeutic agents—Toward an understanding of their favorable tolerability. *Amino Acids* **19**, 133–149.

Rogawski, M. A., and Wenk, G. L. (2003). The neuropharmacological basis for the use of memantine in the treatment of Alzheimer's disease. *CNS Drug Rev* **9**, 275–308.

Sacco, R. L., DeRosa, J. T., Haley, E. C., Jr., Levin, B., Ordronneau, P., Phillips, S. J., Rundek, T., Snipes, R. G., and Thompson, J. L. (2001). Glycine antagonist in neuroprotection for patients with acute stroke: GAIN Americas: A randomized controlled trial. *JAMA* **285**, 1719–1728.

Selkoe, D. J. (2001). Alzheimer's disease: genes, proteins, and therapy. *Physiol Rev* **81**, 741–766.

Simon R. P., Swan J. H, Griffiths T., and Meldrum B. S. (1984). Blockade of *N*-methyl-D-aspartate receptors may protect against ischemic damage in the brain. *Science* **226**, 850–852.

Simon, D. K, Prusky, G. T, O'Leary, D. D, and Constantine-Paton, M. (1992). *N*-methyl-D-aspartate receptor antagonists disrupt the formation of a mammalian neural map. *Proc Natl Acad Sci USA* **89**, 10593–10597.

Sobolevsky, A. I., Koshelev, S. G., and Khodorov, B. I. (1998). Interaction of memantine and amantadine with agonist-unbound NMDA-receptor channels in acutely isolated rat hippocampal neurons. *J Physiol (Lond)* **512**, 47–60.

Takahashi, H., Shin, Y., Cho, S. J., Zago, W. M., Nakamura, T., Gu. Z., Ma, Y. L., FuruKawa, H., Liddington, R., Zhang, D., Tong, G., Chen, H.-S. V., and Lipton, S. A. (2007). Hypoxia enhances S-nitrosylation-Mediated NMDA receptor inhibition via a thiol oxygen sensor motif. *Neuron* **53**, 53–64.

Tanila, H., Minkevicine R., and Banjeree, P. (2003). Behavioral effects of subchronic memantine treatment in APP/PS1 double mutant mice modeling Alzheimer's disease. *J Neurochem* **85**, 42.

Tariot, P. N., Farlow, M. R., Grossberg, G. T., Graham, S. M., McDonald, S., and Gergel, I. (2004). Memantine treatment in patients with moderate to severe Alzheimer disease already receiving donepezil: A randomized controlled trial. *JAMA* **291**, 317–324.

Tominack, R. L., and Hayden, F. G. (1987). Rimantadine hydrochloride and amantadine hydrochloride use in influenza A virus infections. *Infect Dis Clin North Am* **1**, 459–478.

Topper, R., Gehrmann, J., Banati, R., Schwarz, M., Block, F., Noth, J., and Kreutzberg, G. W. (1995). Rapid appearance of beta-amyloid precursor protein immunoreactivity in glial cells following excitotoxic brain injury. *Acta Neuropathol (Berl)* **89**, 23–28.

Vorwerk, C. K., Lipton, S. A., Zurakowski, D., Hyman, B. T., Sabel, B. A., and Dreyer, E. B. (1996). Chronic low-dose glutamate is toxic to retinal ganglion cells. Toxicity blocked by memantine. *Invest Ophthalmol Vis Sci* **37**, 1618–1624.

Wesemann, W., Sturn, G., and Fünfgeld, E. W. (1980). Distribution and metabolism of the potential anti-parkinson drug memantine in the human. *J Neural Transm* **16**, 143–148.

Wesemann, W., Sontag, K.-H., and Maj, J. (1983). Zur Pharmakidynamik und Pharmakokinetik des Memantin. *Arzneim-Forsch/Drug Res* **33**, 1122–1134.

Winblad, B., and Poritis, N. (1999). Memantine in severe dementia: Results of the 9M-Best Study (Benefit and efficacy in severely demented patients during treatment with memantine). *Int J Geriatr Psychiatry* **14**, 135–146.

Wollmuth, L. P., and Sobolevsky, A. I. (2004). Structure and gating of the glutamate receptor ion channel. *Trends Neurosci* **27**, 321–328.

Wollmuth, L. P., Kuner, T., and Sakmann, B. (1998a). Adjacent asparagines in the NR2-subunit of the NMDA receptor channel control the voltage-dependent block by extracellular Mg^{2+}. *J Physiol* **506**, 13–32.

Wollmuth, L. P., Kuner, T., and Sakmann, B. (1998b). Intracellular Mg^{2+} interacts with structural determinants of the narrow constriction contributed by the NR1-subunit in the NMDA receptor channel. *J Physiol* **506**, 33–52.

Wong, E. H., and Kemp, J. A. (1991). Sites for antagonism on the *N*-methyl-D-aspartate receptor channel complex. *Annu Rev Pharmacol Toxicol* **31**, 401–425.

Wu, J., Anwyl, R., and Rowan, M. J. (1995). beta-Amyloid-(1–40) increases long-term potentiation in rat hippocampus in vitro. *Eur J Pharmacol* **284**, R1–R3.

Yun, H. Y., Gonzalez-Zulueta, M., Dawson, V. L., and Dawson, T. M. (1998). Nitric oxide mediates *N*-methyl-D-aspartate receptor-induced activation of p21ras. *Proc Natl Acad Sci USA* **95**, 5773–5778.

Zeevalk, G. D., and Nicklas, W. J. (1992). Evidence that the loss of the voltage dependent Mg^{2+} block at the *N*-methyl-D-aspartate receptor underlies receptor activation during inhibition of neuronal metabolism. *J Neurochem* **59**, 1211–1220.

Zukin, R.S., and Bennett, M. V. (1995). Alternatively spliced isoforms of the NMDARI receptor subunit. *Trends Neurosci* **18**, 306–313.

Zurakowski, D., Vorwerk, C. K., Gorla, M., Kanellopoulos, A. J., Chaturvedi, N., Grosskreutz, C. L., Lipton, S. A., and Dreyer, E. B. (1998). Nitrate therapy may retard glaucomatous optic neuropathy, perhaps through modulation of glutamate receptors. *Vision Res* **38**, 1489–1494.

CHAPTER 19

BRIDGING BENCH TO CLINIC: ROLES OF ANIMAL MODELS FOR POST-GENOMICS DRUG DISCOVERY ON METABOLIC DISEASES

*Chi-Wai Wong and Ling Chen**

19.1 INTRODUCTION

While human diseases may have various origins and forms—including viral or bacterial infectious diseases; inherited disorders such as muscular dystrophy; metabolic diseases such as obesity; diabetes; aging-associated diseases such as Alzheimer's; and cancers—many of these diseases can be modeled in animals by means of infectious agents, genetic manipulation, diet or drug-induced regiments, and surgery or xenograft explants. These animal models have offered important insights into the molecular mechanism of human diseases, served as proof-of-concept studies to validate gene targets for drug discovery, and, ultimately, provided indispensable platforms for evaluating the efficacy and safety of drugs in development. Decisions on whether to further or halt the development of a candidate drug often hinges on

*Corresponding author: chen_ling@gibh.ac.cn

Drug Discovery Research: New Frontiers in the Post-Genomic Era, Edited by Ziwei Huang

comprehensive analysis of its efficacy in proven animal models that faithfully predict the therapeutic value in treating the corresponding human disease, its pharmacodynamic/pharmacokinetic characteristics, and, importantly, its short- and long-term safety profiles.

The genomic era has brought forth tremendous knowledge on the gene sequences and genomic organizations of humans, rodents, and various other animals. An increasing amount of information is being obtained on the tissue, developmental, and disease-related expression patterns of candidate genes. Based on these data, one can easily identify the corresponding orthologue and homologue of a particular target gene of interest from primates to rodents. Combined with advances in techniques such as conventional and conditional transgenics or knockouts, it is relatively straightforward to create rodent models of human monogenic disorders through manipulating the expression levels of those equivalent genes in rodents, while creating models of human diseases of a multifactorial nature still remains a challenge. Nonetheless, these genetic models can be combined with diet-/drug-induced regimens and/or surgery/xenograft explants to further mimic the complex interaction of gene function and the environment, create animal disease models that more closely parallel that of complex human diseases, and allow more predictive evaluation on the effectiveness of newly developed drugs.

Besides demonstrating efficacies in animal disease models, effective drugs are also required to have desirable characteristics in absorption, distribution, metabolism, excretion, and pharmacokinetic (ADME/PK) profiles. Advances in techniques such as genechip/microarray for studying gene expression or proteomics for protein–protein interactions have presented novel testing paradigms in understanding the mechanisms of diseases as well as evaluating the ADME/PK profiles of drugs in both *in vitro* and *in vivo* rodent models. These data from ADME/PK studies of candidate drugs are then utilized for better drug design through further medicinal chemistry work and additional efficacy studies, forming an essential feed-forward loop in the drug discovery process.

Although rodent models of human diseases have provided a broader spectrum of human disease models to choose from for drug testing, the differences in primate and rodent physiology—in particular, the regulation of metabolic pathways—have generated some uncertainty in relying on only rodent models for drug development decisions (Ktorza et al., 1997). Nonhuman primates are closer to humans in terms of physiology; thus, an increasing number of nonhuman primate models are gaining popularity among pharmaceutical companies for testing developmental track drugs for their efficacy and safety profiles. In the following sections, we will attempt to highlight the impact of recent advances in genomics, proteomics, and pharmacogenomics on generating highly predictive animal models in both rodents and nonhuman primates, with emphasis on how differences in physiology affect the outcomes of studying metabolic diseases and decisions on drug discovery.

19.2 RODENT MODELS OF HUMAN DISEASES

Years of traditional breeding, crossing, and selecting desirable phenotypes from different strains have yielded a number of mouse lines that show phenotypic similarity to human diseases. Mapping and crossing of these lines reveal the genetic markers or loci that are linked to the phenotypes. For instance, the obese and insulin-resistant strains of *ob/ob* and *db/db* mice and *fa/fa* mutations in Kolestsky, ZDF, and Zucker rats are shown to be defective in leptin and its receptor (Chen et al., 1996; Chua et al., 1996; Takaya et al., 1996; Tartaglia et al., 1995; Wu-Peng et al., 1997; Zhang et al., 1994). Leptin is an important adipocyte-derived cytokine or adipokine that regulates appetite and energy metabolism. Defects in this system—in either the adipokine itself or its receptor—leads to obese and insulin-resistant phenotypes. Although genetic defects in the leptin system are rare in humans, rodents provide a platform in which antiobesity or antidiabetic drugs can be tested for their effects on body weight, blood glucose, and insulin levels as well as sensitivity to glucose or insulin challenges. Other spontaneous mutants used as obesity and diabetes models include tubby mice, yellow $A^{vy/-}$ mice with ectopic expression of the agouti gene, and growth-hormone-deficient dwarf rats (*dw/dw*) (Kleyn et al., 1996; Clark et al., 1996; Yen et al., 1994). This type of traditional cloning and mapping of mutant genes associated with a particular phenotype, although useful, requires large-scale breeding and crossing to identify phenotype-associated loci and eventually the associated gene. Yet other, more direct testing of candidate target genes of human disease by reverse genetics is carried out through genetic manipulations in mice to either knock in or knock out those genes to study the associated disease phenotypes.

19.2.1 Genetic Deletion Models

Even though the technique of generating mouse embryonic stem cells had been established before the genomic era, the advances in sequencing and mapping the human and mouse genomes have greatly assisted the process of targeted mutagenesis—in particular, conditional deletion of target genes by the Cre-loxP recombinase system (Branda and Dymecki, 2004; Yu and Bradley, 2001). While conventional target deletions have been fruitful in yielding useful and relevant information on the function of genes (Table 19.1), due to the essential function of some of these genes in early development, many of these targeted deletions have resulted in embryonic lethal phenotypes, preventing the evaluation of their functions in adults and in a tissue-specific fashion. The utilization of the Cre-loxP recombinase system, combined with the knowledge gained from studying promoter and enhancer elements that govern the tissue and developmental stage-specific expression of genes, allows the design of targeting vectors for conditional expression of Cre recombinase that would be expressed in desirable tissues during specific developmental stages. Therefore, targeted homologous recombination can be tailored to excise the targeted gene of interest in a controlled manner, generating deletion mutants that yield more informative phenotypes.

An example of utilizing conditional knockout to clarify the function of a gene in adult animals due to embryonic homozygous lethality involves a well-documented drug discovery target in diabetes mellitus, peroxisome proliferator-activated receptor gamma (PPARγ) (Berger et al., 2005). PPARγ is a nuclear hormone receptor, a class of transcription factor in which activity is modulated by small molecular ligands (Lazar, 2005). Nuclear hormone receptors also include other well-established drug

TABLE 19.1 Selected Examples of Gene Deletions Used in Obesity and Diabetes Research

Gene	Phenotypes	References
Pro-opiomelanocortin (*Pomc1*)	Obese, IR[a]	Yaswen et al. (1999)
Melanocortin-3 receptor (*Mc3r*)	Obese, IR	Chen et al. (2000)
Melanocortin-4 receptor (*Mc4r*)	Obese, IR	Huszar et al. (1997)
NPY1 receptor (*Npy1r*)	Obese[b]	Kushi et al. (1998)
NPY2 receptor (*Npy2r*)	Obese[b]	Naveilhan et al. (1999)
NPY5 receptor (*Npy5r*)	Obese[b]	Marsh et al. (1998)
5-Hydroxytryptamine 2C receptor (*Htr2c*)	Obese, IR	Nonogaki et al. (1998)
Bombesin receptor subtype-3 (*Brs3*)	Obese	Ohki-Hamazaki et al. (1997)
Peroxisome proliferator-activated receptor α (*Ppara*)	Obese	Costet et al. (1998)
Steroid receptor coactivator 1 (*Src1*)	Obese[c]	Picard et al. (2002)
Glucocorticoid receptor interacting protein 1 (*Grip1*)	Lean[d]	Picard et al. (2002)
Receptor interacting protein 140 (*Rip140*)	Lean[d]	Leonardsson et al. (2004)
Estrogen receptor α (*Esr1*)	Obese	Heine et al. (2000)
Estrogen=related receptor α (*Esrra*)	Lean[d]	Luo et al. (2003)

[a]IR, insulin resistance.
[b]Late-onset obesity.
[c]Obesity due to high-fat diet.
[d]Resistant to high-fat-diet-induced obesity.

discovery targets such as peroxisome proliferator-activated receptor alpha (PPARα) for dyslipidemia, estrogen receptor (ER) for osteoporosis, and glucocorticoid receptor (GR) for anti-inflammation (Staels, 2005). The activity of PPARγ is modulated by endogenous ligands such as prostanglin J_2 (PGJ_2) and synthetic drugs such as rosiglitazone that belong to the thiazolindinedione (TZD) class of molecules for the treatment of type II diabetes (Forman et al., 1995). Since a point mutation in PPARγ leads to diabetes and hypertension in humans and TZD stimulates the activity of PPARγ and alleviates the symptoms of diabetes, one would simply expect that a deletion of the PPARγ gene would prime the animal for the development of diabetes (Deeb et al., 1998). A homozygous deletion of PPARγ results in embryonic lethality; however, a heterozygous deletion of PPARγ actually partially protects the animal from developing diabetes (Kubota et al., 1999; Miles et al., 2000).

PPARγ is expressed in multiple tissues, more abundantly in adipose than in muscle, the primary organ responsible for glucose disposal upon insulin stimulation. The paradoxical observation from heterozygous mice has led to the hypothesis that differential regulation of PPARγ activity in these insulin-responsive tissues is required to confer protection from developing diabetes. By selecting appropriate targeting vectors that would result in tissue-specific homozygous deletion of PPARγ, different strains of adipose- or muscle-specific knockouts have revealed the complex relationship of PPARγ and glucose metabolism (He et al., 2003; Hevener et al., 2003). Adipose-specific PPARγ knockout causes adipose hypertrophy, elevated levels of plasma-free fatty acids and triglyceride, and decreased levels of adipokines leptin and adiponectin, showing some of the characteristics of metabolic syndrome X. Despite causing insulin resistance in adipose and the liver, the blood glucose level, glucose and insulin tolerance, and insulin-stimulated muscle glucose uptake remain normal. On the other hand, muscle–specific PPARγ knockout causes serve muscle glucose intolerance and insulin resistance with an 80% reduction in insulin-stimulated glucose uptake. Importantly, the animals are not responsive to TZD treatment, strongly suggesting the primary action of TZD is through regulating the activity of PPARγ in the muscle, paving ways for medicinal modification of TZD derivatives for improved efficacies (Argmann et al., 2005).

19.2.2 Transgenic and Viral-Mediated Overexpression Models

While knockout studies have been useful in identifying the necessary functions of genes, transgenic models and viral-mediated overexpression are of equal value in building predictive animal models for drug development. By choosing the appropriate conventional or conditional transgenic vector, it is possible to link the elevated expression level of candidate genes associated with human diseases to the cause of the corresponding diseases in rodents (Table 19.2). Moreover, through viral-mediated overexpression in combination with conventional knockout, one can ameliorate the disease condition caused by gene knockout, providing additional proof-of-concept evidence.

Chronic and subacute inflammation is associated with obesity, insulin resistance, and diabetes (Bullo et al., 2003; Haffner, 2003; Shoelson et al., 2003). In diabetic patients, elevations of pro-inflammatory markers, particularly those under the control of inflammatory transcription activator NF-κB, are found in adipose and liver tissues. The activity of NF-κB is activated by the inhibitor IκB, which is in turn inhibited by its negative regulating kinases, IKKα and IKKβ. Cai et al. (2004)

TABLE 19.2 Selected Examples of Transgenic and Adenovirus-Mediated Overexpression Used in Obesity and Diabetes Research

Gene	Method	Phenotypes	References
Peroxisome proliferator-activated receptor δ (*Ppard*)	Muscle transgenic	IS[a], Lean[b]	Wang et al. (2004)
Angiopoietin-related growth factor (*Angptl6*)	Adenovirus	IS, Lean[b]	Oike et al. (2005)
Retinol binding protein-4 (*Rbp4*)	Muscle transgenic	IR[c]	Yang et al. (2005)
Forkhead transcription factor C2 (*Foxc2*)	Adipose transgenic	Lean[b]	Kim et al. (2005)
Glucose transporter 4 (*Glut4*)	Adipose transgenic	IS	Shepherd et al. (1993)
Agouti-related transcript (*Agrt*)	Transgenic	Obese	Graham et al. (1997)

[a]IS, insulin-sensitive.
[b]Resistant to high-fat-diet-induced obesity.
[c]IR, insulin-resistant.

generated a hepatic-specific transgenic model in which constitutively active IKKβ is overexpressed in the liver to induce a state of subacute chronic inflammation. These mice develop hyperglycemia with severe hepatic insulin resistance and moderate systemic insulin resistance, including effects in the muscle. Using a similar strategy, the same group of researchers tested the effect of muscle-specific expression of IKKβ and only observed profound muscle wasting resembling clinical cachexia. These tissue-specific transgenic studies provided strong evidence that chronic inflammation in the liver can cause insulin resistance and that anti-inflammatory drugs may be of therapeutic potential.

An alternative to generating transgenic lines is through adenovirus-mediated studies, although these studies show a limited range of overexpression. Adenovirus carrying the desirable gene of interest can be easily engineered by commercially available vectors and helper cell lines. Tail-vein injection of recombinant adenovirus typically leads to overexpression of genes in the liver. Previous work on conventional knockout of insulin signal negative regulator protein tyrosine phosphatase 1B (PTP1B) showed that these mice were hypersensitive to insulin (Elchebly et al., 1999). However, controversies arise as to the primary tissue of PTP1B negative action. By using adenovirus to re-express PTP1B in the liver, Haj and co-workers observed marked attenuation of insulin sensitivity in these animals, thus demonstrating that the liver is the primary site of action for the peripheral effect of PTP1B and suggesting that a liver-specific PTP1B inhibitor may be of clinical value in restoring insulin sensitivity in diabetic patients (Haj et al., 2005; Harley and Levens, 2003).

Adenovirus-mediated *in vivo* gene delivery is even more commonly used to investigate the functions of secreted proteins and constitutes an important aspect of the target validation process. Oike et al. (2005) demonstrated the potential therapeutic value of angiopoietin-related growth factor (ANGPTL6) as an antiobesity/antidiabetic agent through a combination of gene knockout, transgenic overexpression, and adenovirus-mediated gene delivery. Specifically, they found that ANGPTL6 knockout mice displayed an obese phenotype with insulin resistance. In contrast, ANGPTL6 transgenic mice were lean and resistant to diet-induced obesity. Importantly, when mice that had already gained weight on a high-fat diet were treated with an adenovirus driving the expression of ANGPTL6, they lost some of the weight gained and displayed improved insulin sensitivity, strongly suggesting that ANGPTL6 may be a novel therapeutic target in obesity/diabetes.

19.2.3 Diet- or Drug-Induced Models

Besides genetic manipulation by deletion or transgenic, an even broader range of animal models for testing the effectiveness of therapeutic compounds can be built through administration of special diet or drug regimens. For example, rodents can be induced to develop non-insulin-dependent diabetes mellitus (NIDDM) or type I diabetes through injection of streptozotocin (STZ), which works primarily through impairing the glucose-dependent insulin secretion from pancreatic beta islet cells (Tsuji et al., 1988). This model can be used to study the changes of a gene expression profile by microarray in beta islet cells or other insulin-responsive tissues such as the liver, adipose, and muscle for mechanistic studies or target discovery purposes as the animal begins to lose its ability to secrete a sufficient amount of insulin to maintain euglycemia (Fan et al., 2003; Kume et al., 2005).

In addition to drug-induced models, more physiologically relevant studies through altering animal diet composition can also help gain important insights into mechanistic studies. Obesity is an important risk factor in humans for developing type II diabetes. Mice become overweight when put on a high-fat and/or high-carbohydrate diet and develop symptoms of metabolic diseases such as hyperinsulinemia and hyperlipidemia (Corsetti et al., 2000; Luo et al., 1998; Wang et al., 2001). This strategy is particularly informative when combined with genetic manipulation and drug-induced models.

Glucagon-like peptide-1 (GLP-1) is a gastric peptide hormone that functions to enhance the glucose-dependent insulin secretion from pancreatic beta islet cells (Stoffers, 2004). The half-life of GLP-1 is relatively short due to its cleavage by the dipeptidyl peptidase IV (DPP-IV) (Chen et al., 2003). Conarello et al. (2003) studied the effects of DPP-IV knockout on a high-fat diet. They found that DPP-IV knockout mice were refractory to weight gain and hyperinsulinemia compared to wild-type mice. With an elevated level of GLP-1, these mice were also protected against STZ-induced loss of beta cell mass and hyperglycemia, highlighting the potential of a GLP-1 analogue or DPP-IV inhibitor to be effective in enhancing glucose-dependent insulin secretion for managing diabetes (Nielsen, 2005).

Another intriguing example of combining a diet-induced regimen with that of genetic manipulation involves conditional knockout of IKKβ in either hepatocyte or myeloid cells to address the effect of modulating inflammatory pathway in different tissue would result in distinctive outcomes. Arkan et al. (2005) found out that mice with liver-specific knockout of IKKβ developed insulin resistance in muscle and fat in response to a high-fat diet despite retaining liver insulin responsiveness. On the other hand, myeloid-specific knockout mice remained insulin-sensitive and protected against insulin resistance on a high-fat diet. Together with the IKKβ transgenics mentioned above, these lines of evidence suggest that specific anti-inflammatory drugs may be of therapeutic value in treating metabolic diseases.

19.3 ANIMAL MODELS FOR PHARMACOLOGICAL AND TOXICOLOGICAL STUDIES

While these animal models can be utilized to establish the proof-of-concept studies and examine the efficacies of drugs under development, other important parameters such as ADME/PK and safety profiles that constitute the overall therapeutic value of drugs are being evaluated in pharmacological and toxicological studies (Theil et al., 2003). The advances in genomics, proteomics, and pharmacogenomics have brought forth improved *in vitro* assays and *in silico* predictions to learn about the absorption, distribution, metabolism, excretion, and pharmacokinetic profiles even before drugs are administered in animals for *in vivo* studies. These techniques if appropriately applied can help eliminate drug candidates that may pose problems early in the drug development process. However, to truly analyze the ADME/PK profiles, it is essential to use indicative animal models. Primarily, these pharmacodynamic/pharmacogenetic studies are conducted using rodent models with preferred studies in larger animals, including canine and nonhuman primates.

19.3.1 Humanized Mice for Drug Metabolism Studies

Rodents such as mice and rats have been utilized as standards for learning the ADME/PK profiles of leading drug candidates for years. Nonetheless, due to the species difference between primates and rodents in xenobiotic sensors responsible for regulating the expression of drug-metabolizing enzymes, it is noteworthy to conduct at least some ADME/PK studies in nonhuman primates. Before engaging in these kinds of extensive and expensive studies, one can utilize a more humanlike mouse in terms of drug metabolism for such studies.

The majority of drug metabolism takes place in the liver where transcription factors including pregnane X receptor (PXR) and constitutive androstane receptor (CAR), both nuclear hormone receptors, regulate the expression of phase I cytochrome P450 oxygenases (CYPs), phase II UDP-glucuronosyltransferases (UGTs), and phase III transporters such as multidrug resistance-associated proteins (MRPs) and organic anion transporting polypeptide 2 (OATP2) for elimination and detoxification in response to xenobiotics or drugs (Bock and Kohle, 2004; Sonoda et al., 2003). The activities of these nuclear receptors are regulated by binding to structurally distinct xenobiotics through their respective ligand-binding domains.

Due to the differences in the primary sequence of the ligand-binding domain of PXR between humans and mice, drugs recognized by human PXR, such as rifampicin, are not necessarily recognized by mouse PXR; therefore, the predictive value of using only mice as a model system to learn about drug metabolism in humans is less than ideal (Jones et al., 2000). In addition, the enhancer elements that are recognized by PXR in controlling Cyp3A gene expression are only partially conserved between humans and mice, leading to differential action of Cyp3A genes in response to xenobiotics. In order to create a mouse model that recapitulates the specificity and regulatory pattern to that of a human, a line of PXR homozygous knockout was first created and then followed by replacing the deleted mouse gene with that of human PXR (Xie et al., 2000). These humanized mice can then recognize the plethora of drug molecules that would lead to undesirable activation of the Cyp3A gene. Thus, it is now possible to learn about the potential regulation of CYP genes that may be the basis for adverse drug–drug interactions.

19.3.2 Pharmacodynamic/Pharmacokinetic Studies

Besides understanding the regulation of metabolizing enzymes and transporters by potential drug candidates by humanized mice, it can also be used to learn about the pharmacodynamic and pharmacokinetic profiles. Depending on the optimal treatment regimen, a desirable drug may be long- or short-lived once taken. It is therefore important to determine the kinetics and distribution of drugs in circulation and at target tissues, analyze the metabolite derivatives and excretion rate of drugs, obtain the maximal tolerance dose for short-term safety evaluation, and establish the optimal dosing regimens to define the therapeutic window (Li, 2004.).

Therefore, by first selecting the appropriate animal model for testing efficacy, pharmacodynamic parameters measured like the concentration that produces the half-maximal effect, the duration and intensity of responses can then be conducted.

Another key parameter for a desirable drug is bioavailablity. Bioavailability is the amount of exposure measured as free concentration in plasma—the amount of drugs not bound to plasma protein. Bioavailability is often determined by the rates of absorption, extraction, elimination, and variability due to genetic and environmental factors. Bioavailability data including the maximum concentration (C_{max}), time to maximum concentration (t_{max}) in single or multiple-dose studies, area under curve (AUC), peak-trough fluctuation (PTF), and time required to reach steady state can sometimes be extended to calculate the potential safety risk where plasma free concentration is considered to be relevant to compound safety. In general, there is a 30-fold safety multiple between therapeutic activity and concentration causing QT prolongation, a predictive risk of ventricular tachycardia.

19.4 NONHUMAN PRIMATE MODELS FOR METABOLIC DISEASES AND DRUG DISCOVERY

Despite the progress made in creating more humanized rodents to produce more predictive disease models, the differences in physiology and metabolism in particular have raised concern about the relevancy of data obtained from some rodent models. Noticeably, the metabolic regulatory pathways governing the biosynthesis of cholesterol and bile acids in the liver are not entirely conserved between primates and rodents. In humans, one of the major risk factors for the development of atherosclerosis is an elevated circulating level of cholesterol that correlates with high dietary cholesterol and fat intake (Goldstein and Brown, 1990). However, mice and rats on a high-cholesterol diet are resistant to hypercholesterolemia (Horton et al., 1995).

Liver conversion of cholesterol into bile acids is through the rate-limiting step enzyme cytochrome P450 (CYP) 7A1 in which expression is governed by a network of nuclear hormone receptors in response to their respective ligands (Goodwin et al., 2000; Lu et al., 2000; Russell, 1999). CYP7A1 expression is feedback-controlled by the pathway endproduct bile acids through binding to a farnesoid X receptor (FXR). Together with heterodimer partner 9-*cis* retinoic acid receptor RXR, FXR activates its target gene small heterodimer partner (SHP), which in turn acts as a repressor of positive factor liver receptor homolog 1 (LRH1) on the expression of the CYP7A1 gene. In mice and rats, CYP7A1 is induced by oxysterol metabolites of cholesterol, which bind to and activate liver X receptor (LXR). LXR binds to the CYP7A1 gene through a response element that is not conserved in humans. Instead, activation of LXR represses human CYP7A1 expression through inducing the expression of SHP (Goodwin et al., 2002). Thus, there is no substitute for testing the efficacy and safety of metabolic disease drugs in nonhuman primate models, especially those that are designed for treating hypercholesterolemia.

Nonhuman primate models for metabolic diseases have been utilized successfully to provide more informative mechanism studies and predictive efficacy studies for cholesterol-lowering and antidiabetic drugs. Barbara Hansen et al. have established a colony of obese rhesus monkeys at the Obesity and Diabetes Research Center. These monkeys first develop fasting hyperinsulinemia, hyperglycemia, and insulin resistance and then eventually become diabetic with retinopathy and nephropathy upon aging, displaying many characteristics similar to that of the human condition (Ortmeyer et al., 2000). When these monkeys were treated with the antidiabetic drug TZD, the compromised glycogen synthase activity in diabetic muscle was restored close to the normal level, contributing to improved insulin sensitivity and indicating that these animals respond to antidiabetic drugs in a similar way to humans.

In addition to testing antidiabetic drugs, this model has also been shown to be valuable in demonstrating the effectiveness of a novel peroxisome proliferator-activated receptor δ (PPARδ) ligand. Oliver et al. first observed that the synthetic PPARδ ligand can stimulate the expression of LXR and some LXR-regulated genes *in vitro* and speculated that this PPARδ ligand may regulate cholesterol biosynthesis at least partly through LXR (Oliver et al., 2001). They tested this idea in the obese monkey model and observed a dose-dependent decrease in small-dense low density lipoprotein, fasting triglycerides, and insulin and an increase in high-density lipoprotein, dramatically improving the serum profile of these monkeys to a lower risk category for cardiovascular disease. These cases strongly indicate the predictive value of such a nonhuman primate model in assessing drug efficacy.

19.5 CONCLUSION AND FUTURE PROSPECTS

With the ingenious use of various genetic manipulations such as conditional knockins or knockouts with an appropriate genetic strain of rodents in combination with diet- or drug-induced regimens, there is a wide range of possibilities for testing the functions of genes. Once the therapeutic potential of a target has been established by these animal models, lead candidate compounds can also be tested in these models to study their on-target and off-target effects. A compound designed to modulate the activity of a particular target should lose its efficacy in the background of the appropriate target knockout. Residual activity may then suggest off-target effects, possibly through acting on similar family members of the intended target or through unanticipated mechanisms. To further evaluate if these types of compounds still possess therapeutic value, it is then desirable to test the compound in more sophisticated models. The future directions may thus depend on the generation of even more desirable conditional compound knockins and knockouts of multiple genes to tailor the timing and duration of genetic manipulations. One possibility is to combine the current techniques with the siRNA technology that would allow modification of a combination of genes as opposed to a single genetic knockout. Once these animal models have been established to reflect the pathological condition of human diseases, they become highly valued in the drug development process for learning the efficacy of developing compounds, potential off-target effects, and ADME/toxicity profiles.

REFERENCES

Argmann, C. A., Cock, T. A., and Auwerx, J. (2005). Peroxisome proliferator-activated receptor gamma: The more the merrier? *Eur J Clin Invest* **35**(2), 82–92; discussion 80.

Arkan, M. C., Hevener, A. L., Greten, F. R., Maeda, S., Li, Z. W., Long, J. M., Wynshaw-Boris, A., Poli, G., Olefsky, J., and Karin, M. (2005). IKK-b links inflammation to obesity-induced insulin resistance. *Nat Med* **11**(2), 191–198.

Berger, J. P., Akiyama, T. E., and Meinke, P. T. (2005). PPARs: Therapeutic targets for metabolic disease. *Trends Pharmacol Sci* **26**(5), 244–251.

Bock, K. W., and Kohle, C. (2004). Coordinate regulation of drug metabolism by xenobiotic nuclear receptors: UGTs acting together with CYPs and glucuronide transporters. *Drug Metab Rev* **36**(3–4), 595–615.

Branda, C. S., and Dymecki, S. M. (2004). Talking about a revolution: The impact of site-specific recombinases on genetic analyses in mice. *Dev Cell* **6**(1), 7–28.

Bullo, M., Garcia-Lorda, P., Megias, I., and Salas-Salvado, J. (2003). Systemic inflammation, adipose tissue tumor necrosis factor, and leptin expression. *Obes Res* **11**(4), 525–531.

Cai, D., Frantz, J. D., Tawa, N. E., Jr., Melendez, P. A., Oh, B. C., Lidov, H. G., Hasselgren, P. O., Frontera, W. R., Lee, J., Glass, D. J., and Shoelson, S. E. (2004). IKKbeta/NF-kappaB activation causes severe muscle wasting in mice. *Cell* **119**(2), 285–298.

Chen, A. S., Marsh, D. J., Trumbauer, M. E., Frazier, E. G., Guan, X. M., Yu, H., Rosenblum, C. I., Vongs, A., Feng, Y., Cao, L., Metzger, J. M., Strack, A. M., Camacho, R. E., Mellin, T. N., Nunes, C. N., Min, W., Fisher, J., Gopal-Truter, S., MacIntyre, D. E., Chen, H. Y., and Van der Ploeg, L. H. (2000). Inactivation of the mouse melanocortin-3 receptor results in increased fat mass and reduced lean body mass. *Nat Genet* **26**(1), 97–102.

Chen, H., Charlat, O., Tartaglia, L. A., Woolf, E. A., Weng, X., Ellis, S. J., Lakey, N. D., Culpepper, J., Moore, K. J., Breitbart, R. E., Duyk, G. M., Tepper, R. I., and Morgenstern, J. P. (1996). Evidence that the diabetes gene encodes the leptin receptor: Identification of a mutation in the leptin receptor gene in db/db mice. *Cell* **84**(3), 491–495.

Chen, W. T., Kelly, T., and Ghersi, G. (2003). DPPIV, seprase, and related serine peptidases in multiple cellular functions. *Curr Top Dev Biol* **54**, 207–232.

Chua, S. C., Jr., Chung, W. K., Wu-Peng, X. S., Zhang, Y., Liu, S. M., Tartaglia, L., and Leibel, R. L. (1996). Phenotypes of mouse diabetes and rat fatty due to mutations in the OB (leptin) receptor. *Science* **271**(5251), 994–996.

Clark, R. G., Mortensen, D. L., Carlsson, L. M., Carlsson, B., Carmignac, D., and Robinson, I. C. (1996). The obese growth hormone (GH)-deficient dwarf rat: Body fat responses to patterned delivery of GH and insulin-like growth factor-I. *Endocrinology* **137**(5), 1904–1912.

Conarello, S. L., Li, Z., Ronan, J., Roy, R. S., Zhu, L., Jiang, G., Liu, F., Woods, J., Zycband, E., Moller, D. E., Thornberry, N. A., Zhang, B. B. (2003). Mice lacking dipeptidyl peptidase IV are protected against obesity and insulin resistance. PNAS. *Proc Natl Acad Sci USA* **100**(11), 9825–9830.

Corsetti, J. P., Sparks, J. D., Peterson, R. G., Smith, R. L., and Sparks, C. E. (2000). Effect of dietary fat on the development of non-insulin dependent diabetes mellitus in obese Zucker diabetic fatty male and female rats. *Atherosclerosis* **148**(2), 231–241.

Costet, P., Legendre, C. More, J., Edgar, A., Galtier, P., and Pineau, T. (1998). Peroxisome proliferator-activated receptor alpha-isoform deficiency leads to progressive dyslipidemia with sexually dimorphic obesity and steatosis. *J Biol Chem.* **273**(45), 29577–29585.

Deeb, S. S., Fajas, L., Nemoto, M., Pihlajamaki, J., Mykkanen, L., Kuusisto, J., Laakso, M., Fujimoto, W., and Auwerx, J. (1998). A Pro12Ala substitution in PPARgamma2 associated with decreased receptor activity, lower body mass index and improved insulin sensitivity. *Nat Genet.* **20**(3), 284–287.

Elchebly, M., Payette, P., Michaliszyn, E., Cromlish, W., Collins, S., Loy, A. L., Normandin, D., Cheng, A., Himms-Hagen, J., Chan, C. C., Ramachandran, C., Gresser, M. J., Tremblay, M. L., and Kennedy, B. P. (1999). Increased insulin sensitivity and obesity resistance in mice lacking the protein tyrosine phosphatase-1B gene. *Science.* **283**(5407), 1544–1548.

Fan, Q., Shike, T., Shigihara, T., Tanimoto, M., Gohda, T., Makita, Y., Wang, L. N., Horikoshi, S., and Tomino, Y. (2003). Gene expression profile in diabetic KK/Ta mice. *Kidney Int* **64**(6), 1978–1985.

Forman, B. M., Tontonoz, P., Chen, J., Brun, R. P., Spiegelman, B. M., and Evans, R. M. (1995). 15-Deoxy-delta 12, 14-prostaglandin J2 is a ligand for the adipocyte determination factor PPAR gamma. *Cell* **83**(5), 803–812.

Goldstein, J. L., and Brown, M. S. (1990). Regulation of the mevalonate pathway. *Nature* **343**, 425–430.

Goodwin, B., Jones, S. A., Price, R. R., Watson, M. A., McKee, D. D., Moore, L. B., Galardi, C., Wilson, J. G., Lewis, M. C., Roth, M. E., Maloney, P. R., Willson, T. M., and Kliewer, S. A. (2000). A regulatory cascade of the nuclear receptors FXR, SHP-1, and LRH-1 represses bile acid biosynthesis. *Mol Cell* **6**, 517–526.

Goodwin, B., Redinbo, M. R., and Kliewer, S. A. (2002). Regulation of cyp3a gene transcription by the pregnane x receptor. *Annu Rev Pharmacol Toxicol* **42**, 1–23.

Goodwin, B., Watson, M. A., Kim, H., Miao, J., Kemper, J. K., and Kliewer, S. A. (2003). Differential regulation of rat and human *CYP7A1* by the nuclear oxysterol receptor liver X receptor-α. *Mol Endocrinol* **17**(3), 386–394.

Graham, M., Shutter, J. R., Sarmiento, U., Sarosi, I., and Stark, K. L. (1997). Overexpression of Agrt leads to obesity in transgenic mice. *Nat Genet* **17**(3), 273–274.

Haffner, S. M. (2003). Pre-diabetes, insulin resistance, inflammation and CVD risk. *Diabetes Res Clin Pract* **61** (Suppl 1); S9–S18.

Haj, F. G., Zabolotny, J. M., Kim, Y. B., Kahn, B. B., and Neel, B. G. (2005). Liver-specific protein-tyrosine phosphatase 1B (PTP1B) re-expression alters glucose homeostasis of PTP1B-/-mice. *J Biol Chem* **280**(15), 15038–15046.

Harley, E. A., and Levens, N. (2003). Protein tyrosine phosphatase 1B inhibitors for the treatment of type 2 diabetes and obesity: Recent advances. *Curr Opin Investig Drugs* **4**(10), 1179–1189.

He, W, Barak, Y., Hevener, A., Olson, P., Liao, D., Le, J., Nelson, M., Ong, E., Olefsky, J. M., and Evans, R. M. (2003). Adipose-specific peroxisome proliferator-activated receptor gamma knockout causes insulin resistance in fat and liver but not in muscle. *Proc Natl Acad Sci USA* **100**(26), 15712–15717.

Heine, P. A., Taylor, J. A., Iwamoto, G. A., Lubahn, D. B., and Cooke, P. S. (2000). Increased adipose tissue in male and female estrogen receptor-alpha knockout mice. *Proc Natl Acad Sci USA* **97**(23), 12729–12734.

Hevener, A. L., He, W., Barak, Y., Le, J., Bandyopadhyay, G., Olson, P., Wilkes, J., Evans, R. M., and Olefsky, J. (2003). Muscle-specific Pparg deletion causes insulin resistance. *Nat Med.* **9**(12), 1491–1497.

Horton, J. D., Cuthbert, J. A., and Spady, D. K. (1995). Regulation of hepatic 7 α-hydroxylase expression and response to dietary cholesterol in the rat and hamster. *J Biol Chem* **270**, 5381–5387.

Huszar, D., Lynch, C. A., Fairchild-Huntress, V., Dunmore, J. H., Fang, Q., Berkemeier, L. R., Gu, W., Kesterson, R. A., Boston, B. A., Cone, R. D., Smith, F. J., Campfield, L. A., Burn, P., and Lee, F. (1997). Targeted disruption of the melanocortin-4 receptor results in obesity in mice. *Cell* **88**(1), 131–141.

Jones, S. A., Moore, L. B., Shenk, J. L., Wisely, G. B., Hamilton, G. A., McKee, D. D., Tomkinson, N. C., LeCluyse, E. L., Lambert, M. H., Willson, T. M., Kliewer, S. A., and Moore, J. T. (2000). The pregnane X receptor: A promiscuous xenobiotic receptor that has diverged during evolution. *Mol Endocrinol* **14**(1), 27–39.

Kim, J. K., Kim, H. J., Park, S. Y., Cederberg, A., Westergren, R., Nilsson, D., Higashimori, T., Cho, Y. R., Liu, Z. X., Dong, J., Cline, G. W., Enerback, S., and Shulman, G. I. (2005). Adipocyte-specific overexpression of FOXC2 prevents diet-induced increases in intramuscular fatty acyl CoA and insulin resistance. *Diabetes* **54**(6), 1657–1663.

Kleyn, P. W., Fan, W, Kovats, S. G., Lee, J. J., Pulido, J. C., Wu, Y., Berkemeier, L. R., Misumi, D. J., Holmgren, L., Charlat, O., Woolf, E. A., Tayber, O., Brody, T., Shu, P., Hawkins, F., Kennedy, B., Baldini, L., Ebeling, C., Alperin, G. D., Deeds, J., Lakey, N. D., Culpepper, J., Chen, H., Glucksmann-Kuis, M. A., Carlson, G. A., Duyk, G. M., and Moore, K. J. (1996). Identification and characterization of the mouse obesity gene tubby: a member of a novel gene family. *Cell* **85**(2), 281–290.

Ktorza, A., Bernard, C., Parent, V., Penicaud, L., Froguel, P., Lathrop, M., and Gauguier, D. (1997). Are animal models of diabetes relevant to the study of the genetics of non-insulin-dependent diabetes in humans? *Diabetes Metab* **23** (Suppl 2), 38–46.

Kubota, N., Terauchi, Y., Miki, H., Tamemoto, H., Yamauchi, T., Komeda, K., Satoh, S., Nakano, R., Ishii, C., Sugiyama, T., Eto, K., Tsubamoto, Y., Okuno, A., Murakami, K., Sekihara, H., Hasegawa, G., Naito, M., Toyoshima, Y., Tanaka, S., Shiota, K., Kitamura, T., Fujita, T., Ezaki, O., Aizawa, S., Kadowaki, T., et al. (1999). PPAR gamma mediates high-fat diet-induced adipocyte hypertrophy and insulin resistance. *Mol Cell* **4**(4), 597–609.

Kume, E., Aruga, C., Ishizuka, Y., Takahashi, K., Miwa, S., Itoh, M., Fujimura, H., Toriumi, W., Kitamura, K., and Doi, K. (2005). Gene expression profiling in streptozotocin treated mouse liver using DNA microarray. *Exp Toxicol Pathol* **56**(4–5), 235–244.

Kushi, A., Sasai, H., Koizumi, H., Takeda, N., Yokoyama, M., and Nakamura, M. Obesity and mild hyperinsulinemia found in neuropeptide Y-Y1 receptor-deficient mice. (1998). *Proc Natl Acad Sci USA* **95**(26), 15659–15664.

Lazar, M. A. (2005). PPAR gamma, 10 years later. *Biochimie* **87**(1), 9–13.

Leonardsson, G., Steel, J. H., Christian, M., Pocock, V., Milligan, S., Bell, J., So, P. W., Medina-Gomez, G., Vidal-Puig, A., White, R., and Parker, M. G. (2004). Nuclear receptor corepressor RIP140 regulates fat accumulation. *Proc Natl Acad Sci USA* **101**(22), 8437–8442.

Li, A. P. (2004). A comprehensive approach for drug safety assessment. *Chem Biol Interact* **150**(1), 27–33.

Lu, T. T., Makishima, M., Repa, J. J., Schoonjans, K., Kerr, T. A., Auwerx, J., and Mangelsdorf, D. J. (2000) Molecular basis for feedback regulation of bile acid synthesis by nuclear receptors. *Mol Cell* **6**, 507–515.

Luo, J., Quan, J., Tsai, J., Hobensack, C. K., Sullivan, C., Hector, R., and Reaven, G. M. (1998). Nongenetic mouse models of non-insulin-dependent diabetes mellitus. *Metabolism* **47**(6), 663–668.

Luo, J., Sladek, R., Carrier, J., Bader, J. A., Richard, D., and Giguere, V. (2003). Reduced fat mass in mice lacking orphan nuclear receptor estrogen-related receptor alpha. *Mol Cell Biol* **23**(22), 7947–7956.

Marsh, D. J., Hollopeter, G., Kafer, K. E., and Palmiter, R. D. (1998). Role of the Y5 neuropeptide Y receptor in feeding and obesity. *Nat Med.* **4**(6), 718–721.

Miles, P. D., Barak, Y., He, W., Evans R. M., and Olefsky, J. M. (2000). Improved insulin-sensitivity in mice heterozygous for PPAR-gamma deficiency. *J Clin Invest.* **105**(3), 287–292.

Munzberg, H., Bjornholm, M., Bates, S. H., and Myers, M. G., Jr. (2005). Leptin receptor action and mechanisms of leptin resistance. *Cell Mol Life Sci* **62**(6), 642–652.

Naveilhan, P., Hassani, H., Canals, J. M., Ekstrand, A. J., Larefalk, A., Chhajlani, V., Arenas, E., Gedda, K., Svensson, L., Thoren, P., and Ernfors, P. (1999). Normal feeding behavior, body weight and leptin response require the neuropeptide Y Y2 receptor. *Nat Med* **5**(10), 1188–1193.

Nielsen, L. L. (2005). Incretin mimetics and DPP-IV inhibitors for the treatment of type 2 diabetes. *Drug Discov Today* **10**(10), 703–710.

Nonogaki, K., Strack, A. M., Dallman, M. F., and Tecott, L. H. (1998). Leptin-independent hyperphagia and type 2 diabetes in mice with a mutated serotonin 5-HT2C receptor gene. *Nat Med* **4**(10), 1152–1156.

Ohki-Hamazaki, H., Watase, K., Yamamoto, K., Ogura, H., Yamano, M., Yamada, K., Maeno, H., Imaki, J., Kikuyama, S., Wada, E., and Wada, K. (1997). Mice lacking bombesin receptor subtype-3 develop metabolic defects and obesity. *Nature* **390**(6656), 165–169.

Oike, Y., Akao, M., Yasunaga, K., Yamauchi, T., Morisada, T., Ito, Y., Urano, T., Kimura, Y., Kubota, Y., Maekawa, H., Miyamoto, T., Miyata, K., Matsumoto, S., Sakai, J., Nakagata, N., Takeya, M., Koseki, H., Ogawa, Y., Kadowaki, T., and Suda, T. (2005). Angiopoietin-related growth factor antagonizes obesity and insulin resistance. *Nat Med* **11**(4), 400–408.

Oliver, W. R., Jr., Shenk, J. L., Snaith, M. R., Russell, C. S., Plunket, K. D., Bodkin, N. L., Lewis, M. C., Winegar, D. A., Sznaidman, M. L., Lambert, M. H., Xu, H. E., Sternbach, D. D., Kliewer, S. A., Hansen, B. C., and Willson, T. M. (2001). A selective peroxisome proliferator-activated receptor delta agonist promotes reverse cholesterol transport. *Proc Natl Acad Sci USA* **98**(9), 5306–5311.

Ortmeyer, H. K., Bodkin, N. L., Haney, J., Yoshioka, S., Horikoshi, H., and Hansen, B. C. (2000). A thiazolidinedione improves *in vivo* insulin action on skeletal muscle glycogen synthase in insulin-resistant monkeys. *Int J Exp Diabetes Res* **1**(3), 195–202.

Picard, F., Gehin, M., Annicotte, J., Rocchi, S., Champy, M. F., O'Malley, B. W., Chambon, P., and Auwerx, J. (2002). SRC-1 and TIF2 control energy balance between white and brown adipose tissues. *Cell* **111**(7), 931–941.

Roberts, S. A. (2003). Drug metabolism and pharmacokinetics in drug discovery. *Curr Opin Drug Discov Dev* **6**(1), 66–80.

Russell, D. W. (1999). Nuclear orphan receptors control cholesterol catabolism. *Cell* **97**, 539–542.

Shepherd, P. R., Gnudi, L., Tozzo, E., Yang, H., Leach, F., and Kahn, B. B. (1993). Adipose cell hyperplasia and enhanced glucose disposal in transgenic mice overexpressing GLUT4 selectively in adipose tissue. *J Biol Chem* **268**(30), 22243–22246.

Shoelson, S. E., Lee, J., and Yuan, M. (2003). Inflammation and the IKK beta/I kappa B/NF-kappa B axis in obesity- and diet-induced insulin resistance. *Int J Obes Relat Metab Disord* **27**(Suppl 3), S49–S52.

Sonoda, J., Rosenfeld, J. M., Xu, L., Evans, R. M., and Xie, W. (2003). A nuclear receptor-mediated xenobiotic response and its implication in drug metabolism and host protection. *Curr Drug Metab* **4**(1), 59–72.

Staels, B. (2005). PPARgamma and atherosclerosis. *Curr Med Res Opin* **21**(Suppl 1), S13–S20.

Stoffers, D. A. (2004). The development of beta-cell mass: Recent progress and potential role of GLP-1. *Horm Metab Res* **36**(11–12), 811–821.

Takaya, K., Ogawa, Y., Hiraoka, J., Hosoda, K., Yamori, Y., Nakao, K., and Koletsky, R. J. (1996). Nonsense mutation of leptin receptor in the obese spontaneously hypertensive Koletsky rat. *Nat Genet* **14**(2), 130–131.

Tartaglia, L. A., Dembski, M., Weng, X., Deng, N., Culpepper, J., Devos, R., Richards, G. J., Campfield, L. A., Clark, F. T., Deeds, J., Muir, C., Sanker, S., Moriarty, A., Moore, K. J., Smutko, J. S., Mays, G. G., Wool, E. A., Monroe, C. A., and Tepper, R. I. (1995). Identification and expression cloning of a leptin receptor, OB-R. *Cell* **83**(7), 1263–1271.

Theil, F. P., Guentert, T. W., Haddad, S., and Poulin, P. (2003). Utility of physiologically based pharmacokinetic models to drug development and rational drug discovery candidate selection. *Toxicol Lett* **138**(1–2), 29–49.

Tsuji, K., Taminato, T., Usami, M., Ishida, H., Kitano, N., Fukumoto, H., Koh, G., Kurose, T., Yamada, Y., Yano, H., et al. (1988). Characteristic features of insulin secretion in the streptozotocin-induced NIDDM rat model. *Metabolism* **37**(11), 1040–1044.

Tucker, G. T. (1994). Clinical implications of genetic polymorphism in drug metabolism. *J Pharm Pharmacol* **46**(Suppl 1), 417–424.

Wang, P. R., Guo, Q., Ippolito, M., Wu, M., Milot, D., Ventre, J., Doebber, T., Wright, S. D., and Chao, Y. S. (2001). High fat fed hamster, a unique animal model for treatment of diabetic dyslipidemia with peroxisome proliferator activated receptor alpha selective agonists. *Eur J Pharmacol* **427**(3), 285–293.

Wang, Y. X., Zhang, C. L., Yu, R. T., Cho, H. K., Nelson, M. C., Bayuga-Ocampo, C. R., Ham, J., Kang, H., and Evans, R. M. (2004). Regulation of muscle fiber type and running endurance by PPARdelta. *PLoS Biol* **2**(10), 294.

Wu-Peng, X. S., Chua, S. C. Jr., Okada, N., Liu, S. M., Nicolson, M., and Leibel, R. L. (1997). Phenotype of the obese Koletsky (f) rat due to Tyr763Stop mutation in the extracellular domain of the leptin receptor (Lepr): Evidence for deficient plasma-to-CSF transport of leptin in both the Zucker and Koletsky obese rat. *Diabetes* **46**(3), 513–518.

Xie, W., Barwick, J. L., Downes, M., Blumberg, B., Simon, C. M., Nelson, M. C., Neuschwander-Tetri, B. A., Brunt, E. M., Guzelian, P. S., and Evans, R. M. (2000). Humanized xenobiotic response in mice expressing nuclear receptor SXR. *Nature* **406**(6794), 435–439.

Yang, Q., Graham, T. E., Mody, N., Preitner, F., Peroni, O. D., Zabolotny, J. M., Kotani, K., Quadro, L., and Kahn, B. B. (2005). Serum retinol binding protein 4 contributes to insulin resistance in obesity and type 2 diabetes. *Nature* **436** (7049), 356–362.

Yaswen, L., Diehl, N., Brennan, M. B., and Hochgeschwender, U. (1999). Obesity in the mouse model of pro-opiomelanocortin deficiency responds to peripheral melanocortin. *Nat Med* **5**(9), 1066–1070.

Yen, T. T., Gill, A. M., Frigeri, L. G., Barsh, G. S., Wolff, G. L. (1994). Obesity, diabetes, and neoplasia in yellow A(vy)/– mice: Ectopic expression of the agouti gene. *FASEB J* **8**(8), 479–488.

Yu, Y., and Bradley, A. (2001). Engineering chromosomal rearrangements in mice. *Nat Rev Genet* **2**(10), 780–790.

Zhang, Y., Proenca, R., Maffei, M., Barone, M., Leopold, L., and Friedman, J. M. (1994). Positional cloning of the mouse obese gene and its human homologue. *Nature* **372**(6505), 425–432.

CHAPTER 20

STEM CELL RESEARCH AND APPLICATIONS FOR HUMAN THERAPIES

Prithi Rajan, Kook In Park, Vaclav Ourednik, Jean Pyo Lee, Jamie Imitola, Franz-Joseph Mueller, Yang D. Teng, and Evan Snyder*

20.1 INTRODUCTION

While the presence of a pluripotent or multipotent "stem" cell in embryonic and adult animals is conceptually intuitive, a concerted study of these cells as a discipline has acquired prominence only since the early 1990s (Gage et al., 1995; Snyder and Macklis, 1995). It is now widely accepted that stem cells of varying potentials exist in adult organisms ranging from plants to invertebrate and vertebrate animals and are present to an even greater extent in developing and young organisms. Stem cells reside in several tissues in the adult and embryonic stages of mammalian development and have the potential to give rise to several mature lineages (Blau et al., 2001). This complex field is rapidly emerging largely due to the promise it holds for cellular therapies in humans. In this chapter we will attempt to review some salient aspects of the field in general and demonstrate research and therapeutic uses of stem cells with some specific examples.

*Corresponding author: esnyder@burnham.org

Drug Discovery Research: New Frontiers in the Post-Genomic Era, Edited by Ziwei Huang

Embryonic stem cells (ESCs) are isolated from the inner cell mass of a developing embryo, are pluripotent, and have the unbridled capacity to proliferate and differentiate into almost all the cell types that populate the adult organism. Conversely, somatic stem cells are present to varying degrees in young and adult tissues and are thought to have a more restricted potential in terms of the types of fates into which they can differentiate. The best-described of the somatic stem cells include neural stem cells (NSCs), which may be isolated from both adult and embryonic brains; hematopoietic stem cells (HSCs), which are isolated from adult bone marrow; and mesenchymal stem cells (MSCs), which are also isolated from adult bone marrow. Stem cells have also been described from adult adipose tissues (adipose-derived stem cells, ADSCs), umbilical cord blood, adult muscle, and liver, among other tissues. While somatic stem cells are more restricted in their differentiation potential compared to their embryonic counterparts, the plasticity of embryonic stem cells presents as a double-edged sword because they do not form tumors and/or differentiate into unexpected lineages as readily when used in transplantation therapies. Adult stem cells also have fewer political encumbrances in terms of limits on the generation and practical use of human stem cell lines.

While it is possible to harvest HSCs, MSCs, and ADSCs with relative ease from adult humans, ESCs and NSCs present more of a challenge due to size and anatomical considerations. Thus, cell lines have been created from mouse and human NSCs that have been created either by immortalization with genes such as myc or by propagation with selected growth factors. NSCs have been selected from fetal human tissue by a battery of surface antibodies, including CD133 and 5E12 (Uchida et al., 2000), and NSC lines have been derived by immortalization with myc (Villa et al., 2000) and selected by culture techniques and growth factors (Snyder, unpublished data). Similarly, several lines of ESCs have been created and characterized to varying extents (Hoffman and Carpenter, 2005). A registry of hESCs currently eligible for federally funded research is available on http://stemcells.nih.gov/research/registry/. There are currently 21 hESCs on this list, which have been derived in universities and institutes all over the globe. A concerted effort is currently being made to characterize these ES cells and establish culture conditions, including selection and identification criteria that may serve as a benchmark in global stem cell efforts. The results of the characterization of some of these cell lines are presented in the Web site http://stemcells.nih.gov/research/nihresearch/, where selected cell lines have been characterized and the results made public. Similar efforts are in progress in other labs where techniques such as several variations of microarrays (several variations), massively parallel signature sequencing (MPSS), and proteomics are being used in an effort to sift out the various cell populations that compose the complex hESC cultures. Some of these will be discussed in greater detail below. Other groups have also independently created hESC lines, including the 17 lines described in Cowan et al. (2004).

The obvious *in vivo* application for stem cells is transplantation of stem or progenitor cells that lead to "cell replacement therapies." Ameliorative effects could potentially be the result of one or all of these events: (1) appropriate differentiation of the stem cells to mature cells that are functionally and/or phenotypically similar to the cell compromised or lost; (2) trophic/paracrine effects provided by the transplanted cells to the injured tissue; and (3) fusion of transplanted cells to the injured cells.

Although hESCs hold the most therapeutic value because of their pluripotency, our lack of knowledge regarding control of their differentiation and their propensity to promote tumors after transplantation is a major issue that needs to be addressed before they can be serious contenders in human therapeutics. In addition to transplantation, the homing properties of stem cells may prove advantageous in the use of these cells for the delivery of therapeutic agents to compromised tissues in the body, especially tumors (Aboody et al., 2000); Lee et al., 2003; Nakamizo et al., 2005). The latter use may be more tangible as the cells could be engineered in such a manner that they are dispensed with once the therapeutic is delivered.

While the most obvious translation of the potential of stem cells into realizable and immediate benefit to humans is in cell replacement therapies with the endpoint of short- and long-term amelioration of disease phenotypes, there are other equally important and interesting applications of these remarkable cells. Important *in vitro* uses of stem cells include the use of differentiating and mature differentiated phenotypes at several stages of the drug discovery process. In addition to these two applications, the dynamic multipotentiality of stem cells can also be used to establish *in vitro* paradigms that represent diseased states (particularly of developmental disorders) so that the mechanisms of onset and reversal of disease may be studied at the molecular and cellular levels. The insights thus obtained may be used to design therapies for these illnesses. In addition, at a purely basic level of scientific curiosity, the mechanisms of developmentally related events such as fate choice, cell proliferation, and cell motility may be studied in these dynamic cultures. An understanding of the mechanisms of development will drive the manipulation and exploitation of stem cells for novel cell-replacement therapies.

20.2 DEFINITION OF STEM CELLS

The enormous potential of stem cells arises from their plasticity. The use of stem cells in all pertinent applications will be greatly aided by a precise definition of their state in the continuum of "multipotent cell" to "committed precursor." Cell surface markers, in particular, will greatly aid in the isolation of particular populations of stem cells when used in conjunction with magnetic bead technology and fluorescence-activated cell sorting (FACS); these populations of stem cells may be used in both *in vivo* and *in vitro* applications. Cytoplasmic markers could largely be used for the identification of cell types. At the present time there is a dearth of markers that precisely define stem cells and their progeny across the field. Although work in the field is progressing by the absence or presence of groups of markers thought to identify particular groups of cells, the lack of comprehensive sets of markers (or a single marker) defining specific stem cells, precursors, and mature phenotypes leads to a lack of precision, which is a serious limitation in the field.

Several techniques have been used in attempts to define unique markers for embryonic and somatic stem cells markers and the mature phenotypes that they yield. Some of the earliest publications attempting to define "stemness" genes—that is, genes that are unique to the stem cell state irrespective of the source of the stem cells—met with muted success (Ivanova et al., 2002; Ramalho-Santos et al., 2002).

These and other similar efforts suggest that the study of differential gene expression alone is not sufficient to define markers of particular stem cells or restricted lineages, because there doesn't appear to be a clear demarcation of specific markers for different cell types at the level of gene expression. Other markers related to stem cells including Oct4, nanog, nestin, and musashi have been isolated using techniques such as invertebrate genetics, expression cloning, generation of antibodies to whole cells and cell lysates, differential display, and activation of proteins during certain cellular events to define markers of particular cell types or processes (Mitsui et al., 2003; Chambers et al., 2003; Lendahl et al., 1990; Sakakibara et al., 2002). Using subtractive cDNA technology, Brandenberger et al. (2004b) have compared differences in gene expression in a population of undifferentiated hESCs with three populations of cells differentiated from hESCs and have attempted a characterization of differences in levels of expression (not activation) of signaling proteins and related transcription factors. Such "transcriptome" analyses will possibly lead to the definition of cohorts of markers that define particular fates or may be useful for the characterization of populations of cells. The HSC and MSC fields have relied largely on the "CD" surface markers gleaned over time by defining markers of cells as they differentiate into various lineages in the blood (Shizuru et al., 2005; Pittenger and Martin, 2004). Markers such as CD133, which has been used to characterize HSCs as well as NSCs (Uchida et al., 2000; Shizuru et al., 2005), may be a general marker of stem cells. Stem cells have also been isolated based on their position in a fluorescence activated cell sorting (FACS) profile, where a small group of cells exclude dyes with greater efficiency than the rest of the population of cells within that particular tissue. These cells, called "side population" cells, have now been characterized from various tissues including blood, bone marrow, breast, skin, and eye and appear to coincide with the appearance of membrane channels of the ATP binding cassette (ABC) family that includes the MDR and BCRP1 receptors. The stemlike properties of this population are still undergoing characterization (Jonker et al., 2005; Camargo et al., 2005; Uchida et al., 2004). The ongoing characterization of ESCs and the various somatic stem cells will continue to broaden our definition of each kind of cell and perhaps improve our methods of assessing the use of these cells in clinical and research projects.

Other novel identifying events, perhaps more dynamic, are also being explored as markers of particular states of stem cells. Using gene-trapping technology, Scheel et al. (2005) studied differentiating neural stem cells and identified known and novel markers of differentiation. Such data will yield clues as to the dynamic process of differentiation in stem cells and perhaps yield markers, which define several stages of differentiation before the mature phenotype is reached. Since the final arbiter of fate choice is transcription of specific genes that dictate fate choice, of particular interest is the transcriptional state of the cell, the ability of the DNA to bind to proteins such as histones and transcription factors, and the ease with which the DNA can be transcribed—that is, the epigenetic state of the genome. Several clever variations of array technology are being developed where the methylation states of the DNA and histones and the availability of DNA to bind to transcription factors can be detected [reviewed in van Steensel (2005)]. The use of these techniques could possibly lead to novel insights into the "transcriptome" of proliferating and differentiating stem cells.

This includes the ChIP-chip (chromosome immunoprecipitation) that can analyze the sites in the genome that are bound by particular proteins. Alternatively, the presence of specific protein complexes in the cell could be considered a marker, for instance, if it relates specifically to the activation of a transcription factor required for a particular fate. In the example presented in Fig. 20.3, the complex of FRAP-Stat3 would serve as a marker of glial differentiation in similar paradigms, since Stat3 has been shown to be required for glial differentiation by at least two independent groups (Rajan et al., 2003; Bonni et al., 1997). However, Stat3 activation alone cannot serve as a marker for glial differentiation since the neuropoietic cytokines activate STAT proteins in several tissues and under diverse conditions (Aaronson and Horvath, 2002). Such complexes could be detected by colocalization of signal by immunofluorescence or by fluorescence resonance energy transfer (FRET) in combination with bead technology or FACS, depending on the application in question. Stable RNA molecules such as miRNAs have recently been shown to regulate cell cycling in stem cells (Hatfield et al., 2005) and regulate the differentiation of cardiomyocytes (Zhao et al., 2005).

More global efforts at signaling pathway analysis have also been recently attempted. In an example of trying to model the global signaling state of a hippocampal neuron, Ma'ayan et al. (2005) have used a combination of published literature along with computer modeling to propose networks of signals arising from the ligand–receptor interaction at the cell surface to transcription factor activation in the nucleus and have described regulatory motifs along the cellular network. While this study has taken only the chemical reactions into consideration without spatial and temporal criteria, it provides an initial look at a complex cellular network and could develop into more mature definitions of networks of plastic versus mature cells. Others have used microarray approaches to globally detect genes that are inactive and active in hESCs with respect to particular transcription factors and postulate possible feed-forward and feedback loops at the gene expression level (Boyer et al., 2005). It is very possible that a precise description of a stem cell as it traverses its path to differentiation may be defined by combinations of two or more kinds of markers, "static" and "dynamic," some of which are described above. Given the relatively nebulous nature of the markers presently being used to define stem cells and their derivatives, the only stringent definition of a stem cell and progenitor remains a functional one. While it is unrealistic, or sometimes impossible, to perform "clonal assays" to prove that the cell in question is a stem/progenitor cell because it is able to give rise to all the requisite fates, there is no equivalent alternative at the present time. The evolution of markers precise enough to identify the position of the cell in the continuum of "stemness" through "mature fate" will greatly enhance the efficiency of the experimental and therapeutic avenues being pursued (Table 20.1).

While several approaches are being used to define markers as described above, existing knowledge is being used to characterize cultures currently being used in laboratories. This is a particular issue for hESCs, which, due to their enormous plasticity, are notoriously heterogeneous due to variations introduced by passaging of cultures, minor variations in culture protocols that exist in different laboratories. Various techniques characterizing the expression profiles of different stem cells have been used. Expression microarrays have been used to determine the similarities and differences between hESCs (Bhattacharya et al., 2004; Yang et al., 2005). Description

TABLE 20.1 Current Markers Used to Identify ESCs, NSCs, and MSCs[a]

ESCs	Nanog
	Sox 2
	Oct 4
	Rex 1
	CD 9
	SSEA3
	SSEA 4
	Telomerase
	Alkaline phosphatase
NSCs	Nestin
	Sox 1
	Musashi
	CD 133
	PSA-NCAM
	5E12
MSCs	CD 44
	CD 105
	CD 71
	CD 29
	CD 45 negative
	CD 73
	CD 90
	p75

[a]This is a parsimonious list of markers for some human stem cells. While more extensive lists are in use, the most commonly used markers for these three varieties of human stem cells are listed. Because no single marker is unambiguously specific (i.e., there may be overlap in immunoreactivity between different stages of differentiation or even between cells from different organ systems), designation of a given stem cell type should rest only the use of panels of multple markers (not simply one) complemented by functional assays.

of the gene expression profiles of the ESC lines HES3 and HES4 has been done by serial analysis of gene expression (SAGE) to compare between these cell lines and also to compare them with mouse ESC lines and cancer cells (Richards et al., 2004). Massively parallel signature sequencing (MPSS) has also been used to confirm some of the data generated by other techniques (Brandenberger et al., 2004a). MSCs and HSCs have been characterized largely using FACS to determine the surface markers present on the cells. NSCs have also been characterized using FACS, again by the description of their surface markers (Uchida et al., 2000).

A conundrum in the stem cell field is the fact that due to the limited numbers of stem cells present in an embryo or adult animal, stem cells have to be expanded in culture to generate enough numbers of cells for experimental and clinical manipulations. However, due to the inherent plasticity of the cells, the stem cell cultures that result are entirely a product of the culture conditions. This must always be considered while making interpretations about the plasticity, or lack thereof, of stem cells, particularly relating to somatic stem cells. An excellent example of this is the culture of mesenchymal stem cells. When cultured in medium containing 10% serum as is common in the field, these cells differentiate into adipocytes, chondrocytes, and osteoblasts (Pittenger et al., 1999). However, when cultured under alternate culture conditions, including low serum and regulated density, these cells can also be coaxed into ectodermal fates, such as neurons, and endodermal fates, such as endothelium (Jiang et al., 2002; Gregory et al., 2005; Rajan and Mackay, unpublished data). Thus, the differentiation potential of stem cells is performed on stem cells that have undergone proliferation in specific media, and it is possible that the conditions used for, and thus the signals received during, proliferation of the stem cells dictate the restrictions of the differentiation potential in the resulting culture. Similarly, neural stem cells cultured in serum and/or as neurospheres have a greater propensity to glial fates when compared to neural stem cells maintained as monolayers and in defined culture media—that is, without serum (Fig. 20.1). A corollary to this is that

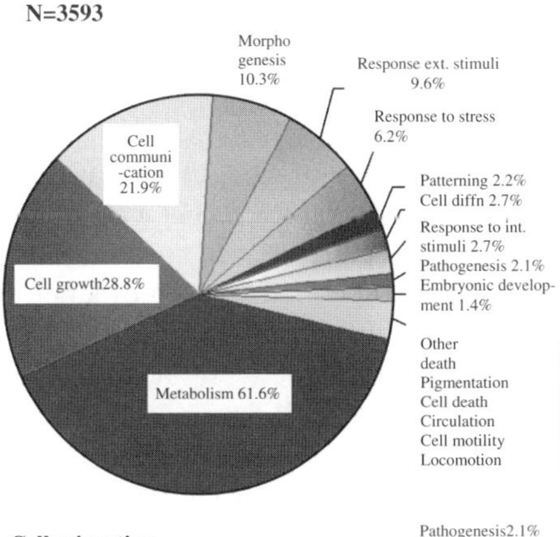

Neuronal fate

gene ID	Undif C17.2	Dif C17.2	Nsphere LCB	ESC LCB	HSC LCB	C17.2 LCB	
Mm.42227	M	A	-	-	-	E	PHOX2b
Mm.34537	M	A	-	-	-	E	C/EBPα
Mm.4639	P	P	-	-	-	E	C/EBPδ
Mm.2924	P	A	-	-	-	E	PDGF Ra
Mm.3400	P	P	-	-	E	-	Disheveled
Mm.4390	M	M	-	-			Numb
Mm.4451	P	P	-	-	-	-	HES-1
Mm.2657	P	P	-	-	-	-	Engrailed-1

Neuronal signaling

gene ID	Undif C17.2	Dif C17.2	Nsphere LCB	ESC LCB	HSC LCB	C17.2 LCB	
Mm.57234	M	A	-	-	-	E	Neuropeptide Y R6
Mm.57040	M	A	-	-	-	E	Noradrenalin
Mm.8053	M	A	-	-	-	E	Serotonin
Mm.16745	M	A	-	-	-	E	Oxytocin-neuropht.:

Cell migration

gene ID	Undif C17.2	Dif C17.2	Nsphere LCB	ESC LCB	HSC LCB	C17.2 LCB	
Mm.16234	M	A	-	-	-	E	fibronection receptor alpha
Mm.4974	M	A	-	E	-	E	NCAM
Mm.19065	M	A	-	E	-	E	Integrin binding pt GRP1
Mm.3730	M	A	-	E	-	E	semiphorin H
Mm.26934	M	A	-	-	-	E	Lectin, galactose-binding
Mm.26852	M	A	-	E	-	E	MHC

Cell signaling

gene ID	Undif C17.2	Dif C17.2	Nsphere LCB	ESC LCB	HSC LCB	C17.2 LCB	
Mm.2770	M	A	-	-	-	E	IGF 1
Mm.578	P	A	-	-	-	E	IGF binding protein 5
Mm.21373	M	A	-	-	-	E	IGF binding protein 6
Mm.2020	M	A	-	E	E	E	LIM-1
Mm.2924	P	A	-	-	-	E	PDGF Ra
Mm.9154	M	A	-	-	-	E	TGFb
Mm.14455	P	P	-	-	-	E	TGFb
Mm.1853	M	A	-	-	-	E	Ganglioside express. 2

Key
P = present expression
M = marginal expression
A = absent expression
E = expressed dCHIP call
- = absent dChip call

Figure 20.1 Sixty-seven percent of NSC genes are lost in neurospheres. This figure, taken from Parker et al. (2005), shows that there are substantial qualitative differences in NSCs that are grown in different conditions. Here comparisons were made between cells grown as neurospheres and monolayers, and differences are seen in genes involved in cell signaling, fate choice, and cell migration.

the culture of stem cells isolated from various tissues such as bone marrow, adipose tissue, and umbilical cord under the same culture conditions, such as that used for the culture of MSCs, will possibly result in the isolation and culture of a similar stem cell from these seemingly unrelated tissues (Strem et al., 2005; Li et al., 2005a). Such problems are only magnified when dealing with embryonic stem cells, which are even more plastic and less defined. The creation of disparate ESC lines from the same starting cell line due to variations in culture conditions extant in independent laboratories is an issue that is being addressed in a concerted manner by the International Stem Cell Initiative (Andrews et al., 2005). Another caveat in the field is that most of the experiments defining plasticity of particular stem cells are performed *in vitro* with cultured stem cells. Although the differentiation potential of some stem cells such as HSCs and ESCs have been confirmed *in vivo*, the limits of multipotentiality of stem cells *in vivo*, and especially *in situ*, remain to be defined. All these observations serve to reiterate the continuing need for monitoring the populations of embryonic and somatic stem cells that are in culture, and possibly *in vivo*, by markers that need to be more comprehensive and stringent in defining the cell types being used or derived.

The culture and expansion of stem cells is evolving into an independent discipline, both at the level of small-scale cultures for research and on a large scale for potential clinical applications. The challenge of stem cell culture is to achieve the dual goals of maintaining the cells in an undifferentiated state while maximizing for yield. More often than not, dense cultures tend to differentiate spontaneously. Three practical considerations impinge on the design of a culture method. (1) A recent comprehensive review by Hoffman and Carpenter (2005) gives an idea of the intricacies of hESC culture. Traditionally, hESCs have been maintained on mouse fibroblasts that serve as "feeder" layers and promote maintenance of the undifferentiated state. MSCs are routinely cultured in serum. Due to the implications of nonhuman pathogens associated with cells to be used in transplantation, the culture conditions for all stem cells destined to be used for human therapies will preferably be "animal origin free"—that is, free of all undefined reagents of animal origin, including serum, usually fetal bovine serum (FBS), and feeder cells. (2) Although FBS is a rich source of growth, survival, and cell adhesion factors, it is essentially an undefined cocktail of proteins containing large amounts of albumin. Again, culture of stem cells in defined medium will permit better control of the culture and a more reproducible maintenance of the undifferentiated and differentiated states. (3) However, the creation of appropriate culture conditions by the addition of the requisite individual growth factors that promote proliferation and survival of stem, differentiating, and differentiated cultures, along with the appropriate adhesion molecules, can be prohibitively expensive. This is particularly true when large-scale production of cells needs to be achieved for patient and drug screening applications. Another practical consideration is that suspension cultures are preferred over adherent cultures for the large-scale production of cells, because they are more feasible in terms of expenditure of reagents and manpower. Several stem cell cultures, including MSCs, NSCs, and ESCs, are best maintained in adherent cultures. Creative methods are being devised in order to adapt these cells to more suspension-like culture methods, including the use of beads as "suspended substrata" on which cells may grow (Schroeder et al., 2005; Sen et al., 2004).

20.3 USES OF STEM CELLS IN THERAPEUTIC APPLICATIONS

The most obvious use of stem cells is in replacement therapies where compromised tissues are replaced with functional equivalents. This had been envisioned as the goal in several disease processes, most commonly in diseases where a specific known cell type is compromised, such as diabetes, spinal cord injury, myocardial infarctions, Parkinson's disease (PD), and some blood cancers reviewed in (Daley et al., 2003; Shizuru et al., 2005). While initial impetus for the field of therapeutic transplantation was given by bone marrow transplantations for blood-related malignancies (Shizuru et al., 2005), it is now being contemplated for several of the somatic stem cells—including NSCs and MSCs and, in the longer term, ESCs as well—when the processes of uncontrolled differentiation and tumorigenesis in these latter cells is understood better. Amelioration of disease phenotypes in animal models have been attempted with various stem cells including hESCs, NSCs, HSCs, and MSCs (Kim et al., 2002; Yang et al., 2002; Eglitis and Mezey, 1997; Zhao et al., 2002). We will discuss the use of NSCs in transplantation with specific reference to PD.

20.3.1 NSCs: Parkinson's Disease

NSCs have been used with varying degrees of success in transplantation therapies in the central nervous system (CNS) (Snyder et al., 2004; Picard-Riera et al., 2004). The impact of NSCs in directly rescuing endangered host neurons was first evinced in a series of experiments in aged rodents in which the nigrostriatal system was impaired (Fig. 20.2). PD is a degenerative disorder characterized by a loss of midbrain dopamine (DA) neurons with a subsequent reduction in striatal DA (Rosenthal, 1998). The disease, in addition to incapacitating many thousands of patients, has also long served as a model for testing neural cell replacement strategies. Transplantation therapy for this CNS disorder has a long history [for review see (Dunnett (1999)]. It was the neural disease that was first treated clinically by neural transplantation, using primary tissue from human fetal ventral mesencephalon to replace DA-expressing cells (Lindvall et al., 1990ab). Indeed, it was in this disease that the limitations of fetal tissue grafts in not only rodent and primate models of PD (Mehta et al., 1998), but also in clinical trials (Lindvall et al., 1990a,b), was first recognized. These limitations include (a) short graft survival and limited integration of the grafts and (b) the possibility of unregulated DA production in improper regions, leading adversely to dyskinesias. Given PD's storied role in the development of cellular therapies, it is appropriate that a model of this disease should have also played a pivotal role in revealing a little-suspected but powerful therapeutic action that NSCs may play in preserving degenerating host cells by some heretofore-unheralded mechanisms that are nevertheless inherent to stem cell biology.

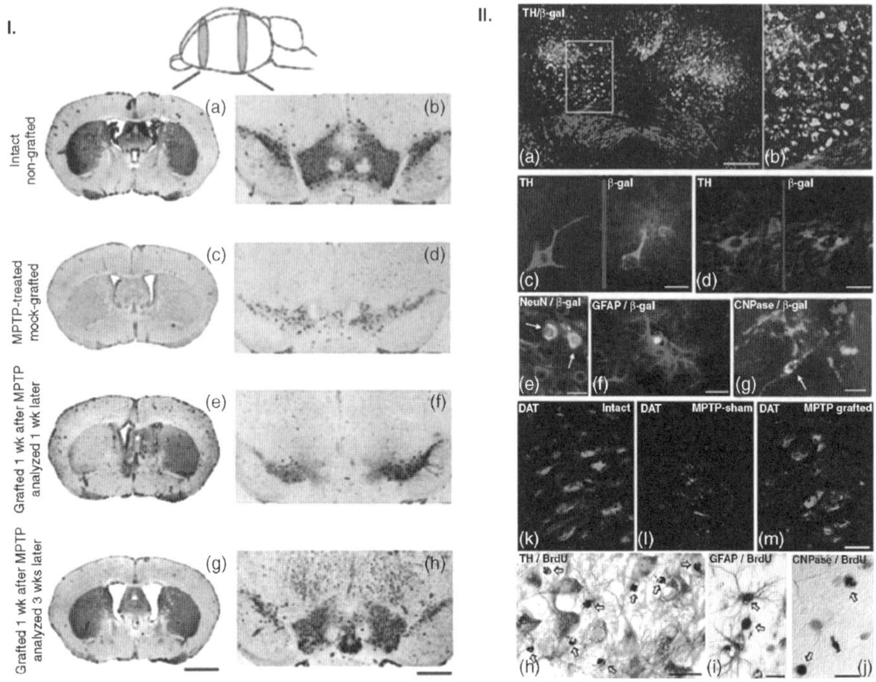

Figure 20.2 NSCs possess an inherent mechanism for rescuing dysfunctional neurons. Shown here is evidence from the effects of NSCs in the restoration of mesencephalic dopaminergic function. [Modified from Ourednik et al. (2002)]. **I.** TH expression in mesencephalon and striatum of aged mice following MPTP lesioning and unilateral NSC engraftment into the substantia nigra/ventral tegmental area (SN/VTA). A model that emulates the slow dysfunction of aging dopaminergic neurons in substantia nigra (SN) was generated by giving aged mice repeated high doses of MPTP. Schematic on top indicates the levels of the analyzed transverse sections along the rostrocaudal axis of the mouse brain. Representative coronal sections through the striatum are presented in the left column (**A**, **C**, **E**, **G**) and through the SN/VTA area in the right column (**B**, **D**, **F**, **H**). **A** and **B**. Immunodetection of TH (black cells) shows the normal distribution of DA-producing TH$^+$ neurons in coronal sections in the intact SN/VTA (**B**) and their projections to the striatum (**A**). **C** and **D**. Within 1 week, MPTP treatment caused extensive and permanent bilateral loss of TH immunoreactivity in both the mesostriatal nuclei (**C**) and the striatum (**D**), which lasted for a lifetime. Shown in this example, and matching the time point in **g** and **h**, is the situation in a mock-grafted animal 4 weeks after MPTP treatment. **E** and **F**. Unilateral (right side) stereotactic injection of NSCs into the nigra is associated, within 1 week after grafting, with substantial recovery of TH synthesis within the ipsilateral DA nuclei (**F**) and their ipsilateral striatal projections (**E**). By 3 weeks posttransplant, however (**G**, **H**), the asymmetric distribution of TH expression disappeared, giving rise to TH immunoreactivity in the midbrain (**H**) and striatum (**G**) of both hemispheres that approached the immunoreactivity of intact controls (**A**, **B**) and gave the appearance of mesostriatal restoration. Similar observations were made when NSCs were injected 4 weeks after MPTP treatment (not shown). Bars: 2 mm (left), 1 mm (right). Note the ectopically placed TH$^+$ cells in **H**. These are analyzed in greater detail, along with the entire SN, in **II**. **II.** Immunohistochemical analyses of TH, DAT, and BrdU-positive cells in MPTP-treated and grafted mouse brains. The presumption was initially that the NSCs had replaced the

dysfunctional TH neurons. However, examination of the reconstituted SN with dual βgal (green) and TH (red) ICC showed that 90% of the TH^+ cells in the SN were host-derived cells that had been rescued [**a c**], and only 10% were donor-derived [**d**]. Most NSC-derived TH^+ cells were actually just above the SN ectopically (blocked area in **a**, enlarged in **b**). These photomicrographs were taken from immunostained brain sections from aged mice exposed to MPTP, transplanted 1 week later with NSCs, and sacrificed after 3 weeks. The following combinations of markers were evaluated: TH (red) with βgal (green) [**a–d**]; NeuN (red) with βgal (green) [**e**]; GFAP (red) with βgal (green) [**f**]; CNPase (green) with βgal (red) [**g**]; TH (brown) and BrdU (black) [**k**]; GFAP (brown) with BrdU (black) [**l**]; CNPase (brown) with BrdU (black) [**m**]. Anti-DAT-stained areas are revealed in green in the SN of intact [**h**], mock-grafted [**i**], and NSC-grafted [**j**] brains. Three different fluorescence filters specific for Alexa Fluor 488 (green), Texas Red (red), and a double-filter for both types of fluorochromes (yellow) were used to visualize specific antibody binding: **c**, **d**, and **h–j** are single-filter exposures; **a**, **b**, and **e–g** are double-filter exposures. **a** shows a low-power overview of the SN/VTA of both hemispheres. The majority of TH^+ cells (red cells in **a**) within the nigra are actually of host origin (~90%), with a much smaller proportion being donor-derived (green cells, ~10%) (representative close-up of such a donor-derived TH^+ cell in **d**). Although a significant proportion of NSCs did differentiate into TH^+ neurons, many of these actually resided ectopically, dorsal to the SN (boxed area in **a**, enlarged in **b**; high-power view of donor-derived (green) cell that was also TH^+ (red) in **c**), where the ratio of donor-to-host cells was inverted: ~90% donor-derived compared with ~10% host-derived. Note the almost complete absence of a green βgal-specific signal in the SN^+VTA, whereas ectopically, many of the TH^+ cells were double-labeled and thus NSC-derived (appearing yellow orange in higher power under a red/green double-filter in panel **b**). **e–g**. NSC-derived non-TH neurons ($NeuN^+$) [**e**, arrow], astrocytes ($GFAP^+$) [**f**], and oligodendrocytes ($CNPase^+$) [**g**, arrow] were also seen, both within the mesencephalic nuclei and dorsal to them. **h–j**. Any proliferative $BrdU^+$ cells after MPTP insult and/or grafting were confined to glial cells, whereas the TH^+ neurons [k] were $BrdU^-$. This finding suggested that the reappearance of TH^+ host cells was not the result of neurogenesis but rather the recovery of extant host TH^+ neurons. Bars: ~10 μm [**a**]; 20 μm [**c**, **d**, **e**]; 30 μm [**f**]; 10 μm [**g**]; 20 μm [**h–j**]; 25 μm, [**k**]; 10 μm, [**l**]; and 20 μm [**m**]. **k–m**. The green DAT-specific signal in **j** suggests that the reconstituted mesencephalic nuclei in the NSC-grafted mice (as in **H**) were functional DA neurons comparable to those seen in intact nuclei [**h**] but not in MPTP-lesioned sham-engrafted controls [**i**]. This further suggests that the TH^+ mesostriatal DA neurons affected by MPTP are, indeed, functionally impaired. (Note that sham-grafted animals [**i**] contain only punctate residual DAT staining within their dysfunctional fibers, whereas DAT staining in normal [**h**] and, similarly, in engrafted [**j**] animals was normally and robustly distributed both within processes and throughout their cell bodies.) See color plates.

In addition to "replacing" the injured tissue, neural stem cells also appear to engage the host in a dynamic series of ongoing reciprocal interactions, each instructing the other. Under instruction from exogenous NSCs, the injured host nervous system also contributes to its own repair. The impact of NSCs in directly rescuing endangered host neurons was first evinced in a series of experiments in aged rodents in which the nigrostriatal system was impaired (Fig. 20.2). In the hope that NSCs might spontaneously differentiate into DA neurons when implanted into a DA-depleted region of the CNS, unmanipulated murine NSCs were implanted unilaterally into the substantia nigra of aged mice that been exposed systemically to high-dose MPTP (1-methyl-4-phenyl-1,2,3,6-tetrahydropyridine), a neurotoxin that produces a persistent impairment of mesencephalic DA neurons and their striatal projections (Ourednik et al., 2002). The NSCs not only migrated from their point of implantation and integrated extensively within both hemispheres, but also were, indeed, associated with a dramatic reconstitution of DA function throughout the mesostriatal system. While there was spontaneous conversion of a subpopulation of donor NSCs into dopaminergic neurons in DA-depleted areas, contributing to nigral reconstitution, the majority ($\sim$80–90%) of dopaminergic neurons in the "reconstituted midbrain" were actually host cells that had been "rescued" by factors produced constitutively by the NSCs with which they were juxtaposed and that had not, themselves, become neurons. These "chaperone cells" constitutively produced substantial amounts of neurosupportive agents. One such prominent molecule among many in this case was glial-cell-line-derived neurotrophic factors (GDNF), a factor known to be neuroprotective of ventrally located neurons (including DA neurons and spinal ventral horn cells). A similar observation is beginning to emerge from the implantation of human NSCs into the MPTP-lesioned subhuman primate model of PD.

A sense for the extent of the cross-talk also became evident when examining rodent models of hypoxia-ischemia (HI), a common cause of neurological disability in adults and children. HI causes much of its damage from extensive loss of cerebral parenchyma and of the cells and connections that reside there. When NSCs are implanted into these regions of extensive degeneration (particularly when transiently supported by biodegradable scaffolds), robust reciprocal interactions ensue spontaneously between the exogenous implant and the injured host brain that result in substantial reconstitution of parenchyma and anatomical connections as well as reduction of parenchymal loss, secondary cell loss, inflammation, and scarring. Similar results are observed in the hemi-resectioned adult rodent spinal cord in which evidence of an upregulated host neuronal regenerative response is noted, resulting in significant functional improvement. Indeed, the ability of engrafted NSCs to exert a protective and regenerative influence on degenerating host neural systems by virtue of their intrinsic expression of trophic factors is being observed in an increasing number of conditions. For example, the implantation of murine and human NSCs into the spinal cords of the SOD1 transgenic mouse model of ALS, a disease characterized by virulent motor neuron degeneration, has been pivotal in protecting these ventral horn cells from death and in preserving motor and respiratory function, blunting disease progression, and extending life. NSCs can similarly protect other neuron pools, promote motor axonal outgrowth following traumatic spinal cord injury, preserve infarcted regions of cerebrum, induce vascularization of reconstituted regions of cortical parenchyma, and inhibit inflammation and scarring following traumatic or ischemic insult.

20.3.2 MSCs: Cardiac Illness

MSCs are relatively robust stem cells that are isolated from bone marrow, where they are present at a frequency of about 0.001% nucleated cells and of about a tenth as much as HSCs. While MSCs are usually cultured in FBS and resemble fibroblast cultures in appearance, unlike fibroblasts they have the capacity to differentiate into fates in the mesodermal lineage, including chondrocytes, adipocytes, and osteoblasts (Pittenger et al., 1999). Two properties of these cells make them attractive candidates for cellular therapies: their apparent lack of immunogenecity after transplantation and their potential to help in remodeling cardiac tissue after myocardial infarction (MI). While allogeneic MSCs have been used in animal models (rodent and porcine) with good functional and histological recovery, autologous bone marrow (presumably including several populations of cells in addition to MSCs) has also been used in the clinic with promising results (Silva et al., 2004). While MSCs have been injected into the site of cardiac ischemia resulting from MI, they have also been administered systemically and have been shown to home to the site of injury (Kraitchman et al., 2005). Subsequent histological examination of the site of transplantation indicates some incorporation and survival of the transplanted cells. However, the extent of recovery of the heart tissue is not concomitant with the numbers of surviving cells, suggesting that there is remodeling of the host cells and the extracellular matrix indicating possible paracrine effects originating from the transplanted cells. This is similar to the trend seen with the NSC transplantations. MSCs also appear to block T-cell responses in the host immune system (Aggarwal and Pittenger, 2005), thus leading to their possible use as adjunct therapy in transplantation of other tissues, including bone marrow. In addition, MSCs have been used with some success in therapies related to bone replacement (Caplan, 2005). Similar work is in progress with ADSCs, which also appear to ameliorate symptoms of MI and have the added advantage of ease of isolation of populations of interest from adipose tissue for autologous transplantation into the same patient.

20.3.3 Stem Cells as Anti-Cancer Agents

Tangential to the above discussion, NSCs and MSCs have been used to deliver therapeutics to tumors, particularly gliomas (Aboody et al., 2000; Lee et al., 2003; Nakamizo et al., 2005). Aboody et al. (2000) first reported that modified NSCs introduced into the parenchyma or cerebral ventricle as well as the systemic circulation could migrate over great distances to sites of intracranial pathology, as modeled by a glioma in rodent hosts and could position themselves in direct juxtaposition to glioma cells migrating away from the tumor bulk to invade normal tissue. This ability to track invading tumor cells signified a potentially powerful way to treat a phenomenon notorious to primary gliomas that has made their management so vexing. The same group observed reduction of tumor bulk and improved host survival with the use of genetically modified NSCs. Combinations of NSCs from different sources have been shown to exhibit the same gliomatropic effect in experimental rodent brain tumor models and to effectively reduce tumor bulk and prolong host survival (Ehtesham et al., 2002, 2004; Barresi et al., 2003). Factors released and expressed by the glioma cells themselves, by the tumor stroma (composed of adjacent reactive astrocytes, microglia, and oligodendrocytes) and by tumor-derived endothelium as well as by the damaged surrounding normal brain itself all contribute to NSC gliomatropism. Some of these agents, such as stem cell factor (SCF) and monocyte chemoattractant protein-1 (MCP1), have been identified (Erlandsson et al., 2004; Widera et al., 2004; Sun et al., 2004), yet others are still to be characterized and their role in NSC gliomatropism defined (Werbowetski et al., 2004). Expression of SDF-1α by tumor-derived endothelium serves to attract the migration of NSCs (Allport et al., 2004; Fears et al., 2004). Blocking SDF-1α/CXCR4 interactions also prevents gliomatropic migration of NSCs (Ehtesham et al., 2004). SDF-1α/CXCR4 interactions appear to be pivotal as well to the gliomatropism of circulating adult hematopoietic progenitor cells (Tabatabai et al., 2005).

Exploiting the unique tropism of NSCs for gliomas, several groups have now confirmed the therapeutic efficacy of using genetically armed NSCs to target neoplasms *in vivo* in a variety of murine brain tumor models through the delivery of a variety of growth-regulating and anti-glioma gene products. Aboody et al. (2000) demonstrated *in vivo* efficacy of murine NSCs transduced with the gene for the enzyme cytosine deaminase (CDA). The enzyme converts the nontoxic pro-drug 5-fluorocytosine (5-FC) into the nucleoside analogue 5-fluorouracil (5-FU), which is then incorporated into the DNA of the neoplastic cell, causing chain termination and cell death. Tumor-bearing mice inoculated with CDA-expressing NSCs and given 5-FC demonstrated dramatic reduction of the intracranial tumor burden. This finding was subsequently corroborated in a different murine tumor model using a different NSC line (Barresi et al., 2003). Another approach utilizes NSCs as engraftable, mobile, gliomatropic viral packaging lines (Lynch et al., 1999). One study reported effective killing of tumor and escaping micro-deposits in a murine host by using murine NSCs to release replication-conditional HSV TK (Herrlinger et al., 2000), hence overcoming the typical low transduction frequency encountered in HSV glioma gene therapy by directing delivery of the virus to the intended cellular targets. This targeted cytotoxic effect, mediated by conversion of the pro-drug ganciclovir by TK into ganciclovir phosphate, is greatly amplified by virtue of the "bystander effect" (Li et al., 2005b). Gliomatropic NSCs engineered to be a viral packaging cell line for adenoviral-based vectors also show similar efficacy (Arnhold et al., 2003). NSCs have also been used to deliver gene products that show therapeutic effects against gliomas such as IL4, IL12, and TRAIL (Benedetti et al., 2000; Ehtesham et al., 2002; Walczak et al., 1999) and a potent anti-angiogenic compound called endostatin (Bjerkvig et al., 2003).

It should be noted that although the tropism of stem cells for cancer was first unveiled by observing the behavior of NSCs, this phenomenon has also been observed in MSCs. MSCs engineered to express interferon-β (IFN-β) when injected into the carotid artery of brain-tumor-bearing mice also appeared to migrate in a gliomatropic fashion and destroyed the tumor via direct cytotoxicity, resulting in prolonged survival of the hosts (Nakamizo et al., 2005). Others have shown that MSCs overexpressing the immunomodulatory cytokine IL-2 migrate to the contralateral tumor-bearing hemisphere via the corpus callosum, helping promote tumor destruction (Nakamura et al., 2004). Lee et al. (2003) also suggested that cells derived from MSCs, which have some characteristics of NSCs, have the capacity to migrate toward an injury or glioma in the brain. Furthermore, MSCs administered systemically appeared to localize to prostate and breast cancers metastatic within the periphery. The question of which type of stem cell is best suited for which type of tumor within which region will need to be determined empirically. At present, we favor the view that stem cells derived from the lineage-of-origin of the cancer are best suited for "hunting it down" and eradicating it.

A brief spate of papers introduced the possibility of somatic cells differentiating into lineages removed from that of their origin—for instance, NSCs differentiating into skeletal muscle and hematopoietic lineages (Bjornson et al., 1999; Galli et al., 2000). Similarly, MSCs differentiate into mesodermal fates such as chondrocytes, adipocytes, and osteocytes as expected (Pittenger et al., 1999), but can also be induced to differentiate into ectodermal fates such as neurons and smooth muscle and endodermal fates such as liver. Whole bone marrow and MSCs have been shown to remain viable when transplanted into the CNS in models of spinal cord and ischemic injury and are thought to provide beneficial effects (Kopen et al., 1999; Chopp and Li, 2002; Akiyama et al., 2002; Hofstetter et al., 2002). While this "transdifferentiation" is an intriguing biological question in its own right, it may also provide alternative therapeutic options, especially in the case of somatic stem cells that are challenging to propagate, such as HSCs, or where the tissue in which they reside is relatively inaccessible, such as NSCs in the brain. However, it is also possible that the starting population of stem cells may not be as homogeneous as originally thought—that is, that the MSCs had some neural progenitors that were carried over from the bone marrow, or that some of the cells scored positive for a particular mature phenotype were scored aberrantly due to experimental artifacts (Daley et al., 2003). In addition, the occurrence of cell fusion of stem cells with other mature cells in the vicinity *in vitro* (Ying et al., 2002; Terada et al., 2002) and *in vivo* (Weimann et al., 2003a,b) introduces the possibility that some of these transdifferentiation events could be a result of cell fusion. In spite of all these caveats, the possibility of transdifferentiation of somatic cells merits further scrutiny because it has not been conclusively disproved.

In summary, the science of transplantation has made significant progress in the past decade, but several of the details remain to be investigated and conditions standardized. While transplantation of adult stem cells appears to be a more mature discipline than that of ESCs at the present time, mechanistic details of related phenomena will greatly enhance the efficacy of the treatment. These phenomena include incorporation and survival of transplanted cells in the compromised tissue, homing of transplanted cells to injured tissue, differentiation of stem cells to pertinent host tissue if any, paracrine/trophic effects of transplanted cells on host tissue (with or without differentiation), fraction of transplanted cells that are retained in the tissue of interest and location, and survival of remaining cells. The method of delivery of the cells may be determined somewhat by the disease being treated, but introduction of cells may be at the site of the injury or systemic. In either scenario, the cells could be delivered with several refinements that enhance survival and perhaps differentiation/incorporation after transplantation, such as addition of growth factors that promote survival, genetic manipulation of the transplanted cells that may allow prolonged expression of genes of interest and may permit control of the transplanted cells in the patient, and, particularly in the case of ESCs, transplantation of cells that have been partially differentiated into appropriate lineages. Perhaps the site of transplantation may be pretreated to be favorable to receive and enhance the efficiency of the graft. Finally, knowledge of the precise mechanism(s) by which symptoms of the illness are reversed—that is, differentiation of transplanted cells and integration of host and transplanted tissues, trophic and paracrine functions of the transplanted cells, and cell fusion—will aid in the design of treatment protocols and in the possible use of combination therapies that include transplantation.

20.4 USES OF STEM CELLS IN RESEARCH APPLICATIONS

20.4.1 Experimental Systems to Study Basic Biology

Since the functional definition of a stem cell lies in its capacity to differentiate into several fates, the obvious follow-up to that question is the determination of the mechanism of these differentiation events. Stem cells provide a tractable system for a study of dynamic signaling processes that occur during fate choice and other events important in mammalian development, such as cell proliferation, differentiation, cell motility, cell survival, senescence, cell death, and formation of functional three-dimensional tissues. A wealth of literature exists describing (a) some of the molecular mechanisms of the complex differentiation phenotypes presented by stem cells and (b) the use of these cells to model three-dimensional cultures that resemble tissues and to model disease states so that the mechanism of diseased phenotypes may be studied. Three-dimensional tissues reminiscent of liver, capillaries, and blood have been described where stem or progenitor cells were cultured in three-dimensional matrices, including those composed of collagen (Imamura et al., 2004), matrigel or scaffolds composed of lactic acid/glycolic acid, and self-assembling peptide matrices (Semino et al., 2003). Ishikawa et al. (1996) studied hESCs, which were deficient for the NFκB precursor, and showed that NFκB is important in inflammatory responses involving p50 and RelA complexes. It is also conceivable that diseases caused by mutations in single genes, such as Lesch–Nyhan's disease, could be modeled in hESCs. With the advent of recent successes in cloning of hESCs, it is conceivable that the generation of mutant lines could expedite mechanistic studies of certain disorders, especially developmental disorders. The regulation of proliferation and differentiation of stem cells may also borrow a few lessons from research progressing in the field of cancer, such as the observation that dicer miRNA circumvents the G1-S boundary, thus allowing stem cells to proliferate in an environment that generally does not promote proliferation, a fact that may be pertinent to cancer (Hatfield et al., 2005). Alternatively, miRNA have been shown to be involved in lymphoma, gastric, and lung cancers (Croce and

Calin, 2005) and in the differentiation of cardiac and fat tissue (Esau et al., 2004; Zhao et al., 2005). The use of stem cells as *in vitro* models to study the mechanism of action of such observations is feasible, especially when differentiation-/development-related events are involved.

We have studied glial differentiation of NSCs in relative detail and will discuss some of these findings below. NSCs differentiate into cells characteristic of the CNS—namely, neurons, astrocytes, and oligodendrocytes—upon withdrawal of the mitogen bFGF (Johe et al., 1996). The addition of specific factors also causes the directed differentiation of NSCs into CNS and non-CNS fates. While the addition of platelet-derived growth factor (PDGF) enhances neuronal differentiation, triiodothyronine (T3) causes oligodendrocyte differentiation, and ciliary neurotrophic factor (CNTF) along with the other gp130 family of cytokines is a robust inducer of the astrocytic fate (Johe et al., 1996; Bonni et al., 1997; Rajan and McKay, 1998). While the neuropoietic cytokines are robust inducers of glial differentiation of neural stem cells, other factors such as addition of bone morphogenetic proteins (BMPs), serum, and (to a much lesser extent) epidermal growth factor (EGF) also cause glial differentiation. Physical parameters are also important: NSC cultures that are maintained and passaged at high density are biased toward a glial fate. BMPs are a large family of soluble factors that have several important functions during development and are involved in the differentiation of several cell fates, including neural crest, cortical neurons, generation of germ cells, lung, tooth, kidney, cartilage, and bone, as well as in the formation of dorsalizing gradients during early differentiation (Liem et al., 1995, 1997; Nguyen et al., 2000; Sela-Donenfeld and Kalcheim, 1999; Weaver et al., 1999; Pizette and Niswander, 2000). There is only one major cytoplasmic signaling pathway described for BMPs, which seems paradoxical in light of the various BMP-mediated effects described. One mechanism by which a single factor may have such pleiotropic effects is by acting in conjunction with other factors present in the local environment. Specific cellular fates may be caused by the interaction of the signaling intermediates of two or more cytoplasmic pathways, leading to different transcriptional profiles and thus different phenotypes. Although this idea is conceptually simple, the details of the molecular interactions and their specific outcomes are not clearly defined in most cases. The complex differentiation phenotype that BMP exerts on NSCs was studied as a model system of this phenomenon.

BMPs cause differentiation into three fates in NSCs in our cultures: neurons, smooth muscle, and glia. The culture conditions used in this study do not support the survival of neurons, and thus the mechanism of differentiation of that lineage was not pursued. However, the differentiation of NSCs into smooth muscle and glia is mutually exclusive and can be regulated in culture by density. Dense cultures treated with BMP yield glia as a result of Stat3 activation, and sparse cultures differentiate into smooth muscle as a result of SMAD activation. BMP-mediated Stat3 activation occurs in the presence of a prior signal initiated by cell–cell contact in dense cultures and is blocked by the drug rapamycin. In dense cultures, the glial fate is dominant and occurs at the expense of smooth muscle (Rajan et al., 2003). In a developmental context, the increase to threshold levels of a differentiation signal by the concerted action of two or more pathways provides a model for the generation of various phenotypes by intersecting gradients of factors. This example shows the involvement of a clinically relevant molecule in regulation of fate choice. It is thus not surprising that rapamycin analogues have been used on occasion in the treatment of glial tumors (Galanis et al., 2005). Also, even though FRAP and Stat3 are relatively ubiquitous proteins, the complex of these two proteins could be considered a marker of glial differentiation in stem cells (Fig. 20.3).

Recent data suggest that acetylation is possible on lysine 685 present on Stat3. In overexpression studies, it was suggested that acetylation is required for optimal activation of Stat3 in addition to phosphorylation (Yuan et al., 2005). It has also been shown that lysine 685 is a substrate for the CBP/p300 protein that forms a complex with Stat3 (Wang et al., 2005). Although the physiological significance of this observation has already been called into question (O'Shea et al., 2005), it is an interesting additional level of control that may be relevant to the glial differentiation of stem cells. The gp130 family of cytokines including CNTF, LIF, oncostatin M, and IL6 cause the activation of STAT proteins in several cell types including liver, kidney, neurons, astrocytes, and neural stem cells. However, the outcome of STAT activation in these various cell types is distinctive. Also, neural stem cells isolated from rats of different ages exhibit varying capacities for astrocytic differentiation. NSCs isolated

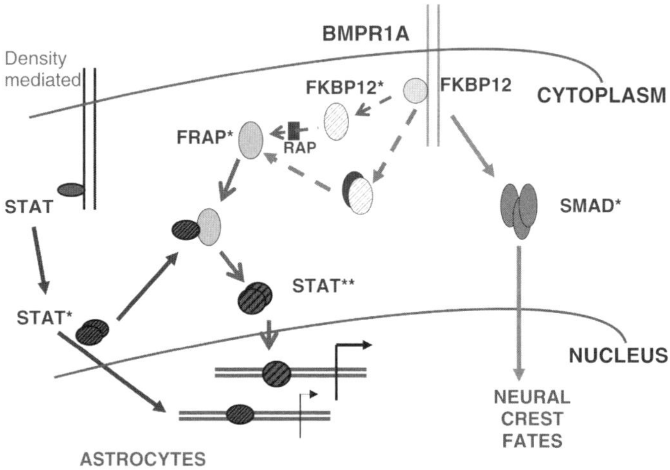

Figure 20.3 Schematic of intracellular signals activated by BMP4 in neural stem cells This model is based on the data presented in Rajan et al. (2003). BMP4 causes the activation of at least two signaling pathways in neural stem cells. Upon ligand–receptor binding, the BMPR receptor complex releases FKBP12 and activates SMAD proteins, resulting in smooth muscle differentiation. The released FKBP12 may bind with rapamycin to inhibit FRAP or may act in some modified form to activate FRAP. FRAP then catalyzes serine phosphorylation (S*) of STAT to augment its prior activation by tyrosine phosphorylation (Y*) by another density-mediated signal. High cell density acts to promote basal STAT activation and DNA binding by an unknown signaling mechanism. This enhanced activation of STAT (Y*S*) causes efficient glial differentiation. Levels of activated STAT proteins in the cell receiving the BMP signal dictate whether the STAT or SMAD signal acquires precedence in the fate choice between smooth muscle and glia. See color plates.

from E12 rat embryos do not respond to LIF by differentiating into glia, while those isolated from E15 embryos do (Molne et al., 2000). The differences in outcomes could possibly be due to the epigenetic state of the cell receiving the signal, as has been suggested by Song and Ghosh (2004). There is also increasing evidence that genes such as bmi1 (which is one of the Polycomb Group of genes, an "oncogene" that regulates cell-cycle-related genes such as INK4a and p16ARF) are involved in the maintenance, proliferation, and differentiation states of hematopoietic and neural stem cells, while bmi1 forms part of a complex called the polycomb repressive complex 1 (PRC1), thought to maintain stable maintenance of gene expression by regulation of epigenetic chromatin modifications (Valk-Lingbeek et al., 2004). Mice deficient in bmi1 display deficits in neural stem cells and cerebellar neurons (Leung et al., 2004). Interestingly, they also show an increase in astrocytes (Zencak et al., 2005). These observations, though fragmented, strongly suggest that epigenetic modifications play an important role in fate choice of glia and, in a broader sense, in the decision of a cell to remain stem-like or to differentiate.

Epigenetic modification includes processes such as DNA methylation, histone modification, and ATP-driven chromatin remodeling. These processes function cooperatively to establish and maintain active or inactive chromatin states in cellular development. In general, acetylation of core histones and methylation of K4 of histone H3 correlate with transcriptional active ("open") chromatin state, whereas deacetylation of core histones and methylation of K9 of histone H3 correlate with transcriptional repressed ("closed") chromatin state. The proteins that mediate these interactions include DNA cytosine methyltransferases, histone acetyl transferases, histone deactylases, histone methyltransferases, and other specialized proteins that mediate the interaction of proteins with nucleosomal DNA. The precise function of these proteins in plasticity of stem cells is not clear. It appears that the expression of Oct-4 is required for the maintenance of pluripotency, self–renewal, and survival (Kehler et al., 2004). A global study of hESC promoters that bind the three markers Oct-4, Nanog, and Sox2 yielded clues as to genes that remain inactive when bound to these transcription factors and others that appear to be activated (Boyer et al., 2005). Follow-up studies may delineate the epigenetic reasons for this observation, including a possible mechanism for the role of these three factors in global regulation of "stemness" (genes of promoters that are activated) and phenotypic commitment (genes of promoters that are repressed) in ESCs. While it is generally agreed that global transcriptional, translational, and epigenetic characteristics of ESCs are unique, the factors that orchestrate these changes are still elusive. In a preliminary experiment to identify these factors, it was determined that characteristics of ESCs could be conferred on somatic cells by fusion (Cowan et al., 2005). These authors showed that the nucleus from the mature fusion partner could be appreciably "reprogrammed" such that it acquired some of the epigenetic features resembling an undifferentiated hESC. It appears that cytoplasmic factors play an important role in the initiation and maintenance of the epigenetic state of the nucleus.

Thus, as expected, there appears to be an intimate relationship between cytoplasmic and nuclear signals. Another facet of regulation acquiring importance is stable, nontranslated RNA molecules called micro RNAs (miRNAs), which lead to reduced translation of specific proteins by either mRNA degradation or inhibition of translation (Alvarez-Garcia and Miska, 2005). Although small interfering RNAs (siRNAs) were first discovered in *C. elegans* (Lee et al., 1993) and have since been used extensively as research reagents that inhibit eukaryotic—including mammalian—gene expression, functionally similar miRNAs have been isolated from mammalian cells, and two have been shown to be involved in cardiac muscle and adipocyte differentiation. Zhao et al. (2005) have shown that miR-1 targets a transcription factor hand-1 that controls the proliferation of ventricular cardiomyocytes during development. Similarly, miR-147 has been shown to regulate adipocyte differentiation, possibly by inhibiting translation of ERK5 (Esau et al., 2004). On a more global level, mutation of the enzyme dicer, which is part of the cellular machinery responsible for the generation of miRNAs, causes a G1-S block in stem cells in *Drosophila*, implying that miRNAs are essential for the proliferation of stem cells in their native environment (Hatfield et al., 2005).

The initial use of mouse ESCs was in the creation of "knock-out" and "knock-in" mice, which have contributed immensely to the study and understanding of mammalian developmental biology. The past decade has seen a revolution in this field with the isolation of murine, rodent, and human stem cells from a variety of sources, including blastocysts, the brain, and bone marrow. In spite of the obvious relevance of the human stem cells to the medical sciences, important discoveries about the basic biology of the stem cells themselves and some of the associated developmental phenomena in mammals are continuing to be made in the rodent and murine systems, and to a lesser extent in invertebrates and lower vertebrates such as *Drosophila* and Zebrafish. All these studies—molecular, cellular, systems biology, and clinical—will contribute to the gestalt of knowledge about stem cells and will aid immensely in the successful exploitation and manipulation of this powerful system to our advantage.

20.4.2 Creation of Applied Systems for Drug Discovery

Stem cells are potentially useful in several stages of the drug discovery process, including target discovery and validation, secondary screens of candidate molecules, and toxicological studies of potential drugs. Alternatively, stem cells could be used as *in vitro* experimental systems to determine the precise mechanism of action of a candidate drug. Using genetic manipulation technology, ESC lines may be created that express specifically mutated proteins or lines that lack specific genes, either of which may be used to study the mechanism of action of particular targets or as a platform for screening drugs that reverse a particular phenotype. Alternatively, "humanized" mice may be produced that express the human version of the gene of interest, which may then be used for some *in vivo* testing of drug action and ADME-TOX (administration, distribution, metabolism, excretion, and toxicity of candidate drug molecules). Mouse ESCs are already being used by companies in drug screening applications (McNeish, 2004). The creation of person-specific and disease-specific hESCs would also aid drug discovery efforts. The former has potential for autologous cellular therapies as well as for pharmacogenomic studies where genomic profiles are correlated with the most effective therapeutic regimen. Disease-specific hESCs could lead to the design of experiment systems for target validation and subsequently to screening platforms that may be used for secondary screens of compounds of interest.

In addition to the advantages that stem cells present in the areas of drug discovery, they also represent a potentially limitless source of "normal," untransformed, diploid cells. This is especially relevant to human systems, because most primary human cells do not propagate beyond a few generations in culture. This property of stem cells could potentially be exploited for the generation of large quantities of cells needed to generate screening platforms for high-throughput screening of chemical libraries. Secondly, drug discovery platforms that would otherwise be challenging to generate on a large scale, such as platforms that employ neuronal cells, could be derived from stem cells or precursors that have been differentiated *in vitro*. Some groups have been successful in the generation of neurons from hESCs (Reubinoff et al., 2001; Schuldiner et al., 2001), which suggests that this may be possible in the future. Even specific types of neurons, such as dopaminergic or serotonergic, could possibly be derived by *in vitro* stem cell differentiation: The former has particular relevance to Parkinson's disease, and the latter has relevance to psychiatric illness. Finally, stem cells and their derivatives could be used for toxicity studies of candidate drugs. They would provide a preferable alternative to the transformed cell lines on which these studies are currently being performed. In addition, they could potentially be differentiated into liver cells, which would provide an ideal platform for toxicity studies to be performed upon *in vitro*, and for which a viable alternative does not exist.

20.5 PROJECTIONS AND CONCLUSIONS

While the enormous potential presented by stem cells exists without a doubt, there are scientific and other pragmatic problems that need to be addressed before this potential is realized. Procedures need to be developed for the reproducible scale-up of all kinds of human stem cells, with precise definition of the resultant cells. The cells could be undifferentiated stem cells (for possible use in transplantation and drug screening assays) or differentiated sufficiently along a particular lineage (as in hESCs differentiated into neuronal precursors of a specific kind for use in transplantation) as dictated by the end use. The scale-up will have to take into account the following: optimization of culture of stem cells in suspension or adherent cultures; culture systems that use defined media and are devoid of animal-derived components; cost-effective culture of undifferentiated stem cells that may then be used as such or further differentiated into fully or partially differentiated cultures; and panels of markers sufficient to characterize the desired endproducts, be they a single population of undifferentiated stem cells or a complex population of cells that have been partially differentiated cells along a particular lineage. Also, the ability of the stem cells to remain stable in their phenotypes through the rigors of the drug screening process will have to be ensured. Understanding the molecular intracellular signals that mediate the proliferation, survival, differentiation, senescence, and motility of stem cells will be integral to these efforts and to the ability to manipulate these cells with precision and reproducibility. When used as cellular therapeutics, the ability to control these cells post-transplantation will be an additional desirable manipulation. The ultimate use of these cells to better the human condition, both directly in the clinic as cellular therapeutics or as surrogates in the pharmaceutical industry, will be dictated by our understanding of these complex cells themselves and by the reconciliation of philosophical differences pertaining to ethical issues surrounding embryonic stem cells, particularly in the United States.

In spite of these considerable hurdles, scientists worldwide are tackling this problem with interest and determination, due to both the interesting scientific challenge it poses and the significant clinical gains it presents. In addition to appreciable efforts in the academic quarter, several companies are investing substantial amounts of manpower and resources to achieve practical usability of these cells. These include Amgen, Cell Genesys, Chemicon, Cytori Therapeutics, Geron, Invitrogen, NeuralStem, Novocell, Osiris Therapeutics, Pfizer, ReNeuron, StemCells, Inc., and Viacell, to name a few. While transplantation of bone marrow as an adjunct for cancer therapies is not uncommon, neural, mesenchymal, and hematopoietic stem cells are increasingly being used in clinically relevant and experimental scenarios. Although initial neural stem cell trials were related to Parkinson's disease, stem cell transplantation is now being thought of as a viable option for replacement therapies in spinal cord injuries. Companies such as Geron, NeuralStem, and StemCells, Inc., have invested substantial resources into determining the practical use of these cells in animal models (Vogel, 2005; Cummings et al., 2005; Keirstead et al., 2005) and hope to progress to clinical trials. Transplantation of whole bone marrow and selected cell populations, including mesenchymal stem cells, has been revisited several times in therapies relating to heart cell replacement (Pittenger and Martin, 2004; Shim and Wong, 2004), and Osiris Therapeutics has initiated clinical trials with human MSCs. Geron is also anticipating the release of cultures derived from hESCs to be used for toxicity studies of candidate drugs (Sinha, 2005). The nature and complexity of the challenge presented by stem cells makes it an ideal discipline for active and extensive collaborations between the academic and corporate scientific establishments at national and international levels.

ACKNOWLEDGMENTS

We would like to acknowledge the A-T Children's Project, NIH, and the Michael J. Fox Foundation, who have supported some of the experiments described here.

REFERENCES

Aaronson, D. S., and Horvath, C. M. (2002). A road map for those who don't know JAK-STAT. *Science* **296**, 1653–1655.

Aboody, K. S., Brown, A., Rainov, N. G., Bower, K. A., Liu, S., Yang, W., Small, J. E., Herrlinger, U., Ourednik, V., Black, P. M., et al. (2000). Neural stem cells display extensive tropism for pathology in adult brain: Evidence from intracranial gliomas. *Proc Natl Acad Sci USA* **97**, 12846–12851.

Aggarwal, S., and Pittenger, M. F. (2005). Human mesenchymal stem cells modulate allogeneic immune cell responses. *Blood* **105**, 1815–1822. Epub 2004 Oct 1819.

Akiyama, Y., Radtke, C., and Kocsis, J. D. (2002). Remyelination of the rat spinal cord by transplantation of identified bone marrow stromal cells. *J Neurosci* **22**, 6623–6630.

Allport, J. R., Shinde Patil, V. R., and Weissleder, R. (2004). Murine neuronal progenitor cells are preferentially recruited to tumor vasculature via alpha4-integrin and SDF-1alpha-dependent mechanisms. *Cancer Biol Ther* **3**, 838–844. Epub 2004 Sept 2018.

Alvarez-Garcia, I., and Miska, E. A. (2005). MicroRNA functions in animal development and human disease. *Development* **132**, 4653–4662.

Andrews, P. W., Benvenisty, N., McKay, R., Pera, M. F., Rossant, J., Semb, H., and Stacey, G. N. (2005). The International Stem Cell Initiative: Toward benchmarks for human embryonic stem cell research. *Nat Biotechnol* **23**, 795–797.

Arnhold, S., Hilgers, M., Lenartz, D., Semkova, I., Kochanek, S., Voges, J., Andressen, C., and Addicks, K. (2003). Neural precursor cells as carriers for a gene therapeutical approach in tumor therapy. *Cell Transplant* **12**, 827–837.

Barresi, V., Belluardo, N., Sipione, S., Mudo, G., Cattaneo, E., and Condorelli, D. F. (2003). Transplantation of prodrug-converting neural progenitor cells for brain tumor therapy. *Cancer Gene Ther* **10**, 396–402.

Benedetti, S., Pirola, B., Pollo, B., Magrassi, L., Bruzzone, M. G., Rigamonti, D., Galli, R., Selleri, S., Di Meco, F., De Fraja, C., et al. (2000). Gene therapy of experimental brain tumors using neural progenitor cells. *Nat Med* **6**, 447–450.

Bhattacharya, B., Miura, T., Brandenberger, R., Mejido, J., Luo, Y., Yang, A. X., Joshi, B. H., Ginis, I., Thies, R. S., Amit, M., et al. (2004). Gene expression in human embryonic stem cell lines: unique molecular signature. *Blood* **103**, 2956–2964. Epub 2003 Dec 2930.

Bjerkvig, R., Read, T. A., Vajkoczy, P., Aebischer, P., Pralong, W., Platt, S., Melvik, J. E., Hagen, A., and Dornish, M. (2003). Cell therapy using encapsulated cells producing endostatin. *Acta Neurochir Suppl* **88**, 137–141.

Bjornson, C. R., Rietze, R. L., Reynolds, B. A., Magli, M. C., and Vescovi, A. L. (1999). Turning brain into blood: A hematopoietic fate adopted by adult neural stem cells *in vivo*. *Science* **283**, 534–537.

Blau, H. M., Brazelton, T. R., and Weimann, J. M. (2001). The evolving concept of a stem cell: entity or function? *Cell* **105**, 829–841.

Bonni, A., Sun, Y., Nadal-Vicens, M., Bhatt, A., Frank, D. A., Rozovsky, I., Stahl, N., Yancopoulos, G. D., and Greenberg, M. E. (1997). Regulation of gliogenesis in the central nervous system by the JAK-STAT signaling pathway. *Science* **278**, 477–483.

Boyer, L. A., Lee, T. I., Cole, M. F., Johnstone, S. E., Levine, S. S., Zucker, J. P., Guenther, M. G., Kumar, R. M., Murray, H. L., Jenner, R. G., et al. (2005). Core transcriptional regulatory circuitry in human embryonic stem cells. *Cell* **122**, 947–956.

Brandenberger, R., Khrebtukova, I., Thies, R. S., Miura, T., Jingli, C., Puri, R., Vasicek, T., Lebkowski, J., and Rao, M. (2004a). MPSS profiling of human embryonic stem cells. *BMC Dev Biol* **4**, 10.

Brandenberger, R., Wei, H., Zhang, S., Lei, S., Murage, J., Fisk, G. J., Li, Y., Xu, C., Fang, R., Guegler, K., et al. (2004b). Transcriptome characterization elucidates signaling networks that control human ES cell growth and differentiation. *Nat Biotechnol* **22**, 707–716. Epub 2004 May 2016.

Camargo, F. D., Chambers, S. M., Drew, E., McNagny, K. M., and Goodell, M. A. (2005). Hematopoietic stem cells do not engraft with absolute efficiencies. *Blood* **4**, 4.

Caplan, A. I. (2005). Review: Mesenchymal stem cells: Cell-based reconstructive therapy in orthopedics. *Tissue Eng* **11**, 1198–1211.

Chambers, I., Colby, D., Robertson, M., Nichols, J., Lee, S., Tweedie, S., and Smith, A. (2003). Functional expression cloning of Nanog, a pluripotency sustaining factor in embryonic stem cells. *Cell* **113**, 643–655.

Chopp, M., and Li, Y. (2002). Treatment of neural injury with marrow stromal cells. *Lancet Neurol* **1**, 92–100.

Cowan, C. A., Atienza, J., Melton, D. A., and Eggan, K. (2005). Nuclear reprogramming of somatic cells after fusion with human embryonic stem cells. *Science* **309**, 1369–1373.

Cowan, C. A., Klimanskaya, I., McMahon, J., Atienza, J., Witmyer, J., Zucker, J. P., Wang, S., Morton, C. C., McMahon, A. P., Powers, D., and Melton, D. A. (2004). Derivation of embryonic stem-cell lines from human blastocysts. *N Engl J Med* **350**, 1353–1356. Epub 2004 Mar 1353.

Croce, C. M., and Calin, G. A. (2005). miRNAs, cancer, and stem cell division. *Cell* **122**, 6–7.

Cummings, B. J., Uchida, N., Tamaki, S. J., Salazar, D. L., Hooshmand, M., Summers, R., Gage, F. H., and Anderson, A. J. (2005). Human neural stem cells differentiate and promote locomotor recovery in spinal cord-injured mice. *Proc Natl Acad Sci USA* **102**, 14069–14074. Epub 12005 Sept 14019.

Daley, G. Q., Goodell, M. A., and Snyder, E. Y. (2003). Realistic prospects for stem cell therapeutics. *Hematology (Am Soc Hematol Educ Program)*, 398–418.

Dunnett, S. B. (1999). Repair of the damaged brain. The Alfred Meyer Memorial Lecture 1998. *Neuropathol Appl Neurobiol* **25**, 351–362.

Eglitis, M. A., and Mezey, E. (1997). Hematopoietic cells differentiate into both microglia and macroglia in the brains of adult mice. *Proc Natl Acad Sci USA* **94**, 4080–4085.

Ehtesham, M., Kabos, P., Gutierrez, M. A., Chung, N. H., Griffith, T. S., Black, K. L., and Yu, J. S. (2002). Induction of glioblastoma apoptosis using neural stem cell-mediated delivery of tumor necrosis factor-related apoptosis-inducing ligand. *Cancer Res* **62**, 7170–7174.

Ehtesham, M., Yuan, X., Kabos, P., Chung, N. H., Liu, G., Akasaki, Y., Black, K. L., and Yu, J. S. (2004). Glioma tropic neural stem cells consist of astrocytic precursors and their migratory capacity is mediated by CXCR4. *Neoplasia* **6**, 287–293.

Erlandsson, A., Larsson, J., and Forsberg-Nilsson, K. (2004). Stem cell factor is a chemoattractant and a survival factor for CNS stem cells. *Exp Cell Res* **301**, 201–210.

Esau, C., Kang, X., Peralta, E., Hanson, E., Marcusson, E. G., Ravichandran, L. V., Sun, Y., Koo, S., Perera, R. J., Jain, R., et al. (2004). MicroRNA-143 regulates adipocyte differentiation. *J Biol Chem* **279**, 52361–52365. Epub 52004 Oct 52325.

Fears, C. Y., Sontheimer, H. W., Bullard, D. C., and Gladson, C. L. (2004). Could labeled neuronal progenitor cells be used to target glioma tumor endothelium? *Cancer Biol Ther* **3**, 845–846. Epub 2004 Sept 2026.

Gage, F. H., Ray, J., and Fisher, L. J. (1995). Isolation, characterization, and use of stem cells from the CNS. *Annu Rev Neurosci* **18**, 159–192.

Galanis, E., Buckner, J. C., Maurer, M. J., Kreisberg, J. I., Ballman, K., Boni, J., Peralba, J. M., Jenkins, R. B., Dakhil, S. R., Morton, R. F., et al. (2005). Phase II trial of temsirolimus (CCI-779) in recurrent glioblastoma multiforme: A North Central Cancer Treatment Group Study. *J Clin Oncol* **23**, 5294–5304. Epub 2005 Jul 5295.

Galli, R., Borello, U., Gritti, A., Minasi, M. G., Bjornson, C., Coletta, M., Mora, M., De Angelis, M. G., Fiocco, R., Cossu, G., and Vescovi, A. L. (2000). Skeletal myogenic potential of human and mouse neural stem cells. *Nat Neurosci* **3**, 986–991.

Gregory, C. A., Ylostalo, J., and Prockop, D. J. (2005). Adult bone marrow stem/progenitor cells (MSCs) are preconditioned by microenvironmental "niches" in culture: A two-stage hypothesis for regulation of MSC fate. *Sci STKE 2005*, pe37.

Hatfield, S. D., Shcherbata, H. R., Fischer, K. A., Nakahara, K., Carthew, R. W., and Ruohola-Baker, H. (2005). Stem cell division is regulated by the microRNA pathway. *Nature* **435**, 974–978. Epub 2005 Jun 2008.

Herrlinger, U., Woiciechowski, C., Sena-Esteves, M., Aboody, K. S., Jacobs, A. H., Rainov, N. G., Snyder, E. Y., and Breakefield, X. O. (2000). Neural precursor cells for delivery of replication-conditional HSV-1 vectors to intracerebral gliomas. *Mol Ther* **1**, 347–357.

Hoffman, L. M., and Carpenter, M. K. (2005). Characterization and culture of human embryonic stem cells. *Nat Biotechnol* **23**, 699–708.

Hofstetter, C. P., Schwarz, E. J., Hess, D., Widenfalk, J., El Manira, A., Prockop, D. J., and Olson, L. (2002). Marrow stromal cells form guiding strands in the injured spinal cord and promote recovery. *Proc Natl Acad Sci USA* **99**, 2199–2204.

Imamura, T., Cui, L., Teng, R., Johkura, K., Okouchi, Y., Asanuma, K., Ogiwara, N., and Sasaki, K. (2004). Embryonic stem cell-derived embryoid bodies in three-dimensional culture system form hepatocyte-like cells *in vitro* and *in vivo*. *Tissue Eng* **10**, 1716–1724.

Ishikawa, H., Ryseck, R. P., and Bravo, R. (1996). Characterization of ES cells deficient for the p105 precursor (NF-kappa B1): Role of p50 NLS. *Oncogene* **13**, 255–263.

Ivanova, N. B., Dimos, J. T., Schaniel, C., Hackney, J. A., Moore, K. A., and Lemischka, I. R. (2002). A stem cell molecular signature. *Science* **298**, 601–604. Epub 2002 Sept 2012.

Jiang, Y., Jahagirdar, B. N., Reinhardt, R. L., Schwartz, R. E., Keene, C. D., Ortiz-Gonzalez, X. R., Reyes, M., Lenvik, T., Lund, T., Blackstad, M., et al. (2002). Pluripotency of mesenchymal stem cells derived from adult marrow. *Nature* **418**, 41–49. Epub 2002 Jun 2020.

Johe, K. K., Hazel, T. G., Muller, T., Dugich-Djordjevic, M. M., and McKay, R. D. (1996). Single factors direct the differentiation of stem cells from the fetal and adult central nervous system. *Genes Dev* **10**, 3129–3140.

Jonker, J. W., Freeman, J., Bolscher, E., Musters, S., Alvi, A. J., Titley, I., Schinkel, A. H., and Dale, T. C. (2005). Contribution of the ABC transporters Bcrp1 and Mdr1a/1b to the side population phenotype in mammary gland and bone marrow of mice. *Stem Cells* **23**, 1059–1065. Epub 2005 Jul 1057.

Kehler, J., Tolkunova, E., Koschorz, B., Pesce, M., Gentile, L., Boiani, M., Lomeli, H., Nagy, A., McLaughlin, K. J., Scholer, H. R., and Tomilin, A. (2004). Oct4 is required for primordial germ cell survival. *EMBO Rep* **5**, 1078–1083.

Keirstead, H. S., Nistor, G., Bernal, G., Totoiu, M., Cloutier, F., Sharp, K., and Steward, O. (2005). Human embryonic stem cell-derived oligodendrocyte progenitor cell transplants remyelinate and restore locomotion after spinal cord injury. *J Neurosci* **25**, 4694–4705.

Kim, J. H., Auerbach, J. M., Rodriguez-Gomez, J. A., Velasco, I., Gavin, D., Lumelsky, N., Lee, S. H., Nguyen, J., Sanchez-Pernaute, R., Bankiewicz, K., and McKay, R. (2002). Dopamine neurons derived from embryonic stem cells function in an animal model of Parkinson's disease. *Nature* **418**, 50–56. Epub 2002 Jun 2020.

Kopen, G. C., Prockop, D. J., and Phinney, D. G. (1999). Marrow stromal cells migrate throughout forebrain and cerebellum, and they differentiate into astrocytes after injection into neonatal mouse brains. *Proc Natl Acad Sci USA* **96**, 10711–10716.

Kraitchman, D. L., Tatsumi, M., Gilson, W. D., Ishimori, T., Kedziorek, D., Walczak, P., Segars, W. P., Chen, H. H., Fritzges, D., Izbudak, I., et al. (2005). Dynamic imaging of allogeneic mesenchymal stem cells trafficking to myocardial infarction. *Circulation* **112**, 1451–1461. Epub 2005 Aug 1429.

Lee, J., Elkahloun, A. G., Messina, S. A., Ferrari, N., Xi, D., Smith, C. L., Cooper, R., Jr., Albert, P. S., and Fine, H. A. (2003). Cellular and genetic characterization of human adult bone marrow derived neural stem-like cells: A potential antiglioma cellular vector. *Cancer Res* **63**, 8877–8889.

Lee, R. C., Feinbaum, R. L., and Ambros, V. (1993). The C. elegans heterochronic gene lin-4 encodes small RNAs with antisense complementarity to lin-14. *Cell* **75**, 843–854.

Lendahl, U., Zimmerman, L. B., and McKay, R. D. (1990). CNS stem cells express a new class of intermediate filament protein. *Cell* **60**, 585–595.

Leung, C., Lingbeek, M., Shakhova, O., Liu, J., Tanger, E., Saremaslani, P., Van Lohuizen, M., and Marino, S. (2004). Bmi1 is essential for cerebellar development and is overexpressed in human medulloblastomas. *Nature* **428**, 337–341.

Li, C. D., Zhang, W. Y., Li, H. L., Jiang, X. X., Zhang, Y., Tang, P., and Mao, N. (2005a). Isolation and Identification of a Multilineage Potential Mesenchymal Cell from Human Placenta. *Placenta* **17**, 17.

Li, S., Tokuyama, T., Yamamoto, J., Koide, M., Yokota, N., and Namba, H. (2005b). Bystander effect-mediated gene therapy of gliomas using genetically engineered neural stem cells. *Cancer Gene Ther* **12**, 600–607.

Liem, K. F., Jr., Tremml, G., Roelink, H., and Jessell, T. M. (1995). Dorsal differentiation of neural plate cells induced by BMP-mediated signals from epidermal ectoderm. *Cell* **82**, 969–979.

Liem, K. F., Jr., Tremml, G., and Jessell, T. M. (1997). A role for the roof plate and its resident TGFbeta-related proteins in neuronal patterning in the dorsal spinal cord. *Cell* **91**, 127–138.

Lindvall, O., Brundin, P., Widner, H., Rehncrona, S., Gustavii, B., Frackowiak, R., Leenders, K. L., Sawle, G., Rothwell, J. C., Marsden, C. D., et al. (1990a). Grafts of fetal dopamine neurons survive and improve motor function in Parkinson's disease. *Science* **247**, 574–577.

Lindvall, O., Rehncrona, S., Brundin, P., Gustavii, B., Astedt, B., Widner, H., Lindholm, T., Bjorklund, A., Leenders, K. L., Rothwell, J. C., et al. (1990b). Neural transplantation in Parkinson's disease: the Swedish experience. *Prog Brain Res* **82**, 729–734.

Lynch, W. P., Sharpe, A. H., and Snyder, E. Y. (1999). Neural stem cells as engraftable packaging lines can mediate gene delivery to microglia: Evidence from studying retroviral env-related neurodegeneration. *J Virol* **73**, 6841–6851.

Ma'ayan, A., Jenkins, S. L., Neves, S., Hasseldine, A., Grace, E., Dubin-Thaler, B., Eungdamrong, N. J., Weng, G., Ram, P. T., Rice, J. J., et al. (2005). Formation of regulatory patterns during signal propagation in a Mammalian cellular network. *Science* **309**, 1078–1083.

McNeish, J. (2004). Embryonic stem cells in drug discovery. *Nat Rev Drug Discov* **3**, 70–80.

Mehta, V., Hong, M., Spears, J., and Mendez, I. (1998). Enhancement of graft survival and sensorimotor behavioral recovery in rats undergoing transplantation with dopaminergic cells exposed to glial cell line-derived neurotrophic factor. *J Neurosurg* **88**, 1088–1095.

Mitsui, K., Tokuzawa, Y., Itoh, H., Segawa, K., Murakami, M., Takahashi, K., Maruyama, M., Maeda, M., and Yamanaka, S. (2003). The homeoprotein Nanog is required for maintenance of pluripotency in mouse epiblast and ES cells. *Cell* **113**, 631–642.

Molne, M., Studer, L., Tabar, V., Ting, Y. T., Eiden, M. V., and McKay, R. D. (2000). Early cortical precursors do not undergo LIF-mediated astrocytic differentiation. *J Neurosci Res* **59**, 301–311.

Nakamizo, A., Marini, F., Amano, T., Khan, A., Studeny, M., Gumin, J., Chen, J., Hentschel, S., Vecil, G., Dembinski, J., et al. (2005). Human bone marrow-derived mesenchymal stem cells in the treatment of gliomas. *Cancer Res* **65**, 3307–3318.

Nakamura, K., Ito, Y., Kawano, Y., Kurozumi, K., Kobune, M., Tsuda, H., Bizen, A., Honmou, O., Niitsu, Y., and Hamada, H. (2004). Antitumor effect of genetically engineered mesenchymal stem cells in a rat glioma model. *Gene Ther* **11**, 1155–1164.

Nguyen, V. H., Trout, J., Connors, S. A., Andermann, P., Weinberg, E., and Mullins, M. C. (2000). Dorsal and intermediate neuronal cell types of the spinal cord are established by a BMP signaling pathway. *Development* **127**, 1209–1220.

O'Shea, J. J., Kanno, Y., Chen, X., and Levy, D. E. (2005). Cell signaling. Stat acetylation—A key facet of cytokine signaling? *Science* **307**, 217–218.

Ourednik, J., Ourednik, V., Lynch, W. P., Schachner, M., and Snyder, E. Y. (2002). Neural stem cells display an inherent mechanism for rescuing dysfunctional neurons. *Nat Biotechnol* **20**, 1103–1110. Epub 2002 Oct 1115.

Parker, M. A., Anderson, J. K., Corliss, D. A., Abraria, V. E., Sidman, R. L., Park, K. I., Teng, Y. D., Cotanche, D. A., and Snyder, E. Y. (2005). Expression profile of an operationally-defined neural stem cell clone. *Exp Neurol* **194**, 320–332.

Picard-Riera, N., Nait-Oumesmar, B., and Baron-Van Evercooren, A. (2004). Endogenous adult neural stem cells: Limits and potential to repair the injured central nervous system. *J Neurosci Res* **76**, 223–231.

Pittenger, M. F., Mackay, A. M., Beck, S. C., Jaiswal, R. K., Douglas, R., Mosca, J. D., Moorman, M. A., Simonetti, D. W., Craig, S., and Marshak, D. R. (1999). Multilineage potential of adult human mesenchymal stem cells. *Science* **284**, 143–147.

Pittenger, M. F., and Martin, B. J. (2004). Mesenchymal stem cells and their potential as cardiac therapeutics. *Circ Res* **95**, 9–20.

Pizette, S., and Niswander, L. (2000). BMPs are required at two steps of limb chondrogenesis: formation of prechondrogenic condensations and their differentiation into chondrocytes. *Dev Biol* **219**, 237–249.

Rajan, P., and McKay, R. D. (1998). Multiple routes to astrocytic differentiation in the CNS. *J Neurosci* **18**, 3620–3629.

Rajan, P., Panchision, D. M., Newell, L. F., and McKay, R. D. (2003). BMPs signal alternately through a SMAD or FRAP-STAT pathway to regulate fate choice in CNS stem cells. *J Cell Biol* **161**, 911–921.

Ramalho-Santos, M., Yoon, S., Matsuzaki, Y., Mulligan, R. C., and Melton, D. A. (2002). "Stemness": Transcriptional profiling of embryonic and adult stem cells. *Science* **298**, 597–600. Epub 2002 Sep 2012.

Reubinoff, B. E., Itsykson, P., Turetsky, T., Pera, M. F., Reinhartz, E., Itzik, A., and Ben-Hur, T. (2001). Neural progenitors from human embryonic stem cells. *Nat Biotechnol* **19**, 1134–1140.

Richards, M., Tan, S. P., Tan, J. H., Chan, W. K., and Bongso, A. (2004). The transcriptome profile of human embryonic stem cells as defined by SAGE. *Stem Cells* **22**, 51–64.

Rosenthal, A. (1998). Auto transplants for Parkinson's disease? *Neuron* **20**, 169–172.

Sakakibara, S., Nakamura, Y., Yoshida, T., Shibata, S., Koike, M., Takano, H., Ueda, S., Uchiyama, Y., Noda, T., and Okano, H. (2002). RNA-binding protein Musashi family: Roles for CNS stem cells and a subpopulation of ependymal cells revealed by targeted disruption and antisense ablation. *Proc Natl Acad Sci USA* **99**, 15194–15199. Epub 12002 Oct 15129.

Scheel, J. R., Ray, J., Gage, F. H., and Barlow, C. (2005). Quantitative analysis of gene expression in living adult neural stem cells by gene trapping. *Nat Methods* **2**, 363–370. Epub 2005 Apr 2021.

Schroeder, M., Niebruegge, S., Werner, A., Willbold, E., Burg, M., Ruediger, M., Field, L. J., Lehmann, J., and Zweigerdt, R. (2005). Differentiation and lineage selection of mouse embryonic stem cells in a stirred bench scale bioreactor with automated process control. *Biotechnol Bioeng* **27**, 27.

Schuldiner, M., Eiges, R., Eden, A., Yanuka, O., Itskovitz-Eldor, J., Goldstein, R. S., and Benvenisty, N. (2001). Induced neuronal differentiation of human embryonic stem cells. *Brain Res* **913**, 201–205.

Sela-Donenfeld, D., and Kalcheim, C. (1999). Regulation of the onset of neural crest migration by coordinated activity of BMP4 and Noggin in the dorsal neural tube. *Development* **126**, 4749–4762.

Semino, C. E., Merok, J. R., Crane, G. G., Panagiotakos, G., and Zhang, S. (2003). Functional differentiation of hepatocyte-like spheroid structures from putative liver progenitor cells in three-dimensional peptide scaffolds. *Differentiation* **71**, 262–270.

Sen, A., Kallos, M. S., and Behie, L. A. (2004). New tissue dissociation protocol for scaled-up production of neural stem cells in suspension bioreactors. *Tissue Eng* **10**, 904–913.

Shim, W., and Wong, P. (2004). Stem cell cardiomyoplasty: State-of-the-art. *Ann Acad Med Singapore* **33**, 451–460.

Shizuru, J. A., Negrin, R. S., and Weissman, I. L. (2005). Hematopoietic stem and progenitor cells: clinical and preclinical regeneration of the hematolymphoid system. *Annu Rev Med* **56**, 509–538.

Silva, G. V., Perin, E. C., Dohmann, H. F., Borojevic, R., Silva, S. A., Sousa, A. L., Assad, J. A., Vaughn, W. K., Mesquita, C. T., Belem, L., et al. (2004). Catheter-based transendocardial delivery of autologous bone-marrow-derived mononuclear cells in patients listed for heart transplantation. *Tex Heart Inst J* **31**, 214–219.

Sinha, G. (2005). Cell biology. Human embryonic stem cells may be toxicology's new best friends. *Science* **308**, 1538.

Snyder, E. Y., Daley, G. Q., and Goodell, M. (2004). Taking stock and planning for the next decade: realistic prospects for stem cell therapies for the nervous system. *J Neurosci Res* **76**, 157–168.

Snyder, E. Y., and Macklis, J. D. (1995). Multipotent neural progenitor or stem-like cells may be uniquely suited for therapy for some neurodegenerative conditions. *Clin Neurosci* **3**, 310–316.

Song, M. R., and Ghosh, A. (2004). FGF2-induced chromatin remodeling regulates CNTF-mediated gene expression and astrocyte differentiation. *Nat Neurosci* **7**, 229–235. Epub 2004 Feb 2008.

Strem, B. M., Hicok, K. C., Zhu, M., Wulur, I., Alfonso, Z., Schreiber, R. E., Fraser, J. K., and Hedrick, M. H. (2005). Multipotential differentiation of adipose tissue-derived stem cells. *Keio J Med* **54**, 132–141.

Sun, L., Lee, J., and Fine, H. A. (2004). Neuronally expressed stem cell factor induces neural stem cell migration to areas of brain injury. *J Clin Invest* **113**, 1364–1374.

Tabatabai, G., Bahr, O., Mohle, R., Eyupoglu, I. Y., Boehmler, A. M., Wischhusen, J., Rieger, J., Blumcke, I., Weller, M., and Wick, W. (2005). Lessons from the bone marrow: how malignant glioma cells attract adult haematopoietic progenitor cells. *Brain* **128**, 2200–2211. Epub 2005 Jun 2209.

Terada, N., Hamazaki, T., Oka, M., Hoki, M., Mastalerz, D. M., Nakano, Y., Meyer, E. M., Morel, L., Petersen, B. E., and Scott, E. W. (2002). Bone marrow cells adopt the phenotype of other cells by spontaneous cell fusion. *Nature* **416**, 542–545. Epub 2002 Mar 2013.

Uchida, N., Buck, D. W., He, D., Reitsma, M. J., Masek, M., Phan, T. V., Tsukamoto, A. S., Gage, F. H., and Weissman, I. L. (2000). Direct isolation of human central nervous system stem cells. *Proc Natl Acad Sci USA* **97**, 14720–14725.

Uchida, N., Dykstra, B., Lyons, K., Leung, F., Kristiansen, M., and Eaves, C. (2004). ABC transporter activities of murine hematopoietic stem cells vary according to their developmental and activation status. *Blood* **103**, 4487–4495. Epub 2004 Feb 4426.

Valk-Lingbeek, M. E., Bruggeman, S. W., and van Lohuizen, M. (2004). Stem cells and cancer; the polycomb connection. *Cell* **118**, 409–418.

van Steensel, B. (2005). Mapping of genetic and epigenetic regulatory networks using microarrays. *Nat Genet 37 Suppl*, S18–S24.

Villa et al. (2000).

Vogel, G. (2005). Cell biology. Ready or not? Human ES cells head toward the clinic. *Science* **308**, 1534–1538.

Walczak, H., Miller, R. E., Ariail, K., Gliniak, B., Griffith, T. S., Kubin, M., Chin, W., Jones, J., Woodward, A., Le, T., et al. (1999). Tumoricidal activity of tumor necrosis factor-related apoptosis-inducing ligand *in vivo*. *Nat Med* **5**, 157–163.

Wang, R., Cherukuri, P., and Luo, J. (2005). Activation of Stat3 sequence-specific DNA binding and transcription by p300/CREB-binding protein-mediated acetylation. *J Biol Chem* **280**, 11528–11534. Epub 12005 Jan 11513.

Weaver, M., Yingling, J. M., Dunn, N. R., Bellusci, S., and Hogan, B. L. (1999). Bmp signaling regulates proximal–distal differentiation of endoderm in mouse lung development. *Development* **126**, 4005–4015.

Weimann, J. M., Charlton, C. A., Brazelton, T. R., Hackman, R. C., and Blau, H. M. (2003a). Contribution of transplanted bone marrow cells to Purkinje neurons in human adult brains. *Proc Natl Acad Sci USA* **100**, 2088–2093. Epub 2003 Feb 2087.

Weimann, J. M., Johansson, C. B., Trejo, A., and Blau, H. M. (2003b). Stable reprogrammed heterokaryons form spontaneously in Purkinje neurons after bone marrow transplant. *Nat Cell Biol* **5**, 959–966. Epub 2003 Oct 2015.

Werbowetski, T., Bjerkvig, R., and Del Maestro, R. F. (2004). Evidence for a secreted chemorepellent that directs glioma cell invasion. *J Neurobiol* **60**, 71–88.

Widera, D., Holtkamp, W., Entschladen, F., Niggemann, B., Zanker, K., Kaltschmidt, B., and Kaltschmidt, C. (2004). MCP-1 induces migration of adult neural stem cells. *Eur J Cell Biol* **83**, 381–387.

Yang, A. X., Mejido, J., Luo, Y., Zeng, X., Schwartz, C., Wu, T., Thies, R. S., Bhattacharya, B., Han, J., Freed, B., et al. (2005). Development of a focused microarray to assess human embryonic stem cell differentiation. *Stem Cells Dev* **14**, 270–284.

Yang, M., Stull, N. D., Berk, M. A., Snyder, E. Y., and Iacovitti, L. (2002). Neural stem cells spontaneously express dopaminergic traits after transplantation into the intact or 6-hydroxydopamine-lesioned rat. *Exp Neurol* **177**, 50–60.

Ying, Q. L., Nichols, J., Evans, E. P., and Smith, A. G. (2002). Changing potency by spontaneous fusion. *Nature* **416**, 545–548. Epub 2002 Mar 2013.

Yuan, Z. L., Guan, Y. J., Chatterjee, D., and Chin, Y. E. (2005). Stat3 dimerization regulated by reversible acetylation of a single lysine residue. *Science* **307**, 269–273.

Zencak, D., Lingbeek, M., Kostic, C., Tekaya, M., Tanger, E., Hornfeld, D., Jaquet, M., Munier, F. L., Schorderet, D. F., van Lohuizen, M., and Arsenijevic, Y. (2005). Bmi1 loss produces an increase in astroglial cells and a decrease in neural stem cell population and proliferation. *J Neurosci* **25**, 5774–5783.

Zhao, L. R., Duan, W. M., Reyes, M., Keene, C. D., Verfaillie, C. M., and Low, W. C. (2002). Human bone marrow stem cells exhibit neural phenotypes and ameliorate neurological deficits after grafting into the ischemic brain of rats. *Exp Neurol* **174**, 11–20.

Zhao, Y., Samal, E., and Srivastava, D. (2005). Serum response factor regulates a muscle-specific microRNA that targets Hand2 during cardiogenesis. *Nature* **436**, 214–220.

INDEX

Drug Discovery Research: New Frontiers in the Post-Genomic Era, Edited by Ziwei Huang